Smart Innovation, Systems and Technologies

Volume 237

The Smart Innovation, Systems and Technologies book series encompasses the topics of knowledge, intelligence, innovation and sustainability. The aim of the series is to make available a platform for the publication of books on all aspects of single and multi-disciplinary research on these themes in order to make the latest results available in a readily-accessible form. Volumes on interdisciplinary research combining two or more of these areas is particularly sought.

The series covers systems and paradigms that employ knowledge and intelligence in a broad sense. Its scope is systems having embedded knowledge and intelligence, which may be applied to the solution of world problems in industry, the environment and the community. It also focusses on the knowledge-transfer methodologies and innovation strategies employed to make this happen effectively. The combination of intelligent systems tools and a broad range of applications introduces a need for a synergy of disciplines from science, technology, business and the humanities. The series will include conference proceedings, edited collections, monographs, handbooks, reference books, and other relevant types of book in areas of science and technology where smart systems and technologies can offer innovative solutions.

High quality content is an essential feature for all book proposals accepted for the series. It is expected that editors of all accepted volumes will ensure that contributions are subjected to an appropriate level of reviewing process and adhere to KES quality principles.

Indexed by SCOPUS, EI Compendex, INSPEC, WTI Frankfurt eG, zbMATH, Japanese Science and Technology Agency (JST), SCImago, DBLP.

All books published in the series are submitted for consideration in Web of Science.

More information about this series at http://www.springer.com/series/8767

Mohamed Ben Ahmed ·
Horia-Nicolai L. Teodorescu · Tomader Mazri ·
Parthasarathy Subashini ·
Anouar Abdelhakim Boudhir
Editors

Networking, Intelligent Systems and Security

Proceedings of NISS 2021

Springer

Editors
Mohamed Ben Ahmed
Faculty of Sciences and Techniques
of Tangier
Abdelmalek Essaadi University
Tangier, Morocco

Tomader Mazri
National School of Applied Sciences
Ibn Tofail University
Kénitra, Morocco

Anouar Abdelhakim Boudhir
Department of Computer Sciences
Faculty of Sciences and Techniques
of Tangier
Abdelmalek Essaadi University
Tangier, Morocco

Horia-Nicolai L. Teodorescu
Technical University of Iasi
Iași, Romania

Parthasarathy Subashini
Department of Computer Science
Avinashilingam University
Coimbatore, Tamil Nadu, India

ISSN 2190-3018 ISSN 2190-3026 (electronic)
Smart Innovation, Systems and Technologies
ISBN 978-981-16-3636-3 ISBN 978-981-16-3637-0 (eBook)
https://doi.org/10.1007/978-981-16-3637-0

This Springer imprint is published by the registered company Springer Nature Singapore Pte Ltd.
The registered company address is: 152 Beach Road, #21-01/04 Gateway East, Singapore 189721, Singapore

Committee

Conference Chair

Tomader Mazri, ENSA, Ibn Tofail University, Kenitra, Morocco

Conference General Chairs

Mohamed Ben Ahmed, FST, Tangier UAE University, Morocco
Anouar Abdelhakim Boudhir, FST, Tangier UAE University, Morocco
Bernadetta Kwintiana Ane, University of Stuttgart, Germany

Conference Technical Programme Committee Chair

Wassila Mtalaa, Luxembourg Institute of Science and Technology, Luxembourg

Keynote and Panels Chair

Domingos Santos, Polytechnic Institute Castelo Branco, Portugal

Publications Chair

İsmail Rakıp Karaşo, Karabuk University

Special Issues Chair

Senthil Kumar, Hindustan College of Arts and Science, India

Local Organizing Committee

Tomader Mazri, ENSA, UIT, Morocco
Hassan Mharzi, ENSA, UIT, Morocco
Mohamed Nabil Srifi, ENSA, UIT, Morocco
Tarik Jarou, ENSA, UIT, Morocco
Benbrahim Mohammed, ENSA, UIT, Morocco
Abderrahim Bajjit, ENSA, UIT, Morocco
Rachid Elgouri, ENSA, UIT, Morocco
Imane Sahmi, ENSA, UIT, Morocco
Loubna El Amrani, ENSA, UIT, Morocco

Technical Programme Committee

Ismail Rakip Karas, Karabuk University, Türkiye
Abdel-Badeeh M. Salem, Ain Shams University, Egypt
Abderrahim Ghadi, FSTT, UAE, Morocco
Accorsi Riccardo, Bologna University, Italy
Aftab Ahmed Khan, Karakoram International University, Pakistan
Ahmad S. Almogren, King Saud University, Saudi Arabia
Ahmed Kadhim Hussein, Babylon University, Iraq
Alabdulkarim Lamya, King Saud University, Saudi Arabia
Alghamdi Jarallah, Prince Sultan University, Saudi Arabia
Ali Jamali, Universiti Teknologi Malaysia
Alias Abdul Rahman, Universiti Teknologi Malaysia
Anabtawi Mahasen, Al-Quds University, Palestine
Anton Yudhana, Universitas Ahmad Dahlan, Indonesia
Arioua Mounir, UAE, Morocco
Assaghir Zainab, Lebanese University, Lebanon
Astitou Abdelali, UAE, Morocco
Aydın Üstün, Kocaeli University, Türkiye
Aziz Mahboub, FSTT, UAE, Morocco
Barış Kazar, Oracle, USA
Bataev Vladimir, Zaz Ventures, Switzerland
Behnam Alizadehashrafi, Tabriz Islamic Art University, Iran
Behnam Atazadeh, University of Melbourne, Australia

Preface

In an age of explosive worldwide growth of electronic data storage and communications, effective protection of information has become a critical requirement.

With the exponential growth of wireless communications, Internet of Things, and cloud computing, and the increasingly dominant roles played by electronic commerce in every major industry, safeguarding the information in storage and travelling over the communication networks is increasingly becoming the most critical and contentious challenges for the technology innovators.

This trend opens up significant research activity for academics and their partners (industrialists, governments, civil society, etc.) in order to establish essential and intelligent bases for developing the active areas of networking, intelligent systems and security.

This edited book aims to present scientific research and engineering applications for the construction of intelligent systems and their various innovative applications and services. The book also aims to provide an integrated view of the problems to researchers, engineers, practitioners and to outline new topics in networks and security.

This edition is the result of work accepted and presented at the Fourth International Conference on Networks, Intelligent Systems and Security (NISS 2021) held on April, 1–2, 2020, in Kenitra, Morocco. It brings together original research, work carried out and proposed architectures on the main themes of the conference.

The goal of this book edition is constructing and building the basics and essentials researches, innovations and applications that can help on the growth of the future next generation of networks and intelligent systems.

We would like to acknowledge and thank Springer Nature staff for their support, guidance and for the edition of this book.

Finally, we wish to express our sincere thanks to Prof. Robert J. Howlett, Mr. Aninda Bose and Ms. Sharmila Mary Panner Selvam for their kind support and help to promote and develop research.

Tangier, Morocco Mohamed Ben Ahmed
Iași, Romania Horia-Nicolai L. Teodorescu
Kénitra, Morocco Tomader Mazri
Coimbatore, India Parthasarathy Subashini
Tangier, Morocco Anouar Abdelhakim Boudhir

Contents

Network and Cloud Technologies

Smart Security

Smart Digital Learning

Smart Information Systems

About the Editors

Mohamed Ben Ahmed is an associate professor of computer sciences at Abdelmalek Essaâdi University, Morocco; he received the Ph.D. degree in computer sciences and telecommunications, in 2010 from Abdelmalek Essaâdi University. His researches are about smart and sustainable cities, data mining and routing in wireless sensor networks. He is currently a supervisor of several thesis and an investigator in several international research projects about smart cities. He is the author of more than fifty papers published in international journals and conferences. He is the co-editor of Springer *Innovations in Smart Cities Applications* book. He is a chair and a committee member of several international conferences.

Prof. Horia-Nicolai L. Teodorescu teaches intelligent systems at Gheorghe Asachi Technical University of Iasi and language technology at Alexandru Ioan Cuza University of Iasi; in addition, he is the director of the Institute of Computer Science of the Romanian Academy. He served for extended periods as a visiting and invited professor at Swiss Federal Institute of Technology, Lausanne, University of South Florida, Tampa, and Kyushu Institute of Technology and FLSI, Iizuka, Japan, among others. He served as a co-director of postgraduate and doctoral studies in Lausanne and in University of Leon, Spain. He also served as a vice-rector of Gheorghe Asachi Technical University of Iasi. When he was included in the Romanian Academy, he was the youngest member of the learned body. Dr. Teodorescu occupied several positions in national and international societies and institutions, including member of the independent expert group and vice-chair of the group for Computer Science of NATO. Dr. Teodorescu served as a member of the editorial boards of several major journals issued by publishers as IEEE, Francis & Taylor, Elsevier, and the Romanian Academy. He authored about 250 conference and journal papers and more than 25 books; he holds 24 national and international patents.

Prof. Tomader Mazri received her HDR degree in Networks and Telecommunication from Ibn Tofail University, Ph.D. in Microelectronics and Telecommunication from Sidi Mohamed Ben Abdellah University and INPT of Rabat, Master's in Microelectronics and Telecommunication Systems, and Bachelor's in Telecommunication from the Cadi Ayyad University. She is currently a professor at the National School of Applied Sciences of Kenitra, a permanent member of Electrical and Telecommunications Engineering Laboratory, and an author and a co-author of 15 articles journals, 40 articles in international conferences, 3 chapters, and 5 books. Her major research interests are on microwave systems for mobile and radar, smart antennas, and mobile network security.

Parthasarathy Subashini has also received Ph.D. in Computer Science in 2009 from Avinashilingam University for Women, Tamil Nadu, India. From 1994, she is working as a professor in the Computer Science Department of Avinashilingam University. Concurrently, she contributed to several fields of mathematics, especially nature-inspired computing. She has authored or co-authored 4 books, 6 book chapters, 1 monograph, 145 papers, including IEEE, Springer's in various international, national journals, and conferences. She has held positions as a reviewer, chairpersons for different peer-reviewed journals. Under her supervision, she has ten research projects of worth more than 2.7 crores from various funding agencies like Defence Research and Development Organization, Department of Science and Technology, SERB, and University Grants Commission. She has visited many countries for various knowledge sharing events. As a member of IEEE, IEEE Computational Intelligence Society, and IEEE Computer Society of India, she extended her contribution as IEEE Chair for Women in Computing under IEEE Computer Society of India Council in the year 2015–2016.

Anouar Abdelhakim Boudhir is currently an associate professor at the Faculty of Sciences and Technique of Tangier. Actually, he is the president of the Mediterranean Association of Sciences and Technologies. He is an adviser at the Moroccan union against dropping out of school. He received the HDR degree from Abdelmalek Essaadi University; he is the co-author of several papers published in IEEExplorer, ACM, and in high indexed journals and conference. He co-edited a several books published on Springer series, and he is a co-founder of a series of international conferences (Smart health17, SCIS'16, SCA18, SCA19, SCA20, NISS18, NISS19, NISS20, DATA21) till 2016. His supervise several thesis about artificial intelligence, security, and E-healthcare. His key research relates to ad hoc networks, VANETS, WSN, IoT, big data, computer healthcare applications, and security applications.

Artificial Intelligence for Sustainability

Detection of Human Activities in Wildlands to Prevent the Occurrence of Wildfires Using Deep Learning and Remote Sensing

Ayoub Jadouli and Chaker El Amrani

Abstract Human activities in wildland are responsible for the largest part of wildfire cases. This paper presents a work that uses deep learning on remote sensing images to detect human activity in wildlands to prevent fire occurrence that can be caused by humans. Human activities can be presented as any human interaction with wildlands, and it can be roads, cars, vehicles, homes, human shapes, agricultural lands, golfs, airplanes, or any other human proof of existence or objects in wild lands. Conventional neural network is used to classify the images. For that, we used three approaches, in which one is the object detection and scene classification approach, the second is land class approach where two classes of lands can be considered which are wildlands with human interactions and wildland without human interaction. The third approach is more general and includes three classes that are urban lands, pure wildlands, and wildlands with human activities. The results show that it is possible to detect human activities in wildlands using the models presented in this paper. The second approach can be considered the most successful even if it is the simpler.

1 Introduction

1.1 Wildfire and Machine Learning

Machine learning (ML) is the term for techniques that allow the machine to find a way to resolve problems without being specially programmed for that. ML approaches are used in the data science context, relating: data size, computational requirements, generalizability, and interpretability of data. In the last two decades, there is a big increase in using ML methods in wildfire fields. There are three main types of ML methods:

A. Jadouli (✉) · C. El Amrani
LIST Lab, Faculty of Sciences and Techniques of Tangier, Abdelmalek Essaâdi University, Tangier, Morocco
e-mail: ajadouli@uae.ac.ma

M. Ben Ahmed et al. (eds.), *Networking, Intelligent Systems and Security*,
Smart Innovation, Systems and Technologies 237,
https://doi.org/10.1007/978-981-16-3637-0_1

- **Supervised ML**: The goal is learning a parametrized function or model that maps a known input (i.e., predictor variables) to a known output (or target variables). So, an algorithm is used to learn the parameters of that function using examples. Supervised learning can solve two types of problems, and it can be a classification problem when the target variables are categorical or a regression problem when the target variables are continuous. There is a lot of methods that can be categorized as supervised ML: Naive Bayes (NB), decision trees (DT), classification and regression tree (CART), random forest (RF), deep neural network (DNN), Gaussian processes (GP), artificial neural networks (ANN), genetic algorithms (GA), recurrent neural network (RNN), maximum entropy (MAXENT), regression trees (BRT), random forest (RF), K-nearest neighbor (KNN), support vector machines (SVM) [Hearst, Dumais, Osuna,Platt, Scholkopf, 1998], and K-SVM. Supervised ML can be used in these fields (fire spread/burn area prediction, fire occurrence, fire severity, smoke prediction, climate change, fuels characterization, fire detection, and fire mapping) [1].
- **Unsupervised Learning**: It is used when the target variables are not available, and generally, the goal is understanding the patterns and discovering the output, dimensionality reduction, or clustering. The relationships or patterns are extracted from the data without any guidance as to the right answer. A lot of methods can be considered in that field (K-means clustering (KM), self-organizing maps SOM, autoencoders, Gaussian mixture models (GMM), iterative self-organizing DATA algorithm (ISODATA), hidden Markov models (HMM), density-based spatial clustering of applications with noise (DBSCAN)), T-distributed stochastic neighbor embedding (t-SNE), random forest (RF), boosted regression trees (BRT) [Freund, Shapire, 1995], maximum entropy (MaxEnt), principal component analysis (PCA), and factor analysis). Unsupervised ML can be used for fire detection, fire mapping, burned area prediction, fire weather prediction landscape controls on fire, fire susceptibility, and fire spread/burn area prediction [1].
- **Agent-Based Learning**: A single or a group of autonomous agents interact with the environment following specific rules of behavior. Agent-based learning can be used for optimization and for decision making. The next algorithms can be considered as agent-based: genetic algorithms (GA), Monte Carlo tree search (MCTS), Asynchronous Advantage Actor-Critic (A3C), deep Q-network (DQN), and reinforcement learning (RL) [Sutton, Barto, 1998]. The agent-based learning can be useful for optimizing fire simulators, fire spread and growth, fuel treatment, planning and policy, and wildfire response [1].

1.2 Deep Learning and Remote Sensing

In the last decade, deep learning models can be considered as the most successful ML methods. It is considered to be an artificial neural network that involves multiple hidden layers [2]. Because of the large successful use of these methods by big companies in production, the research interest in the field has increased, and more and more applications have been used to solve a large scale of problems including remote sensing problems.

Remote sensing is a technique that uses reflected or emitted electromagnetic energy to get information about the earth's land and water surfaces, obtaining quantitative measurements and estimations of geo-bio-physical variables. That is possible because every material in the scene has a special interaction with electromagnetic radiation that can be emitted, reflected, or absorbed by these materials depending on their shapes and their molecular composition. With the increase of the spatial resolution of satellite images created by merging their data with information collected at a higher resolution, it is possible to achieve resolution up to 25-cm [3].

1.3 Problem

Wildfires are mostly caused by humans. So, to prevent the occurrence of fire, the monitoring of human activities in wildlands is an essential task. A lot of technical challenges are involved in that field. Deep learning applied to high-resolution spatial images can be considered to solve a part of the problem. So, this work focuses on the usage of convolutional neural network (CNN) to detect human activities in wildlands based on remote sensing images. The results of this work will be used in future work to achieve a prediction of fire occurrence that can be caused by humans. The details can be found in Sect. 3.

2 Related Work

Deep neural networks are the most successful methods used to solve problems linked to the interpretation of remote sensing data. So, a lot of researchers are interested in these methods:

Kadhim et al. [4] presented useful models for satellite image classification that are based on convolutional neural networks, the features that are used to classify the image extracted by using four pre-trained CNN models: Resnet50, GoogleNet, VGG19, and AlexNet. The Resnet50 model achieves a better result than other models for all the datasets they used. Varshney and Debvrat [5] used a convolutional neural network and fused SWIR and VNIR multi-resolution data to achieve pixel-wise classification through semantic segmentation, giving a good result with a high snow-and-cloud F1 score. They found that their DNN was better than the traditional methods and was able to learn spectro-contextual information, which can help in the semantic segmentation of spatial data. Long et al. [6] propose a new object localization framework, which can be divided into three processes: region proposal, classification, and accurate object localization process. They found that the dimension-reduction model performs better than the retrained and fine-tuned models and the detection precision of the combined CNN model is much higher than that of any single model. Wang et al. [7] used mixed spectral characteristics and CNN to propose a remote sensing recognition method for the detection of landslides and obtained an accuracy of 98.98 and 97.69%.

With the success of CNN in image recognition, more studies are now interested to find the best and appropriate model that can be used to solve a specific problem. This is the case of Zhong et al. [8] proposing RSNet a remote sensing DNN framework. Their goal is to automatically search and find the appropriate network architecture for image recognition tasks based on high-resolution remote sensing images (HRS).

3 Proposed Solution and Methods

3.1 *Architecture and Implementation of the Design*

Our research focuses on the detection of human activities on wildlands using CNN on satellite high-resolution Images. Even if this subject is very general, it is very much related to our project because human activities are the primary cause of the wildfire. The goal is to use the result of the models to be able to predict the areas where the occurrence of fire in wildlands will start with the help of weather data (See Fig. 1). We proposed three approaches to solve this problem:

- A simple CNN model trained by UC Merced dataset with five classes of 21 classes as output. And based on the 21 classes, a conclusion can be made.
- A simple CNN model for a simple classification used by two classes (wildland with human activities and pure wildlands)
- ResNet50 pre-trained model with transfer learning to output with three classes (urban lands, wildlands with human interactions, and pure wildland without human interaction) (See Fig. 2).

3.2 *Conventional Neural Network Model*

Convolutional neural networks (CNN) are a type of deep neural network which have one or more convolutional layers. This network can be very useful when there is a need to detect a specific pattern in data and make sense of them. CNN is very useful for image analyses and matches the requirement of our study.

In our case, we have used the same sequential model for the **first approach** with five classes and the **second approach** with two classes (See Fig. 3). Where ten layers can be found described in Table 1 for the first approach and Table 2 for the second approach.

The models are built thanks to Python 3 and Keras library, and the training and tests are made with the parallelization on the top of NVIDIA GeForce 840M single GPU device, with 384 CUDA cores, 1029 MHz in frequency, and 2048 MB DDR3.

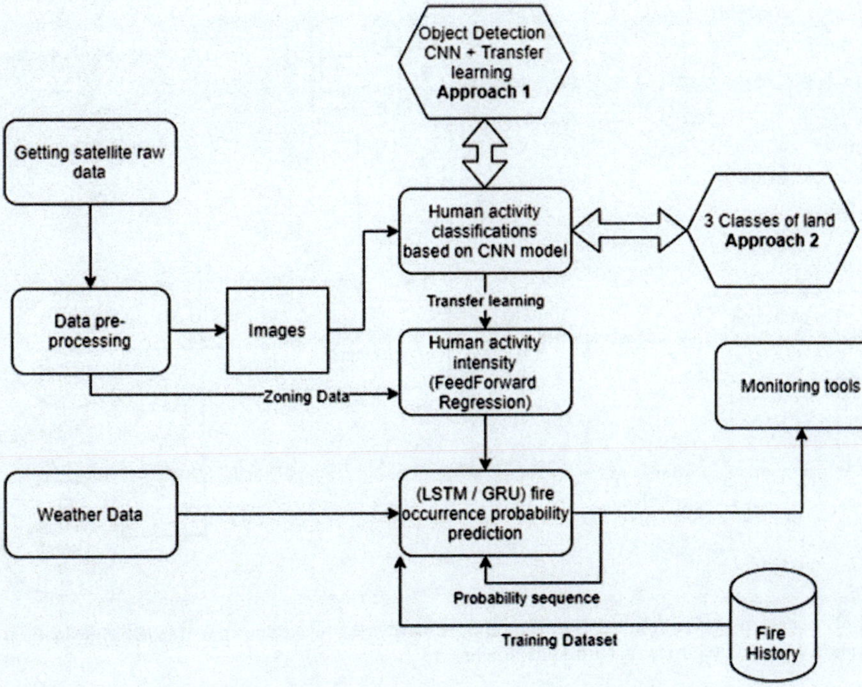

Fig. 1 Process of wildfire prediction using DL CNN and LSTM/GRU based on human activity detection

3.3 Transfer Learning

Transfer learning is a technique where information learned in a primary model of machine learning is transferred to a secondary machine learning model to achieve a new task. This technique is used in the **third** approach for the classification of three land classes (pure wildland, wildland with human activities, and urban lands) See Fig. 2.

In this model, we try to use the pre-trained model of RestNet50 without the top layer "resnet50_weights_tf_dim_ordering_tf_kernels_notop.h5" with transfer learning to classify the three classes that we work on based on the pre-trained network (Table 3; Fig. 4).

3.4 Datasets

This study uses UC Merced dataset as the principal source of data because it is widely utilized in land use cases of studies and showed good results in machine learning classification problems.

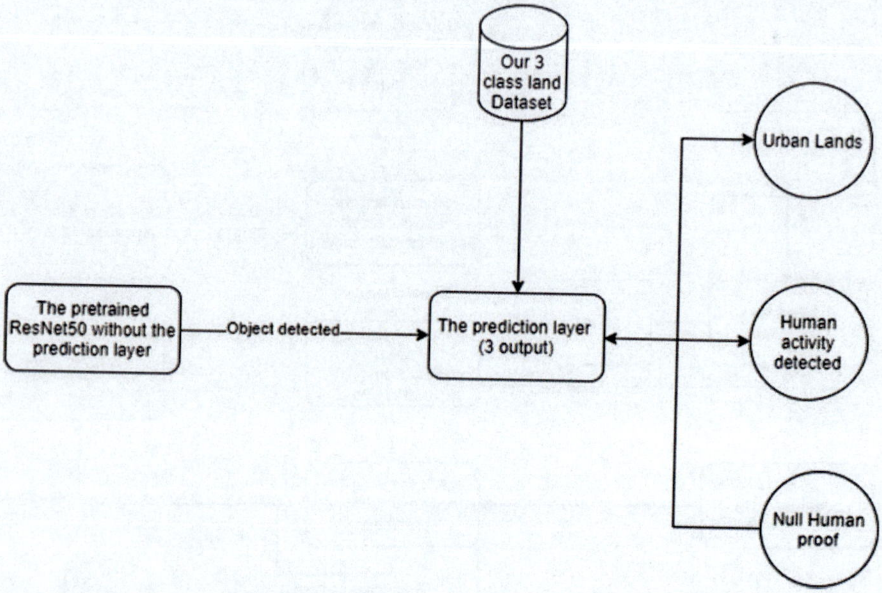

Fig. 2 Diagram shows how ResNet50 pretreated network is used with transfer learning methods to classify three categories of wildlands images

UC Merced land use dataset is introduced by Yang and Newsam [9]. It is a land use image dataset with 21 classes, each class has 100 images, and each image measures 256 × 256 pixels, with a spatial resolution of 0.3 m per pixel. The images are extracted from United States Geological Survey National Map Urban Area Imagery. The images were manually cropped to extract various urban areas around the United States [9].

We have subdivided the dataset into three other datasets to match our case study.

In the first dataset, we have extracted five classes that can be linked to wildlands which are as follows: forest because it is the main study area, freeways because they are mostly built in wildland and are proof of human activities in wildlands, golf course because the images match the forest images and we need our model to find the differences, river because sometimes it looks like a freeway or a road and we need our model to find the differences, and finally, sparse residential areas because we found buildings that are mostly built near wildlands and that can be considered as another proof of human activities in wildlands (See Fig. 6).

In the second dataset, we have split the first dataset into two classes which are Human Activity Wildland (wildland or images that look like wildlands with human activity proofs) and Pure Wildland (clean forest and rivers) (See Fig. 7).

The third UC Merced dataset images are subdivided into three classes which are Urban Land (images of urban areas), Human Activity Wildland (wildlands or wildland like images where a trace of human activities can be found), and Pure Wildland (images of wildlands with no trace of human activity) (See Fig. 8). The

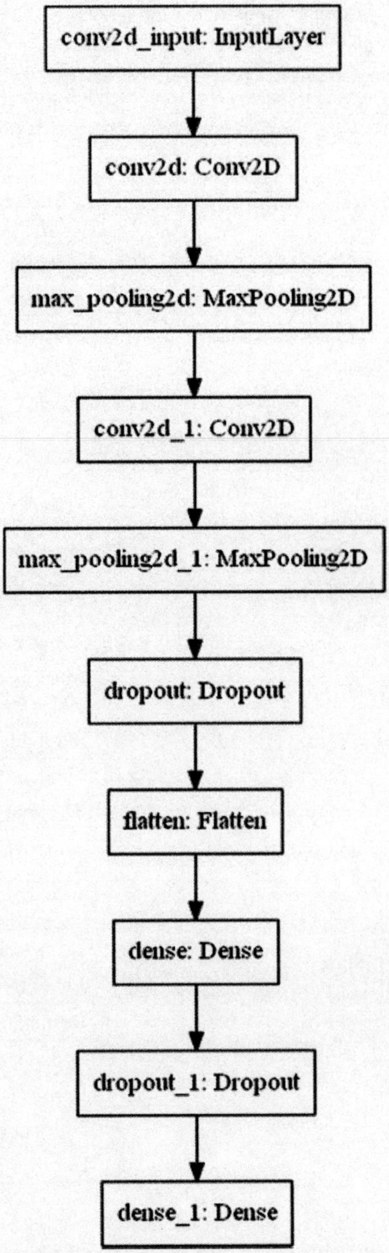

Fig. 3 Approach 1 and approach 2 model's layers

Table 1 Approach 1 model's details

Layer (type)	Output shape	Param #
conv2d (Conv2D)	(None, 254, 254, 32)	896
max_pooling2d (MaxPooling2D)	(None, 127, 127, 32)	0
conv2d_1 (Conv2D)	(None, 125, 125, 32)	9248
Max_pooling2d_1 (MaxPooling2D)	(None, 62, 62, 32)	0
Dropout (Dropout)	(None, 62, 62, 32)	0
Flatten (Flatten)	(None, 123,008)	0
Dropout_1 (Dropout)	(None, 128)	15,745,152
Dropout_1 (Dropout)	(None, 128)	0
Dense_1 (Dense)	(None, 5)	645

Table 2 Approach 2 model's details

Layer (type)	Output shape	Param #
conv2d (Conv2D)	(None, 254, 254, 32)	896
max_pooling2d (MaxPooling2D)	(None, 127, 127, 32)	0
conv2d_1 (Conv2D)	(None, 125, 125, 32)	9248
max_pooling2d_1 (MaxPooling2D)	(None, 62, 62, 32)	0
Dropout (Dropout)	(None, 62, 62, 32)	0
Flatten (Flatten)	(None, 123,008)	0
Dropout_1 (Dropout)	(None, 128)	15,745,152
Dropout_1 (Dropout)	(None, 128)	0
Dense_1 (Dense)	(None, 2)	645

Table 3 Approach 3 model's details

Layer (type)	Output shape	Param #
Resnet50 (Functional)	(None, 2048)	89,623,587,712
Dropout (Dropout)	(None, 2048)	0
Dense (Dense)	(None, 128)	262,272
Dropout (Dropout)	(None, 128)	0
Dense_1 (Dense)	(None, 3)	387

goal of this dataset is to have a global class type that can match any image obtained using visual bands remote sensing (Fig. 5).

For the third approach, we have used the pre-trained model of ResNet50 (trained using ImageNet dataset). The retained model can be downloaded at this link: https://www.kaggle.com/keras/resnet50

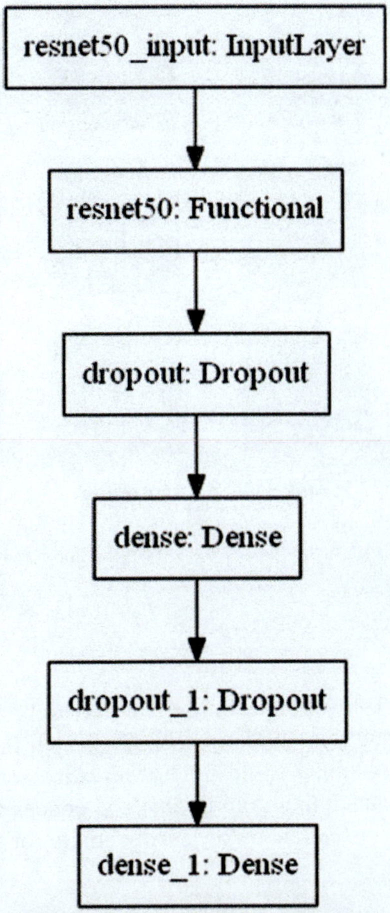

Fig. 4 Approach 3 model's layers

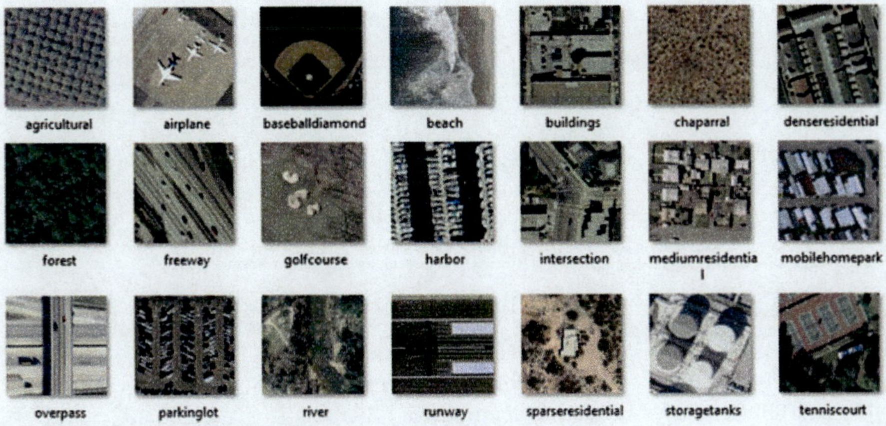

Fig. 5 Sample images from UC merced land dataset

Fig. 6 Sample images from the approach 1 dataset

3.5 *Results*

To avoid overfitting, we have split the data into two parts, namely the training set and the validation set. We watch the results based on validation accuracy that we obtained based on the validation set to evaluate the model's performance. We also watch the training time and epochs which can be considered as batches or sets that are fed one by one, sequentially for training to avoid saturation of the memory.

3.5.1 Approach 1

See Table 4; Fig. 9.

3.5.2 Approach 2

See Table 5; Fig. 10.

3.5.3 Approach 3

With the third approach, we have obtained the result of training in 100 s with the same environment and a validation accuracy of 63%.

Fig. 7 Sample images from the approach 2 dataset

3.6 *Discussions*

Both the approaches one and two have a maximum accuracy with 50 epochs: 70% for the fst approach and 75% for the second approach which means that 50 epochs are enough to train the model.

The results show that the second approach performs better than the first approach in both accuracy and time which prove that using a maximum of two classes is better than using multiple classes.

The accuracy of 75% seems to be low compared to the accuracy in other researches and studies, but the goal of this research is not to achieve the best possible accuracy but just to prove that we can use deep learning and remote sensing images to detect human activities in wildlands. With 75% accuracy applied to large areas that have multiple images, we can effectively detect human activities.

With the third approach, the training time is faster than the two previous approaches, but the accuracy is inferior. We can explain the low accuracy because

Fig. 8 Sample images from the approach 3 dataset

Table 4 Approach 1 results details based on training validation accuracy and training time

Epoch	5	10	20	50	100
Time (s)	123.77	244.19	489.50	1200.34	2406.10
Acc	0.48	0.60	0.54	0.70	0.52

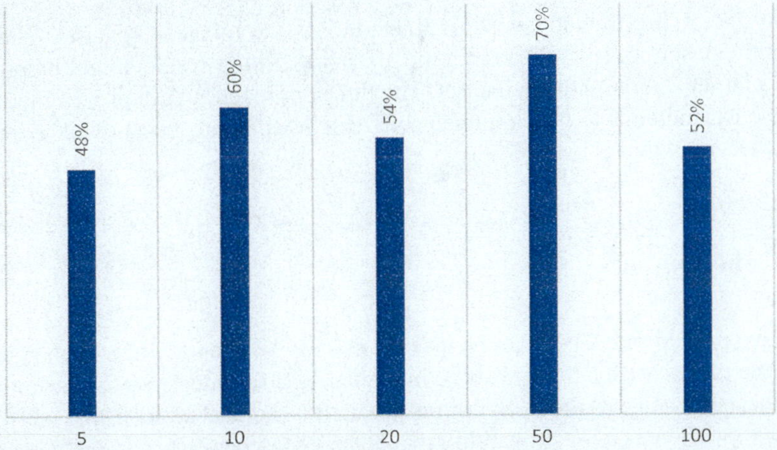

Fig. 9 Approach 1 validation accuracy by number of epochs

Table 5 Approach 2 results details based on training validation accuracy and training time

Epoch	5	10	20	50	100
Time (s)	50.86	103.16	201.76	495.31	992.35
Acc (%)	60	65	55	75	75

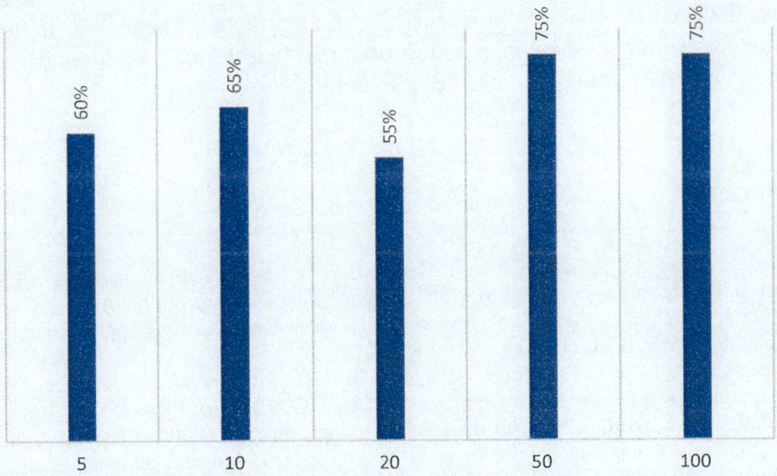

Fig. 10 Approach 2 validation accuracy by number of epochs

the ResNet50 pre-trained layers are trained using the ImageNet dataset. Which is a general dataset not spatialized in remote sensing. Better results might be produced with a larger remote sensing dataset. The three approaches have proven that we can use the DL methods to watch human activities in wildlands because all results have a test accuracy that is over 63%.

4 Conclusion

We have proved that CNN can be used to classify wildland with human activities. So while there is still a lot of work to implement the full ideas, we can be optimistic nonetheless when we think of the results of the work. The idea may seem very intuitive, so there is a big probability that other researchers are working on the same problem even if we cannot find any work that used the same ideas in the field of our research.

The detection of human activities in wildland may be used for other purposes if we can obtain a better accuracy to watch wildlands. But for our purpose, more than 70% is enough to detect human activities in wildland because the same areas have a lot of images and the purpose is to calculate the probabilities of fire occurrence with the help of weather data and the history of fire.

A larger dataset may be introduced in future works with better DL models to increase the accuracy and let our research be more efficient. The hope is to help the professionals better do their jobs and in doing so help reduce wildfires caused by humans by increasing the efficiency of the monitoring.

References

1. Jain, P., Coogan, S.C.P., Subramanian, S.G., Crowley, M., Taylor, S., Flannigan, M.D.: A review of machine learning applications in wildfire science and management (2020)
2. LeCun, Y., Bottou, L., Bengio, Y., Haffner, P.: Gradient-based learning applied to document recognition. Proc. IEEE (1998)
3. Gomez-Chova, L., Tuia, D., Moser, G., Camps-Valls, G.: Multimodal classification of remote sensing images: a review and future directions. Proc. IEEE **103**(9), 1560–1584 (2015)
4. Kadhim, M.A., Abed, M.H.: Convolutional neural network for satellite image classification. In: Studies in Computational Intelligence, vol. 830, Issue January. Springer International Publishing (2020)
5. Varshney, D.: Convolutional Neural Networks to Detect Clouds and Snow in Optical Images (2019). http://library.itc.utwente.nl/papers_2019/msc/gfm/varshney.pdf
6. Long, Y., Gong, Y., Xiao, Z., Liu, Q.: Accurate object localization in remote sensing images based on convolutional neural networks. IEEE Trans. Geosci. Remote Sens. **55**(5), 2486–2498 (2017)
7. Wang, Y., Wang, X., Jian, J.: Remote sensing landslide recognition based on convolutional neural network. Mathematical Problems in Engineering (2019)
8. Wang, J., Zhong, Y., Zheng, Z., Ma, A., Zhang, L.: RSNet: the search for remote sensing deep neural networks in recognition tasks. IEEE Trans. Geosci. Remote Sens. (2020)

9. Yang, Y., Newsam, S.: Bag-of-visual-words and spatial extensions for land-use classification. In: GIS: Proceedings of the ACM International Symposium on Advances in Geographic Information Systems (2010)
10. Waghmare, B., Suryawanshi, M.: A review- remote sensing. Int. J. Eng. Res. Appl. **07**(06), 52–54 (2017)
11. Li, T., Shen, H., Yuan, Q., Zhang, L.: Deep learning for ground-level PM2.5 prediction from satellite remote sensing data. In: International Geoscience and Remote Sensing Symposium (IGARSS), 2018-July (November), 7581–7584 (2018)
12. Tondewad, M.P.S., Dale, M.M.P.: Remote sensing image registration methodology: review and discussion. Procedia Comput. Sci. **171**, 2390–2399 (2020)
13. Xu, C., Zhao, B.: Satellite image spoofing: Creating remote sensing dataset with generative adversarial networks. Leibniz Int. Proc. Inf. LIPIcs **114**(67), 1–6 (2018)
14. Zhang, L., Xia, G. S., Wu, T., Lin, L., Tai, X.C.: Deep learning for remote sensing image understanding. J. Sens. 2016 (2015)
15. Rodríguez-Puerta, F., Alonso Ponce, R., Pérez-Rodríguez, F., Águeda, B., Martín-García, S., Martínez-Rodrigo, R., Lizarralde, I.: Comparison of machine learning algorithms for wildland-urban interface fuelbreak planning integrating ALS and UAV-Borne LiDAR data and multispectral images. Drones **4**(2), 21 (2020)
16. Li, Y., Zhang, H., Xue, X., Jiang, Y., Shen, Q.: Deep learning for remote sensing image classification: a survey. Wiley Interdisc. Rev. Data Min. Knowl. Discov. **8**(6), 1–17 (2018)
17. Khelifi, L., Mignotte, M.: Deep learning for change detection in remote sensing images: comprehensive review and meta-analysis. IEEE Access **8**(Cd), 126385–126400 (2020)
18. Alshehhi, R., Marpu, P.R., Woon, W.L., Mura, M.D.: Simultaneous extraction of roads and buildings in remote sensing imagery with convolutional neural networks. ISPRS J. Photogramm. Remote. Sens. **130**(April), 139–149 (2017)
19. de Lima, R.P., Marfurt, K.: Convolutional neural network for remote-sensing scene classification: Transfer learning analysis. Remote Sens. **12**(1) (2020)
20. Liu, X., Han, F., Ghazali, K.H., Mohamed, I.I., Zhao, Y.: A review of convolutional neural networks in remote sensing image. In: ACM International Conference Proceeding Series, Part F1479 (July), 263–267 (2019)
21. Goodfellow, I.: 10—Slides—Sequence Modeling: Recurrent and Recursive Nets (2016). http://www.deeplearningbook.org/
22. Semlali, B.-E.B., Amrani, C.E., Ortiz, G.: Adopting the Hadoop architecture to process satellite pollution big data. Int. J. Technol. Eng. Stud. **5**(2), 30–39 (2019)

The Evolution of the Traffic Congestion Prediction and AI Application

Badr-Eddine Soussi Niaimi, Mohammed Bouhorma, and Hassan Zili

Abstract During the past years, there were so many researches focusing on traffic prediction and ways to resolve future traffic congestion; at the very beginning, the goal was to build a mechanism capable of predicting the traffic for short-term; meanwhile, others did focus on the traffic prediction using different perspectives and methods, in order to obtain better and more precise results. The main aim was to come up with enhancements to the accuracy and precision of the outcomes and get a longer-term vision, also build a prediction's system for the traffic jams and solve them by taking preventive measures (Bolshinsky and Freidman in Traffic flow forecast survey 2012, [1]) basing on artificial intelligence decisions with the given predictions. There are many algorithms; some of them are using statistical physics methods; others use genetic algorithms… the common goal was to achieve a kind of framework that will allow us to move forward and backward in time to have a practical and effective traffic prediction. In addition to moving forward and backward in time, the application of the new framework allows us to locate future traffic jams (congestions). This paper reviews the evolution of the existing traffic prediction's approaches and the edge given by AI to make the best decisions; we will focus on the model-driven and data-driven approaches. We start by analyzing all advantages and disadvantages of each approach to reach our goal in order to pursue the best approaches for the best output possible.

1 Introduction

Nowadays, we are noticing that our cities are becoming overpopulated very fast, which leads to a greater number of vehicles as well as a considerable number of deaths caused by traffic accidents. Therefore, our cities need to become smarter in order to deal with the risks that come with these evolutions. As a matter of fact, becoming smarter requires a lot of improvements to be made in the related sectors. In

B.-E. Soussi Niaimi (✉) · M. Bouhorma · H. Zili
Faculty of Sciences and Techniques of Tangier, Abdelmalek Essaâdi University, Tangier, Morocco
e-mail: bsoussiniaimi@uae.ac.ma

M. Ben Ahmed et al. (eds.), *Networking, Intelligent Systems and Security*,
Smart Innovation, Systems and Technologies 237,
https://doi.org/10.1007/978-981-16-3637-0_2

the hope of reducing the number of incidents and the time/money waste, also having a better monitoring of our cities' roads as well as implementing the best preventive measures to the infrastructure to have the optimal structure possible. Therefore, building features that allow us to control our infrastructure should be our number one priority to overcome the dangers we are facing every day on our roads. In other words, taking our road's management to the next level, using all that we have today; technologies, frameworks and sources of data that we can gather. Furthermore, exploiting the advantages of the traffic congestion prediction algorithms will save a lot of human lives as well as the time and money, to have a brighter and smarter future. However, at this moment, to precisely reroute the right amount of traffic can be developed in the future [2].

Regarding the high speed of the evolution in the transportation sector, the use of these algorithms became crucial to keep up with the impact that affects our cities, given that our cities are becoming bigger and more crowded than ever. Moreover, applying other concepts such as artificial intelligence (AI) and big data… seems to be an obligation to have an edge in the future, because the traffic jams are causing a huge time/money loss nowadays.

Moreover, in Morocco, there were more than 3700 deaths and over 130,00 injuries in 1 year (2017) caused by road accidents (89,375) [3], alongside with the occurrence of so many traffic jams over the years in the populated areas. As well as, during the special occasions (Sport events, holidays …), we cannot help noticing that the accidents' counter is increasing rapidly year after another with more than 10% between 2016 and 2017. As we know, many road accidents are caused by the traffic congestion, the road capacity and management, as well as, the excess speed of the vehicles while traveling and not respecting the traffic signs and the road's marks. We should concentrate our efforts to reduce these accidents, given that traffic prediction algorithms can prevent future congestion. As a result, we practically will have the ability to reduce the number of accidents and save lives, time and money. Furthermore, making traveling through the road is easier, safer and faster.

We will discuss in this paper the different approaches of traffic prediction approaches. Moreover, how to exploit those results using AI to make enhancements to the current roads and the new ones. Furthermore, we will shed light on some relevant projects in order to have an accurate overview of the utilities of these predictions in real life simulated situations. Also, we will answer some questions such as how can we predict short/long-term traffic dynamics using the real-time inputs? What are the required tools and algorithms in order to achieve the best traffic management?

The data have a huge impact on the output results, when it comes to transportation research, the old traffic models are not data driven. As a result, handling modern traffic data seems to be out of hand; in order to analyze modern traffic data from multiple sources to cover an enormous network, the data source could be retrieved from sensors or analyzing the driving behavior and extracting patterns basing on the trajectory data, as well as, transit schedule and airports, etc.

What do we need? A technology that allows us to teleport in order to save all our problems, but this is unlikely to happen, also we do not really need it, what we need is an improvement of what we have, a technology breakthrough to enhance our vehicles and our road's infrastructure. In the matter of facts, the existing road's element can be extremely powerful and efficient with a little adaptation by using mathematics, information technologies and all the available sources of information. Because nowadays, we are living the peak of communication evolution and AI breakout, regarding all the inventions happening in almost every sector, information became available with a huge mass, more that we can handle. Therefore, processing those data is the biggest challenge and extracting the desired information and making decisions is the ultimate goal to achieve the best travel experience, we do have all the requirements to move forward and go to the next step, as well as, making the biggest revolution in traffic management and road's infrastructure.

2 Method

In contemplation of pointing out the advantages and the weakness of each approach, we conducted a systematic literature review [4], as we know every existing approach has its own strengths as well as its limitations or weaknesses; in our detailed review, we will focus on the strength spots and the possibility to combine multiple approach in order to find a way to overcome the limitations that come with the existing approaches.

The first goal was to compare the existing approaches to come out with the best one of them, but after conducting a global review of every approach, we realized that every single approach is unique in its own way. As a result, the approaches cannot be compared because they handle a different aspect or area of expertise. Therefore, to achieve our goal, which is building the ultimate traffic congestion prediction mechanism, we should combine multiple approaches, but before that, we have to analyze the weaknesses and the strengths to choose what will work the best for us.

3 Data-Driven Approach

The data-driven approach is a relatively new approach, because it is a data-based approach. Therefore, we will shed the light in this part on the most common source of data used in this approach which are the weather, the traffic intensity, road sensors, GPS, social media, etc., because of the evolutions happening around the world regarding the communication tools, sharing data became easier and faster, as well as, available to everyone around the globe. Thanks to the smartphones and their new amazing features nowadays, it becomes easier to collect traffic's information basing on the publicly shared location and the integrated GPS along with mobile network that can give us a the approximated coordinate of the phone also the vehicle estimated location; as a result [5], we can also extract the traveling speed and the accuracy of the data, and the Wi-Fi also can be used to enhance the accuracy of the previous methods; in the final analysis, we can say the data are not an issue because of the various ways that we can use in order collect the desired specimen, but the quality and the accuracy of these data are crucial to have an accurate output or prediction; in our case, the first step to handle the incoming data is the storage problem because of the enormous size of the inputs. Thanks to the appearance of the new storage technologies with the capacity to handle huge amount of data in an efficient way (big data), also the evolutions happened to the computer capacities to process a huge amount of inputs in short duration; because of all the previous breakouts, it was time for data-driven approach to rise; using this approach, we are capable now of finding the link between the traffic's conditions and in incoming information, and we use this link in order to predict future traffic's jam and preventing it.

3.1 Real-Time Data Sources

As stated before, the main source of data is the vehicles GPS, road sensors, surveillance system and phone's location mostly combining GPS, mobile network and Wi-Fi to enhance the accuracy. As well as, historic data, there are other useful information that we could use, such as the taxis and bus station and trajectory that can be integrated to the collected data, to have a wide vision and more accurate results, the timing and quality of these data are crucial for the outcome of any data-driven approach; in order to locate every vehicle on the grid in real time, we mostly corporate all the previous sources of data to enhance the accuracy of the coordinates.

Many methods can be used such static observation using traffic surveillance to locate all vehicles using sensors and camera, but it requires a lot of data processing to end up with high-quality results. On the other hand, there is the route observation using GPS-enabled vehicles and phone's coordinate on the road, which allow us to get more information about the targeted vehicle such the current speed and exact position in real-time without the necessity to process a huge amount of data. The vehicles' position is not the only data required as input to the data-driven approach the weather reports (because the rain only can have a huge impact on the traffic [6]), the road's state, the special events, the current date and time (holidays, working days and hours) have a huge impact on the traffic flow and the vehicle travel's time.

3.2 Gravity Model

In order to focus on the mobility patterns, the gravity model was created inspired by Newton's law of gravitation [7], the gravity model is commonly used in public transportation management system [8], geography [9], social economics [10] and telecommunication [11], and the gravity model is defined with the following equation:

$$T_{i,J} = \frac{x_i^{\propto} x_j^{\beta}}{f(d_{i,j})},$$

Originally, $T_{i,j}$ is the bilateral trade volume between the two locations i and j; in our case, it is the volume of the people flow between the two given locations in direct proportion of the population size, and $f(d_{i,j})$ is the function of the distance $d_{i,j}$ between them [12, 13], and the model measures the travel costs between the two given locations in the urban area.

However, the model considers the travel between the source the destination which is the same for both directions. In reality, this is far from being accurate. Also, the model needs in advance to estimate some parameters based on the empirical data [14, 15].

4 Model-Driven Approach

This approach is mostly used for the long-term traffic prediction. Therefore, the goal is most likely changing the infrastructure, because we will be modeling months or years of future traffic regardless of the real-time information [16], and we can use the long-term predictions in order to simulate events such as conferences, concerts and football matches. Regarding the last technological evolutions that happened in the last few years, we are capable of analyzing the current traffic conditions and handling multiple factors and parameters in order to end up with an accurate prediction (hours in the future). It is mostly used for road evaluation that helps to determine the best signals to use, the speed limits, junctions, round bounds, lanes … it is mostly used throughout simulators to see the behaviors according to the giving model.

There are many simulators that use this model in order to give real time and future traffic information and conditions; we will shed the light on the following ones:

4.1 VISTA (Visual Interactive System for Transport Algorithms)

It is a very powerful framework that combines multiple transportations analysis tools, the framework unified the data interface in order to build a user interface that could be exploited using many programming languages, with the capacity to be ran over a network in many OS. The user interface to VISTA functions as a geographic information system (GIS) with zooming panning, with the possibility to execute queries to retrieve the desired data. Model can also be used to obtain traffic estimation and prediction in roads with partially loop detector coverage [17]. It was developed in 2000 by Ziliaskopoulos and Waller [18].

4.2 Dynamic Network Assignment for Management of Information to Travelers (DynaMIT)

The project is based on a dynamic traffic assignment (DTA) system 1 for estimation of network conditions, real-time traffic prediction and generation of drivers' guidance developed at MIT's Intelligent Transportation Systems Laboratory [18].

In order to work properly, the system needs real-time information and offline ones. The offline information is the representation of the network topology using a set of links, nodes and loading elements [19]. As well as the traveler's socioeconomic data such as gender, age, vehicle ownership, income and the reason for the trip that we can get using polls and questionnaires. For the real-time information, those are mostly the road sensors and cameras data, also the traffic control strategies and the incident properties such as the coordinates, the date, time, the expected results on the traffic flow and the road capacity. After the integration of the previous data to the system, the system will be capable of providing prediction, estimation and travel information [18], as well as flow speed, link density and driver characteristics (travel time, route choice and departure time) [20]

5 AI Application

Every city across the world is using or planning to use AI in order to build the optimal traffic management system, a system that is capable of solving complex issues, real time and future ones, as we can notice that most of the real-time approaches can solve most of the problems. However, in real life situations and in case of complex traffic jams, there will be some side-effects, and the same goes with the long-term traffic prediction because of the unpredictable changes that could cause anomalies that makes the accuracy of the model questionable. Therefore, we are far away from having a perfect traffic prediction model. The main goal of having an accurate prediction is to take conclusive and efficient decisions in any situation, because each model has its own flaws, as well as its strengths, there is where the role of artificial intelligence comes in. By harnessing the capability of the AI in hand, we will be able to give solutions in real-time. As well as, suggestions with the goal to improve the roads infrastructure to avoid future problems. As a result, achieving an advanced traffic management system that can also provide to the traveler some valuable information such as:

The best transportation mode to use.
The best route to take.
The predicted traffic jams.
The best parking spots and its availability.
The services provided along the roads.

This information can be also used to manipulate or even solve congestion and control the flow in order to have a better travel experience for everyone as well as avoiding traffic incidents and all the losses that come with it.

If we want an overview of the system, we should have three important levels. The first level is the data collection (raw data). The second level is the data analysis and interpretation; at this level, we could use any approach or even combine the model-driven and data-driven approach to have more accuracy and precision. And the last level is the decision making and traffic control basing on the previous level output, improving the final result depends on the quality of every level, from data collection and analysis of the decision making and the control actions.

In this part, we will focus on the decision making of traffic control actions management. Currently, the method used is the signal plan (SP) selection. The SP is selected after computing offline data in a library; all the SP of the network is a result of an optimization program executed on a predefined traffic flow structure. The selected SP is based on a set of values representing traffic flow situations. Therefore, there is no insurance that the selected set is appropriate for the actual situation. As a result, our choice will be based on the similarities [21], the selected SP is supposed to be the best match for the current traffic flow situation. By analyzing the output results of trying multiple combinations of signal plans in order to choose the most suitable one.

6 Discussion

The goal of our review is to build a new perspective on the existing approaches of the traffic congestion prediction. Instead of comparing the strengths and the weaknesses, we concluded that combining the strengths is the way to achieve the ultimate intelligent transport system. A system whose capable of predicting the future traffic congestions, as well as, the capability of proposing solution and handling real-time changes with high efficiency. Furthermore, AI usage is a crucial part to accomplish our goal, while it is challenging to harmonize the data-driven approach to work along with the model-driven approach. Because the foundation of every approach is completely different. Hence, the data-driven approach consists of analyzing data's masses in real-time to come out with predictions. On the other hand, the model-driven approach is mostly based on the historical data to propose changes for the current infrastructure. In order to reach our goal, we must use the model-driven approach to build an efficient infrastructure and then comes the data-driven approach to handle the variable factor of the road and the traffic flow. We will also need an AI capable of making decisions in any critical situation. By combining these approaches, we will be empowered by an efficient source of information, so we can notify the drivers of the road's state, the best route to take, the possible traffic jams and the existing ones, as well as, the optimal transportation mode and all the existing services provided along the roads.

7 Related Works

The cognitive traffic management system (CTMS): Basing on the Internet of Things, the development and deployment of smart traffic lights are possible. The e-government system proposes smart traffic lights as replacements for the current traditional ones, by taking advantages proposed by the cloud computing power and the unlimited scaling opportunity that we have nowadays, the build of the smart traffic lights is possible using the wireless technologies in order to interconnect different things without human interaction.

The real-time traffic management system (RTMS): In order to adapt with the rapid growth with the population of India, and the unbalance between the roads and the vehicles' number, the proposed solution is based on real-time monitoring composed of mobile units. These units work alongside with small networks of road-side units, the goal is to calculate dynamically the time for each traffic light, and the data have to be processed in real time in order to update dynamically each traffic light time accordingly.

8 Proof of Concept

The new system will cover all the bases and provide futuristic features with advanced capabilities. The first and the most important part is the long-term prediction; in order to construct the road model, this step allows us to detect the exact points and locations that required the intervention, which leads up to reduce the cost and the time required to set up the features. The new system contains much more features than any other existing or proposed ones; we will shed the light on them in the next paragraph. The new features allow us to control the traffic flow in the most efficient way, along with a very high accuracy. Furthermore, the cost of the installations will be at its minimum comparing to other projects thanks to our road's model that allows us to have a deeper understanding of the roads flaws and the possible enhancements.

The new system will include many imbedded elements that allow us to manage the road and shape-shift the infrastructure depending on the need and to perfectly adapt to any given situation. Also, the road smart signs and the road marks will be placed strategically to avoid any possible congestion, the preventive measure will allow us to meticulously be ready for any traffic flow and control it with high efficiency thanks to the integrated system, and every part of the system's units is connected to the same network to exchange information and roads state. Empowered by AI, the system will be capable of making critical decisions and solve complex road's situation with the optimal way possible.

9 Proposed Architecture

The first and the most important part is the data analysis to extract information about the future traffic hints. Using the model driving approach, we managed to move forward in time and predict the future traffic of the selected area. For the first trial, we moved 24 h in time in order to verify the accuracy of our model. Afterward, we moved 2 months, then 2 years. The congestion points were consistent in few roads, which makes the road selection for our smart features easier and more accurate in matter of result and traffic congestion wise.

The location of each smart sign is selected based on the road's model, the goal is to prevent the congestions and redirect the traffic flow when it is needed; each feature (smart traffic light, smart sign and smart road mark) is a stationary microcomputer connecter to the main server by a transceiver.

In order to solve traffic congestion problems, we should start by establishing a new road model, a new architecture for our road, giving the capability to our road to adapt itself to the traffic flow. We propose a smart road, a shape shifting road, in other words, a road that can change its properties depending on the situation, and the challenge is to make it efficient, economize the costs, because building a smart road occupied with a lot of smart features such as:

- **Smart signs**: Electrical signs that can be changed by analyzing the traffic flow, as well as, the short-term prediction in order to avoid any future congestions.
- **Smart traffic light**: The duration of the traffic light depends on the current traffic situation [22], but the real goal is to avoid any future traffic jam.
- **Smart road marking**: The road can change its lines in order to relieve the congestion on the congested direction; the road marking should be based on the historical data to be set on the road and the real-time data to get prediction thus handling the real-time changes.

Every part of our system is enhanced with AI capability in order to make decisions based on the current situation, according to the given predictions (Fig. 1).

Processing the historical data consists of analyzing the traffic road situation for the past years, and pointing out all the previous traffic jam situations, the causes (Sport events, holidays, schools, weather…). By processing all those data, we will be able to locate all the possible congestion points, which are the locations that are known for congestions with a constant frequency. In order to locate the congestion points, we used these historical data: Weather reports, emergency calls, police/public reports, incident reports, traffic sensors, traffic cameras, congestion reports, social media of interest, transit schedules and status.

After processing all previous sources of information in order to extract patterns, we were able to locate the congestion points shown below (Fig. 2).

In the previous figure, we can see the roads in red which are the most known for congestion. Therefore, they have the highest priority for the smart features' installation, but it is not the only criteria in order to pick the right, the impact of the new

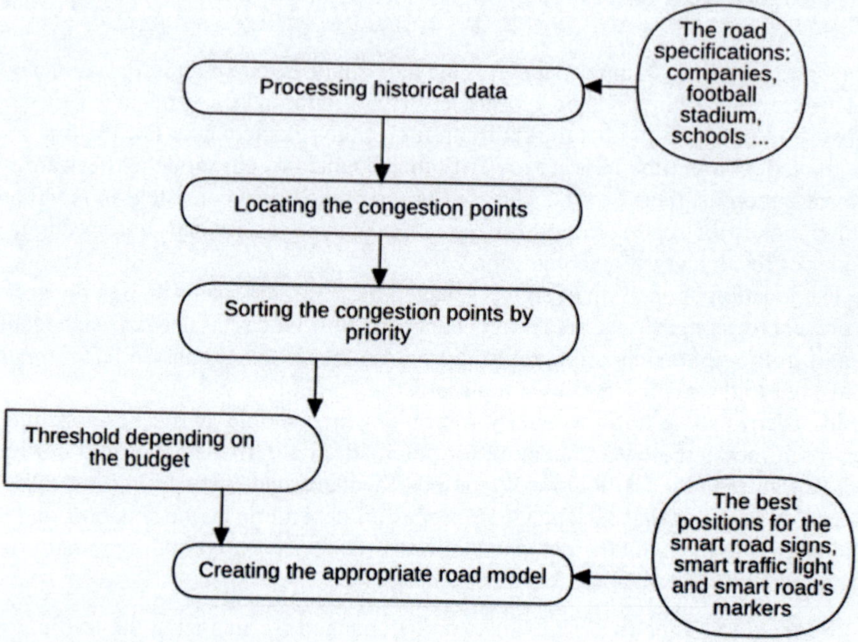

Fig. 1 Process to set up the road's model

Fig. 2 Example of historical data processing results for roads

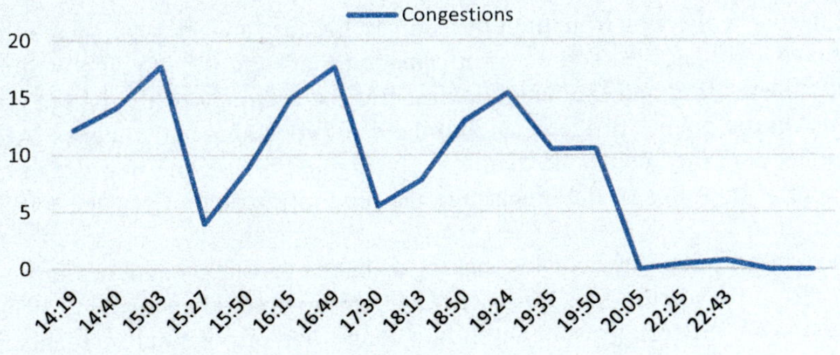

Fig. 3 Daily traffic congestion for the selected area

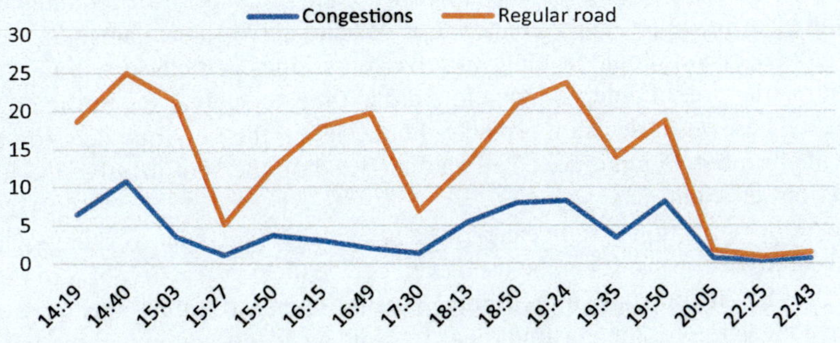

Fig. 4 Daily traffic congestions for the same area after setting up the smart features

installation on the other road should be considered as well, the cost of the installations and efficiency of the road's features regarding in possible event (Fig. 3).

After running a traffic road simulation on a road in order to observe the congestion variation during a typical day (without any special events), we can notice that during periods of time the congestion reached the pick. Those results are obtained before the addition of the smart features (Fig. 4).

The figure above shows us the same road congestions statistics during the same day, the blue line displays the congestion after the integration of the smart signs, traffic light and smart markers, we were able to reduce the traffic congestions by 50.66%, and the percentage can be much higher if we made the surrounding roads smart as well.

10 Conclusion and Future Research

In this paper, we presented different approaches of predictions. As well, we will be focusing on the output and the inputs of each method used by those approaches. Moreover, the impact of the input data in the result regardless of the approach used.

After the data collection from multiple sources, because nowadays, there are so many real times and historical source of information thanks to the ongoing evolution in the communication and the technologies. But the accuracy of the output depends directly on the quality of the input and the data processing and analysis. After the data collection, then comes the data's analyzing and processing in order to make a sense out of the mass of data regarding the road infrastructure and the traffic flow concerned.

The second step is the decision making using the prediction and the output information from the appropriate approach. Using those predictions, we will be able to take actions in order to prevent any future congestion or potential accidents, moreover, by doing the aftermath of each action and the consequences in the short and long term, we will have a clear path ahead. Because some changes should be made to the road itself, by changing the current infrastructure to make it better and smarter, in addition of having a better chance to avoid and solve future jams. As well as, some other decision should be made to solve congestions, with the constraints of the road infrastructure. In order to move forward in time, to analyze the traffic flow and choose the best set of decision possible. Furthermore, the real-time decision based on the real-time data input collected and analyzed on the spot to solve the instant congestions efficiently.

The traffic prediction system can be used in so many ways, such as changing roads infrastructure decisions. It can be used even before the construction of a road, with the goal of having an edge in the future. When it comes to traffic jams, as well as, solving congestion for the existing roads to avoid accidents and giving the travelers the best experience possible. Also, these approaches can be applied to reduce time of an ambulance to reach a certain destination with the most efficient way and the optimal time and save life as a result, or simply allowing for a regular traveler to travel safer, faster and more comfortable, and having the best traveling experience possible.

References

1. Bolshinsky, E., Freidman, R.: Traffic Flow Forecast Survey. Technion—Computer Science Department, Tech. Rep. (2012)
2. Matthews, S.E.: How Google Tracks Traffic. Connectivist (2013)
3. Ministry of Equipment, Transport, Logistics and Water (Roads Management) of Morocco (2017)
4. vom Brocke, J., Simons, A., Riemer, K., Niehaves, B., Plattfaut, R., Cleven, A.: Standing on the Shoulders of Giants: Challenges and Recommendations of Literature Search in Information Systems Research (2015)
5. Barbosa, H., Barthelemy, M., Ghoshal, G., James, C.R., Lenormand, M., Louail, T., Menezes, R., Ramasco, J.J., Simini, F., Tomasini, M.: Human mobility: models and applications. Phys. Rep. **734**, 1–74 (2018)
6. Saberi, K.M., Bertini, R.L.: Empirical Analysis of the Effects of Rain on Measured Freeway Traffic Parameters. Portland State University, Department of Civil and Environmental Engineering, Portland (2009)

7. Zipf, G.K.: The p1p2/d hypothesis: on the intercity movement of persons. Am. Sociol. Rev. **11**(6), 677–686 (1946)
8. Jung, W.S.: Gravity model in the korean highway. **81**(4), 48005 (2008)
9. Feynman, R.: The Brownian movement. Feynman Lect. Phys. **1**, 41–51 (1964)
10. Matyas, L.: Proper econometric specification of the gravity model. World Econ. **20**(3), 363–368 (1997)
11. Kong, X., Xu, Z., Shen, G., Wang, J., Yang, Q., Zhang, B.: Urban traffic congestion estimation and pre- diction based on floating car trajectory data. Futur. Gener. Comput. Syst. **61**, 97–107 (2016)
12. Anderson, J.E.: The gravity model. Nber Work. Papers **19**(3), 979–981 (2011)
13. Barth´elemy, M.: Spatial networks. Phys. Rep. **499**(1), 1–101 (2011)
14. Lenormand, M., Bassolas, A., Ramasco, J.J.: Sys- tematic comparison of trip distribution laws and mod- els. J. Transp. Geogr. **51**, 158–169 (2016)
15. Simini, F., Gonz´alez, M.C., Maritan, A., Baraba´si, A.L.: A universal model for mobility and migration patterns. Nature **484**(7392), 96–100 (2012)
16. INRIX.: Who We Are. INRIX Inc. (2014)
17. Lopes, J.: Traffic prediction for unplanned events on highways (2011)
18. Ziliaskopoulos, A.K., Waller, S.: An Internet-based geographic information system that integrates data, models and users for transportation applications. Transp. Res. Part C: Emerg. Technol. **8**(1–6), 427–444 (2000)
19. Ben-akiva, M., Bierlaire, M., Koutsopoulos, H., Mishalani, R.: DynaMIT: a simulation-based system for traffic prediction. DACCORD Short Term Forecasting Workshop, pp. 1–12 (1998)
20. Milkovits, M., Huang, E., Antoniou, C., Ben-Akiva, M., Lopes, J.A.: DynaMIT 2.0: the next generation real-time dynamic traffic assignment system. In: 2010 Second International Conference on Advances in System Simulation, pp. 45–51 (2010)
21. Li, Q., Zheng, Y., Xie, X., Chen, Y., Liu, W., Ma, W.Y.: Mining user similarity based on location history. In: ACM Sigspatial International Conference on Advances in Geographic Information Systems, page 34. ACM (2008)
22. Wheatley, M.: Big Data Traffic Jam: Smarter Lights, Happy Drivers. Silicon ANGLE (2013)

Tomato Plant Disease Detection and Classification Using Convolutional Neural Network Architectures Technologies

Djalal Rafik Hammou and Mechab Boubaker

Abstract Agriculture is efficient from an economic and industrial point of view. The majority of countries are trying to be self-sufficient to be able to feed their people. But unfortunately, several states are suffering enormously and are unable to join the standing up to satisfy their populations in sufficient quantities. Despite technological advances in scientific research and advances in genetics to improve the quality and quantity of agricultural products, today we find people who die of death. In addition to famines caused by wars and ethnic conflicts and above all plant diseases that can devastate entire crops and have harmful consequences for agricultural production. With the advancement of artificial intelligence and vision from computers, solutions have brought to many problems. Smartphone applications based on deep learning using convolutionary neural network for deep learning can detect and classify plant diseases according to their types. Thanks to these processes, many farmers have solved their harvesting problems (plant diseases) and considerably improved their yield and the quality of the harvest. In our article, we propose to study the plant disease (tomato) using the PlantVillage [1] database with 18,162 images for 9 diseased classes and one seine class. The use of CNN architectures DenseNet169 [2] and InceptionV3 [3] made it possible to detect and classify the various diseases of the tomato plant. We used transfer learning technology with a batch-size of 32 as well as the RMSprop and Adam optimizers. We, therefore, opted for a range of 80% for learning and 20% for the test with a period number of 100. We evaluated our results based on five criteria (number of parameters, top accuracy, accuracy, top loss, score) with an accuracy of 100%.

D. R. Hammou (✉)
Faculty of Exact Science, EEDIS, Department of Computer Sciences, Djillali Liabes University, BP 89, 22000 Sidi Bel Abbes, Algeria
e-mail: r_hammou@esi.dz

M. Boubaker
Faculty of Exact Science, LSPS, Department of Probability and Statistics, Djillali Liabes University, BP 89, 22000 Sidi Bel Abbes, Algeria

M. Ben Ahmed et al. (eds.), *Networking, Intelligent Systems and Security*,
Smart Innovation, Systems and Technologies 237,
https://doi.org/10.1007/978-981-16-3637-0_3

1 Introduction

Plants represent an enormous economic stake in the world of industrialized agriculture. The development of agriculture has enabled many countries to combat famine and eradication from the face of the earth. Many necessitous states are suffering, and their people have not eaten anything. Despite technological progress and advanced scientific research that had increased production and improved yields, there are still people dying of hunger. Agriculture is a large and rich field. Since the dawn of time, human beings have produced their food by cultivating the land. Agriculture offers a very varied diversification for human food such as cereals (wheat, rice, corn, barley, starch, etc.), fruits (banana, apple, strawberry, kiwi, pear, etc.), and vegetables (potatoes, tomatoes, carrot, zucchini, onion, etc.).

Plants are the staple of our diet and represent a large field of research. They are living organisms and consist of complex plant cells. They are part of the eukaryotic culture. There are several specialties in the field of plants (medicinal plants, botanical plants, etc.). The plant classification depends on different criteria such as climate, temperature, size, type of stem, and geographical area. We can find plants of the polar regions, high mountains, tropical, cold, and hot (classification according to climate). We cannot determine the exact number of plant varieties, but a scientific study in 2015 determined that more than 400,000 plant species existed [4]. The weak point of plants is diseases that can kill them or demolish an entire crop. Among the efficient diseases that attack vegetables and fruit trees: early blight, anthracnose, blight, mildew, moniliasis, Mosaic Virus, blossom end necrosis, phytophthora, rust, and virosis.

In our article, we are interested in tomato plants (vegetables). The name of tomato comes from the word Solanum Lycopersicum [5], and it is a vegetable in the classification of the agricultural world. It is part of the family Solanaceae that originally comes from South America in the northwest (Peru, Ecuador). It was cultivated for the first time in Mexico. It is famous all over the world, and it has become a staple in our daily life. It is a specific plant cultivated in the Mediterranean countries such as Algeria, Morocco, Tunisia, Spain, Italy, etc.

Algeria is among the most efficient countries in the production and export of tomatoes. During the 2017–2018 agricultural season (see Table 1), the annual production reached 2.91 million tons, 1.37 million tons for household consumption, and 1.54 million tons for industrial processing (see Fig. 1) [6].

The structure of our article is represented according to the following plant. Section 1 gives a general introduction to the importance of agriculture and the cultivation of plants. Section 2 describes a literature review on machine learning, and deep learning techniques in the detection and classification of plant diseases. Section 3 is devoted to the choices of CNN architectures. Section 4 presents the strategy to follow for the deployment of CNN deep learning architecture. Section 5 gives a general idea of the hardware and software tools to be used for the experiments. Section 6 describes the results obtained from the experiments on the PlantVillage [1] database. Finally, the last section describes the conclusion and future research perspectives.

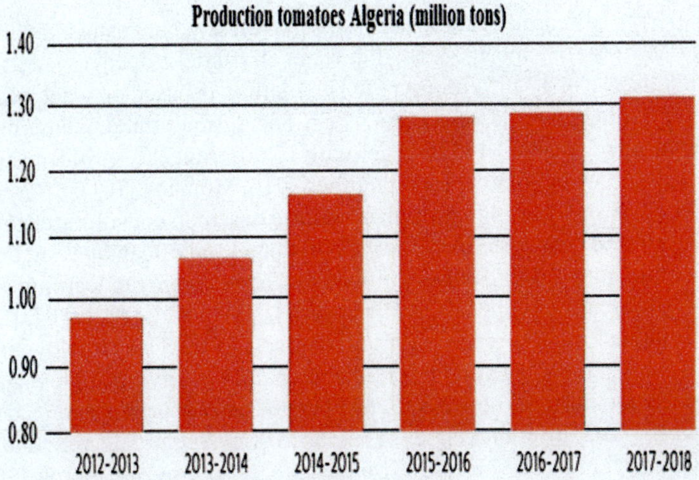

Fig. 1 Tomato production in Algeria [6]

2 Related Work

In 2012, Hanssen et al. [5] described the tomato plant in detail from its origin to its implementation in the Mediterranean region. They explain the different diseases that tomato production can affect and the solutions adopted to deal with this kind of disease. In December 2013, Akhtar et al. [7] implemented a three-part method. First, the segmentation to locate the diseased region of the plant. Then it extracts the segmented region image to be able to code the features. Then these characteristics are classified according to the type of disease. They obtained an accuracy of 94.45% by comparing with the techniques of state of the art (K-nearest neighbor (KNN), Naïve Bayes classifier, support vector machine (SVM), decision tree classifier (DTC), recurrent neural networks (RNN)). In December 2015, Kawasaki et al. [8] proposed an innovative method based on convolutional neural networks (CNN) with a custom architecture. The experiments were performed on a cucumber image database with a total of 800 images. They used a cross-validation (fourfold) strategy by classifying the plants

Table 1 The 2017–2018 Algerian annual yield of tomatoes in the different wilayas of the country concerning industrial production and household consumption [6]

Wilaya	Household consumption (tons)	Wilaya	Industrila processing (tons)
Biskra	233,000	Skikda	465,000
Mostaganem	133,000	ElTarf	350,000
Tipaza	106,000	Guelma	206,000
Ain Defla	73,000	Ain Defla	168,000

into two classes (diseased cucumber class, seine class). The results gave an average accuracy of 94.9%. In June 2016, Sladojevic et al. [9] developed a system for identifying plant disease of 13 different types. The method is based on a deep convolution network with the help of a Caffe framework. The agricultural database used for the experiments, which contain 4483 images with 15 different classes of fruit. The results reached an accuracy of 96.30%. In September 2016, Mohanty [10] proposed a system for classification and recognition of plant disease based on convolutional neuron networks. They tested their system on a corpus of images (54,306 images) with two types of CNN architectures: AlexNet and GoogleNet. They employed learning and testing strategy of different rates ([80–20%], [60–40%], [50–50%], [40–60%], [20–80%]), and they obtained a good result with an accuracy of 99.34%. In November 2016, Nachtigall et al. [11] used a system for detecting and classifying apple plant diseases using convolutional neural networks. They carried out experiments on a database of 1450 images with 5 different classes. They used AlexNet architecture and achieved 97.30% accuracy. In December 2017, Lu et al. [12] proposed an approach to solving the problem of plant pathology of plant disease (rice stalk leaf). The CNN architecture used in the experiments is AlexNet. They used a database of 500 rice stem plant images with 10 disease classes. Finally, they were able to obtain an accuracy of 95.48%. In July 2017, Wang et al. [13] submitted an idea of detecting diseases in apple plants using deep learning technology. They proceeded to use the following CNN architectures: VGG16, VGG19, InceptionV3, ResNet50. The operation of the experiments is done with a rate of 80% for learning and 20% for testing. They used the technology of transfer learning. The PlantVillage database was used, with an image count of 2086 for 4 classes of apple plant disease. The best result was obtained with the VGG16 architecture for an accuracy of 90.40%. In 2018, Rangarajan et al. [14] proposed a system to improve the property and quantity of tomato production by trying to detect plant diseases. The system involves using deep and convolutional neural networks. They experimented with 6 classes of diseased tomatoes and a seine from the PlanteVillage database (number of images is 13 262). The CNN architectures deployed for the tests are AlexNet and VGG16, with a result of 97.49% accuracy. In September 2018, Khandelwal et al. [15] implemented an approach for classification and visual inspection of the identification of plant diseases in general. They used a large database (PlanteVillage, which contains 86 198 images) of 57 classes from 25 different cultures with diseased plants and seines. The approach is based on deep learning technology using CNN architectures (InceptionV3 and ResNet50). They used transfer learning with different rates for learning and testing ([80–20%], [60–40%], [40–60%], [20–80%]) as well as a batch-size of 25 and 25 epochs. They reached an accuracy of 99.374%. In February 2020, Maeda-Gutiérrez [16] proposed a method, which consists of using 5 CNN deep learning architectures (AlexNet, GoogleNet, InceptionV3, ResNet18, ResNet34) for the classification of tomato plant disease. They carried out a learning rate of 80 and 20% for the test. They also used the learning transfer with the following hyper-parameters: batch-size of 32, 30 epochs. They used the PlantVillage database (tomato plant with 9 different disease classes and one seine class) with 18 160 images. The results are evaluated based on five criteria (accuracy, precision, sensitivity, specificity, F-score) with an accuracy of 99.72%.

Aproch proposed: our approach is based on the following points with our modest contribution from our article:

- First, we will study the methods used in the literature of machine learning and deep learning for the detection and classification of plant diseases.
- We will use specific and particular convolutional neural network architectures for this type of problem.
- Next, we will test our approach on a corpus of images.
- We will evaluate our results obtained according to adequate parameters (accuracy, number of parameters, top accuracy, top loss, score).
- We will establish a comparative table of our approach with those of state of the art.
- Finally, we will end with a conclusion and research perspectives.

3 CNN Architecture

We have chosen two CNN architectures for our approach:

3.1 DenseNet169

Huang et al. [2] invented the DenseNet architecture based on convolutional neural networks. The specificity of this architecture is that each layer is connected directly to all the other layers. The DenseNet contains L (L + 1) 2 direct connection and is an enhanced version of the ResNet [3] network. The difference between the two is that the DenseNet architecture contains fewer parameters, and it computes faster than the ResNet architecture. The DenseNet network has certain advantages: such as the principle of reuse of features and alleviates the gradient problem. The DenseNet architecture has been evaluated in benchmark object recognition competitions (CIFAR100, ImageNet, SVHN, CIFAR-10) and achieved significant results with other architectures in the bibliographic literature. Among the variants of this architecture is DenseNet169. It has a depth of 169 layers and an input image of 224 × 224 pixels.

3.2 InceptionV3

Over the years, InceptionV3 architecture has emerged as a result of several researchers. It is built on the basis in the article by Szegedy et al. [17] in 2015. They designed the inception module (network complex). Its design depends on the depth and width of the network. For InceptionV3 to emerge, it was necessary to go through InceptionV1 and InceptionV2. Inception V1 (GoogleNet) was developed in the ImageNet visual recognition competition (ILSVRC14) [18]. GoogleNet is a deep 22-layer network, and it uses size convolution filters (1×1, 3×3, and 5×5) as input. The trick was to use a 1×1 size filter before 3×3 and 5×5 because 1×1 convolutions are much less expensive (computation time) than 5×5 convolutions. InceptionV2 and InceptionV3 created by Szegedy and Vanhoucke [19] in 2016. InceptionV2 has the particularity of factoring the 5×5 convolution product into two 3×3 convolution products. It has a significant impact on the computation time. This improvement is important (a 5 × 5 convolution is more costly than a 3 × 3 convolution). InceptionV3 used the InceptionV2 architecture with upgrades in addition to the RMSProp optimizer, 7×7 convolution factorization, and BatchNorm in the auxiliary classifier. InceptionV3 is a deep 48-layer architecture with an input image of 299 × 299 pixels.

4 Deployment Strategy for the Deep Learning Architecture

We have adopted a strategy for training and testing CNN architectures. The goal is to optimize the neural network and avoid the problem of over fighting by using mathematical methods with the following criteria:

4.1 Data-Collection

The data collection consists of preparing the dataset for the neural network. Consider the technical characteristics of the CNN architecture. The database must be large enough for CNN to function correctly. The size of the input image must be compatible with the input of the neural network. The image database used in our experiments is PlantVillage [1]. It contains over 18,000 plant images and is the best dataset.

4.2 Transfert Learning

Learning transfer is an optimized mathematical method of machine learning. It is a method that consists of using pre-trained weights from the ImageNet [20] database (it is a database that contains over 1.2 million images with over 1000 different object classes) in another CNN architecture to solve a well-defined problem. The interest of the process is in using the information and knowledge gained from ImageNet to feed it back into another network to solve another classification problem.

4.3 Data Augmentation

It is a mathematical technique that increases the dimension of the image database. The process consists of using mathematical operations (such as rotations (rotate the image 90°, 180°, 270°), translations (scaling), changing the image size, decreasing the clarity of the picture, blur the image with different degrees, change the color and effect of the image, geometric transformation of the picture, flip the picture horizontally or vertically, etc.).

4.4 Fine-Tuning

Fine-tuning is a technique for optimizing convolutional neural networks. The principle is to modify the last layer of the neural network or to add an intermediate layer before the output. The goal is to adapt the new classification problem so that it can have good accuracy.

5 Hardware and Software Tool

To test our approach on CNN architectures, we used the material which is described in Table 2.

6 Result and Discussion

Dataset:

PlantVillage [1] is a plant image database that contains pictures of healthy and diseased plants. It is dedicated to agriculture so that they can get an idea of the type of

Table 2 Hardware and software characteristics

Hard and soft	Technical characteristics
Processor (CPU)	Intel (R) Xeon (R) @ 2.20 GHz.
Graphic card (GPU)	GeForce GTX 1080 X, 8Gb
Memory (RAM)	25 GB
Operating system	Windows 8, 64 bits
Programming language	Pyhton 3.6
Architecture	keras 2.3

Fig. 2 Pictures of tomato plants from the PlantVillage database [1]

Table 3 Characteristic of hyper-parameters

Hyper-parameters	Values
Batch-size	32
Epochs	100
Optimizer	Adam
Learning rate	0.0001
Beta 1	0.9
Beta 2	0.999
Epsilon	1e−08
Number class	10

disease in the plant. It contains 54,309 images of 14 different fruit and vegetable plants (Strawberry, Tomato, Soybean, Potato, Peach, Apple, Squash, Blueberry, Raspberry, Pepper, Orange, Corn, Grape, Cherry). The database contains images of 26 diseases (4 bacterial, 2 virals, 17 fungal, 1 mite, and 2 molds (oomycetes)). There are also 12 species of seine plant images, making a total of 38 classes. A digital camera type (Sony DSC - Rx100/13 20.2 megapixels) was used to take the photos from the database at Land Grant University in the USA.

In our article, we are interested in the tomato plant (healthy and sick). The PlantVillage [1] database contains 10 classes of tomato plants (see Fig. 2) with a total of 18,162 images. The different classes of tomatoes are *Alternaria solani*, *Septoria lycopersici*, *Corynespora cassiicola*, *Fulvia fulva*, *Xanthomonas campestris* pv. Vesicatoria, *Phytophthora infestans*, Tomato Yello Leaf Curl Virus, Tomato Mosaic Virus, *Tetranychus urticae*, healthy (see Table 4).

Concerning the hyper-parameters, they are described in Table 3.

Regarding the dataset partitioning method, we used a rate of 80% for training and 20% for evaluation using the cross-validation method. The strategy for dividing the dataset is to separate the dataset into three parts (training, validation, test). Since the

Table 4 The different characteristics of tomato plant diseases (PlantVillage database) [1]

Name classe	Nb images	Fungi	Bacteria	Mold	Virus	Mite	Healthy
Tomato bacterial spot	2127	–	Xanthomonas campestris pv. Vesicatoria	–	–	–	–
Tomato early blight	1000	*Alternaria solani*	–	–	–	–	–
Tomato healthy	1592	–	–	–	–	–	Healthy
Tomato late blight	1910	–	–	*Phytophthora infestans*	–	–	–
Tomato leaf mold	952	*Fulvia fulva*	–	–	–	–	–
Tomato septoria leaf spot	1771	*Septoria lycopersici*	–	–	–	–	–
Tomato spider mites	1676	–	–	–	–	*Tetranychus urticae*	–
Tomato target spot	1404	*Corynespora cassiicola*	–	–	–	–	–
Tomato mosaic virus	373	–	–	–	Tomato mosaic virus	–	–
Tomato yellow leaf curl virus	5357	–	–	–	Tomato yello leaf curl virus	–	–
Total	18,162	–	–	–	–	–	–

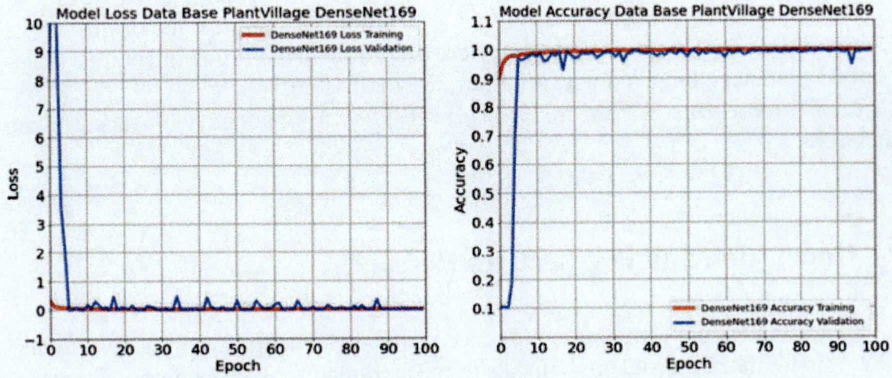

Fig. 3 Result of the experiments with the DenseNet169 architecture for loss and accuracy

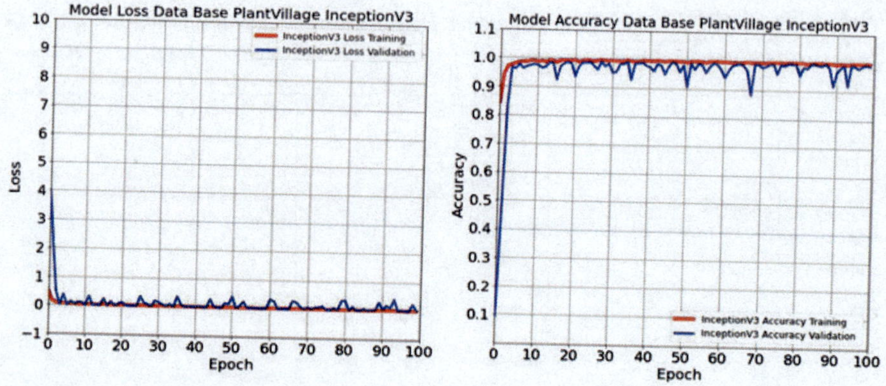

Fig. 4 Result of the experiments with the InceptionV3 architecture for loss and accuracy

Table 5 Comparison table of the results of the experiments the tomatoes database of PlantVillage of the different CNN architecture

Architecture	Parameters	Top accuracy (%)	Accuracy (%)	Top loss	Score
DenseNet169	12,659,530	99.80	100	1.2665 e−07	0.0178
InceptionV3	21,823,274	99.68	100	3.5565 e−05	0.0002

database contains 18 162 images of plants, we took 11,627 for training, 2903 for validation, and 3632 for testing.

The experiments on the tomato plant image database (PlantVillage) gave good results. We were able to obtain an accuracy of 100% with the DenseNet169 architecture (see Fig. 3) and the same thing with the InceptionV3 architecture (see Fig. 4).

Table 5 reflects the comparison the evaluation of the results of the CNN DenseNet169 and InceptionV3 architecture according to the following points: Number of parameters, top accuracy, accuracy, top loss, score.

Table 6 represents a comparison of the results we obtained with those of the literature.

7 Conclusion and Perspectives Research

Computer vision and artificial intelligence technology (deep learning) have solved a lot of plant disease problems. Thanks to the architectures of convolutional neuron networks (CNN), detection, and classification have become accessible. Farmers who use deep learning applications on SmartPhone in remote areas can now detect the type of plant disease and provide solutions that can improve the productivity of their crops. The results of our experiments on tomato plant disease reached an accuracy of 100%. We plan to use autoencoder architecture (such as U-Net) for visualization

Table 6 Comparison chart between different methods

Author	Year	Nbr of class	Nbr of images	CNN architecture	Accuracy (%)
Kawasaki et al. [8]	2015	3	800	Customized	94.90
Sladojevic et al. [9]	2016	15	4483	CaffeNet	96.30
Mohanty et al. [10]	2016	38	54,306	AlexNet GooglexNet	99.34
Nachtigall et al. [11]	2016	5	1450	AlexNet	97.30
Lu et al. [12]	2017	10	500	AlexNet	95.48
Wang et al. [13]	2017	4	2086	VGG16 VGG19 InceptionV3 ResNet50	90.40
Rangarajan et al. [14]	2018	7	13,262	AlexNet VGG16	97.49
Khandelwal et al. [15]	2018	57	86,198	InceptionV3 ResNet50	99.37
Maeda-Gutiérrez [16]	2020	10	18,160	AlexNet GoogleNet InceptionV3 ResNet18 ResNet34	99.72
Our aproch.	2020	10	18,162	DenseNet169 InceptionV3	100

of plant leaves, which can improve the detection and segmentation of the diseased region and facilitate classification work.

Acknowledgements I sincerely thank Doctor Mechab Boubaker from the University of Djillali Liabes of Sidi Bel Abbes for encouraging and supporting me throughout this work and also for supporting me in hard times because it is thanks to him that I was able to do this work.

References

1. Hughes, D., Salathé, M.: An open access repository of images on plant health to enable the development of mobile disease diagnostics through machine learning and crowdsourcing. arXiv preprint arXiv:1511.08060 (2015): n. pag
2. Huang, G., Liu, Z., K. Q. Weinberger, and L. van der Maaten, "Densely connected convolutional networks," in 2016 IEEE Conference on Computer Vision and Pattern Recognition (CVPR), Jun 2017, pp. 4700–4708
3. He, K., Zhang, X., Ren, S., Sun, J.: Deep residual learning for image recognition. In: Proceedings of the IEEE Conference on Computer Vision and Pattern Recognition, Las Vegas, NV, USA, 26 June–1 July 2016, pp. 770–778 (2016)

4. Bachman, S.: State of the World's Plants Report. Royal Botanic Gardens, Kew, p. 7/84 (2016) (ISBN 978-1-84246-628-5)
5. Hanssen, I.M., Lapidot, M.: Major tomato viruses in the Mediterranean basin. In: Loebenstein, G., Lecoq, H. (eds.) Advances in Virus Research, vol. 84, pp. 31–66. Academic Press, San Diego (2012)
6. Market developments in Fruit and Vegetables Algeria [https://meys.eu/media/1327/market-developments-in-fruit-and-vegetables-algeria.pdf], MEYS Emerging Markets Research
7. Akhtar, A., Khanum, A., Khan, S.A., Shaukat, A.: Automated plant disease analysis (APDA): performance comparison of machine learning techniques. In: Proceedings of the 11th International Conference on Frontiers of Information Technology, pp. 60–65 (2013)
8. Kawasaki, Y., Uga, H., Kagiwada, S., Iyatomi, H.: Basic study of automated diagnosis of viral plant diseases using convolutional neural networks. In: Advances in Visual Computing: 11th International Symposium, ISVC 2015, Las Vegas, NV, USA, December 14–16, 2015. Proceedings, Part II, 638–645 (2015)
9. Sladojevic, S., Arsenovic, M., Anderla, A., Culibrk, D., Stefanovic, D.: Deep neural networks based recognition of plant diseases by leaf image classification. Comput. Intell, Neurosci (2016)
10. Mohanty, S.P., Hughes, D.P., Salathé, M.: Using deep learning for image-based plant disease detection. Front. Plant Sci. **7**, 1419 (2016)
11. Nachtigall, L.G., Araujo, R.M., Nachtigall, G.R.: Classification of apple tree disorders using convolutional neural networks. In: Proceedings of the 2016 IEEE 28th International Conference on Tools with Artificial Intelligence (ICTAI), pp. 472–476. San Jose, CA 6–8 November 2016
12. Lu, Y., Yi, S., Zeng, N., Liu, Y., Zhang, Y.: Identification of rice diseases using deep convolutional neural networks. Neurocomputing **267**, 378–384 (2017)
13. Wang, G., Sun, Y., Wang, J.: Automatic image-based plant disease severity estimation using deep learning. Comput. Intell. Neurosci. 2917536 (2017)
14. Rangarajan, A.K., Purushothaman, R., Ramesh, A.: Tomato crop disease classification using pre-trained deep learning algorithm. Procedia Comput. Sci. **133**, 1040–1047 (2018)
15. Khandelwal, I., Raman, S.: Analysis of transfer and residual learning for detecting plant diseases using images of leaves. Computational Intelligence: Theories. Applications and Future Directions-Volume II, pp. 295–306. Springer, Singapore (2019)
16. Maeda-Gutiérrez, V., Galván-Tejada, C.E., Zanella-Calzada, L.A., Celaya-Padilla, J.M., Galván-Tejada, J.I., Gamboa-Rosales, H., Luna-García, H., Magallanes-Quintanar, R., Guerrero Méndez, C.A., Olvera-Olvera, C.A.: Comparison of convolutional neural network architectures for classification of tomato plant diseases. Appl. Sci. **10**, 1245 (2020)
17. Szegedy, C., Liu, W., Jia, Y., Sermanet, P., Reed, S., Anguelov, D., Erhan, D., Vanhoucke, V., Rabinovich, A.: Going deeper with convolutions. In: 2015 IEEE Conference on Computer Vision and Pattern Recognition (CVPR), pp. 1–9 (2015)
18. Simonyan, K., Zisserman, A.: Very deep convolutional networks for large-scale image recognition. CoRR, vol. abs/1409.1556 (2014)
19. Szegedy, C., Vanhoucke, V., Ioffe, S., Shlens, J., Wojna, Z.: Rethinking the inception architecture for computer vision. In: 2016 IEEE Conference on Computer Vision and Pattern Recognition (CVPR), pp. 2818–2826 (2016)
20. Krizhevsky, A., Sutskever, I., Hinton, G.E.: ImageNet classification with deep convolutional neural networks. CACM (2017)

Generative and Autoencoder Models for Large-Scale Mutivariate Unsupervised Anomaly Detection

Nabila Ounasser, Maryem Rhanoui, Mounia Mikram, and Bouchra El Asri

Abstract Anomaly detection is a major problem that has been well studied in various fields of research and fields of application. In this paper, we present several methods that can be built on existing deep learning solutions for unsupervised anomaly detection, so that outliers can be separated from normal data in an efficient manner. We focus on approaches that use generative adversarial networks (GAN) and autoencoders for anomaly detection. By using these deep anomaly detection techniques, we can overcome the problem that we need to have a large-scale anomaly data in the learning phase of a detection system. So, we compared various methods of machine based and deep learning anomaly detection with its application in various fields. This article used seven available datasets. We report the results on anomaly detection datasets, using performance metrics, and discuss their performance on finding clustered and low density anomalies.

1 Introduction

Anomaly detection is an important and classic topic of artificial intelligence that has been used in a wide range of applications. It consists of determining normal and abnormal values when the datasets converge to one-class (normal) due to insufficient sample size of the other class (abnormal). Models are typically based on large amounts of labeled data to automate detection. Insufficient labeled data and high labeling effort limit the power of these approaches.

N. Ounasser (✉) · M. Rhanoui · B. E. Asri
IMS Team, ADMIR Laboratory, Rabat IT Center ENSIAS, Mohammed V University in Rabat, Rabat, Morocco

M. Rhanoui · M. Mikram
Meridian Team, LYRICA Laboratory, School of Information Sciences, Rabat, Morocco

M. Mikram
Faculty of Sciences, LRIT Laboratory, Rabat IT Center, Mohammed V University in Rabat, Rabat, Morocco

M. Ben Ahmed et al. (eds.), *Networking, Intelligent Systems and Security*, Smart Innovation, Systems and Technologies 237,
https://doi.org/10.1007/978-981-16-3637-0_4

While it is a problem widely studied in various communities including data mining, machine learning, computer vision, and statistics, there are still some challenges that require advanced approaches. In recent years, deep learning enabled anomaly detection has emerged as a critical direction toward addressing these challenges.

Generative models [12] are used in various domains such as Person Identification [10], Image Synthesis [8], Image Generation (WGAN) [1], Face Aging [4, 17], etc.

Autoencoders are a very interesting group of neural network architectures with many applications in computer vision, natural language processing, and other fields. Applications of autoencoders include also compression, recommender systems, and anomaly detection.

We aim to provide a comparative study of the research on deep anomaly detection. We have grouped existing techniques into different categories based on the underlying approach adopted by each technique. For each category, we have identified key assumptions, which are used by the techniques to separate between normal and anomalous behavior. When we apply a technique to a particular domain, these assumptions can be used as guidelines to assess the effectiveness of the technique in that domain. Further, we identify the advantages and disadvantages of the techniques. Then, we report results, using performance metrics, and we discuss their performance to decide who is the most efficient in anomalies detection.

2 Background and Context: Anomaly Detection

An anomaly is an observation which deviates so much from other observations as to arouse suspicion, it was generated by a different mechanism [13]. Anomalies are also referred to as abnormalities, deviants, or outliers in the data mining and statistics literature. Anomaly detection is the problem of determining if a given point lies in a low-density region [20].

Anomalies or outliers are extreme values that differ from other observations on data. They may be due to variability in a measure, experimental errors or novelty.

Thus, the detection of anomalies highlights the importance of the diversity of the fields it covers and the advantageous results it brings. In particular, anomaly detection is used to identify fraudulent banking transactions. In this context, companies in the banking sector, for example, try to identify abnormal customer behavior, detect fake cards, etc. In addition, anomaly detection also applies in the detection of network intrusions. In fact, cyber attacks are currently on the rise. These attacks mainly target information theft and system malfunctions. Generally, the detection of these attacks can be achieved through the control and monitoring of atypical behaviors of all information system entities. In addition, web-based anomaly detection applications are used to detect malicious and malicious users, including spammers and scammers who publish false news, such as false recommendations in e-commerce sites such as Amazon, Wish, etc.

3 Anomaly Detection Techniques

This section illustrates the different anomaly detection techniques (supervised, unsupervised, semi-supervised), with a focus on unsupervised detection. This approach is the most flexible method that does not require any labeled data. Usually supervised models allow labeled data, which is not always available, hence the use of unsupervised models (Fig. 1).

3.1 Unsupervised Machine Learning for Anomaly Detection

Anomalies detection is the process of identifying outliers. It is based on the assumption that the behavior of the intruder that generates an anomaly is significantly different from normal or legitimate behavior.

Unsupervised anomaly detection includes several approaches, and we can categorize these approaches as:

Linear Models A linear model is specified as a linear combination of features. Based on the training data, the learning process calculates a weight for each entity to train a model that can predict or estimate the target value. We will explain two algorithms that are part of this approach; PCA and OC-SVM.

Kernel PCA [21] is an anomaly detection method based on kernel PCA [25] and reconstruction error. The method consists of assigning scores to the points based on their projection in the space generated by the core PCA. The greater the reconstruction error of these projections, the more likely the points are to be anomalies.

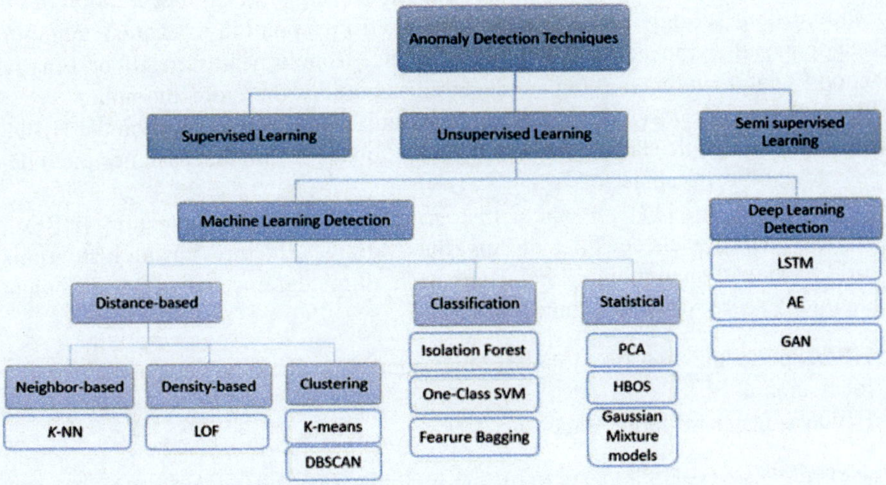

Fig. 1 Anomaly detection techniques

OC-SVM: One-Class Support Vector Machines Schoelkopf et al. [24] present an anomaly detection method based on SVMs, particularly the one-class SVM. This method estimates the support vertor SV of a distribution by identifying the regions in the input space where most of the cases occur. In this purpose, data are projected nonlinearly in a feature space and separated of their origin by a margin as wide as possible. All data points outside this range are considered anomalies.

Proximity Proximity models observe the spatial proximity of each object in the data space. If the proximity of an object differs significantly from the proximity of other objects, it is considered an anomaly. For this approach, we will look at three algorithms: LOF, K-NN, and HBOS.

LOF: Local Outlier Factor Breunig et al. [5] propose LOF, an anomaly detection algorithm. LOF is the most widely known algorithm for detecting local anomalies and has introduced the concept of local anomalies.

The LOF score is therefore essentially a ratio of local density. This means that normal instances, whose density is as high as the density of their neighbors, obtain a score of about 1.0. The anomalies, which have a low local density, will have a higher score. At this point, we also see why this algorithm is local: it is based only on its direct neighborhood, and the score is a ratio mainly based on the k neighbors only.

It is important to note that, in anomaly detection tasks, when local anomalies are not of interest, this algorithm can generate many false alarms.

K-NN: k-Nearest Neighbors The K-NN method can be summarized as follows:

1. For each record in the dataset, the k-nearest neighbors must be selected
2. An anomaly score is calculated using these k neighbors and this according to two possibilities: either the distance to a single K-th nearest neighbor or the average distance of all k-th nearest neighbors.

In Ramaswamy et al. [22], a new formula for scoring distance-based anomalies is proposed. This scoring is based on the distance from a point in its nearest neighbor. Each point is then ranked according to its distance from its nearest neighbor. Finally, the top % points in this ranking are declared as outliers, therefore anomalies.

HBOS: Histogram-based Outlier Score The outlier value score based on the HBOS histogram is a simple statistical anomaly detection algorithm that assumes the independence of the variables.

Goldstein et al. [11] present a histogram-based anomaly detection (HBOS) method, which models densities of univariate variables (features) using histograms with a fixed or dynamic bin width. Thereafter, all histograms are used to calculate an anomaly score for each instance of data.

- HBOS performs well on global anomaly detection problems but cannot detect local anomalies.
- HBOS is faster for larger datasets.

Nevertheless, HBOS appears to be less effective on problems of local outliers.

Outlier Ensembles and Combination Frameworks Outlier ensembles and combination frameworks consist of combining the results of different models in order to create a more robust model. We will detail in this subsection; isolation forest and feature bagging.

Isolation Forest Isolation forest [16] explicitly identifies anomalies instead of profiling normal data points. Isolation forest, like any other method of tree "Set", is built on the basis of decision trees. In these trees, partitions are created by first randomly selecting an element, then selecting a random split value between the minimum and maximum values of the selected function.

As with other anomaly detection methods, an anomaly score is required to make decisions.

Feature Bagging Feature bagging is a method that consists of using several learning algorithms to achieve the best predictive performance that could come from any learning algorithm used alone. Lazarevic et al. [14], through tests on synthetic and real datasets, have found that the combination of several methods gives better results than each algorithm used separately, and this on datasets with: different degrees of contamination, different sizes, different dimensions, benefiting from different output combinations and the diversity of individual predictions.

Clustering Clustering models classify data into different clusters and count points that are not part of any of the clusters known as outliers. We mention here: K-means and DBSCAN.

K-means Syarif et al. [26] present a benchmark between the k-means algorithm, as well as three other variants (improved k-means, k-medoids, EM clustering). K-means is a clustering method used for the automatic detection of similar data instances. K-means starts by randomly defining k centroids.

- Methods based on the K-means algorithm are relatively fast
- On the other hand, they have a high rate of FP (False Positive)

DBSCAN: density-based spatial clustering of applications with noise DBSCAN [9],? is a clustering algorithm that detects clusters of arbitrary shapes and sizes, relying on a notion of cluster density: clusters are high-density regions in space, separated by areas of low density. It is not necessary to specify parameters that are generally difficult to define a priori, such as the number of k clusters, unlike K-means. It addresses different clustering and correlation analysis methods for unsupervised anomaly detection. For the DBSCAN algorithm, the following points were discussed :

- DBSCAN provides better results in small spaces, because high-dimensional spaces are usually rare, making it difficult to distinguish between high and low density.
- DBSCAN has certain parameters that limit its performance; in particular, two parameters that define the notion of cluster density: the minimum number of models that define a cluster and the maximum neighboring distance within models (Table 1).

Table 1 Models synthesis

Approach	Model	Strengths	Weaknesses
Linear Models	PCA	Suitable for large data, sensitive to noise	Can not model complex data distributions
	OC-SVM	Do not make assumptions about data distribution can characterize a complex boundary	When clusters become more complex, performance decreases
Proximity models	LOF	Easy to use (only one parameter *k*)	Based in the calculation only on the nearest neighbors
	K-NN	Simple and intuitive memory-based adapts to new training data	As the dataset grows, the efficiency and speed of the algorithm declines poor performance in unbalanced data
	HBOS	Faster than clustering and nearest neighbors models	Less efficient on local outlier problems
Outlier ensembles	Isolation forest	Efficient for large and high dimensionality dataset slow temporal complexity	Does not handle categorical data
	Feature bagging	Valid for different degrees of contamination, sizes and dimensions	Sensitive to the size of sampled datasets
Clustering models	K-means	Easy to implement, fast	Requires numerical data requires number of cluster *k* when clusters become more complex, performance decreases
	DBSCAN	No need to set cluster number *k*	Some parameters limit performance

3.2 *Unsupervised Deep Learning for Anomaly Detection*

Autoencoder Anomaly Detection Autoencoders are deep neural networks used to reproduce the input at the output layer; i.e., the number of neurons in the output layer is exactly the same as the number of neurons in the layer entry.

The architecture of the autoencoders may vary depending on the network applied (LSTM, CNN, etc.). A deep autoencoder is made up of two symmetrical deep arrays

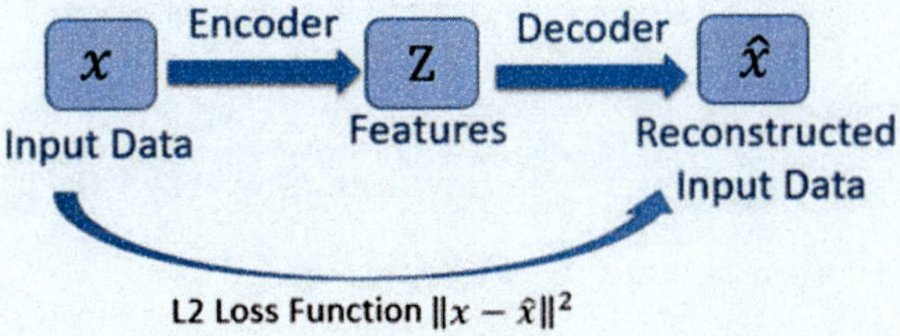

Fig. 2 Autoencoder: loss function

used to reproduce the input at the output layer. One network takes care of the encoding of the network and the second of decoding (Fig. 2).

Deep Autoencoding Gaussian Mixture Model (DAGMM) proposed by Zong et al. [27] is a deep learning framework that addresses the challenges of unsupervised anomaly detection from several aspects. This paper is based on a critique of existing methods based on deep autoencoding. First of all, the authors state the weakness of compression networks in anomaly detection, as it is difficult to make significant modifications to the well-trained deep autoencoder to facilitate subsequent density estimation tasks. Second, they find that anomaly detection performance can be improved by relying on the mutual work of compression and estimation networks. First, with the regularization introduced by the estimation network, deep autoencoder in the compression network learned by the end-to-end training can reduce the reconstruction error as low as the error of its pre-processed counterpart. This can be achieved only by performing end-to-end training with deep autoencoding. Second, with the well learned low-dimensional representations of the compression network, the estimation network is capable of making significant density estimates.

Chen et al. [6] For unsupervised anomaly detection tasks, the GMAA is a model that aims to jointly optimize dimensionality reduction and density estimation. In this paper, the authors' attention was focused on the subject of confidentiality. In this new approach which aims at improving model performance, we aggregate the parameters of the local training phase on clients to obtain knowledge from more private data. In this way, confidentiality is properly protected. This work is inspired by the work we discussed before. Therefore, this paper presents a federated deep autocoded Gaussian federated mixture model (DAGMM) to improve the performance of DAGMM caused by a limited amount of data.

Matsumoto et al. [19] This paper presents a detection method of chronic gastritis (an anomaly in the medical field) from gastric radiographic images. Among the constraints mentioned in this article and that traditional methods of anomaly detection cannot overcome is the distribution of normal and abnormal data in the dataset. The number of non-gastritis images is much higher than the number of gastritis images. To cope with this problem, the authors of this article propose the DAGMM as a new approach to detect chronic gastritis with high accuracy. DAGMM allows also the detection of chronic gastritis using images other than gastritis. Moreover, as mentioned above, the DAGMM differs from other models by the simultaneous learning of dimensionality reduction and density estimation.

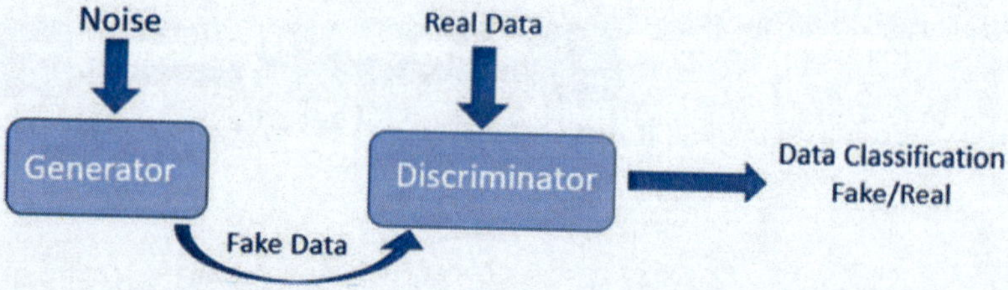

Fig. 3 GANs: generator + discriminator

GAN-Based Anomaly Detection In addition to the different approaches mentioned, mainly in the machine learning domain, there are also other anomaly detection methods that prefer the use of neural networks, with deep learning, and in particular the GAN model.

Generative adversarial networks (GAN) is a powerful member of the neural network family. It is used for unsupervised deep learning. It is made up of two competing models, a generator and a discriminator. The generator takes care of creating the realistic synthetic samples from the noise, the z-latent space, and the discriminator is designed to distinguish between a real sample and a synthetic sample (Fig. 3).
AnoGAN proposed by Schlegl et al. [23]is the firstly proposed method using GAN for anomaly detection. It is a deep convolutional generative adversarial network (DCGAN) that has been trained in synthetics data then it can detect anomalies in new images. The objective of AnoGAN is the standard GAN exploitation, formed on positive samples, to learn a mapping of the representation of the latent space z to the real sample G (z), and the goal is to use this representation learned to map new samples to latent space.
BiGANs Donahue et al. [7] present BiGAN that extends the classic GAN architecture by adding a third component: The encoder, which learns to map from data space x to latent space z. The objective of the generator remains the same, while the objective of the discriminator is modified to classify between a real sample and a synthetic sample and in addition between a real coding, i.e., given by the coder, and a synthetic coding, i.e., a sample of the latent space z.
DOPING The introduction of the generative antagonist network (GAN) allowed the generation of realistic synthetic samples, which were used to expand the training sets. In this paper [15], they focused on unsupervised anomaly detection and proposed a new generative data augmentation framework optimized for this task. Using a GAN variant known as the contradictory auto-analyzer (CAA), they imposed a distribution on the latent space of the dataset and systematically sample the latent space to generate artificial samples. This method is the first data augmentation technique focused on improving the performance of unsupervised anomaly detection.
GANomaly Ackay et al. [2, 3] introduce an anomaly detection model, GANomaly, including a conditional generative contradictory network that "jointly learns the generation of a high-dimensional image space and the inference of latent space." The GANomaly model is different from AnoGAN and BiGANs because it compares

the encoding of images in latent space rather than the distribution of images. The generator network in this model uses sub-networks encoder-decoder-encoder.
GAAL Liu et al. [18] present a new model that brings together GAN and active learning strategy. The aim is to train the generator G to generate anomalies that will serve as an input to the discriminator D, together with the real data to train him to differentiate between normal data and anomalies in an unsupervised context.

4 Datasets Description and Performance Evaluation

4.1 Datasets

Experimentally, we perform experiments on synthetic and real datasets. Several aspects are taken into consideration in the choice of these datasets. First, the nature of the unlabeled data. In addition, the influx of data from a multitude of sources requires the development of a proactive approach that takes into account the volume, variety, and velocity of the data.

The choice of method and model depends completely on the context, the intended objective, the available data, and their properties. In this study, we investigate the problem of anomaly detection in its global sense. We are going to try to process diversified datasets, each dataset represents a specific domain. Our study therefore concerns all data and information production sectors, regardless of the type of anomaly: failure, defect, fraud, intrusion ...

We used for this study anomaly detection several available datasets. These datasets used are all labeled, but the labels is only be used in the evaluation phase, where comparison measures between the predicted labels and the real labels is applied (Table 2).

Table 2 Datasets descriptions

Dataset	Speciality	Size	Dimension	Contamination
Credit cards	Bank fraud, financial crime	7074	14	4.8
KDDCup99	Intrusion/cyber security	4,898,431	41	20
SpamBase	Intrusion/cyber security	4207	57	39.9
Waveform	Sport	5000	21	2.9
Annthyroid	Medicine	7129	21	7.4
WDBC	Medicine	367	30	2.7
OneCluster	–	1000	2	2

4.2 Results and Discussion

The table below lists the models and the categories to which they belong, the datasets and their characteristics, i.e., specialty, size, dimension, and contamination rate, and finally, the measurement metrics chosen for the evaluation of the models (Table 3).

The goal of this study is to be able to detect anomalies using unlabelled datasets. To do so, we used several methods: detection using machine learning algorithms (one-class SVM, LOF, isolation forest, and K-means) and deep learning SO-GAAL, MO-GAAL, and DAGMM approaches.

In this section, we will evaluate these elaborated methods by comparing the performance of several techniques allowing the detection of anomalies.

To evaluate the models, we have used the metrics of AUC, precision, F1 score, and recall. This combination of measures is widely used in classification cases and allows a fair comparison and a correct evaluation.

We applied the different algorithms on the seven datasets. From the table above, several observations can be obtained:

While in general, DAGMM, MO-GAAL, and SO-GAAL demonstrate superior performance to machine learning methods in terms of F1 score on all datasets. Especially on KDDCup99, DAGMM achieves a 14 and 10% improvement in F1 score compared to other methods. OC-SVM, K-means and isolation forest suffer from poor performance on most datasets. For these machine learning models, the curse of dimensionality could be the main reason that limits their performance. For LOF, although it performs reasonably well on many datasets, the deep learning models outperform it. For example, at DAGMM, the latent representation and the reconstruction error are jointly taken into account in the energy modeling.

One of the main axes that affects method performance is the contamination rate, so contaminated training data negatively affects detection accuracy. In order to achieve better detection accuracy, it is important to form a model with high quality data, i.e., clean or keep the contamination rate as low as possible. When the contamination rate does not exceed 2% of the mean accuracy, the recall and F1 score decreases for all methods except GAAL (SO-GAAL and MO-GAAL). During this time, we observe that the DAGMM is more sensitive to the contamination rate, we notice that it maintains a good detection accuracy with a high contamination rate.

In addition, the size of the datasets is an essential factor affecting the performance of the methods. For MO-GAAL, as the number of dimensions increases, superior results are more easily obtained.

In particular, MO-GAAL is better than SO-GAAL. SO-GAAL does not perform well on some datasets. It depends if the generator stops the training before falling into the problem of mode collapse. This demonstrates the need for several generators with different objectives, which can provide more user-friendly and stable results. MO-GAAL directly generates informative potential outliers.

In summary, our experimental results show that the GAN and DAGMM models suggest a promising direction for the detection of anomalies on large and complex datasets. On the one hand, this is due to the strategy of the GAAL models, which

Table 3 Models evaluation

Model	Category	Dataset	AUC	Precision	F1	Recall
DAGMM	AE	WDBC	0.9201	0.932	0.937	0.942
		Annthyroid	0.9013	0.9297	0.944	0.936
		KddCup	0.9746	0.9683	0.961	0.954
		SpamBase	0.9236	0.9370	0.938	0.937
		Credit card	0.8777	0.8478	0.765	0.804
		Waveform	0.8611	0.8762	0.833	0.820
		Onecluster	0.4981	0.5078	0.5078	0.4983
SO-GAAL	GAN	WDBC	0.981	0.9411	0.9486	0.9561
		Annthyroid	0.6269	0.6371	0.6147	0.6248
		KddCup99	0.691	0.6804	0.6501	0.6822
		SpamBase	0.645	0.6311	0.6389	0.6577
		Credit card	0.7066	0.7234	0.7109	0.7278
		Waveform	0.8581	0.8302	0.8356	0.8411
		Onecluster	0.9984	0.9640	0.9412	0.9715
MO-GAAL	GAN	WDBC	0.9885	0.9714	0.9681	0.9704
		Annthyroid	0.6972	0.7002	0.7212	0.7326
		KddCup99	0.7688	0.7717	0.7703	0.7656
		SpamBase	0.6864	0.6745	0.6945	0.6812
		Credit card	0.5682	0.5504	0.5324	0.5579
		Waveform	0.8526	0.8479	0.8456	0.8681
		Onecluster	0.9994	0.9811	0.9808	0.9739
Isolation forest	Classification	WDBC	0.0545	1	0.028	0.0545
		Annthyroid	0.1495	0.9981	0.0808	0.1495
		KddCup99	0.4113	0.9868	0.2598	0.4113
		SpamBase	0.5338	0.6679	0.4445	0.5338
		Credit card	0.2111	1	0.1180	0.2111
		Waveform	0.0581	1	0.0299	0.0581
		Onecluster	0.040	1	0.0204	0.040
LOF	Density based	WDBC	0.9155	0.9939	0.9188	0.9549
		Annthyroid	0.8509	0.9311	0.9058	0.9183
		KddCup99	0.7657	0.8096	0.9205	0.8615
		SpamBase	0.5583	0.5882	0.8815	0.7056
		Credit card	0.8398	0.9080	0.9135	0.9107
		Waveform	0.8911	0.9790	0.9073	0.9418
		Onecluster	0.9155	0.9939	0.9788	0.9549

(continued)

Table 3 (continued)

Model	Category	Dataset	AUC	Precision	F1	Recall
KMeans	Clustering	WDBC	0.9046	0.9969	0.9048	0.9486
		Annthyroid	0.3494	0.9474	0.3142	0.4719
		KddCup99	0.2083	0	0	0
		SpamBase	0.460	1	0.0107	0.0212
		Credit card	0.1508	0.9036	0.0566	0.1065
		Waveform	0.5193	0.9694	0.5214	0.6781
		Onecluster	0.4640	0.9805	0.4622	0.6283
One-Class SVM	Classification	WDBC	0.4741	0.9457	0.4874	0.6433
		Annthyroid	0.5176	0.928	0.5095	0.6615
		KddCup99	0.4868	0.7870	0.4822	0.5980
		SpamBase	0.3530	0.4534	0.3776	0.4120
		Credit card	0.1055	0	0	0
		Waveform	0.5086	0.97977	0.9043	0.6659
		Onecluster	0.4940	0.9740	0.4969	0.6581

do not require the definition of a scoring threshold to separate normal data from anomalies, and the architecture of the sub-models, generator G and discriminator D, which give the possibility to set different parameters in order to obtain the optimal result: activation function, number of layers and neurons, input and output of each model, optimizer as well as the number of generators. On the other hand, the end-to-end learned DAGMM achieves the highest accuracy on public reference datasets and provides a promising alternative for unsupervised anomaly detection.

Among the constraints we faced, it is in the data collection phase that we do not find valid databases for anomaly detection. When dealing with datasets that contain a high contamination rate, we will converge to binary classification instead of anomaly detection. As discussed, anomaly detection aims to distinguish between "normal' and "abnormal" observations. Anomalous observations should be rare, and this also implies that the dataset should be out of balance. Unlike classification, class labels are meant to be balanced so that all classes have almost equal importance. Also, GAN and AE are powerful models that require high performance materials which are not always available.

5 Conclusion

In this article, we have compared various machine and deep learning methods for anomaly detection along with its application across various domains. This paper has used seven available datasets.

In the experimental study, we have tested four machine learning models and three deep learning models. One of our findings is that, with respect to performance metrics, DAGMM, SO-GAAL, and MO-GAAL were the best performers. They had demonstrated superior performance over state-of-the-art techniques on public benchmark datasets with up to over 10% improvement on the performance metrics and suggests a promising direction for unsupervised anomaly detection on multidimensional datasets.

Deep learning-based anomaly detection is still active research, and a possible future work would be to extend and update this article as more sophisticated techniques are proposed.

References

1. Adler, J., Lunz, S.: Banach wasserstein gan. In: Advances in Neural Information Processing Systems, pp. 6754–6763 (2018)
2. Akçay, S., Abarghouei, A.A., Breckon, T.P.: Ganomaly: semi-supervised anomaly detection via adversarial training. In: ACCV (2018)
3. Akçay, S., Atapour-Abarghouei, A., Breckon, T.P.: Skip-ganomaly: Skip connected and adversarially trained encoder-decoder anomaly detection (2019). arXiv preprint arXiv:1901.08954
4. Antipov, G., Baccouche, M., Dugelay, J.L.: Face aging with conditional generative adversarial networks. In: 2017 IEEE International Conference on Image Processing (ICIP), pp. 2089–2093. IEEE (2017)
5. Breunig, M.M., Kriegel, H.P., Ng, R.T., Sander, J.: Lof: identifying density-based local outliers. In: ACM Sigmod Record, vol. 29, pp. 93–104. ACM (2000)
6. Chen, Y., Zhang, J., Yeo, C.K.: Network anomaly detection using federated deep autoencoding gaussian mixture model. In: International Conference on Machine Learning for Networking, pp. 1–14. Springer (2019)
7. Donahue, J., Krähenbühl, P., Darrell, T.: Adversarial feature learning (2016). arXiv preprint arXiv:1605.09782
8. Dong, H., Liang, X., Gong, K., Lai, H., Zhu, J., Yin, J.: Soft-gated warping-gan for pose-guided person image synthesis. In: Advances in Neural Information Processing Systems, pp. 474–484 (2018)
9. Ester, M., Kriegel, H.P., Sander, J., Xu, X., et al.: A density-based algorithm for discovering clusters in large spatial databases with noise. Kdd. **96**, 226–231 (1996)
10. Ge, Y., Li, Z., Zhao, H., Yin, G., Yi, S., Wang, X., et al.: Fd-gan: pose-guided feature distilling gan for robust person re-identification. In: Advances in Neural Information Processing Systems, pp. 1222–1233 (2018)
11. Goldstein, M., Dengel, A.: Histogram-based outlier score (hbos): a fast unsupervised anomaly detection algorithm. In: Poster and Demo Track of the 35th German Conference on Artificial Intelligence, pp. 59–63 (2012)
12. Goodfellow, I., Pouget-Abadie, J., Mirza, M., Xu, B., Warde-Farley, D., Ozair, S., Courville, A., Bengio, Y.: Generative adversarial nets. In: Advances in Neural Information Processing Systems, pp. 2672–2680 (2014)
13. Hawkins, D.M.: Identification of Outliers, vol. 11. Springer (1980)
14. Lazarevic, A., Kumar, V.: Feature bagging for outlier detection. In: Proceedings of the Eleventh ACM SIGKDD International Conference on Knowledge Discovery in Data Mining, pp. 157–166. ACM (2005)
15. Lim, S.K., Loo, Y., Tran, N.T., Cheung, N.M., Roig, G., Elovici, Y.: Doping: Generative data augmentation for unsupervised anomaly detection with gan. In: 2018 IEEE International Conference on Data Mining (ICDM), pp. 1122–1127. IEEE (2018)

16. Liu, F.T., Ting, K.M., Zhou, Z.H.: Isolation-based anomaly detection. ACM Trans. Knowl. Discov. Data (TKDD) **6**(1), 3 (2012)
17. Liu, S., Sun, Y., Zhu, D., Bao, R., Wang, W., Shu, X., Yan, S.: Face aging with contextual generative adversarial nets. In: Proceedings of the 25th ACM International Conference on Multimedia, pp. 82–90. ACM (2017)
18. Liu, Y., Li, Z., Zhou, C., Jiang, Y., Sun, J., Wang, M., He, X.: Generative adversarial active learning for unsupervised outlier detection. IEEE Trans. Knowl, Data Eng (2019)
19. Matsumoto, M., Saito, N., Ogawa, T., Haseyama, M.: Chronic gastritis detection from gastric x-ray images via deep autoencoding gaussian mixture models. In: 2019 IEEE 1st Global Conference on Life Sciences and Technologies (LifeTech), pp. 231–232. IEEE (2019)
20. Menon, A.K., Williamson, R.C.: A loss framework for calibrated anomaly detection. In: Proceedings of the 32nd International Conference on Neural Information Processing Systems, pp. 1494–1504. Curran Associates Inc. (2018)
21. Mika, S., Schölkopf, B., Smola, A.J., Müller, K.R., Scholz, M., Rätsch, G.: Kernel pca and de-noising in feature spaces. In: Advances in Neural Information Processing Systems, pp. 536–542 (1999)
22. Ramaswamy, S., Rastogi, R., Shim, K.: Efficient algorithms for mining outliers from large data sets. In: ACM Sigmod Record, vol. 29, pp. 427–438. ACM (2000)
23. Schlegl, T., Seeböck, P., Waldstein, S.M., Schmidt-Erfurth, U., Langs, G.: Unsupervised anomaly detection with generative adversarial networks to guide marker discovery. In: International Conference on Information Processing in Medical Imaging, pp. 146–157. Springer (2017)
24. Schölkopf, B., Platt, J.C., Shawe-Taylor, J., Smola, A.J., Williamson, R.C.: Estimating the support of a high-dimensional distribution. Neural Comput. **13**(7), 1443–1471 (2001)
25. Schölkopf, B., Smola, A., Müller, K.R.: Kernel principal component analysis. In: International Conference on Artificial Neural Networks, pp. 583–588. Springer (1997)
26. Syarif, I., Prugel-Bennett, A., Wills, G.: Unsupervised clustering approach for network anomaly detection. In: International Conference on Networked Digital Technologies, pp. 135–145. Springer (2012)
27. Zong, B., Song, Q., Min, M.R., Cheng, W., Lumezanu, C., Cho, D., Chen, H.: Deep autoencoding gaussian mixture model for unsupervised anomaly detection. In: International Conference on Learning Representations (2018)

Automatic Spatio-Temporal Deep Learning-Based Approach for Cardiac Cine MRI Segmentation

Abderazzak Ammar, Omar Bouattane, and Mohamed Youssfi

Abstract In the present paper, we suggest an automatic spatio-temporal aware, deep learning-based method for cardiac segmentation from short-axis cine magnetic resonance imaging MRI. This aims to help in automatically quantifying cardiac clinical indices as an essential step towards cardiovascular diseases diagnosis. Our method is based on a lightweight Unet variant with the incorporation of a 2D convolutional long short-term memory (LSTM) recurrent neural network based layer. The 2D convolutional LSTM-based layer is a good fit for dealing with the sequential aspect of cine MRI 3D spatial volumes, by capturing potential correlations between consecutive slices along the long-axis. Experiments have been conducted on a dataset publically available from the ACDC-2017 challenge. The challenge's segmentation contest focuses on the evaluation of segmentation performances for three main cardiac structures: left, right ventricles cavities (LVC and RVC respectively) as well as left ventricle myocardium (LVM). The suggested segmentation network is fed with cardiac cine MRI sequences with variable spatial dimensions, leveraging a multi-scale context. With less overhead on preprocessing and no postprocessing steps, our model has accomplished near state-of-the-art performances, with an average dice overlap of 0.914 for the three cardiac structures on the test set, alongside good correlation coefficients and limits of agreement for clinical indices compared to their ground truth counterparts.

1 Introduction

The World Health Organization (WHO) repeatedly reports its concern about increasing cardiovascular diseases threats counting amongst the leading cause of death globally [16]. Cardiovascular diseases have attracted the attention of researchers in an attempt to early identifying heart diseases and predicting cardiac dysfunction. As it is generally admitted by the cardiologists community, this goes necessarily by quantifying ventricular volumes, masses and ejection fractions (EF) also called clinical

A. Ammar (✉) · O. Bouattane · M. Youssfi
ENSET Mohamedia, Hassan II University, Casablanca, Morocco

M. Ben Ahmed et al. (eds.), *Networking, Intelligent Systems and Security*,
Smart Innovation, Systems and Technologies 237,
https://doi.org/10.1007/978-981-16-3637-0_5

parameters or indices. On the other hand, cardiac cine MRI among other modalities is now recognized as one of the favorite tools for cardiac function analysis. Generally acquired as 3D spatial volumes evolving over time from diastole to systole then back to diastole, cine MRI sequences present 2D short-axis images at each slice level aggregated on the long-axis as a third spatial dimension and the temporal dimension or frame index. According to the cardiologists, evaluation of cardiac indices at only two frames: End Diastole and End Systole respectively (ED and ES) are sufficient for a reliable cardiac analysis. Given the spatial resolutions, calculation of these volumetric-based parameters could be achieved by first delineating cardiac chambers cavities and walls boundaries. However, manual delineation by experts of such contours is tedious and time consuming; this is why a fully automatic cardiac segmentation is highly sought after.

Over time, researchers have attempted to perform the cardiac segmentation task by adopting one of two main approaches: image-based methods such as thresholding, clustering and deformable models with no prior knowledge, or model-based methods such as statistical shape models, active appearance models and deformable models with prior knowledge. Recently, with the advances in deep learning techniques, particularly with convolutional neural networks (CNNs) [9] and fully convolutional networks (FCNs) [10], it turned out to be a most promising tools for high performances in segmentation tasks. CNNs advocate weight sharing and reduced connectivity to only a restricted receptive field to leverage spatial relationships within images.

Most of the research methods dealing with medical images segmentation relied on the FCN use, either solely or in combination with other methods. However, many of them attempted the combination of FCN-based models with recurrent neural networks (RNNs) architectures [1, 3, 12] as it is the case for our suggested model.

In the following section, we present the dataset, the related data preparation steps along with detailed description of our segmentation method. Subsequently, the segmentation metrics, loss functions, training settings and hyperparameters tuning are presented. Discussions and results presentation along with comparisons with the state-of-the-art methods are dealt with thereafter. Finally, a conclusion section summarizes the work in this paper and gives indications on future research work to further enhance accomplished results.

2 Materials and Methods

2.1 *Dataset Presentation*

The dataset on which our experiments have been conducted has been made publically available by the Automated Cardiac Diagnosis Challenge ACDC-2017 organizers. It comprises 150 real clinical exams of different patients evenly divided in five pathological classes: NOR (normal), MINF (previous myocardial infarction), DCM (dilated cardiomyopathy), HCM (hypertrophic cardiomyopathy) and RV (abnormal right ventricle). The dataset has been acquired by means of cine MRI short-axis slices with two MRI scanners of different magnetic strengths (1.5–3.0 T). The cine MRI short-axis slices go through the long-axis from base (upper slice) to apex (lower slice), each slice is of 5–8 mm thickness, 5 or 10 mm inter-slice gap and 1.37–1.68 mm^2/px for spatial resolution [2]. The dataset was divided into two separate subsets: the train set with 100 cases (20 for each pathological category) and a test set with 50 cases (10 for each pathological category). For each patient, the 3D spatial volumes at the two crucial instants ED and ES were provided separately. For the training set, images alongside their respective manually annotated ground truth GT masks, drawn by two clinical experts, were also provided for training purposes. For the test set, only cine MRI images were provided while their GTs counterparts were kept private for evaluation and participant methods ranking purposes.

2.2 Data Preprocessing

From the provided cine MRI sequences of the ACDC-2017 dataset, there are noticeable differences in both images spatial dimensions and intensity distributions. While CNN-based classification-oriented applications need to standardize spatial dimensions to a common size, this is not mandatory for FCN architectures such as Unet. We thus, choose to keep the original dimensions for two main reasons: this offers a multi-scale context in the learning process and mainly because, as we are planing the use of an LSTM-based RNN module, we need to proceed by handling one patient volume at the time where the sequential character makes sense. A small adjustment though has been carried out on images spatial dimensions which consisted in aligning down to the closest multiple of 32px for both height H and width W. On the other hand, before feeding the segmentation network, images intensities need to be normalized, we choose to operate on a per slice-based normalization:

$$X_{i,j}^{\text{prep}} = \frac{X_{i,j} - X_{\min}}{X_{\max} - X_{\min}} \tag{1}$$

where $X_{i,j}$ is image intensity at pixel (i, j), $X_{\min}$, $X_{\max}$ are minimum and maximum intensities of image X, respectively, given the assumption of independancy and identical distribution iid of image intensity.

2.3 Data Augmentation

As it is a common practice, to cope with training data scarcity leading to overfitting models, we resort to the use of a data augmentation technique as a means of regularization. Based on the provided 100 patients data in the training set, we proceeded to create 100 other virtual patients. This is achieved by small shifts both horizontally and vertically, small rotations and small zooms on the original training set images. Input images and their GT counterparts need to be jointly transformed.

2.4 2D Convolutional LSTM Layer

RNNs are basically designed to inherently exhibit a temporal dynamic behavior. Thus, they are a good fit for learning-based models for sequential data handling. Early RNNs implementations, also called vanilla RNN, faced the well-known vanishing and/or exploding gradient issues; thus, training such architecture-based models is harder. Long short-term memory (LSTM)-based RNN [6], one of the most popular implementations, precisely comes with the idea to overcome these issues by implementing gates controlling the contributions to the cell memory state $\mathcal{C}_t$. As such, an LSTM unit features the ability to maintain its cell memory state $\mathcal{C}_t$—in a learning-based way—from relevant contributions of previous observations throughout sequential inputs, while being able to discard irrelevant informations too. A 2D convolutional-based extension to the 1D LSTM unit has been suggested by [14] and allows FCNs-like architectures to benefit from this structure while preserving spatial correlations. Figure 1 shows an 2D convolutional LSTM cell's architecture with peephole [4] and the following equations (Eq. 2) summarize its working principle.

$$
\begin{aligned}
i_t &= \sigma(W_{ix} * \mathcal{X}_t + W_{ih} * \mathcal{H}_{t-1} + W_{ic} \odot \mathcal{C}_{t-1} + b_i) \\
f_t &= \sigma(W_{fx} * \mathcal{X}_t + W_{fh} * \mathcal{H}_{t-1} + W_{fc} \odot \mathcal{C}_{t-1} + b_f) \\
\tilde{\mathcal{C}}_t &= \tanh(W_{cx} * \mathcal{X}_t + W_{ch} * \mathcal{H}_{t-1} + b_c) \\
\mathcal{C}_t &= f_t \odot \mathcal{C}_{t-1} + i_t \odot \tilde{\mathcal{C}}_t \\
o_t &= \sigma(W_{ox} * \mathcal{X}_t + W_{oh} * \mathcal{H}_{t-1} + W_{oc} \odot \mathcal{C}_t + b_o) \\
\mathcal{H}_t &= o_t \odot \tanh(\mathcal{C}_t)
\end{aligned}
\tag{2}
$$

where subscript t denotes the current time step or frame. W_{ij}, b_i learnable convolutional weights between input j and output i before activation and related bias, respectively. In our application, the sequential aspect is not of a temporal nature but is rather sought after between consecutive slices along the long-axis. Indeed, the cardiac structures should presumably show some kind of shape's variability pattern along the long-axis, in that they are getting smaller starting from the base towards the apex while keeping some shape similarity. However, this is not evenly true in the same way for all structures especially for the RV structure and particularly for pathological cases.

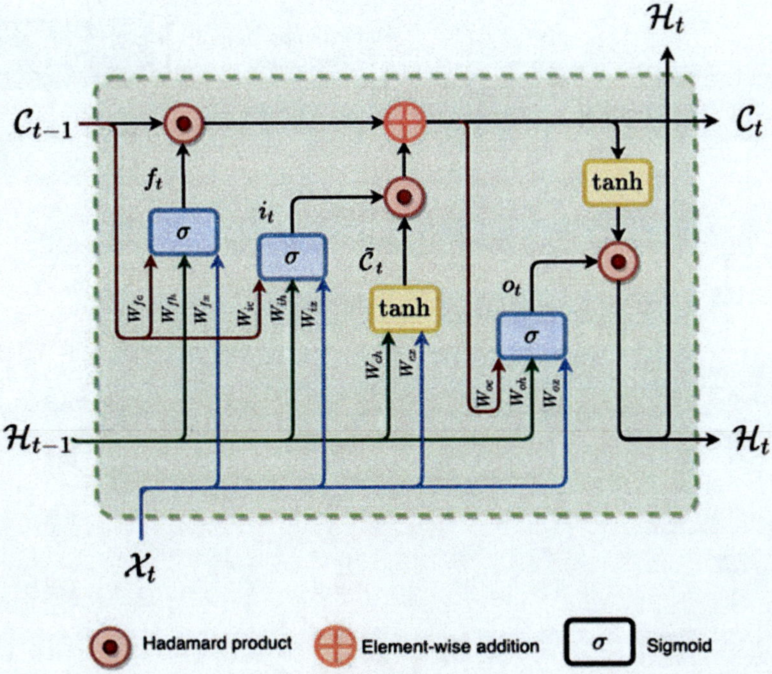

Fig. 1 2D convolutional LSTM cell with peephole: i_t, f_t and o_t are input, forget and output gates activations, respectively. $\mathcal{X}_t$, $\hat{\mathcal{C}}_t$, $\mathcal{C}_t$, and $\mathcal{H}_t$ are input, cell input, cell memory state and hidden state, respectively

2.5 *Unet-Based Segmentation Network*

The segmentation network is a lightweight variant of the well-known Unet architecture [13]. It is fed with images from the cine MRI sequences at their original dimensions slightly cropped at the lowest dimension: min (H, W) aligned down to 32px. As it is shown in Fig. 2, the Unet-based segmentation network follows an encoder/decoder pattern. A contracting path (on the left, in Fig. 2) chains up with a series of encoding blocks, starting from input down to a bottleneck. Each encoding block comprises two stacked subblocks 3 ×3 kernel convolutional layers, followed by a dropout DP layer as a means of regularization [5], a batch normalization layer BN [7] then a rectified linear unit (Relu) activation layer [11]. At each level in the contracting path, the number of feature maps increases (×2), while the spatial dimensions decrease (/2) by a prior downsampling (D) block, through 2×2 strided maxpooling layer. In The expanding path [in the middle of Fig. 2], at each ascending level, upsampling blocks (U) increase the spatial dimensions (×2) by means of transposed convolutions or deconvolutions [17], while feature maps number is reduced by the half. The particularity of the Unet architecture is the reuse of earlier feature maps of the contracting path at their corresponding level in the expanding path

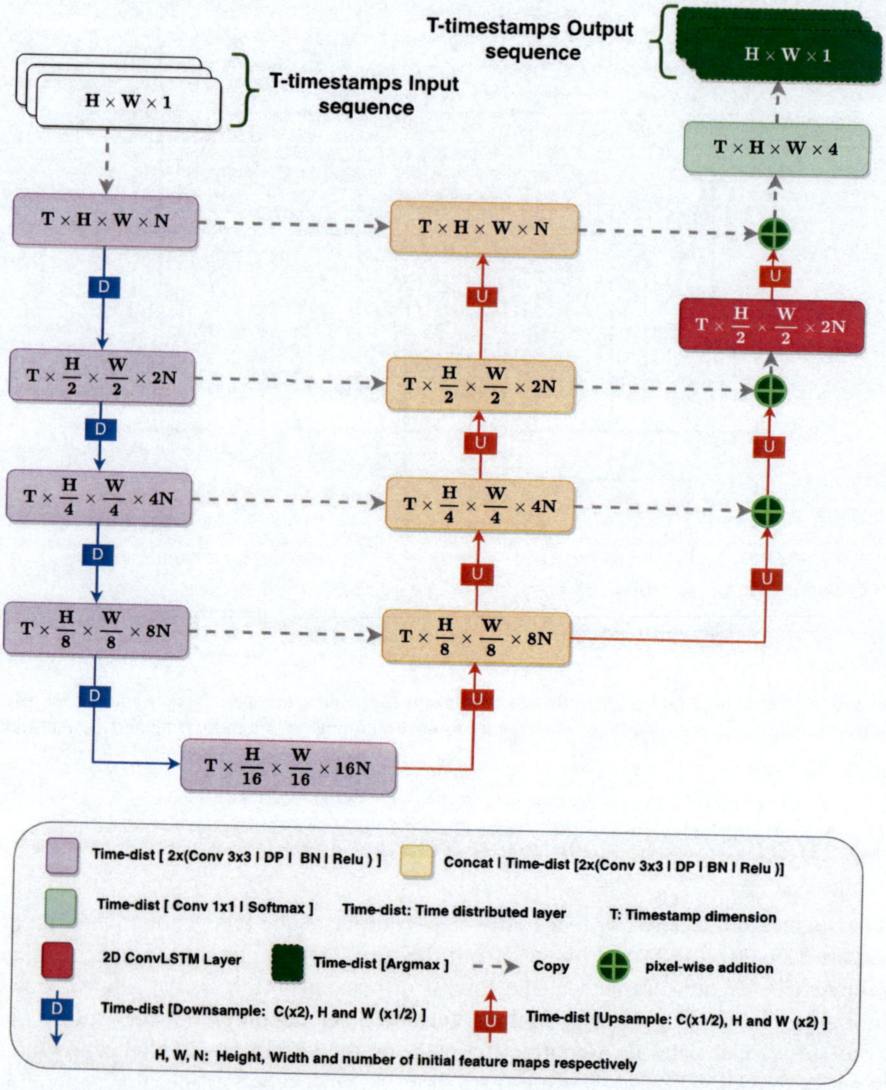

Fig. 2 Segmentation network architecture

where spatial dimensions match. This is achieved by simple channel-wise concatenation operations. As shown on the right of Fig. 2, our proposed architecture is a Unet variant where the output is constructed by aggregating all the expanding path levels outputs by upsampling with appropriate projections to perform pixel-wise additions. The aggregating output path ends with a four outputs softmax layer to predict pixel-wise class for the background, RVC, LVM and LVC as a raw one-hot encoded four valued vector. In the training phase, this is enough to guide the learning process; however in the inference phase, we need to retrieve the one channel mask to compare against the GT counterpart; this is achieved simply by an argmax operator applied to the softmax outputs. We choose to introduce the 2D convolutional LSTM layer in the middle of the aggregating path to keep the overall architecture as lightweight as possible while keeping a solid enough contribution of the 2D convolutional LSTM layer in the learning process. It is noteworthy that because of the 2D nature of the convolutional blocks in the construction of the Unet derived architecture and as this is fed with temporal sequences of images, these blocks need to be wrapped within time distributed layers referred to as Time-dist in Fig. 2.

3 Experiments and Model Training

3.1 Segmentation Evaluation Metrics

Let $\mathcal{C}_a$ the predicted or automatic contour delineating the object boundary in the image to segment, $\mathcal{C}_m$ its ground truth counterpart and $\mathcal{A}_a$, $\mathcal{A}_m$ sets of pixels enclosed by these contours, respectively. In the following, we recall the definitions of two well-known segmentation evaluation metrics:

Hausdorff Distance (HD) This is a symmetric distance between $\mathcal{C}_a$ and $\mathcal{C}_m$:

$$\mathcal{H}(\mathcal{C}_a, \mathcal{C}_m) = \max\left(\max_{i\in\mathcal{C}_a}\left(\min_{j\in\mathcal{C}_m} d(i,j)\right), \max_{j\in\mathcal{C}_m}\left(\min_{i\in\mathcal{C}_a} d(i,j)\right)\right) \quad (3)$$

where i and j are pixels of $\mathcal{C}_a$ and $\mathcal{C}_m$ respectively and $d(i, j)$ the distance between i and j. Low values of HD indicate both contours are much closer to each other.

Dice Overlap Index Measures the overlap ratio between $\mathcal{A}_a$ and $\mathcal{A}_m$. Ranging from 0 to 1, high dice values imply a good match:

$$\text{Dice} = \frac{2 \times |\mathcal{A}_m \cap \mathcal{A}_a|}{|\mathcal{A}_m| + |\mathcal{A}_a|} \quad (4)$$

3.2 Loss Functions

Supervised learning-based models training is achieved by the minimization of a suitable loss function. In our suggested method, which is actually acting as a pixel-wise classification model to achieve a semantic segmentation, we choose a tandem based on a dice overlap related and a crossentropy-based terms.

Cross Entropy Loss The categorical or multi-class crossentropy loss is defined as:

$$-\sum_{c=1}^{C} y(c,x)\log\left(\hat{y}(c,x)\right) \tag{5}$$

at the pixel level. C denotes the number of classes, y is the ground truth one-hot encoded label vector for pixel x and

$$\hat{y}(c,x) = \frac{e^{a(c,x)}}{\sum_{i=1}^{C} e^{a(i,x)}} \tag{6}$$

its estimated softmax score counterpart applied to activation functions a. As the ground truth is one-hot encoded, only the positive label is retained and the sum of C terms reduces to only one term. The crossentropy for the whole image sample yields:

$$\mathcal{L}_{ce} = -\sum_{x\in\Omega} \log\left(\hat{y}(p,x)\right)\big|_{p = \arg\max_{c}\,(y(c,x))} \tag{7}$$

where Ω is the image spatial domain.

Dice Overlap Based Loss The dice overlap index we recalled the definition above as a performance metric could also serve for the definition of a loss function. Seen as a metric, a good segmentation is achieved by maximizing the dice overlap index, or similarly by minimizing the deducted loss function:

$$\mathcal{L}_{\text{dice}} = -\log(\text{dice}) \tag{8}$$

3.3 Total Loss

We choose as a total loss function for training our segmentation network a combination of the above mentioned individual loss terms (Eqs. 7 and 8) plus an L_2 based weights decay penalty as a regularization term.

$$\mathcal{L}_{\text{tot}} = \alpha\mathcal{L}_{ce} + \beta\mathcal{L}_{\text{dice}} + \gamma W^2 \tag{9}$$

where W represents network weights. We choose to set both the crossentropy and dice-based loss terms contribution weights to 1, the L_2 based regularization contribution weight γ is adjusted to 2×10^{-4} (see 3.4).

3.4 Training Hyperparameters Tuning

Our experiments have been conducted with the following hyperparameters:

- # of initial filters: $N = 16$, 2.061 M as total number of learnable parameters.
- Dropout [15]-based regularization probability: 0.1.
- L_2 based regularization: $\gamma = 2 \times 10^{-4}$.
- Relu activation function [11].
- Adam optimizer [8], learning rate = 1×10^{-4}
- Number of iterations: 100 epochs, with variable batch size depending on the number of slices in the patient's volume.
- five fold stratified cross-validation.

Figure 3 shows the evolution, over training epochs, of both total loss and heart overall dice overlap training curves.

4 Results and Discussions

After training the suggested model in a five fold stratified cross-validation way and before going in the inference phase on the test set, we first gather validation results and try to analyze them. It is noteworthy that the achieved segmentation results are raw predictions, without any postprocessing actions.

4.1 Validation Results

From Fig. 4 it can be seen that

1. Figure 4a, LVC presented the highest dice score among the three structures, ED frames scores are higher than ES ones. Except for the HCM pathology at ES frames where there is a substancial dispersion, most of the obtained scores are less spread and keep a median above 0.95.
2. Figure 4b While RVC dice scores at ED frames are better than ES frames ones as for the LVC, the distributions are noticeably more spread, especially for ES frames.
3. Figure 4c At the opposite of the previous two cavities, LVM wall segmentation tends to report good performances in the ES frames, while globally its scores are the lowerest among the three structures.

This can be explained by the fact that the LVC component presents the most regular shape close to a circular one along the long-axis, except for the very early basal slices. The RVC component however presents the most larger variability in shape from basal to apical slices. LVM suffers from relying on two boundaries delineation (endocardium and epicardium), which is responsible for prediction cumulative errors.

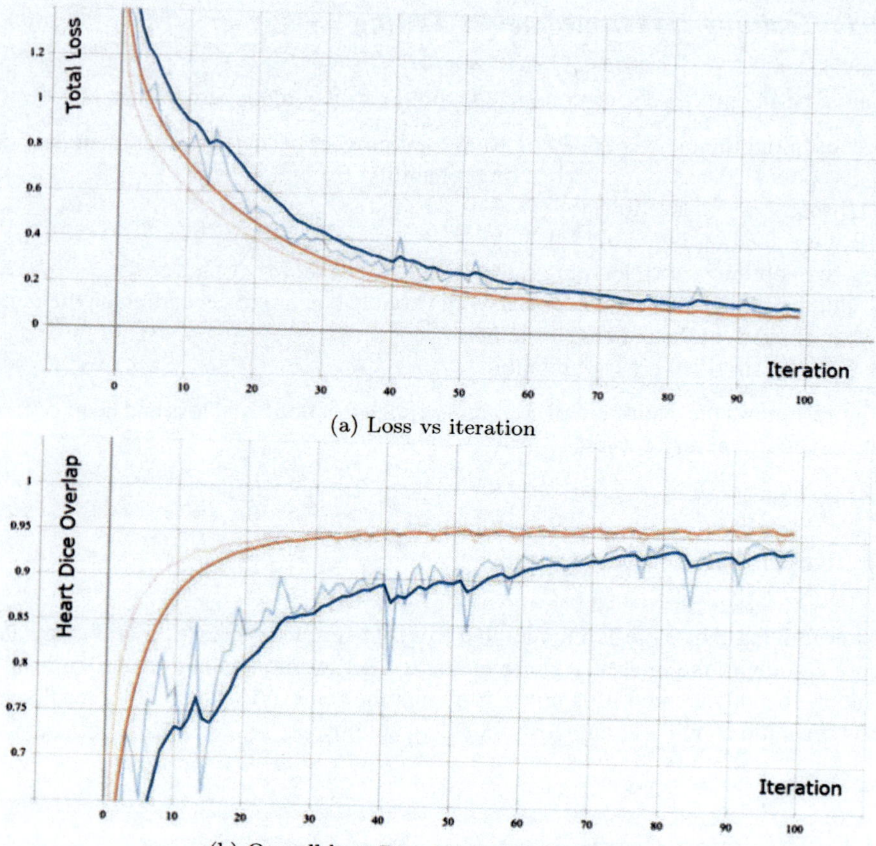

(a) Loss vs iteration

(b) Overall hart Dice overlap vs iteration.

Fig. 3 Training and validation curves, in orange (training) and in blue (validation)

Finally, due to the shrinking state of the heart at the end of systole phase ES, the LVC and RVC structures see their predictions performances decrease, while the LVM performance is rather increasing as the cumulated errors gets minimized with small structures.

4.2 *Test Results*

Our segmentation results on the test set (unseen data), along with clinical indices are reported in Tables 1, 2 and 3 for LVC, RVC and LVM structures, respectively. Compared to the top ranking participant methods on the same challenge's test set, our method achieved rather good results while being lightweight in that it requires a few parameters. Highlighted results indicate either first or second rank. This agrees with

Fig. 4 Dice overlap index validation results

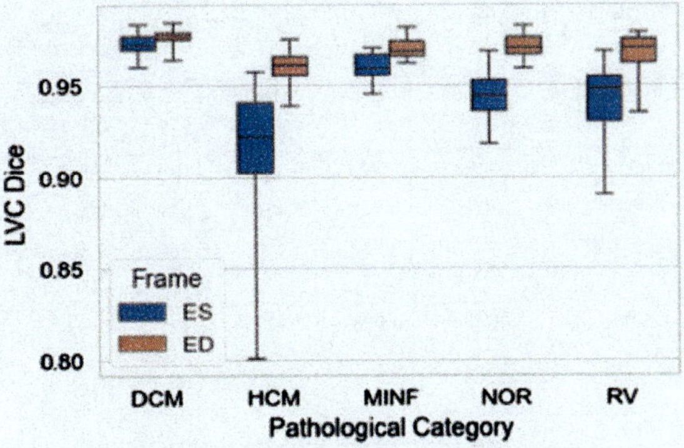

(a) LVC dice vs pathological category

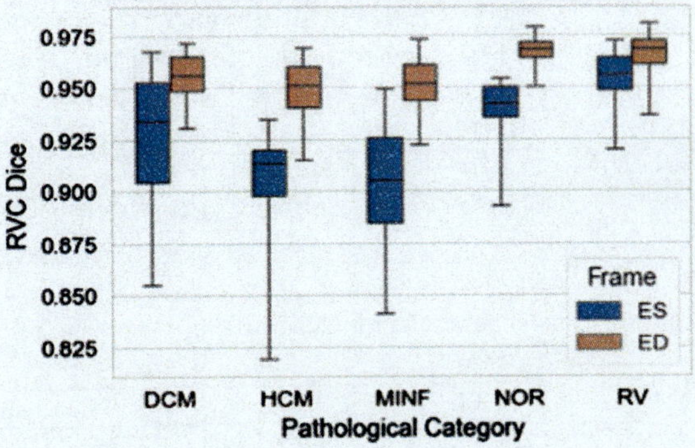

(b) RVC dice vs pathological category

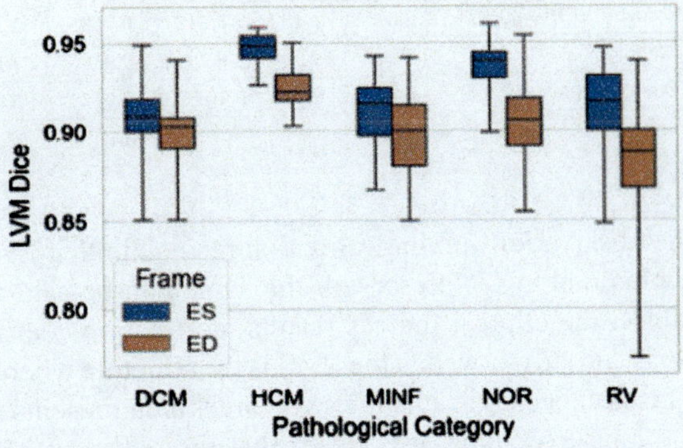

(c) LVM dice vs pathological category

Table 1 Challenge results for LVC structure (on the test set)

Method	LVC									
	DICE		HD (mm)		EF (%)			Vol. ED (ml)		
	ED	ES	ED	ES	Corr.	Bias	Std.	Corr.	Bias	Std.
Simantiris 2020	**0.967**	**0.928**	6.366	7.573	**0.993**	**−0.360**	**2.689**	**0.998**	2.032	**4.611**
Isensee 2018	**0.967**	**0.928**	**5.476**	**6.921**	0.991	0.49	2.965	0.997	1.53	5.736
Zotti 2019	0.964	0.912	6.18	8.386	0.99	−0.476	3.114	0.997	3.746	5.146
Painchaud 2019	0.961	0.911	6.152	8.278	0.99	−0.48	3.17	0.997	3.824	5.215
Ours	**0.966**	**0.928**	7.429	8.150	**0.993**	−0.740	**2.689**	0.995	**−0.030**	7.816

Table 2 Challenge results for RVC structure (on the test set)

Method	RVC									
	DICE		HD (mm)		EF (%)			Vol. ED (ml)		
	ED	ES	ED	ES	Corr.	Bias	Std.	Corr.	Bias	Std.
Isensee 2018	**0.951**	**0.904**	**8.205**	**11.655**	**0.91**	−3.75	**5.647**	**0.992**	**0.9**	**8.577**
Simantiris 2020	0.936	0.889	13.289	14.367	0.894	−1.292	6.063	0.990	0.906	9.735
Baldeon 2020	0.936	0.884	10.183	12.234	0.899	−2.118	5.711	0.989	3.55	10.024
Zotti 2019	0.934	0.885	11.052	12.65	0.869	**−0.872**	6.76	0.986	2.372	11.531
Ours	0.924	0.871	10.982	13.465	0.846	−2.770	7.740	0.955	−6.040	20.321

Table 3 Challenge results for LVM structure (on the test set)

Method	LVM									
	DICE		HD (mm)		Vol. ES (ml)			Mass ED (g)		
	ED	ES	ED	ES	Corr.	Bias	Std.	Corr.	Bias	Std.
Isensee 2018	**0.904**	**0.923**	**7.014**	**7.328**	**0.988**	−1.984	**8.335**	0.987	−2.547	8.28
Simantiris 2020	0.891	0.904	8.264	9.575	0.983	−2.134	10.113	**0.992**	−2.904	6.460
Baldeon 2020	0.873	0.895	8.197	8.318	**0.988**	−1.79	8.575	0.989	−2.1	7.908
Zotti 2019	0.886	0.902	9.586	9.291	0.98	**1.16**	10.877	0.986	**−1.827**	8.605
Ours	0.890	**0.906**	9.321	10.029	0.972	5.420	12.735	0.980	2.080	10.199

the observations on the validation results, in that LVC, RVC and LVM dice overlap scores ranking is preserved, the same can be said for HD metric. From the same tables, the clinical indices results: correlation coefficients and limits of agreement (bias and std) show that the RVC is the structure where the network is less performing. This is expected as it is the structure which presents the high shape variability along the long-axis; thus, it is likely that the recurrent LSTM-based convolutional layer captures less relevant correlations in the related input sequences. An example of a successful segmentation is shown in Fig. 5.

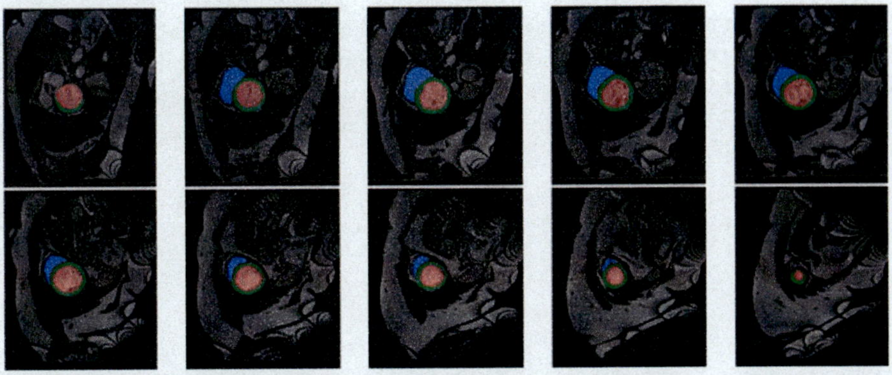

(a) LVC Dice: 0.97, HD: 7.34 mm - RVC Dice: 0.96, HD: 6.91 mm - LVM Dice: 0.88, HD: 11.56 mm

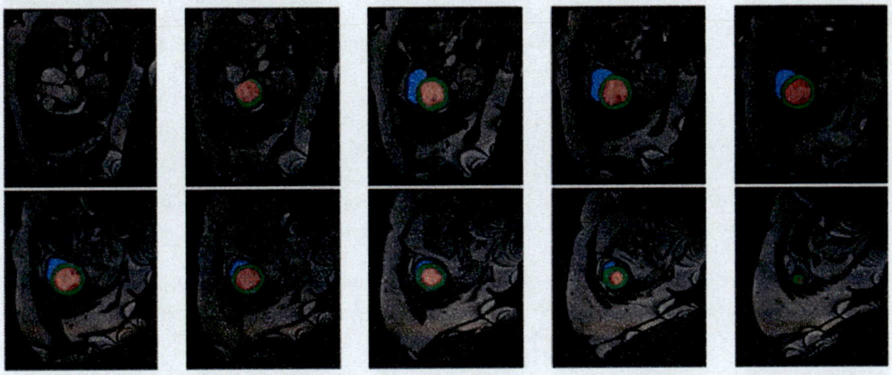

(b) LVC Dice: 0.97, HD: 8.37 mm - RVC Dice: 0.93, HD: 14.57 mm - LVM Dice: 0.89, HD: 11.96 mm

Fig. 5 Example of a successful volume segmentation from the test set **a** ED frame, **b** ES frame. Showing images from basal (top left) to apical (bottom right) slices for each frame. In overlay are predicted masks annotations in red, green and blue for LVC, LVM and RVC, respectively

5 Conclusion

In this paper, we suggested an automatic spatio-temporal deep learning-based approach for cardiac cine MRI segmentation. This has been implemented by incorporating a convolutional LSTM layer into a lightweight Unet variant to capture potential correlations between consecutive slices along the long-axis. The suggested model has been trained, validated and tested on a public dataset provided by the ACDC-2017 challenge and achieved an average dice overlap score of 0.947, 0.898, 0.899 and 0.914 for LVC, RVC, LVM and overall heart bi-ventricle chambers respectively on the challenge's segmentation contest. Given the context of medical images segmentation challenging task, our method has rather achieved good performances and even outperformed, for some metrics, the state-of-the-art participant methods to the challenge. Our method could benefit from further postprocessing operations to refine the obtained predicted masks, seeking for coupling with other established methods, adding a multi-scale approach to the architecture. These are some of the directions we will head to, in future work, to extend the suggested model and enhance the obtained results.

References

1. Alom, M.Z., Hasan, M., Yakopcic, C., Taha, T.M., Asari, V.K.: Recurrent residual convolutional neural network based on U-Net (R2U-Net) for medical image segmentation (2018). arXiv:1802.06955
2. Bernard, O., Lalande, A., Zotti, C., Cervenansky, F., Yang, X., Heng, P.A., Cetin, I., Lekadir, K., Camara, O., Gonzalez Ballester, M.A., Sanroma, G., Napel, S., Petersen, S., Tziritas, G., Grinias, E., Khened, M., Kollerathu, V.A., Krishnamurthi, G., Rohe, M.M., Pennec, X., Sermesant, M., Isensee, F., Jager, P., Maier-Hein, K.H., Full, P.M., Wolf, I., Engelhardt, S., Baum- gartner, C.F., Koch, L.M., Wolterink, J.M., Isgum, I., Jang, Y., Hong, Y., Patravali, J., Jain, S., Humbert, O., Jodoin, P.M.: Deep learning techniques for automatic MRI cardiac multi-structures segmentation and diagnosis: is the problem solved? IEEE Trans Med Imaging **37**, 2514–2525 (2018). https://doi.org/10.1109/TMI.2018.2837502
3. Chakravarty, A., Sivaswamy, J.: RACE-Net: a recurrent neural network for biomedical image segmentation. IEEE J. Biomed. Health Inform. 23, 1151–1162 (2019). doi: 10.1109/JBHI.2018.2852635
4. Gers, F., Schmidhuber, J.: Recurrent nets that time and count. In: Proceedings of the IEEE-INNS-ENNS International Joint Conference on Neural Networks. IJCNN 2000. Neural Computing: New Challenges and Perspectives for the New Millennium, pp. 189–194, vol.3. IEEE (2000). https://doi.org/10.1109/IJCNN.2000.861302
5. Hinton, G.E., Srivastava, N., Krizhevsky, A., Sutskever, I., Salakhutdinov, R.R.: Improving neural networks by preventing co-adaptation of feature detectors, 1–18 (2012). arXiv:1207.0580
6. Hochreiter, S., Schmidhuber, J.: Long short-term memory. Neural Comput. 9, 1735–1780 (1997). doi: 10.1162/neco.1997.9.8.1735
7. Ioffe, S., Szegedy, C.: Batch normalization: accelerating deep network training by reducing internal covariate shift. In: 32nd International Conference on Machine Learning, ICML 2015 1, 448–456 (2015). arXiv:1502.03167
8. Kingma, D.P., Ba, J.: Adam: a method for stochastic optimization. In: 3rd International Conference on Learning Representations, ICLR 2015—Conference Track Proceedings abs/1412.6 (2014). arXiv:1412.6980
9. Lecun, Y., Bengio, Y., Hinton, G.: Deep learning. Nature **521**, 436–444 (2015). DOI 10.1038/nature14539, arXiv:1807.07987
10. Long, J., Shelhamer, E., Darrell, T.: Fully convolutional networks for semantic segmentation. IEEE Trans. Pattern Anal. Mach. Intell. 39, 640–651 (2014). doi: 10.1109/TPAMI.2016.2572683
11. Nair, V., Hinton, G.E.: Rectified linear units improve restricted Boltzmann machines. In: ICML 2010—Proceedings, 27th International Conference on Machine Learning, 807–814 (2010). URL https://icml.cc/Conferences/2010/papers/432.pdf
12. Poudel, R.P.K., Lamata, P., Montana, G.: Recurrent fully convolutional neural networks for multi-slice MRI cardiac segmentation. In: Lecture Notes in Computer Science (including subseries Lecture Notes in Artificial Intelligence and Lecture Notes in Bioinformatics), volume 10129 LNCS, pp. 83–94 (2017). DOI 10.1007/978-3-319-52280-7_8
13. Ronneberger, O., Fischer, P., Brox, T.: U-Net: convolutional networks for biomedical image segmentation **9351**, 234–241 (2015). https://doi.org/10.1007/978-3-319-24574-4_28

14. Shi, X., Chen, Z., Wang, H., Yeung, D.Y., Wong, W.K., Woo, W.C.: Convolutional LSTM network: a machine learning approach for precipitation nowcasting. In: Advances in Neural Information Processing Systems, pp. 802–810 (2015). arXiv:1506.04214
15. Srivastava, N., Hinton, G., Krizhevsky, A., Sutskever, I., Salakhutdinov, R.:. Dropout: a simple way to prevent neural networks from overfitting. J. Mach. Learn. Res. **15**, 1929–1958 (2014). URL http://jmlr.org/papers/v15/srivastava14a.html
16. World Health Organization.: Cardiovascular diseases (CVDs) (2017). URL https://www.who.int/en/news-room/fact-sheets/detail/cardiovascular-diseases-(cvds)
17. Zeiler, M.D., Krishnan, D., Taylor, G.W., Fergus, R.: Deconvolutional networks. In: Proceedings of the IEEE Computer Society Conference on Computer Vision and Pattern Recognition 2528–2535 (2010). https://doi.org/10.1109/CVPR.2010.5539957

Skin Detection Based on Convolutional Neural Network

Yamina Bordjiba, Chemesse Ennehar Bencheriet, and Zahia Mabrek

Abstract Skin detection is an essential step in many human–machine interaction systems such as e-learning, security, communication… etc., it consists of extracting regions containing the skin in a digital image. This problem has become the subject of considerable research in the scientific community where a variety of approaches has been proposed in the literature; however, few recent reviews exist. Our principal goal in this paper is to extract skin regions using a Convolutional neural network called LeNet5. Our framework is divided into three main parts: At first, a deep learning is performed to Lenet5 network using 3354 positive examples and 5590 negative examples from SFA dataset, then and after a preprocessing of each arbitrary image the trained network will classify image pixels into skin/non-skin. Lastly, a thresholding and prost-processing of classified regions is carried out. The tests were carried out on images of variable complexity: indoor, outdoor, variable lighting, simple and complex background. The results obtained are very encouraging, we show the qualitative and quantitative results obtained on SFA and BAO datasets.

1 Introduction

Skin is one of the most important parts of the human body, so it is logical to consider it as the main element to be detected in many artificial vision systems operating on human beings such as medicine for disease detection and recognition, security for intrusion detection, people identification, facial recognition, gesture analysis, hand tracking, etc.

Although considered an easy and simple task to be performed by the human, the recognition of human skin remains an operation of high complexity for the machine

Y. Bordjiba · Z. Mabrek
LabStic Laboratory, 8 Mai 1945- Guelma University, BP 401, Guelma, Algeria
e-mail: bordjiba.yamina@univ-guelma.dz

C. E. Bencheriet (✉)
Laig Laboratory, 8 Mai 1945- Guelma University, BP 401, Guelma, Algeria
e-mail: bencheriet.chemesseennehar@univ-guelma.dz

M. Ben Ahmed et al. (eds.), *Networking, Intelligent Systems and Security*,
Smart Innovation, Systems and Technologies 237,
https://doi.org/10.1007/978-981-16-3637-0_6

despite the technological progress of the sensors and processors used, for several reasons such as lighting and shooting conditions of the captured image, background variation (indoor/outdoor), skin color variation (different ethnicity), etc....

The main objective of our work is to design a model with a deep learning architecture, and to implement a convolutional neural network model for skin detection, for these we propose an approach based on LeNet 5 network.

Our contribution is divided into three main parts: At first, a deep learning is performed to Lenet5 network using 3354 positive examples and 5590 negative examples from SFA dataset, then and after a preprocessing of each arbitrary image the trained network will classify image pixels into skin/non-skin. Lastly, a thresholding and prost-processing of classified regions are carried out.

The remainder of this paper is structured as follows: Sect. 2 gives the development of principal steps of our proposed framework. Section 3 provides the experimental results using two different datasets and Sect. 4 concludes the paper with discussions and future research directions.

2 Related Work

Skin detection is a difficult problem and has become the subject of considerable study, to improve the skin detection process [1], but this requires a high rate of accuracy due to the noise and complexity of the images. In this context, the research community is divided into two parts: Conventional research and deep learning-based research [2]. Conventional methods can be divided into different categories. They can be based on pixel classification [3, 4] or region segmentation [5, 6], while other studies have selected a hybrid of two or more methods. Among researches based on region segmentation, authors of [7] propose a technique purely based on region for skin color detection, they cluster similarly colored pixels, based on color and spatial distance. First, they use a basic skin color classifier, then, they extract and classify regions called superpixel. Finally, a soothing procedure with CRF (Conditional Random Field) is applied to improve result. This proposed method reaches 91.17% true positive rate and 13.12% false-positive rate. Authors indicate that skin color detection has to be based on regions rather than pixels.

Many studies have also investigated the effects of color space selection [8, 9]; they confirm that RGB color space is not the best one for this task. In [10], authors use Cb-Cr color space and extract Skin regions using the Gaussian skin color model. The likelihood ratio method is used to create a binary mask. To design skin color model, they also use a combination of two different databases to encompass larger skin tones. For performance evaluation, a total of 165 facial images from the Caltech database were randomly selected; the achieved accuracy is about 95%.

Color spaces have been widely used in skin detection. In [11], the authors present a comparative study of skin detection in two color spaces HSV and YCbCr. The detection result is based on the selection of a threshold value. The authors concluded that HSV-based detection is the most appropriate for simple images with a uniform background. However, the YCbCr color space is more effective and efficient to be applied for complex color images with uneven illumination.

The authors of [12] propose to model skin color pixels with three statistical functions. They also propose a method to eliminate the correlation between skin chrominance information. For this method's tests, they used the COMPAQ skin data set for the training and testing stages, with different color spaces. The accuracy achieved was 88%, which represents, according to the authors, an improvement over previous statistical method.

Many researchers have used neural networks to detect skin color, and recently deep learning methods have been widely used and have achieved successful performance for different problems of classification in computer vision. However, there are few researches on human skin detection based on deep learning (especially convolutional neural networks) and they limited their studies to diagnosing skin lesions, disorders and cancers only [13].

In [13], authors propose a sequential deep model to identify the regions of the skin appearing on the image. This model is inspired by the VGGNet network, and contains modifications to treat finer grades of microstructures commonly present in skin texture. For their experiments, they used two datasets: Skin Texture Dataset and FSD dataset, and compared their results with conventional texture-based techniques. Based on the overall accuracy, they claim to obtain superior results.

Kim et al. [14] Realize one of the most interesting work in skin detection using deep learning, where they propose two networks based on well-known architectures, one based on VGGNet, and the second based on the Network in Network (NiN) architecture. For both, they used two training strategies; one based on full image training, and the other based on patch training. Their experiences have shown that NiN-based architectures provide generally better performance than VGGNet-based architectures. They also concluded that full image-based training is more resistant to illumination and color variations, in contrast to the patch-based method, which learns the skin texture very well, allowing it to reject skin-colored background when it has different texture from the skin.

3 Proposed Method

The aim of this work is to propose a new approach to skin detection based on deep learning. The detection will be done in two steps. the first is a learning phase of the CNN, once its weights are found, they are used in the second phase which is the segmentation, based on patches; where the input image has to be pre-processed, then it is divided into overlapping patches obtained by a sliding window. These patches are classified as skin or non-skin by the CNN already trained in the first phase. Finally, a post-processing stage is applied. The global architecture of our skin detection system is illustrated in Fig. 1.

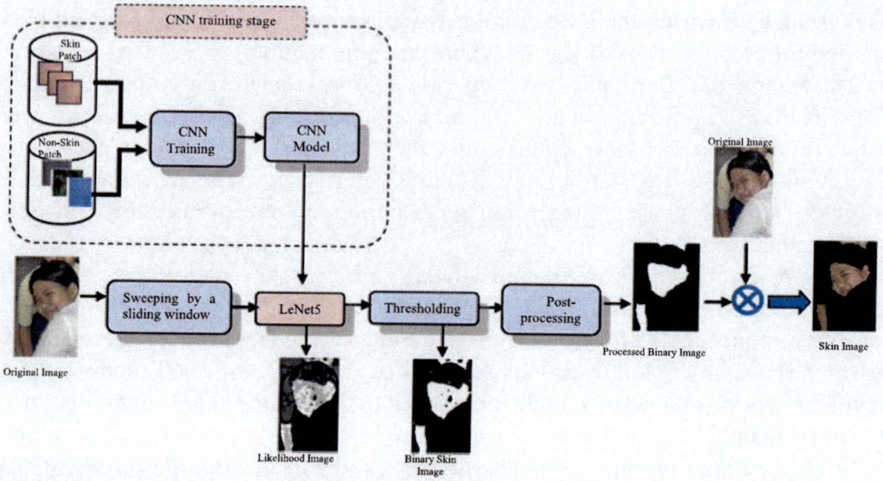

Fig. 1 Global architecture of our skin detection system

3.1 The Used CNN Architecture

Recently, "convolutional neural networks" (CNNs) have emerged as the most popular approach to classification and computer vision problem and several convolutional neural network architectures were proposed in literature. One of the first successful CNNs was LeNet by LeCun [15], which was used to identify handwritten numbers on checks at most banks in the United States. Consisting of two convolutional layers, two maximum grouping layers, and two fully connected layers for classification, it has about 60,000 parameters, most of which are in the last two layers.

Later, LeNet-5 architecture (one of the multiple models proposed in [15]) was used for handwritten character recognition [16]. It obtained a raw error rate of 0.7% out of 10,000 test examples. As illustrated in Fig. 2 and Table 1, the network defined the basic components of CNN, but according to the hardware of the time, it required high computational power. This makes it unable to be as popular and used as other algorithms (such as SVM), which could obtain similar or even better results. One of the main reasons for choosing "Lenet-5" is its simplicity, and this feature allows us to preserve as many characteristics as possible, because the large number of layers

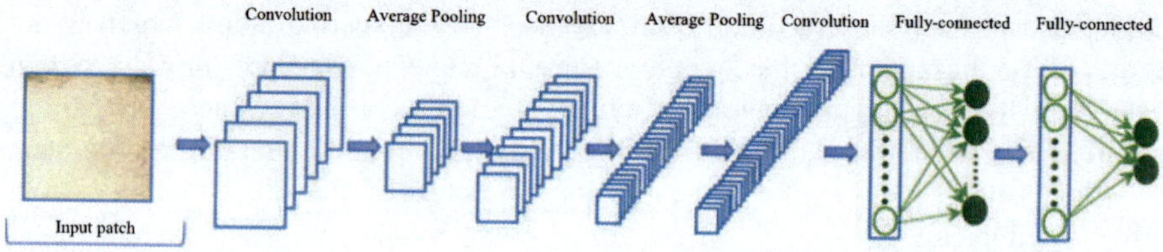

Fig. 2 Used CNN architecture

Table 1 [12–16] A detailed description of different layers of used CNN

	Convolution	Average pooling	Convolution	Average pooling	Convolution	Fully connected	Fully connected
Input size	17 × 17 × 3	17 × 17 × 6	16 × 16 × 6	12 × 12 × 16	6 × 6 × 16	480	84
Kernel size	5 × 5	5 × 5	5 × 5	2 × 2	5 × 5	–	–
Stride	1	1	1	2	1	–	–
Pad	Same	Valid	Valid	Valid	Valid	–	–
# filters	6	–	16	–	120	–	–
Output size	17 × 17 × 6	16 × 16 × 6	12 × 12 × 16	6 × 6 × 16	2 × 2 × 120	84	2

in our experience destroys the basic characteristics, which are color and texture. The database contains small sample sizes, which makes the use of a large number of convolution or pooling layers unnecessary or bad.

LeNet5 model has been successfully used in different application areas, such as facial expression recognition [17], vehicle-assisted driving [18], traffic sign recognition [19] and medical application like sleep apnea detection [20] …etc.

3.2 Training Stage

The training phase is a very important step; it is carried out to determine the best weights of all CNN layers. Our network is trained with image patches of positive and negative examples where inputs are skin/non-skin patches and outputs correspond to the label of these patches, sized 17 × 17, are manually extracted by us from the training images of the database (Fig. 3). Training is achieved by optimizing a loss function using the stochastic gradient descent approach (the Adam's optimizer). The loss function in our case is simply cross entropy. Finally, a low learning rate is set at 0.001 to form our CNN.

3.3 Segmentation Stage

Our proposed skin detection system is actually a segmentation algorithm, which consists of scanning the entire input image through a 17 × 17 window and a 1:16 delay step.

Then, each of these thumbnails is classified by the CNN network, previously formed, so each pixel of these thumbnails is replaced by its probability of belonging to the skin or non-skin class. The obtained result is grayscale probability image (Fig. 3) then a thresholding is applied to obtain a skin binary image (Fig. 3).

In order to clean the binary image resulting from noise, we applied morphological operators as post-processing: the closure is used to eliminate small black holes (Fig. 4), and the aperture is used to eliminate small white segments of the image (Fig. 5).

The last step is the displaying of skin image performed by a simple multiplication between the original image and the binary image, to give as a result an RGB image with only skin detected regions.

It is necessary to note that we foresee a pre-processing phase to improve the quality of images that are too dark or too light, because the lighting can seriously affect the color of the skin.

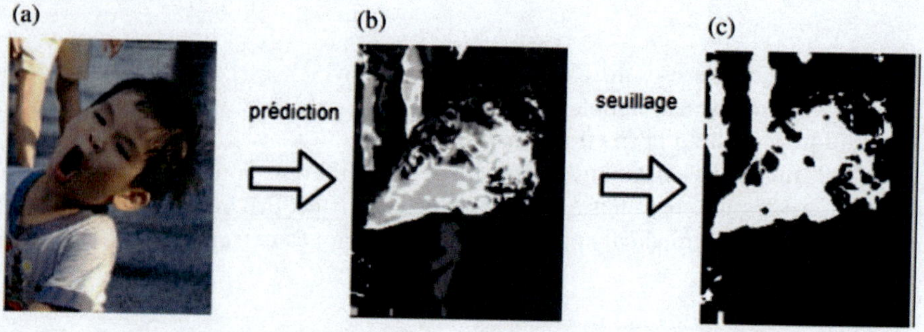

Fig. 3 Segmentation stage **a** original image, **b** likelihood image, **c** binary image

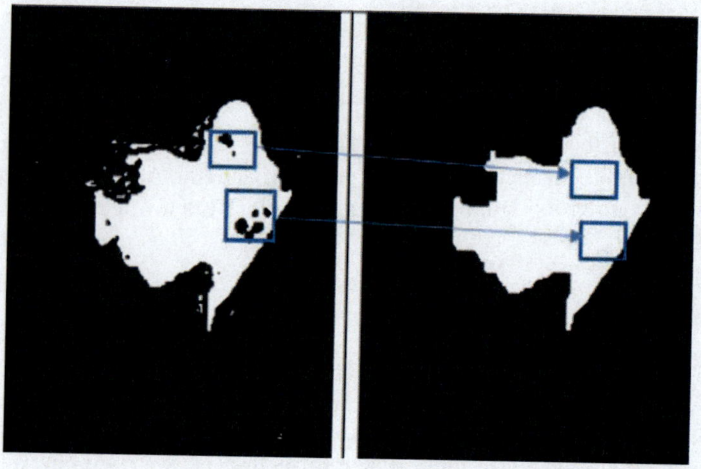

Fig. 4 Application of closure to eliminate small black holes

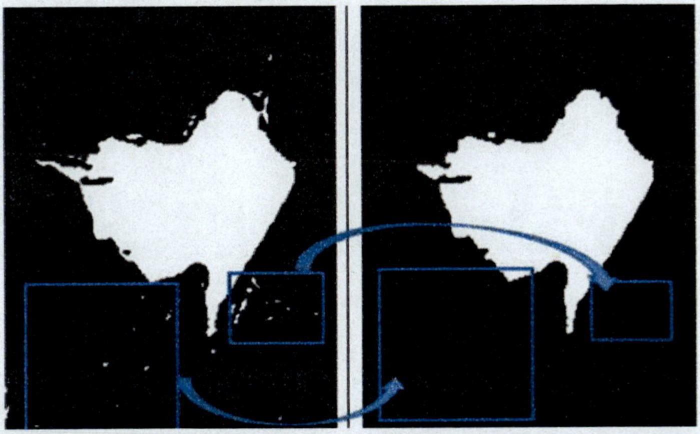

Fig. 5 Application of aperture to eliminate small white segments

4 Results and Discussion

4.1 Dataset and Training

This section reports the result of skin detection using LeNet5 Convolutional Neural Network. In the training stage, we used the SFA dataset [21] that was constructed on the basis of face images from FERET dataset [22] (876 images) and AR dataset [23] (242 images) databases, from which skin and non-skin samples were retrieved differently from 18 different scales (Fig. 6).

The dataset contains over 3354 manually labeled skin images and over 5590 non-skin images. The dataset is divided into 80% for training and 20% for testing (validation).

The testing phase is a crucial step in the evaluation of training of the CNN network. It consists of evaluating the network on a complete scene (indoor/outdoor) without any conditions on the shots. We select images from both SFA [13] (Fig. 7) and BAO

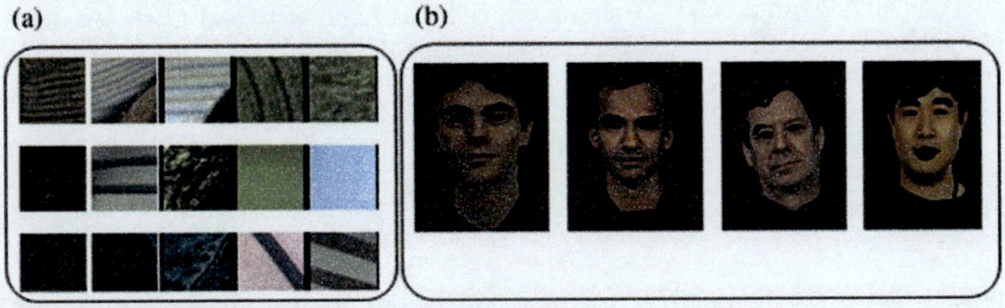

Fig. 6 SFA dataset used for training. **a** Non-skin examples. **b** Skin examples

Fig. 7 Examples from SFA database used in testing stage

Fig. 8 Examples from SFA database used in testing stage

[24] datasets (Fig. 8). Different lighting conditions and complex scenes make these datasets suitable for evaluating our skin detection system.

4.2 Experiments and Discussion

For quantitative analysis of the obtained results, accuracy and error rate was used, shown by the accuracy rates called respectively training accuracy (train-accuracy) and testing or validation accuracy (val-accuracy), and the error rates called respectively training loss (train-loss) and testing or validation loss (val-loss). Figure 9 shown the results obtained with precision training rate of 93%. Figures 10 and 11 shows some tests performed on SFA and BAO datasets where the precision tests obtained are consecutively 96 and 95%.

5 Conclusion

Our principal goal in this paper is to extract skin regions using a Convolutional neural network called LeNet5. Our framework is divided into three main parts: At first, a deep learning is performed to Lenet5 network using 3354 positive examples and

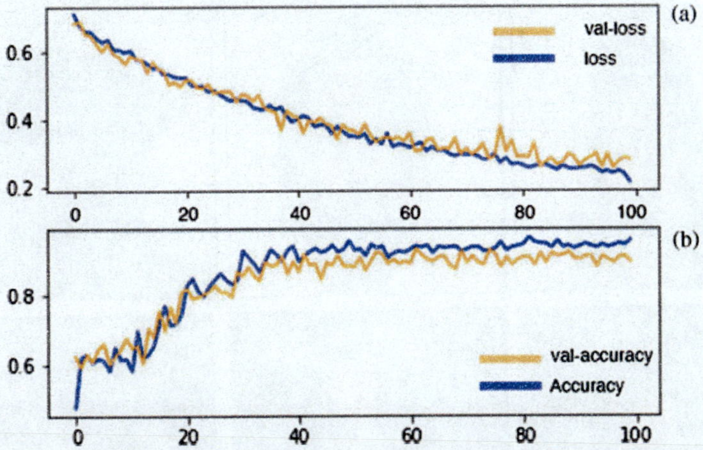

Fig. 9 Training results. **a** Training and validation loss. **b** Training and validation accuracy

Fig. 10 Tests on BAO dataset. **a** Original image. **b** Skin image results

5590 negative examples from SFA dataset, then and after a preprocessing of each arbitrary image the trained network will classify image pixels into skin/non-skin. Lastly, a thresholding and post-processing of classified regions are carried out. The tests were carried out on images of variable complexity: indoor, outdoor, variable lighting, simple and complex background. The results obtained are very encouraging, we show the qualitative and quantitative results obtained on SFA and BAO datasets where the precision tests obtained are consecutively 96 and 95%.

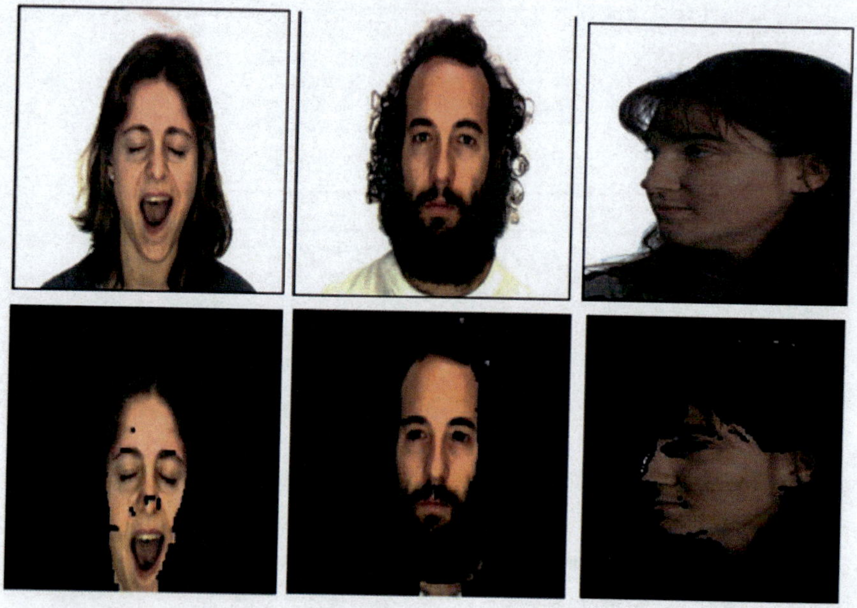

Fig. 11 Tests on SFA dataset. **a** Original image. b Skin image results

Acknowledgements The work described herein was partially supported by 8 Mai 1945 University and PRFU project through the grant number C00L07UN240120200001. The authors thank the staff of LAIG laboratory, who provided financial support.

References

1. Naji, S., Jalab, H.A., Kareem, S.A.: A survey on skin detection in colored images. Artif. Intell. Rev. **52**, 1041–1087 (2019). https://doi.org/10.1007/s10462-018-9664-9
2. Zuo, H., Fan, H., Blasch, E., Ling, H.: Combining convolutional and recurrent neural networks for human skin detection. IEEE Sig. Process. Lett. **24**, 289–293 (2017). https://doi.org/10.1109/LSP.2017.2654803
3. Zarit, B.D., Super, B.J., Quek, F.K.H.: Comparison of five color models in skin pixel classification. In: Proceedings International Workshop on Recognition, Analysis, and Tracking of Faces and Gestures in Real-Time Systems. In Conjunction with ICCV'99 (Cat. No. PR00378). pp. 58–63 (1999). https://doi.org/10.1109/RATFG.1999.799224
4. Phung, S.L., Bouzerdoum, A., Chai, D.: Skin segmentation using color pixel classification: analysis and comparison. IEEE Trans. Pattern Anal. Mach. Intell. **27**, 148–154 (2005). https://doi.org/10.1109/TPAMI.2005.17
5. Ashwini, A., Murugan, S.: Automatic skin tumour segmentation using prioritized patch based region—a novel comparative technique. IETE J. Res. **1,** 12 (2020). https://doi.org/10.1080/03772063.2020.1808091
6. Li, B., Xue, X., Fan, J.: A robust incremental learning framework for accurate skin region segmentation in color images. Pattern Recogn. **40**, 3621–3632 (2007). https://doi.org/10.1016/j.patcog.2007.04.018

7. Poudel, R.P., Nait-Charif, H., Zhang, J.J., Liu, D.: Region-based skin color detection. In: VISAPP (1) VISAPP 2012-Proceedings of the International Conference on Computer Vision Theory and Applications 1, pp. 301–306. VISAPP (2012)
8. Kolkur, S., Kalbande, D., Shimpi, P., Bapat, C., Jatakia, J.: Human skin detection using RGB, HSV and YCbCr Color Models. In: Presented at the International Conference on Communication and Signal Processing 2016 (ICCASP 2016) (2016). https://doi.org/10.2991/iccasp-16.2017.51
9. Brancati, N., De Pietro, G., Frucci, M., Gallo, L.: Human skin detection through correlation rules between the YCb and YCr subspaces based on dynamic color clustering. Comput. Vis. Image Underst. **155**, 33–42 (2017). https://doi.org/10.1016/j.cviu.2016.12.001
10. Verma, A., Raj, S.A., Midya, A., Chakraborty, J.: Face detection using skin color modeling and geometric feature. In: 2014 International Conference on Informatics, Electronics Vision (ICIEV). pp. 1–6 (2014). https://doi.org/10.1109/ICIEV.2014.6850755
11. Shaik, K.B., Ganesan, P., Kalist, V., Sathish, B.S., Jenitha, J.M.M.: Comparative study of skin color detection and segmentation in HSV and YCbCr color space. Procedia Comput. Sci. **57**, 41–48 (2015)
12. Nadian-Ghomsheh, A.: Pixel-based skin detection based on statistical models. J. Telecommun. Electron. Comput. Eng. (JTEC) **8**, 7–14 (2016)
13. Oghaz, M.M.D., Argyriou, V., Monekosso, D., Remagnino, P.: Skin identification using deep convolutional neural network. In: Bebis, G., Boyle, R., Parvin, B., Koracin, D., Ushizima, D., Chai, S., Sueda, S., Lin, X., Lu, A., Thalmann, D., Wang, C., Xu, P. (eds.) Advances in Visual Computing, pp. 181–193. Springer International Publishing, Cham (2019). https://doi.org/10.1007/978-3-030-33720-9_14
14. Kim, Y., Hwang, I., Cho, N.I.: Convolutional neural networks and training strategies for skin detection. In: 2017 IEEE International Conference on Image Processing (ICIP), pp. 3919–3923 (2017). https://doi.org/10.1109/ICIP.2017.8297017
15. Lecun, Y., Jackel, L.D., Bottou, L., Cartes, C., Denker, J.S., Drucker, H., Müller, U., Säckinger, E., Simard, P., Vapnik, V., et al.: Learning algorithms for classification: a comparison on handwritten digit recognition. In: Neural Networks: The Statistical Mechanics Perspective, pp. 261–276. World Scientific (1995)
16. Lecun, Y., Bottou, L., Bengio, Y., Haffner, P.: Gradient-based learning applied to document recognition. Proc. IEEE **86**, 2278–2324 (1998). https://doi.org/10.1109/5.726791
17. Wang, G., Gong, J.: Facial expression recognition based on improved LeNet-5 CNN. In: 2019 Chinese Control and Decision Conference (CCDC), pp. 5655–5660 (2019). https://doi.org/10.1109/CCDC.2019.8832535
18. Zhang, C.-W., Yang, M.-Y., Zeng, H.-J., Wen, J.-P.: Pedestrian detection based on improved LeNet-5 convolutional neural network. J. Algorithms Comput. Technol. **13**, 1748302619873601 (2019). https://doi.org/10.1177/1748302619873601
19. Zhang, C., Yue, X., Wang, R., Li, N., Ding, Y.: Study on traffic sign recognition by optimized Lenet-5 algorithm. Int. J. Patt. Recogn. Artif. Intell. **34**, 2055003 (2019). https://doi.org/10.1142/S0218001420550034
20. Wang, T., Lu, C., Shen, G., Hong, F.: Sleep apnea detection from a single-lead ECG signal with automatic feature-extraction through a modified LeNet-5 convolutional neural network. PeerJ**7**, e7731 (2019) https://doi.org/10.7717/peerj.7731
21. Casati, J.P.B., Moraes, D.R., Rodrigues, E.L.L.: SFA: a human skin image database based on FERET and AR facial images. In: IX workshop de Visao Computational, Rio de Janeiro (2013)
22. Phillips, P.J., Moon, H., Rizvi, S.A., Rauss, P.J.: The FERET evaluation methodology for face-recognition algorithms. IEEE Trans. Pattern Anal. Mach. Intell. **22**, 1090–1104 (2000). https://doi.org/10.1109/34.879790
23. Martinez, A., Benavente, R.: The AR face database. Tech. Rep. 24 CVC Technical Report. (1998)
24. Wang, X., Xu, H., Wang, H., Li, H.: Robust real-time face detection with skin color detection and the modified census transform. In: 2008 International Conference on Information and Automation, pp. 590–595 (2008). https://doi.org/10.1109/ICINFA.2008.4608068

CRAN: An Hybrid CNN-RNN Attention-Based Model for Arabic Machine Translation

Nouhaila Bensalah, Habib Ayad, Abdellah Adib, and Abdelhamid Ibn El Farouk

Abstract Machine Translation (MT) is one of the challenging tasks in the field of Natural Language Processing (NLP). The Convolutional Neural Network (CNN)-based approaches and Recurrent Neural Network (RNN)-based techniques have shown different capabilities in representing a piece of text. In this work, an hybrid CNN-RNN attention-based neural network is proposed. During training, Adam optimizer algorithm is used, and then, a popular regularization technique named dropout is applied in order to prevent some learning problems such as overfitting. The experiment results show the impact of our proposed system on the performance of Arabic machine translation.

1 Introduction

MT is an intricate process that uses a computer application to translate text or speech or even capture from one natural language to another [2]. Many approaches from traditional rule-based approaches to the recent neural methods have been applied since the introduction of MT [4, 7, 8, 25]. Due to the excellent performance that achieves Deep Learning (DL) on difficult problems such as question answering [3, 6], sentiment analysis [3, 9], and visual object recognition [14, 20] for a small amount of steps, Google has investigated the use of DL to develop its own MT system. In the same context, Linguee team have developed DeepL based on CNNs that support a various number of languages such as French, Spanish, English, and

N. Bensalah (✉) · H. Ayad · A. Adib
Team Networks, Telecoms & Multimedia, University of Hassan II Casablanca, Casablanca 20000, Morocco
e-mail: nouhaila.bensalah@etu.fstm.ac.ma

A. Adib
e-mail: abdellah.adib@fstm.ac.ma

A. Ibn El Farouk
Teaching, Languages and Cultures Laboratory Mohammedia, Mohammedia, Morocco

M. Ben Ahmed et al. (eds.), *Networking, Intelligent Systems and Security*,
Smart Innovation, Systems and Technologies 237,
https://doi.org/10.1007/978-981-16-3637-0_7

others. MT systems based on DL often use a sequence-to-sequence model in order to map between the input and the target sequences directly.

The whole sequence-to-sequence process is described by Sutskever et al. in [25]. In short, the first step is to compute a representation of the source sentence using an encoder which can be Long Short Term Memory (LSTM), or Gated Recurrent Unit (GRU). In view to extract the relevant features from that encoder, an attention module can be used. Finally, the obtained vectors are transferred to the decoder which generates the output sentence. The aim of this research is to exploit the full advantages of CNN and RNN in order to map the input sentence to a low dimensional vector sequence. Specifically, in the first stage, a conventional CNN is used. Hence, the encoding of the input sentence will be processed in parallel to properly manipulate and optimize GPU hardware during training. And due to the extensive attention that has gained RNN in recent years, an improved architecture of RNN, namely Bidirectional GRU (BiGRU), is performed on the same input sequence. Finally, a mechanism of self-attention is applied to merge the features generated by both BiGRU and CNN, and then, the obtained vectors will be utilized as inputs to a GRU layer in order to generate the translation of the input sentence. The used attention mechanism could be considered as an instance of the widely known attention mechanism [4], as well as its recent variants; i.e. the self attention [21] and the inner attention [11]. In this case, the attention mechanism is performed within the same input sentence rather than the alignment of the output and the input sentences. The remainder of this paper is organized as follows. Section 2 describes the proposed model for implementing the Arabic MT system. Section 3 details the experimental setup and results. Finally, the conclusion is summarized in Sect. 4.

2 CRAN Arabic MT Model

Selecting the best features, from an input sequence, lies at the core of any MT system. Most of the state-of-the-art MT systems employ neural network-based approaches such as the CNN and the RNN-based architectures. In spite of their easy deployment and generally their capabilities in representing a piece of text, they represent some disadvantages [19]. In the CNN-based approaches, each source sentence is presented as a matrix by concatenating the embedding vector sequence as columns. Then, the CNN is applied to identify the most influential features about the input sentence. Nonetheless, these techniques could uniquely learn regional features, and it is not straightforward to handle with the long-term dependency between the features extracted from the source sentence. On the other hand, employing the (GRU or LSTM)-based approaches allow the model to generate effective sentences representation using temporal features since they capture the long-term dependencies between the words of a source sentence. However, these approaches represent some weaknesses, most significantly, their inability to distinguish between the words that contribute to the selection of the best features. This is due to the fact that they manipulate each word in a source sentence equally. Since the RNN and the CNN can complement each other for MT task, various solutions have been proposed [1, 17]. Most of the existing methods that combine these two models focus on applying the LSTM or GRU on the top of the CNNs. Consequently, they could not be applied directly on the source sentence, and hence, some features will be lost. In order to incorporate the full strength of these two groups of architectures, we present in this paper a novel architecture based on the use of the CNN and BiGRU architectures applied both on the input data. The proposed model is depicted in Fig. 1, and it is summarized as following:

1. The input sentence is preprocessed and then decomposed into words; each one is represented as a fixed-dimension vector using FastText model [10]. The obtained vectors are concatenated to generate a fixed-size matrix.
2. Several convolutional filters are applied on the resulting matrix. Each convolution filter has a view over the entire source sequence, from which it picks features. To extract the maximum value for each region determined by the filter, a max-pooling layer is used.
3. In order to deal with the long-term dependency problem and extract the temporal features from the same input sentence, a BiGRU is applied on the whole input sentence.
4. An attention mechanism is then performed with the objective of merging the useful temporal features and the regional ones obtained, respectively, by the BiGRU layer and the CNN model acting on the whole input sequence.
5. To generate the output sentence from the obtained vector sequence, a GRU layer is used, and a Softmax layer is then applied to generate the translation of the input sequence.

Hereafter, we will detail the different layers through which the whole process passes.

2.1 Input Layer

This is the first layer in our model, and it is used to represent each word in a sentence as a vector of real values. First, the input sentence is decomposed into words. Then, to obtain the same length for all the input sentences, a padding technique is performed on the sentences which are short (length $< n$) where n is the maximum length of the source sentences. Then, every word is embedded as $\mathbf{g} = (g_1, \ldots, g_n)$ where g_i ($i \in (1, 2, .., n)$) represents a column in the embedding matrix. Finally, the obtained vectors, called the embedding vector sequence, will be fed into the BiGRU layer and the CNN model.

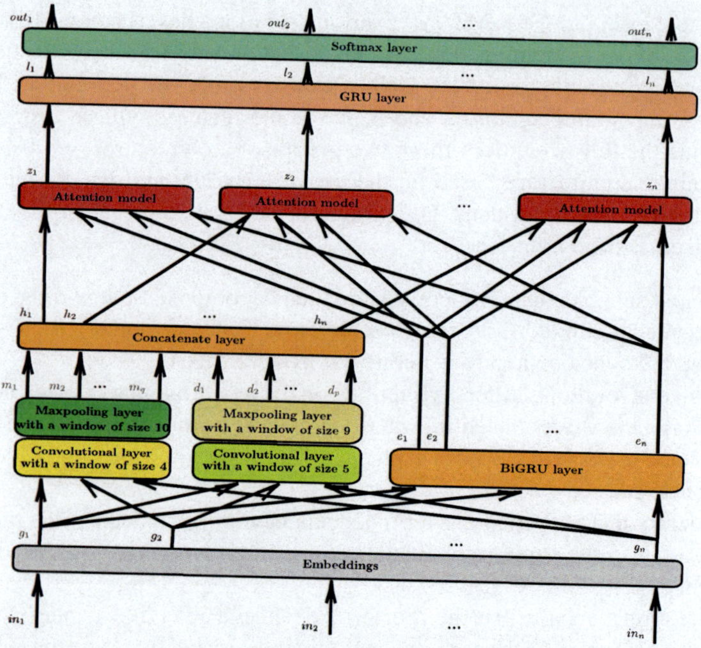

Fig. 1 Block diagram of the overall architecture of the proposed approach

2.2 *Conventional CNN*

The CNN architecture was developed by LeCun et al. [20] and has risen to prominence as a state of the art in MT. In this study, a conventional CNN is investigated to extract the most influential features from the embedding vector sequence. Generally, a conventional CNN model consists of the following layers:

- **Convolutional layer**: utilizes a set of filters (kernels) to convert the embedding vector sequence **g** into feature maps.
- **Nonlinearity**: between convolutional layers, an activation function, such as tan h or ReLU which represents, respectively, a tangent hyperbolic and rectified Linear Unit functions, is applied to the obtained feature maps to introduce nonlinearity into the network. Without this operation, the network would hence struggle with complex data. In this paper, ReLU was adopted, which is defined as:

$$f(x) = \max(0, x) \tag{1}$$

- **Pooling layer**: Its main role is to reduce the amount of parameters and computation in the network by decreasing the feature maps size. Two common methods used in the pooling operation are:

  - Average pooling: outputs the average value in a region determined by the filter.
  - Maximum pooling (or Max pooling): The output is the maximum value over a region processed by the considered filter.

  In this paper, we used the max pooling to preserve the largest activation in the feature maps.
- **Dropout layer**: its role is to randomly drop units (along with their connections) from the neural network during training to avoid overfitting.

2.3 *BiGRU Layer*

Given the embedding vector sequence $\mathbf{g} = (g_1, \ldots, g_n)$, a standard RNN [15] generates the hidden vector sequence $\mathbf{e} = (e_1, \ldots, e_n)$. The main objective of RNN is to capture the temporal features from the source sentence. The output of the network is calculated by iterating the following equation from $t = 1$ to $t = n$:

$$e_t = f(W_{ge} g_t + W_{ee} e_{t-1} + b_e) \tag{2}$$

where the W terms are weight matrices, i.e., for example, W_{ge} denotes the input-hidden weight matrix, the b_e denotes the hidden bias vector, and f is the hidden layer activation function. To avoid the issue of the vanishing gradient that penalizes the standard RNN, GRU [12] is proposed to store the input information without a memory unit. A single GRU cell is illustrated in Fig. 2, and is defined as follows:

$$r_t = \sigma(W_{gr} g_t + W_{er} e_{t-1} + b_r) \tag{3}$$

$$u_t = \sigma(W_{gu} g_t + W_{eu} e_{t-1} + b_u) \tag{4}$$

$$ee_t = \tan h(W_{ge} g_t + W_{ee}(r_t \odot e_{t-1}) + b_e) \tag{5}$$

$$e_t = (1 - u_t) \odot e_{t-1} + u_t \odot ee_t \tag{6}$$

where $\tan h$ is the tangent hyperbolic function, σ the element-wise sigmoid activation function, $\odot$ is the element-wise Hadamard product, r and u are, respectively, the reset gate and update gate, all have the same size as the hidden vector e and the b terms are bias vectors, i.e., for example, b_u denotes the update bias vector.

In MT, the GRU architecture is motivated by two main reasons. First, such architecture has been shown to represent the sequential data by taking into account the previous data. Second, it is better at exploiting and capturing long-range context due to its gates that decide which data will be transferred to the output. However, the GRU is only able to make use of the amount of information seen by the hidden states at the previous steps. In order to exploit the future information as well, bidirectional RNNs

Fig. 2 Architecture of GRU

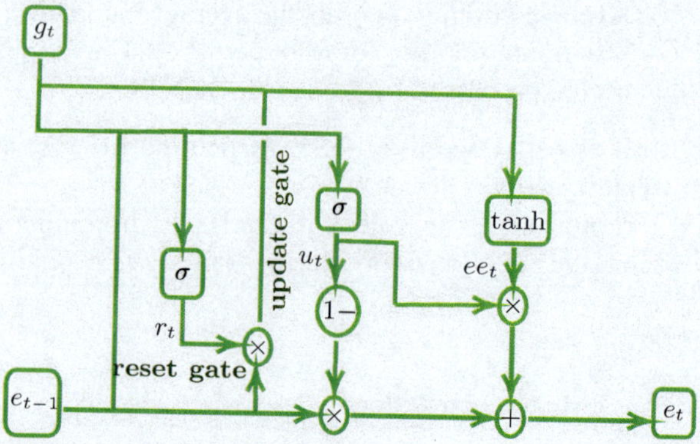

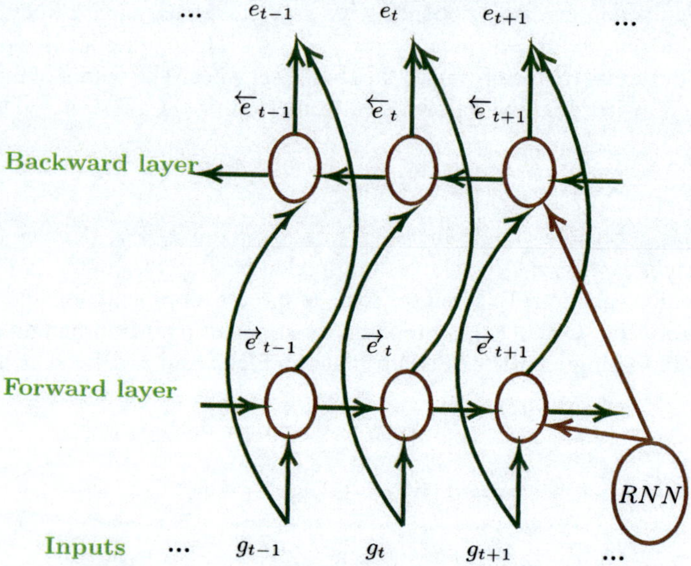

Fig. 3 Bidirectional RNN

(BRNNs) [24] are introduced. They process the data in two opposite directions with two separate hidden layers as illustrated in Fig. 3.

In this case, the forward hidden vector sequence $\overrightarrow{e}_t$ and the backward hidden vector sequence $\overleftarrow{e}_t$ are computed by iterating the backward layer from $t = 1$ to $t = n$ and the forward layer from $t = 1$ to $t = n$.

$$\overrightarrow{e}_t = \tan h(W_{g\overrightarrow{e}} g_t + U_{\overrightarrow{e}\overrightarrow{e}} e_{t-1} + b_{\overrightarrow{e}}) \tag{7}$$

$$\overleftarrow{e}_t = \tan h(W_{g\overleftarrow{e}} g_t + U_{\overleftarrow{e}\overleftarrow{e}} e_{t-1} + b_{\overleftarrow{e}}) \tag{8}$$

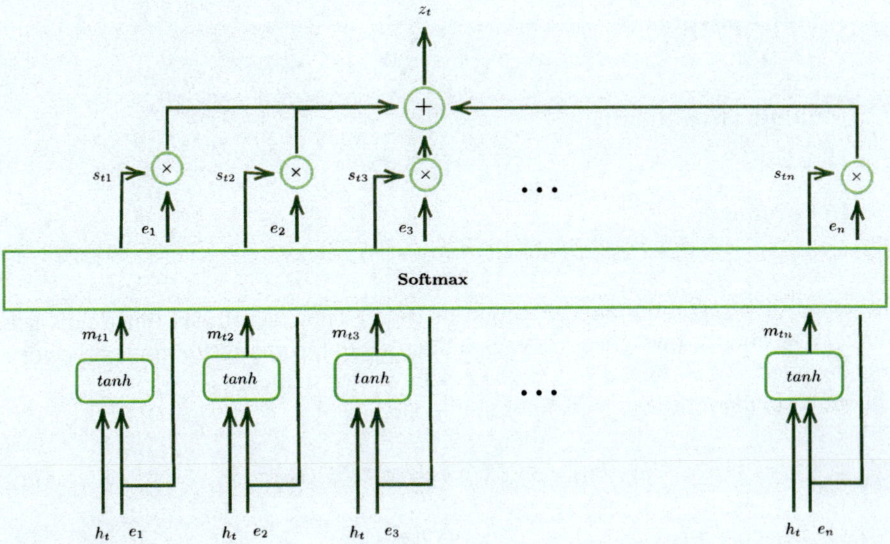

Fig. 4 Attention mechanism

In this way, a sentence could be represented as $\mathbf{e} = (e_1, e_2, .., e_n)$ where $e_t = [\overrightarrow{e}_t, \overleftarrow{e}_t]$. Combining BRNN with GRU gives bidirectional GRU [18] that could exploit long-range context in both input directions.

2.4 Attention Layer

The attention mechanism is one of the key components of our architecture. In this context, there are several works that can help to locate the useful words from the input sequence [4, 22]. Motivated by the ability of the CNN model to capture regional syntax of words and the capacity of the BiGRU model to extract temporal features of words, we aim to use the output vector sequence generated by the CNN, $\mathbf{h} = (h_1, h_2, \ldots, h_n)$, and the hidden vector sequence calculated by the BiGRU layer $\mathbf{e} = (e_1, e_2, \ldots, e_n)$ during the attention mechanism. In the proposed approach, at each step t ($t \in (1, 2, \ldots, n)$), a unique vector h_t and the hidden vector sequence $\mathbf{e} = (e_1, e_2, \ldots, e_n)$ are used to calculate the context vector z_t. The detail of the proposed attention mechanism computation process is illustrated in Fig. 4.
In short, the first step is to measure the similarity denoted as m_{tj} between the hidden vector e_j ($j \in (1, 2, ..n)$) generated by the BiGRU layer and the vector h_t ($t \in (1, 2, ..n)$) produced by the CNN. Three different methods could be used to calculate m_{tj}:

1. **Additive attention**:

$$m_{tj} = W_a^{\intercal} \tan h(W_e e_j + U_h h_t) \tag{9}$$

2. **Multiplicative attention**:

$$m_{tj} = e_j^{\intercal} W_m h_t \tag{10}$$

3. **Dot product**:

$$m_{tj} = e_j^{\intercal} h_t \tag{11}$$

where: W_a, W_e, U_h, W_m are the weight matrices. Then, a Softmax function is used to get a normalized weight s_{tj} of each hidden state e_j. Finally, the context vector z_t is computed as a weighted sum of the hidden vector sequence **e**.
In our case, the following equations are iterated from $j = 1$ to $j = n$:

$$m_{tj} = W_a^{\intercal} \tan h(W_e e_j + U_h h_t) \tag{12}$$

$$s_{tj} = \frac{\exp(m_{tj})}{\sum_{k=1}^{n} \exp(m_{tk}))} \tag{13}$$

The context vector sequence $\mathbf{z} = (z_1, z_2, \ldots, z_n)$ is generated by iterating the following equation from $t = 1$ to $t = n$:

$$z_t = \sum_{j=1}^{n} s_{tj} e_j \tag{14}$$

2.5 The Output Layer

In summary, the goal of our model is to map the input sentence to a fixed sized vector sequence $\mathbf{z} = (z_1, z_2, \ldots, z_n)$ using the CNN-BiGRU and a mechanism of attention. Then, a GRU layer is applied on the obtained vector sequence. Finally, we add a fully connected output layer with a Softmax activation function which gives, at each time step, the probability distribution across all the unique words in the target language. The predicted word at each time step is selected as the one with the highest probability.

2.6 Training and Inference

The Arabic MT process involves two main stages: training and inference. During the training stage, features extraction is performed after the built of the training Arabic-English sentences. It aims at providing a useful representation of an input sentence in such a way that it can be understandable by the model. Then, a CNN-BiGRU with a mechanism of attention followed by a GRU layer are applied on these obtained features. Finally, a Softmax layer is performed with the objective of optimizing the parameters of the neural network by comparing the model outputs with the target sequences (what we should achieve). After the training is done, the Arabic MT model is built and can be used to translate an input sentence with any help from the target sentences. The output sequence is generated word by word using the Softmax layer. Its main role during inference stage is to generate at each time step the probability distribution across all unique words in the target language.

3 Results and Evaluation

The experiments were conducted over our own Arabic-English corpus. It contains a total of 266,513 words in Arabic and 410,423 ones in English, and the amount of unique words was set to 23,159 words in Arabic and 8323 ones in English. The database was divided randomly into a training set, a validation set, and a testing set. 20,800 sentences for both Arabic and English languages were used for training, 600 sentences for validation and 580 for testing.

To build our corpora, we select 19,000 sentences from the UN dataset from the Web site.[1] In order to improve the performance of our model, we have used two other datasets. First, we select manually the best English-Arabic sentences from the Web site[2] which contains blog-posts, tweets in many languages. Finally, we have used the sentences in the English-Arabic pair which can be found in this Web site.[3]

In the following, we present a series of experiments for Arabic MT analysis to understand the practical utility of the proposed approach. As an evaluation metric, we compute the BLEU score [23], the GLEU score [27], and the WER score [26] which are the most commonly used in MT task.

3.1 Hyperparameters Choices

In this part, we investigated the impact of each hyperparameter on Arabic MT system to select the best ones to use during training. The performance of Arabic MT was first evaluated by varying the batch size to 32, 64, 128, and 256. Table 1 shows the BLEU, GLEU, and WER scores for different batch sizes.
From Table 1, we can see that the minimal WER score (WER score = 0.449) is reached using 64 as batch size. Examining the results, it seems that using a too low or too high batch size does not result in better performance.

[1] http://opus.nlpl.eu/.

[2] http://www.cs.cmu.edu.

[3] http://www.manythings.org.

Table 1 Arabic MT performance for different batch sizes

	Batch size	32	64	128	256
BLEU score	1-gram	0.571	0.575	0.512	0.479
	2-gram	0.577	0.578	0.518	0.470
	3-gram	0.592	0.590	0.523	0.473
	4-gram	0.603	0.602	0.527	0.478
GLEU score	1–4 gram	0.445	0.463	0.371	0.317
	1–3 gram	0.471	0.487	0.404	0.352
	1–2 gram	0.513	0.526	0.451	0.424
WER score		0.473	0.449	0.529	0.567

Table 2 Arabic MT performance with respect to the length of the input sentences

	Sentence length	10	20	30
BLEU score	1-gram	0.485	0.529	0.575
	2-gram	0.483	0.526	0.578
	3-gram	0.484	0.536	0.590
	4-gram	0.489	0.538	0.602
GLEU score	1 to 4 gram	0.324	0.401	0.463
	1–3 gram	0.369	0.426	0.487
	1–2 gram	0.445	0.472	0.526
WER score		0.589	0.546	0.449

Then, we evaluated the performance of Arabic MT by varying the sentence length to 10, 20, and 30 which is the maximum length of the source sentences. Table 2 illustrates the influence of the sentence length on the BLEU, GLEU, and WER scores.
The reported results show clearly that we get better performance as the sentence length increases. In our work, the optimal value of the sentence length is set to 30.

The performance of Arabic MT was evaluated for different optimization techniques. The results are reported in Table 3.

We can clearly observe that, globally, the best WER score has been reached for Adam optimizer.

Next, we increase the number of units in the BiGRU and the output layers so as to compare the performance of Arabic MT achievable by our model in this case. Table 4 reports the results obtained using different values of units.
We can notice that changing the number of units affects the performance of Arabic MT and the best results are obtained using 400 units.

In the following, we analyze the impact of increasing the number of layers on the Arabic MT quality. The results are reported in Table 5.

Table 3 Arabic MT performance for different optimization techniques

	Optimization technique	SGD	RMSprop	Adam
BLEU score	1-gram	0.263	0.573	0.575
	2-gram	0.225	0.580	0.578
	3-gram	0.223	0.597	0.590
	4-gram	0.248	0.610	0.602
GLEU score	1–4 gram	0.123	0.460	0.463
	1–3 gram	0.151	0.483	0.487
	1–2 gram	0.194	0.522	0.526
WER score		0.758	0.455	0.449

Table 4 Arabic MT performance for different number of units

	Number of units	100	200	300	400
BLEU score	1-gram	0.507	0.501	0.547	0.575
	2-gram	0.493	0.492	0.546	0.578
	3-gram	0.492	0.493	0.556	0.590
	4-gram	0.491	0.496	0.562	0.602
GLEU score	1–4 gram	0.348	0.342	0.406	0.463
	1–3 gram	0.384	0.377	0.437	0.487
	1–2 gram	0.437	0.429	0.482	0.526
WER score		0.529	0.542	0.497	0.449

Table 5 Arabic MT performance for different numbers of layers

	Number of layers	1	2	3
BLEU score	1-gram	0.568	0.575	0.564
	2-gram	0.574	0.578	0.569
	3-gram	0.588	0.590	0.582
	4-gram	0.598	0.602	0.593
GLEU score	1–4 gram	0.448	0.463	0.452
	1–3 gram	0.474	0.487	0.476
	1–2 gram	0.514	0.526	0.514
WER score		0.465	0.449	0.462

Table 6 Arabic MT performance for different CNN filter sizes

	Size	2	3	4	5
BLEU score	1-gram	0.567	0.569	0.577	0.573
	2-gram	0.569	0.572	0.579	0.575
	3-gram	0.579	0.584	0.591	0.587
	4-gram	0.591	0.594	0.602	0.599
GLEU score	1–4 gram	0.450	0.455	0.462	0.455
	1–3 gram	0.476	0.479	0.487	0.480
	1–2 gram	0.516	0.519	0.526	0.521
WER score		0.458	0.452	0.448	0.454

The reported results in Table 5 show that using a too low or too high number of layers does not result in better performance. For next experiments, we choose the number of layers to be 2.

Finally, based on a manual tuning, we initialize the learning rate with a value of 0.001, and with the beginning of overfitting, we start to multiply this value by 0.4 at each 2 epochs until it falls below 10^{-7}. If the performance of the model on the validation set stops to grow, the early stopping technique based on the validation accuracy is performed. It aims to stop the training process after 5 epochs. More details of these techniques and other tips to reach better training process are described in [16].

The performance of Arabic MT was also evaluated by varying the size of CNN filters to 2, 3, 4, and 5. Table 6 shows the BLEU, GLEU, and WER scores for different CNN filter sizes.

From Table 6, we can see that the best Arabic MT scores come from the CNN filter of size 4 and 5 which have been concatenated in this work.

3.2 *The Impact of Different RNN Variants on Arabic MT*

To study the performance of the Arabic MT under different RNNs, four different combinations of RNNs have been used:

1. BiLSTM for the encoding process (discussed in 2.3) and GRU in the output layer.
2. BiGRU for the encoding process (discussed in 2.3) and LSTM in the output layer.
3. BiLSTM for the encoding process (discussed in 2.3) and LSTM in the output layer.
4. BiGRU for the encoding process (discussed in 2.3) and GRU in the output layer.

It is clear from Table 7 that the combination 4, which is the proposed approach, gives better performance in terms of BLEU, GLEU, and WER scores. Furthermore, one of the attractive characteristics of our model is its ability to train faster than the combinations 1, 2, and 3 (Total Time = 610 s).

Table 7 Arabic MT performance for different combinations of RNNs

	Combination	1	2	3	4
BLEU score	1-gram	0.544	0.380	0.344	0.575
	2-gram	0.549	0.357	0.328	0.578
	3-gram	0.564	0.359	0.336	0.590
	4-gram	0.577	0.371	0.358	0.602
GLEU score	1–4 gram	0.414	0.202	0.173	0.463
	1–3 gram	0.423	0.239	0.207	0.487
	1–2 gram	0.467	0.295	0.258	0.523
WER score		0.511	0.663	0.704	0.449
Total time (s)		1270	2027	1908	610

Table 8 Comparison between the Arabic MT performance of RCAN and our approach

		RCAN	Our approach
BLEU score	1-gram	0.535	0.575
	2-gram	0.545	0.578
	3-gram	0.555	0.590
	4-gram	0.565	0.602
GLEU score	1–4 gram	0.407	0.463
	1–3 gram	0.434	0.487
	1–2 gram	0.476	0.523
WER score		0.511	0.449

3.3 *Comparison with RCAN Model*

In this part, another model denoted as RCAN is proposed for comparison. In this case, the context vector z_t $(t \in (1, 2, .., n))$ is calculated using the vector sequence $\mathbf{h} = (h_1, h_2, \ldots, h_n)$ and the hidden vector e_t.

Table 8 illustrates the results of Arabic MT using the proposed architecture and compares them to the results of RCAN.

It can be seen, from Table 8, that our approach achieved relatively ideal overall performance using our corpus and improved the performance by 6.2% in terms of WER score. These findings may be explained by the use of the temporal vector sequence generated by the BiGRU, instead of the regional vector sequence produced by CNN, to calculate the context vector. In this case, the model becomes able to automatically search for parts of a source sentence that are relevant to predict a target word.

Table 9 Comparison with state-of-the art works using our own corpus

	[4]	[13]	Our approach
BLEU score	0.493	0.485	0.575
WER score	0.492	0.515	0.449

3.4 Comparison with Previous Related Works and Qualitative Evaluation

Because this work is inspired by the approaches proposed by Bahdanau et al. [4] and Cho et al. [13], the performance of Arabic MT is evaluated in terms of BLEU, GLEU, and WER scores reached using our model and these works. Table 9 summarizes the obtained results for Arabic MT task on our corpus with the considering literature works.

We can clearly observe from Table 9 that in all the cases the best performance is achieved using our approach with a limited vocabulary. This is likely due to the fact that our model does not encode the whole input sentence into a single vector. Instead, it focus on the relevant words of the source sentence during the encoding process. As an example, consider this source sentence from the test set:

جميلة في سبتمبر باريس لطيفة في نوفمبر ، لكنها عادة

Our model translated this sentence into:
paris is pleasant during November but it is usually beautiful in September
The truth is:
paris is nice during November but it is usually beautiful in September
The proposed approach correctly translated the source sentence, but it replaced *nice* with *pleasant*.
Let us consider another sentence from the test set:

هذه كتبي

Our model translated this sentence into:
these books are my books
The truth is:
these books belong to me
These qualitative observations demonstrates that the proposed approach does not translate the input sentence as the truth, but instead it preserves the original meaning of the source sentence.

4 Conclusion

In this paper, we proposed the use of both CNN and BiGRU with the mechanism of attention system for the task of MT between English and Arabic texts. The motivation for introducing such a system is to improve the performance of Arabic MT by capturing the most influential words in the input sentences using our corpora. In this context, we described first how the used corpus is produced. A comparative performance analysis of the hyperparameters is performed. As expected, the experimental results show that the proposed method is capable of providing satisfactory performance for Arabic MT. As part of future work, we aim to use saliency to visualize and understand neural models in NLP [5].

References

1. Alayba, A.M., Palade, V., England, M., Iqbal, R.: A combined cnn and lstm model for arabic sentiment analysis. In: International Cross-Domain Conference for Machine Learning and Knowledge Extraction, pp. 179–191 (2018)
2. Alqudsi, A., Omar, N., Shaker, K.: Arabic machine translation: a survey. Artif. Intell. Rev. **42**(4), 549–572 (2014)
3. Antoun, W., Baly, F., Hajj, H.M.: Arabert: transformer-based model for arabic language understanding (2020) . *CoRR abs/2003.00104*
4. Bahdanau, D., Cho, K., Bengio, Y.: Neural machine translation by jointly learning to align and translate. In: Bengio, Y., LeCun, Y. (eds) 3rd International Conference on Learning Representations, ICLR (2015)
5. Bastings, J., Filippova, K.: The elephant in the interpretability room: why use attention as explanation when we have saliency methods? In: Proceedings of the Third BlackboxNLP Workshop on Analyzing and Interpreting Neural Networks for NLP, pp. 149–155. Association for Computational Linguistics (2020)
6. Bensalah, N., Ayad, H., Adib, A., el farouk, A.I.: Combining word and character embeddings in Arabic Chatbots. In: Advanced Intelligent Systems for Sustainable Development, AI2SD'2020, Tangier, Morocco (2020)
7. Bensalah, N., Ayad, H., Adib, A., Farouk, A.I.E.: LSTM or GRU for Arabic machine translation? Why not both! In: International Conference on Innovation and New Trends in Information Technology, INTIS 2019, Tangier, Morocco, Dec 20–21 (2019)
8. Bensalah, N., Ayad, H., Adib, A., Farouk, A.I.E.: Arabic machine translation based on the combination of word embedding techniques. In: Intelligent Systems in Big Data, Semantic Web and Machine Learning (2020)
9. Bensalah, N., Ayad, H., Adib, A., Farouk, A.I.E.: Arabic sentiment analysis based on 1-D convolutional neural network. In: International Conference on Smart City Applications, SCA20 (2020)
10. Bojanowski, P., Grave, E., Joulin, A., Mikolov, T.: Enriching word vectors with subword information. Trans. Assoc. Comput. Linguist., 135–146 (2017)
11. Cheng, J., Dong, L., Lapata, M.: Long short-term memory-networks for machine reading. In: Su, J., Carreras, X., Duh, K. (eds) Proceedings of the 2016 Conference on Empirical Methods in Natural Language Processing, EMNLP, pp. 551–561 (2016)
12. Cho, K., van Merrienboer, B., Bahdanau, D., Bengio, Y.: On the properties of neural machine translation: encoder-decoder approaches. In: Proceedings of SSST@EMNLP 2014, Eighth Workshop on Syntax, Semantics and Structure in Statistical Translation, pp. 103–111. Association for Computational Linguistics (2014)
13. Cho, K., Van Merriënboer, B., Gulcehre, C., Bahdanau, D., Bougares, F., Schwenk, H., Bengio, Y.: Learning phrase representations using rnn encoder-decoder for statistical machine translation (2014). *arXiv preprint* arXiv:1406.1078
14. Ciresan, D.C., Meier, U., Schmidhuber, J.: Multi-column deep neural networks for image classification. In: 2012 IEEE Conference on Computer Vision and Pattern Recognition, Providence, pp. 3642–3649. IEEE Computer Society (2012)

15. Elman, J.L.: Finding structure in time. Cognit. Sci. **14**(2), 179–211 (1990)
16. Feurer, M., Hutter, F.: Hyperparameter optimization. In: Automated Machine Learning, pp. 3–33. Springer (2019)
17. Gehring, J., Auli, M., Grangier, D., Dauphin, Y.N.: A convolutional encoder model for neural machine translation. In: Proceedings of the 55th Annual Meeting of the Association for Computational Linguistics, ACL 2017, pp. 123–135. Association for Computational Linguistics (2017)
18. Graves, A., Schmidhuber, J.: Framewise phoneme classification with bidirectional lstm and other neural network architectures. Neural Netw. **18**(5–6), 602–610 (2005)
19. Guo, L., Zhang, D., Wang, L., Wang, H., Cui, B.: Cran: a hybrid CNN-RNN attention-based model for text classification. In: International Conference on Conceptual Modeling, pp. 571–585. Springer (2018)
20. LeCun, Y., Bottou, L., Bengio, Y., Haffner, P., et al.: Gradient-based learning applied to document recognition. Proc. IEEE **86**(11), 2278–2324 (1998)
21. Lin, Z., Feng, M., dos Santos, C.N., Yu, M., Xiang, B., Zhou, B., Bengio, Y.: A structured self-attentive sentence embedding. In: 5th International Conference on Learning Representations, ICLR (2017)
22. Luong, M., Pham, H., Manning, C.D.: Effective Approaches to Attention-based Neural Machine Translation. *CoRR abs/1508.04025* (2015)
23. Papineni, K., Roukos, S., Ward, T., Zhu, W.-J.: Bleu: a method for automatic evaluation of machine translation. In: Proceedings of the 40th Annual Meeting on Association for Computational Linguistics, pp. 311–318. Association for Computational Linguistics (2002)
24. Schuster, M., Paliwal, K.K.: Bidirectional recurrent neural networks. IEEE Trans. Sig. Process. **45**(11), 2673–2681 (1997)
25. Sutskever, I., Vinyals, O., Le, Q.: Sequence to sequence learning with neural networks. Advances in NIPS (2014)
26. Wang, Y.-Y., Acero, A., Chelba, C.: Is word error rate a good indicator for spoken language understanding accuracy. In: 2003 IEEE Workshop on Automatic Speech Recognition and Understanding (IEEE Cat. No. 03EX721), pp. 577–582. IEEE (2003)
27. Wu, Y., Schuster, M., Chen, Z., Le, Q.V., Norouzi, M., Macherey, W., Krikun, M., Cao, Y., Gao, Q., Macherey, K., Klingner, J., Shah, A., Johnson, M., Liu, X., Kaiser, L., Gouws, S., Kato, Y., Kudo, T., Kazawa, H., Stevens, K., Kurian, G., Patil, N., Wang, W., Young, C., Smith, J., Riesa, J., Rudnick, A., Vinyals, O., Corrado, G., Hughes, M., Dean, J.: Google's neural machine translation system: bridging the gap between human and machine translation. *CoRR abs/1609.08144* (2016)

Impact of the CNN Patch Size in the Writer Identification

Abdelillah Semma, Yaâcoub Hannad, and Mohamed El Youssfi El Kettani

Abstract Writer identification remains a very interesting challenge where many researchers have tried to find the various parameters which can help to find the right writer of a handwritten text. The introduction of deep learning has made it possible to achieve unprecedented records in the field. However, the question to ask, what size of patch to use to train a CNN model in order to have the best performance? In our paper, we try to find an answer to this question by investigating the results of the use of several patch sizes for a Resnet34 model and two languages Arabic and French from the LAMIS-MSHD dataset.

1 Introduction

Writing remains one of the great foundations of human civilization for communication and the transmission of knowledge. Indeed, many objects that are around us are presented in the form of writing: signs, products, newspapers, books, forms ...

Allowing the machine to read and capture more information will surely help in the process of identifying the authors of handwritten documents in a faster and more efficient manner. Indeed, with the advent of new information technologies, and with the increase in the power of machines, the automation of processing operations (editing, searching, and archiving) seems inevitable. Therefore, a system that enables the machine to understand human handwriting is needed.

Writer recognition is used to identify the author or writer of a handwritten text using a database containing samples of training writing. This challenge is not easy

A. Semma (✉) · M. E. Y. El Kettani
Ibn Tofail University, Kenitra, Morocco
e-mail: abdelillah.semma@uit.ac.ma

M. E. Y. El Kettani
e-mail: elkettani@univibntofail.ac.ma

Y. Hannad
National Institute of Statistics and Applied Economics, Rabat, Morocco
e-mail: y.hannad@insea.ac.ma

M. Ben Ahmed et al. (eds.), *Networking, Intelligent Systems and Security*,
Smart Innovation, Systems and Technologies 237,
https://doi.org/10.1007/978-981-16-3637-0_8

because a person's writing style depends on several factors such as their mental and physical state, their pen, the position of the paper, and the writer at the time of writing and the font size, which can vary according to the need.

The identification of the writers of handwritten documents touches several fields such as criminology which in some cases seeks to identify the exact person who produced a handwritten document. Writer identification also helps to recognize the name of the author of a book whose writer is not known. For a long time, human experts has tried to guess the writer of a manuscript manually, something which is not easy, that is why researchers have tried to design automatic systems for identifying writers.

For the writer identification, we proceed generally by three main steps: The pre-processing phase to prepare the handwritten images to the processing phase. The feature extraction phase that allows the extraction of the characteristics of images or parts of images in vectors. The last phase is that of classification where one seeks to calculate the distance between the test features and those of training in order to know the minimum distance which corresponds to the images of the requested author.

In writer recognition, we distinguish between two types: writer identification and writer retrieval. In the process of writer identification, the system must find the right writer of a handwritten document through a training database. While in the process of writer retrieval, we must find all handwritten documents similar to a test document.

The key highlights of our study are:

- Study of the performance induced by different patch sizes.
- Investigation of the probable relationship between the language of the dataset and the performance of diverse patch sizes.
- Study of the impact of the choice of the patch size on the rate of attribution of test patches to the right writer.
- Test the performance of several patch sizes by conducting experiments on two Arabic and French datasets.

The content of our paper is presented in four sections. In the following section, we present the old research works that have been interested in the writer identification, and we focus on the works that have used deep learning. In Sect. 3, we explain the methodology adopted as well as the dataset and the CNN used. The representation of the tests performed will be in Sect. 4. Finally, we end with a brief conclusion.

2 Related Work

Among the earliest work in the field of offline writer identification is that of [18] who employed the multichannel Gabor filtering technique to identify 1000 test scripts from 40 writers. He obtained an accuracy of 96.0%. Srihari et al. [20] tried to identify the writings of 1500 people in the USA by taking the overall characteristics of their writings such as line separation, slant, and character shapes. Other works have focused on the use of descriptors based on LBP [12], LPQ [4], GLCM [5], OBIF [16], or HOG [11]. While other researchers have proposed systems based on codebook-based graphemes [3] or codebook-based small fragments [19].

AlexNet [15] success in ImageNet competition of large-scale visual recognition challenge (LSVRC) in 2012 allowed the entry of the era of deep learning in the field of image recognition. Thus, in 2015, [10] used patches of 56 × 56 to train a neorones network comprising three convolutional layers and two others fully connected. The study achieved an accuracy of 99.5% in the ICDAR-2011 database, 88.5% in ICDAR-2013, and 98.9% in CVL.

Xing and Qia [21] tested a deepWriter with several patches sizes 227 × 227, 131 × 131, and 113 × 113. Their experiments carried out in IAM and HWDB datasets achieved an accuracy: 99.01% on 301 IAM writers and 93.85% on 300 HWDB writers.

Yang et al. [22] proposed a deep CNN DCNN with patches size 96 × 96 which gave results with an accuracy of 95.72% on NLPR handwriting database (Chinese text) and 98.51% on NLPR (English text).

Christlein et al. [6] used a GMM encoding vector of a CNN layer to identify the writers of the ICDAR13 and CVL datasets and exemplar SVM for classification. The same author published another study [7] in which he uses Resnet as CNN and the cluster indices of the clustered SIFT descriptors of each image as the targets and the SIFT 32 × 32 patches as input data. The VLAD encoding of activations of the penultimate layer was considered as local feature descriptors. Classification with an SVM exemplar gave an accuracy of 88.9% on Historical-WI and 84.1% on CLaMM16. In [8], the same author conducted experiments in three datasets KHATT, CVL, and ICDAR13 where he used a Resnet-34 with 32 × 32 patches and the encoding of the penultimate layer as the local descriptor.

Rehman et al. [17] extract local descriptors from CNN activations features. They used QUWI as a test database. The width of each image is resized to 1250 pixels with respect for the width/height ratio. The CNN used is Alexnet with patches of 227×227 and a data-augmentation of the sharped, contoured, sharped contours, and the negatives form version of these patches. They conduct their experiments using the outputs of each of the seven layers to achieve scores of 92.78% on English, 92.20% on Arabic, and 88.11% on Arabic and English.

As we have seen in these works, several sizes of patches were used. So, the question to ask is what is the best patch size to train a CNN and have higher identification rates.

We will try to answer this question in this paper by testing several patch sizes for a Resnet-34 and a bilingual dataset LAMIS-MSHD [9].

3 Proposed Approach

In this section, we present the methodology adopted to verify the impact of patch sizes on the performance of convolutional networks in the field of writer identification.

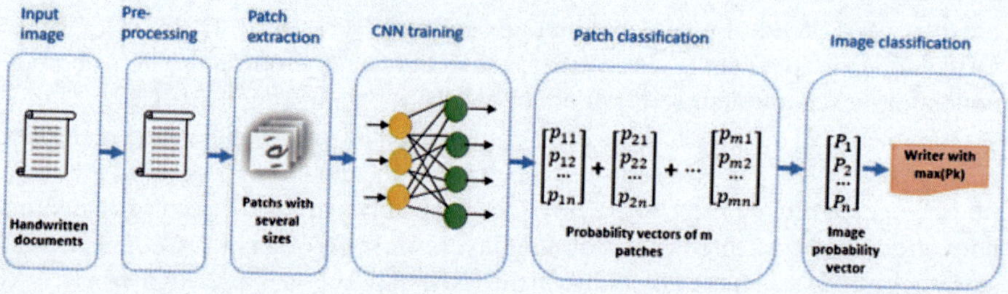

Fig. 1 CNN classification

This methodology is based on the following points:

- Apply a preprocessing step to different images
- Extract random patches of size 32×32 and save the locations of the patch centers to take them into consideration when extracting other patch sizes.
- Train the convolutional neural network (CNN) on the patches extracted from the training images.
- Classify the test images by passing the test patches through the CNN to predict the corresponding writer.

Figure 1 represents the main steps of the methodology adopted.

3.1 Preprocessing

Before segmenting an image and extracting the CNN input patches, a preprocessing phase is necessary. In this phase, all the images are converted to gray mode. Then, we try to correct the skew at the level of each image using the skew algorithm developed by Abecassis [1].

3.2 Extraction of Patches

Since our current study is based on the types or sizes of patches, then we opt for seven main sizes (32 × 32, 64 × 64, 100 × 100, 125 × 125, 150 × 150, 175 × 175, and 200 × 200).

As we know, the CNN input patches can have several sizes in terms of their width and height, and each type of CNN has a minimum size that must absolutely be respected, and this depends on the types of layers contained in the CNN. For example, a max-pooling or average-pooling layer of pool size of 2 × 2 implies that the size of the inputs of the next layer will be that of the previous layer divided by two. Knowing that the size of the input of the last convolutional layer must not be less than (1,1). In our study, we use a Resnet-34 which requires a minimum size of 32×32.

So we opted for patches of size greater than or equal to 32×32. In our study, we limit ourselves to square patches where the width is equal to the height.

The patches were extracted randomly the first time for each dataset with the condition that the center of each patch is a black pixel (containing text). The centers of each image are then saved to a file.

To be able to make a fair comparison of the performance resulting from the use of each patch size, we extracted the different patch sizes from the same centers saved during the first extraction.

We took 400 patches from each image, which gave us 1800 training patches, 200 validation patches, and 400 test patches for each writer.

3.3 CNN Training

In our study, we employ a convolution network which proved its worth in the ImageNet competition of ILSVRC 2015 by taking the first place. The residual networks are known by residual blocks which allow using the skip connection technique to skip 2 or 3 layers and thus save the identity of the input of the block. In Resnet-34, the residual block allows the skip connections of two layers. In the original version of ResNet-34 [14], the addition of the identity is done before the application of the activation function, but in the modified version [13] and which is used in our study, the activation function is applied before the addition of identity.

We trained the CNN with batch sizes ranging from 500 for the patch size 32×32 to 20 for the patch size 200×200. For the value of the learning rate, we start with a value of 0.001 which would be divided by 10 after the 6th, 9th and 12th epoch.

3.4 Database

Our study was carried out in the bilingual LAMIS-MSHD database. This dataset contains 1300 signatures, 600 handwritten documents in Arabic, 600 others in French, and 21,000 digits. To design this database, 100 Algerians including 57% female and 43% male of different age, and level of education were invited to complete 1300 forms. The main purpose of the database is to provide scientific researchers with a multi-script platform. The average line height is approximately 139 pixels and 127 for the Arabic and French version, respectively.

To train our CNN we took 75% of the data, for the validation phase, we took 8% and the rest 17% for the test phase which corresponds to one image per writer.

We can see some samples of the LAMIS-MSHD dataset in Fig. 2

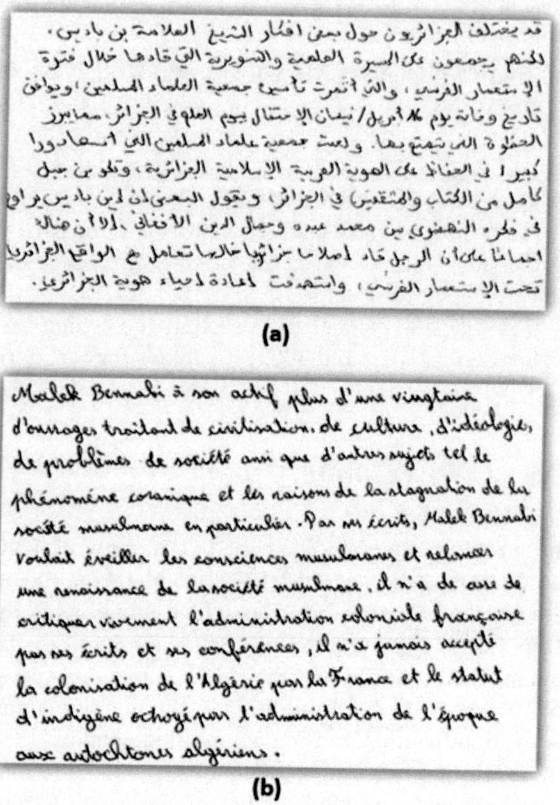

Fig. 2 Handwriting samples from Lamis-MSHD dataset **a** Arabic sample, **b** French sample

4 Experiments and Results

In this section, we present the results of the experiments carried out. We start with a presentation of the values obtained from the accuracy and loss of the training patches, and then, we continue with the results obtained in the test images, after we present the accuracy and loss of the patches test followed by a description of the various results obtained.

4.1 Training Patches

As can be seen in Fig. 3 which describes the evolution of accuracy with epochs and patch sizes, more and more the patch size is increased more and more CNN converges faster. The CNN trained by 200 × 200 patches of the Arabic version of the Lamis-MSHD dataset for example reached from the first epoch an accuracy of 60.23% and ended with an accuracy of 99.70% at the end of the 12th epoch. Unlike the small

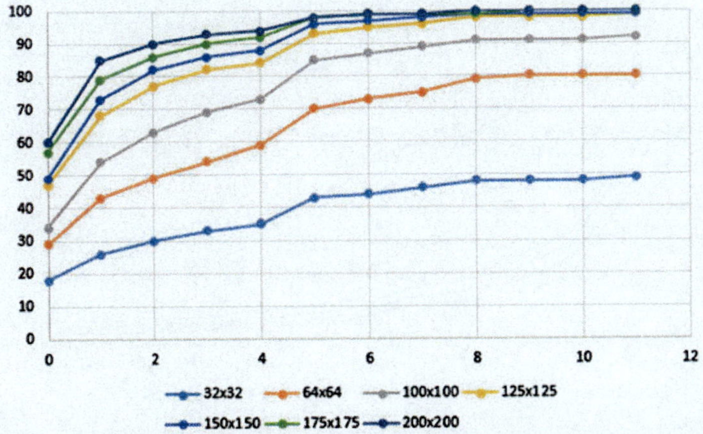

Fig. 3 Patch training accuracy for different patch size of Lamis-MSHD Arabic database

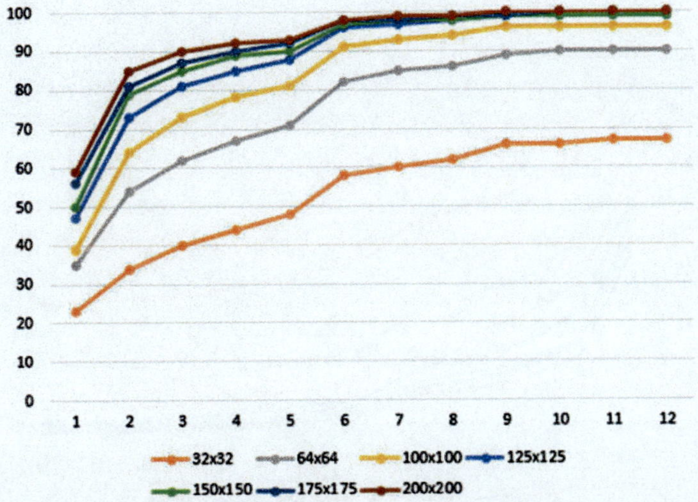

Fig. 4 Patch training accuracy for different patch size of Lamis-MSHD French database

32×32 patches which reached 18.28% at the first epoch and ended up 48.70% at the 14th epoch.

The evolution of the accuracy compared to the epochs and different patch sizes of the French version of the Lamis-MSHD dataset which is represented by Fig. 4 looks like that described previously for the Arabic version of the Lamis-MSHD.

Since the CNN accuracy converges faster for large patches, then the best values for the loss parameter are those recorded for large patches. The same observation can be shared between the Arabic and French version of the LAMIS-MSHD dataset as can be seen in Figs. 5 and 6.

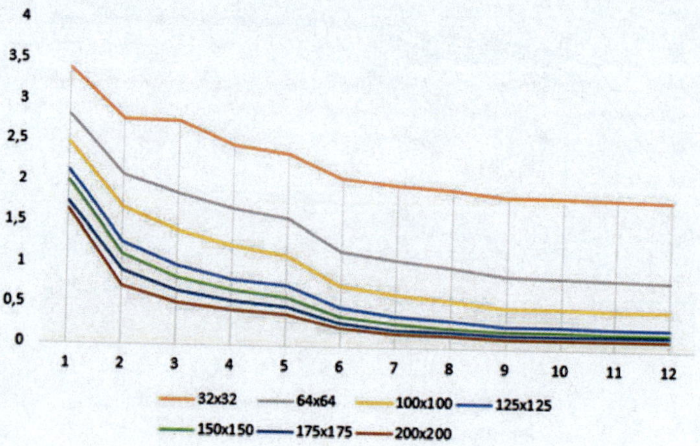

Fig. 5 Patch training loss for different patch size of Lamis-MSHD Arabic database

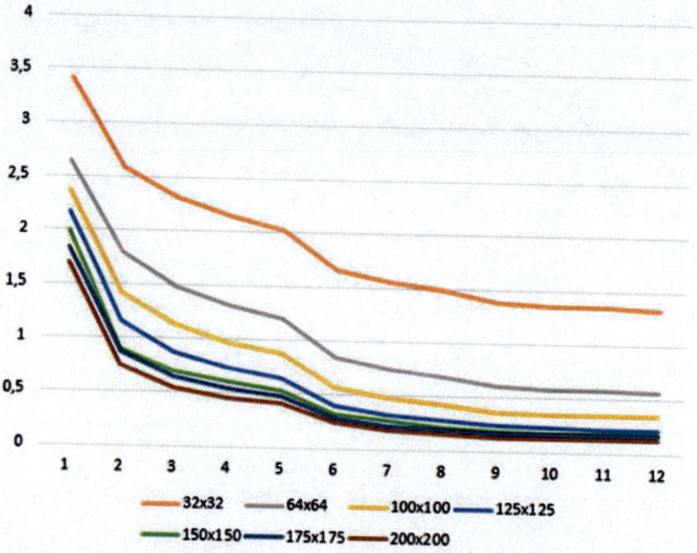

Fig. 6 Patch training loss for different patch size of Lamis-MSHD French database

4.2 Test Patches

After having trained our CNN on the training patches, we proceed to the test phase. In this phase, we extract the test patches in the same way as in the training phase with 400 patches per image. This phase allows us to provide us with three main values:

Table 1 Top-1 classification of test images

Patch size	Lamis-MSHD Arabic	Lamis-MSHD French
32 × 32	93	**98**
64 × 64	97	**98**
100 × 100	99	**98**
125 × 125	99	**98**
150 × 150	**100**	**98**
175 × 175	99	91
200 × 200	40	68

- Top-1 ranking rate for identifying the right writer for test images presented in Table 1.
- The percentage of test patches that have been assigned to the real writer is presented by Fig. 7.
- The average probability that the last fully connected layer gives to a test patch in the classification vector for the correct writer's box (see Table 2).

As we can see, the best performance for the Arabic version of the LAMIS-MSHD dataset corresponds to that of the patches of size 150 × 150 where the top-1 ranking rate is 100% followed by the patches of size 125 × 125 and 175 × 175 with 99%. While for the French version the best performance regarding the image-level classification rate is recorded for patch sizes less than or equal to 150 × 150 with a score of 98%.

The second remark concerns the large sizes 200 × 200 where the classification rate records very low values with 40 and 68% for the Arabic and French version, respectively. This shows that for very large patch sizes, the performance of CNN deteriorates rapidly.

In addition, if we look at the values relating to the probability of assigning test patches to the right writer and the percentage of test patches that were assigned to the good writer, we can see that the best scores are recorded for patches of size 125 × 125 for the French version and 150 × 150 for the Arabic version.

As can be seen, the best performance for the Arabic version corresponds to a patch size of 150 × 150 which is close to the average line height of the Arabic version which is around 139 pixels. Likewise, the correct values for the French version correspond to the patch size 125 × 125 which is very close to the average height of the French version of the LAMIS-MSHD dataset which is approximately 127 pixels.

Another observation can be deduced, it is that in the various tests carried out we can say that most of the values recorded for the French version of the LAMIS-MSHD dataset are significantly better than those relating to the Arabic version, especially for the value of the percentage of test patches attributed to the good writer and the value of the probability of assigning a test patch to its real writer. This may be due to the complexity of the Arabic language compared to the French language (see Table 2 and Fig. 7).

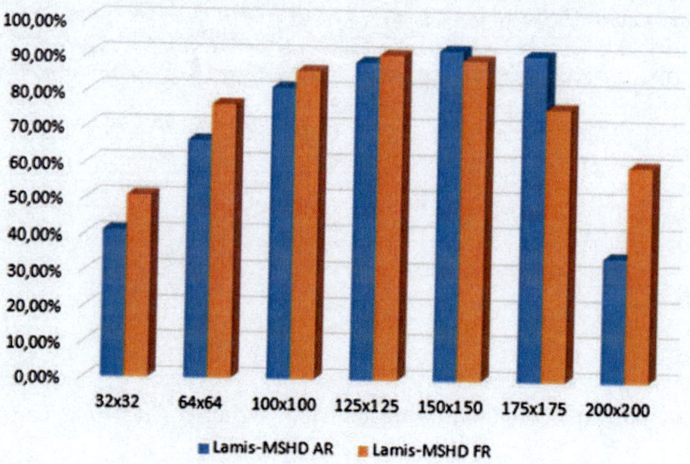

Fig. 7 Percentage of test patches assigned to the right writer

Table 2 Probability of assigning a test patch to the true writer

Patch size	Lamis-MSHD Arabic	Lamis-MSHD French
32 × 32	29.43	40.11
64 × 64	57.19	70.11
100 × 100	75.11	82
125 × 125	84.75	**87.89**
150 × 150	**89.24**	86.95
175 × 175	88.33	73.61
200 × 200	34.18	58.60

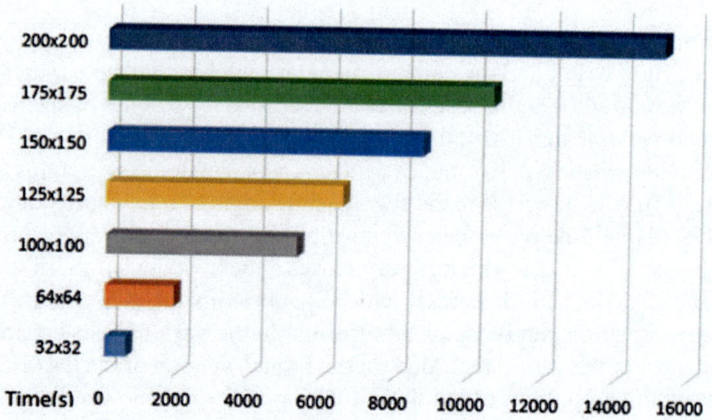

Fig. 8 Average training time of CNN during an epoch

Although the very good scores are recorded for patch sizes between 125 × 125 and 150 × 150, but the execution and training time of the CNN seems to be much higher for these patch sizes with average times going up to 145 min per epoch against 10 min for patches of size 32 × 32 (See Fig. 8). This shows that we certainly gain in the performance of CNN, but we lose in terms of execution time. So if we have very large databases like KHATT which contains 1000 writers or QUWI [2] which contains 1017 writers and to train a Resnet-34 with patch sizes of 32 × 32 we will have on average a week of training (for 4 million training patches), while for patches of size 150, for example, the CNN must train for 14 weeks and with more powerful machines. So, for larger datasets, we must resize the images and train the CNN with small-sized patches.

5 Conclusion

In this paper, we tried to verify the impact of the choice of the size of the patches on the performance of convolutional networks in the field of writer identification. The best scores were recorded for square patches that have dimensions closer to the average line height of the dataset manuscripts. Certainly, the study cannot give an absolute answer about the good size of patches to train all CNN, because we did not test all types of CNN nor all sizes of patches. But, the study offered an answer among others to the question raised in the abstract: What size of patch to use to train a CNN model in order to have the best performance?

The study can be improved by investigating the effect of image resizing on the performance of CNN in the field of writer identification and with the testing of several types of convolutional networks.

References

1. Abecassis, F.: Opencv-morphological skeleton. Retrieved from Félix Abecassis Projects and Experiments: http://felix.abecassis.me/2011/09/opencv-morphological-skeleton/geological mapping at Cuprite Nevada: a rule-based system. Int. J. Remote Sens. **31**, 7 (2011)
2. Al Maadeed, S., Ayouby, W., Hassaïne, A., Mohamad Aljaam, J.: Quwi: an arabic and english handwriting dataset for offline writer identification. In: 2012 International Conference on Frontiers in Handwriting Recognition, pages 746–751. IEEE (2012)
3. Bensefia, A., Paquet, T., Heutte, L.: A writer identification and verification system. Pattern Recogn. Lett. **26**(13), 2080–2092 (2005)
4. Bertolini, D., Oliveira, L.S., Justino, E., Sabourin, R.: Texture-based descriptors for writer identification and verification. Expert Syst. Appl. **40**(6), 2069–2080 (2013)
5. Chawki, D., Labiba, S.M.: A texture based approach for arabic writer identification and verification. In: 2010 International Conference on Machine and Web Intelligence, pages 115–120. IEEE (2010)
6. Christlein, V., Bernecker, D., Maier, A., Angelopoulou, E.: Offline writer identification using convolutional neural network activation features. In: German Conference on Pattern Recognition, pages 540–552. Springer (2015)

7. Christlein, V., Gropp, M., Fiel, S., Maier, A.: Unsupervised feature learning for writer identification and writer retrieval. In: 2017 14th IAPR International Conference on Document Analysis and Recognition (ICDAR), volume 1, pages 991–997. IEEE (2017)
8. Christlein, V., Maier, A.: Encoding cnn activations for writer recognition. In: 2018 13th IAPR International Workshop on Document Analysis Systems (DAS), pages 169–174. IEEE (2018)
9. Djeddi, C., Gattal, A., Souici-Meslati, L., Siddiqi, I., Chibani, Y., El Abed, H.:. Lamis-mshd: a multi-script offline handwriting database. In: 2014 14th International Conference on Frontiers in Handwriting Recognition, pages 93–97. IEEE (2014)
10. Fiel, S., Sablatnig, R.: Writer identification and retrieval using a convolutional neural network. In: International Conference on Computer Analysis of Images and Patterns, pages 26–37. Springer (2015)
11. Hannad, Y., Siddiqi, I., Djeddi, C., El-Kettani, M.E.Y.: Improving arabic writer identification using score-level fusion of textural descriptors. IET Biometr. **8**(3), 221–229 (2019)
12. Hannad, Y., Siddiqi, I., El Kettani, M.E.Y.: Writer identification using texture descriptors of handwritten fragments. Expert Syst. Appl. **47**, 14–22 (2016)
13. He, K., Zhang, X., Ren, S., Sun, J.: Deep residual learning for image recognition. In: Proceedings of the IEEE Conference on Computer Vision and Pattern Recognition, pages 770–778 (2016)
14. He, K., Zhang, X., Ren, S., Sun, J.: Identity mappings in deep residual networks. In: European Conference on Computer Vision, pages 630–645. Springer (2016)
15. Hinton, G.E., Krizhevsky, A., Sutskever, I.: Imagenet classification with deep convolutional neural networks. Adv. Neural Inf. Process. Syst. **25**, 1106–1114 (2012)
16. Newell, A.J., Griffin, L.D.: Writer identification using oriented basic image features and the delta encoding. Pattern Recognit. **47**(6), 2255–2265 (2014)
17. Rehman, A., Naz, S., Razzak, M.I., Hameed, I.A.: Automatic visual features for writer identification: a deep learning approach. IEEE Access **7**, 17149–17157 (2019)
18. Said, H.E.S., Tan, T.N., Baker, K.D.: Personal identification based on handwriting. Pattern Recognit. **33**(1), 149–160 (2000)
19. Siddiqi, I., Vincent, N.: Writer identification in handwritten documents. In: Ninth International Conference on Document Analysis and Recognition (ICDAR 2007), volume 1, pages 108–112. IEEE (2007)
20. Srihari, S.N., Cha, S.-H., Arora, H., Lee, S.: Individuality of handwriting. J. Forensic Sci. **47**(4), 856–872 (2002)
21. Xing, L., Qiao, Y.: Deepwriter: a multi-stream deep cnn for text-independent writer identification. In: 2016 15th International Conference on Frontiers in Handwriting Recognition (ICFHR), pages 584–589. IEEE (2016)
22. Yang, W., Jin, L., Liu, M.: Deepwriterid: an end-to-end online text-independent writer identification system. IEEE Intell. Syst. **31**(2), 45–53 (2016)

Network and Cloud Technologies

Optimization of a Multi-criteria Cognitive Radio User Through Autonomous Learning

Naouel Seghiri, Mohammed Zakarya Baba-Ahmed, Badr Benmammar, and Nadhir Houari

Abstract Dynamic and optimal management of radio spectrum congestion is becoming a major problem in networking. Various factors can cause damage and interference between different users of the same radio spectrum. Cognitive radio provides an ideal and balanced solution to these types of problems (overload and congestion in the spectrum). The cognitive radio concept is based on the flexible use of any available frequency band of the radio spectrum that could be detected. In the world of cognitive radio, we distinguish two categories of networks, namely the primary ones, which have priority and control over access to the radio spectrum, and the secondary ones, called cognitive radio networks, which allocate the spectrum dynamically. In this paper, we focus on the dynamic management of the radio spectrum based on a multi-criteria algorithm to ensure the quality of service (QoS) of the utilization by secondary users. Our approach is to use a multi-agent system based on autonomous learning and focused on a competitive cognitive environment. In this paper, we evaluate the secondary user's performance in an ideal environment of cognitive radio systems; we use the multi-agent platform called Java Agent Development (JADE), in which we implement a program that applies the multi-criteria TOPSIS algorithm to choose the best primary user (PU) among several PUs detected in the radio spectrum. Another paper allows scalability over 100 primary users evaluating four different types of technologies, namely voice, email, file transfer and video conferencing, and a comparison at the end of the convergence time for the latter technology with results from another paper.

N. Seghiri · B. Benmammar
Laboratory of Telecommunication of Tlemcen (LTT), Aboubekr Belkaid University, 13000 Tlemcen, Algeria

M. Z. Baba-Ahmed (✉)
Laboratory of Telecommunication of Tlemcen (LTT), Hassiba Ben Bouali University, 02000 Chlef, Algeria
e-mail: m.babaahmed@univ-chlef.dz

N. Houari
Laboratory of Telecommunication of Tlemcen (LTT), ZSoft Consulting, 75010 Paris, France
e-mail: nadhir.houari@protonmail.com

M. Ben Ahmed et al. (eds.), *Networking, Intelligent Systems and Security*,
Smart Innovation, Systems and Technologies 237,
https://doi.org/10.1007/978-981-16-3637-0_9

1 Introduction

In the last decade, the number of wireless devices has exceeded the world's population. Billions of devices causing a lot of unused spectrum [1]. A big challenge is to manage and share the allocated spectrum [2]. Conventional radio systems have not been able to manage these gaps in the radio spectrum. In contrast, intelligent radio systems, such as cognitive radio systems, manage the spectrum better.

Cognitive radio was officially introduced in 1998 by Joseph Mitola via a seminar at Royal Inst. of Technology in Stockholm and later published in an article written by Mitola and Maguire [3]. A cognitive radio is a programmable radio for automatic detection of available channels and their flexible use in the radio spectrum [4]. By combining the two systems, traditional and cognitive, we obtain a spectrum with two types of users, the primary users who have priority and control over the allocation of their radio spectrum, and the secondary users who dynamically rent a portion of the spectrum from the primary users. This is referred to as autonomy.

Autonomous computing is not considered as a new technology, but rather a new holistic, goal-oriented approach to computer system design that holds promise for the development of large-scale distributed systems [5]. Autonomous computing, as the name suggests, is a way of designing mechanisms to protect software and hardware, whether internal or external, in such a way that they can anticipate threats or automatically restore their function in the event of unforeseen tampering. It was first introduced by IBM, and research on autonomic agents and multi-agent systems is heavily inspired by it [6].

A system of multi-agent is a grouping of agents where each has its own capabilities. It allows us to build complex systems consisting of different interacting intelligent agents [7]. Each agent can adopt certain behaviors based on local information to maximize the overall performance of the system [8].

In this paper, we describe a solution for dynamic spectrum allocation in a multi-agent environment. Here, a multi-criteria decision analysis algorithm is used to find out the ideal allocation for a secondary user in a radio spectrum with multiple primary users. We have chosen the Technique for Order of Preference by Similarity to Ideal Solution (TOPSIS) algorithm, which consists in choosing the alternative with the shortest geometric distance to the ideal solution and the longest distance to the anti-ideal solution.

2 Related Work

A cognitive radio terminal can interact with its radio environment to adapt to it, detect free frequencies and exploit them. It will have the capabilities to efficiently manage all radio resources. Current research on cognitive radio is mainly focused on the improvement of detection, analysis and decision techniques [9]. Several approaches are proposed to optimize it.

2.1 Bayesian Nonparametric Approaches for CR Optimization

The Bayesian approach is based on a random model that represents the importance of the anterior distribution to generate the posterior distribution using Bayes' theorem. In [10], the authors proposed the NOnparametric Bayesian channEls cLustering (NOBEL) scheme. It allows quantifying channels and identifying multi-channel CRN quality of service levels. In NOBEL, SU observes the channels and extracts the characteristics of PU's channels. Then NOBEL exploits these characteristics and models them using infinite Gaussian mixture model and Gibbs collapsed sampling method. NOBEL helps SUs find the optimal channel that meets these requirements.

2.2 Reinforcement Learning-Based Approaches for CR Optimization

Reinforcement learning is one of the most important machine learning techniques in which desirable behaviors are rewarded and/or undesirable behaviors are sanctioned. The paper [11] lists the most recent spectrum allocation algorithms which uses reinforcement learning techniques in CR networks. There are six algorithms including Q-learning, improved Q-learning, deep Q-networks, on-policy RL, policy gradient and actor-critic learning automata. The algorithms' focal points and impediments are analyzed in their specific practical application scenarios.

2.3 SVM-Based Approaches for CR Optimization

Support vector machine (SVM) is a very efficient machine learning algorithm for classification problems. In the paper [12], the authors proposed to use and evaluate SVM-based approaches to appropriately classify free channels in the licensed frequency bands available in a cognitive radio network, i.e., from the best to the least optimal characteristics for a secondary SU user to choose the best channel.

2.4 ANN-Based Approaches for CR Optimization

An artificial neural network ANN is a computer system based on the way the human brain works to learn; it consists of a large set of artificially interconnected neurons. Researchers of cognitive networks have tried to integrate ANN-based techniques to dynamically access the spectrum. The authors of [13] proposed a spectrum detection scheme that uses a neural network to determine whether a PU channel is free or busy.

The proposed scheme uses likelihood ratio testing statistics and energy detection to train the neural network.

2.5 Game Theoretic Approaches for CR Optimization

Game theory is a mathematical model that has gained considerable importance in scientific research due to its efficiency and accuracy in modeling individual behavior. The authors in [13] proposed two new approaches based on game theory to model cooperative spectrum sensing in a cognitive radio environment. The first scenario is an evolutionary game model, where SUs have the right to choose whether to cooperate in spectrum detection or not. The second scenario is the Stackelberg game, where the fusion center (FC) can intervene in the cooperation process to allocate payments to SUs in return for their participation in the detection.

3 TOPSIS Method

TOPSIS is a multi-criteria analysis method for decision support. It was introduced in 1981 by Yoon and Hwang [14]. The TOPSIS main idea is about the geometric distance from both the ideal and the anti-ideal solution, i.e., the most appropriate solution is the one with the smallest distance from the ideal solution and the larger distance from the anti-ideal solution [14]. In our work, we implemented this method to calculate the ideal choice based on the multiple criteria imposed by our secondary user. The keyword TOPSIS stands for Technique for Order Preference by Similarity to Ideal Solution [15].

3.1 Ideal and Anti-ideal Solution

Ideal solution: $A^* = \{g_1{}^*, \ldots, g_j{}^*, \ldots, g_n{}^*\}$.
With $g_j{}^*$ the best value for the jth criterion among all the shares.

Anti-ideal solution: $A' = \{g_1{}', \ldots, g_j{}', \ldots, g_n{}'\}$.
With $g_j{}'$ the worst value for the jth criterion among all the actions.

3.2 Decision Matrix

TOPSIS assumes that we have m options (alternatives) and n attributes/criteria, and we have the score of each alternative with regard to each criterion.

Let x_{ij}, x_{ij} score of option i in relation to criterion j. We have a matrix $D = (x_{ij})$ matrix of $n \times m$. Let J be the set of benefit criteria or attributes (more is better). Let J' be the set of negative criteria or attributes (less is more) [16].

$$D = \begin{matrix} & C_1 & C_2 & \cdots & C_n \\ A_1 & x_{11} & x_{12} & \cdots & x_{1n} \\ A_2 & x_{21} & x_{22} & \cdots & x_{2n} \\ \vdots & \vdots & \vdots & \ddots & \vdots \\ A_m & x_{m1} & x_{m2} & \cdots & x_{mn} \end{matrix}$$

3.3 The Six Steps of the TOPSIS Algorithm

Step1: Development of the normalized decision matrix

$$r_{ij} = \frac{x_{ij}}{\sqrt{\sum_{i=1}^{m} x_{ij}^2}} \qquad (1)$$

$$i = 1 \ldots m \qquad j = 1 \ldots n$$

Step2: Development of the weighted normalized decision matrix

$$v_{ij} = w_j * r_{ij}$$
$$i = 1 \ldots m \qquad j = 1 \ldots n \qquad (2)$$

$$V = \begin{bmatrix} v_{11} & \cdots & v_{1n} \\ \vdots & \ddots & \vdots \\ v_{m1} & \cdots & v_{mn} \end{bmatrix} == \begin{bmatrix} w_1 \,.\, r_{11} & \cdots & w_n \,.\, r_{1n} \\ \vdots & \ddots & \vdots \\ w_1 \,.\, r_{m1} & \cdots & w_n \,.\, r_{mn} \end{bmatrix}$$

Step3: Calculate ideal and negative-ideal solutions

$$V_j^+ = \{(\max(v_{ij})|j \in J_1), (\min(v_{ij})|j \in J_2)\} \qquad (3)$$

$$V_j^- = \{(\min(v_{ij})|j \in J_1), (\max(v_{ij})|j \in J_2)\} \qquad (4)$$

J1: set of benefit criteria.
J2: set of cost criteria.

Step4: Determine the separation measure

$$S_i^+ = \sqrt{\sum_{j=1}^{n}\left(V_j^+ - V_{ij}\right)} \quad i = 1\ldots m, j = 1\ldots n \tag{5}$$

$$S_i^- = \sqrt{\sum_{j=1}^{n}\left(V_j^- - V_{ij}\right)} \quad i = 1\ldots m, j = 1\ldots n \tag{6}$$

Step5: Determine the relative closeness to the ideal solution

$$P_i^* = \frac{S_i^-}{S_i^- + S_i^+} \quad 0 < P_i^* < 1 \tag{7}$$

Step6: Ranking the preference order.

- Choose the action with the highest similarity index (choice problem).
- Rank the shares in descending order of similarity indexes (ranking problem) [17].

4 Proposed Approach

In our approach, there are two types of users, primary users (PUs) and secondary users (SUs). We have defined a negotiation strategy, one-to-many strategy, where a secondary user (SU) initiates the negotiation with multiple primary users (PUs). In our case study, there are ten PUs, as shown in Fig. 1.

The SUs have several specific requirements, such as number of channels, bandwidth, technology and price. At the beginning of the negotiation, the SU sends the first hello request to all PUs. The goal of this first request is to find out which PUs are available; when available we mean that all PUs that have at least a minimum number of channels and a minimum bandwidth and the required technology or newer and the price of better are required by the SU. Once a PU acknowledges the request, it responds with affirmative response if it has at least the minimum requirements of the SU, or a negative response if it does not have at least one of the requirements of the SU.

Once the SU has the list of the PUs that respond to its needs, this is where our work comes in which is to find the best PU among all PUs. We have chosen to perform this task using the TOPSIS multi-criteria algorithm. We give the list of PUs that responded with an acknowledgement and their critters (number of channels, bandwidth, technology…) as input, and as output we expect the best ideal PU, which best answers our SU needs.

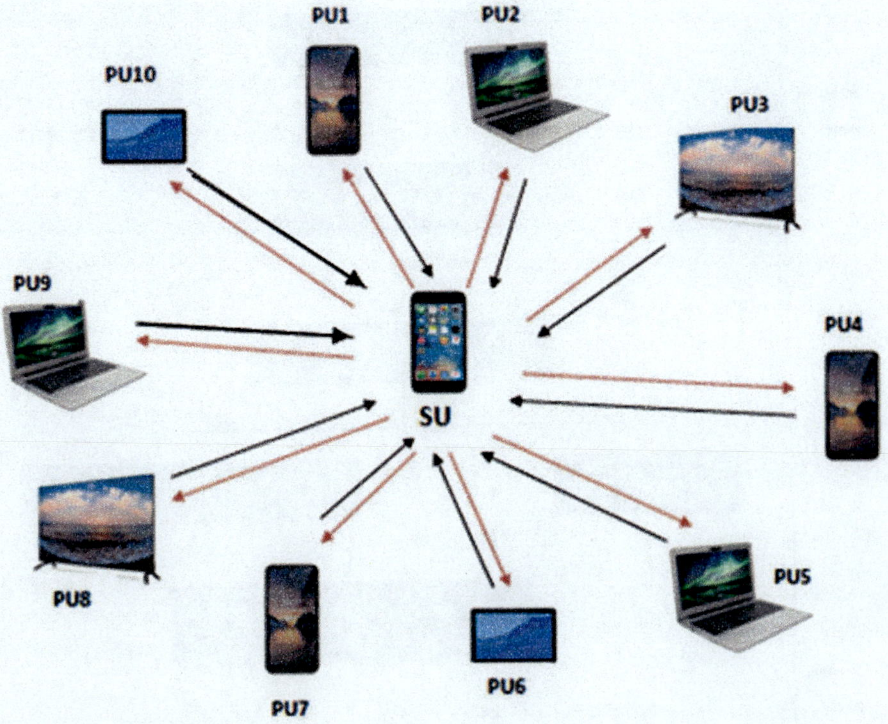

Fig. 1 Proposed scenario

4.1 Flowchart and Objective Functions of the TOPSIS Algorithm

The flowchart represents the execution steps of our application. First the detection phase—the SU detects the environment; once it detects a free part of the spectrum, it broadcasts the minimum number of required channels to all PUs. Second the decision phase—SU must select a single PU based mainly on the number of channels. The PUs receive the broadcasted request with the required number of channels, the PUs that meet this requirement send an acknowledgement that contains their information, such as the exact number of channels available, the technology used, etc... On the other hand, the PU, which does not have the required number of channels, rejects the request. So, it is a matter of which PU is most ideal for the SU. All this is illustrated in the flowchart (Fig. 2).

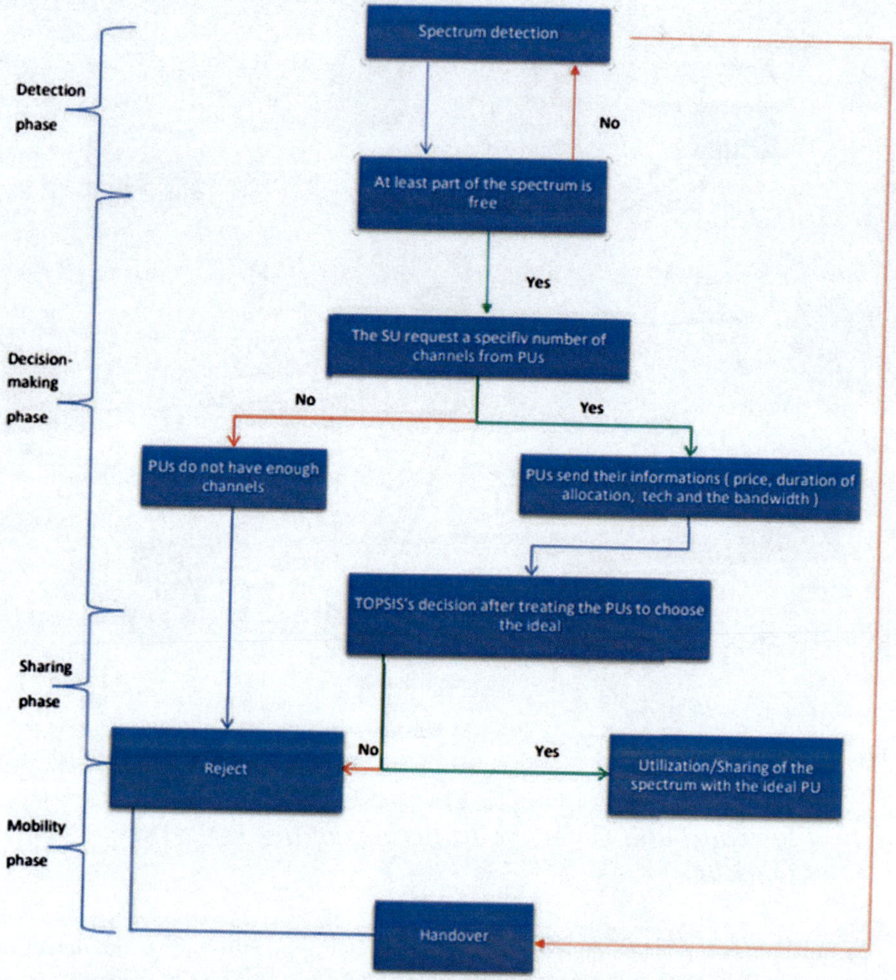

Fig. 2 Flowchart and objective functions of the TOPSIS algorithm

5 JADE Simulation

The simulation was performed under Apache NetBeans IDE 12.0 (Integrated Development Environment) using the JADE platform, which contains all the components for controlling control the SMAs, which are explained in more detail below:

In this first part of the simulation, we decided to define a cognitive agent for the secondary user named SU and ten primary user agents named PU1 to PU10, recognized by the same SU. This SU agent will communicate with the ten PUs simultaneously until it finds a PU that is compatible with its requirements (Fig. 3).

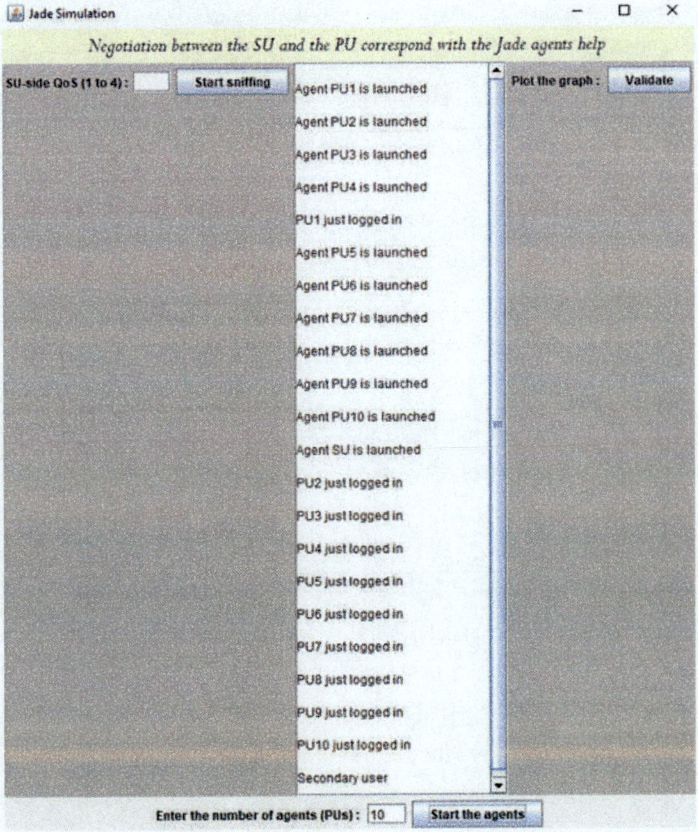

Fig. 3 JADE simulation interface of our approach

5.1 Presentation of the Proposed Approach

Our goal in this work is to improve our autonomous learning system by integrating an algorithm that helps us choose the best primary user based on multiple criteria. We chose the TOPSIS algorithm because it is simple, flexible and fast to find the ideal choice. In what follows, we will present our simulation scenario in which we implement our flowchart for the scenario of a SU communicating with ten PUs.

First, the SU requests three channels to ensure the QoS of the file transfer and therefore sends requests to all detected PUs. The PUs that have the required number of channels inform with an ACL message that contains the number of requested channels and important information about the price of the allocation, the technology, the allocated time and the bandwidth to be used. Otherwise, the PU that does not have the required number of channels rejects the request. In this example, we have eight PUs responding positively with different proposals ranging from the price of the

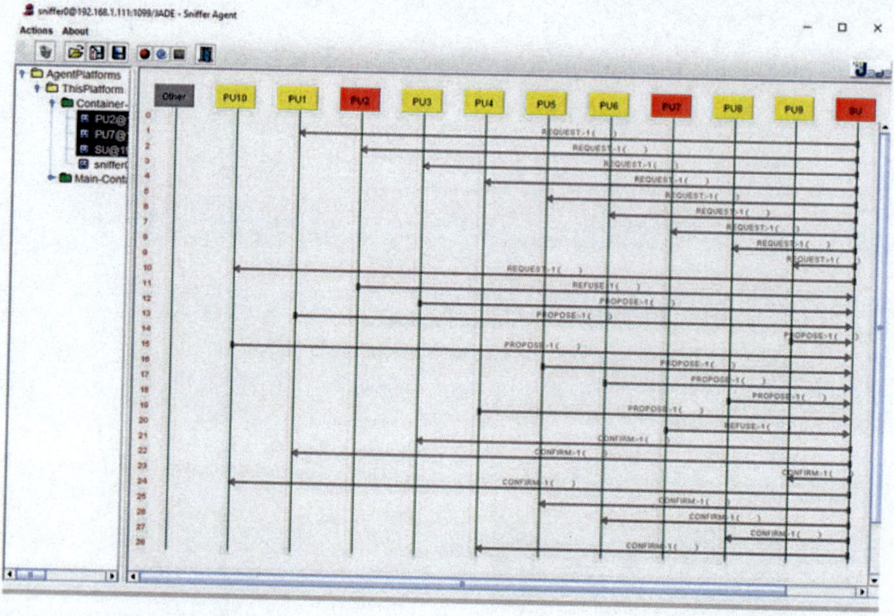

Fig. 4 Negotiation between SU and the ten PUs

allocation to the technology and bandwidth used and two PUs responding negatively, PU2 and PU7 (they do not have the required number of channels) (Fig. 4).

5.2 Negotiation Between Secondary and Primary Users

The TOPSIS algorithm and the choice of the best PU

The multiple positive PU responses confuse the SU, and it cannot decide which of them is optimal. Precisely, the context of the example is to rank the following PUs that form the choice of the SU (PU1, PU3, PU4, PU5, PU6, PU8, PU9, PU10) using the TOPSIS algorithm and based on the four criteria listed below:

- Allocation time
- The technology used
- The bandwidth
- The Price

Data

The first step is to decide a uniform scale of measurement of the levels (scores) to be assigned to each criterion relative to the corresponding alternative (PU) by defining numerical values (1–8) generally in ascending order and the linguistic meaning of each level (from "Not interesting at all" to "Perfectly interesting"). These values are used to measure both positive (favorable) and negative (unfavorable) criteria.

The Alternatives X Criteria data matrix is determined by assigning each alternative the level of each of its attributes based on the previously defined scale.

- For positive criteria (time, technology, bandwidth), the higher the score, the more positive (favorable) the criterion.
- For the negative criterion (price), the higher the score, the more negative (unfavorable) the criterion.

For each criterion, a weighting is assigned (a weight that reflects the importance of the criterion in our final choice). The weights must be defined so that their sum is equal to 1 and are usually defined in %. Even if the weights are not between 0 and 1, they can always be reduced to the interval [0, 1] by simply dividing each weight by the sum of all the weights. The following weights are assigned to the four criteria in order:

- Allocation time: 0.25.
- Technology: 0.25.
- Bandwidth: 0.2.
- Price: 0.3.

Also giving us an interval of these four criteria as follows:

- Allocation time: [between 1 and 24] h.
- Technology: [3G, 3.5G, 3.75G, 4G, 4.5G, 5G].
- Bandwidth: [144 Mbps–10 Gbps].
- Price [from: 120 up to 300] DA per hour.

After the simulation, we found the results as shown in Table 1.

Figure 5 shows the result of the best and worst primary users sharing the spectrum with the secondary user, among the ten primary users with different criteria.

In conclusion, here is the ranking in descending order of the eight PUs of the most satisfactory at least in terms of quality of service for file transfer which is given as follows:

1. PU1 the most favorable
2. PU6
3. PU8
4. PU9
5. PU10
6. PU5
7. PU3
8. PU4 the least favorable.

We notice that PU2 and PU7 do not have required number of channels to share the spectrum with our secondary user.

Table 1 PU classified after the use of the TOPSIS algorithm

Alternative name	Criteria values
PU1	Channel number = 6, Price = 153, Allocated time (h) = 3, Tech = 5G, Bd = 10,788.112
PU6	Channel number = 4, Price = 132, Allocated time (h) = 6, Tech = 4G, Bd = 6602.962
PU8	Channel number = 3, Price = 238, Allocated time (h) = 13, Tech = 3.75G, Bd = 13.385
PU9	Channel number = 4, Price = 155, Allocated time (h) = 14, Tech = 3G, Bd = 0.7641
PU10	Channel number = 7, Price = 271, Allocated time (h) = 7, Tech = 3.75G, Bd = 5.0514
PU5	Channel number = 3, Price = 220, Allocated time (h) = 7, Tech = 3G, Bd = 0.635
PU3	Channel number = 8, Price = 253, Allocated time (h) = 3, Tech = 3.75G, Bd = 11.056
PU4	Channel number = 6, Price = 201, Allocated time (h) = 4, Tech = 4G, Bd = 66.441

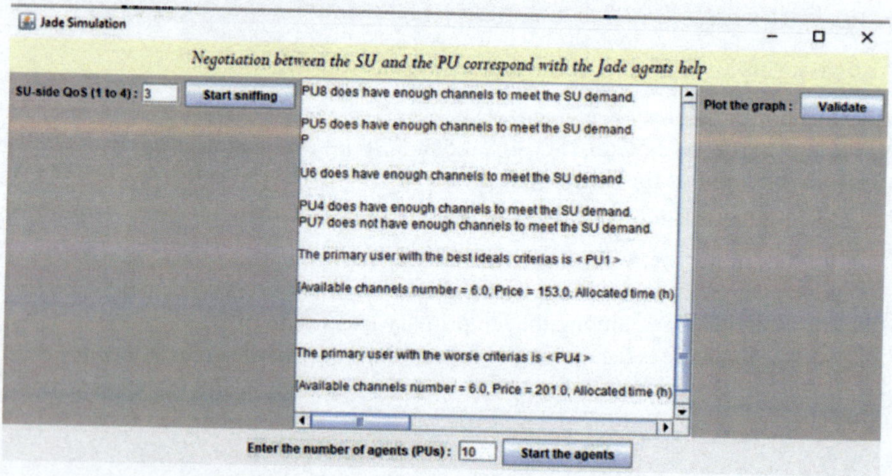

Fig. 5 Simulation results displayed on the console of the Java program

6 Results and Discussion

To further strengthen our study, we have opted for a phased study by upscaling for deeper and better-quality learning and using four QoS for four different technologies, namely voice, email, file transfer and video conferencing for a secondary user communicating with multiple PUs.

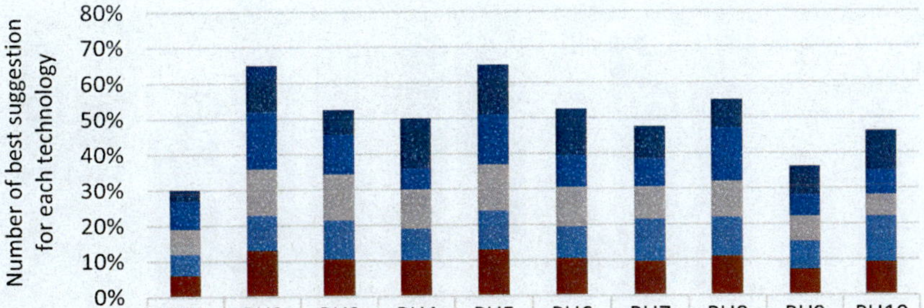

	PU1	PU2	PU3	PU4	PU5	PU6	PU7	PU8	PU9	PU10
Video conference	3%	13%	7%	14%	14%	13%	9%	8%	8%	11%
File Transfer	8%	16%	11%	6%	14%	9%	8%	15%	6%	7%
E-mail	7%	13%	13%	11%	13%	11%	9%	10%	7%	6%
voice	6%	10%	11%	9%	11%	9%	12%	11%	8%	13%
TOTAL	6%	13%	11%	10%	13%	11%	10%	11%	7%	9%

Fig. 6 Best suggestion results for SU choosing between ten PUs out of 100 communication attempts

The scaling is done by 100 communication trials of SU with ten PUs requesting firstly one channel for voice, secondly two channels for email, thirdly three channels for file transfer and fourthly four channels for video conferencing to find out which PU is the most optimal. Figure 6 shows the best proposal results between the SU and the ten PUs for the four technologies.

A comparison between technologies showed that PU4 and PU5 are better for video conferencing, while PU2 is better for file transfer; PU2, PU3 and PU5 are better for email; and PU10 is better for voice. For a global view of all technologies, PU2 and PU5 are the best.

Figure 7 represents a ranking of the PUs for 100 negotiation attempts of a SU with ten PUs compared to the different technologies.

Now we come to another contribution, namely the convergence time. Figure 8 shows the average convergence time for 100 communication attempts between a SU and ten PUs.

Figure 8 shows us the average convergence time required between the SU and the ten PUs to share the spectrum with. One of them for video conferencing, PU1 has the best time at 55.84 ms, while PU6 is the last with 96 ms; despite this, all users have a convergence time <150 ms (a time required by the QoS of the technology used for video conferencing), a value that ensures the QoS required in the literature [18].

And Fig. 9 shows us the comparison of our contribution with the article [19] in terms of average convergence time, and we find that the best convergence time of our contribution is much better and optimal compared to their convergence time which is fixed at 138.9 ms.

The result obtained allows us to say that our cognitive system is optimized, precisely:

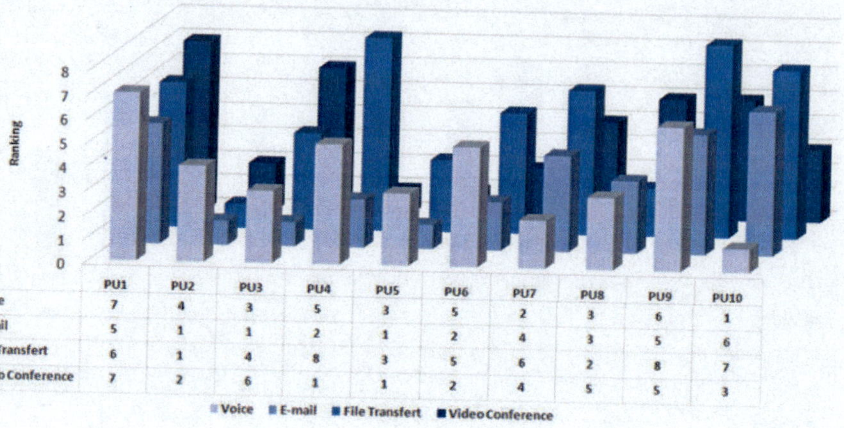

	PU1	PU2	PU3	PU4	PU5	PU6	PU7	PU8	PU9	PU10
Voice	7	4	3	5	3	5	2	3	6	1
E-mail	5	1	1	2	1	2	4	3	5	6
File Transfert	6	1	4	8	3	5	6	2	8	7
Video Conference	7	2	6	1	1	2	4	5	5	3

Fig. 7 Best suggestion ranking for SU choosing between 10 PUs out of 100 communication attempts

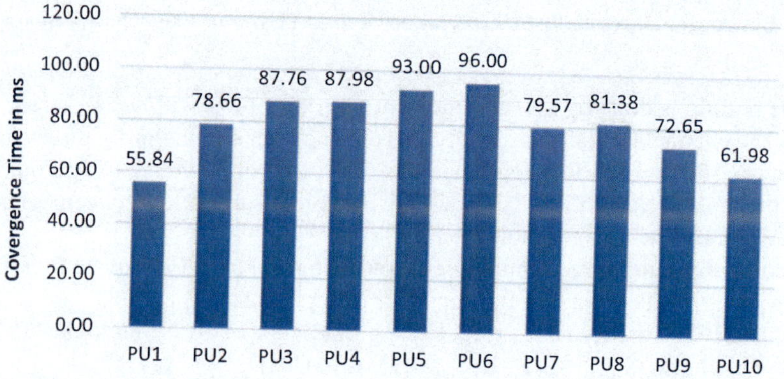

Fig. 8 Average of the convergence time over 100 communication attempts between the SU and the different PUs in millisecond

1. The cognitive engine has been strengthened with more suggestion data and a ranking from the best to the worst PUs, which allows the secondary user to optimize the planning and decision making in the cognitive cycle, also to make suggestions to other SUs in case of collaboration.
2. A huge win for convergence time optimized for different technologies and especially for video conferencing, which allows us to extend our work in the future for real-time technologies.

Last but not least, the results we found show us that the TOPSIS algorithm is indispensable for negotiation in a multi-agent system to ensure optimal quality of service in a cognitive radio environment.

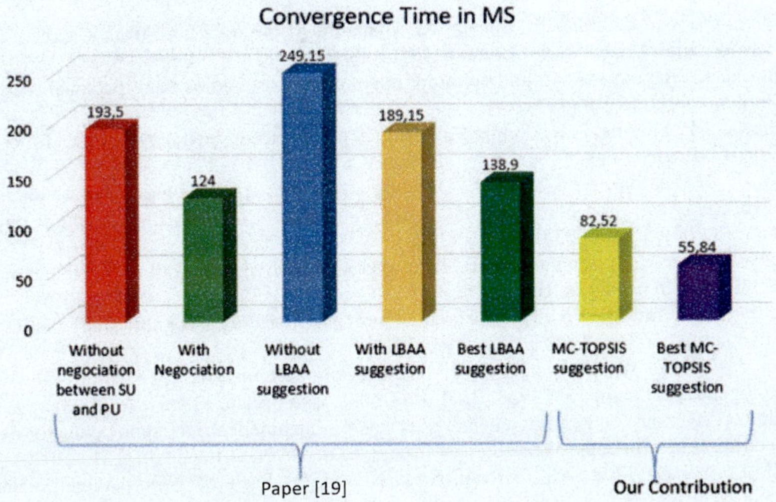

Fig. 9 Comparison of the average convergence time with other work

7 Conclusion

In this paper, we have studied the autonomous behavior of cognitive agents through multi-agent systems and see their impact on intelligent networks (cognitive radio) based on the requirements of the secondary user on multiple criteria to choose the best offer among those of the primary users.

This new contribution allowed us to develop an algorithm based on multiple criteria to illustrate the communication of the secondary user with the primary users by selecting the best offer to allocate a part of the spectrum from them with different requirements. Subsequently, the simulation results using the platform JADE, which is remarkably close to the ideal case of cognitive radio, proved to be conclusive with a significant gain in quality of service in cognitive radio systems. The system can efficiently exploit the spectrum in an opportunistic and reliable manner. Finally, the results of our new approach are better and optimized than those found in the literature.

As a perspective, we hope to further strengthen our system based on security in cognitive radio systems and using our new multi-criteria approach, to test the scalability of the system and to implement it on a system based on real signals.

References

1. Tomar, G., Bagwari, A., Kanti, J.: Introduction to Cognitive Radio Networks and Applications, pp. 124–133. CRC Press (2017)

2. Song, M., Xin, C., Zhao, Y., Cheng, X.: Dynamic spectrum access: from cognitive radio to network radio. IEEE Wirel. Commun. **19**(1), 23–29 (2012)
3. Mitola, J., Maguire, G.: Cognitive radio: making software radios more personal. IEEE Pers. Commun. **6**(4), 13–18 (1999)
4. Jaiswal, M., Sharma, A.K., Singh, V.: A survey on spectrum sensing techniques for cognitive radio. In: Proceedings of the Conference on ACCS, pp. 1–14 (2013)
5. Lin, P., MacArthur, A., Leaney, J.: Defining autonomic computing: a software engineering perspective. In: Proceedings of the 2005 Australian Software Engineering Conference (ASWEC'05), 1530-0803/05. IEEE (2005)
6. Baba-Ahmed, M.Z., Benmammar, B., Bendimered, F.T.M: Spectrum allocation for autonomous cognitive radio networks. IJACT: Int. J. Adv. Comput. Technol. **7**(2), 48–59 (2015)
7. Amraoui, A.: Towards a multi-agent architecture for opportunistic cognitive radio. Ph.D. thesis, University Abou bekr Belkaid Tlemcen (2015)
8. van der Hoek, W., Wooldridge, M.: Multi-agent systems. In: Foundations of Artificial Intelligence, vol. 3, pp. 887–928. Bradford Books, Cambridge, MA, USA (2008)
9. Kaur, A., Kumar, K.: A comprehensive survey on machine learning approaches for dynamic spectrum access in cognitive radio networks. J. Exp. Theor. Artif. Intell., 1–40 (2020)
10. Ali, A., Ahmed, M.E., Ali, F., Tran, N.H., Niyato, D., Pack, S.: NOn-parametric Bayesian channEls cLustering (NOBEL) scheme for wireless multimedia cognitive radio networks. IEEE J. Sel. Areas Commun. **37**(10), 2293–2305 (2019)
11. Wang, Y., Ye, Z., Wan, P., Zhao, J.: A survey of dynamic spectrum allocation based on reinforcement learning algorithms in cognitive radio networks. Artif. Intell. Rev. **51**(3), 493–506 (2019)
12. Sarmiento, D.A.L., Viveros, L.J.H., Trujillo, E.R.: SVM and ANFIS as channel selection models for the spectrum decision stage in cognitive radio networks. Contemp. Eng. Sci. **10**(10), 475–502 (2017)
13. Patel, D.K., Lopez-Benitez, M., Soni, B., Garcia-Fernandez, A.F.: Artificial neural network design for improved spectrum sensing in cognitive radio. Wirel. Netw. **26**(8), 6155–6174 (2020)
14. Benmammar, B.: Resource allocation in a cognitive radio network using JADE. Research Report in Telecommunications, Tlemcen University (2015)
15. Loganathan, J., Latchoumi, T.P., Janakiraman, S., Parthiban, L.: A novel multi-criteria channel decision in co-operative cognitive radio network using E-TOPSIS. In: Proceedings of the International Conference on Informatics and Analytics, pp. 1–6 (2016)
16. Bhatia, M., Kumar, K.: Network selection in cognitive radio enabled wireless body area networks. Digit. Commun. Netw. **6**(1), 75–85 (2020)
17. Beg, I., Rashid, T.: Multi-criteria trapezoidal valued intuitionistic fuzzy decision making with Choquet integral based TOPSIS. Opsearch **51**(1), 98–129 (2014)
18. Szigeti, T., Hattingh, C., Barton, R., Briley, Jr., K.: End-to-End QoS Network Design: Quality of Service for Rich-Media & Cloud Networks. Cisco Press (2013)
19. Baba-Ahmed, M.Z., et al.: Self-management of autonomous agents dedicated to cognitive radio networks. In: International Conference in Artificial Intelligence in Renewable Energetic Systems. pp. 372–380. Springer, Cham (2019)

MmRPL: QoS Aware Routing for Internet of Multimedia Things

Hadjer Bouzebiba and Oussama Hadj Abdelkader

Abstract This paper provides an improved version of the routing protocol for low power and lossy networks (RPL), called multimedia RPL (MmRPL). This protocol is proposed as a solution for some restrictions in the RPL storing mode. RPL consumes much energy when the network size increases which may degrade its performance and cause a degradation of the network performance, resulting in an inconsistency in the use of the RPL protocol in constrained internet of multimedia things (IoMT) networks. IoMT applications can be very demanding in terms of quality of service (QoS) requirements such as minimum delay, reduced rate of control overhead and low-energy consumption. The proposed extension overcomes the memory overload challenge and improves the QoS requirements in IoMT networks. The obtained simulation results proved that the proposed algorithm outperforms the standard RPL and another extension of RPL, in terms of control-plane overhead, end-to-end delay and energy consumption.

1 Introduction

The emerging Internet of Things technology (IoT) consists in large networks of small-sized low-power embedded devices that can support various applications. The Internet of multimedia things (IoMT) is an extension of the IoT, considered as an interconnection of multimedia objects in which they are able to acquire, process and display multimedia contents from and to the real world using specific input and output devices [1]. With the growth of this type of smart devices, the tendency of IoMT applications has recently became much more important by getting involved

H. Bouzebiba (✉)
STIC Laboratory, University of Tlemcen, 13000 Tlemcen, Algeria
e-mail: hadjer.bouzebiba@univ-tlemcen.dz

O. Hadj Abdelkader
Faculty of Engineering, SYSTEC - Research Center for Systems and Technologies, University of Porto and Institute for Systems and Robotics, 4200-465 Porto, Portugal
e-mail: hadjabdelkader@fe.up.pt

M. Ben Ahmed et al. (eds.), *Networking, Intelligent Systems and Security*,
Smart Innovation, Systems and Technologies 237,
https://doi.org/10.1007/978-981-16-3637-0_10

in many fields such as smart homes, smart health, smart vehicles, smart cities ... etc. This diversity may bring new stringent requirements than the IoT environment, mostly due to the increase of multimedia content in the network.

For communication purposes, IoT devices use some standardized routing protocols to manage the data transfer within their network. The routing protocol should be reliable, energy efficient and mainly scalable. As mentioned in [2], a good routing protocol in this context should extend to satisfy the requirements of large network size and densities in spite of the resource constraints of wireless sensor networks (WSNs). The internet engineering task force working group (IETF) Routing over Low power and Lossy networks (ROLL) have proposed a routing protocol named Routing Protocol for Low power and Lossy Networks (RPL) [3]. RPL meets the specific requirements of low power and lossy networks (LLN). For this latter, RPL has rapidly became the standard routing protocol for IoT. RPL uses a tree approach to organize the network into a tree structure which allows the data to flow in both upward and downward directions. In the downward direction, the traffic flow from the root toward its associated nodes which spoils the scalability of RPL significantly [4]. Moreover, in storing mode, each RPL router has to store routes for destinations in its sub-DODAG [5]. This feature generates a limiting factor in RPL, which is the amount of available memory to store the neighbors in the routing table. Thus, each close node to the root is obliged to store the routing state for almost the entire destination oriented acyclic graph (DODAG), which can be challenging for resource-constrained devices [6].

To the best of our knowledge, the standard RPL does not deal with the case where a parent node cannot accept to add a new downward route when its routing table is full, this problem can happen in scalable networks such as smart cities. When the new scalable nodes want to join network, it will consume buffer memory, while exchanging the control messages besides of the necessary amount of energy taken during this process. This will negatively impact on the network performance in terms of delay and energy consumption without the establishment of any proper routing.

In this paper, we propose an RPL extension named MmRPL to tackle the problem of insufficient storage memory and to optimize the protocol for IoMT networks. The proposed MmRPL reduces the amount of control-plane overhead by checking the memory of nodes once having a new connection in the network. The remainder of this paper is structured as follows. Section 2 provides an overview of RPL protocol. Section 3 presents a description of the problem statement. Section 4 summarizes the related work. In Section 5, the proposed solution is discussed. In Section 6, the performance of MmRPL is evaluated and the results are discussed in comparison with RPL. The final section concludes the paper and gives some hints about future works.

2 RPL Overview

The RPL [3, 7] represents a distance vector routing protocol, in which a DODAG based on sets of metrics is constructed. The final destination node in the DODAG is called the root (as Low Power and Lossy Border Routers (LBR)). The latter acts like a bridge between the LLN and Internet. RPL supports different types of traffic : Multipoint to Point (M2P), Point to Multipoint (P2M) and Point to Point (P2P). Each node in the DODAG is characterized by a rank value, which represents its distance towards the root node calculated according to an objective function. The OF [8, 9] determines the rank of each node based on one or more metrics and selects the optimal route in a DODAG. The DODAG is constructed from the root by broadcasting ICMPv6 control message (the DODAG Information Object (DIO)) to its neighborhood. This kind of messages contains some configuration parameters (such as the DODAG roots identity, routing metrics, as well as the rank) needed to build the topology. Once DIO message is received by neighboring nodes, the node joining the DODAG will: (1) add the sender prefix to its candidate parent list; (2) calculate its rank; (3) select the closest node to the root from its candidate parents which acts as its next hop (preferred parent) toward the root and (4) update the received DIO message with its own rank and repeat the same procedure until all the network's nodes will have an upward route toward the root.

After the construction of the upward route, another type of ICMPv6 control messages named as destination advertisement object (DAO) is used to build the downward routes. DAO messages are unicasted to the node's preferred parent for nodes, which have already joined the DODAG and want to advertise one or more reachable destination prefixes including their own. RPL affords two operation modes for downward routes: the storing and the non-storing mode. The storing mode requires for each parent that receives a DAO message from one of its children, to store its prefix (DAO sender address) in its routing table as a next hop prefix. Besides, the DAO receiver can optionally acknowledges the DAO sender using the DAO-ACK message (DAO Acknowledgement). The parent by its turn forwards the received DAO to its own preferred parent repeating the same process until the DAO achieves the DODAG root. The RPL's network structure and construction have been illustrated in Fig. 1.

3 Related Work

Since RPL has been considered as an open standard by IETF community, many enhancements have been proposed to improve its performance [4, 11, 12].

Authors in [11], proposed an improved version of RPL protocol named Enhanced-RPL that treats the problem of unreachable destination caused by storage limitations of certain node's preferred parent. In RPL, when a child node wants to announce its prefix in the downward routing, it should unicast a DAO message to its preferred parent, but when the parent's routing table is full, it will not accept any additional nodes. This can happen in scalable networks when the number of nodes increases. As a solution to this problem, in the proposed Enhanced-RPL, a list of candidate parents was proposed to the child node that has lost the chance to announce itself to its preferred parent.

Authors in [12] suggested an extension of RPL named MERPL which reduces the memory consumption by improving the storing mode scalability. In this extension,

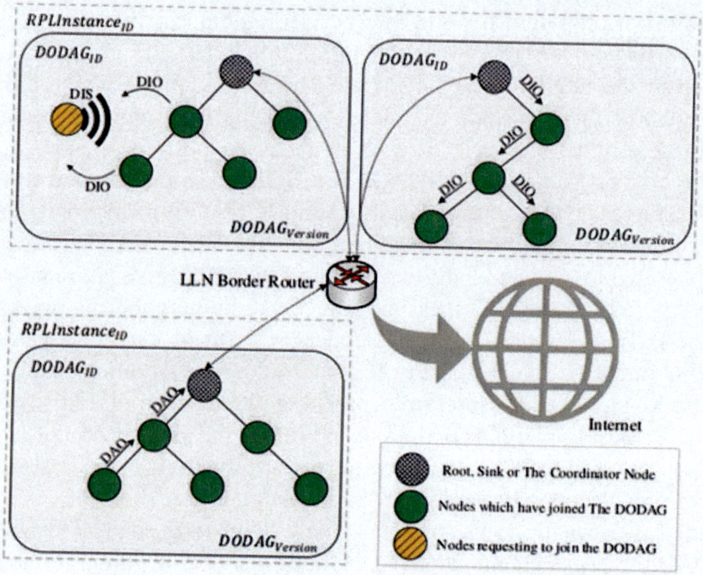

Fig. 1 RPL structure [10]

whenever a node reaches a predefined number of N in its routing table, it must delegate one of its children (which has the highest number of routing table entries) to play the role of a storing node. Then, the delegating node (parent) must delete all entries that are related to the child (new storing node) from its routing table. After that the new reachable destinations selected by the child will be advertised to the root in a DAO message.

Further, various optimizations of the RPL protocol which take the QoS as routing metrics have been studied, such as [13–15] which aimed to satisfy a specific QoS. For example in [13], authors proposed a compliant OF which allows the RPL to support the multi-instance approach, called OFQS. The OFQS has focused on multi-objective metrics taking in consideration the following metrics: delay, remaining energy of the nodes' battery and quality of link. In [14], the authors measured the RPL performance in terms of the QoS parameters such as the packet delivery ratio, the network convergence time, the remaining energy, the latency and the total amount of overhead. In the same context, in [16], the authors choosed the fuzzy systems in order to design a new set of contextual OFs for RPL. They specified a set of sensitive functions "Delivery Quality and Context" (DQCA-OF), by combining alternately the metrics: *ETX*, energy consumed and number of hops (HC). Thus, a set of OFs can be dynamically chosen according to the QoS requirements of the IoT application. Moreover, in [17], authors proposed an enhanced version of the RPL routing protocol by providing a new QoS aware OF based on free bandwidth named as FreeBW-RPL. The proposed OF works on distributing the traffic load in the network by taking, in a dynamic way, multiple paths for routing the heavy amount of traffic, while creating an energy balanced routing paths. Another work which aims to perform on a multi-path parent selection is the on-demand selection (ODeSe) algorithm suggested in [18]. This algorithm focused on the dynamic conditions at the packet forwarding time. This work implements the packet automatic repeat request, replication and elimination and overhearing (PAREO) [19] functions in order to improve both of reliability and energy efficiency.

A particular RPL improvement has been proposed in [20] called energy efficient optimal parent selection (EEOPS–RPL), in which authors have used the firefly optimization algorithm in order to extend the lifespan of the IoT network. This algorithm calculates the current location of firefly (each node is considered as a firefly), attraction of the fireflies, random function, velocity and the global best values in the network. Thus, during the data transmission, the distance is viewed as the movement parameter for choosing the optimum parent in the DODAG. The firefly algorithm offers a fast convergence, while choosing the optimal parent, reduces the packet loss during the route establishment and expands the lifespan of the entire network.

In any routing protocol, scalability represents an important feature bearing a direct impact on network's performance and reliability. In this regard, RPL did not specify any action to do when a node's routing table is full and still receiving some solicitation (DIS or DAO messages) from new nodes or from an already joined node. However, the aforementioned research did not take in consideration the unnecessary route establishment and reduction of overhead in IoMT networks. In addition, the routing protocol should account for the performance of the multimedia communications within IoMT network in terms of end-to-end delay, overhead and rate of packets delivered. Moreover, we argue that the problem of scaling with network density has not been sufficiently analyzed, especially the optimization of the total network overhead in order to extend the RPL protocol to large data stream routing such as multimedia data routing. The protocol MmRPL proposed in this paper is dedicated for these issues especially the overloaded case of an IoMT node.

4 Problem Statement

IoMT's applications require the satisfaction of many QoS constraints such as minimum amount of overhead, limited delay and energy. However, in the storing mode of the RPL protocol, nodes may exhaust their memory easily since each node is required to store the routing information about its sub-DODAG which may lead to a storage limitation of neighbors and routing tables. In consequences, this latter represents a more serious problem especially in a scalable network. Generally, nodes that are around the root are more likely to run out their memory, especially in large networks such as IoMT; as presented in [12] these nodes may not possess the required large memory resources. This issue figures as a challenging especially for the resource constrained devices. Besides, RPL does not provide any action to take when a parent rejects to install a new downward route (i.e., case of an overloaded routing table). These issues can impact negatively on the QoS of the IoMT network.

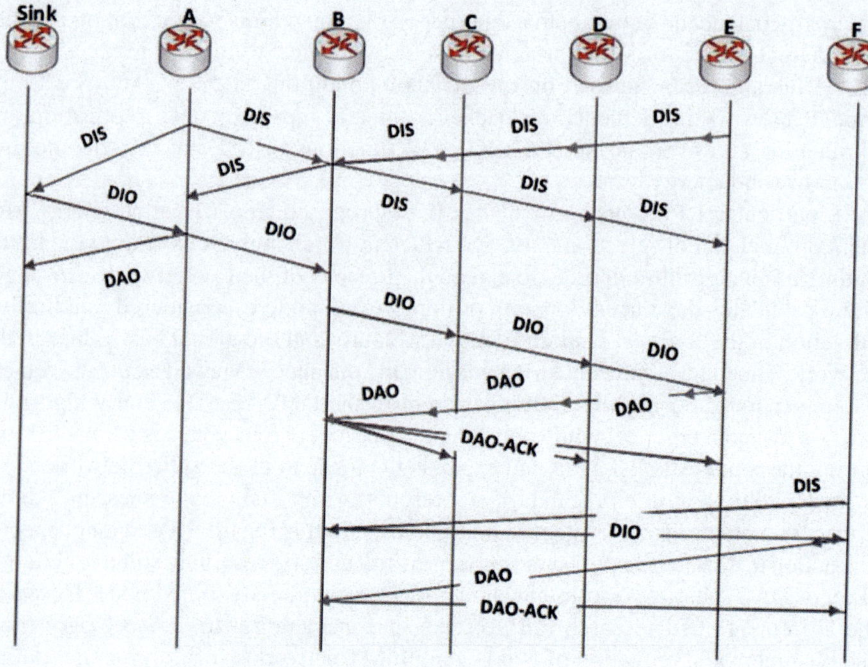

Fig. 2 RPL's memory saturation case in downward route

Another remaining problem that will obstruct the process, of RPL, is the amount of overhead exchanged in the network.

4.1 *Case Study*

In this subsection, we have defined a sample case study of a scalable network with limited memory storage and resources constrained nodes. In many cases, some of the node's preferred parent runs out of its routing table storage such as the node B depicted in Fig. 2 which has a memory capacity of three routing entries per node. After that some new nodes (such as node F) want to join the network, at that time the preferred parent (node B) will try to add its routing entry for a specific target rendering. Consequently, the announced target will be unreachable and the entire packet destined for it will be lost. This will also increase the energy consumption. In order to overcome this issue, we illustrate in Fig. 3 the same network with a proposed solution which aims to stop sending the control messages and the unnecessary energy loss. Once node F solicits node B (overloaded memory node), this later stops interacting with any new node till it succeeds to manage its routing table or the node F finds another reachable parent node.

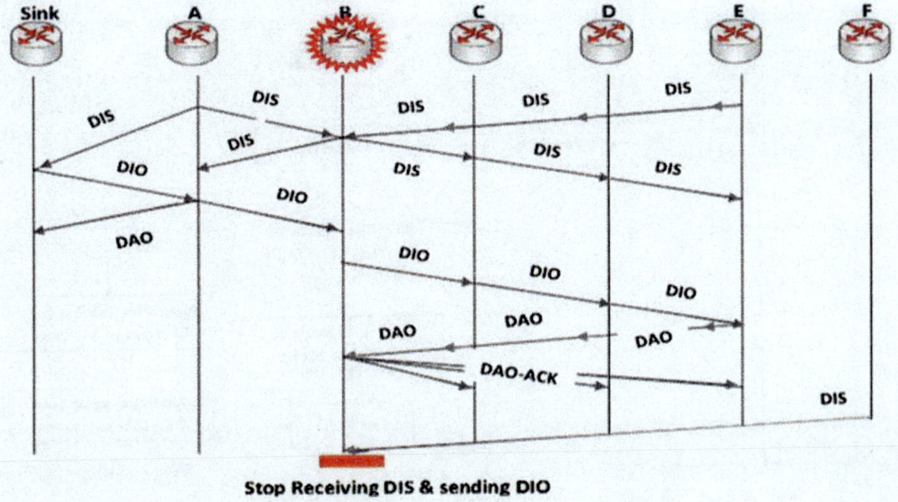

Fig. 3 The proposed solution by MmRPL

5 The Proposed Solution

Our proposed MmRPL takes into account the scalability of the network in the case of memory saturation of a new node, while reducing the total overhead. The main objective focuses on the economy of the number of control messages knowing that it will also reduce the amount of energy which reflects more benefits for a such constrained network as IoMT. Our purpose is to avoid the unnecessary messages in the case of saturation, as depicted in Fig, 4, once the parent node receives a DIS control message, it will check the memory of its neighboring table. If there is much space to let the new node join, the RPL's process will continue. Otherwise, in the case of overloaded memory, the proposed scheme will prohibit the parent node from exchanging any control messages which saves its energy until this latter manages to clear some space in its routing table. Alternatively, the new node will find its new way to another reachable parent. The use of MmRPL would reflect positively on the performance of the IoMT network.

6 Performance Evaluation and Discussion

In this section, we explore the performance of MmRPL in scalable network in terms of control overhead, energy consumption and end to end delay.

We evaluate the MmRPL performance using Cooja emulator [21] which allows us to test the binary code that could run on real $TelosB$ WSN nodes, besides, it offers for us a full flexibility to evaluate radio environments and topologies. We have fixed

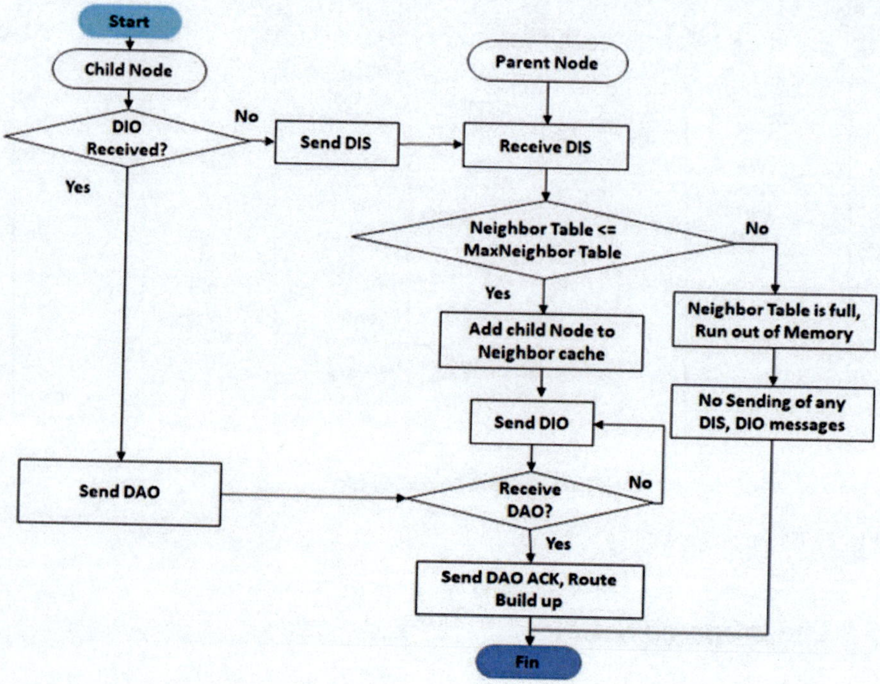

Fig. 4 Diagram of the proposed solution for the RPL memory saturation case in downward route

the routing table entries N to 3 in order to easily clarify the saturation case. Moreover, ContikiMAC [22] is employed as the underlying duty cycle MAC layer. Regarding the radio propagation model, we used the Unit Disk Graph Model: Distance Loss [23]. Each node sends an application data packet every minute to the root.

All the log files (trace files) of all experiments are analysed by a Perl script in order to extract the statistical results. The main simulation parameters are summarized in Table 1.

6.1 Performance Metrics

- **Average Power Consumption (APC)** In order to compute the energy consumption of the network, we use a novel technique of the software named Powertrace which is a tool available in Contiki [24]. This tool tracks the power state by estimating the energy consumption for CPU processing, packet transmission and listening. The power trace mechanism provides CPU, low-power mode (lpm), radio listen and radio transmit values which gives the total time consumed in each power trace interval. Moreover, we take the average percent radio on time of all the nodes in the whole network setup. The average power consumption (APC) is calculated

Table 1 Simulation parameters

Parameter	Value
Simulation time	500s
Number of nodes	[5–20]
Objective function	OF0, MRHOF
Routing metric	Overhead, Energy consupmtion
DIO, DAO, DIS packet size	30 bytes
Data packet size	(50, 100) Bytes
Send interval	1 pckt/min
Ranges of nodes	Tx: 50 m , Interference: 100 m
Radio duty cycle	ContikiMAC
Radio propagation model	Unit disk graph model

according to Eq. 2.

$$\text{APC} = \frac{\sum_{i=0}^{n}(\text{LPM} + \text{CPU} + \text{RadioListen} + \text{RadioTransmit})}{n} \tag{1}$$

where n represents the number of nodes in the network.

- **End-to-end delay** The end-to-end delay (Packet Delay) is another evaluation metric used for measuring the total time taken by packets in order to be successfully delivered from node to sink.
- **Total Overhead** The Overhead in RPL is calculated by the Eq. 2.

$$\text{Overhead} = \sum(\text{DIS} + \text{DIO} + \text{DAO} + \text{DAOACK}) \tag{2}$$

6.2 Results Discussion

In Figs. 5a–c and 6a–c, we plot our simulation results after improving the RPL protocol by reducing the amount of total overhead circulating in the network during both of the construction and the routing process.

We notice that as long as new nodes join the network (the number of nodes increases) the performance of MmRPL is showing on by reducing the amount of overhead around 2000 packets when the network reaches 20 nodes as shows Fig. 5b. By consequence, the amount of energy consumed by MmRPL network is less than the one consumed by the conventional RPL as depict all of Fig. 6. At the meantime, MmRPL results in the same benefits of the total amount of overhead in Fig. 5c when we tried to expand the simulation time to 14450 seconds in order to observe the variation of the overhead. As well, we have noticed a stabilization of the average end to end delay of MmRPL, which takes less delay than the standard RPL, as depict

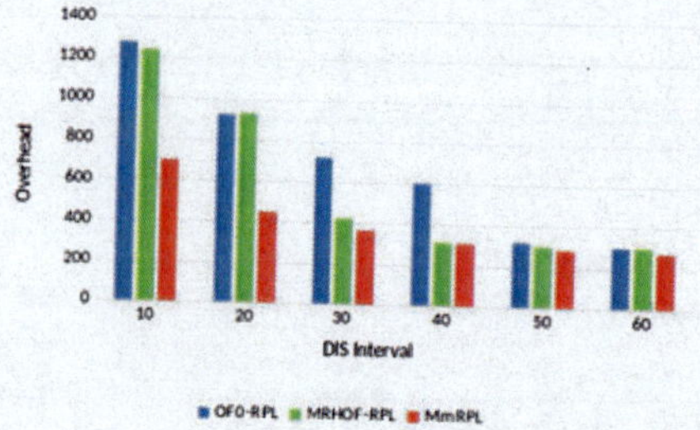

(a) Control Overhead vs. DIS interval.

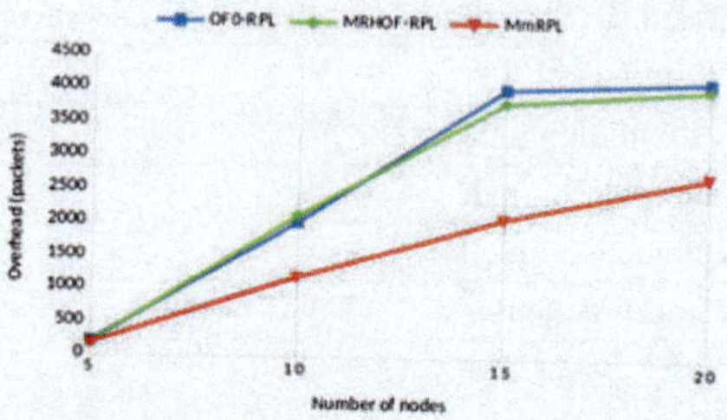

(b) Control Overhead vs. Network size.

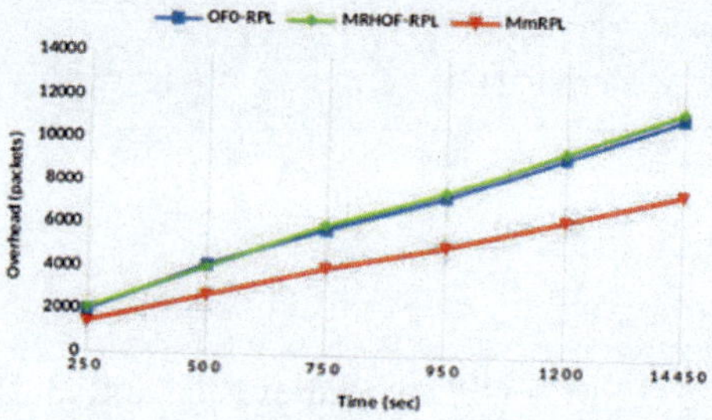

(c) Control Overhead vs. Simulation time.

Fig. 5 Overhead results

Fig. 6. In Fig. 6, we plot the radio transmit power of the saturated node, preferred parent represented as node 4 in Fig. 3 against time where our proposed MmRPL outperforms the conventional RPL.

6.3 *Comparative Results*

This subsection investigates some further analysis about our work by comparing it with another RPL enhancement based on a combination of multiple sinks support named MRRD+ (Multiple, RSSI, Rank and dynamic) [25] as illustrated in Table 2.

The first analysis highlighted the impact of the number of nodes on the total number of control packets. After comparing our simulation parameters (Data packet size: 50–100 Bytes; Radio propagation model: Unit Disk Graph Model; Simulation Time: 500 s) with the ones used for the MRRD+ in [25], (Data packet size: 30 Bytes; Radio propagation model: MRM with random behavior; Simulation Time: 300 s), respectively, we have noticed that when the network density is equal to 20 nodes, the MmRPL shows an improvement in terms of the number of control packets. In another finding, the MmRPL revealed remarkable results better than (MRRD+)-1S, (MRRD+)-2S, (MRRD+)-3S and (MRRD+)-4S in terms of the average end-to-end delay. Specifically, in 20 nodes, MRRD+ delay varies from (20–45 ms), while our

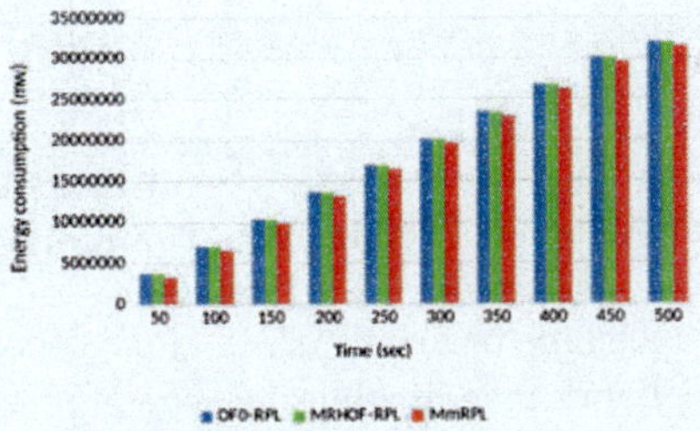

(a) Energy consumption of the network vs. Simulation Time.

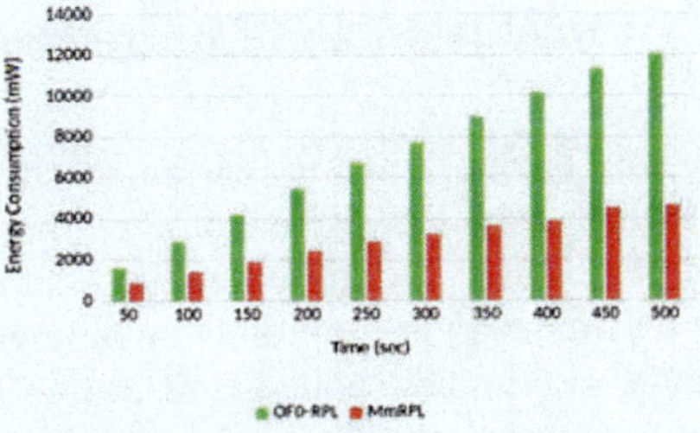

(b) Energy consumption of the saturated Node (Preferred parent) vs. Time.

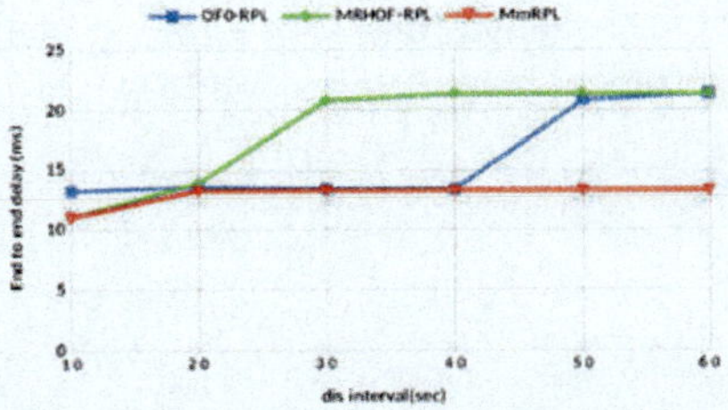

(c) Average End to End Delay vs. DIS interval.

Fig. 6 Energy consumption results

Table 2 Comparative results between MmRPL and MRRD+

Routing protocol	Performance parameter	
	Average end to end delay (ms)	
MmRPL	Network size = 20	12
MRRD+ (1S, 2S, 3S, 4S)		20–45
Routing protocol	Performance parameter	
	Number of control packets	
MmRPL	Network size = 20	1800
MRRD+ (1S, 2S, 3S, 4S)		≈2200

MmRPL keeps a stable delay consumption for 12 ms, knowing that the DIS sending interval varies between 20 and 60 DIS message per second.

Even in the presence of existing differences between our simulation parameters and the comparative work's ones, especially in what concerns the mobility, the comparison of our work results against the aforementioned work have been very promising. Thus, the results show that the proposed MmRPL would represent a good solution for the memory problem, and it reduces the total amount of overhead in the network and satisfies the QoS requirements of the IoMT networks.

7 Conclusion and Future Work

In this paper, we proposed an extension of the RPL routing protocol named as MmRPL which satisfies some of the QoS of the IoMT networks. The proposed MmRPL allows to minimize the total control messages once detecting an event of node's memory saturation by avoiding the exchange of unnecessary messages specially when a new node wants to join the network by soliciting the saturated node in its vicinity. In this situation, reducing the amount of overhead will help us to reduce the energy consumption during the exchange process. Besides, the MmRPL optimizes another IoMT QoS which is the end to end delay taken from the DODAG construction till the end of the routing process. In future, we will study the validation of the proposed MmRPL protocol in real embedded IoMT system and test it in large networks.

References

1. Floris, A., Atzori, L.: Quality of experience in the multimedia internet of things: definition and practical use-cases. In: 2015 IEEE International Conference on Communication Workshop (ICCW), pp. 1747–1752. IEEE (2015)
2. Kim, E., Kaspar, D., Gomez, C., Bormann, C.: Problem statement and requirements for 6lowpan routing. In: Draft-IETF-6LoWPAN-routing-requirements-04, IETF Internet Draft (Work in Progress) (2009)
3. Winter, T.: RPL: IPv6 routing protocol for low-power and lossy networks (2012)
4. Kiraly, C., Istomin, T., Iova, O., Picco, G.P.:. D-RPL: overcoming memory limitations in RPL point-to-multipoint routing. In 2015 IEEE 40th Conference on Local Computer Networks (LCN), pp. 157–160. IEEE (2015)
5. Clausen, T., Herberg, U., Philipp, M.: A critical evaluation of the ipv6 routing protocol for low power and lossy networks (RPL). In: 2011 IEEE 7th International Conference on Wireless and Mobile Computing, Networking and Communications (WiMob), pp. 365–372. IEEE (2011)
6. Iova, O., Picco, P., Istomin, T., Kiraly, C.: RPL: the routing standard for the internet of things... or is it? IEEE Commun. Mag. **54**(12), 16–22 (2016)
7. Gaddour, O., Koubâa, A.: RPL in a nutshell: a survey. Comput. Netw. **56**(14), 3163–3178 (2012)
8. Thubert, P.: Objective function zero for the routing protocol for low-power and lossy networks (RPL) (2012)
9. Gnawali, O.: The minimum rank with hysteresis objective function (2012)
10. Safaei, B., Hosseini Monazzah, A.M., Shahroodi, T., Ejlali, A.: Objective function: a key contributor in internet of things primitive properties. In: 2018 Real-Time and Embedded Systems and Technologies (RTEST), pp. 39–46. IEEE (2018)
11. Ghaleb, B., Al-Dubai, A., Ekonomou, E., Wadhaj, I.: A new enhanced RPL based routing for internet of things. In: 2017 IEEE International Conference on Communications Workshops (ICC Workshops), pp. 595–600. IEEE (2017)
12. Gan, W., Shi, Z., Zhang, C., Sun, L., Ionescu, D.: MERPL: a more memory-efficient storing mode in RPL. In: 2013 19th IEEE International Conference on Networks (ICON), pp. 1–5. IEEE (2013)
13. Nassar, J., Berthomé, M., Dubrulle, J., Gouvy, N., Mitton, N., Quoitin, B.: Multiple instances QoS routing in RPL: application to smart grids. Sensors **18**(8), 2472 (2018)

14. Joseph Charles , A.S., Kalavathi, P.: QoS measurement of RPL using cooja simulator and wireshark network analyser (2018)
15. Zier, A., Abouaissa, A., Lorenz, P.: E-RPL: A routing protocol for IoT networks. In: 2018 IEEE Global Communications Conference (GLOBECOM), pp. 1–6. IEEE (2018)
16. Da Silva Araújo, H., Rodrigues, J.J.P.C., De Al Rabelo, R., De C Sousa, N., Filho, C.C.L.S.J., Sobral, J,V.V., et al.: A proposal for IoT dynamic routes selection based on contextual information. Sensors (Basel), **18**(2), 353 (2018)
17. Bouzebiba, H., Lehsaini, M.; FreeBW-RPL: a new RPL protocol objective function for internet of multimedia things. Wirel. Pers. Commun. 1–21 (2020)
18. Jenschke, T.L., Koutsiamanis, R.-A., Papadopoulos, G.Z., Montavont, N.: ODeSe: on-demand selection for multi-path RPL networks. Ad Hoc Netw. 102431 (2021)
19. Koutsiamanis, R.-A., Papadopoulos, G.Z., Jenschke, T.L., Thubert, P., Montavont, N.: Meet the PAREO functions: towards reliable and available wireless networks. In: ICC 2020-2020 IEEE International Conference on Communications (ICC), pp. 1–7. IEEE (2020)
20. Sennan, S., Somula, R., Luhach, A.K., Deverajan, G.G., Alnumay, W., Jhanjhi, N.Z., Ghosh, U., Sharma, P.: Energy efficient optimal parent selection based routing protocol for internet of things using firefly optimization algorithm. Trans. Emerging Telecommun. Technol. e4171 (2020)
21. Osterlind, F., Dunkels, A., Eriksson, J., Finne, N., Voigt, T.: Cross-level sensor network simulation with COOJA. In: Proceedings. 2006 31st IEEE Conference on Local Computer Networks, pp. 641–648 (2006)
22. Dunkels, A.: The ContikiMAC radio duty cycling protocol (2011)
23. Clark, B.N., Colbourn, C.J., Johnson, D.S.: Unit disk graphs. Discrete Math. **86**(1–3), 165–177 (1990)
24. Dunkels, A., Eriksson, J., Finne, N., Tsiftes, N.: Network-level power profiling for low-power wireless networks. Powertrace (2011)
25. Wang, J., Chalhoub, G.: Mobility support enhancement for RPL with multiple sinks. Ann. Telecommun. **74**(5), 311–324 (2019)

Channel Estimation in Massive MIMO Systems for Spatially Correlated Channels with Pilot Contamination

Mohamed Boulouird, Jamal Amadid, Abdelhamid Riadi, and Moha M'Rabet Hassani

Abstract This work treats multi-cell (M-C) multi-user (M-U) massive MIMO (M-MIMO) systems taking into consideration pilot contamination (PC), where Rayleigh fading channels are correlated in the spatial domain. An appropriate exponential correlation (EC) is using as an approximation model for uniform-linear arrays (Un-LA). The statistics of minimum mean square error (MMSE), element-wise MMSE (EW-MMSE), approximate MMSE (Approx.MMSE), and least-squares (LS) estimators are evaluated and analyzed. The Approx.MMSE estimator uses an imperfect covariance matrix (CM), which relies on sample CM to estimate a true CM provided by the MMSE. Analytical NMSE formulas for idealistic and realistic CMs are presented and interpreted. An analytical normalized mean square error (NMSE) formula is also given for EW-MMSE.

1 Introduction

The M-MIMO technology offers such enhancement in spectral efficiency (S-E) and energy efficiency (E-E) using spatial multiplexing (S-M) and a high channel gain, respectively [1, 2]. The literature deals with two families of scenarios for a M-MIMO systems. The first scenario where the channels are independent [3–5], in which the

M. Boulouird (✉)
Smart Systems and Applications (SSA) Group, National School of Applied Sciences of Marrakesh (ENSA-M), Cadi Ayyad University, Marrakesh, Morocco
e-mail: m.boulouird@uca.ac.ma

J. Amadid · A. Riadi · M. M. Hassani
Instrumentation, Signals and Physical Systems (I2SP) Group, Faculty of Sciences Semlalia, Cadi Ayyad University, Marrakesh, Morocco
e-mail: jamal.amadid@edu.uca.ac.ma

A. Riadi
e-mail: abdelhamid.riadi@edu.uca.ac.ma

M. M. Hassani
e-mail: hassani@uca.ac.ma

M. Ben Ahmed et al. (eds.), *Networking, Intelligent Systems and Security*,
Smart Innovation, Systems and Technologies 237,
https://doi.org/10.1007/978-981-16-3637-0_11

expression of channel coefficient is composed of two factors, namely large-scale fading (LSF) and small-scale fading (SSF), the first is including both path loss and shadowing, while the second factor is including statistics variations of the channels. In the second scenario, the channels are spatially correlated (SC) [6–11]. Where the channel CM describes the correlation between antennas in spatial domain. In reality, in certain cases channels are dependents (i.e., SC) [6, 12, 13]. That is to say, in certain measure the element of the channels is correlated. The property of the channel CMs is to make sense of the spatial correlation among antennas. Where the off-diagonal elements of CMs usually are nonzero. We consider in our studies SC channels for a square system M-C M-U M-MIMO under PC.

1.1 Related Works

In [14], the authors have tackled SC channels and their greatness over a system that performs with M-MIMO technology since SC channels describe more practical channels and draw a real propagation environment. They present up-to-date results regarding the core boundaries of M-MIMO, which are not primarily a matter of pilot contamination rather the capacity to get statistics of the channels. Hence, this result gives rise to an updated version of M-MIMO, namely M-MIMO 2.0.

The authors in [15] have proposed an approximate model for SC channels for two arrangements known in the literature as the Un-LA and uniform circular array employing the Laplacian distribution. A metric for evaluating the performance has been proposed by which they evaluate the performance of the proposed model. Furthermore, under this metric, the proposed model works best in small-angle spreading situations.

The authors in [16] discussed the SC channel in the up-link stage, where the BS has a large number of antennas. Besides, they prolong the LSF concept to SC channels. For which they have developed a signal-to-interference ratio plus noise formulation that relies only on SSF factors.

In [6], the authors have dealt with different distributions that describe correlation among channels to investigate and evaluate the influence of spatial correlation using either Gaussian, uniform, and Laplacian distribution. The Gaussian distribution is known as the local scattering model, whereas the uniform distribution is known as the one-ring model. Besides, they consider that the system operated under a higher PC level. Furthermore, in each scenario (i.e., for each distribution), they have analyzed the channel estimation quality using three estimators, namely LS, MMSE, and EW-MMSE employing the MSE metric. This evaluation is performed using the effective signal-to-noise ratio, where the best performance is achieved using the local scattering model (i.e., Gaussian distribution).

1.2 Contributions

The contributions of this work are given in twofold as follows :

1. This work introduced the concept of spatially correlated channels (i.e., that model a more realistic channel environment) on ULA using the exponential correlation model and the influence of this concept on channel estimation.
2. An EW-MMSE channel estimator is proposed, which performs similarly to MMSE with lower complexity and shown results better than the Approx.MMSE even if the number of realizations is large.

1.3 Work Scheduling

The rest sections of this work are arranged as follows: In Sect. 2, the system model is defined including the expression of the received signal and the setup used for CMs. In Sect. 3, the NMSE expressions are given and analyzed for all used channel estimator. In Sect. 4, the simulation results are given, in which the efficiency of our proposed method is evaluated and compared to existing methods. We finished our work by giving a conclusion in Sect. 5.

2 System Model

Our studies treat a square M-C scheme, where each cell has a BS in the corner and K user with single antenna in the cell-edge area [17], as depicted in Fig. 1.

All channels are considered as correlated Rayleigh fading (CRF) channels. The h_{jlk} exemplify vector of the channel from the k-th user in the j-th cell to the N antennas at l-th BS, which define as $\mathbf{h}_{jlk} = [h_{jlk1}, h_{jlk2}, \ldots, h_{jlkN}]^{\mathrm{T}} \sim \mathbb{CN}(0_N, \mathbf{R}_{jlk})$. Where $\mathbf{R}_{jlk} \in \mathbf{C}^{\mathbf{N}\times\mathbf{N}}$ display a positive semi-defined channel CM. It is remarkable to underlining that $\mathbf{R}_{ijk}$ is not an identity matrix, which characterizes macroscopic effects including path loss in various directions and channel correlation in the spatial domain.

The EC model is studied in our work for a Un-LA referring to Bjornson work [17], with which we modeled the correlation between contiguous antennas. The inter-antenna correlation can be represented by

$$[\mathbf{R}]_{m,n} = \beta r^{|n-m|} e^{i\varphi(n-m)} 10^{(\nu_n+\nu_m)//20} \tag{1}$$

β, φ and r are respectively, the LSF coefficient, angle of arrival, correlation co-efficient/factor. While $\nu_1, \ldots, \nu_M \sim \mathcal{CN}(0, \sigma^2)$ afford independent LSF variations through the array. Otherwise, the users in all cells send their Uplink (UP) pilot sequence (PS) simultaneously. The PSs used in each cell are duplicated in all other

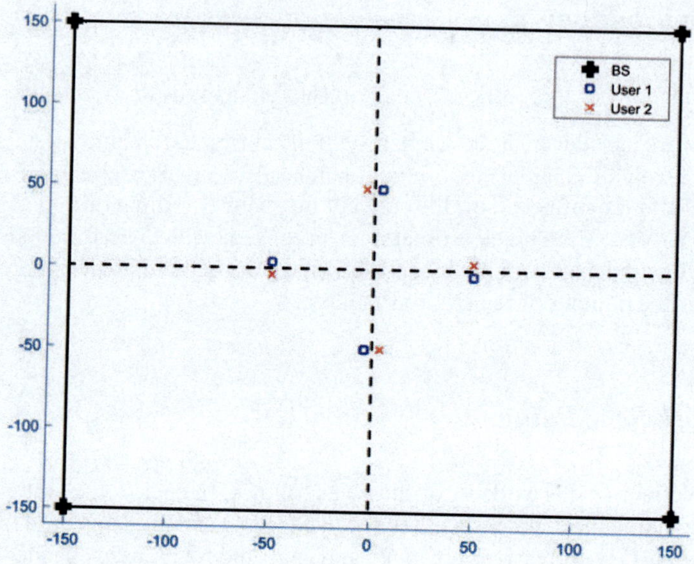

Fig. 1 Square M-C setup

cells (i.e., leading to the PC problem). That is to say, the frequency factor is one. $\phi_k \in \mathbb{C}^\tau$ is the PS sent by k-th user, where τ is the length of each PS, ϕ_k have as property the following equation $\phi_k^H \phi_k = 1, \forall k$. For all K users in each cell, the global matrix $\tau \times K$ of the PS becomes $\Phi^H \Phi = \mathbb{I}_N, \forall k$, where $\Phi = [\phi_1, \phi_2, \ldots, \phi_K]$. The received UP signal Y_j at j-th BS can be defined as

$$Y_j = \sqrt{q} \sum_{l=1}^{L} H_{jl} \ \Phi^H + W_j \tag{2}$$

where q represents UP transmit power (TP) and $W_j^{N \times \tau}$ is a noise matrix, each element of $W_j^{N \times \tau}$ follow a $\mathcal{CN}(0, 1)$.

3 Channel Estimation

3.1 LS Channel Estimation

In general, an estimate for the channel vector h_{jjk} at the j-th BS using LS channel estimator through adequate statistics [18] is defined as follows

$$\hat{h}_{jjk}^{\mathrm{LS}} = \theta_{jk} = \frac{1}{\sqrt{q}} Y_j \phi_k = \sum_{l=1}^{L} h_{jlk} + w_{jk} \tag{3}$$

Here the vector w_{jk} is the multiplication between the noise matrix and PS of k-th user divided by square root of pilot power as shown in the following equation

$$w_{jk} = \frac{1}{\sqrt{q}} W_j \phi_k \sim \mathcal{CN}(0_N, \frac{1}{q}\mathbb{I}_N) \tag{4}$$

where $\hat{h}_{jjk}^{\mathrm{LS}} \sim \mathcal{CN}(0_N, \Psi_{jk})$, Ψ_{jk} represent the covariance of LS channel estimate expressed as $\Psi_{jk} = \sum_{l=1}^{L} \mathbf{R}_{jlk} + \frac{1}{q}\mathbf{I}_N$. While the error of estimation is represented by $\tilde{h}_{jjk}^{\mathrm{LS}} = h_{jjk} - \hat{h}_{jjk}^{\mathrm{LS}}$ which follows $\mathcal{CN}(0_N, \Psi_{jk} - \mathbf{R}_{jjk})$. It is dependent of both $\hat{h}_{jjk}^{\mathrm{LS}}$ and the CM of $\hat{h}_{jjk}^{\mathrm{LS}}$; where $\tilde{h}_{jjk}^{\mathrm{LS}}$ is determined by $\mathrm{Cov}(\hat{h}_{jjk}^{\mathrm{LS}}, \tilde{h}_{jjk}^{\mathrm{LS}}) = \Psi_{jk} - \mathbf{R}_{jjk}$.

Therefore, the NMSE is computed in accordance with each antenna using LS estimator, and it is formulated as follows

$$\varepsilon_{jk}^{\mathrm{LS}} = \frac{1}{N}\mathbb{E}\{\|\hat{h}_{jjk}^{\mathrm{LS}} - h_{jjk}\|^2\} = \frac{1}{N} Tr[\Psi_{jk} - \mathbf{R}_{jjk}] \tag{5}$$

We can notice that the NMSE of LS estimator does not rely on any previously statistics of the channel (i.e., LSF coefficients). Besides, LS estimator is a linear estimator that has low complexity and large NMSE compared to MMSE estimator [5]. In the rest of this subsection, some notes for LS channel estimation are given as follows:

Note 1: If each element of $\mathbf{h}_{jjk}$ for all cells, BSs and all users in each cell follows $\mathcal{CN}(0, 1)$. Hence, the CM in this case is a diagonal matrix represented as follows $\Psi_{jk} = \psi_{jk}\mathbf{I}_N = (\sum_{l=1}^{L} \beta_{jlk} + \frac{1}{q})\mathbf{I}_N$ and $\mathbf{R}_{jjk} = \beta_{jjk}\mathbf{I}_N$. Thus, $\varepsilon_{jk}^{\mathrm{LS}} = \psi_{jk} - \beta_{jjk}$ and $\beta_{jjk} = \frac{1}{N} Tr[\Psi_{jk}]$.

Note 2: if $\mathbf{h}_{jjk} \sim \mathcal{CN}(0, 1)$, q tends to infinity and under PC, the NMSE of LS estimator is: $\varepsilon_{jk}^{\mathrm{LS}} \longrightarrow \sum_{l=1}^{L} \beta_{jlk}$ when $q \longrightarrow \infty$.

Note 3: According to Eq. (3), we can easily remark that $\hat{h}_{jjk}^{\mathrm{LS}} = \hat{h}_{jlk}^{\mathrm{LS}}$ (i.e., the vectors are parallel). Thus, the BS cannot separate these channels (the channels from the users that have the same PSs).

3.2 MMSE Channel Estimation

The MMSE channel estimation is an estimator which relies on statistics of the channel and is among Bayesian estimators category [18]. To estimate the channel $\mathbf{h}_{jjk}$, the MMSE estimator is determined by

$$\hat{\mathbf{h}}_{jjk}^{\text{MMSE}} = \frac{\mathbf{R}_{jjk}}{\Psi_{jk}}\theta_{jk} = \mathbf{R}_{jjk}\Psi_{jk}^{-1}\theta_{jk} \tag{6}$$

For the Gaussian model, MMSE channel estimator has a special property represented by the independence between the vector estimate $\hat{\mathbf{h}}_{jjk}^{\text{MMSE}}$ and the estimation error $\tilde{\mathbf{h}}_{jjk}^{\text{MMSE}} = \mathbf{h}_{jjk} - \hat{\mathbf{h}}_{jjk}^{\text{MMSE}}$, which are randomly distributed vectors as $\hat{\mathbf{h}}_{jjk}^{\text{MMSE}} \sim \mathcal{CN}(\mathbf{0}_N, \mathbf{R}_{jjk}\Psi_{jk}^{-1}\mathbf{R}_{jjk})$ and $\tilde{\mathbf{h}}_{jjk}^{\text{MMSE}} \sim \mathcal{CN}(\mathbf{0}_N, \mathbf{R}_{jjk}(\mathbf{I}_N - \Psi_{jk}^{-1}\mathbf{R}_{jjk}))$. The estimated vector $\hat{\mathbf{h}}_{jjk}^{\text{MMSE}}$ is not correlated with the received vector (de-spread vector) θ_{jk}. Which means $\text{Cov}(\hat{\mathbf{h}}_{jjk}^{\text{MMSE}}, \theta_{jk}) = 0$. As a result, θ_{jk} is not dependent of both estimated vector and estimation error vector.

Therefore, we compute the NMSE in accordance with each antenna using MMSE estimator. Thus, the following result is obtained

$$\varepsilon_{jk}^{\text{MMSE}} = \frac{1}{N}\mathbb{E}\{\|\hat{h}_{jjk}^{\text{MMSE}} - h_{jjk}\|^2\} = \frac{1}{N}Tr[\mathbf{R}_{jjk} - \mathbf{R}_{jjk}\Psi_{jk}^{-1}\mathbf{R}_{jjk}] \tag{7}$$

We present some notes for MMSE channel estimation as follows.

Note 1: If each element of $\mathbf{h}_{jjk}$ for all cells, BSs and all users in each cell follows $\mathcal{CN}(0, 1)$. Therefore, $\Psi_{jk} = \psi_{jk}\mathbf{I}_N = (\sum_{l=1}^{L}\beta_{jlk} + \frac{1}{q})\mathbf{I}_N$ and $\mathbf{R}_{jjk} = \beta_{jjk}\mathbf{I}_N$. Thus, $\varepsilon_{jk}^{\text{MMSE}} = \beta_{jjk} - \frac{\beta_{jjk}^2}{\psi_{jk}}$ and $\beta_{jjk} = \frac{1}{N}Tr[\Psi_{jk}]$.

Note 2: if $\mathbf{h}_{jjk} \sim \mathcal{CN}(0, 1)$, q tends to infinity and Under PC, the NMSE of MMSE estimator is $\varepsilon_{jk}^{\text{MMSE}} \longrightarrow \beta_{jjk}(1 - \frac{\beta_{jjk}}{\sum_{l=1}^{L}\beta_{jlk}})$ when $q \longrightarrow \infty$.

Note 3: According to Eq. (6), and if $\mathbf{R}_{jjk}$ is invertible, we can conclude that $\hat{\mathbf{h}}_{jlk}^{\text{MMSE}} = \mathbf{R}_{jlk}\mathbf{R}_{jjk}^{-1}\hat{\mathbf{h}}_{jjk}^{\text{MMSE}}$. In the reference [17], they have been shown that if $\mathbf{R}_{jlk}$, $\forall l$ are linearly not dependant, which means $\mathbf{R}_{jjk} = \alpha\mathbf{R}_{jlk}$, $\forall l \neq j$ where α is a real number. Consequently, the channels vector are not parallel (i.e., the BS can separate users sent the same PS). When $\mathbf{h}_{jjk} \sim \mathcal{CN}(0, 1)$, according to Eq. (6), we remark that $\hat{\mathbf{h}}_{jlk}^{\text{MMSE}} = \frac{\beta_{jlk}}{\beta_{jjk}}\hat{\mathbf{h}}_{jjk}^{\text{MMSE}}$. The BS cannot separate the users used the same PS because the channel vectors are parallel differed by a factor $\frac{\beta_{jlk}}{\beta_{jjk}}$.

3.3 *Approximate MMSE Channel Estimator*

Generally, to get the CMs $\mathbf{R}_{jlk}, \forall l, k$ is a difficult process that it requires the estimation of large matrices. In this part, we showed a straightforward and efficient approach to this issue. According to Eq. (6), we can estimate Ψ_{jk} which consists of the sum of $\mathbf{R}_{jlk}$ plus identity matrix $\mathbf{I}_N$ times the inverse of q. Thus, we can denote the estimated of Ψ_{jk} by $\hat{\Psi}_{jk}$, and swapping Ψ_{jk} with $\hat{\Psi}_{jk}$ in Eq. (6). A traditional way to this estimation issue is to approximate the CM including the sample CM [7]. To estimate Ψ_{jk}, we can remark that $\mathbb{E}[\theta_{jk}\theta_{jk}^H] = \Psi_{jk}$. To approximate Ψ_{jk}, we can use the sample CM below

$$\hat{\Psi}_{jk} = \frac{1}{N_\Psi}\sum_{n=1}^{N_\Psi} \theta_{jk}(n)\theta_{jk}^H(n), \quad n = 1, 2, \ldots, N_\Psi \tag{8}$$

where N_Ψ represents the number of observations (NoO). In addition, $\mathbb{E}[\hat{\Psi}_{jk}] = \Psi_{jk}$. Note that the convergence of the sample CM is proportional to the NoO of the de-spread vector $\theta_{jk}(n)$.

$$\lim_{N_\Psi \to +\infty} \hat{\Psi}_{jk} = \frac{1}{N_\Psi}\sum_{n=1}^{N_\Psi} \theta_{jk}(n)\theta_{jk}^H(n) \to \Psi_{jk} \tag{9}$$

We show that if $N_\Psi \to +\infty \Rightarrow \hat{\Psi}_{jk} \to \Psi_{jk}$. During this study, we look to the channels as ergodic. The covariance of the i-th column of $\hat{\Psi}_{jk}$ is given by

$$\mathrm{Cov}([\hat{\Psi}_{jk}]_i) = \frac{1}{N_\Psi}[\Psi_{jk}]_i[\Psi_{jk}]_i^{\mathrm{T}} \tag{10}$$

where $[\Psi_{jk}]_i$ is the i-th column of Ψ_{jk}. If $\mathbf{h}_{jjk} \sim \mathcal{CN}(0, 1)$, thus $[\Psi_{jk}]_i = [\psi_{jk}0]^{\mathrm{T}}$. Hence,

$$\mathrm{Cov}([\hat{\Psi}_{jk}]_i) = \frac{1}{N_\Psi}\psi_{jk}^2\mathbf{I}_N \tag{11}$$

The presence of error in each element of $\hat{\Psi}_{jk}$ leads to losing its eigenstructure, which makes each eigenvalue and eigenvector non-aligned with that of Ψ_{jk}. While the MMSE estimator is taking into account the eigenstructure of Ψ_{jk} to achieve efficient channel estimate. Therefore, it has a high effect on the Approx.MMSE performance. Thus, to get over these problems, a convex combination scheme in [7, 8, 19] is given to estimate Ψ_{jk} (CM)

$$\hat{\Psi}_{jk}(\eta) = \eta\hat{\Psi}_{jk} + (1-\eta)\hat{\Psi}_{jk}^{\mathrm{diag}} \tag{12}$$

This convex combination is a type of regularization that transforms $\hat{\Psi}_{jk}$ into a full-rank (F-R) matrix for all η values in the interval [0,1[. Even if when $N_\Psi < N$, the off-diagonal of $\hat{\Psi}_{jk}$ be under-estimated.

The authors in [6] estimate $\mathbf{R_{jlk}}$ $\forall j, l, k$ for each user in a specific pilot phase. In our work, we are not considering the individual $\mathbf{R_{jlk}}$, but we considering the estimation of Ψ_{jk} and performance evaluation under certain parameters. From Eqs. (6), (8), and (12), $\hat{\Psi}_{jk}$ is considered as the actual CM. We can give the appro.MMSE formula for the channel vector $\hat{\mathbf{h}}_{jjk}$ as follows

$$\hat{\mathbf{h}}_{jjk}^{\mathrm{Approx.MMSE}} = \frac{\mathbf{R}_{jjk}}{\hat{\Psi}_{jk}(\eta)}\theta_{jk} = \mathbf{R}_{jjk}\hat{\Psi}_{jk}(\eta)^{-1}\theta_{jk} \tag{13}$$

We suppose that θ_{jk} is independent of $\hat{\Psi}_{jk}$, that is to say $\hat{\Psi}_{jk}$ is not estimated using θ_{jk}. In addition, N_{Ψ} is high enough to generate a great estimate of Ψ_{jk}. The expression of NMSE for each BS antenna using appro.MMSE is as follows

$$\begin{aligned} \varepsilon_{jk}^{\text{Approx.MMSE}} &= \frac{1}{N}\mathbb{E}\{\|\hat{h}_{jjk}^{\text{Approx.MMSE}} - h_{jjk}\|^2\} \\ &= \frac{1}{N}Tr[\mathbf{R}_{jjk} - \mathbf{R}_{jjk}\Psi_{jk}(\eta)^{-1}\mathbf{R}_{jjk}] \end{aligned} \tag{14}$$

We noticed that when N_{Ψ} tends to infinity, the NMSE of Approx.MMSE estimator and MMSE estimator are the same. The factor η is also chosen so as to the NMSE is minimized. The authors in [19] have been given an optimization for this problem of choosing η which minimizes the NMSE with proof.

3.4 EW-MMSE Estimator

The EW-MMSE estimator has low computational compared to the MMSE estimator, which is also called a diagonalized MMSE estimator. This estimator relies on the diagonal of the CM, where the off-diagonal of the CM is zero. Therefore, the EW-MMSE is avoiding the correlation between CM elements. In some work, the EW-MMSE is used as an alternative estimator [19, 20]. The estimate of $\mathbf{h}_{jjk}$ using EW-MMSE estimator is determined by

$$\hat{\mathbf{h}}_{jjk}^{\text{EW-MMSE}} = \frac{\mathbf{D}_{jjk}}{\mathbf{\Omega}_{jk}}\theta_{jk} = \mathbf{D}_{jjk}\mathbf{\Omega}_{jk}^{-1}\theta_{jk} \tag{15}$$

where $\mathbf{D}_{jjk} \in \mathbb{C}^{N\times N}$ and $\mathbf{\Omega}_{jk} \in \mathbb{C}^{N\times N}$ are the diagonal matrices of $\mathbf{R}_{jjk}$ and Ψ_{jk}, respectively. The estimated vector $\hat{\mathbf{h}}_{jjk}^{\text{EW-MMSE}}$ and the estimation error $\tilde{\mathbf{h}}_{jjk}^{\text{EW-MMSE}} = \mathbf{h}_{jjk} - \hat{\mathbf{h}}_{jjk}^{\text{EW-MMSE}}$ are random vectors distributed as $\hat{\mathbf{h}}_{jjk}^{\text{EW-MMSE}} \sim \mathcal{CN}(\mathbf{0}_N, \Gamma_{jjk})$ and $\tilde{\mathbf{h}}_{jjk}^{\text{EW-MMSE}} \sim \mathcal{CN}(\mathbf{0}_N, \tilde{\Upsilon}_{jjk})$, respectively.
where $\Gamma_{jjk} = q\tau\mathbf{D}_{jjk}\mathbf{\Omega}_{jk}(\mathbf{\Psi_{jk}})^{-1}\mathbf{\Omega}_{jk}\mathbf{D}_{jjk}$ and $\tilde{\Upsilon}_{jjk} = \mathbf{R}_{jjk} - \mathbf{R}_{jjk}\mathbf{\Omega}_{jk}\mathbf{D}v - \mathbf{D}_{jjk}\mathbf{\Omega}_{jk}\mathbf{R}_{jjk} + \Gamma_{jjk}$.

An important remark is that the vector estimate $\tilde{\mathbf{h}}_{jjk}^{\text{EW-MMSE}}$ and the estimation error $\tilde{\mathbf{h}}_{jjk}^{\text{EW-MMSE}}$ are correlated (is not the case in the MMSE estimator) which can be computed as

$$\begin{aligned} \text{Cov}(\tilde{\mathbf{h}}_{jjk}^{\text{EW-MMSE}}, \tilde{\mathbf{h}}_{jjk}^{\text{EW-MMSE}}) &= \mathbf{D}_{jjk}\mathbf{\Omega}_{jk}^{-1}\mathbf{R}_{jjk} - \Gamma_{jk} \\ &= \mathbf{D}_{jjk}\mathbf{\Omega}_{jk}^{-1}\mathbf{R}_{jjk} - \mathbf{D}_{jjk}\mathbf{\Omega}_{jk}(\mathbf{\Psi_{jk}})^{-1}\mathbf{\Omega}_{jk}\mathbf{D}_{jjk} \end{aligned} \tag{16}$$

The expression of NMSE for each BS antenna of the proposed EW-MMSE estimator is as follows

$$\begin{aligned}\varepsilon_{jk}^{\text{EW-MMSE}} &= \frac{1}{N}\mathbb{E}\{\|\hat{\mathbf{h}}_{jjk}^{\text{EW-MMSE}} - \mathbf{h}_{jjk}\|^2\} \\ &= \frac{1}{N}Tr\Big[\mathbf{R}_{jjk} - \mathbf{R}_{jjk}\boldsymbol{\Omega}_{jk}^{-1}\mathbf{D}_{jjk} - \mathbf{D}_{jjk}\boldsymbol{\Omega}_{jk}^{-1}\mathbf{R}_{jjk} + \\ &\mathbf{D}_{jjk}\boldsymbol{\Omega}_{jk}^{-1}\boldsymbol{\Psi}_{\mathbf{jk}}\boldsymbol{\Omega}_{jk}^{-1}\mathbf{D}_{jjk}\Big],\end{aligned} \tag{17}$$

where $\chi_{jk} = \mathbf{R}_{jjk}\boldsymbol{\Omega}_{jk}^{-1}\mathbf{D}_{jjk}$ and $\chi_{jk}^{\mathrm{T}} = \mathbf{D}_{jjk}\boldsymbol{\Omega}_{jk}^{-1}\mathbf{R}_{jjk}$.

The new NMSE expression of the EW-MMSE estimator is as follows

$$\varepsilon_{jk}^{\text{EW-MMSE}} = \frac{1}{N}\mathrm{Tr}[\mathbf{R}_{jjk} - \chi_{jk} - \chi_{jk}^{\mathrm{T}} + \mathbf{D}_{jjk}\boldsymbol{\Omega}_{jk}^{-1}\boldsymbol{\Psi}_{\mathbf{jk}}\boldsymbol{\Omega}_{jk}^{-1}\mathbf{D}_{jjk}] \tag{18}$$

After computing the NMSE of all methods used in this work, we have defined the following lemma.

Lemma 1 *We consider that the estimated channel vector* $\hat{\mathbf{h}}_{jjk}$ *which is defined by* $\hat{\mathbf{h}}_{jjk} = \Lambda_{jk}\theta_{jk}$. *In this case, the NMSE for each antenna is defined as follows:*

$$\begin{aligned}\frac{1}{N}\mathbb{E}\{\|h_{jjk} - \Lambda_{jk}\theta_{jk}\|^2\} = \frac{1}{N}Tr[(\mathbf{I}_N - \Lambda_{jk} - \Lambda_{jk}^{\mathrm{T}})\mathbf{R}_{jjk}] \\ + \frac{1}{N}Tr[(\Lambda_{jk}\boldsymbol{\Psi}_{\mathbf{jk}}\Lambda_{jk}^{\mathrm{T}})]\end{aligned} \tag{19}$$

where Λ_{jk} *is a square matrix given by:*

$$\Lambda_{jk} = \begin{cases}\mathbf{I}_N, & \textit{LS estimator.} \\ \mathbf{R}_{jjk}\boldsymbol{\Psi}_{jk}^{-1} & \textit{MMSE estimator.} \\ \mathbf{R}_{jjk}\hat{\boldsymbol{\Psi}}_{jk}(\eta)^{-1} & \textit{Approx.MMSE estimator.} \\ \mathbf{D}_{jjk}(\boldsymbol{\Omega}_{jk})^{-1} & \textit{Proposed EW-MMSE estimator.}\end{cases} \tag{20}$$

Demonstration: For the demonstration of NMSE given in Eq. (19), we can easily get it by direct computation.

4 Simulation Results

During this section, simulation results are presented to prove our theoretical result. We have used an appropriate model in which the channels are correlated in spatial domain. Therefore, an EC model is adopted with a factor of correlation $r = 0.5$ [17]. The standard deviation of LSF σ is fixed to 4. We adopt in our study that $L = 4$ cells,

$K = 2$ users in each cell. While $\tau_b = 200$ is the coherence block and $q = 100\,\text{mW}$ is the power of transmission for each device. We can convert q to dBm, which gives $q = 20\,\text{dBm}$.

In all the simulation results, we have considered that $N = 100$ and $\eta = 0.5$ except the result shown in Figs. 2 and 3, respectively. The NMSE of each antenna can be written for each estimator as follows $\mathbb{E}\{\|\mathbf{h}_{jjk} - \Lambda_{jk}\theta_{jk}\|^2\}/NTr[\mathbf{R}_{jjk}]$. In addition, all estimators performances are presented in terms of NMSE.

In Fig. 2, we present the NMSE versus the number of BS antennas. From this figure, the LS estimator presents the worst performance compared to the others. NMSE of all used estimators decreases with increasing the number of BS antennas. It is very important to underline that when the number of BS antennas grows, the Approx.MMSE and EW-MMSE performances get closer to that of MMSE. However, the EW-MMSE presents a better performance compared to Approx.MMSE.

Figure 3 shows the NMSE performance against the RFc η. From this figure, the MMSE, LS, and EW-MMSE estimators have constant NMSE values; meaning that they do not rely on RFc η. The LS estimator has the largest NMSE compared to the others, but it is important to underline that the LS estimator does not require previous knowledge of channel statistics. Unlike LS, the MMSE estimator requires prior statistics information of the channel. The Approx.MMSE performance is relying on the value of the factor of η. When η varies from 0 to 0.4, the NMSE of Approx.MMSE is very close to that of MMSE. We can say that they have similar NMSE values. In this interval, the effect of off-diagonal elements is not important in comparison with the diagonal elements. This is resulting from small values of

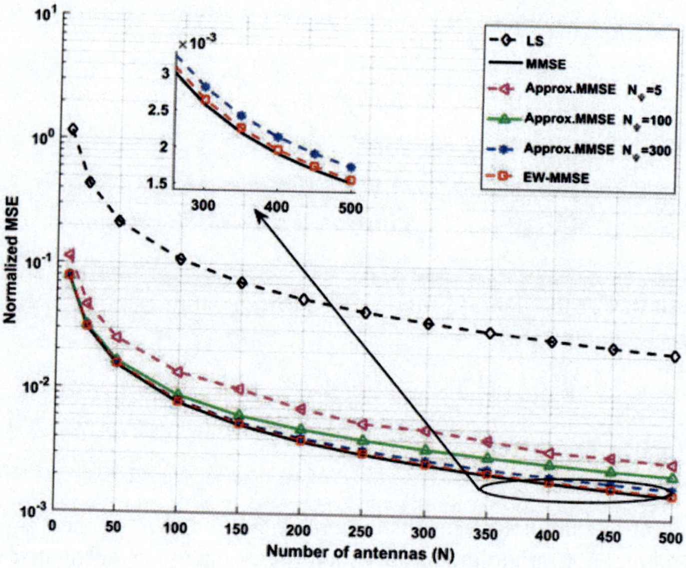

Fig. 2 NMSE in dependance of the number of BS antennas, N

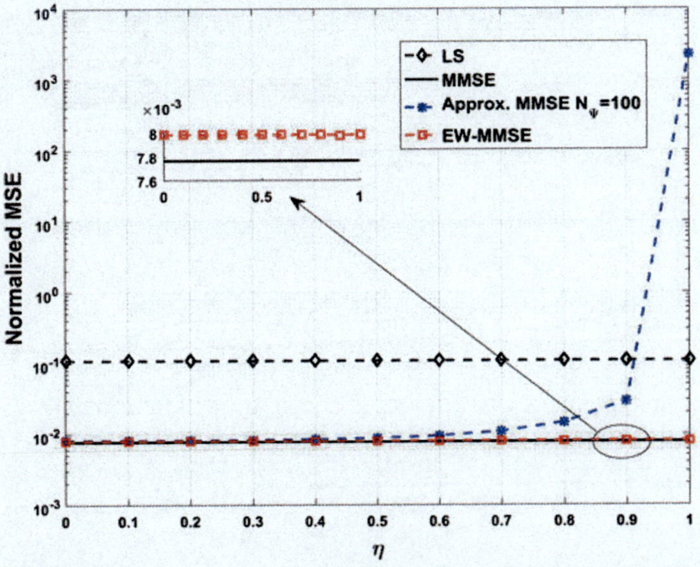

Fig. 3 NMSE in dependence of the Regularization Factor (RFc), η

the factor η. When η varies from 0.5 to 0.9, the performance of Approx.MMSE is lower compared with the MMSE estimator, but it is yet surpassing the performance of LS. When η exceeding 0.9, the Approx.MMSE performance is worse compared to LS and MMSE estimators. For the reason that the CM is not a F-R matrix when η exceeding 0.9. On the other side, the EW-MMSE performance is better compared to the performances of LS and Approx.MMSE. The EW-MMSE performance is very close to the MMSE estimator for all RFc η values. It is seen as a low-complexity estimator compared to MMSE.

In Fig. 4, we present the NMSE versus the NoO N_Ψ, to show the performances of each estimator under the NoO N_Ψ. The LS estimator is the worse one, whereas the MMSE Approx.MMSE and EW-MMSE estimators present higher quality performance than LS. The MMSE estimator supposes that all channel statistics are perfectly known. From the Fig. 4, we can notice that the NMSE for LS, MMSE, and EW-MMSE estimators remained the same for all N_Ψ values; it implies that the performance of LS, MMSE, and EW-MMSE estimators does not rely on N_Ψ. However, the Approx.MMSE estimator is approaching the MMSE estimator when the NoO N_Ψ increases. It implies that the Approx.MMSE depends on N_Ψ. In addition, we can say that the NoO and the error (the errors between the items of the sample CM and the true CM) are inversely proportional, when N_Ψ increases. The errors between the items of the sample CM and the items of the true CM decrease. Nevertheless, the EW-MMSE estimator is very nearly to MMSE performance, and it presents a better performance compared to Approx.MMSE and LS estimators.

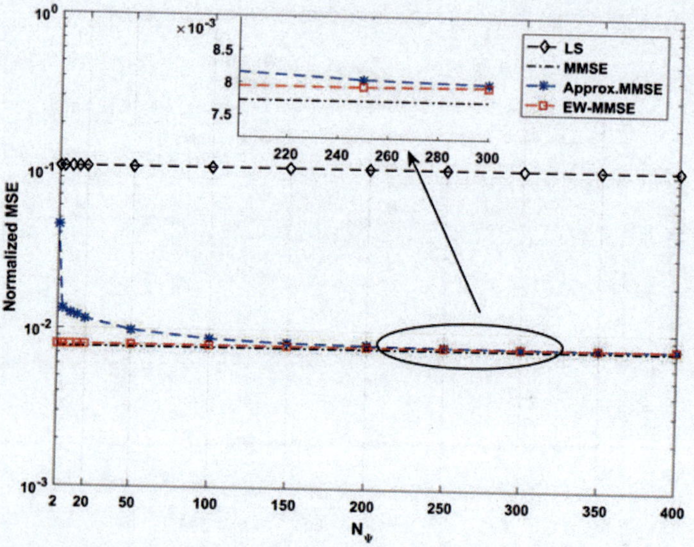

Fig. 4 NMSE versus the NoO, N_Ψ

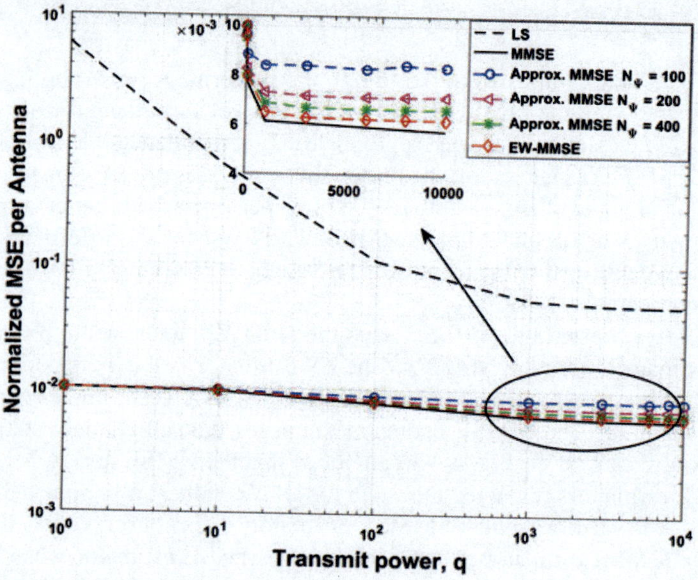

Fig. 5 NMSE versus the transmit pilot power, q

In Fig. 5, we present the NMSE against the UP power q. From this figure, it is clearly that the LS estimator provides a lowest NMSE values. While, with increasing the UP Transmit Power, the EW-MMSE and MMSE are nearly the same. On the other hand, the performance of Approx.MMSE is relying on both the UP Transmit Power q and the NoO N_Ψ. For small values of q (from 0 to 100), the NMSE of Approx.MMSE is close to MMSE and EW-MMSE estimators. But regarding q ($q > 100$), the MMSE and EW-MMSE estimator performances become better than Approx.MMSE. Increasing N_Ψ, the performance of Approx.MMSE approaches to the MMSE estimator. The Approx.MMSE presents a lower performance compared to the EW-MMSE and MMSE estimators. Consequently, the EW-MMSE estimator gives a better performance than Approx.MMSE for all q and N_Ψ values.

5 Conslusion

This paper has suggested a straightforward and powerful channel estimator in terms of NMSE performance. The Approx.MMSE estimator has substituted the covariance matrix of the MMSE estimator through a sample CM. It has presented NMSE results approaching the MMSE estimator with an increasing number of samples. While, the worst performance has provided using LS estimator. Nevertheless, the EW-MMSE has provided better performance than Approx.MMSE. The NMSE results are almost the same like MMSE estimator with lower complexity.

References

1. Marzetta, T.L.: Noncooperative cellular wireless with unlimited numbers of base station antennas. IEEE Trans. Wirel. Commun. **9**(11), 3590–3600 (2010)
2. Larsson, E.G., Edfors, O., Tufvesson, F., Marzetta, T.L.: Massive MIMO for next generation wireless systems. IEEE Commun. Mag. **52**(2), 186–195 (2014)
3. Khansefid, A., Minn, H.: On channel estimation for massive MIMO with pilot contamination. IEEE Commun. Lett. **19**(9), 1660–1663 (2015)
4. De Figueiredo, F.A.P., Cardoso, F.A.C.M., Moerman, I., Fraidenraich, G.: Channel estimation for massive MIMO TDD systems assuming pilot contamination and frequency selective fading. IEEE Access **5**, 17733–17741 (2017)
5. De Figueiredo, F.A.P., Cardoso, F.A.C.M., Moerman, I., Fraidenraich, G.: Channel estimation for massive MIMO TDD systems assuming pilot contamination and flat fading. EURASIP J. Wirel. Commun. Netw. **2018**(1), 1–10 (2018)
6. Mandal, B.K., Pramanik, A.: Channel estimation in massive MIMO with spatial channel correlation matrix. In: Intelligent Computing Techniques for Smart Energy Systems, pp. 377–385. Springer (2020)
7. de Figueiredo, F.A.P., Lemes, D.A.M., Dias, C.F., Fraidenraich, G.: Massive MIMO channel estimation considering pilot contamination and spatially correlated channels. Electron. Lett. **56**(8), 410–413 (2020)
8. Björnson, E., Sanguinetti, L., Debbah, M.: Massive MIMO with imperfect channel covariance information. In: 2016 50th Asilomar Conference on Signals, Systems and Computers, pp. 974–978. IEEE (2016)

9. Filippou, M., Gesbert, D., Yin, H.: Decontaminating pilots in cognitive massive MIMO networks. In: 2012 International Symposium on Wireless Communication Systems (ISWCS), pp. 816–820. IEEE (2012)
10. Adhikary, A., Nam, J., Ahn, J.-Y., Caire, G.: Joint spatial division and multiplexing-the large-scale array regime. IEEE Trans. Inf. Theory **59**(10), 6441–6463 (2013)
11. Yin, H., Gesbert, D., Filippou, M., Liu, Y.: A coordinated approach to channel estimation in large-scale multiple-antenna systems. IEEE J. Sel. Areas Commun. **31**(2), 264–273 (2013)
12. Gao, X., Edfors, O., Rusek, F., Tufvesson, F.: Massive MIMO performance evaluation based on measured propagation data. IEEE Trans. Wirel. Commun. **14**(7), 3899–3911 (2015)
13. Özdogan, Ö., Björnson, E., Larsson, E.G.: Massive MIMO with spatially correlated Rician fading channels. IEEE Trans. Commun. **67**(5), 3234–3250 (2019)
14. Sanguinetti, L., Björnson, E., Hoydis, J.: Toward massive MIMO 2.0: understanding spatial correlation, interference suppression, and pilot contamination. IEEE Trans. Commun. **68**(1), 232–257 (2019)
15. Forenza, A., Love, D.J., Heath, R.W.: Simplified spatial correlation models for clustered MIMO channels with different array configurations. IEEE Trans. Veh. Technol. **56**(4), 1924–1934 (2007)
16. Adhikary, A., Ashikhmin, A.: Uplink massive MIMOfor channels with spatial correlation. In 2018 IEEE Global Communications Conference (GLOBECOM), pp. 1–6. IEEE (2018)
17. Björnson, E., Hoydis, Jakob, Sanguinetti, L.: Massive MIMO has unlimited capacity. IEEE Trans. Wirel. Commun. **17**(1), 574–590 (2017)
18. Sengijpta, S.K.: Fundamentals of Statistical Signal Processing: Estimation Theory (1995)
19. Shariati, N., Björnson, E., Bengtsson, M., Debbah, M.: Low-complexity polynomial channel estimation in large-scale MIMO with arbitrary statistics. IEEE J. Sel. Topics Signal Process. **8**(5), 815–830 (2014)
20. Björnson, E., Hoydis, J., Sanguinetti, L.: Massive MIMO networks: Spectral, energy, and hardware efficiency. Found. Trends Signal Process. **11**(3–4), 154–655 (2017)

On Channel Estimation of Uplink TDD Massive MIMO Systems Through Different Pilot Structures

Jamal Amadid, Mohamed Boulouird, Abdelhamid Riadi, and Moha M'Rabet Hassani

Abstract This work is considered as a comparative study in which the quality of channel estimation (CE) in massive multiple-input multiple-output (M-MIMO) systems is studied by operating at Uplink (UL) phase according to a time division duplex (TDD) scheme using commonly known channel estimators existing in the literature. The least squares (LS) and minimum mean square error (MMSE) channel estimators are investigated with three categories of pilots, namely regular pilots (RPs), time-superimposed (or superimposed) pilots and staggered pilots (StP). Two patterns of frequency reuse (FR) per category are used. The simulation results showed that by increasing the number of BS antennas with a fixed number of symbols dedicated to the UL phase and vice versa, the normalized mean square error (NMSE) of the LS and MMSE estimators using the superimposed pilot (SuP) or StP is asymptotically approaches the NMSE of the LS and MMSE estimators using the RP, respectively. An asymptotic behavior is studied for two different FR scenarios.

1 Introduction

M-MIMO cellular networks rely on a large number of antennas (NoA) at the base stations (BS) to serve a large number of users. M-MIMO technology has attracted considerable interest as a candidate for future cellular systems [1, 2]. Respecting

J. Amadid · A. Riadi · M. M. Hassani
Instrumentation, Signals and Physical Systems (I2SP) Group, Faculty of Sciences Semlalia, Cadi Ayyad University, Marrakesh, Morocco
e-mail: jamal.amadid@edu.uca.ac.ma

A. Riadi
e-mail: abdelhamid.riadi@edu.uca.ac.ma

M. M. Hassani
e-mail: hassani@uca.ac.ma

M. Boulouird (✉)
Smart Systems and Applications (SSA) Group, National School of Applied Sciences of Marrakesh (ENSA-M), Cadi Ayyad University, Marrakesh, Morocco
e-mail: m.boulouird@uca.ac.ma

M. Ben Ahmed et al. (eds.), *Networking, Intelligent Systems and Security*, Smart Innovation, Systems and Technologies 237,
https://doi.org/10.1007/978-981-16-3637-0_12

the NoA at the BS, these systems offer a major enhancement in the UL stage, in the same way improving the energy efficiency (EE) and spectral efficiency (SE), when the accurate channel state information (CSI) is convinced to be obtainable at reception [3–5]. By using linear processing at the BS [6, 7], the throughput has been increased under advantageous spreading conditions [8]. The previously mentioned advantages of a large MIMO cellular network depend on the presumption that such BS has access to reliable CSI. The CE process in both multiplexing modes (i.e., TDD and frequency division duplex (FDD) is performed by involving orthogonal training sequences. Whereas for the CE phase in M-MIMO systems, the FDD mode is considered as an impractical use [4, 9]. While TDD mode is largely applied and became the most promising for M-MIMO. Since we want to build a network that can perform successfully under any form of the propagation environment, in TDD mode, the CSI is available at the BS when the pilot sequence and data are received at the BS. While the channel reciprocity scenario is always used [10], and the CE can be made with more accuracy.

Despite the advantage provided by TDD mode for M-MIMO, the M-MIMO system has constraint of pilot contamination (PC) resulting from the duplication of the same pilot sequences (PS) in contiguous cells, which cannot disappear even if the NoA at the BS reaches infinity. Therefore, PC keeps till now as a choke-point for the large MIMO TDD systems [3, 6, 11].

1.1 Related Works

In the literature, CE regarding the PC problem has been addressed in several works. In [12], the authors rely on the hypothesis that adjacent cells coordinate their transmission signals based on second-order statistics to facilitate the CE process. In the case of non-coordination among the cells, a lot of works focused on mitigating PC in the CE phase. In [13, 14], authors deal with CE concerning the PC problem using singular-value decomposition (SVD) and semi-blind channel estimate to avoid PC. For practical use, the BS has no information regarding channel statistics of the contiguous cells. The authors in [15–17] suggested an estimator based on maximum likelihood (ML) which can afford similar accuracy to that of MMSE without knowing any information about channel statistics of the contiguous cells. To summarize, the previously mentioned literature dealt with the send of pilots followed by payload data (PD) symbols (herein indicated to as RPs or time-multiplexed (TMu) pilots [18–20]). In mobile communication scenarios, the channel coherent time is restricted by the users' mobility. In this case, the RP intended to present the worst performance. Alternatively, to RP, current studies have centered on the SuP channel estimate in the UL M-MIMO [19–22], where SuP is regarded as a supported pilot scheme. By comparing SuP and RP, no added time for services is needed for SuPs. Thus, it can effectively contribute better SE compared to RP [23] which demand added time to accomplish that service. The power allocation strategy across SuPs and PD was studied and evaluated in [24]. In [25], the authors have been introduced the SuP channel estimate in traditional MIMO systems.

In recent years, numerous studies [26–30] have been conducted on M-MIMO systems with SuPs and have concluded that they are efficient in avoiding PC problems. However, the assorted pilot style is subject to co-interference of PD symbols, which frequently restricts its effectiveness especially in the situation of low signal-to-noise ratio (SNR).

1.2 Organization of This Work

The main parts of our study are outlined as follows: First, the system model is presented in Sect. 2. Next, LS performance is assessed for three categories of pilots using NMSE in Sect. 3. Then, the MMSE estimator is discussed for three pilot categories in Sect. 4. After that, the results of the simulation are presented in Sect. 5 in which we affirm our theoretical study. Finally, our final remarks are summarized in Sect. 6.

1.3 Contributions of This work

The main concern of this paper is to study the UL channel estimation for M-MIMO cellular networks. The two major contributions of this work are as follows:

1. Investigate and evaluate the performances of LS and MMSE estimators using either Rer pilots and Sup pilots with different frequency reuse schemes.
2. Introduce an Stg pilot, which considered as a particular case of Sup pilots, and analyze this pilot type under different frequency reuse schemes.

2 System Model

Our model deals with a multi-cell (MC) multi-user (MU) scenario in the UL phase. The TDD mode is used with L cells and K users with a single antenna in each cell with $M \gg K$ (M is the BS antennas). Generally, in communication networks, a band of symbols is assumed in which the channel coherent is considered. In our work, this symbol band is symbolized by C and presumed to be split into two sub-band C_{up} and C_{dl} defined the number of time slots in UL and Downlink, successively. The spreading matrix received at the jth BS symbolized by $Y_j \in \mathbb{C}^{M \times C_{\text{up}}}$, which can be expressed as

$$Y_j = \sum_{l=0}^{L-1} \sum_{k=0}^{K-1} \sqrt{q_{lk}} \mathbf{g}_{jlk} \mathbf{s}_{lk}^T + N_j \tag{1}$$

Here $\mathbf{g}_{jlk} \in \mathbb{C}^{M\times 1}$ represents the channel from user k in the lth cell to the jth BS. While $\mathbf{s}_{lk}$ represents the vector of symbols dispatched by user k in the lth cell, where q_{lk} represents the power by which the symbols $\mathbf{s}_{lk}$ are dispatched. In addition, $N_j \in \mathbb{C}^{M\times C_{\text{up}}}$ represents the noise matrix, where each column of N_j is distributed as $\mathcal{CN}(0; \sigma^2)$. In this paper, we adopt the assumption that the columns are not dependent on each other.

Generally, the channel $\mathbf{g}_{jlk} \sim \mathcal{CN}(0_M; \beta_{jlk}\mathbb{1}_M)$ expressed in function of two coefficients, namely small-scale fading (SSF) and large-scale fading (LSF) coefficients, where the SSF is defined by the quick change over the phase and amplitude of a signal. Its SSF coefficient is regarded as complex normal distribution $\mathcal{CN}(0; 1)$. While the LSF coefficient includes path loss or attenuation of the path as well as log-normal shadowing. Furthermore, for the coherence duration time, the channel $\mathbf{g}_{jlk}$ is assumed to be static, which means that the channel is supposed to be constant over C symbols. While β_{jlk} is supposed to be consistent for a considerably longer duration. The symbol $\mathbf{s}_{lk}$ in Eq. (1) depends on the type of pilot dispatched on the UL. Whenever RP is employed, the pilots are dedicated to some of the components of $\mathbf{s}_{lk}$, and the rest is considered for PD. On the other hand, when the SuPs is employed, all the pilots and PD dispatched alongside each other.

We supposed that there is a synchronization of all pilots. This hypothesis is typically used in the large MIMO research [3, 12, 31]. However, such a device is simple to examine numerically using that hypothesis. In reality, the synchronization of a wide-area network may not be feasible. In this work, the CE quality gained from RP, SuP, and StP on LS and MMSE estimators is evaluated using NMSE.

3 Least Square Channel Estimation

In this section, the performances are studied, evaluated and discussed for the LS channel estimation for three pilot schemes.

3.1 Regular Pilot

The RP has been used in many works in the literature [15–17]. In this category of pilots, each user dispatches a pilot/training sequence of length τ for channel estimates which is followed by PD. The PS used in this subsection are taken/extracted from a unitary matrix $\Psi \in \mathbb{C}^{\tau\times\tau}$ such that $\Psi\Psi^H = \tau I_\tau$, where every PS is represented by a column of this matrix. These PS are orthogonal and shared over r^{RP} cells. Meaning that, at every r^{RP} th cell, the PS ψ_{lk} that is dispatched by user k is re-used in all r^{RP} cells, where r^{RP} is defined as $r^{\text{RP}} = \tau/K$ and K symbolizes the number of user per cell. Hence, the LS channel estimate using RP is formulated as [8, 20, 32]

$$\hat{g}_{jjk}^{\mathrm{RP}^{ls}} = g_{jjk} + \sum_{l \in \mathcal{P}_j(r^{\mathrm{RP}}) \setminus j} \sqrt{\frac{q_{lk}}{q_{jk}}} g_{jlk} + n_{jk} \tag{2}$$

Here $n_{jk} = N_j \psi_{jk} / (\tau \sqrt{q_{lk}})$ and the cells employed the same PS as cell j are referred to as subgroup $\mathcal{P}_j(r^{\mathrm{RP}})$. When employing the RP scheme together with LS channel estimate, the NMSE is formulated as

$$\begin{aligned} \mathcal{NMSE}_{jk}^{\mathrm{RP}^{ls}} &= \frac{\mathbb{E}\{\|\hat{g}_{jjk}^{\mathrm{RP}} - g_{jjk}\|^2\}}{\mathbb{E}\{\|g_{jjk}\|^2\}} \\ &= \frac{1}{\beta_{jjk}} \left(\sum_{l \in \mathcal{P}_j(r^{\mathrm{RP}}) \setminus j} \frac{q_{lk}}{q_{jk}} \beta_{jlk} + \frac{\sigma^2}{\tau q_{jk}} \right) \end{aligned} \tag{3}$$

The NMSE expression in (3) depends on interference from contiguous cells. In other words, from the cells that employ the same PS as cell j (i.e., PC) which occurs when using the same pilot in the previously mentioned subgroup $\mathcal{P}_j(r^{\mathrm{RP}})$.

3.2 Superimposed Pilots

The SuPs are the second category introduced in our work in which the users dispatched pilots accompanying PD with reduced power (i.e., $\mathbf{s}_{lk} = \rho \mathbf{d}_{lk} + \lambda \boldsymbol{\phi}_{lk}$). The two parameters $\lambda^2, \rho^2 > 0$ are the UL transmit power assigned for pilot and PD successively. Under the constraint, $\rho^2 + \lambda^2 = 1$. The LS channel estimate using SuP is formulated as [20]

$$\hat{g}_{jlk}^{\mathrm{SuP}^{ls}} = \sum_{n \in \mathcal{P}_j(r^{\mathrm{SuP}})} \sqrt{\frac{q_{nk}}{q_{lk}}} g_{jnk} + \frac{\rho}{C_u \lambda} \sum_{n=0}^{L-1} \sum_{p=0}^{K-1} \sqrt{\frac{q_{np}}{q_{jk}}} g_{jnp} \mathbf{d}_{np}^T \boldsymbol{\phi}_{lk}^* + \frac{N_j \boldsymbol{\phi}_{lk}^*}{\lambda C_{\mathrm{up}} \sqrt{q_{lk}}} \tag{4}$$

Here $\boldsymbol{\phi}_{lk} \in \mathbb{C}^{C_{\mathrm{up}}}$ and $\mathbf{d}_{lk} \in \mathbb{C}^{C_{\mathrm{up}}}$ are, successively, the pilot and PD symbols dispatched by the user k in the ith cell. In this case, the C_{up} orthogonal SuPs are reused in all r^{SuP} cells. Here $r^{\mathrm{SuP}} = C_{\mathrm{up}}/K$, K symbolizes the number of users in each cell. Besides, the cells employed the same PS as cell j are referred to as subgroup $\mathcal{P}_j(r^{\mathrm{SuP}})$. Furthermore, The PS used in this subsection are extracted from a unitary matrix $\boldsymbol{\Phi} \in \mathbb{C}^{C_{\mathrm{up}} \times C_{\mathrm{up}}}$ such that $\boldsymbol{\Phi}\boldsymbol{\Phi}^H = C_{\mathrm{up}} I_{C_{\mathrm{up}}}$. Hence, $\boldsymbol{\phi}_{lk}^H \boldsymbol{\phi}_{np} = \delta_{lk}\delta_{np}$. When employing the SuP scheme together with LS channel estimate, the NMSE is formulated as

$$\begin{aligned} \mathcal{NMSE}_{jk}^{\mathrm{SuP}^{ls}} &= \frac{\mathbb{E}\{\|\hat{g}_{jjk}^{\mathrm{SuP}} - g_{jjk}\|^2\}}{\mathbb{E}\{\|g_{jjk}\|^2\}} \\ &= \frac{1}{\beta_{jjk}} \left(\sum_{l \in \mathcal{P}_j(r^{\mathrm{SuP}}) \setminus j} \frac{q_{lk}}{q_{jk}} \beta_{jlk} + \frac{\rho^2}{C_u \lambda^2} \sum_{n=0}^{L-1} \sum_{p=0}^{K-1} \frac{q_{np}}{q_{jk}} \beta_{jnp} + \frac{\sigma^2}{\lambda^2 C_u q_{jk}} \right) \end{aligned} \tag{5}$$

The NMSE expression in (5) depends on interference from contiguous cells as in the previous scheme, added to an additional interference term comes from sending pilot alongside PD.

3.3 Staggered Pilots

The StPs are the third category of pilots studied in our work, where users in each cell are staggering their pilot communications. Guaranteed that if the users of a specific cell send UL pilots, users in the rest of $r^{\text{StP}} - 1$ cells send PD [33, 34]. This pilot category is considered as a particular case of the SuP, where the pilot power p_p for this category depends on the length of coherence time C_{up} as well as the length of the PS τ used in the RP cases. The PD power p_d for this category of pilot depends on PD power in the former discussed category as exemplified the equation below

$$\begin{cases} p_p = q\lambda^2 C_{\text{up}}/\tau \\ p_d = q\rho^2 \\ P = \sqrt{\frac{C_{\text{up}}}{\tau}}\text{blkdiag}\{\Phi_0, \ldots, \Phi_{L-1}\} \end{cases} \tag{6}$$

Considered $Y_n \in \mathbb{C}^{M\times\tau}$ as the spread matrix attained the jth BS when the users in the nth cell (where $0 \leqslant n \leqslant r^{\text{StP}}$) sent UL pilots. Remark that the index j has been removed from Y_n for simplicity.

$$Y_n = \sum_{l\in\mathcal{P}_n(r^{\text{SuP}})} \sum_k \sqrt{q_{lk}p_p} g_{jlk}\phi_{nk}^T + \sum_{l\notin\mathcal{P}_n(r^{\text{SuP}})} \sum_k \sqrt{q_{lk}p_d}\, g_{jlk}(\mathbf{d}_{lk}^n)^T + N_n \tag{7}$$

($\mathbf{d}_{lk}^n$) represents the vector of the data symbols dispatched at the nth block by the user k in cell l. The LS channel estimate using StP is formulated as

$$\begin{aligned} \hat{g}_{jnk}^{\text{StP}^{ls}} = &\sum_{l\in\mathcal{P}_n(r^{\text{SuP}})} \sqrt{\frac{q_{lk}}{q_{nk}}} g_{jlk} + \frac{1}{C_u}\sqrt{\frac{p_d}{p_p}} \sum_{l\notin\mathcal{P}_n(r^{\text{SuP}})} \sum_p \sqrt{\frac{q_{lp}}{q_{nk}}} g_{jlp}(\mathbf{d}_{lp}^n)^T\phi_{nk}^* \\ &+ \frac{N_n\phi_{nk}^*}{C_{\text{up}}\sqrt{p_p q_{nk}}} \end{aligned} \tag{8}$$

When employing the StP scheme together with LS channel estimate, the NMSE is formulated as

$$\mathcal{NMSE}_{jk}^{\mathrm{StP}^{ls}} = \frac{\mathbb{E}\{\|\hat{g}_{jjk}^{\mathrm{SuP}} - g_{jjk}\|^2\}}{\mathbb{E}\{\|g_{jjk}\|^2\}}$$
$$= \frac{1}{\beta_{jjk}}\Bigg(\sum_{l \in \mathcal{P}_j(r^{\mathrm{SuP}})\backslash j} \frac{q_{lk}}{q_{jk}} \beta_{jlk} + \frac{p_d}{C_{\mathrm{up}} p_p} \sum_{l \notin \mathcal{P}_j(r^{\mathrm{SuP}})} \sum_{k} \frac{q_{lk}}{q_{jk}} \beta_{jlk} \tag{9}$$
$$+ \frac{\sigma^2}{p_p C_{\mathrm{up}} q_{jk}} \Bigg)$$

As in the case of SuPs, the NMSE expression in (9) depends on interference from contiguous cells, which belongs to the same $\mathcal{P}_j(r^{\mathrm{SuP}})$ subgroup (*As mentioned early the cells employed the same PS as cell j are referred to as subgroup* $\mathcal{P}_j(r^{\mathrm{SuP}})$). An additional interference term comes from dispatching UL data over other cells simultaneously with the sent pilots on $\mathcal{P}_j(r^{\mathrm{SuP}})$ subgroup.

4 MMSE Channel Estimation

In this section, the performances of the MMSE channel estimate for three pilot schemes are studied, evaluated and discussed . As in the previous section (Sec. 3), in which we discussed the LS channel estimate for three categories of pilots. In the same manner, this section discusses the MMSE channel estimate for the three pilot categories discussed in the previous section. We assumed that the same symbols and properties elaborated in the previous section are valid in this section.

4.1 Regular Pilot

As in Sect. 3.1, this subsection evaluates and studies the MMSE channel estimate for a system employing RPs. Assuming that the same symbols and properties elaborated in Sect. 3.1 are valid in this subsection. The RPs have been addressed in several works in the literature[15–17]. Therefore, the MMSE channel estimate using RP is written as follows [8, 20, 32]

$$\theta_{jk}^{\mathrm{RP}} = \hat{g}_{jjk}^{\mathrm{RP}} = g_{jjk} + \sum_{l \in \mathcal{P}_j(r^{\mathrm{RP}})\backslash j} \sqrt{\frac{q_{lk}}{q_{jk}}} g_{jlk} + n_{jk} \tag{10}$$

Here, $n_{jk} = N_j \psi_{jk} / (\tau \sqrt{q_{lk}})$ and the cells employed the same PS as cell j are referred to as subgroup $\mathcal{P}_j(r^{\mathrm{RP}})$, where the RP scheme is used with the MMSE channel estimate. In this case, the MMSE channel coefficient is written as follows

$$\hat{g}_{jjk}^{\mathrm{RP}^{\mathrm{mmse}}} = \frac{\beta_{jjk}}{\left(\sum_{l\in\mathcal{P}_j(r^{\mathrm{RP}})} \frac{q_{lk}}{q_{jk}}\beta_{jlk} + \frac{\sigma^2}{\tau q_{jk}}\right)}\theta_{jk} = \frac{\beta_{jjk}}{\Lambda_{jk}^{\mathrm{RP}}}\theta_{jk} \tag{11}$$

where $\Lambda_{jk}^{\mathrm{RP}} = \left(\sum_{l\in\mathcal{P}_j(r^{\mathrm{RP}})} \frac{q_{lk}}{q_{jk}}\beta_{jlk} + \frac{\sigma^2}{\tau q_{jk}}\right)$. The metric NMSE of the MMSE estimator using the RP is formulated as follows

$$\begin{aligned} \mathcal{NMSE}_{jk}^{\mathrm{RP}^{\mathrm{mmse}}} &= \frac{\mathbb{E}\{\|\hat{g}_{jjk}^{RP^{\mathrm{mmse}}} - g_{jjk}\|^2\}}{\mathbb{E}\{\|g_{jjk}\|^2\}} \\ &= \frac{1}{\Lambda_{jk}^{\mathrm{RP}}}\left(\sum_{l\in\mathcal{P}_j(r^{\mathrm{RP}})\backslash j} \frac{q_{lk}}{q_{jk}}\beta_{jlk} + \frac{\sigma^2}{\tau q_{jk}}\right) \end{aligned} \tag{12}$$

The NMSE formula in (12) relies on interference from neighboring cells meaning that from a cell that uses the same PS as cell j. This happens in our scenario when the same pilot is used in the previously described subgroup $\mathcal{P}_j(r^{\mathrm{RP}})$.

4.2 Superimposed Pilots

As in Sect. 3.2, this subsection investigates and discusses the MMSE channel estimate for a system working under SuPs. Considering that the same symbols and properties elaborated in Sect. 3.2 are valid in this subsection. The SuPs have a large benefit for M-MIMO systems [20, 21] where the pilot and PD are dispatched simultaneously. The MMSE channel estimate using SuP is formulated as [20]

$$\theta_{jk}^{\mathrm{SuP}} = \hat{g}_{jlk}^{\mathrm{SuP}^{ls}} = \sum_{n\in\mathcal{P}_j(r^{\mathrm{SuP}})} \sqrt{\frac{q_{nk}}{q_{lk}}} g_{jnk} + \frac{\rho}{C_u\lambda}\sum_{n=0}^{L-1}\sum_{p=0}^{K-1}\sqrt{\frac{q_{np}}{q_{jk}}} g_{jnp}\mathbf{d}_{np}^T\boldsymbol{\phi}_{lk}^* + \frac{N_j\boldsymbol{\phi}_{lk}^*}{\tau\sqrt{q_{lk}}} \tag{13}$$

where the SuP scheme is used with the MMSE channel estimate. In this case, the MMSE channel coefficient is written as follows

$$\begin{aligned} \hat{g}_{jjk}^{\mathrm{SuP}^{\mathrm{mmse}}} &= \frac{\beta_{jjk}}{\left(\sum_{l\in\mathcal{P}_j(r^{\mathrm{SuP}})} \frac{q_{lk}}{q_{jk}}\beta_{jlk} + \frac{\rho^2}{C_u\lambda^2}\sum_{n=0}^{L-1}\sum_{p=0}^{K-1}\frac{q_{np}}{q_{jk}}\beta_{jnp} + \frac{\sigma^2}{\lambda^2 C_u q_{jk}}\right)}\theta_{jk}^{\mathrm{SuP}} \\ &= \frac{\beta_{jjk}}{\Lambda_{jk}^{\mathrm{SuP}}}\theta_{jk}^{\mathrm{SuP}} \end{aligned} \tag{14}$$

Here $\Lambda_{jk}^{\text{SuP}} = \left(\sum_{l \in \mathcal{P}_j(r^{\text{SuP}})} \frac{q_{lk}}{q_{jk}} \beta_{jlk} + \frac{\rho^2}{C_u \lambda^2} \sum_{n=0}^{L-1} \sum_{p=0}^{K-1} \frac{q_{np}}{q_{jk}} \beta_{jnp} + \frac{\sigma^2}{\lambda^2 C_u q_{jk}} \right)$. When employing the SuP scheme together with MMSE channel estimate, the MSE is formulated as

$$\begin{aligned} \mathcal{NMSE}_{jk}^{\text{SuP}^{\text{mmse}}} &= \frac{\mathbb{E}\{\|\hat{g}_{jjk}^{SuP^{mmse}} - g_{jjk}\|^2\}}{\mathbb{E}\{\|g_{jjk}\|^2\}} \\ &= \frac{1}{\Lambda_{jk}^{\text{SuP}}} \left(\sum_{l \in \mathcal{P}_j(r^{\text{SuP}}) \setminus j} \frac{q_{lk}}{q_{jk}} \beta_{jlk} + \frac{\rho^2}{C_u \lambda^2} \sum_{n=0}^{L-1} \sum_{p=0}^{K-1} \frac{q_{np}}{q_{jk}} \beta_{jnp} + \frac{\sigma^2}{\lambda^2 C_u q_{jk}} \right) \end{aligned} \tag{15}$$

The NMSE expression in (15) depends on interference from contiguous cells as in the previous scheme, added to an additional interference term comes from sending pilot alongside PD.

4.3 Staggered Pilots

As in Sect. 3.3, this subsection introduces StP as a particular case of SuPs. As stated previously in Sect. 3.3, the users in each cell are staggering their pilot communications. Guaranteed that if the users of a specific cell send UL pilots, users in the rest of $r^{\text{StP}} - 1$ cells send PD [33, 34]. We assume that the same symbols and properties elaborated in Sect. 3.3 are valid in this subsection. Hence, the MMSE channel estimate for a system working under StP is expressed in the following form

$$\begin{aligned} \theta_{jk}^{\text{StP}} &= \hat{g}_{jnk}^{\text{StP}^{ls}} \\ &= \sum_{l \in \mathcal{P}_n(r^{\text{SuP}})} \sqrt{\frac{q_{lk}}{q_{nk}}} g_{jlk} + \frac{1}{C_u} \sqrt{\frac{p_d}{p_p}} \sum_{l \notin \mathcal{P}_n(r^{\text{SuP}})} \sum_{p} \sqrt{\frac{q_{lp}}{q_{nk}}} g_{jlp} (\mathbf{d}_{lp}^n)^T \phi_{nk}^* \\ &\quad + \frac{N_n \phi_{nk}^*}{C_{\text{up}} \sqrt{p_p q_{nk}}} \end{aligned} \tag{16}$$

The expression of the MMSE channel coefficient when using StP is written as follows

$$\begin{aligned} \hat{g}_{jjk}^{\text{StP}^{\text{mmse}}} &= \frac{\beta_{jjk}}{\left(\sum_{l \in \mathcal{P}_j(r^{\text{SuP}}) \setminus j} \frac{q_{lk}}{q_{jk}} \beta_{jlk} + \frac{p_d}{C_{\text{up}} p_p} \sum_{l \notin \mathcal{P}_j(r^{\text{SuP}})} \sum_k \frac{q_{lk}}{q_{jk}} \beta_{jlk} + \frac{\sigma^2}{p_p C_{\text{up}} q_{jk}} \right)} \theta_{jk}^{\text{StP}} \\ &= \frac{\beta_{jjk}}{\Lambda_{jk}^{\text{StP}}} \theta_{jk}^{\text{StP}} \end{aligned} \tag{17}$$

where $\Lambda_{jk}^{\text{StP}} = \left(\sum_{l \in \mathcal{P}_j(r^{\text{SuP}}) \setminus j} \frac{q_{lk}}{q_{jk}} \beta_{jlk} + \frac{p_d}{C_{\text{up}} p_p} \sum_{l \notin \mathcal{P}_j(r^{\text{SuP}})} \sum_k \frac{q_{lk}}{q_{jk}} \beta_{jlk} + \frac{\sigma^2}{p_p C_{\text{up}} q_{jk}} \right)$

When using the StP scheme with MMSE channel estimate, the NMSE is formulated as follows

$$\mathcal{NMSE}_{jk}^{\text{StP}^{ls}} = \frac{\mathbb{E}\{\| \hat{g}_{jjk}^{\text{SuP}} - g_{jjk} \|^2\}}{\mathbb{E}\{\| g_{jjk} \|^2\}}$$
$$= \frac{1}{\Lambda_{jk}^{\text{StP}}} \Bigg(\sum_{l \in \mathcal{P}_j(r^{\text{SuP}}) \setminus j} \frac{q_{lk}}{q_{jk}} \beta_{jlk} + \frac{p_d}{C_{\text{up}} p_p} \sum_{l \notin \mathcal{P}_j(r^{\text{SuP}})} \sum_{k} \frac{q_{lk}}{q_{jk}} \beta_{jlk} \tag{18}$$
$$+ \frac{\sigma^2}{p_p C_{\text{up}} q_{jk}} \Bigg)$$

As in the case of SuPs, the NMSE expression in (18) depends on interference from contiguous cells which belongs to the same $\mathcal{P}_j(r^{\text{SuP}})$ subgroup (*As mentioned early the cells employed the same PS as cell j are referred to as subgroup* $\mathcal{P}_j(r^{\text{SuP}})$). An additional interference term comes from dispatching UL data over other cells simultaneously with the sent pilots on $\mathcal{P}_j(r^{\text{SuP}})$ subgroup.

5 Simulation Results

Simulation results are provided in this section to validate our theoretical analysis given in the previous sections. This section aims to evaluate and compare the performances of LS and MMSE channel estimates using the NMSE metric. For a system using $L = 91$ cells (five tiers of cells) and $K = 10$ users per cell for all pilot categories aforementioned. Users are distributed across the cells. With the aim of studying the PC effect, we assume that the users are at a distance greater than 100 m from the BS, where the shadowing effect is taken into consideration, which is usually assumed to obtain from tall buildings. We analyze the performance of LS and MMSE for pilot categories discussed in previous sections under two FR schemes ($r = 3, r = 7$). The SNR value for the UL phase is fixed to 10 dB.

Figure 1 shows the NMSE in dependence on C_{up}, which presents the number of symbols used in the UL phase. The number of BS M antennas is fixed at 100 in all simulation except where M is varied. As C_{up} increases, the performance provided through the LS estimator using SuPs and StPs in both FR cases ($r^{\text{SuP}} = 3$, $r^{\text{StP}} = 3$; $r^{\text{SuP}} = 7$, $r^{\text{StP}} = 7$) is asymptotically closed to the performance provided from RPs in both FR cases ($r^{\text{RP}} = 3$, $r^{\text{RP}} = 7$), respectively. In addition, system performance is improved by using FR equal to 7 (which is visualized in the NMSE values for all pilot categories). Noted that the effect of FR is a major factor in the performance of SuP and StPs (i.e., overcome the NMSE gap between SuP and StPs) since as FR increases. The performance obtained with SuPs is close to that obtained with StPs (similar behavior).

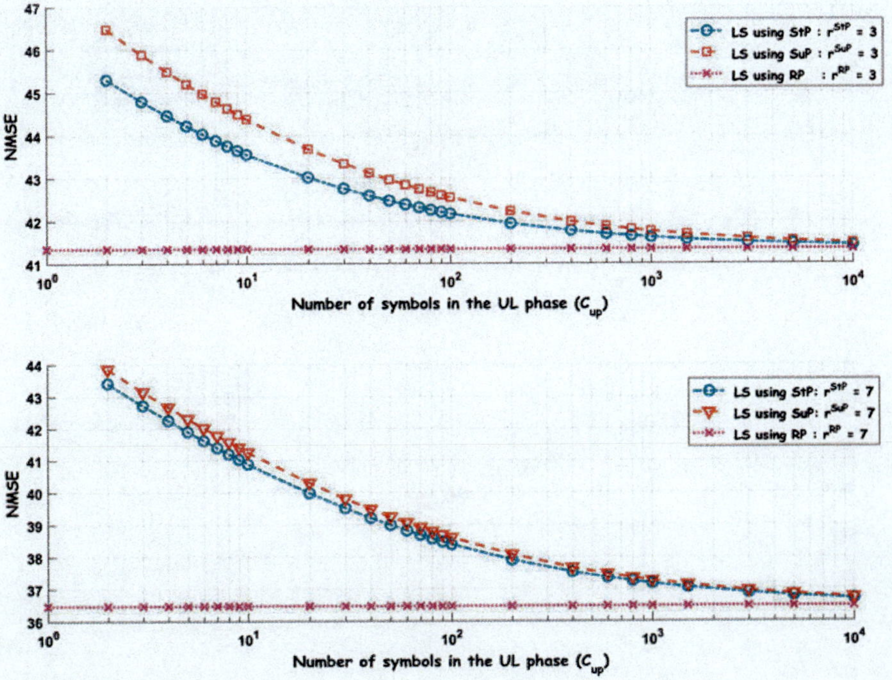

Fig. 1 NMSE in dependence on the number of symbols in the UL C_{up} for the LS estimator using three different pilot categories and considering two cases of FR

Figure 2 shows the NMSE in dependence on C_{up} for the MMSE estimator under different FR. As C_{up} increases, the performance afforded by the MMSE estimator using SuP and StPs in the two FR cases ($r^{\mathrm{SuP}} = 3$, $r^{\mathrm{StP}} = 3$; $r^{\mathrm{SuP}} = 7$, $r^{\mathrm{StP}} = 7$) is asymptotically closed to the performance afforded by the RPs in the two FR cases ($r^{\mathrm{RP}} = 3$, $r^{\mathrm{RP}} = 7$) respectively. It is worth noting that the performance of the MMSE estimator is better than that of the LS estimator. Furthermore, the system performance is improved with a FR of 7 compared to the case of 3. Besides, note that the impact of FR is crucial for the performance of the SuP and StPs, where the difference between the NMSE of the SuP and StPs using the MMSE estimator is relatively small compared to the case of the NMSE of the SuP and StPs using LS.

Figure 3 shows the NMSE versus M. The performances of the LS estimator are presented for three categories of pilots under two FR values. The number of symbols C_{up} in the UL phase is fixed at 35 in all simulations except where C_{up} is varied. For the case where $r^{\mathrm{SuP}} = r^{\mathrm{StP}} = 3$, a large gap is given between the NMSE of the SuP and StPs for small values of M. While in the case of $r^{\mathrm{SuP}} = r^{\mathrm{StP}} = 7$, this gap is relatively narrow. As M increases, this gap becomes quite narrow and the NMSE of the SuP and StPs asymptotically approaches to the NMSE of the RPs for both FR scenarios.

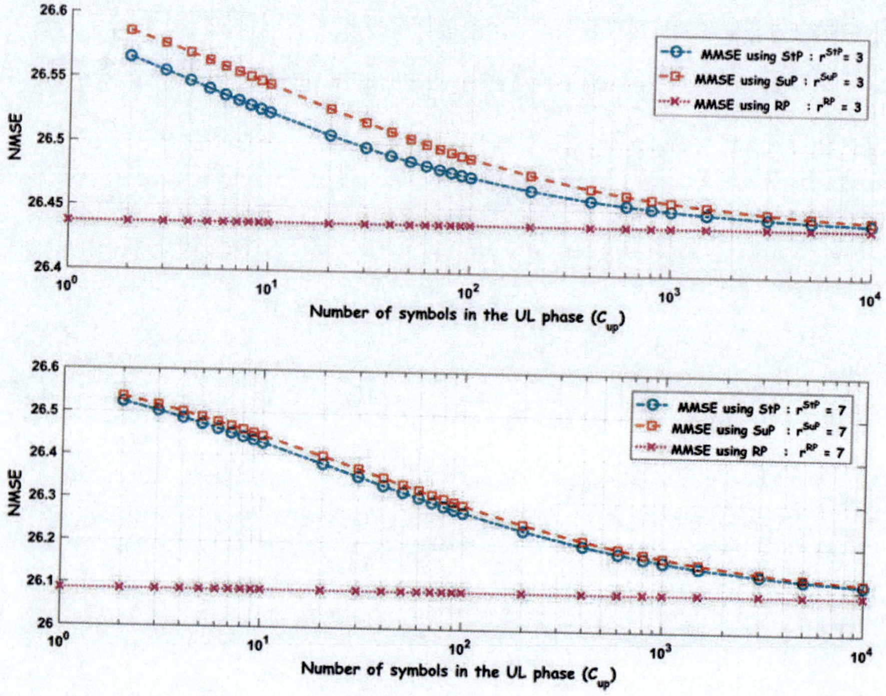

Fig. 2 NMSE in dependence on the number of symbols in the UL C_{up} for the MMSE estimator using three different pilot categories and considering two cases of FR

Figure 4 shows the NMSE in dependence on M. The performances of the MMSE estimator are presented for three categories of pilots under two FR values. It is obvious that the performance of the MMSE estimator is better than that of the LS estimator (by comparing the results provided in Figs. 3 and 4). For the case where $r^{SuP} = r^{StP} = 3$, a large gap is given between the NMSE of the SuP and StPs for small values of M emphasizing. This difference is less than that provided by LS under the same conditions. Whereas in the case of $r^{SuP} = r^{StP} = 7$, this gap is relatively narrow. As M increases, this gap became rather narrow and the NMSE of the SuP and StPs is asymptotically approaches the NMSE of the RPs for both FR scenarios.

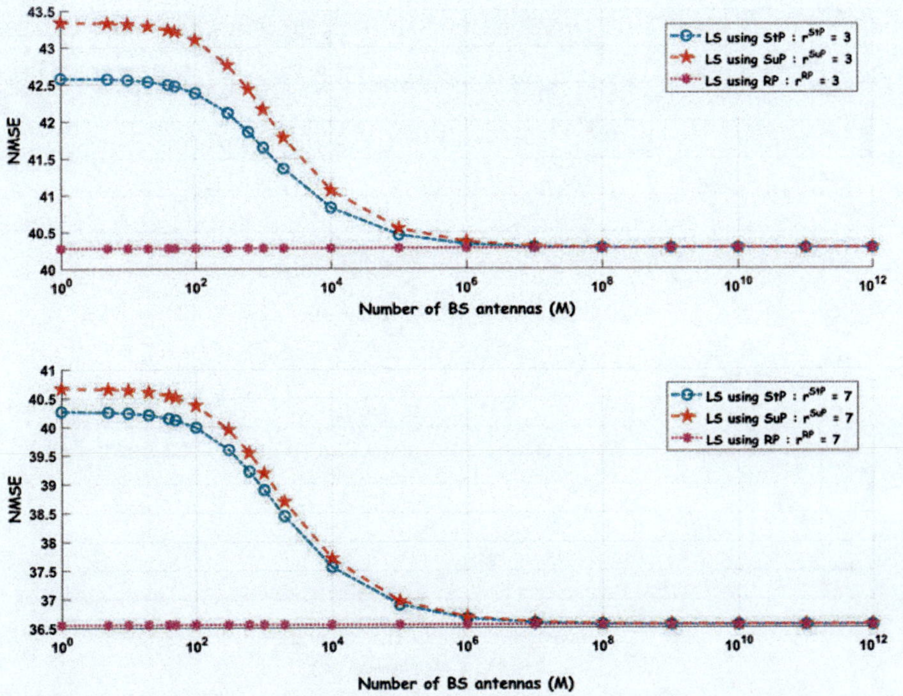

Fig. 3 NMSE in dependence on the NoA M at the BS for the LS estimator using three different pilot categories and considering two cases of FR

6 Conclusion

In this work, we have studied and analyzed the quality of CE for the M-MIMO system in the UL phase. The TDD scheme is operated for three categories of pilots. We have assessed CE quality employing the LS and MMSE channel estimators for regular, SuP, and StPs for two different FR scenarios. We have shown that when the number of symbols dedicated to the UL phase increases, an asymptotic behavior using LS and MMSE estimators with staggered and SuPs is observed. Wherein their NMSE approaches the NMSE of LS and MMSE estimators that employ RP pilots as the number of symbols C_{up} dedicated to the UL phase increases. Furthermore, we also studied the performance of our system under the NoA at the BS, where an identical asymptotic behavior or curve shape is obtained. We have also studied the impact of FR, where we have concluded that the performance is improving by using FR of 7. While a very small gap in terms of the NMSE is obtained between staggered and SuPs by using FR of 7 where this gap is very narrow using the MMSE estimator in comparison to LS estimator.

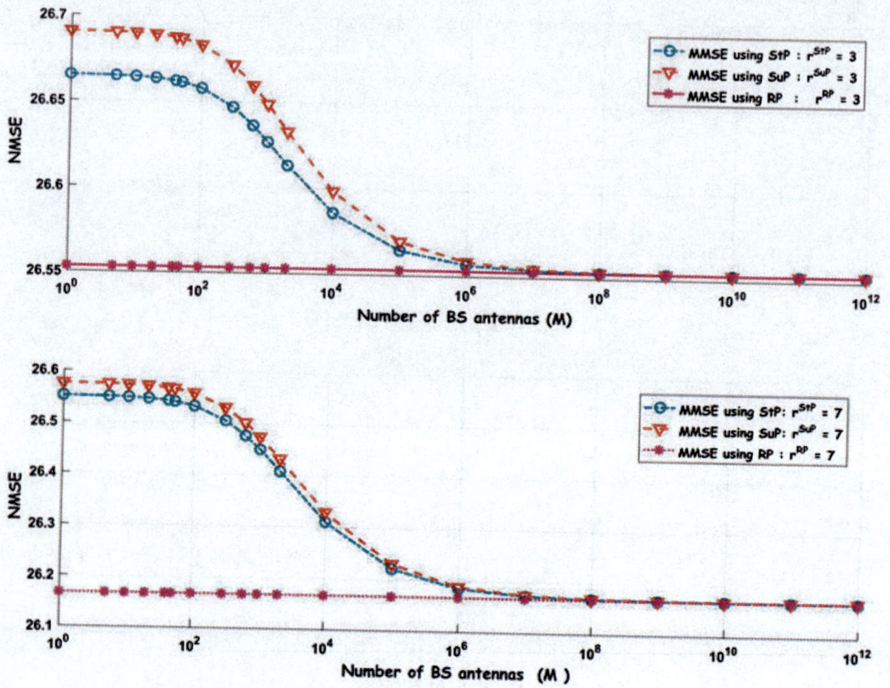

Fig. 4 NMSE in dependence on the NoA M at the BS for the MMSE estimator using three different pilot categories and considering two cases of FR

References

1. Boccardi, F., Heath, R.W., Lozano, A., Marzetta, T.L., Popovski, P.: Five disruptive technology directions for 5g. IEEE Commun. Mag. **52**(2), 74–80 (2014)
2. Osseiran, A., Boccardi, F., Braun, V., Kusume, K., Marsch, P., Maternia, M., Queseth, O., Schellmann, M., Schotten, H., Taoka, H., et al.: Scenarios for 5g mobile and wireless communications: the vision of the metis project. IEEE Commun. Mag. **52**(5), 26–35 (2014)
3. Ngo, H.Q., Larsson, E.G., Marzetta, T.L.: Energy and spectral efficiency of very large multiuser MIMO systems. IEEE Trans. Commun. **61**(4), 1436–1449 (2013)
4. Lu, L., Li, G.Y., Swindlehurst, A.L., Ashikhmin, A., Zhang, R.: An overview of massive MIMO: benefits and challenges. IEEE J. Sel. Topics Signal Process. **8**(5), 742–758 (2014)
5. Rusek, F., Persson, D., Lau, B.K., Larsson, E.G., Marzetta, T.L., Edfors, O., Tufvesson, F.: Scaling up MIMO: Opportunities and challenges with very large arrays. IEEE Signal Process. Mag. **30**(1), 40–60 (2012)
6. Hoydis, J., Ten Brink, S., Debbah, M.: Massive MIMO in the UL/DL of cellular networks: How many antennas do we need? IEEE J. Sel. Areas Commun. **31**(2), 160–171 (2013)
7. Yang, H., Marzetta, T.L.: Performance of conjugate and zero-forcing beamforming in large-scale antenna systems. IEEE J. Sel. Areas Commun. **31**(2), 172–179 (2013)
8. Marzetta, T.L.: Noncooperative cellular wireless with unlimited numbers of base station antennas. IEEE Trans. Wirel. Commun. **9**(11), 3590–3600 (2010)
9. Björnson, E., Larsson, E.G., Marzetta, T.L.: Massive MIMO: ten myths and one critical question. IEEE Commun. Mag. **54**(2), 114–123 (2016)

10. Paulraj, A.J., Ng, B.C.: Space-time modems for wireless personal communications. IEEE Pers. Commun. **5**(1), 36–48 (1998)
11. Larsson, E.G., Edfors, O., Tufvesson, F., Marzetta, T.L.: Massive MIMO for next generation wireless systems. IEEE Commun. Mag. **52**(2), 186–195 (2014)
12. Yin, H., Gesbert, D., Filippou, M., Liu, Y.: A coordinated approach to channel estimation in large-scale multiple-antenna systems. IEEE J. Sel. Areas Commun. **31**(2), 264–273 (2013)
13. Ngo, H.Q., Larsson, E.G., EVD-based channel estimation in multicell multiuser MIMO systems with very large antenna arrays. In: 2012 IEEE International Conference on Acoustics, Speech and Signal Processing (ICASSP), pp. 3249–3252. IEEE (2012)
14. Guo, K., Guo, Y., Ascheid, G.: On the performance of EVD-based channel estimations in mu-massive-MIMO systems. In: 2013 IEEE 24th Annual International Symposium on Personal, Indoor, and Mobile Radio Communications (PIMRC), pp. 1376–1380. IEEE (2013)
15. Khansefid, A., Minn, H.: On channel estimation for massive MIMO with pilot contamination. IEEE Commun. Lett. **19**(9), 1660–1663 (2015)
16. de Figueiredo, F.A.P., Cardoso, F.A.C.M., Moerman, I., Fraidenraich, G.: Channel estimation for massive MIMO TDD systems assuming pilot contamination and flat fading. EURASIP J. Wirel. Commun. Netw. **2018**(1), 1–10 (2018)
17. de Figueiredo, F.A.P., Cardoso, F.A.C.M., Moerman, I., Fraidenraich, G.: Channel estimation for massive MIMO TDD systems assuming pilot contamination and frequency selective fading. IEEE Access **5**, 17733–17741 (2017)
18. Guo, C., Li, J., Zhang, H.: On superimposed pilot for channel estimation in massive MIMO uplink. Phys. Commun. **25**, 483–491 (2017)
19. Upadhya, K., Vorobyov, S.A., Vehkapera, M.: Downlink performance of superimposed pilots in massive MIMO systems in the presence of pilot contamination. In: 2016 IEEE Global Conference on Signal and Information Processing (GlobalSIP), pp. 665–669. IEEE (2016)
20. Upadhya, K., Vorobyov, S.A., Vehkapera, M.: Superimposed pilots are superior for mitigating pilot contamination in massive MIMO. IEEE Trans. Signal Process. **65**(11), 2917–2932 (2017)
21. Zhang, H., Pan, D., Cui, H., Gao, F.: Superimposed training for channel estimation of OFDM modulated amplify-and-forward relay networks. Science China Inf. Sci. **56**(10), 1–12 (2013)
22. Li, J., Zhang, H., Li, D., Chen, H.: On the performance of wireless-energy-transfer-enabled massive MIMO systems with superimposed pilot-aided channel estimation. IEEE Access **3**, 2014–2027 (2015)
23. Zhou, G.T., Viberg, M., McKelvey, T.: A first-order statistical method for channel estimation. IEEE Signal Process. Lett. **10**(3), 57–60 (2003)
24. Huang, W.-C., Li, C.-P., Li, H.-J.: On the power allocation and system capacity of OFDM systems using superimposed training schemes. IEEE Trans. Veh. Technol. **58**(4), 1731–1740 (2008)
25. Dai, X., Zhang, H., Li, D.: Linearly time-varying channel estimation for MIMO/OFDM systems using superimposed training. IEEE Trans. Commun. **58**(2), 681–693 (2010)
26. Zhang, H., Gao, S., Li, D., Chen, H., Yang, L.: On superimposed pilot for channel estimation in multicell multiuser MIMO uplink: large system analysis. IEEE Trans. Veh. Technol. **65**(3), 1492–1505 (2015)
27. Upadhya, K., Vorobyov, S.A., Vehkapera, M.: Superimposed pilots: an alternative pilot structure to mitigate pilot contamination in massive MIMO. In: 2016 IEEE International Conference on Acoustics, Speech and Signal Processing (ICASSP), pp. 3366–3370. IEEE (2016)
28. Li, F., Wang, H., Ying, M., Zhang, W., Lu, J.: Channel estimations based on superimposed pilots for massive MIMO uplink systems. In: 2016 8th International Conference on Wireless Communications & Signal Processing (WCSP), pp. 1–5. IEEE (2016)
29. Die, H., He, L., Wang, X.: Semi-blind pilot decontamination for massive MIMO systems. IEEE Trans. Wirel. Commun. **15**(1), 525–536 (2015)
30. Wen, C.-K., Jin, S., Wong, K.-K., Chen, J.-C., Ting, P.: Channel estimation for massive MIMO using Gaussian-mixture Bayesian learning. IEEE Trans. Wirel. Commun. **14**(3), 1356–1368 (2014)

31. Björnson, E., Hoydis, J., Kountouris, M., Debbah, M.: Massive MIMO systems with non-ideal hardware: Energy efficiency, estimation, and capacity limits. IEEE Trans. Inf. Theory **60**(11), 7112–7139 (2014)
32. Fisher, R.A.: On the mathematical foundations of theoretical statistics. Philos. Trans. R. Soci. Lond. Ser. Containing Pap. Math. Phys. Char. **222**(594–604), 309–368 (1922)
33. Kong, D., Daiming, Q., Luo, K., Jiang, T.: Channel estimation under staggered frame structure for massive MIMO system. IEEE Trans. Wirel. Commun. **15**(2), 1469–1479 (2015)
34. Mahyiddin, W.A.W.M., Martin, P.A., Smith, P.J.: Performance of synchronized and unsynchronized pilots in finite massive MIMO systems. IEEE Trans. Wirel. Commun. **14**(12), 6763–6776 (2015)

NarrowBand-IoT and eMTC Towards Massive MTC: Performance Evaluation and Comparison for 5G mMTC

Adil Abou El Hassan, Abdelmalek El Mehdi, and Mohammed Saber

Abstract Nowadays, the design of 5G wireless network should consider the Internet of Things (IoT) among the main orientations. The emerging IoT applications need new requirements other than throughput to support a massive deployment of devices for massive machine-type communication (mMTC). Therefore, more importance is accorded to coverage, latency, power consumption and connection density. To this purpose, the third generation partnership project (3GPP) has introduced two novel cellular IoT technologies enabling mMTC, known as NarrowBand IoT (NB-IoT) and enhanced MTC (eMTC). This paper provides an overview of NB-IoT and eMTC technologies and a complete performance evaluation of these technologies against the 5G mMTC requirements is presented. The performance evaluation results show that these requirements can be met but under certain conditions regarding the system configuration and deployment. At the end, a comparative analysis of the performance of both technologies is conducted mainly to determine the limits and suitable use cases of each technology.

1 Introduction

Internet of Things (IoT) is seen as a driving force behind recent improvements in wireless communication technologies such as third generation partnership project (3GPP) long-term evolution advanced (LTE-A) and 5G New Radio (NR) to meet the expected requirements of various massive machine-type communication (mMTC) applications. The mMTC introduce a new communication era where billions of

A. Abou El Hassan (✉) · A. El Mehdi · M. Saber
SmartICT Lab, National School of Applied Sciences,
Mohammed First University Oujda, Oujda, Morocco
e-mail: a.abouelhassan@ump.ac.ma

A. El Mehdi
e-mail: a.elmehdi@ump.ac.ma

M. Saber
e-mail: m.saber@ump.ac.ma

M. Ben Ahmed et al. (eds.), *Networking, Intelligent Systems and Security*,
Smart Innovation, Systems and Technologies 237,
https://doi.org/10.1007/978-981-16-3637-0_13

devices, such as remote indoor or outdoor sensors, will need to communicate with each other, while connected to the cloud-based system.

The purpose of 5G system design is to cover three categories of use cases: enhanced mobile broadband (eMBB), massive machine-type communication (mMTC), as well as ultra reliable low-latency communication (uRLLC) [1].

The benefit of 5G system is the flexibility of its structure, which allows the use of a common integrated system to cover many use cases, by using a new concept which is network slicing based on SDN (Software-Defined Networking) and NFV (Network Function Virtualization) technologies [2].

3GPP has introduced two low-power wide area (LPWA) technologies for IoT in Release 13 (Rel-13): NarrowBand IoT (NB-IoT) and enhanced machine-type communication (eMTC) which were designed to coexist seamlessly with existing LTE systems. The 3GPP Rel-13 core specifications for NB-IoT and eMTC were finalized in June 2016 [3, 4], whereas Rel-14 and Rel-15 enhancements were completed, respectively, in June 2017 and June 2018 [3, 4]. About the Rel-16 enhancements, they are underway and scheduled for completion in 2020 [1]. In Rel-15, 3GPP has defined in its work five requirements of 5G mMTC in terms of coverage, throughput, latency, battery life and connection density [5].

The aim of this paper is to determine the system configuration and deployment required for NB-IoT and eMTC technologies in order to fully meet the 5G mMTC requirements. In addition, a comparative analysis is performed of the performances of NB-IoT and eMTC technologies against the 5G mMTC requirements, in order to determine the limits and suitable use cases of each technology.

The remainder of the paper is organized as follows. Section 2 presents the related works. In Sect. 3, overviews of both NB-IoT and eMTC technologies are provided. This is followed, in Sect. 4, by a complete performance evaluation of NB-IoT and eMTC technologies against 5G mMTC requirements in terms of coverage, throughput, latency, battery lifetime and connection density. In addition, the enhancements provided by the recent 3GPP releases are also discussed. A comparative analysis of the performances evaluated of NB-IoT and eMTC technologies is presented in Sect. 5 in order to specify the limits and suitable use cases of each technology. Finally, Sect. 6 concludes the paper.

2 Related Works

Many papers address 3GPP LPWA technologies including NB-IoT and eMTC and non-3GPP LPWA technologies such as LoRa and Sigfox. El Soussi et al. [6] propose an analytical model and implement NB-IoT and eMTC modules in discrete-event network simulator NS-3, in order to evaluate only battery life, latency and connection density. Whereas Jörke et al. [7] present typical IoT smart city use cases such as waste management and water metering to evaluate only throughput, latency and battery life of NB-IoT and eMTC. Pennacchioni et al. [8] analyze the performance of NB-IoT in a massive MTC scenario focusing on only the evaluation of coverage and connection density, by choosing a smart metering system placed in a dense urban scenario as a case study. However, Liberg et al. [9] focus on NB-IoT technology only but provide a performance evaluation against 5G mMTC requirements. On the other hand, Krug and O'Nils [10] compare the delay and energy consumption of data transfer covering a various IoT communication technologies such as Bluetooth, WiFi, LoRa, Sigfox and NB-IoT. However to our knowledge, there is no paper covering the evaluation of the performances of NB-IoT and eMTC technologies against 5G mMTC requirements, as well as the comparative analysis of these performances. This motivated us to perform a comparative analysis of the evaluated performances of NB-IoT and eMTC technologies against 5G mMTC requirements, in order to highlight the use cases of each technology.

3 Overview of Cellular IoT Technologies: NB-IoT and eMTC

3.1 Narrowband IoT: NB-IoT

The 3GPP design aims for Rel-13 were low cost and low-complexity devices, long battery life and coverage enhancement. For this purpose, two power saving techniques have been implemented to reduce power consumption of device: Power saving mode (PSM) and extended discontinuous reception (eDRX) introduced in Rel-12 and Rel-13, respectively, [7, 11]. The bandwidth occupied by the NB-IoT carrier is 180 kHz corresponding to an one physical resource block (PRB) of 12 subcarriers in an LTE system [11]. There are three operation modes to deploy NB-IoT: as a stand-alone carrier, in guard-band of an LTE carrier and in-band within an LTE carrier [11, 12].

In order to coexist with LTE system, NB-IoT uses orthogonal frequency division multiple access (OFDMA) in downlink with the identical subcarrier spacing of 15 kHz and frame structure as LTE [11]. Whereas NB-IoT uses in uplink single-carrier frequency division multiple access (SC-FDMA) and two numerologies which use 15 kHz and 3.75 kHz subcarrier spacings with 0.5 ms and 2 ms slot durations, respectively, [11]. The restricted QPSK and BPSK modulation schemes are used in downlink and uplink by NB-IoT device with a single antenna [3, 11]. Also, NB-IoT defines three coverage enhancement (CE) levels in a cell: CE-0, CE-1 and CE-2 corresponding to the maximum coupling loss (MCL) of 144 dB, 154 dB and 164 dB, respectively, [8].

Two device categories Cat-NB1 and Cat-NB2 are defined by NB-IoT which correspond to the device categories introduced in Rel-13 and Rel-14, respectively. The maximum transport block size (TBS) supported in uplink by Cat-NB1 is only 1000 bits compared to 2536 bits for Cat-NB2. For downlink, the maximum TBS supported by Cat-NB1 is only 680 bits compared to 2536 bits for Cat-NB2 [3].

The signals and channels used in downlink (DL) are as follows: Narrowband primary synchronization signal (NPSS), narrowband secondary synchronization signal

(NSSS), narrowband reference signal (NRS), narrowband physical broadcast channel (NPBCH), narrowband physical downlink shared channel (NPDSCH) and narrowband physical downlink control channel (NPDCCH). NPDCCH is used to transmit downlink control information (DCI) for uplink, downlink and paging scheduling [3, 11].

Whereas only one signal and two channels are used in uplink (UL): Demodulation reference signal (DMRS), narrowband physical uplink shared channel (NPUSCH) and narrowband physical random access channel (NPRACH). Two formats are used for NPUSCH which are: Format 1 (F1) and Format 2 (F2). NPUSCH F1 is used by the user equipment (UE) to carry uplink user's data to the evolved Node B (eNB), whereas NPUSCH F2 is used to carry uplink control information (UCI) which are the DL hybrid automated repeat request acknowledgement (HARQ-ACK) and negative ACK (HARQ-NACK) [11].

For cell access, the UE must first synchronize with the eNB using NPSS and NSSS signals to achieve time and frequency synchronization with the network and cell identification. Then, it receives narrowband master information block (MIB-NB) and system information block 1 (SIB1-NB) carried by NPBCH and NPDSCH, respectively, from eNB to access the system [11].

3.2 Enhanced Machine-Type Communication: eMTC

The overall time structure of the eMTC frame is also identical to that of the LTE frame described in Sect. 3.1. eMTC reuses an identical numerology as LTE, OFDMA and SC-FDMA are used in downlink and uplink, respectively, with subcarrier spacing of 15 kHz [12]. The eMTC transmissions are limited to a narrowband size of 6 PRBs corresponding to 1.4 MHz including guardbands. As the LTE system has a bandwidth from 1.4 to 20 MHz, a number of non-overlapping narrowbands (NBs) can be used if the LTE bandwidth exceeds 1.4 MHz [4]. Up to Rel-14, eMTC device uses QPSK and 16-QAM modulation schemes with a single antenna for downlink and uplink. Whereas support for 64-QAM in downlink has been introduced in Rel-15 [4].

Two device categories are defined by eMTC: Cat-M1 and Cat-M2 corresponding to device categories introduced in Rel-13 and Rel-14, respectively. Cat-M1 has only an maximum channel bandwidth of 1.4 MHz compared to 5 MHz for Cat-M2 [4]. In addition, Cat-M2 supports a larger TBS of 6968 bits and 4008 bits in uplink and downlink, respectively, compared to 2984 bits in both downlink and uplink for Cat-M1 [4].

The following channels and signals are reused by eMTC in downlink: Physical downlink shared channel (PDSCH), physical broadcast channel (PBCH), primary synchronization signal (PSS), secondary synchronization signal (SSS), positioning reference signal (PRS) and cell-specific reference signal (CRS). MTC physical downlink control channel (MPDCCH) is the new control channel which has the role of carrying DCI for uplink, downlink and paging scheduling [4, 12].

Whereas for uplink, the following signals and channels are reused: Demodulation reference signal (DMRS), sounding reference signal (SRS), physical uplink shared channel (PUSCH), physical random access channel (PRACH) and physical uplink control channel (PUCCH) which conveys UCI [4, 12].

For cell access, the UE uses the PSS/SSS signals to synchronize with the eNB, and PBCH which carries the master information block (MIB). After decoding the MIB and then the new system information block for reduced bandwidth UEs (SIB1-BR) carried by PDSCH, the UE initiates the random access procedure using PRACH to access the system [12].

4 NB-IoT and eMTC Performance Evaluation

4.1 *Coverage*

The MCL is a common measure to define the level of coverage a system can support. It is depending on the maximum transmitter power (P_{TX}), the required signal-to-interference-and-noise ratio (SINR), the receiver noise figure (NF) and the signal bandwidth (BW) [13]:

$$\mathrm{MCL} = P_{\mathrm{TX}} - (\mathrm{SINR} + \mathrm{NF} + N_0 + 10\log_{10}(\mathrm{BW})) \tag{1}$$

where N_0 is the thermal noise density which is a constant equal −174 dBm/Hz.

Based on the simulation assumptions given in Table 1 according to [14] and using (1) to calculate MCL, Tables 2 and 3 show the NB-IoT and eMTC channel coverage, respectively, to achieve the MCL of 164 dB which corresponds to the 5G mMTC coverage requirement to be supported [5].

Tables 2 and 3 also indicate the required acquisition time and block error rate (BLER) associated with each channel to achieve the targeted MCL of 164 dB. From the acquisition times shown in Tables 2 and 3, we note that to achieve the MCL of 164 dB at the appropriate BLER, it is necessary to use the time repetition technique for the simulated channels.

Table 1 Simulation and system model parameters

Parameter	Value
System bandwidth	10 MHz
Channel model	Tapped delay line (TDL-iii/NLOS)
Doppler spread	2 Hz
NB-IoT mode of operation	Guard-band
eNB Rx/Tx	4/2 and 4/4 only for NPSS/NSSS transmissions
Device Rx/Tx	1/1

Table 2 Downlink and uplink coverage of NB-IoT

Assumptions for simulation	Downlink physical channel			Uplink physical channel		
	NPBCH	NPDCCH	NPDSCH	NPRACH	NPUSCH F1	NPUSCH F2
TBS (Bits)	24	23	680	–	1000	1
Acquisition time (ms)	1280	512	1280	205	2048	32
BLER (%)	10	1	10	1	10	1
Max transmit power (dBm)	46	46	46	23	23	23
Transmit power/Carrier (dBm)	35	35	35	23	23	23
Noise figure NF (dB)	7	7	7	5	5	5
Channel bandwidth (kHz)	180	180	180	3.75	15	15
Required SINR (dB)	−14.5	−16.7	−14.7	−8.5	−13.8	−13.8
MCL (dB)	163.95	166.15	164.15	164.76	164	164

Table 3 Downlink and uplink coverage of eMTC

Assumptions for simulation	Downlink physical channel			Uplink physical channel		
	PBCH	MPDCCH	PDSCH	PRACH	PUSCH	PUCCH
TBS (Bits)	24	18	328	–	712	1
Acquisition time (ms)	800	256	768	64	1536	64
BLER (%)	10	1	2	1	2	1
Max transmit power (dBm)	46	46	46	23	23	23
Transmit power/Carrier (dBm)	39.2	36.8	36.8	23	23	23
Noise figure NF (dB)	7	7	7	5	5	5
Channel bandwidth (kHz)	945	1080	1080	1048.75	30	180
Required SINR (dB)	−17.5	−20.8	−20.5	−32.9	−16.8	−26
MCL (dB)	163.95	164.27	163.97	164.7	164	165.45

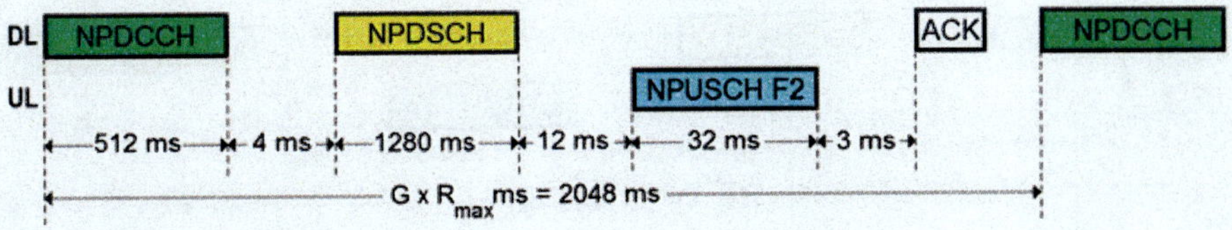

Fig. 1 NPDSCH scheduling cycle ($R_{\max} = 512$; $G = 4$) at the MCL

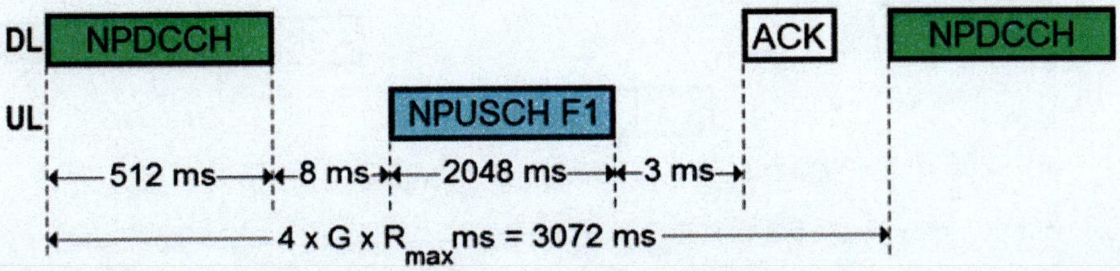

Fig. 2 NPUSCH $F1$ scheduling cycle ($R_{\max} = 512$; $G = 1.5$) at the MCL

4.2 *Throughput*

The downlink and uplink throughputs of NB-IoT are obtained according to the NPDSCH and NPUSCH F1 transmission time intervals issued from NPDSCH and NPUSCH F1 scheduling cycles, respectively, and using the simulation assumptions shown in Tables 1 and 2. While the downlink and uplink throughputs of eMTC are determined based on the PDSCH and PUSCH transmission time intervals issued from PDSCH and PUSCH scheduling cycles respectively and the simulation assumptions given in Tables 1 and 3. The MAC-layer throughput (THP) is calculated with the following formula:

$$\text{THP} = \frac{(1 - \text{BLER})(\text{TBS} - \text{OH})}{\text{PDCCH Period}} \tag{2}$$

where PDCCH period is the period of physical downlink control channel of NB-IoT and eMTC that are NPDCCH and MPDCCH, respectively, and OH is the overhead size in bits corresponding to the radio protocol stack.

Figure 1 depicts NPDSCH scheduling cycle of NB-IoT according to [14], where the NPDCCH user-specific search space is configured with a maximum repetition factor $R_{\max}$ of 512 and a relative starting subframe periodicity G of 4.

Based on BLER and TBS given in Table 2 and using an overhead (OH) of 5 bytes, a MAC-layer THP in downlink of 281 bps is achieved according to the formula (2).

The NPUSCH F1 scheduling cycle depicted in Fig. 2 corresponds to scheduling of NPUSCH F1 transmission once every fourth scheduling cycle according to [14], which ensures a MAC-layer THP in uplink of 281 bps according to the formula (2) and based on BLER and TBS given in Table 2 and an overhead (OH) of 5 bytes.

Figure 3 depicts the PDSCH scheduling cycle of eMTC which corresponds to scheduling of PDSCH transmission once every third scheduling cycle, where the

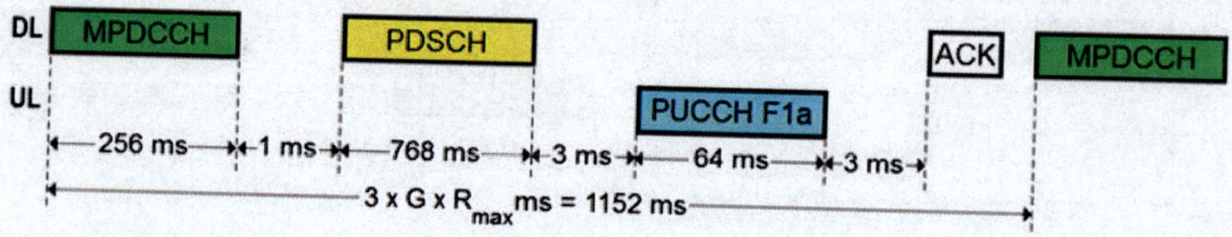

Fig. 3 PDSCH scheduling cycle ($R_{\max} = 256$; $G = 1.5$) at the MCL

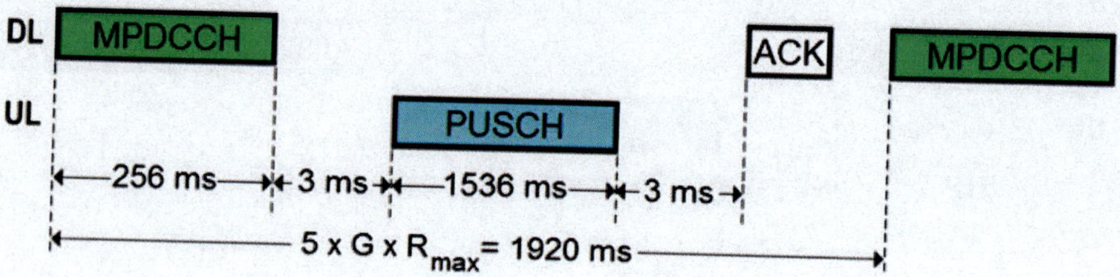

Fig. 4 PUSCH scheduling cycle ($R_{\max} = 256$; $G = 1.5$) at the MCL

MPDCCH user-specific search space is configured with $R_{\max}$ of 256 and a relative starting subframe periodicity G of 1.5 according to [14]. Whereas the PUSCH scheduling cycle depicted in Fig. 4 corresponds to scheduling of PUSCH transmission once every fifth scheduling cycle according to [14].

From BLER and TBS indicated in Table 3 and the use of an overhead (OH) of 5 bytes, the MAC-layer throughputs obtained in downlink and uplink are 245 bps and 343 bps respectively according to the formula (2).

As part of 3GPP Rel-15, 5G mMTC requires that downlink and uplink troughputs supported at the MCL of 164 dB must be at least 160 bps [5]. As can be seen, the MAC-layer throughputs of both NB-IoT and eMTC technologies meet the 5G mMTC requirement.

It should be noted that the BLER targets associated with each NB-IoT and eMTC channel require the acquisition times shown in Tables 2 and 3, respectively. Therefore, the throughput levels of NB-IoT and eMTC can be further improved by using the new Cat-NB2 and Cat-M2 device categories, respectively, which support a larger TBS in downlink and uplink and also enhanced HARQ processes.

4.3 *Latency*

The latency should be evaluated for the following procedures: Radio resource control (RRC) resume procedure and early data transmission (EDT) procedure that has been introduced in Rel-15 and allowing the device to terminate the transmission of small data packets earlier in RRC-idle mode. Figures 5 and 6 depict the data and signaling flows corresponding to the RRC Resume and EDT procedures respectively that are used by NB-IoT. Whereas the data and signalling flows corresponding to the

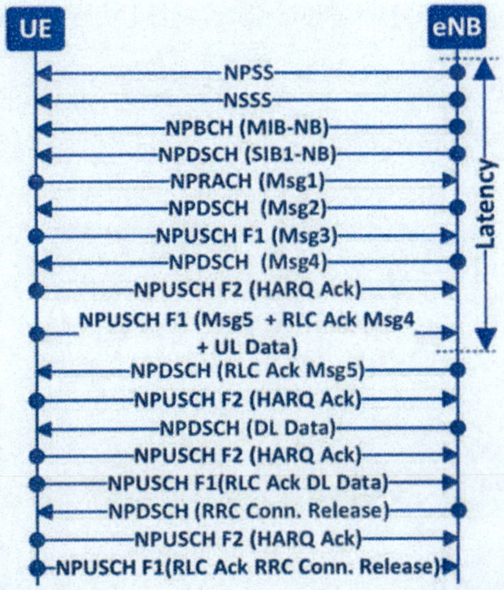

Fig. 5 NB-IoT RRC resume procedure

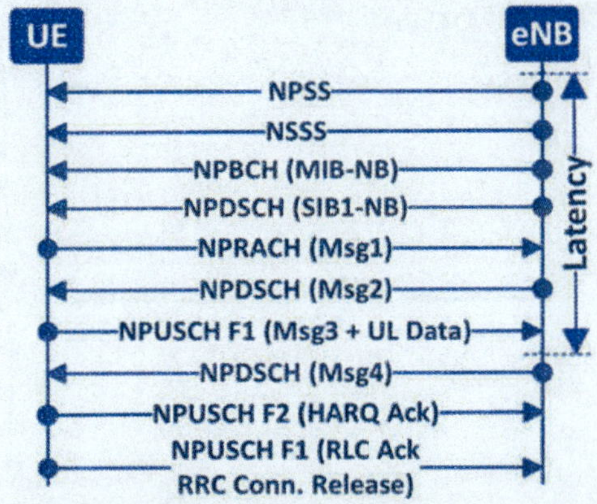

Fig. 6 NB-IoT EDT procedure

RRC Resume and EDT procedures used by eMTC are illustrated in Figs. 7 and 8, respectively.

The latency evaluation is based on the same radio related assumptions and the system model given in Table 1, whereas the packet sizes used and the latency evaluation results of NB-IoT and eMTC at the MCL of 164 dB are shown in Tables 4 and 5, respectively, according to [14].

As can be seen from Tables 4 and 5, the 5G mMTC target of 10 s latency at the MCL of 164 dB defined in 3GPP Rel-15 [5] is met by NB-IoT and eMTC technologies, for both RRC Resume and EDT procedure. However, the best latencies of 5.8 and 5 seconds obtained by NB-IoT and eMTC, respectively, using the EDT procedure are mainly due to the multiplexing of the user data with Message 3 on the dedicated traffic channel, as shown in Figs. 6 and 8, respectively.

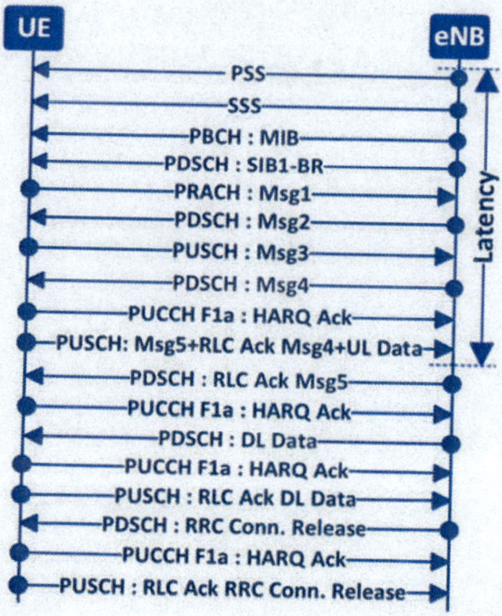

Fig. 7 eMTC RRC resume procedure

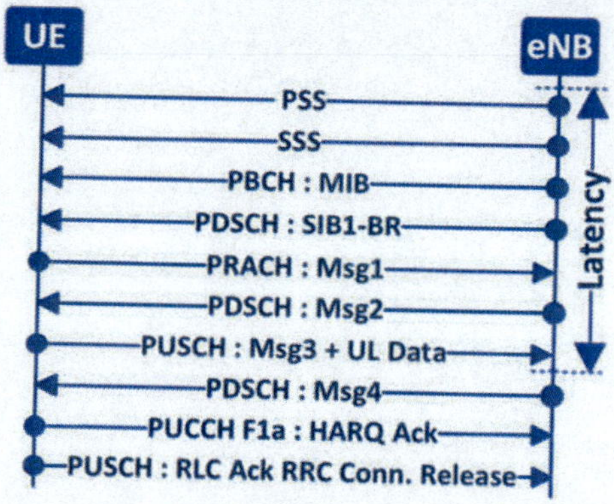

Fig. 8 eMTC EDT procedure

4.4 Battery Life

The RRC resume procedure is used in battery life evaluation instead of the EDT procedure since EDT procedure does not support uplink TBS larger than 1000 bits which requires long transmission times. The packet flows used to evaluate battery life of NB-IoT and eMTC are the same as shown in Figs. 5 and 7, respectively, where DL data corresponds to the application acknowledgment regarding receipt of UL report by the eNB. Four levels of device power consumption are defined, including transmission (P_{TX}), reception (P_{Rx}), Idle-Light sleep (P_{ILS}) corresponding to device in RRC-Idle mode or RRC-Connected mode but not actively transmitting or receiving, whereas Idle-Deep sleep (P_{IDS}) corresponds to power saving mode.

Table 4 Packet sizes and results of NB-IoT latency evaluation

RRC Resume procedure		EDT procedure	
Random access response: Msg2	7 bytes	Random access response: Msg2	7 bytes
RRC Conn. Resume request: Msg3	11 bytes	RRC Conn. Resume request: Msg3 + UL report	11 + 105 bytes
RRC Conn. Resume: Msg4	19 bytes	RRC Conn. Release: Msg4	24 bytes
RRC Conn. Resume complete: Msg5 + RLC Ack Msg4 + UL report	22 + 200 bytes		
RRC Conn. Release	17 bytes		
Latency	9 s	Latency	5.8 s

Table 5 Packet sizes and evaluation results of eMTC latency

RRC resume procedure		EDT procedure	
Random access response: Msg2	7 bytes	Random access response: Msg2	7 bytes
RRC Conn. Resume request: Msg3	7 bytes	RRC Conn. Resume request: Msg3 + UL report	11 + 105 bytes
RRC Conn. Resume: Msg4	19 bytes	RRC Conn. Release: Msg4	25 bytes
RRC Conn. Resume complete: Msg5 + RLC Ack Msg4 + UL report	22 + 200 bytes		
RRC Conn. Release	18 bytes		
Latency	7.7 s	Latency	5 s

The battery life in years is calculated using the following formula according to [13]:

$$\text{Battery life [years]} = \frac{\text{Battery energy capacity}}{365 \times \frac{E_{\text{day}}}{3600}} \tag{3}$$

Where E_{day} is the device energy consumed per day in Joule and calculated as follows :

$$E_{\text{day}} = [(P_{\text{TX}} \times T_{\text{TX}} + P_{\text{Rx}} \times T_{\text{Rx}} + P_{\text{ILS}} \times T_{\text{ILS}}) \times N_{\text{rep}}] + (P_{\text{IDS}} \times 3600 \times 24) \tag{4}$$

Table 6 Simulation and system model parameters for battery life evaluation

Parameter	Value
LTE system bandwidth	10 MHz
Channel model and Doppler spread	Rayleigh fading ETU—1 Hz
eNB power and antennas configuration	NB-IoT: 46 dBm (Guard-band, In-band)—2Tx/2Rx 43 dBm (Stand-alone)—1Tx/2Rx
	eMTC: 46 dBm—2Tx/2Rx
Device power and antennas configuration	23 dBm—1Tx/1Rx

Table 7 Traffic model and device power consumption

Message format	
UL report	200 bytes
DL application acknowledgment	20 bytes
Report periodicity	Once every 24 h
Device power consumption levels	
Transmission and reception power consumption	P_{TX}: 500 mW—P_{Rx}: 80 mW
Idle mode power consumption	P_{ILS}: 3 mW—P_{IDS}: 0.015 mW

As for T_{TX}, T_{Rx} and T_{ILS}, they correspond to overall times in seconds for transmission, reception and Idle-Light sleep, respectively, according to packet flows of NB-IoT and eMTC shown in Figs. 5 and 7, respectively. While N_{rep} corresponds to the number of uplink reports per day.

The simulation and system model parameters used to evaluate the battery life of NB-IoT and eMTC are given in Table 6 according to [15, 16]. While the assumed traffic model according to Rel-14 scenario and device power consumption levels used are given in Table 7.

Based on the transmission times of the signals and downlink and uplink channels given in [15] and using the formulas (3) and (4) with the simulation assumptions given in Table 7 and a 5 Wh battery, the evaluated battery lifes of NB-IoT to achieve the MCL of 164 dB in in-band, guard-band and stand-alone operation modes are 11.4, 11.6 and 11.8 years, respectively. Whereas the evaluated battery life of eMTC to achieve the MCL of 164 dB is 8.8 years according to the assumed transmission times given in [16].

The 5G mMTC requires battery life beyond 10 years at the MCL of 164 dB, supposing an energy storage capacity of 5Wh [5]. Therefore, NB-IoT achieves the targeted battery life in all operations modes. However, eMTC does not fulfill the 5G mMTC targeted battery life. In order to significantly increase eMTC battery life, the number of base station receiving antennas should be increased to reduce UE transmission time. Therefore, if the number of base station receiving antennas is 4 instead of only 2, the evaluated battery life is 11.9 years which fulfills the 5G mMTC target according to [14].

To further increase the battery life of NB-IoT and eMTC, the narrowband wake-up signal (NWUS) and MTC WUS signal (MWUS) introduced in 3GPP Rel-15 can be implemented, respectively. Since these signals allow the UE to remain in idle mode until informed to decode NPDCCH/MPDCCH channel for a paging occasion, thereby achieving energy saving.

4.5 *Connection Density*

The 5G mMTC target on connection density which is also part of the international mobile telecommunication targets for 2020 and beyond (IMT-2020), requires the support of one million devices per square kilometer in four different urban macro scenarios [5]. These scenarios are based on two channel models (UMA A) and (UMA B) and two distances of 500 and 1732 m between adjacent cell sites denoted by ISD (inter-site distance) [17].

Based on the simulation assumptions given in Table 8 and the non-full buffer system level simulation to evaluate connection density of NB-IoT and eMTC according to [18], Fig. 9 shows the latency required at 99% reliability to deliver 32 bytes payload as a function of the connection requests intensity (CRI) to be supported, corresponding to the number of connection requests per second, cell and PRB.

It should be noted that the latency shown in Fig. 9 includes the idle mode time to synchronize to the cell and read the MIB-NB/MIB and SIB1-NB/SIB1-BR. Knowing that each UE must submit a connection request to the system periodically, we can calculate the connection density to be supported (CDS) per cell area using the following formula:

$$\mathrm{CDS} = \frac{\mathrm{CRI} \cdot \mathrm{CRP}}{A} \tag{5}$$

Table 8 System level simulation assumptions of urban macro scenarios

Parameter	Value
Frequency band	700 MHz
LTE and eMTC system bandwidths	10–1.4 MHz
Operation mode of NB-IoT	In-band
Cell structure	Hexagonal grid with 3 sectors per size
Pathloss model	UMA A, UMA B
eNB power and antennas configuration	46 dBm—2Tx/2Rx
UE power and antennas configuration	23 dBm—1Tx/1Rx

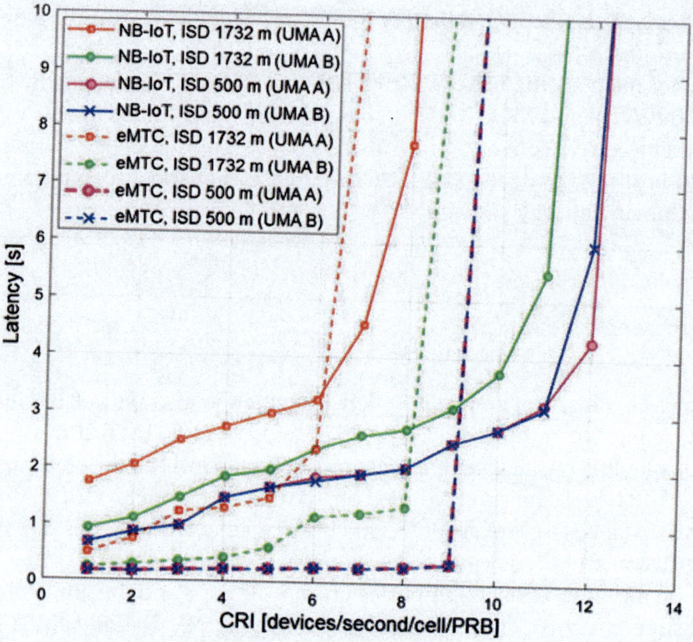

Fig. 9 Intensity of connection requests in relation to latency

where CRP is the periodicity of connection requests given in seconds and the hexagonal cell area A is calculated by: $A = \text{ISD}^2 \cdot \sqrt{3}/6$.

For NB-IoT, to evaluate the connection density per PRB and square kilometer depicted in Fig. 10 and which corresponds to the overall number of devices that successfully transmit a payload of 32 bytes accumulated over two hours, the CDS is obtained from (5) using the CRI values of Fig. 9 and a periodicity of connection requests of two hours. While for eMTC, to evaluate the connection density per narrowband and square kilometer shown in Fig. 11, the CDS is determined from (5) using the CRI values of Fig. 9, a periodicity of connection requests of two hours and a scaling of a factor 6 corresponding to the eMTC narrowband of 6 PRBs.

As can be seen from Fig. 10, in the two scenarios corresponding to the 500 m ISD, more than 1.2 million devices per PRB and square kilometer can be supported by an NB-IoT carrier with a maximum 10 s latency. However, only 94000 and 68000 devices per PRB and square kilometer can be supported using the (UMA B) and (UMA A) channel models, respectively, with an ISD of 1732 m within the 10-s latency limit. Indeed, in the 1732 m ISD scenario, the density of base stations is 12 times lower than with a 500 m ISD. Therefore, this difference in base station density results in differences of up to 18 times between the connection densities relating to scenarios of 500 and 1732 m ISD.

For the 500-m ISD scenario shown in Fig. 11, a single eMTC narrowband can support up to 5.68 million devices within the 10-s latency limit, by addition of 2

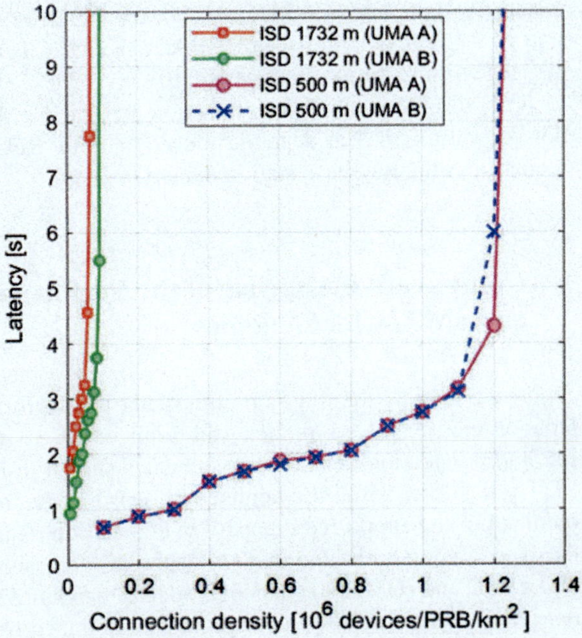

Fig. 10 Connection density of NB-IoT in relation to latency

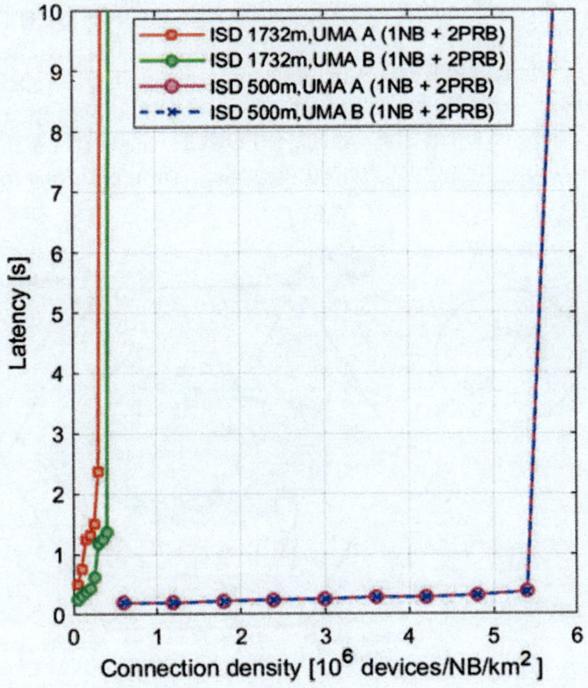

Fig. 11 Connection density of eMTC in relation to latency

further PRBs to transmit PUCCH. For the 1732 m ISD and (UMA B) scenario, cell size is a 12 times larger which explains a eMTC carrier can only support 445,000 devices within a limit of latency of 10 s.

Also, to further improve connection density of eMTC, sub-PRB resource allocation for uplink that has been introduced in 3GPP Rel-15 can be used in the case of a scenario with a low-base station density.

5 Comparative Analysis of the Performance of NB-IoT and eMTC Technologies

Figure 12 depicts the diagram comparing the performance of NB-IoT and eMTC technologies evaluated in Sect. 4 in terms of coverage, throughput, latency, battery life and connection density. The latencies shown in Fig. 12 are that obtained with EDT procedure, while the connection densities are represented by the best values obtained of the supported intensity of connection requests (CRI) from Fig. 9 within the 10-s latency limit, and that correspond to the same urban macro scenario using 500 m ISD and (UMA B) channel model. The 5G mMTC requirement of CRI shown in Fig. 12 corresponds to the targeted CRI obtained from (5) to achieve one million devices per square kilometer for 500 m ISD scenario.

From Tables 2 and 3, it can be seen that NPUSCH F1 and PUSCH channels need the maximum transmission times to reach the coverage target of 164 dB. Thus, for NB-IoT, NPDCCH must be configured with 512 repetitions to achieve the targeted BLER of 1%, while the maximum configurable repetition number for NPDCCH is 2048. Whereas for eMTC, MPDCCH needs to be configured with the maximum configurable repetition number, i.e., 256 repetitions to reach the targeted BLER of

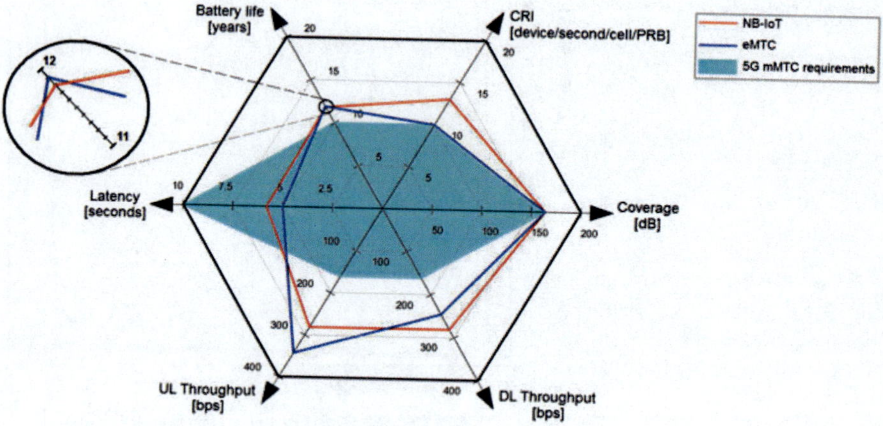

Fig. 12 Performance comparison diagram of NB-IoT and eMTC technologies

1%. Therefore, to support operations in extreme coverage, NB-IoT technology can be considered more efficient than eMTC technology.

As shown in Fig. 12, eMTC can offer significantly higher uplink throughput due to the larger device bandwidth and reduced processing time. In addition, Fig. 12 shows that eMTC performs slightly better than NB-IoT in terms of latency using the EDT procedure. The justification is that NPDCCH and MPDCCH achieve an MCL of 164 dB for a transmission times of 512 ms and 256 ms, respectively, according to Tables 2 and 3. Therefore, eMTC technology is capable of serving IoT applications requiring relatively short response times such as End-Device positioning and voice over LTE (VoLTE).

Figure 12 shows that eMTC is slightly more efficient than NB-IoT in terms of battery life to achieve the MCL of 164 dB, but only if the number of base station receiving antennas is 4 instead of only 2 according to [14]. In fact, the increase of the number of base station receiving antennas improves the uplink throughput, thereby reducing UE transmission time to achieve energy savings.

Figure 12 also indicates that NB-IoT offers a higher connection density than eMTC, which is due to the efficient use of sub-carrier NPUSCH F1 transmissions with a large number of repetitions. Therefore, NB-IoT technology is likely to meet IoT applications requiring a massive number of connected devices, such as smart metering system for gas, electricity and water consumption.

6 Conclusion

To conclude, this paper shows that the five 5G mMTC targets are achieved by both NB-IoT and eMTC technologies. However, the results of performance evaluation show that the performances are achieved except under certain conditions regarding system configuration and deployment, such as the number of repetitions configured for channels transmission, the number of antennas used by the base station and the density of base stations. Regarding the coverage and connection density, NB-IoT offers a better performances than eMTC and precisely for the scenario of a high-base station density with 500 m inter-site distance. While eMTC performs more efficiently than NB-IoT in terms of throughput, latency and battery life. Therefore, NB-IoT can be claimed to be the best performing technology for IoT applications supporting operations in extreme coverage and requiring a massive number of devices. On the other hand, to meet the requirements of IoT applications that need relatively shorter response times, eMTC is the most efficient technology to choose.

References

1. Ghosh, A., Maeder, A., Baker, M., Chandramouli, D.: 5G evolution: a view on 5G cellular technology beyond 3GPP release 15. IEEE Access **7**, 127639–127651 (2019). https://doi.org/10.1109/ACCESS.2019.2939938
2. Barakabitze, A.A., Ahmad, A., Mijumbi, R., Hines, A.: 5G network slicing using SDN and NFV: a survey of taxonomy, architectures and future challenges. Comput. Netw. **167**, 106984 (2020). https://doi.org/10.1016/j.comnet.2019.106984
3. Ratasuk, R., Mangalvedhe, N., Xiong, Z., Robert, M., Bhatoolaul, D.: Enhancements of narrowband IoT in 3GPP Rel-14 and Rel-15. In: 2017 IEEE Conference on Standards for Communications and Networking (CSCN), pp. 60–65. IEEE (2017). https://doi.org/10.1109/CSCN.2017.8088599
4. Ratasuk, R., Mangalvedhe, N., Bhatoolaul, D., Ghosh, A.: LTE-M evolution towards 5G massive MTC. In: 2017 IEEE Globecom Workshops (GC Wkshps), pp. 1–6. IEEE (2018), https://doi.org/10.1109/GLOCOMW.2017.8269112
5. 3GPP: TR 38.913, 5G: Study on scenarios and requirements for next generation access technologies Release 15, version 15.0.0. Technical Report, ETSI. https://www.etsi.org/deliver/etsi_tr/138900_138999/138913/15.00.00_60/tr_138913v150000p.pdf (2018)
6. El Soussi, M., Zand, P., Pasveer, F., Dolmans, G.: Evaluating the performance of eMTC and NB-IoT for smart city applications. In: 2018 IEEE International Conference on Communications (ICC), pp. 1–7. IEEE (2018). https://doi.org/10.1109/ICC.2018.8422799
7. Jörke, P., Falkenberg, R., Wietfeld, C.: Power consumption analysis of NB-IoT and eMTC in challenging smart city environments. In: 2018 IEEE Globecom Workshops, GC Wkshps 2018—Proceedings, pp. 1–6. IEEE (2019). https://doi.org/10.1109/GLOCOMW.2018.8644481
8. Pennacchioni, M., Di Benedette, M., Pecorella, T., Carlini, C., Obino, P.: NB-IoT system deployment for smart metering: evaluation of coverage and capacity performances. In: 2017 AEIT International Annual Conference, pp. 1–6 (2017). https://doi.org/10.23919/AEIT.2017.8240561
9. Liberg, O., Tirronen, T., Wang, Y.P., Bergman, J., Hoglund, A., Khan, T., Medina-Acosta, G.A., Ryden, H., Ratilainen, A., Sandberg, D., Sui, Y.: Narrowband internet of things 5G performance. In: IEEE Vehicular Technology Conference, pp. 1–5. IEEE (2019). https://doi.org/10.1109/VTCFall.2019.8891588
10. Krug, S., O'Nils, M.: Modeling and comparison of delay and energy cost of IoT data transfers. IEEE Access **7**, 58654–58675 (2019). https://doi.org/10.1109/ACCESS.2019.2913703
11. Feltrin, L., Tsoukaneri, G., Condoluci, M., Buratti, C., Mahmoodi, T., Dohler, M., Verdone, R.: Narrowband IoT: a survey on downlink and uplink perspectives. IEEE Wirel. Commun. **26**(1), 78–86 (2019). https://doi.org/10.1109/MWC.2019.1800020
12. Rico-Alvarino, A., Vajapeyam, M., Xu, H., Wang, X., Blankenship, Y., Bergman, J., Tirronen, T., Yavuz, E.: An overview of 3GPP enhancements on machine to machine communications. IEEE Commun. Mag. **54**(6), 14–21 (2016). https://doi.org/10.1109/MCOM.2016.7497761
13. 3GPP: TR 45.820 v13.1.0: Cellular system support for ultra-low complexity and low throughput Internet of Things (CIoT) Release 13. Technical Report, 3GPP. https://www.3gpp.org/ftp/Specs/archive/45_series/45.820/45820-d10.zip (2015).
14. Ericsson: R1-1907398, IMT-2020 self evaluation: mMTC coverage, data rate, latency & battery life. Technical Report, 3GPP TSG-RAN WG1 Meeting #97. https://www.3gpp.org/ftp/TSG_RAN/WG1_RL1/TSGR1_97/Docs/R1-1907398.zip (2019)
15. Ericsson: R1-1705189, Early data transmission for NB-IoT. Technical Report, 3GPP TSG RAN1 Meeting #88bis. https://www.3gpp.org/ftp/TSG_RAN/WG1_RL1/TSGR1_88b/Docs/R1-1705189.zip (2017)
16. Ericsson: R1-1706161, Early data transmission for MTC. Technical Report, 3GPP TSG RAN1 Meeting #88bis. https://www.3gpp.org/ftp/TSG_RAN/WG1_RL1/TSGR1_88b/Docs/R1-1706161.zip (2017)

17. ITU-R: M.2412-0, Guidelines for evaluation of radio interface technologies for IMT-2020. Technical Report, International Telecommunication Union (ITU) (2017). https://www.itu.int/dms_pub/itu-r/opb/rep/R-REP-M.2412-2017-PDF-E.pdf
18. Ericsson: R1-1907399, IMT-2020 self evaluation: mMTC non-full buffer connection density for LTE-MTC and NB-IoT. Technical Report, 3GPP TSG-RAN WG1 Meeting #97. https://www.3gpp.org/ftp/TSG_RAN/WG1_RL1/TSGR1_97/Docs/R1-1907399.zip (2019)

Integrating Business Intelligence with Cloud Computing: State of the Art and Fundamental Concepts

Hind El Ghalbzouri and Jaber El Bouhdidi

Abstract The majority of the problems that organizations are currently facing is the lack of use of cloud computing as shared resources, and it is always looking to become smarter and more flexible by trying to use recent and powerful technologies, such as business intelligence solutions. The business intelligence solution is considered a quick investment and an easy-to-deploy solution. It has become very popular with organizations that process a huge amount of data. Thus, to make this solution more accessible, we thought to use cloud-computing technology to migrate the business intelligence system using frameworks and adapted models to process the data in a way that is efficient. The most important goal is to satisfy users, in terms of security and availability of information. There are many benefits to using cloud BI solution, especially in terms of cost reduction. This paper will address an important definition regarding cloud computing and business intelligence, the importance of each one, and the combination of both, evoking a cloud BI, we will present the management risks to take into account before proceeding to solutions, and the benefits and challenges of cloud computing will also discussed by comparing existing scenarios and their approach. The perspective of our future research will be based on this state of the art that remains an important opening for future contributions.

1 Introduction

Over recent years, augmentation of business application has become very huge, the data and information stored in different business systems are also increasing, the business intelligence has become a trendy technology used in lot of company, and especially the organizations specialized in digital transformation. The business intelligence has evolved rapidly to the level of recent technology like new software

H. El Ghalbzouri (✉) · J. El Bouhdidi
SIGL Laboratory, National School of Applied Sciences Tetuan, Tetuan, Morocco
e-mail: h.elghalbzouri@uae.ac.ma

J. El Bouhdidi
e-mail: jaber.elbouhdidi@uae.ac.ma

M. Ben Ahmed et al. (eds.), *Networking, Intelligent Systems and Security*,
Smart Innovation, Systems and Technologies 237,
https://doi.org/10.1007/978-981-16-3637-0_14

and hardware solutions, using BI process, the organizations become more scalable, intelligent and flexible at the data management level, business intelligence has been historically one of the most resource intensive applications, and it helps the decisions makers to have a clear visibility and make a better decision to take a good strategy to improve their business.

However, at a decline of economic, business or other activities, the majority of organizations find it difficult to make a huge funding investment in technology and human resources, so for this, they always look for implementing a software solutions and options to improve theirs business with a lower costs because the costs can be immensely expensive using recent technologies.

Cloud computing is a model for managing, storing and processing data online via the Internet, it is an economic solution to resolve the problem of high costs, because they are subscription-based witches means that we are paying a monthly rental fee, which is inclusive of all underlying hardware infrastructure and software technology, and this technical advantages attract organizations to migrate theirs BI system to the cloud computing. The latter conceptualized three models services: SAAS—software as service, PAAS platforms as a service, IAAS infrastructure as services, each one of those has advantages and inconvenient, but the most used service in organization currently is the SAAS service. Because of its performance and accessibility via any Web browser, there is no need to install software or to buy any hardware.

Business intelligence contains a set of theories and methodologies that allow to transform a unstructured data into significant and useful information for business and decisions makers, BI offers to users an elastic utilization of storage and networking resources that cloud computing gives a resilient pay-as-you-go manner. The integration of BI system into cloud environment needs to respect a lot of characteristic of each BI component and look for the interdependencies between those, and the deployment of cloud BI presents a lot of challenges at the technical side, conceptual and organizational.

There are many challenges of cloud BI, including ownership, data security and cloud providers confidence or host used for deployment. The cloud BI is the hosted buzzword talked in every industry, and it is the combined power of cloud computing and bi technology. A data center powered by BI technology can provide access to data and monitor business performance. This enables the acquisition of a massive data warehouse with 10 to 100 terabytes of relational databases (RDBMS). This elastic data evolution is what makes cloud BI so powerful, so for that we can provide a framework to help organizations to move their data or system to cloud environment, but like each technology, there is many risks to take into account in terms of security, financial factor, response time and how to choose the best service cloud to insure a good migration with minimizing risks. These issues will be very important to take in account before proceeding to any solution. In this paper, we will discuss and define the existing approach and their future orientations. We will present also the existing frameworks and their implementation aspect related to cloud BI system, afterward we will compare existing scenarios and discussing them in terms of the effectiveness, of each one, in order to optimize the quality of requirements.

2 Overview of Fundamental Concepts

Cloud computing systems deal with large volumes of data using almost limitless computing resources, while data warehouses are multidimensional databases that store huge volumes. Combining cloud computing with business intelligence systems is among the new solution addressed by most of the organization, and in the next section, we will clarify the basic terms and discuss the associated concepts and their adaptation to business intelligence domain.

2.1 Cloud Computing

Cloud computing is described as a new model represented by a pool of systems in which computing infrastructure resources are connected into a network over the Internet, and it offers a scalable infrastructure for management of data resources. Adapting a cloud computing as a solution, the costs may be reduced significantly. The old concept of cloud computing is grid computing, previously grid computing was used of free resources, which mean that all computers connected to a network to solve just a single problem at the same time, so if one system fails, there is a high risk of others to fail. Currently, the extended version of grid computing is named as cloud computing or technology vendors like IBM, Google, Microsoft [1], so this new concept of cloud computing tries to resolve this issue by using all the systems in the network so that if one system fails, another will automatically replace it.

The National Institute of Standards and Technology (NIST) defines cloud computing as "a model for enabling ubiquitous, convenient, on-demand network access to a shared pool of configurable computing resources (e.g., networks, servers, storage, applications and services) that can be rapidly provisioned and released with minimal management effort or service provider interaction" [2].

Cloud computing offers mobility features known as mobile cloud computing. Mobile cloud computing is defined as a "new worldview for portable applications, where the information handling and storage are relocated from the nearby clients to intense and centralized computing platforms situated in the clouds" [3].

Cloud resources can be reassigned according to customer needs. For example, some countries do not allow their user data to be stored outside their borders. In order to achieve this, cloud providers create an infrastructure that can reside in each country that this offers flexibility in terms of the use of several time zones to work with.

2.2 Characteristics' and Deployment Model

According to NIST, cloud computing has four characteristics: on-demand self-service, broad network access, resource pooling, rapid elasticity and measured service.

Regarding the "on-demand self-service" and "broad network access", a consumer can forecast computer capabilities, such as server time and network storage, and can access to functionality over the network using multiples platforms client (e.g., mobile phones, tablets, laptops and workstations).

Cloud infrastructure is characterized by rapid elasticity and resource pooling, for example the IT resource provider is pooled to serve multiple consumers using the multi-tenant model, it is rapidly elastic, the capacities are available, and it is often unlimited and can be scaled to any amount at any time. In addition, cloud infrastructure is a very measured service because the cloud system automatically monitors and optimizes shared resources. It can be monitored and reported on, providing transparency for both the service provider and the consumer of the service being used.

Cloud computing has proven adequate for hosting multiple databases, processing analytic workloads and providing a database as a service. Cloud computing is a new model that resources of its infrastructures are provided as a service on the Internet. All data owners can use all services provided by the cloud computing and outsource their data to him, enjoying a good quality of services.

Cloud computing architecture has three types of services such as software as a service (SaaS), platform as a service (PaaS) and infrastructure as a service (IaaS).

- Infrastructure as a service (IaaS): The IaaS providers offer the physicals or virtual machines to request needs of the costumers and help them to deploy theirs logical solutions, this service is very benefic for the client who needs flexible and security infrastructure with reduced costs, and the costumer with this service pays only for the allocated resource, without worrying for hardware or software maintenance.

With IaaS service, the costumer does not manage or control the underlying cloud infrastructure, and he has just a limited control of certain network component (e.g., host firewalls).

- Platform as a service (PaaS): The PaaS service is most service used for cloud because its facilitates the implementations and testing software solutions, it provides also a necessary resources needed to run an application, and they are automatically allocated so that user does not need to do it manually, also in PaaS service. The consumer does not manage or control the underlying cloud infrastructure, including the network, servers, operating systems or storage, but controls the applications and possibly configuration settings for the application hosting environment.
- Software as a service (SaaS): The SaaS service is described as pay-per-use service, where cloud providers offer to theirs users a complete configured solution (hardware and software), the costumer or organization has just to pay a monthly or annually subscription fee that will depend on the customization of the BI solution

and resources allocated to this company. SaaS can offer full access to a business to implement their BI solution, with a benefits that concern maintenance of software and hardware solutions, the whole implementation is backed by the provider, and costumer does not have to worry about this.

Likewise, the customer does not manage or control the underlying cloud infrastructure, including network, servers, operating systems, storage or even individual application features, but there is an exception that may be possible of the user-specific application configuration [4, 5].

3 Business Intelligence

Business intelligence is a generic term used for combining multiple technologies and processes, it is a set of software tools and hardware solutions, it stores and analyzes data in order to help in decisions making, and business intelligence is both a process and a product. The process consists of methods that organizations use to develop their useful information that can help organizations to survive and predict certainly the behavior of their competitors.

Business intelligence system includes specific components. For example, data warehouses, ETL tools, tools for multidimensional analysis and visualization.

3.1 Data Warehouse

Data warehouse is a database used for reporting and processing data, and it is a central repository of data created for integration of the data from multiple heterogeneous sources that support analytical reporting.

Data warehouses are used to historize data and create reports for management, such as annual and quarterly comparisons.

3.2 ETL

ETL process is responsible for extracting data from multiple heterogeneous sources, its necessary role is transformation of the data from many different formats into a common format, and after that, we load it in a data warehouse.

3.3 Multidimensional Analysis

Online analytical processing (OLAP) can be taken as an example of multidimensional analysis. It allows the analysis of user information from different databases of multiple systems at the same time.

Relational databases are considered like two-dimensional, and OLAP process is multidimensional, which means that users can analyze multidimensional data interactively from multiple perspectives.

3.4 Restitutions

This component is very important, the main objective of the restitution of the data is to communicate a good information for a decision-makers and to show them the results in a clear and effective way, through graphical presentation like dashboards, more and more the data collected well, the decision-maker takes the right decision of their business, and this helps them to communicate good hypotheses and prediction for the future.

3.5 BI Architecture

To process this huge data in a BI system and integrate it through a cloud environment, we need to run a basic architecture based on the BI cloud solution, this architecture contains cloud-compatible components that facilitate interaction between them, and in order to achieve this migration, cloud computing provides an environment based on the use of multiple services separated by layers forming the hardware and software system [6] (Fig. 1).

Data integration: This is related to ETL tools that are needed to transform the data purifying process.

Database: This is related to multidimensional or relational databases.

Data warehousing tools: It is related to a package of applications and tools that allow the maintenance of data warehouse.

Bi tools: This analyzes the data that are stored in data warehouse.

Hardware: It is related to storage and networks on which data will be physically stored.

Software: This refers to everything related to the operating systems and drivers necessary to handle the hardware.

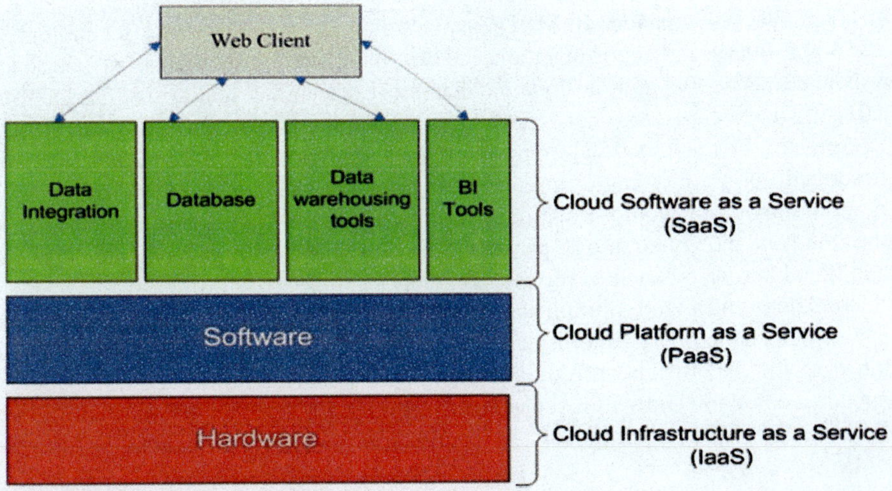

Fig. 1 BI on the cloud architecture

4 Integrating Business Intelligence into Cloud Computing

4.1 Cloud BI

The business intelligence solution based on a cloud computing platform is called "Cloud Business Intelligence". Cloud business intelligence is a revolutionary concept that makes it possible to deploy the BI system to the cloud environment, which is easy to deploy and flexible, also with reduced cost, cloud BI is based on various services offered by the cloud, for example, software as a service business intelligence (SAAS BI), and it is a software of delivery model for business intelligence in which applications are typically deployed outside the corporate firewall hosted in Web site and accessed by a secure Internet connection for the end users [7]. BI cloud technology is sold by providers, with a subscription or pay-per-view basis, instead of using a traditional licensing model with annual maintenance fees.

4.2 Migration of BI to the Cloud

Integrating BI into a cloud environment will solve the problem of technology obsolescence and is an advantage for organizations in term of scalability that will be achieved, no matter how company's data are complex. This integration is not related only for scalability, but also for elasticity and ease of use. By Elasticity that's refer to ability of a BI system to absorb continually information's from new added software [8].

Technologies considered by Dresney Advisory include cloud technologies, including query and reporting tools, OLAP, data mining, analytics, ad hoc analysis and query tools and dashboards. The most researched features of cloud BI in 2018 are dashboards, advanced visualization, data integration and self-service [9].

Some of the features of BI migration to the cloud include the ease of deployment described in the following example: A company that decides at one time to build a new online solution would say helpdesk for its customers. This can be implemented in the cloud and integrated into BI processes in a short time, without having to purchase additional hardware, such as servers, so helpdesk software will run as well.

Implementing a Web solution based on BI system to help decision-making process is also determined by portability and eases of access on any Web browser, that is, the user will visualize the information using various devices; however, it is inside or outside a company and will be constantly informed [10].

Pros and cons of integration of a BI solution into cloud environment are as follows.

Pros:

- Scalability and elasticity;
- Reduced costs;
- Ease of use and access;
- Availability;
- Hardware and software maintenance.

Cons:

- Privacy
- Government regulations (where applied).

As stated before in the cloud computing section of this article, privacy remains an issue. BI solutions are not an exception. The security provided by BI solutions is only at an user interface (UI) level. The data stored on cloud database is exposed to the provider. Government regulations are, in some cases, a barrier in the migration of BI solutions of companies to a cloud infrastructure outside the border. This represents a downside in terms of cloud computing expenses. The cloud providers that are located in the same country with an organization might have higher costs than foreign providers.

4.3 Comparison Between Private and Public Cloud BI

To make a good decision for business in order to migrate their business intelligence system to cloud environment, we must choose best type of cloud computing such us the private, public or hybrid one. The hybrid deployment combines IT services from private and public deployments [11]. In terms to ensure a high security of the data migrated in public cloud, organizations have little control of their resources, because the data are open for public use and accessed via the Internet. It does not require a maintenance or need time changing because the cloud provider is responsible for public cloud: It is a shared cloud that means that we are paying only for what we need, it is used especially in testing Web sites or pay-per-use applications. In public cloud, organizations have little control of their resources, because the data are open for public use and accessed via the Internet. It does not require a maintenance or need time changing because the cloud provider is responsible for it.

The BI infrastructure software platforms available for cloud providers hosting are SAP, RightScale, blue insight (IBM), Web-Sphere, Infromatica and Salesforce.com.

For these platforms, we find those who are public and private one, for example,

- RightScale: It is a public cloud BI, and it is open source and publicly available, for all and not just for organizations. It is a full BI cloud that all business intelligence functions are processed. For example, report generation, online analytical processing, comparative and predictive analysis.
- Blue Insight (IBM): It is a private BI cloud and is not open source. The data management technique that is used by it has more than a petabyte of data storage. It not only supports forecasting, but it is also scalable and flexible.
- Salesforce.com: It is a public BI cloud and is not open source. The data management technique it uses is automated data management. It supports forecasting, but it is not flexible, and it has low scalability.
- Informatica: It is a public BI cloud and is not open source. The data management technique that is used by it is data migration, replication and archiving. It not only supports forecasting, but it is also not flexible, and it has low scalability [12, 13]

This comparison concludes that the best public solution for cloud BI is RightScale, while for private solutions, we can take only those implemented by IBM (Table 1).

Table 1 Characteristics of public, private, hybrid cloud computing

Features	Public cloud	Private cloud	Hybrid cloud
Scalability	Very high	Limited	Very high
Reliability	Moderate	Very high	Medium to high
Security	Totally depends on service providers	High-level security	Secure
Performance	Low to medium	Good	Good
Cost	Free	Depends on resource located	Lower than private cloud
Flexibility	High	Very high	Very high
Maintenance	No maintenance	There is a maintenance	
Time saving	Yes	No	Yes
Pricing	Pay per use	Fixed	Fixed
Examples	Amazon EC2, Google appEngine	VMWARe, Microsoft, KVM, XEN, IBM	IBM, VMWARe, vCloud, Eucalyptus

5 Inspected Scenarios

With the rapid evolution of business intelligence technology and its new concept of cloud integration, many research and studies have been done with implementation of various solutions. Now, it has become difficult for societies to choose the best one. In our case of study, several scenarios based on business intelligence cloud will be analyzed. In this section, we will discuss two of these scenarios used to illustrate the issues discussed previously.

5.1 *Scenario 1: OLAP Cloud BI Framework Based on OPNET Model*

This scenario deals with the hosting of BI systems in the cloud based on OLAP cubes by integrating them on the Web.

The data structures used in the OLAP cube must be converted to XML files based on DTD structures to be compatible with the Web object component (Web cubes), and this solution provides better performance for exploring data in the cloud.

For this, we integrate an OLAP framework comprising the dashboards and the data analytics layer as SAAS model, for the integration of data warehouse and OLTP/DSS databases as PAAS model and for the underlying servers, databases we integrate it as IAAS model.

In our case, we used the OPNET model, and this network can be used to integrate BI and OLAP applications that has been designed in such a way that the load can be evenly distributed to all the relational database management systems (RDBMS) servers in such a way that all RDBMS servers are evenly involved in receiving and processing the OLAP query load. The main architecture of OPNET model is that it comprises two large domains—the BI on the cloud domain and the extranet domain comprising six corporates having 500 OLAP users in each as shown in Figs. 2 and 3

The application clouds are IP network cloud objects comprising application server arrays and database server arrays, connected to a cloud network [14].

In the following figure, the BI framework contains four numbers of Cisco 7609 series layer 3 high and routing switches connecting in such a way that the load can be evenly distributed. The cloud switch 4 is routing all inbound traffic to the servers and sends their responses back to the clients.

The cloud switches 1 and 3 are serving four RDBMS servers, and the cloud switch 2 is serving all the OLAP application servers. An array of five numbers of OLAP application servers and an array of eight numbers of RDBMS servers. The blue dotted lines from each OLAP server are drawn to all the RDBMS servers indicating that each OLAP server will use the services of all the RDBMS servers available in the array to process a database query. The customer's charge is routed to the OLAP application servers using destination preference settings on the client objects configured in the extranet domain [14] (Fig. 4).

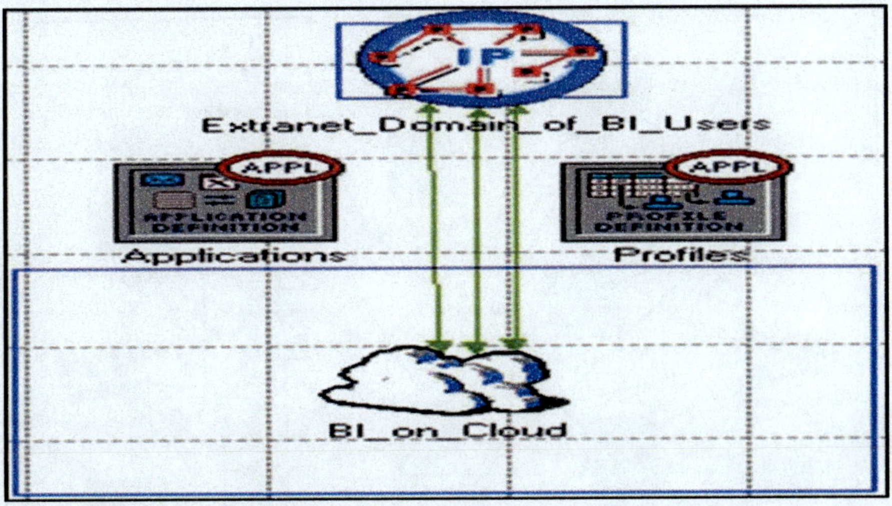

Fig. 2 Architecture of OPNET model [14]

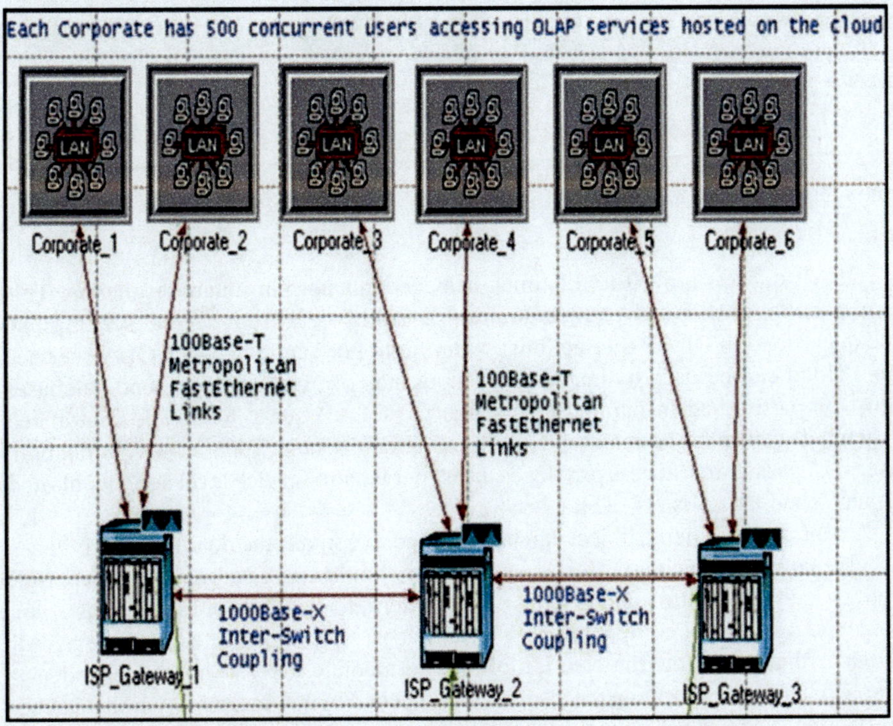

Fig. 3 BI on the cloud architecture [14]

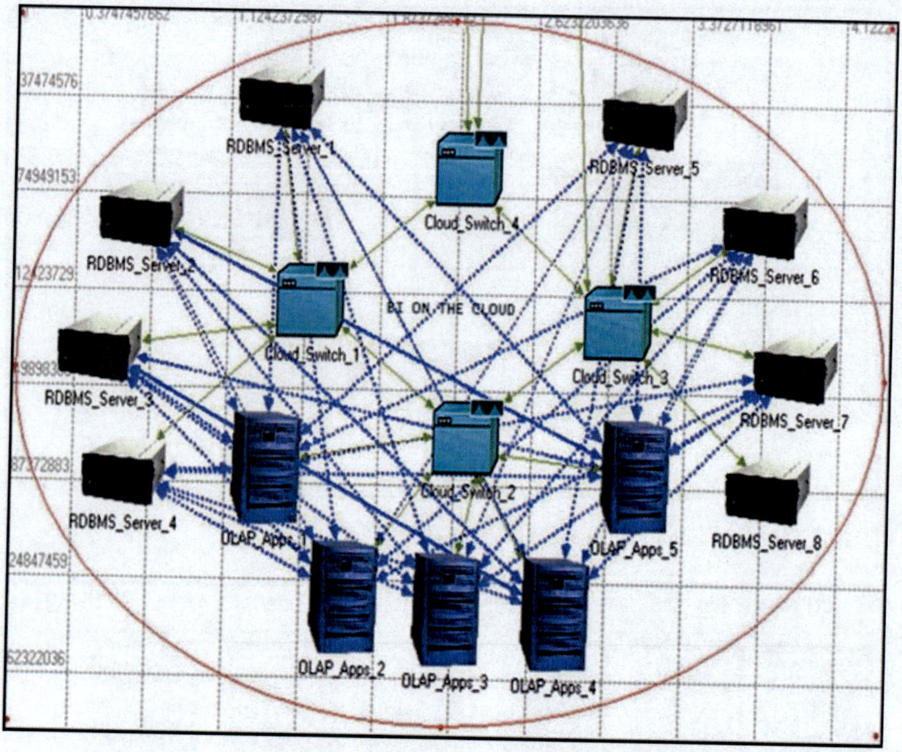

Fig. 4 Extranet domain comprising six corporate shaving 500 OLAP users in each corporate [14]

OLAP queries are 10 to 12 times heavier than normal database queries. This explains that each query extracts multidimensional data from several schemas, so the query load in OLAP transactions is very high. For example, if the OLAP service on a cloud can be used by hundreds of thousands of users, the back-end databases must be partitioned in parallel to manage the OLAP query load. The centralized schema object must be maintained with all tenant details, such as—identification, user IDs, passwords, access privileges, users per tenant, service level agreements and tenant schema details [14, 15].

A centralized schema object can be designed to contain the details and privileges of all tenants on the cloud. The IAAS provider should ensure a privacy and control, both of load distribution and response time pattern, the OLAP application hosted on the cloud may be not compatible with the services, so for this, we can use the SAAS provider that can allow the creation of an intermediate layer to host a dependency graph that helps in dropping the attributes not needed in the finalized XML data cube.

BI and OLAP must have a high level of resources as a multilayer architecture composed of multidimensional OLAP cubes with multiplexed matrices representing the relationships between various business variables. All the cubes send OLAP queries to data warehouses stored in RDBMS servers. The response time size of an OLAP query is typically 10–12 times greater than an ordinary database query.

5.2 *Scenario 2: Model of Cloud BI Framework by Using Multi-cloud Providers*

The approach of this scenario is about the factors, which affect the migration of BI to cloud, so that we adapt organizational requirements and different deployment model to alternative cloud service models, such as the example system shown in Figs. 5 and 6.

This model of framework helps decision-makers to take into account of cloud BI system as well as security, cost and performance.

(1) Bi-user represents the organization's premises where BI system is running before the cloud migration.

In this deployment, the BI-user pushes and extracts the data to cloud environment. Push and pull communications are secured by encrypting the dataflow with the transport layer security/secure sockets layer cryptographic protocols (TLS/SSL).

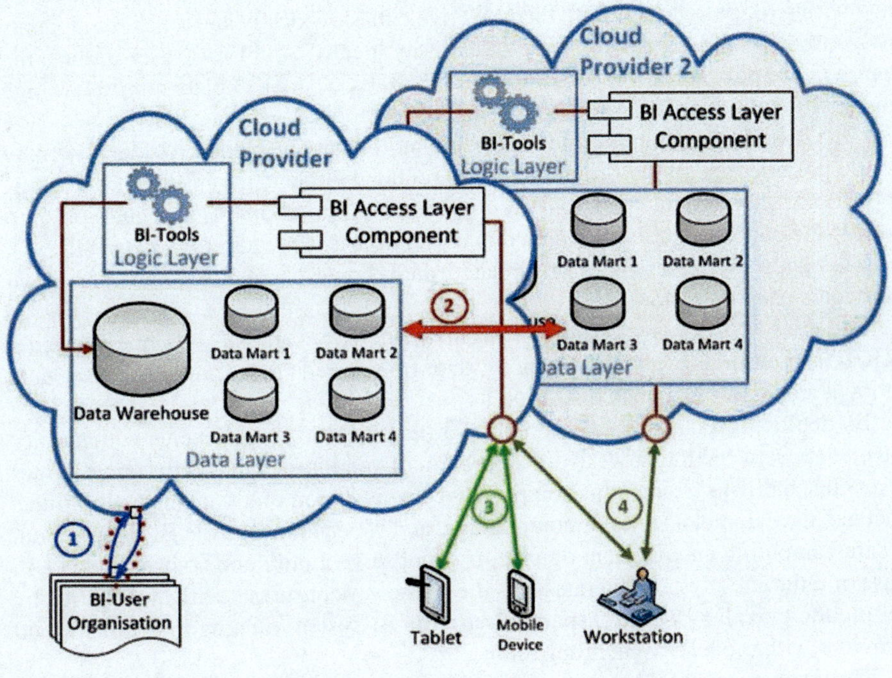

Fig. 5 BI system in a cloud provider [16]

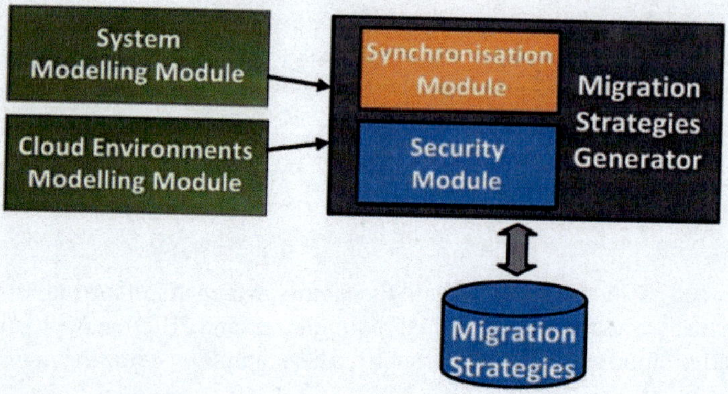

Fig. 6 BI system in a cloud provider [16]

Once the Bi-user transfers its data to cloud premises, the BI tools start to run in cloud environment and analyze all this data which are stored in data warehouse to generate data analysis and report, in order to be accessed by different devices such as the workstation, tablet or mobile phone shown at rounded circle (3) in Fig. 6.

Regarding the trust of the organization to the data transferred, our approach takes a partial migration strategy using more than one cloud provider, to insure security and opting for partial migration that sensitive data stay locally and other components move to the cloud providers, while others stay locally. And for Bi-user pushes, the anonymized data needed to use BI tools on IaaS, PaaS, or SaaS platforms to leverage the additional scalable resources for available BI tools.

This partial migration is done with using more than one cloud provider—namely cloud provider and cloud provider 2—see rounded circle (2) to insure a portability, synchronization module and high security using end-to-end SSL/TLS encryption to secure the communication between cloud premises as shown in Fig. 3 [16] (Fig. 7).

This approach gives users an updated data, however, of the number of cloud providers used to explore it.

For this, we address a data synchronization with a globally unique identifier (GUID) to enforce consistency among data transferred from source to target data storage and harmonize data over time.

By deploying a copy of a BI system to different cloud environments with harmonized data between them, we avoid a problem of vendor lock in and that ameliorates a resilience of Bi systems, this solution gives an isolation of a system so all failures that can have do not attack our components, and we can manage the BI system from a safe computing environment outside, if we observe a failure to control it. Also, in case of failure, BI system tolerates it as the framework ensures it with availability, by letting the overall system transparently use the BI system running in another cloud provider with model of synchronization.

The mechanisms of this framework work as interactions between the data in local premises and cloud environment, and this interaction can be affected by several risks for example:

- The loss of data can happen during the migration of the system BI to cloud environment, because the size of the data to be transferred to cloud environments has implications in terms of the cost of large-scale communications and overall system performance, so this cloud migration framework can recover and save the data to avoid this case, so for this, our framework re-computes the data transferred and compares it with stored one [17].
- Security, we supported by granting access to these data only to users with a user role related to them and the necessary level of authorization.

For the sensitive data we do tokenization method to replace the original data, for example we use the Social Security Number with randomly generated values. The Integration of Business Intelligence into the cloud keep the original format of the data and preserve the functionality running on the cloud premises, and we translate from token to real data in the cloud provider side [18].

6 Comparison of the Scenarios

Discussing about these two scenarios for the integration of BI systems in the cloud, we can see that each of them has its strengths and weaknesses at the same time.

The first scenario is based on the use of OLAP framework based on OPNET which is a structured network model with three different cloud services: SAAS, PAAS and IAAS, which are offered by providers. The use of network modeling OPNET and OLAP framework, OLAP queries are 10 to 12 times heavier than normal database queries. Because each query makes an effort to extract multidimensional data from multiple schemas and the load in OLAP is very high, this implies that when multiple users need to use this service the load on the back-end databases must be balanced and partitioned with schemas in parallel to handle OLAP queries. BI and OLAP must have a high level of resources as a multilayer architecture composed of multi-dimensional OLAP cubes with multiplexed matrices representing the relationships between different variables in the enterprise.

Concerning the second scenario, we discussed a partial migration of the BI system to the cloud, using more than one cloud provider, in this migration, we benefit with a high level of security of the data so that sensitive data stays locally, also using more than one cloud provider that insures performance at the data transfer level, so Bi-user is always informed with updated data over time, however, of the number of cloud providers used to explore the data. But with this solution, some data can be lost while the migrating, so that is because of the implications in terms of costs induced by large-scale communications and overall system performance. That is why our cloud migration framework backs up and recovers data in the event of a disaster to protect against this eventuality, so lost data are still a problem in cloud BI migration at different levels, moreover, that every organization has a level of security that it wants to implement for its solution.

7 Conclusion and Perspective

Cloud computing in recent years became a trend of the majority of organization who uses business intelligence process, and it has a very important role to facilitate the integration and access to the information with a level of performance. The cloud BI solution has been improved with his flexibility of implementation, scalability and high performance of software and hardware business intelligence tools.

In this paper, we discuss the importance of business intelligence for decision-making and the importance to integrate it into the cloud environment, in order to make it flexible to access into the data.

We discussed also about some considerations that we should take into account, and in order to choose a best service for cloud, we define some components and architecture of BI. In additionally, the benefits and inconvenience of cloud BI have been discussed; finally, we compared public, private and hybrid cloud with the characteristics of each, we made a case study of existing solutions, and we compare them with taking into account two important scenarios.

The cloud BI has many benefits in terms of data processing performance, but some challenges still need more researches; for example, security challenges, performance and response time of requests in the OLAP process will be different and much more complex. In the next step of our research, we will develop other application scenario to verify it in the practice. So, this state of the art has significant openings for future contributions, and it is only the beginning of studies on future challenges.

References

1. Kumar, V., Laghari, A.A., Karim, S., Shakir, M., Brohi, A.A.: Comparison of Fog computing & cloud computing. Int. J. Math. Sci. Comput. (2019)
2. Mell, P., Grance, T.: The Nist Definition of Cloud Computing, pp. 800–145. National Institute of Standards and Technology Special Publication (2011)
3. Laghari, A.A., He, H., Shafiq, M., Khan, A.: Assessing effect of Cloud distance on end user's Quality of Experience (QoE). In: 2016 2nd IEEE International Conference on Computer and Communications (ICCC), pp. 500–505. IEEE (2016)
4. http://faculty.winthrop.edu/domanm/csci411/Handouts/NIST.pdf
5. Cloud Computing: An Overview. http://www.jatit.org/volumes/researchpapers/Vol9No1/10Vol9No1.pdf
6. Mohbey, K.K.: The role of big data, cloud computing and IoT to make cities smarte, Jan 2017
7. https://searchbusinessanalytics.techtarget.com/definition/Software-as-a-Service-BI-SaaS-BI
8. Patil, S., Dr. Chavan, R.: Cloud business intelligence: an empirical study. J. Xi'an Univ. Architect. Technol. (2020) (KBC North Maharashtra University, Jalgaon, Maharashtra, India)
9. Bastien, L.: Cloud Business Intelligence 2018: état et tendances du marché Cloud BI', 9 april 2018
10. Tole, A.A.: Cloud computing and business intelligence. Database Syst. J. **V4** (2014)
11. Westner, M., Strahringer, S.: Cloud Computing Adoption. OTH Regensburg, TU Dresden
12. Kasem, M., Hassanein, E.E.: Cloud Business Intelligence Survey. Faculty of Computers and Information, Information Systems Department, Cairo University, Egypt
13. Rao, S., Rao, N., Kumari, K.: Cloud Computing : An Overview. Associate Professor in Computer Science, Nova College of Engineering, Jangareddygudem, India

14. Al-Aqrabi, H., Liu∗, L., Hill, R., Antonopoulos, N.: Cloud BI: Future of business intelligence in the Cloud
15. https://onlinelibrary.wiley.com/doi/abs/https://doi.org/10.1002/cpe.5590
16. Juan-Verdejo, A., Surajbali1, B., Baars2, H., Kemper, H.-G.: Moving Business Intelligence to Cloud Environments. CAS Software A.G, Karlsruhe, Germany
17. https://www.comparethecloud.net/opinions/data-loss-in-the-cloud/
18. https://link.springer.com/chapter/10.1007/978-3-319-12012-6_1

Distributed Architecture for Interoperable Signaling Interlocking

Ikram Abourahim, Mustapha Amghar, and Mohsine Eleuldj

Abstract The interoperability in railway systems and especially in railway signaling interlocking is an issue for mobility need and to master the evolution of technology. Today, traffic of trains need a continuous communication, and the diversity of technologies makes the interoperability difficult. In Europe, some projects are in development to solve the interoperability problem. The European Rail Traffic Management System (ERTMS) is the first project deployed: It aims to establish an exchange of signaling information between interlocking and train. EULYNX is another project interested in standardization of interfaces between field equipment and computer interlocking. In this paper, we propose an architecture of computer interlocking that deal with the interoperability between adjacent calculator through a combination between functional blocks of IEC 61499 standard and service-oriented architecture (SOA). Moreover, the combination is executed on a distributed mode of sub-calculators that compose the calculator of the computer interlocking.

1 Introduction

Railway signaling is the system allowing a fluent mobility of trains and, at the same time, ensuring its security.

The main roles of railway signaling are: the spacing between successive trains and traffic in two opposite directions on the same track between stations; the management of internal movement at the station and the protection of trains from speeding and the risk of derailment.

The research work for this paper is the result of a collaboration between EMI and ONCF.

I. Abourahim (✉) · M. Amghar · M. Eleuldj
Mohammed V University in Rabat, Rabat, Morocco
e-mail: ikramabourahim@research.emi.ac.ma

M. Amghar
e-mail: amghar@emi.ac.ma

M. Eleuldj
e-mail: eleuldj@emi.ac.ma

M. Ben Ahmed et al. (eds.), *Networking, Intelligent Systems and Security*,
Smart Innovation, Systems and Technologies 237,
https://doi.org/10.1007/978-981-16-3637-0_15

To meet those requirements, a set of principles, rules, and logic process put together in the railway signaling system to ensure the safety of trains' mobility by acting on field equipment (points and signals . . .). The command of those equipment and the verification of their statutes are usually done from a signaling machine called interlocking.

Actually, most of infrastructure managers migrate to computer interlocking that allows an easy management of field equipment and gives new functions and services. But this new technology is faced to a lack of homogeneity, and then, a difficulty of communication between interlocking proposed by the various suppliers especially on borders. Also, each modification or upgrade of the infrastructure requires a partial change that may cost more than the total change of the interlocking.

The first project initiated in Europe to deal with interoperability issue is *ERTMS* [1]. This system aims to facilitate trains' mobility from a country to an other without a big investment through a direct transmission of signaling information to the onboard system of the train.

A group of railway infrastructure managers in Europe has carried out a project entitled *EULYNX* [2], since 2014, to standardize the communication protocol between computer interlocking and field equipment independently of their industrial supplier.

An other need of interoperablity is the communication between interlockings on border. Actually, most solutions deployed in different countries are using electrical logic even if they rely on computer interlockings in each side of border. This solution cannot be general but is realized specifically for each case differently.

Unfortunately, there are not enough articles in the literature that study the interoperability between computer interlocking in the signaling railway field due the lack of knowledge synergies between manufacturer competitors since R&D is their respective competitive advantage.

Our paper presents in the first part a review about signaling systems and some existing architecture of computer interlocking. In the second part, we introduce our approach for interoperability of computer interlocking that unifies the architecture and facilitates the communication on borders between stations through *SOA* [3] and *IEC 61499* standard [4]. Moreover, in third part, we explain our proposition of distributed architecture model to combine the interoperability with better processing of the computer interlocking . After that, we analyze results of the execution of our proposed architecture. And finally, we conclude with summary and perspectives of the project.

2 Railway Signaling System: Existing

2.1 Railways Control System

Railway systems, like all transport systems, are in continuous evolution and development. They take advantage of new technologies to improve operations and services.

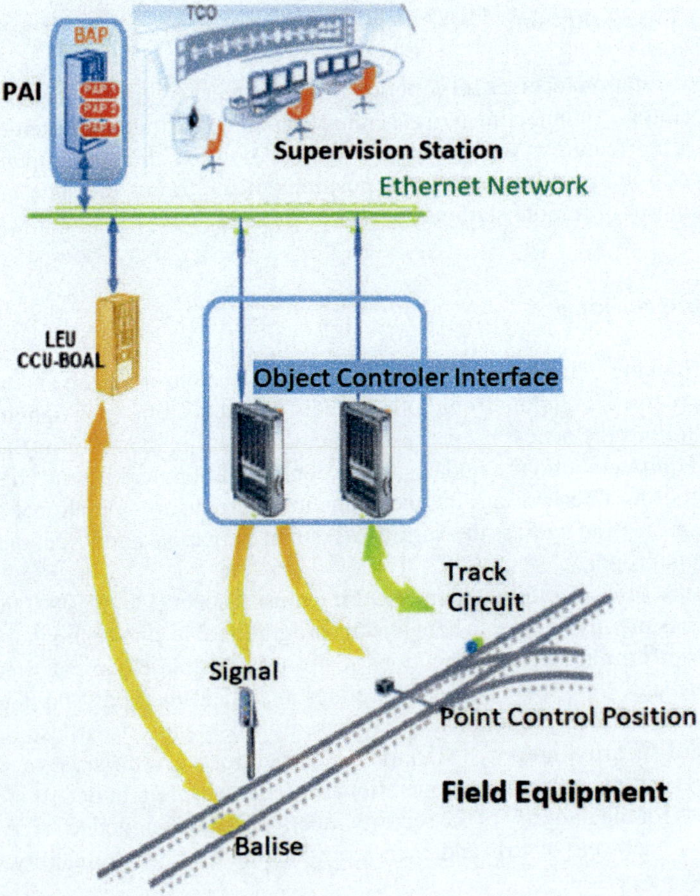

Fig. 1 Railway control system architecture

Safety is a major advantage of rail, particularly for signaling system. Rail traffic safety depends on the reliability of signaling systems especially when it is about an automatic command and control system.

The railway control system allows to continuously supervise, control, and adjust the train operations, ensuring a safe mobility of trains at all times through a continuous communication between interlocking and field equipment.

Being primarily electrical or mechanical, field equipment needs intermediate elements to communicate with computer interlocking called objects controllers. Then, we get a global network ensuring a safe and reliable interaction as shown in Fig. 1.

For high-level process supervisory management, many architectures of signaling system are deployed to allow an interconnection and a continuous exchange between object controllers of all field equipment and the calculator of computer interlocking. This calculator is also operating in an interconnection with the control station.

2.1.1 Control Station

The control station receives information from the interlocking and makes it available to the operator by connecting it to the control device of the railway system. It provides the following features: Visualization of the state of signaling equipment; ordering the desired routes; route control; the positions of the trains; the state of the areas; remote control of remote stations.

2.1.2 Interlocking

The interlocking being the intermediary between the control station and field equipment, it receives its inputs from both systems and also from the neighboring interlocking. It does the necessary processing and calculation, then returns the orders to the field equipment and the updates to the control station, and it sends data back to the neighboring interlocking. Its main function is to ensure operational safety and adequate protection against the various risks that may arise and affect the safety of persons and property.

To ensure safe operation, the interlocking must respect IEC 61508 standard. The latter applies mainly in cases where it is the programmable automaton that is responsible for performing security functions for programmable electrical, electronic or electromechanic systems. IEC 61508 defines analytical methods and development methods for achieving functional safety based on risk analysis. It also determines the levels of security integrity (SIL) to be achieved for a given risk.

The SIL can be defined as an operational safety measure that determines recommendations for the integrity of the security functions to be assigned to safety systems. There are four levels of SIL and each represents an average probability of failure over a 10-year period.

- SIL 4: Very significant impact on the community resulting in a reduction of the danger from 10,000 to 100,000.
- SIL 3: Very important impact on the community and employees reducing the danger from 1000 to 10,000
- SIL 2: Significant protection of installation, production, and employees reducing the danger from 100 to 1000.
- SIL 1: Low protection of installation and production resulting in a reduction in danger of 10–100.

For any type of interlocking, the SIL 4 level is required.

2.1.3 Object Controller

The object controller subsystem consists of several types of hardware cards, its role is to connect the computer interlocking to field equipment: signals, switch engines, etc., and return their state to the interlocking.

The object controller system has two main functions:

- Control communications with the computer switch station.
- Provide the electrical interface to track equipment.

The object controller (OC) receives commands from the central interlocking system (via the transmission system and transmission controller unit) and executes them, converting the software order into the appropriate electrical signal for the field object. It handles trackside objects and returns state information to the central calculator.

2.1.4 Field Equipment

Field equipment refers to different object of trackside which acts locally for the safety of train movement: like signals, ERTMS balises, track circuits, switches, track pedals, etc.

- *Signals*: The signals are essentially used to perform the following functions: stop signals, speed-limiting signals, and directions signals. Each of these functions usually includes an announcement signal and an execution or recall signal.
- *ERTMS balises* Point-to-track transmission transmitter, using magnetic transponder technology. Its main function is to transmit and/or receive signals. The Eurobalise transmits track data to trains in circulation. For that, the Eurobalise is mounted on the track, in the center or on a crossbar between two rails. The data transmitted to the train comes either from the local memory contained in the Eurobalise or from the lateral electronic unit (LEU), which receives the input signals from the lights or the interlocking and selects the appropriate coded telegram to transmit.
- *Track circuits (CdV)*: allows an automatic and continuous detection of the presence or absence of vehicles at all points in a specific section of lane. By therefore, it provides information on the state of occupation of an area that will be used to ensure train spacing, crossing announcements, and electrical immobilization of switches. Its detection principle effectively ensures that the signal is closed entrance to the area not only as soon as an axle enters it, but also when an incident intervenes on the track (broken rail, track shunted by a bar metal closing the track circuit, etc.).
- *Switches*: are a constituent of the railway that allows support and guidance of a train in the a given route during a crossing. The motors are used to move and maintain railway switches in the appropriate position. Switches can be automatically controlled from a station or on the ground by an authorized person.
- *Track pedals*: These devices, also called pedal repeaters (RPds), are located near the track, and they are intended to indicate the presence of a train in a part of the track where they are located. When a train passes, its axles press a pedal and close an electrical circuit. This pedal remains supported until the passage of the last axle.

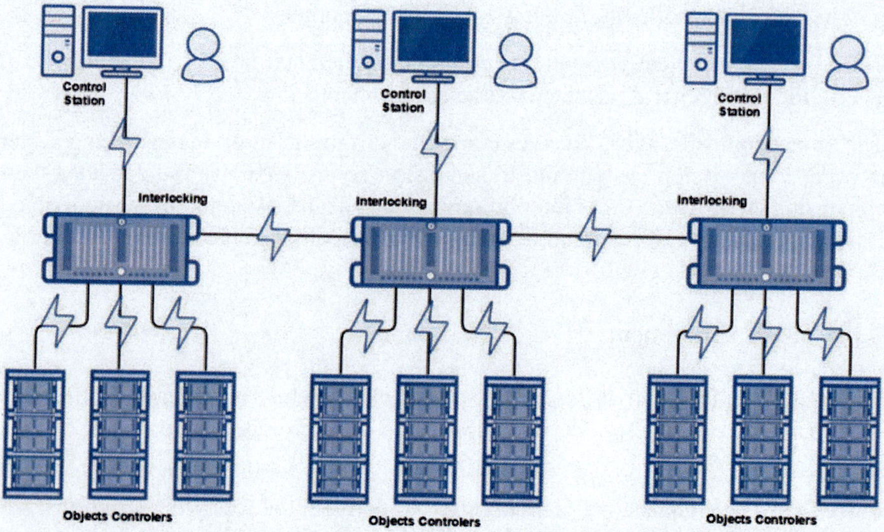

Fig. 2 Decentralized architecture

2.2 *Computer Interlocking Architectures*

As the computer interlocking enables to take advantage of new technologies and the fluidity of communication that emanates from them, many architectures are used actually to ensure a continuous communication between elements of signaling control systems.

There are two types of architecture that are globally implemented: Decentralized and Centralized Architectures.

2.2.1 Decentralized Architecture

In this architecture (Fig. 2), interlocking exchanges data with objects controllers and the station control related to its area of control, and there is a direct link for exchange between adjacent interlocking. This direct link, when it is about different suppliers of computer interlocking, in an electromechanic interface because the protocol of communication is most of the time different and needs the acceptance of suppliers to collaborate. But when we have the same supplier, the serial or Ethernet link is chosen, and the communication is adequate with the context of computer interlocking.

2.2.2 Centralized Architecture

In this architecture (Fig. 3), data from OCs is send to the interlocking managing the area where those objects are located, and all interlocking of a region exchange with the same control station that called central command station.

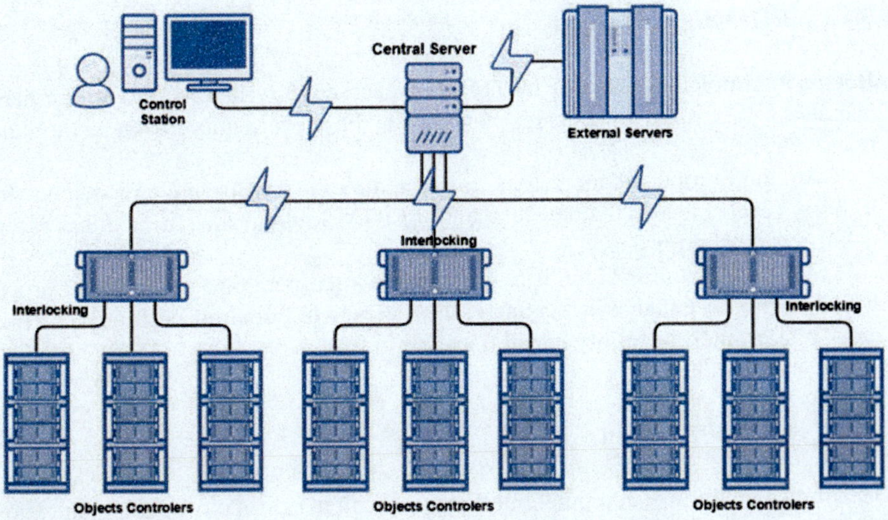

Fig. 3 Centralized architecture

When the command and control are centralized, the communication between adjacent interlocking does not need a direct and specific link, also more functionalities become possible like:

- Tracking Trains: Each train has an identifier that is monitored along the route.
- Programmable List: allows to prepare a route order list for one or many days.
- Automatic Routing: regarding to a number of train and its position a route or itinerary is commanded automatically.

Some of those operations need an interaction with external system through the external server.

3 Interoperability of Computer Interlocking

3.1 The European Initiatives for Railway Signaling Interoperability

Rail transport accompanies openness and free movement between European countries. So, to ensure a security of trains traffic, European infrastructure managers needed to unify their signaling systems in order to establish interoperability through new projects like European Rail Traffic Management System (ERTMS) [1] and EULYNX project [2]

3.1.1 ERTMS

Before the implementation of ERTMS [1], each country had its own traffic management system used for transmission of signaling information between train and track-side.

Then, the European Union (EU), conducted by the European Union Agency for Railway (ERA), leaded the development of ERTMS with principal suppliers of signaling systems in Europe.

The main target of ERTMS is to promote the interoperability of trains in Europe. It aims to greatly enhance safety, increase efficiency of train transports, and enhance cross-border interoperability of rail transport in EU.

3.1.2 EULYNX Project

The implementation of computer interlocking result in a difficulty of communication with field equipment when suppliers are different. So the European community, especially 12 European Infrastructure Managers, leaded an initiative called EULYNX project [2, 5]. This project aims to standardize interfaces and communication protocol between field equipment and computer interlocking.

3.2 *Software Architecture Proposition for Interoperability*

Until now, the problem of interoperability between computer interoperability is not approached specifically in research and literature articles and R&D for industrials is their competitive advantage.

In our approach, we choose to deal with software architecture to meet the challenge of interoperability and homogeneity of signaling interlocking. Then, we rely on two principles of software architecture from the fields of computer science and industrial computing which are the SOA [3] and the functional blocks according to IEC 61499 [4]. This combination has yielded evidence of success regarding global industrial interoperability and the ease of hot upgrading without interrupting production [6].

3.2.1 IEC 61499 Standard

The international standard IEC 61499 [4], dealing with the topic of function blocks for industrial process measurement and control systems, was originally published in 2005 and revised in 2012.

The IEC 61499 standard [4] relies on an execution model driven by the event. This execution model allows a rationalization of the execution of all functions according to a justified order and need.

3.2.2 SOA: Service-Oriented Architecture

Service-oriented architecture [3] (SOA) is a form of mediation architecture that is an application for services (Software components) implemented in interaction model with :

- strong internal consistency using a pivot exchange format, usually XML or JSON.
- loose external couplings using an interoperable interface layer, usually a WS (Web Service).

The service-oriented architecture is a very effective response to the problems that companies face in terms of re-usability, interoperability, and reduction of coupling between the different systems implemented on their information systems. Thus, we distinguish three types of services in the SOA [3] architecture:

- service provider: which provides the information or the service/function.
- service requester: which requests the information or the service/function.
- service repository: the directory of all available services.

4 Distributed Architecture for Interoperable Signaling Interlocking

Some work in railway signaling domain deal with the approach of distributed architecture [7, 8] in different ways independently of interoperability issue.

To perform the proposition of interoperability through functional blocs, we consider a new proposal of distributed architecture for signaling system.

Indeed, functional blocks regarding IEC 61499 standard allow to decompose a system on elementary functions that can be executed in a distributed environment respecting a synchronous logic. So, we choose to keep a central control and supervision and distribute calculation related to interlocking.

Previously, in central architecture, the central calculator is connected directly to object controllers (OC). Each OC is an intermediary between the calculator and the field equipment only for data exchange (Fig. 4).

So, as a distributed configuration, we propose for each station a network of subsystems as shown in Fig. 5:

- Principal functions are executed in the central calculator.
- Functions related to field equipment in borders of station are executed on auxiliary station calculator and only needed information is sent to central calculator. As an example, if we consider a plan station like Fig. 6, we will cut the station into two parts, left and right; then, the equipment of each part is linked to the auxiliary station calculator right or left.
- Functions related to outside area between stations are executed on auxiliary block calculator and only needed information is sent to central calculator.

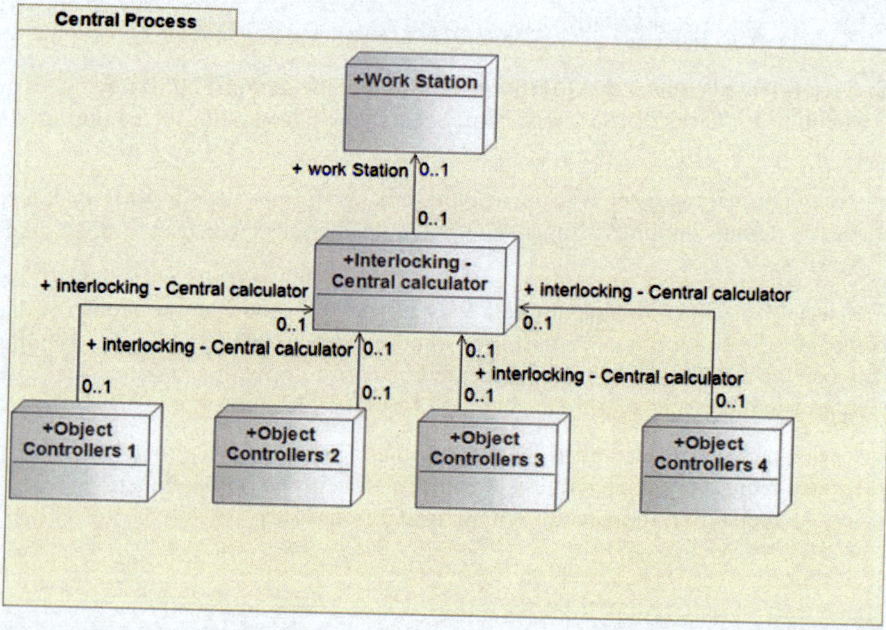

Fig. 4 Central process deployment

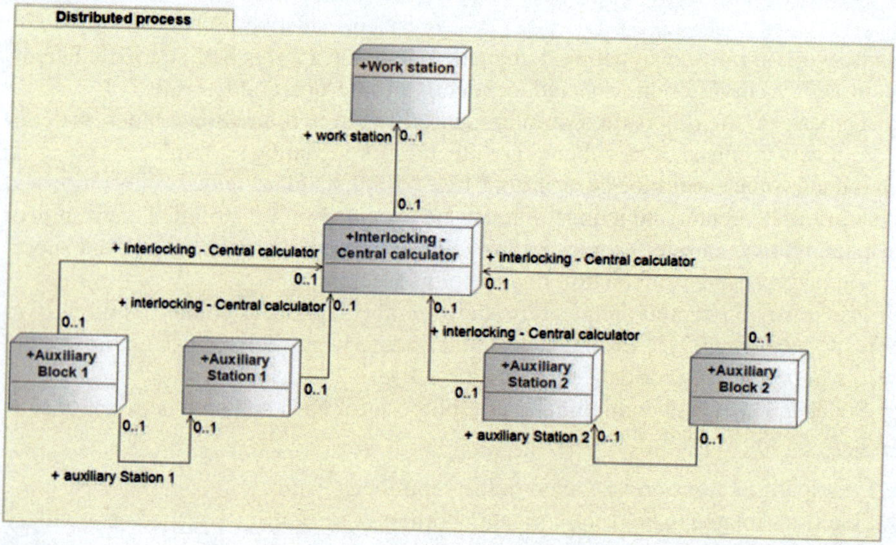

Fig. 5 Distributed process deployment

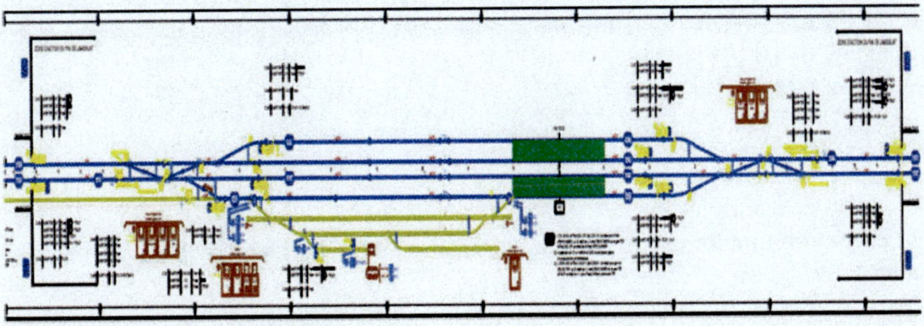

Fig. 6 Station plan

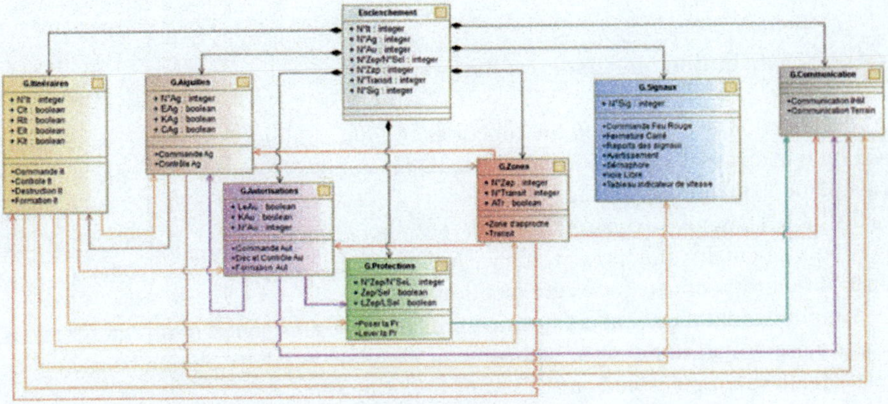

Fig. 7 Functional block diagram Interlocking—SysML

As a result of the choice of software architecture explains in previous part, we made the functional model shown in Fig. 7 and explained in [5].

This model allows to categorize functions of interlocking in families of functional blocks and then their distribution in different calculators of our proposed distributed architecture (Fig. 5).

As centralized architecture is the most architecture used for computer interlocking commissioning around the world, we choose to make a comparison between centralized and distributed architectures.

4.1 Quantitative Calculation Comparison

Regarding to Moroccan principals of railway signaling, we made a new distribution of functions of interlocking. Instead of calculating all the functions in the central calculator, we chose to distribute them between the central and the auxiliary station or the auxiliary block calculators.

Each function is related to *an execution time*. That result on execution time of cycle modeled in (1):

$$t_{\text{exe}} = \sum_{i=1}^{N} t_i \,, \tag{1}$$

t_{exe}: cycle execution time.
t_i: execution time of each function depending on the number of variables and operations.
N: total number of functions.

So reducing the number of functions executed by a calculator allows reducing time of execution in a cycle by each calculator and then ensure .

4.1.1 Distribution on Auxiliary Station

At each station, we separated the functions of field equipment on either side of the station into an auxiliary station calculator that results on having two auxiliary station calculator as mentioned in Fig. 11.

In Table 1, we give a quantitative overview about functions that we keep executing in central calculator and those which we migrate to auxiliary station calculators. This distribution respect categories are mentioned in Fig. 7.

We notice that the numbers mentioned in Table 1 are related to all the functions that we use not all in the same station or in the same auxiliary station because they relate to different possible configurations in the station.

4.1.2 Distribution Auxiliary Block

In auxiliary block, the functions executed are related to signals and areas. Indeed between stations, train traffic management is done automatically through the logical

Table 1 Computing interlocking functions

Categories	Total functions	Central execution	Auxiliary station execution
Itinerary	103	98	5
Point	11	9	2
Area	18	5	13
Signal	195	0	195
Protection	12	9	3
Authorization	56	10	38
Field switch	10	0	10
Total	405	159	246

link between signals and the occupation of the areas that frame them. So all functions are executed in the auxiliary block, and only results are sent to the central calculator for supervision need.

4.2 *Flow Exchange Variables*

For a centralized or distributed architecture, the exchange of the state of variables between functions is essential to enable a global execution of the system with coherence and logical synchronization. Each category of functions has a flow of data exchanged as shown in Fig. 8.

In the case of centralized architecture, all information collected by object controllers is sent to the central calculator. But in the case of distributed architecture, only the results of the auxiliary calculators are sent to central calculator.
As an example, we choose signal category to made a comparison of data flow between central and distributed architectures due to the fact that the functions of this category are calculated entirely in auxiliary calculator.

For internal variables of signal's functions, we can reduce the flow of data from central to object controller when calculation is made in the auxiliary calculator, so the exchange of 56 variables in centralized architecture (Fig. 9) is reduced to 0 variables in distributed architecture (Fig. 10). Also, it allows to reduce data flow to central calculator from 45 variables that are sent from object controllers (Fig. 9) to 23 variables sent from auxiliary calculator (Fig. 10). If we consider a linear model (2) for communication time:

$$t_{\text{com}} = a + b.K_{\text{v}} \quad (2)$$

t_{com}: communication time.
a: latency.
b: debit.
K_{v}: nombre of variables.

So, reducing the number of variables exchanged allows reducing time of communication.

5 Analysis and Discussion of Results

To perform our choice of distributed architecture, we made a simulation through ISaGRAF simulator regarding to the deployment network shown in Fig. 11.
This simulation combined the use of functional blocks respecting the IEC 61499 standard and their execution in a distributed architecture for computer interlocking. We can distinguish in Fig. 3 in the center of the simulator of the central computer or calculator; the simulators that surround it relate to the auxiliary station computers and finally at borders the auxiliary block computers. For the simulation, we were

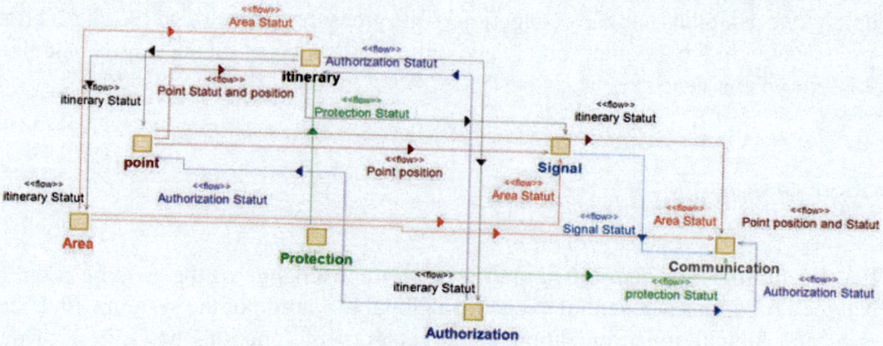

Fig. 8 Exchange diagram

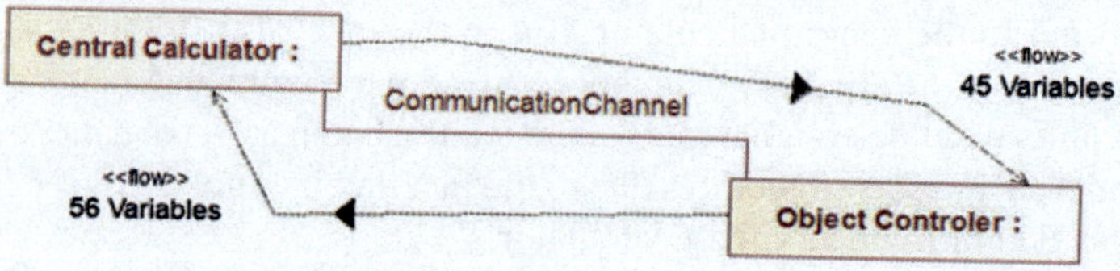

Fig. 9 Flow signal's data—Centralized architecture

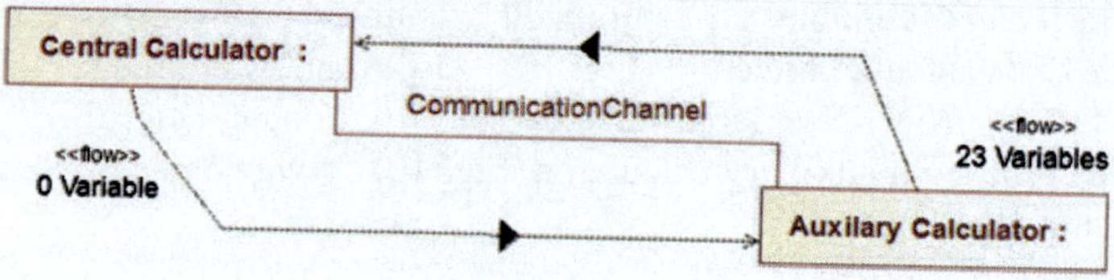

Fig. 10 Flow signal's data—Distributed architecture

content with two auxiliary block, but in reality we can find four or more depending on the extent of the distance between the stations.

The results of simulation ensure, on one side, the equivalent between central and distributed architecture regarding synchronous logic of execution of functions related to the interlocking. On the other side, the distribution of functions' execution allows a reduction at the level of the charge on central calculator and the time of execution cycle (t_{exe}) as well as the decrease in the flow of exchange (t_{com}) between the interlocking and the field equipment.

Fig. 11 Distributed architecture-deployment network

6 Conclusion and Perspectives

Technological change and the need for more speed for train mobility requires support for infrastructure and especially rail signaling systems. Then, the use of computer interlocking for signaling management and the interoperability between interlocking and related equipment becomes an evidence.

Moreover, our proposal takes into account interoperability through elementary functions by the use of functional blocks regarding IEC 61499 standard and also through a distributed architecture of calculators that facilitate the exchange on borders between stations in the same country or between different countries.

A simulation confirmed the relevance of the model using some functions respecting signaling principals of Morocco. An other simulation will be considered in the upcoming work steps having for objective the parameter test of real stations. Through the analysis of different parameters, mainly the respect of the process scheduled in the distributed system architecture model and the exchange security expectations, we can then consider the deployment phase.

References

1. The ERTMS/ETCS signaling system an overview on the standard European interoperable signaling and train control system. http://www.railwaysignalling.eu
2. EULYNX JR EAST Seminar 2016. International Railway Signaling Engineers (IRSE)
3. Newcomer, E., Lomow, G: Understanding SOA with Web Services, 14 décembre 2004
4. Christensen J.H. (Holobloc Inc, Cleveland Heights, OH USA), Strasser, T. (AIT Austrian Institute of Technology, Vienna AT), Valentini, A. (O^3neida Europe, Padova IT), Vyatkin, V. (University of Auckland, NZ), Zoitl, A. (Technical University of Vienna, AT): The IEC 61499 Function Block Standard: Overview of the Second Edition. Presented at ISA Automation Week (2012)
5. Abourahim, I., Amghar, M., Eleuldj, M.: Interoperability of signaling interlocking and its cybersecurity requirements. In: 2020 1st International Conference on Innovative Research in Applied Science, Engineering and Technology (IRASET)
6. Dai, W. Member IEEE, Vyatkin, V., Senior Member IEEE, Christensen, J.H., Dubinin, V.N.: Bridging service-oriented architecture and IEC 61499 for flexibility and interoperability. IEEE Trans. Ind. Inform. **11**(3) (2015)
7. Pascale, A., Varanese, N., Maier, G., Spagnolini, U.: A wireless sensor network architecture for railway signaling, Dip. Elettronica e Informazione, Politecnico di Milano, Italy. In: Proceedings of the 9th Italian Networking Workshop, Courmayeur, 11–13 2012
8. Hassanabadi, H., Moaveni, B., Karimi, M., Moaveni, B.: A comprehensive distributed architecture for railway traffic control using multi-agent systems. Proc. Ins. Mech. Eng. Part F: J. Rail Rapid Trans. **229**(2), 109–124 (2015) (School of Railway Engineering, Iran University of Science and Technology, Narmak, Tehran, Islamic Republic of Iran)

A New Design of an Ant Colony Optimization (ACO) Algorithm for Optimization of Ad Hoc Network

Hala Khankhour, Otman Abdoun, and Jâafar Abouchabaka

Abstract In this paper we have used a new approach of the ACO algorithm to solve the problem of routing data between two nodes, the source to the destination in the AD HOC network, specifically, we have improved a new variable GlobalACO to decrease the cost between the ants (cities), and to better manage the memory management where the ants stored the pheromones. Indeed, we used the BENCHMARK instances to evaluate our new approach and compared them with the other article after we applied this new approach to an AD HOC Network topology. The simulation results of our new approach show convergence and speed with a smaller error rate.

1 Introduction

Since their inception, Mobile wireless sensor networks have enjoyed ever-increasing success within industrial and scientific communities [1], AD Hoc wireless communication networks consist of a large number of mobile sensor nodes that can reposition themselves, get organized in the network, and move to another node to increase the coverage area and reach the destination, and of course, interact with the physical environment [2]. And each node powered by a battery, then the lifespan of a wireless sensor network depends on the lifespan of the energy resources of the sensor nodes, the size of the network; therefore the challenge is to have reliable and fast communication with all these constraints on the Ad Hoc sensor network. In addition to these constraints, the researchers have shown that routing in vehicular networks is an NP-hard problem with several conflicting goals [3, 4]. Therefore, the time taken by an exact method to find an optimal solution is exponential and sometimes inapplicable. For this reason, this challenge can be reduced to an optimization problem to be solved

H. Khankhour (✉) · J. Abouchabaka
Computer Science Department, Science Faculty, University IBN Tofail, Kenitra, Morocco
e-mail: Hala.khankhour@uit.ac.ma

O. Abdoun
Computer Science Department, Faculty Polydisciplinary, University Abdelmalek Essaadi, Larache, Morocco

M. Ben Ahmed et al. (eds.), *Networking, Intelligent Systems and Security*, Smart Innovation, Systems and Technologies 237,
https://doi.org/10.1007/978-981-16-3637-0_16

with approximate methods called metaheuristics in polynomial time, a metaheuristic can be adapted to a specific problem to find high-quality solutions [5] like routing in Ad Hoc big map network, some algorithms are inspired by nature, like Ant Colony Optimization, Genetic Algorithm, and Simulated Annealing [6–8], and others that are not inspired by nature, such as iterative local search and tabu search [9]. In a real Ad Hoc network, it is possible to have many obstacles between two or more nodes. These obstacles can attenuate the transmitted signal and disrupt communication, in addition to the constraint of the battery for each node. It is therefore necessary to find a brief reliable path between a source node and a destination node. To achieve this goal, a new design is presented in this article, about the heuristic routing algorithm; it is based on Ant Colony Optimization (ACO) for Ad Hoc Networks.

Several specific cases of ACO meta-heuristics have been proposed in Dorigo's literature. According to history, the strongest and most efficient systems are the ant system (1991, 1992, 1996), ant colony system (ACS 1997) [10, 11] for more details, see the difference between them [12].

The first ACO routing algorithm was dedicated to wired networks, they have used the proactive routing protocol, relying mainly on ants to find the path, but it does not adapt to changing topologies, so does not apply to ad hoc networks [13]. In 2008, Yu Wan-Jun proposed an ACO-AHR algorithm using the reactive routing protocol [14]. Another algorithm proposed by Abdel-Moniem, The MRAA algorithm based on-demand Distance Vector (AODV) for Ad Hoc Network, the goal is to find the best path in short delay [15]. The MAR-DYMO algorithm proposed by Correia, based on two ACO procedures to be applied to take advantage of routing in Ad Hoc Network [16]. Xueyang Wang proposed a new algorithm called ACO-EG for finding the best path between two nodes based on evolutionary graph theory [17].

In this paper, our strategy is to create a new design algorithm based on the ACO algorithm, we have applied this algorithm on ad hoc networks to find the best path in a short time while avoiding several constraints such as loss of node energy, collisions, loss of packets.

The paper is presented as follows, Sect. 2 will explain the algorithm ACO and the related works, in Sect. 3 will describe our new design algorithm of ACO (GlobalACO), the efficiency of our new design algorithm on BENCHMARK instances, and the Ad hoc network, and finally the conclusion in Sect. 5.

2 Presentation of the Algorithm ACO

Ants are small insects, weigh 1–150 mg, and measure from 0.01 to 3 cm, These social insects form colonies that contain millions of ants (Fig. 1).

the body of the ant is divided into three major parts:

- The head is the support of the antennae (extremely developed sensory receptors) and of the mandibles (members located at the level of the mouth which are in the form of toothed and powerful pincers).

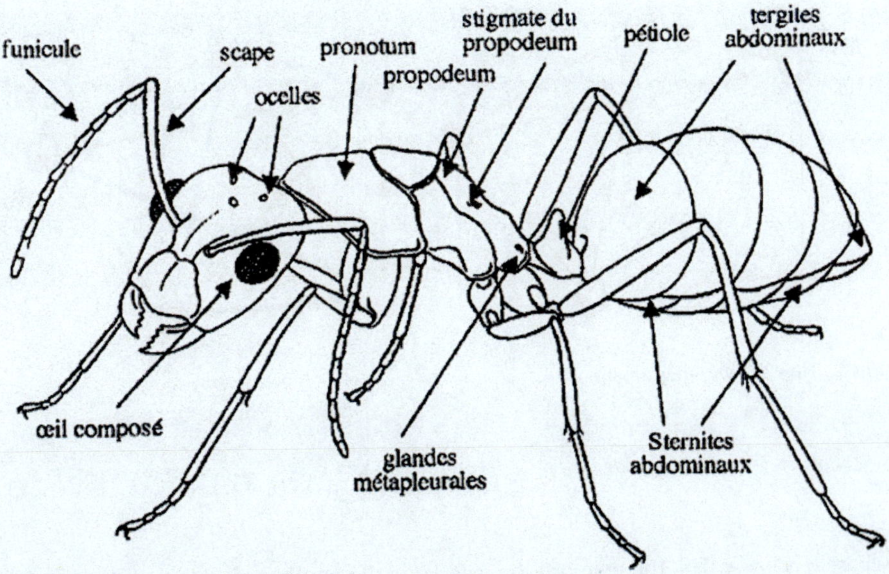

Fig. 1 Description of the ant

- The thorax allows communication between the head and the abdomen, supported by three pairs of very long and very thin legs that allow ants to move in all directions and all possible positions.
- The abdomen contains the entire digestive system and the motor of the blood system [10].

Ant colony optimization is an iterative population-based algorithm where all individuals share a common knowledge that allows them to guide their future choices and to indicate to other individuals directions to follow or on the contrary to avoid.

Strongly inspired by the movement of groups of ants, this method aims to build the best solutions from the elements that have been explored by other individuals. Each time an individual discovers a solution to the problem, good or bad, he enriches the collective knowledge of the colony. So, each time a new individual will have to make choices, he can rely on collective knowledge to weigh his choices [5] (Fig. 2).

To use the natural name, individuals are ants who will move around in search of solutions and who will secrete pheromones to indicate to their fellows whether a path is interesting or not. If a path is found to be heavily pheromized, it will mean that many ants have judged it as part of an interesting solution and that subsequent ants should consider it with interest.

In the literature, the first ACO algorithm to be proposed by Dorigo was the Ant (AS) [10, 11]. After each turn between source and destination, the ants are updated with all the pheromone values traveled. The edges of the graph are the components of the solution, then the update of the pheromones between the cities r and s is as follows (1) [18]:

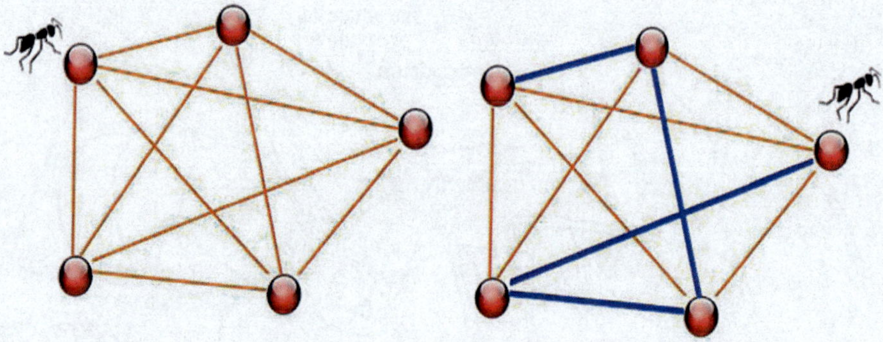

Fig. 2 How the ant finds a path

$$\tau(r,s) \leftarrow (1-\rho)\tau(r,s) + \sum_{k=1}^{m}\left(\Delta\tau(r,s)^k\right) \quad (1)$$

where $0 < \rho < 1$ is the evaporation rate, m is the number of ants and $\Delta\tau(r,s)^k$ is the quantity of pheromone put on edge (r,s) by the k-th oven (2):

$$\Delta\tau(r,s) = \begin{cases} \frac{1}{L_k} & \{if \text{ ant } k \text{ uses } edges\ (\ldots) \text{ in its tour}\} \\ 0 & \text{otherwise} \end{cases} \quad (2)$$

where L_k is the tour length of the k-th ant.

Origo mentioned in his article [19], that ants are initially randomly distributed in cities, so an ant colony algorithm is an iterative population-based algorithm where all individuals share a common knowledge that allows them to guide their future choices and indicate to other individuals directions to follow or on the contrary to avoid. Once the city tour has been completed, an ant k deposits a quantity of pheromone on each edge of its route; then a pheromone update is necessary [20].

In general, the ants are used not to meet on a common path, that's why we used the update of the local pheromones, the rule (2), to encourage the ants to visit edges not yet visited [17].

This helps to mix the cities so that the cities visited at the start of ant visits are visited later in another ant tour.

The pheromone level is updated by applying the local equation update rule (3)

$$\tau(r,s) = (1-\rho).\tau(r,s) + \rho\Delta(r,s)\Delta(r,s) \quad (3)$$

where $\tau(r,s)$ is the quality of pheromone on the edge (r,s) at time t.

ρ: is a parameter governing the decrease of pheromones such that $0 < \rho < 1$.

3 Proposed Approach: New Design of ACO in AD HOC (GlobalACO)

In this part, we will apply the ACO method on Ad Hoc Network to approach the optimal solution of the problem big map network AD Hoc. An ant k placed on the city *i* at instant t will choose the next city *j* according to the visibility *n* of this city and the quantity of pheromones t deposited on the arc connecting these two cities, other algorithmic variants drop the pheromone on the nodes of the network Ad Hoc. The choice of the next city will be made stochastically, with a probability of choosing the city j given by the following algorithm:

Initialization of pheromone tracks;
Initialization pheromone trails;
Place the ants in the source;
Loop as long as the stop criterion has not been reached:

- Build the solutions component by component, crossing the disjunctive graph.
- Use of a heuristic.
- Updating pheromone tracks, the update done globally.

End of the loop.

4 Proposed Simulations and Results

To evaluate our work, as a first step we used the instances of BENCHMARK, to evaluate our algorithm by comparing with another work of Darren M. Chitty [19] just for the best solution, the latter proposed a new PartialACO variant, whose aim is to reduce memory constraints, that is, the PartcialACO variant does not update partial for the best tower. In our article, we added a new variable is GlobalACO, this last makes the global update the best trick; after different tests and configuration runs. The approximate solutions found are In Table 1 we compared the results obtained with Darren M. Chitty (PartialACO) and our result (GlobalACO); knowing that the stopping criterion is 100 iterations; In Table 1 we also note the error of the value of the solution obtained, according to Eq. (4):

$$Er\% = \frac{\text{the solution obtained} - \text{the solution optimal}}{\text{the solution optimal}} \times 100 \tag{4}$$

According to Table 1 we notice that for the pcb442 instance, the error rate for the GlobalACO variant (0.51%) is very small compared to the rate of the PartialACO variant (1.14%), and this is also the case for the d657 instance, the error rate of

Table 1 Comparison between PartialACO and GlobalACO by using TSP instances

Problem		ACO			
Name	Optimal	Result PartialACO	Er%	Result GlobalACO	Er%
pcb442	50,779	51357.8806	1.14	51,038	0.51
d657	48,912	50320.6656	2.88	49,394	0.98
rat783	8806	8997.9708	2.18	8904	1.12
pr1002	259,045	265806.0745	2.61	262,930	1.49
pr2392	378,032	396971.4032	5.01	388,616	2.75

GlobalACO (0.98%) is very small compared to the rate of the PartialACO variant (2.88%), and the same is the case for large cities like for example the instance pr2392, the error rate of the variant GlobalACO (2.79%) is small compared to the rate of the PartialACO variant (5.01%).

The illustrated Fig. 3 shows that the GlobalACO algorithm is closer to the optimal than the PartialACO algorithm.

In Fig. 4, we notice that there is a large distance between the PartialACOEr et GlabalACOEr, this means that our algorithm GlobalACO gave better results compared to the work of Darren M. Chitty PartialACO for the five instances, and it seems that the GlobalACO algorithm converges quickly. After testing our GlobalACO algorithm on TSP instances, we will now apply our algorithm to the AD HOC network.

We suggest more comparative studies between the simulation used (GlobalACO) and other approaches using the genetic algorithm (AG), for example, in the article by

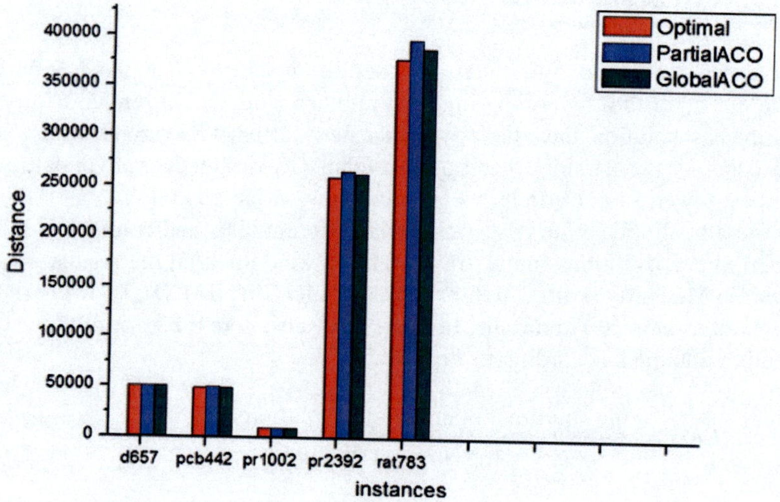

Fig. 3 MTSP comparison with PartialACO and GLOBALACO

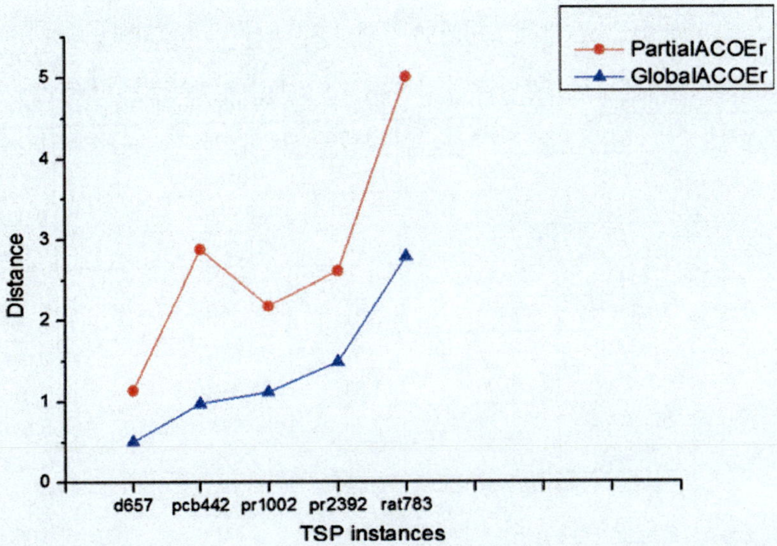

Fig. 4 Comparison error rate between PartialACO and GLOBALACO

Esra'a Alkafaween and Ahmad B. A. Hassanat [21], they proposed a genetic algorithm to produce the offspring using a new mutation operator named "IRGIBNNM", subsequently, they created a new SBM method using three mutation operators, to solve the traveling salesman problem (TSP). This approach is designed on the combination of two mutation operators; random mutation and knowledge-based mutation, the goal is to accelerate the convergence time of the genetic algorithm proposed. Table 2 compares the results obtained by Esra'a Alkafaween and Ahmad B. A. Hassanat and our results (GlobalACO) for the 12 instances.

From Table 2 and Fig. 5, we notice that the error rate is very small for our result of GlobalACO algorithm compared to the result of the New SBM algorithm, especially for the number of cities greater than 40,000, as well as the results obtained by our program are almost the same as those of the literature, this means that the size of the cities will have a great effect on the outcome of the problem.

From Fig. 6, the error rates obtained by our program (GlobalACO) are almost zero, and close to the results of the literature, so our algorithm GlobalACO is powerful for Large Scale TSP Instances.

4.1 Solve the Sensor Network Big Map Using GlobalACO

In this section we applied the GlobalACo algorithm to a sensor array, first, we considered the starting location as the source sensor and the food location as the destination sensor, the ant antennas as the sensor antennas, and the tour as the circuit on the AD

Table 2 Comparison between New SBM and GlobalACO by using 12 TSP instances

Problem		AG		ACO	
Name	Optimal	New SBM	Er%	Result GlobalACO	Er%
eil51	426	428	0.47	426	0
a280	2579	2898	12.37	2582	0.12
bier127	118,282	121,644	2.84	118,285	0.002
kroA100	21,282	21,344	0.29	21,286	0.018
berlin52	7542	7544	0.02	7542	0
kroA200	29,368	30,344	3.32	29,370	0.007
pr152	73,682	74,777	1.49	73,686	0.005
lin318	42,029	470,06	11.84	42,033	0.016
pr226	80,369	82,579	2.75	80,370	0.0012
ch150	6528	6737	3.2	6528	0
st70	675	677	0.29	675	0
rat195	2323	2404	3.49	2325	0.08

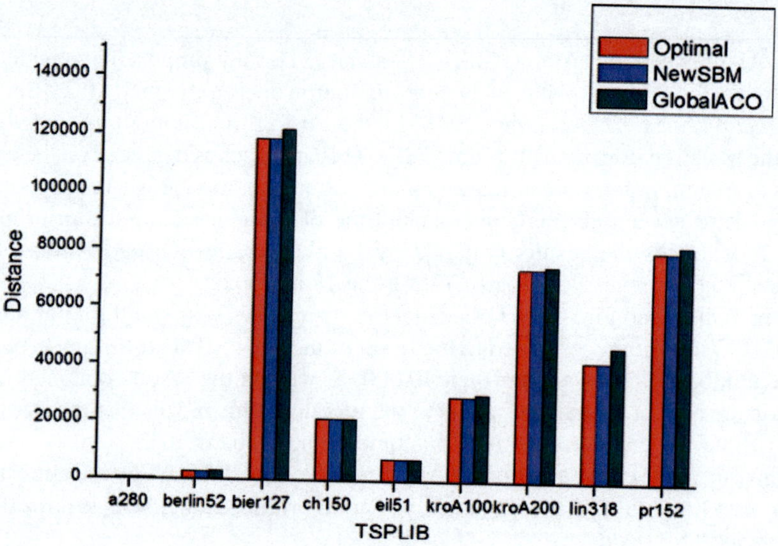

Fig. 5 MTSP comparison with NewSBM and GlobalACO

HOC network, from the source node to the destination node. After several searches, unfortunately almost there is no AD HOC topology to work with, I found a Chang Wook Ahn topology on this article [22].

As shown in Fig. 7 we generated a network topology with 20 nodes and we displayed the results found in Table 2, after several runs we found the total path cost equals 142 just in 0.015 s.

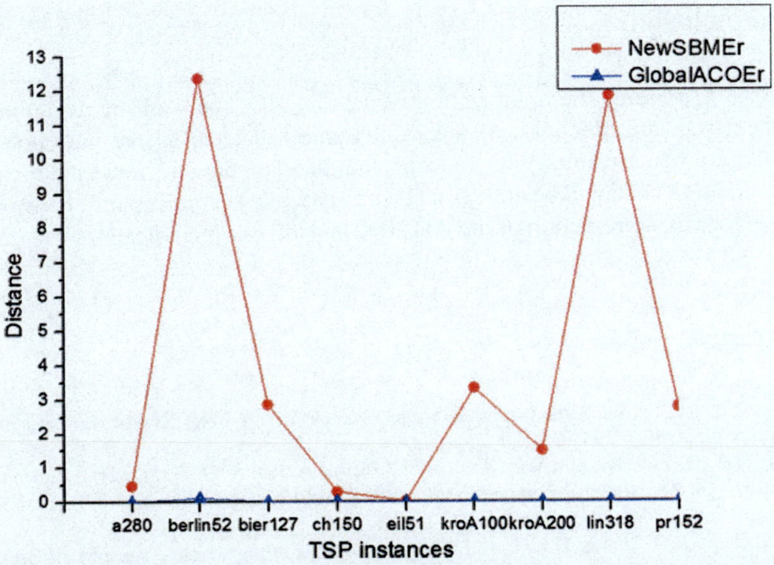

Fig. 6 Comparison error rate between NewSBM and GlobalACO

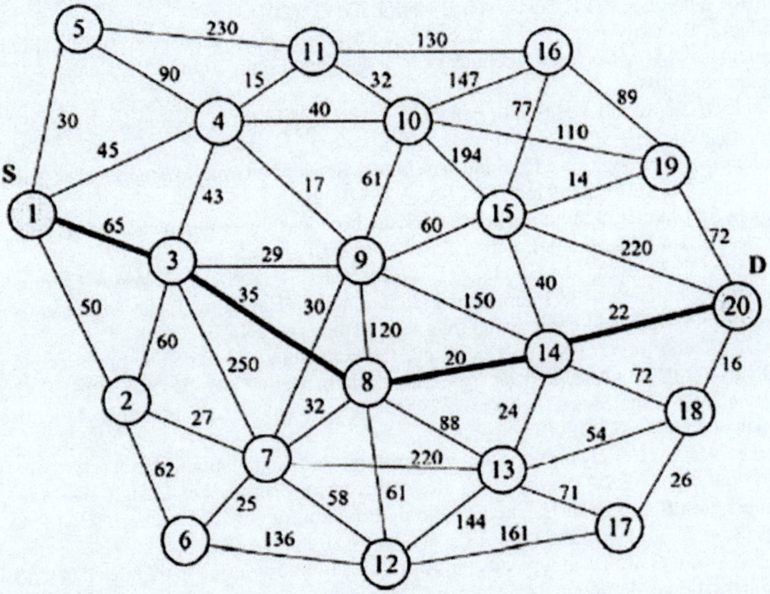

Fig. 7 The topology used in AD HOC by using ACO

5 Conclusion

This article presents the optimization of the AD HOC network by using the ACO metaheuristic, the execution of the results show that the use of the GlobalACO variant gave better results, which means, that the data flow from the source node to the destination node will be done in a faster way while keeping up the energy of each node before the termination of the AD HOC network communication.

References

1. Akyildiz, I.F., Su, W., Sankarasubramaniam, Y., Cayirci, E.: Wireless sensor networks: a survey, Comp. Net. **38**(4), 393–422 (2002)
2. Sagar, S., Javaid, N., Khan, Z. A., Saqib. J., Bibi, A., Bouk, S. H.: Analysis and modeling experiment performance parameters of routing protocols in manets and vanets, IEEE 1lth International Conference, 1867–1871 (2012)
3. Cai Zheng, M., Zhang, D.F., Luo, l.: Minimum hop routing wireless sensor networks based on ensuring of data link reliability. IEEE 5th International Conference on Mobile Ad-hoc and Sensor Networks, pp. 212–217 (2009)
4. Eiza, M.H., Owens, T., Ni, Q., Shi, Q.: Situation-aware QoS routing algorithm for vehicular Ad Hoc networks. IEEE Trans. Veh. Technol. **64**(12) (2015)
5. Hajlaoui, R., Guyennet, H., Moulahi, T.: A Survey on Heuristic-Based Routing Methods in Vehicular Ad-Hoc Network: Technical Challenges and Future Trends. IEEE Sens.S J., **16**(17), September (2016)
6. Alander, J.T.: An indexed bibliography of genetic algorithms in economics, Technical Report Report (2001)
7. Okdem, S., Karaboga, D.: Routing in Wireless Sensor Networks Using an Ant Colony Optimization (ACO) Router Chip. **9**(2), 909–921 (2009)
8. Kumar, S., Mehfuz, S.: Intelligent probabilistic broadcasting in mobile ad hoc network: a PSO approach". J. Reliab. Intell. Environ. **2**, 107–115 (2016)
9. Prajapati, V. K., Jain, M., Chouhan, L.: Tabu Search Algorithm (TSA): A Comprehensive Survey ", Conference 3rd International Conference on Emerging Technologies in Computer Engineering Machine Learning and Internet of Things (ICETCE) (2020)
10. Voss, S.: Book Review: Morco Dorigo and Thomas Stützle: Ant colony optimization (2004) ISBN 0-262-04219-3, MIT Press. Cambridge. Math. Meth. Oper. Res. **63**, 191–192 (2006)
11. Stalling, W.: High-Speed networks: TCP/IP and ATM design principles. Prentice-Hall, Englewood Cliffs, NJ (1998)
12. Sharkey. P.: Ant Colony Optimisation: Algorithms and Applications March 6 (2014)
13. Xiang-quan, Z., Wei, G., Li-jia, G., Ren-ting, L.: A Cross-Layer Design and Ant-Colony Optimization Based Load-Balancing Routing Protocol for Ad Hoc Network (CALRA). Chin. J. Electron.**7**(7), 1199–1208 (2006)
14. Yu, W.J., Zuo, G.M., Li, Q.Q.: Ant colony optimization for routing in mobile ad hoc networks. 7th International Conference on Machine Learning and Cybernetics, pp. 1147–1151 (2008)
15. Abdel-Moniem, A. M., Mohamed, M. H., Hedar, A.R.: An ant colony optimization algorithm for the mobile ad hoc network routing problem based on AODV protocol. In Proceedings of 10th International Conference on Intelligent Systems Design and Applications, pp. 1332–1337 (2010]
16. Correia, S.L.O.B., Celestino, J., Cherkaoui, O.: Mobility-aware ant colony optimization routing for vehicular ad hoc networks. IEEE Wireless Communications and Networking Conference, pp. 1125–1130 (2011)

17. Wang, X., Liu, C., Wang, Y., Huang, C.: Application of Ant Colony Optimized Routing Algorithm Based on Evolving Graph Model In VANETs, 17th International Symposium on Wireless Personal Multimedia Communications (WPMC2014)
18. Chitty, M.D: Applying ACO to large scale TSP instances. Adv. Comput. Intell. Syst. **350**, 104–118 (2017)
19. Rana, H., Thulasiraman, P., Thulasiram, R.K.: MAZACORNET: Mobility Aware Zone based Ant Colony Optimization Routing for VANET, IEEE Congress on Evolutionary Computation June 20–23, pp. 2948-2955, Cancún, México (2013)
20. Tuani A.F., Keedwell E., Collett M.: H-ACO A Heterogeneous Ant Colony Optimisation Approach with Application to the Travelling Salesman Problem. In: Lutton E., Legrand P., Parrend P., Monmarché N., Schoenauer M. (eds.) Artificial Evolution. EA 2017. Lecture Notes in Computer Science, vol 10764. Springer (2018)
21. Alkafaween. E., Hassanat. A.: Improving TSP solutions using GA with a new hybrid mutation based on knowledge and randomness, Computer Science, Neural and Evolutionary Computing (2018)
22. Ahn, C.W., Ramakrishna, R. S.: A Genetic Algorithm for Shortest Path Routing Problem and the Sizing of Populations, IEEE Trans. Evol. Comput. **6**(6) (2002)

Real-Time Distributed Pipeline Architecture for Pedestrians' Trajectories

Kaoutar Bella and Azedine Boulmakoul

Abstract Cities are suffering from traffic accidents. Every one results in significant material or human injuries. According to WHO (World Health Organization), 1.35 million people perish each year as a consequence of road accidents and more end up with serious injuries. One of the most recurrent factors is distracted driving. 16% of pedestrian injuries were triggered by distraction due to phone use, and the amount of pedestrian accidents caused by mobile distraction continues to increase, some writers call Smombie a smartphone zombie. Developing a system to eliminate these incidents, particularly those caused by Smombie, has become a priority for the growth of smart cities. A system that can turn smartphones from being a cause of death to a key player for pedestrians' safety. Therefore, the aim of this paper is to develop a real-time distributed pipeline architecture to capture pedestrians' trajectories. We collect pedestrians' positions in real-time using a GPS tracker mounted in their smart phones. The collected data will be displayed to monitor trajectories and stored for analytical use. To achieve real-time distribution, we are using delta architecture. To enforce this pipeline architecture, we are using open-source technologies such Traccar as GPS tracking Server and Apache Kafka to consume the collected data such as messages, Neo4j to store the increasing data collected for analytical purposes, as we use Spring boot for API development, and finally.

1 Introduction

Road incidents can only lead to tragedies. Whether due to speeding, poor road structure, or due to human error, it must be handled. The implementation of a safety

This work was partially funded by Ministry of Equipment, Transport, Logistics and Water-Kingdom of Morocco, The National Road Safety Agency (NARSA) and National Center for Scientific and Technical Research (CNRST). Road Safety Research Program. An intelligent reactive abductive system and intuitionist fuzzy logical reasoning for dangerousness of driver-pedestrians interactions analysis.

K. Bella · A. Boulmakoul (✉)
LIM/IOS, FSTM, Hassan II University of Casablanca, Casablanca, Morocco

M. Ben Ahmed et al. (eds.), *Networking, Intelligent Systems and Security*,
Smart Innovation, Systems and Technologies 237,
https://doi.org/10.1007/978-981-16-3637-0_17

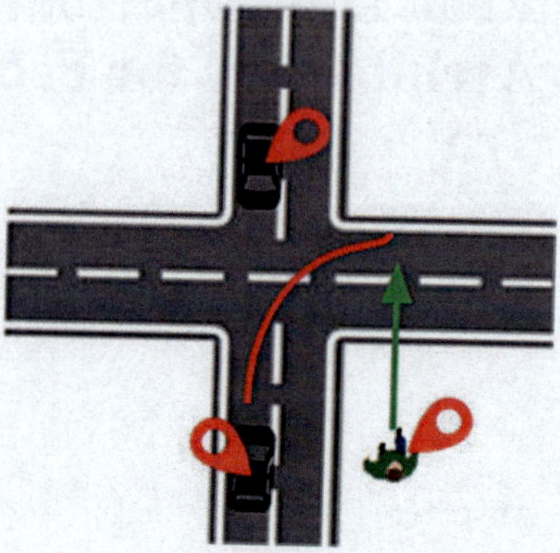

Fig. 1 Estimated intersection collision

system for pedestrians has become a must in presence of the mortality rates that are climbing every year due to injuries [1–3]. As mentioned before, smartphone can be a negative player in this scenario, so in a time driven by technology when it is used for social use case improvement, it is obvious to take advantage of this negative player for our benefit. In this paper, we are using smartphone as GPS trackers to collection pedestrians' locations in order to assemble their trajectory. By knowing each pedestrians and drivers' location we can estimate their next future position and alert if a collision is about to happen (Fig. 1).

Collecting positions for multiple users in real-time, results in big amount of data [4, 5]. This data must be processed for real-time monitoring and stored for analytical uses. However, collecting and handling these massive data presents challenges in how to perform optimized online data analysis. Since speed is very critical in this use case. In order to implement such a complex architecture and to achieve the segments of the reactive manifesto (responsive, resilient, elastic, message-driven) we need a robust architecture with highly scalable frameworks. To collect the locations, we are using Traccar as a GPS tracking server. The collected positions are consumed by a messaging queue. Based on previous work, traditional messaging queues such as ActiveMQ can manage small amounts of information and retain the distribution state of each one, resulting in a lower throughput and no horizontal scale because of the lake of replication concept.

Therefore, we used Apache Kafka. Kafka is a stream-processing platform built by LinkedIn and currently developed by the Apache Software Foundation. Kafka aims to provide low-latency ingestion of large amounts of events. It's highly scalable due to partition replications, providing higher availability too. Now we collect the information, we need to store them for future use. If the right database meets the usage requirement, it can ease and speed up the exploitation of these data which is a key player for the system responsiveness. For our use case where we value data connection for semantics, we used a graph database Neo4j. A graph database is significantly simpler and more expressive than relational one and we won't have to worry about out-of-band processing, such as MapReduce.

Recent developments in information and communication technologies are seen as an essential design vector for smart cities [4, 6–12]. One of these areas is that of mobility in a city. Currently, various cities are considering innovative ways to reduce emissions by increasing active mobility. Considerations related to pedestrian safety are a major challenge for cities. Technologies such as spatial databases, geographic information systems, internet of things, mobile computing, spatial data mining, and. define the fundamental basis for research in the field of urban computing. This work is part of this area and plans to put these technologies at the service of road safety.

The remainder of this article is organized as follows; the next section describes the architecture to connect different components of this pipeline. Section 3 details the implementation of the proposed architecture. A case study, describing how the software solution is tested, is presented in Sect. 4. The article concludes with a discussion and proposes future work in Sect. 5.

2 Architecture

Defining the architecture of the pipeline is a challenging action. The architecture describes how the various elements interact with each other. The goal of this paper is to provide an efficient data flow in order to incorporate real-time data processing. We initially implemented lambda architecture, but due to higher costs when running our jobs, we switched to delta architecture, which is an upgrade of Lambda architecture. It unifies the batch and streaming layers to avoid the gap between the two layers and we won't have to treat data differently [4, 13, 14] (Fig. 2).

In this paper the batch layer is not 100% used, only for formatting and conversion purposes, even though we chose it for future objectives (Fig. 3).

This architecture allows us first to have an exact view of data in real time and a view on analyzed and batched data (Table 1). Traccar Server is the source of our real-time data (locations). Apache Kafka is used for data processing in real time as a message broker. Spring boot for Restful APIs development. And to display real time and batched data for final user, we are using ReactJs as a frontend framework.

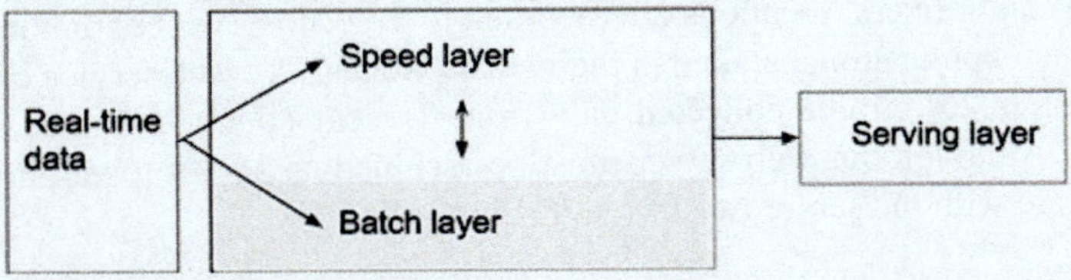

Fig. 2 Delta architecture layers

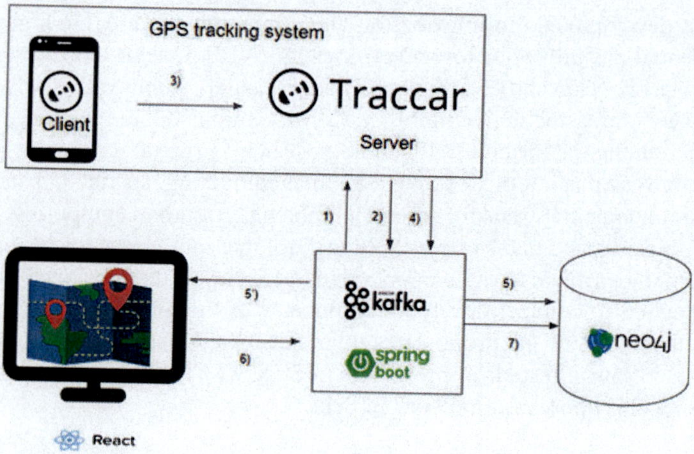

Fig. 3 Pipeline architecture components and data flow

Table 1 Data flow details

Ref	Description
(1)	Http request to Session API to establish connection (session token)
(2)	Reply with cookies
(3)	Position {longitude, latitude}
(4)	Position {longitude, latitude}: Kafka producer
(5)	Sends data to NEO4J after conversion (batch Layer)

2.1 Traccar GPS Tracker

Traccar Server Traccar is an open-source GPS tracking system [15]. It is built of Jetty Java HHTP, Natty network pipeline framework, and MySQL database. For every connection, it creates a pipeline of event handlers. The received messages from GPS devices are formatted and stored in the database (SQL database). In this paper, pedestrians and drivers' locations are recorded to Traccar server continuously from Traccar client application installed in their smartphones. Our web service can access Traccar server, to retrieve collected data from Traccar client; longitude, latitude, and speed. Although our web server must be assigned an access token, in order to communicate with the server by Http API (Figs. 4 and 5).

Based on research, there are various sets of Traccar server API. In our use case, we only need session and position APIs. Our web server sends access token parameters in Session API request, in order to initiate the connection. Traccar server sends in response cookie string to establish trusted connection. The cookie is essential use position and devices APIs. Position API is used to read users locations (longitude and latitude) and speed. We can get users locations in real time with a time difference

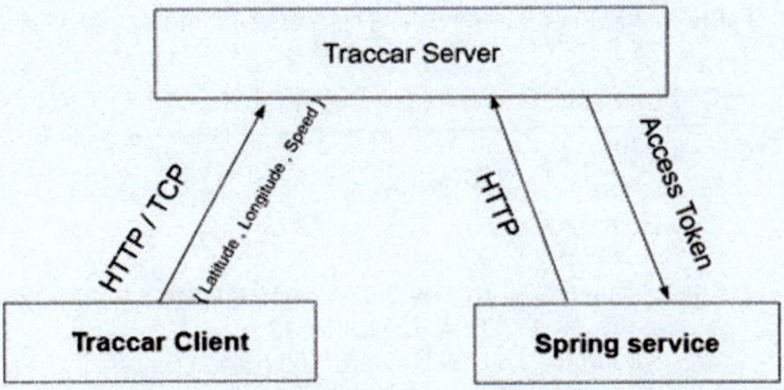

Fig. 4 Communication between Traccar client and our spring boot application

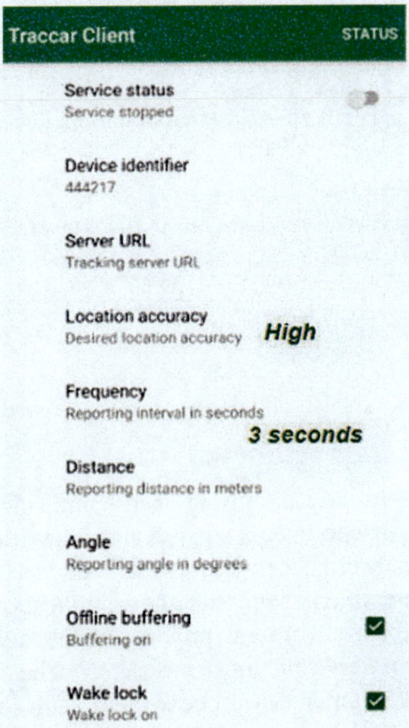

Fig. 5 Android GPS tracking application configuration

of three seconds or less. Our webserver establishes the connection using the access token, and whenever a new client is connected it collects their locations and speed at all time. Using these locations, we can set pedestrians trajectories using Traccar client (Table 2).

Pedestrians In order to keep track of pedestrian's locations, a GPS tracker is needed. In this paper, we are using and Android GPS tracker application developed by Traccar. In order to establish the connection, we must first set some configurations.

Table 2 Parameters pf Traccar HTTP API

HTTP API	Request/Response
Session	**Request:** http://{ServerIP/Domain}/api/session?token=Cy4J8amzLZ1DpzYAw76TpEDfcRWPi5yU **Response:** HTTP/1.1 200 OK Date: Wed, 18 Nov 2020 06:55:10 GMT Server: Jetty(9.4.20.v20190813) Expires: Thu, 19 Nov 2020 00:00:00 GMT Content-Type: application/json Content-Length: 532 Set-Cookie: JSESSIONID = node09ys9frpwk3i51458s79g37dgc5.node0; Path =/ Keep-Alive: timeout = 5, max = 100 Connection: Keep-Alive
Position	**Request:** http://{ServerIP/Domain}/api/positions **Reply:** accuracy: 1500 device Id: 1 fix Time: "2020-11-18T06:21:27.000+0000" id: 19 latitude: 33.584807 longitude: −7.584743 speed: 0

2.2 *Kafka*

Apache Kafka is an open-source distributed streaming platform, licensed under the Apache license [16]. Kafka is written in java and Scala; it implements a publish-subscribe messaging system that is designed to be fast and scalable. Message brokers are middleware that allow applications to exchange messages (events) in cross platforms. In order to provide reliable message storage and guaranteed delivery, message brokers rely on a message queue that stores and orders the messages until the consumer can process them. This strategy prevents the loss of valuable data and enables systems to continue functioning even in the face of the intermittent connectivity or latency issues. Messages are sent by producers, a given message concerns a subject, in Kafka this is called a Topic. The consumer subscribes to one or more topics. He will therefore receive all the messages concerning these subscribed topics. This architecture is an abstraction that provides us developers with a standard of handling the flow of data between an application's services so that we can focus on the core logic.

Configuration Kafka cluster can be composed of multiple brokers. Each broker is identified with an ID and can contain certain topic partitions. In this paper, we are using a single node with 2 brokers as Kafka configuration (replication-factor = 2), so that when a broker is down, another one can serve the data of the topic. This architecture is able to handle more producers. However, this is still a basic

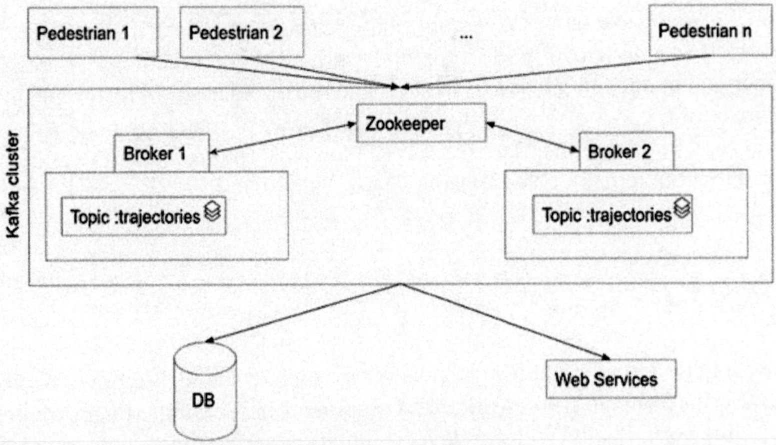

Fig. 6 Kafka cluster consisting of one node and two brokers

configuration and, since Kafka is distributed in nature, a cluster typically consists of multiple nodes with several brokers. The results shown in this paper are based on a two-brokers architecture in order to reveal the first step towards distributing the system. Zookeeper is responsible for managing the load over the nodes (Fig. 6).

2.3 *Spring Boot*

Spring boot is a Java-based framework for building web and enterprise applications. This framework provides a flexible way to configure Java beans and database transactions. It provides also powerful for managing Rest APIs as well as it contains an embedded Servlet Container. We chose It, to abstract the Api service configuration to focus on writing our logic instead of spending time configuring the project and server. It provides a template as a high-level abstraction for sending messages, as well as support for Message-driven POJOs with @KafkaListener annotations and a listener container. With this service, we can create Topics and different instances of producer and consumers very easily.

2.4 *Neo4j Instantiation*

When dealing with a huge amount of data, storing and retrieving these data become a real challenge. In this paper, not only, we are dealing with a lot of data but also our data is highly interconnected. According to previous researches, Cypher is a promising candidate for a standard graph query language. This supports our choice of using a graph database. A graph database saves the data in an object format represented as a node and binds it together with edges (association). It uses Cypher as a query language that allows us to store and retrieve data from the graph database. The syntax of Cypher offers a visual and logical way of matching node patterns and relationships in the graph. Also, we can use the sink connector with Kafka to move data from Kafka topics to Neo4j using Cypher templates.

2.5 React Js

We are using ReactJs for handling view layer for mobile and web application. ReactJs is a JavaScript frontend framework; it is component-based Single Page Applications (SPA) framework. It suits well for data presentation in real time since we can update data without refreshing the page. Its main purpose is being fast and simple. It uses the conception of Virtual Dom for performance optimization. Each DOM object has a corresponding virtual DOM object. When an object is updated, all the DOM changes, which sound incredibly inefficient, but the cost is negligible because the virtual DOM can update so fast. React compares the virtual DOM with its snapshot, which was taken just before the update. By comparing the latest virtual DOM with the pre-update version, ReactJs only updates the one changed. Each semantic data is represented in a higher component. Which will help us structure the project; keep it maintainable and easy to read. In this case, we are using two HOCs: Historical data and real-time data.

3 Architecture Components Implementation

3.1 Processing Real-Time Data Operation

The architecture of this paper consists of several components which are GPS Tracking device system, Message broker, database, and a view platform. Traccar client sends pedestrians locations with a 3 s delay to Traccar server. Kafka producer collects from Traccar server these locations using restful APIs in our Spring boot service and publishes it to the convenient topic. Kafka consumer listens to this topic and each record is sent to Neo4j as nodes. The core concept is to get data from Kafka and dispatch it to Neo4j and controllers. The diagram below explains the flow of Data (Fig. 7).

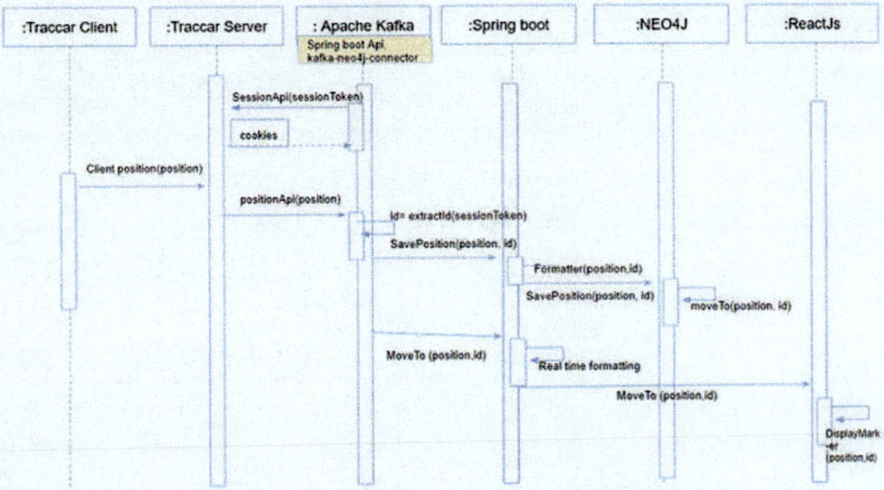

Fig. 7 Traccar sequence diagram for the real-time locations monitoring and historical data

Fig. 8 Graph of pedestrian's locations

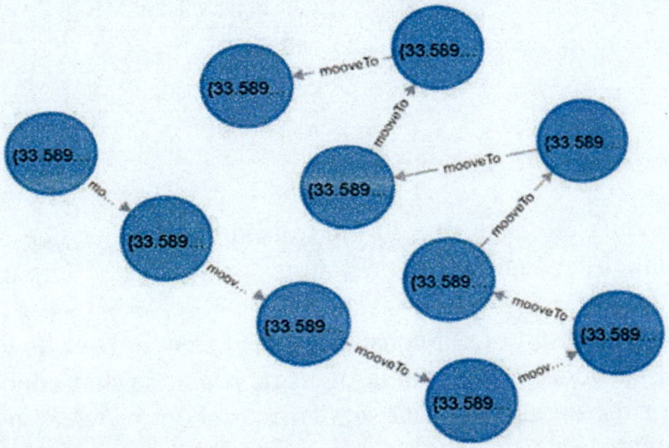

3.2 *Neo4J Instance*

In graph databases, data are presented under graph format. The figure below represents the data saved with hibernate instance, each message is a node under position format {longitude, latitude, speed, userId} related by "move to" connection (Fig. 8).

3.3 *Testing*

We used kafka-*-perf-test library to measure read and write throughput and to stress test the cluster. First testing our producer with 1000 messages. We chose not to

Table 3 Producer test results

Parameter	Result
start.time	2020-11-27 21:38:28:413
end.time	2020-11-27 21:38:29:002
compression	0
message.size	100
batch.size	400
total.data.sent.in.MB	50
MB.sec	0.269
total.data.sent.in.nMsg	1000
nMsg.sec	239.61

Table 4 Consumer test results

Parameter	Result
start.time	2020-11-27 22:40:27:403
end.time	2020-11-27 22:40:28:002
fetch.size	1951429
data.consumed.in.MB	2.2653
MB.sec	3.8700
data.consumed.in.nMs	1001
nMsg.sec	40931.2567

specify message size during tests since our messages are not very large. We've set initial-message-id to generate test data. The results of our producer are as follow (Table 3).

The result of our consumer tests given in the following (Table 4).

According the results, we can conclude that configuration is resilient enough for the number of data set. Traccar client provides us with pedestrians' positions and their speed each three seconds. Although, while testing we realized that some positions are not 100% accurate. And after testing two different trajectories each with approximately 50 positions, 87% of the positions are received: bases on each three seconds we receive a location.

4 Results

In this paper, we have implemented a real-time pipeline for pedestrians' trajectories, using open-source technologies. From our view layer implemented with ReactJs, we retrieve data using REST API from our Spring boot application. Pedestrian's trajectories are displayed on a Google Map using our ReactJs application, as shown in the figure (Fig. 9).

Fig. 9 Pedestrian trajectory

We can see pedestrians' trajectory, but at a certain point where positions are not recorded properly either because the position is not sent properly to Traccar server or delay time. We record accident locations, intersections where accidents happen more often; with this simple peace of information, we can categorize intersections as red zone, orange zone, or green zone. According to the zone type we are going to rise up the collision percentage as shown in Fig. 10. Accidents are yellow dots, and according to the number of accidents we categorize the intersection as dangerous or normal.

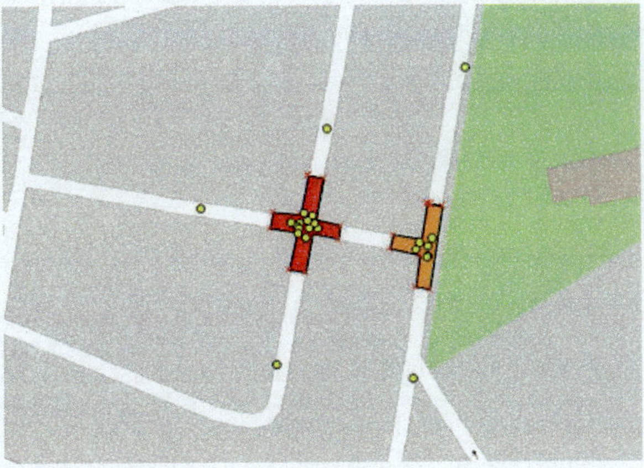

Fig. 10 Intersection categories

5 Conclusion and Future Work

Nowadays, traffic accidents are one of the most serious problems of the transportation system. Pedestrians' safety is treated as a priority to solve in order to upgrade to a smart and safe city ecosystem. In this paper, we present real-time distributed pipeline architecture for pedestrians' trajectories. Our primer concern is pedestrians because it's the weakest component on roads accidents. The main challenge was to define an optimized architecture to provide real-time processed data. Using Traccar GPS tracking server, we are collecting pedestrians' and drivers' positions from Android application installed in their smart phones. Using this data, we can set trajectories and visualize them. For future work we aim, to estimate intersection collisions in order to alert the pedestrian. For more accuracy, we want to record more information besides positions and speed.

References

1. Hussian, R., Sharma, S., Sharma, V.: WSN applications: automated intelligent traffic control system using sensors. Int. J. Soft Comput. Eng. **3**, 77–81 (2013)
2. MarketWatch: Inattention is leading cause of deadly pedestrian accidents in el paso. https://kfoxtv.com/news/local/inattention-leading-cause-ofdeadly-pedestrian-accidents-in-el-paso-say-police, (2019)
3. Tribune, C.: Look up from your phone: Pedestrian deaths have spiked (2019) https://www.chicagotribune.com/news/opinion/editorials/ct-editpedestrian-deaths-rise-20190301-story.html
4. Maguerra, S., Boulmakoul, A., Karim, L., et al.: Towards a reactive system for managing big trajectory data. J. Ambient Intell. Human Comput. **11**, 3895–3906 (2020). https://doi.org/10.1007/s12652-019-01625-3
5. Bull, A., Thomson, I., Pardo, V., Thomas, A., Labarthe, G., Mery, D., Diez, J.P., Cifuentes, L.: Traffic congestion the problem AND how to deal with it. United Nations Publication, Santiago (2004)
6. Atluri, G., Karpatne, A., Kumar, V.: Spatio-temporal data mining: A survey of problems and methods. ACM Comp. Surveys **51**(4), Article No. 83 (2018)
7. Boulmakoul, A., Bouziri, A.E.: Mobile object framework and fuzzy graph modelling to boost HazMat telegeomonitoring. In: Garbolino, E., Tkiouat, M., Yankevich, N., Lachtar, D. (eds.) Transport of dangerous goods. NATO Science for Peace and Security Series C: Environmental Security. Springer, Dordrecht (2012)
8. Das, M., Ghosh, S.K.: Data-Driven Approaches for Spatio-Temporal Analysis: A Survey of the State-of-the-Arts. J. Comput. Sci. Technol. **35**, 665–696 (2020). https://doi.org/10.1007/s11390-020-9349-0
9. D'silva, G.M., Khan, A., Gaurav, J., Bari, S.: Real-time processing of IoT events with historic data using Apache Kafka and Apache Spark with dashing framework, 2017 2nd IEEE International Conference on Recent Trends in Electronics, Information & Communication Technology (RTEICT), Bangalore, pp. 1804–1809 (2017), https://doi.org/10.1109/rteict.2017.8256910
10. Duan, P., Mao, G., Liang, W., Zhang, D.: A unified spatiotemporal model for short-term traffic flow prediction. IEEE Transactions on Intelligent Transportation Systems (2018)
11. Goodchild, M.F.: Citizens as sensors: the world of volunteered geography. GeoJournal, 211–221 (2007)
12. Chen, L., Roy, A.: Event detection from Flickr data through wavelet-based spatial analysis. In CIKM'09, 523–532 (2009)

13. Marz, N., Warren, J.: Big data: principles and best practices of scalable real time data system, ISBN 9781617290343, Manning Publications (2015)
14. Psomakelis, E., Tserpes, K., Zissis, D., Anagnostopoulos, D., Varvarigou, T.: Context agnostic trajectory prediction based on λ-architecture, Future Generation Computer Systems **110**, 531–539 (2020). ISSN 0167-739X, https://doi.org/10.1016/j.future.2019.09.046
15. Watanabe, K., Ochi, M., Okabe, M., Onai, R.: Jasmine: a real-time local-event detection system based on geolocation information propagated to microblogs. In CIKM '11, 2541–2544 (2011)
16. Abdelhaq, H., Sengstock, C., Gertz, M.: Eventweet: Online localized event detection from twitter. VLDB **6**, 12 (2013)

Reconfiguration of the Radial Distribution for Multiple DGs by Using an Improved PSO

Meriem M'dioud, Rachid Bannari, and Ismail Elkafazi

Abstract PSO is one of the famous algorithms that help to find the global solution; in this study, our main objective is to improve the result found by the PSO algorithm to find the optimal reconfiguration by adjusting the inertia weight parameter. In this paper, I select the chaotic inertia weight parameter and the hybrid strategy using the combination between the chaotic inertia weight and the success rate, these kinds of parameters are chosen due to their accuracy, and they give the best solution compared with other types of parameter. To test the performance of this study, I used the IEEE 33 bus in the case of the presence of the DGs, and a comparative study is done to check the reliability and the quality of these two suggested strategies. In the end, it is noticed that the reconfiguration by using the chaotic inertia weight gives a better result than the hybrid strategy and the other studies: reduce losses, improve the voltage profile at each node, and give the solution at a significant time.

1 Introduction

Distributed generation (DG) is an old idea that appears for a long time. But it is still a good technology that provides the electricity at or near where it will be consumed [1]. When it is integrated into the electrical network, the distributed generation can allow reducing the losses at the level of the transmission and distribution network [2]. It is an emerging strategy used to help the electrical production plants to follow client consumption and used to minimize the quantity of electricity that should be produced at power generation plants, besides, help to reduce the environmental impacts that leftover by the electrical production plants [3]. The authors of [4] have used a method to determine the forecasts of the electrical consumption, and based on the electrical quantity requested, in the same direction, the authors of [5] have focused on the

M. M'dioud (✉) · R. Bannari
Laboratory Sciences Engineering ENSA, Ibn Tofail University, Kenitra, Morocco

I. Elkafazi
Laboratory SMARTILAB, Moroccan School of Engineering Sciences, EMSI, Rabat, Morocco

M. Ben Ahmed et al. (eds.), *Networking, Intelligent Systems and Security*,
Smart Innovation, Systems and Technologies 237,
https://doi.org/10.1007/978-981-16-3637-0_18

treatment of the injection of intermittent energy sources. Thus, they have chosen to use some techniques of energy management and check the response to demand, and they have concluded their paper by studying the importance of the smart grids to check a balance between the supply and the demand inter electricity producers and consumers.

However, the authors of [6] have discussed the injection of the intermittent energy generators will give rise to impacts on the reliability of the electrical system (maybe voltage exceed the voltage plane limits), and they have proposed some solution that helps us to insert the intermittent power in electrical systems with a high injection rate. And the authors of [7] have introduced a supervisor algorithm (the predictive control model) to control the energy produced aiming to minimize the total cost of production.

During these last years, the electrical enterprises have been oriented toward new techniques, to enhance, and minimize energy exploitation by searching for suitable reconfigurations, so that these new strategies encourage the reduction of losses. To find the solutions to this type of problem, various studies have been done, such as the authors of [8] have based on the modified shark smell optimization to search for the new reconfiguration of the radial distribution with and without DG aiming to minimize the total power system losses. On the other hand, the authors of [9] have suggested using a simple algorithm for distribution system load with DG, where they have considered the DGs as negative loads. In the article [10], the author gives a comparison between two types of DG (PQ bus, it is mean that the integrated power is considered as a load with a negative sign, the second type is the PV bus is mean that the reactive power injected in the network counts on the expected tension on the bus, but the active power injected is considered constant).

In this vision of the issue, the authors of [11] have used the genetic algorithm to reduce the losses and optimize the voltage at each node after they used the forward–backward method to apply the load flow analysis, aiming to predict the optimum plan for a power distribution network with the presence of multiple DGs. The authors of [12] have discussed in their study the influence resulting from the presence of DG in the network on the losses and the voltage profile of the electrical network. In this vision, [13] have studied the performance analysis of the electrical distribution system with and without DGs. And the authors of [14] have used the prim particle swarm optimization algorithm to search the new reconfiguration and reduce losses of the electrical network with the presence of the DGs.

In this paper, I will use an enhanced and adaptive PSO algorithm based on the adjusting strategy of the inertia weight parameter, this algorithm is chosen thanks to its high speed to meet an optimal and best solution, thereby it easy to implement and it bases on simple equations. And the main goal of this research is to find a significant reduction concerning the value of the losses and a significant amelioration in the voltage profile. For testing the performance and the quality of this proposed method, this paper focused on the IEEE 33 bus with the presence of the DGs, and a comparative study is done to compare the results with other recent studies described in the following paragraph.

2 Related Works

On this side, the authors of [15] have tried to solve this problem by using the particle swarm optimization algorithm focused on the decreasing inertia weight parameter "w" to update the new position and velocity of the particle then have applied the backward forward sweep to define the power flow analysis. The authors of [16] have chosen also the linear decreasing inertia weight by using the sigmoid transformation to restrict the value of the velocities. On another study, the authors of [17] have used the linear decreasing weight by eliminating the wend in the second term. Regarding the authors of [18] have done the comparative analysis to determine the best equation for inertia weight that helps to enhance the quality of the PSO algorithm and to check the reliability and the performance of their paper they focused on five mathematic equations, and they have concluded that the chaotic inertia weight is the best technique to define a result with higher accuracy; however, the random inertia weight technique is best for better efficiency. On the other hand, the authors of [19] have thought to combine the swarm success rate parameter with the chaotic mapping to define the new inertia weight parameter and to validate the performance and the quality of their proposed tool, they examined this new parameter by solving the five functions (Griewank, Rastrigin, Rosenbrock, Schaffer f6, and Sphere), and they concluded that the swarm success rate is a useful tool to improve the performance of any swarm focused on the optimization algorithms. The authors of [8] have used the modified shark smell optimization to reduce the losses and improve the voltage profile this algorithm have the same idea as the particle swarm optimization, and they have concluded that this method helps to find the solution in a significant time and enhance the profile voltage at each node. Therefore, in this study, the main goal is to adjust the inertia weight parameter based on these two best strategies already described in the previous paragraph (the chaotic inertia weight and the swarm success rate combined with the chaotic mapping), aiming to find the optimal reconfiguration of the radial distribution network in the case of the presence of multiple DGs with losses minimized and voltage profile improved. To show the performance of these techniques, it is important to test these suggested techniques on IEEE 33 bus with DGs to compare the solution of this paper with the other studies focused on the particle swarm optimization algorithm.

To study this issue, this paper is divided into five parts. Section 3 introduces the main objective of this study, gives the objective function, and defines the constraints and describes the main steps of the chosen method, and presents the case studies with the presence of DGs. Section 4 discusses and gives an analysis study about the found results and makes the difference between this studies with the other recent works. And in the fifth and the final section, I conclude the research and I present an idea about the future research.

3 Suggested Algorithm

3.1 Problematic

The main cause that encourages several companies to search for some strategies to reduce losses comes from the peak demand it is means when all resources are operating at maximum, this last one gives rise to unnecessary expenses for the electric companies.

When the load increases, these losses increase. This paper has studied the problem using the famous reconfiguration strategy of the electrical network, by the implementation of data using MATLAB software to find the optimal solution using the PSO algorithm to find a new structure of the network with minimum total losses.

Objective Function. As described above, in this paper it is important to reduce the losses, I use the following expression to calculate this last one:

$$Losses = \sum 1 \in S R_1 * I_1^2 \tag{1}$$

With S is the set of the system edges and I_1 is the current of line l, R_1 is the current of line l.

This optimization problem is solved under the following constraints [20].

Constraints.

Kirchhoff's law:

$$I * A = 0 \tag{2}$$

where I: row vector of current of each line of graph and X: incidence matrix of graph.

(Aij = 0 if there are no arcs between i and j; Aij = 1 else);

Tolerance limit:

$$|(Vjn - Vj)|\, Vjn \,| \leq \varepsilon j \max \tag{3}$$

Vjn nominal voltage at node j, Vj is voltage in node j and $\varepsilon\, j\, max$ is tolerance limit at node j [4] (±5% for HTA and + 6%/−10%BT).

Admissible current constraint:

$$I_l \leq I_{l,maxadm} \tag{4}$$

I_l current of line l and $I_{l,maxadm}$.: current maximum admissible of line l.

Radial topology constraint. To have a simple and to keep the security of the electrical network, it is better to choose the radial configuration. It means that on each loop exists an open line. To have this topology, the system should follow these constraints.

Total number of main loops:

$$N_{\text{main loops}} = \left(N_{\text{edge}} - N_{\text{node}}\right) + 1 \tag{5}$$

where N_{edge} is the total edges of the network, N_{node} is the total number of nodes, and $N_{\text{main loops}}$ are the total number of loops in the system.

The total number of sectionalizing switches

$$N_{\text{edge}} = N_{\text{node}} - 1 \tag{6}$$

The total number of tie lines should be equal to the number of main loops in the electrical system.

To study this issue, I break up this problem into two important parts, the first one regarding the radial electrical system with the presence of the DGs. The second one is focused on the Newton–Raphson methods to do the power flow module. This load flow method is chosen to its advantage to a fast convergence rate [21].

3.2 Network with DGs

For this study of reconfiguration with the presence of DGs, I take the case of IEEE 33 bus with tie line is 33–37 as shown in Fig. 1.

Table 6 in appendices gives the line and load data of this network [22]. In this paper, we assume that the DGs data as shown in Table 1:

In this vision, according to the insertion of DG, the power integrated into a node linked to a DG will be modified. And, to update the new value of active and reactive power, I based on the following formulas [24].

$$P = P_{\text{load}} - P_{\text{DG}} \tag{7}$$

$$Q = Q_{\text{load}} - P_{\text{DG}} \tag{8}$$

$$\text{PDG} = a * Q_{\text{DG}} \tag{9}$$

To comprehend how the electrical system with DG works, Fig. 2 makes things easier.

So, the losses in this case become:

$$P_{\text{loss}} = R * \left((P_{load} - P_{DG})^2 + (Q_{load} - (\pm Q_{DG}))^2\right)/V^2 \tag{10}$$

where

R is the line resistance.

P_{loss} is the line losses.

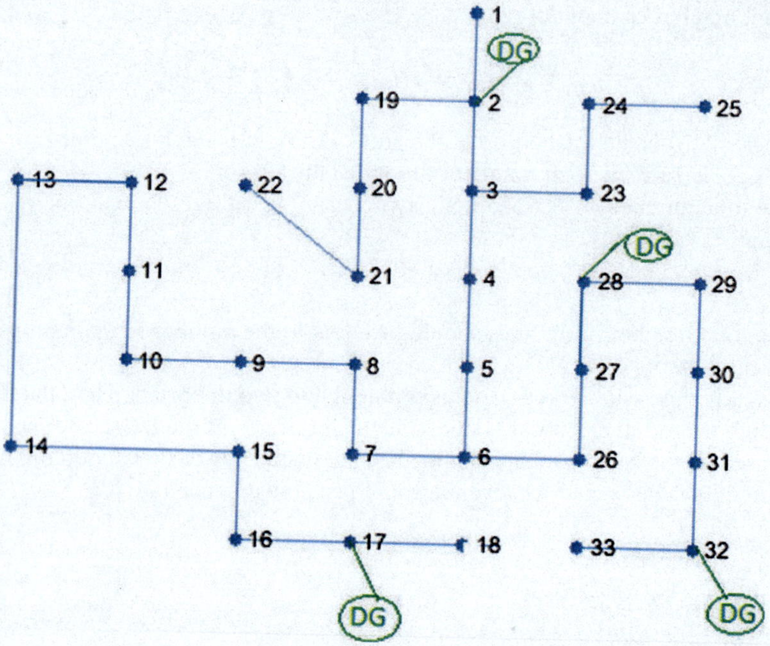

Fig. 1 Network of IEEE 33 bus with DGs

Table 1 DGs data [23]

Location(bus)	Size (MW)	Power factor
28	0.1	0.95
17	0.2	0.95
2	0.14	0.98
32	0.25	0.85

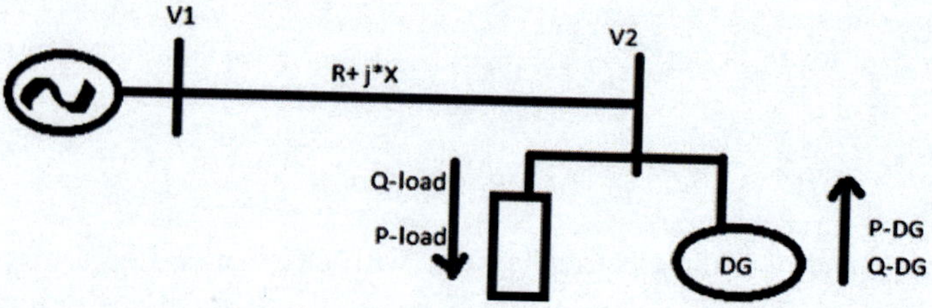

Fig. 2 DG injected as a negative load to the bus

P_{load} active power consumption of load.
Q_{load} reactive power consumption of load.
$(P_{\text{DG}}, Q_{\text{DG}})$ is the active and reactive power output of distributed generation.
a is the power factor of DG.

3.3 PSO Algorithm

PSO is a metaheuristic using to search for the optimal solution, invented by [25]. This optimization method is based on the collaboration of individuals among themselves. In this study, due to the feature of the simple travel rules in the solution space, the particles can gradually find the global minimum. This algorithm follows these steps:

Step 1: In the first step, initialize the number of particles and the number of tie lines by respecting the condition of the system is in radial nature (Table 4).

Step 2: Initialize iteration number (maxiter), inertia coefficient (w_1 and w_2), and acceleration coefficients (C_1 and C_2), the initial velocity of each particle is randomly generated (Table 5 in appendices).

Step 3: Identify the search space for each D dimension (all possible reconfiguration).

Step 4: Apply the Newton–Raphson method [21] to load flow analysis.

Step 5: Define the best value among all pbest values.

Step 6: Find the global best and identify the new tie switches.

Step 7: Update the velocity and new position for each D dimension of the ith particle using the following equation:

Select a random number z in the interval [0, 1] and use a chaotic mapping by using the logistic mapping to set inertia weight coefficient [18]:

$$z(\text{iter} + 1) = 4 * z(\text{iter}) * (1 - z(\text{iter})) \tag{11}$$

or use the success rate [19]:

$$\text{Success}_t^i = \begin{cases} 1 & f(\text{Pbest}_t^i) < f(\text{Pbest}_{t-1}^i) \\ 0 & f(\text{Pbest}_t^i) \geq f(\text{Pbest}_{t-1}^i) \end{cases} \tag{12}$$

$$\text{Succ}_{rate} = \sum_{i=1}^{n} \text{Success}_t^i / n \tag{13}$$

$$\text{z(iter} + 1) = 4 * \text{Succ}_{rate} * (1 - \text{Succ}_{rate}) \tag{14}$$

According to the choice of the inertia weight calculation strategy, adjust and calculate the inertia coefficient by using [25].

$$\text{W} = (\text{w}_1 - \text{w}_2) * (\text{maxiter} - \text{iter})/\text{maxiter} + \text{w}_2 * z(\text{iter} + 1) \tag{15}$$

Now, I use this value of the inertia weight to update the new velocity and of the new position.

Update velocity by using [25]:

$$V_i(\text{iter}+1) = W * V_i(\text{iter}) + C_1 * r_1 * (P_\text{i}(\text{iter}) - X_\text{i}(\text{iter})) + C_2 * r_2 * (G(t) - X_i(\text{iter})) \tag{16}$$

Update position by this equation [25]:

$$X_i(\text{iter}+1) = X_i(\text{iter}) + V_i(\text{iter}+1) \tag{17}$$

Define the new fitness values for the new position [25]

$$\text{Pbest}_\text{t}^\text{t+1} = \begin{cases} \text{Pbest}_\text{i}^\text{t} & f\left(\text{x}_\text{t+1}^\text{i}\right) > \text{Pbest}_\text{t}^\text{i} \\ \text{x}_\text{t+1}^\text{i} & f\left(\text{x}_\text{t+1}^\text{i}\right) \le \text{Pbest}_\text{t}^\text{i} \end{cases} \tag{18}$$

Define the global best by using [25]

$$\text{Gbest} = \min\left\{\text{Pbest}_i^{t+1}\right\} \tag{19}$$

Step 8: Until iter = maxiter, go to step 4. Else print the optimal results.
Step 9_: Display results.

4 Test and Results

4.1 Chaotic Inertia Weight Results

Using the chaotic inertia weight, the set of the tie line becomes 2–13–26–29–33 instead of 33–34-35–36—37 in the base case. Figure 3 shows the new reconfiguration.

In Table 2, a comparative study is done; in this paper, I compare the result of the PSO algorithm by using the chaotic inertia parameter with the study of [24] and the [26] study where the authors solve this same problem by using the minimum spanning tree method.

In Fig. 4, it is clear that the profile voltage is improved compared with the initial case, and the minimum voltage equals to 0.9359 p.u at node 30 instead of 0.8950 p.u at node 18 in the base case. Table 7 in the appendices gives the value of the voltage at each node.

As already described in the previous table, the reconfiguration using the chaotic inertia weight gives the best results, where the losses in this case equal to 0.1005 MW and this value is better than the other study [24] where the losses equal to 0.1241 MW and better than the case of the reconfiguration by using Prim's algorithm [26].

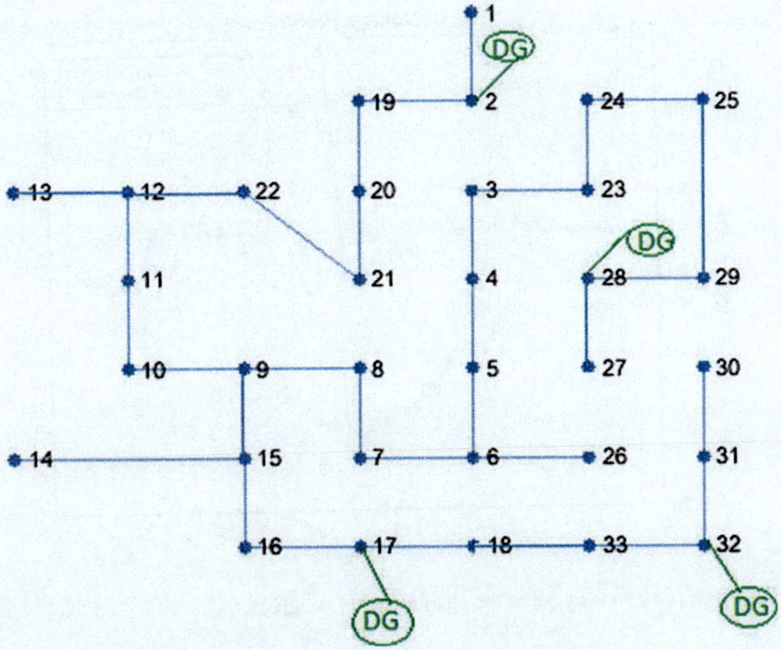

Fig. 3 Network after reconfiguration using the chaotic inertia weight

Table 2 Comparative study

	Base Case with DG	GA with DG [19]	Prim's algorithm with DG [18]	Proposed PSO with DG using the chaotic inertia weight
Tie line	33–34–35–36–37	12–15–18–21–22	12–27–33–34–35	2–13–26–29–33
Node of the minimum voltage	18	–	25	30
Minimum voltage profile (p.u)	0.8950	0.9124	0.95025	0.9359
Total Losses (MW)	0.2465	0.1241	0.1331	0.1005

Concerning the value of the voltage profile at each node, it is noticed that the minimum voltage profile is improved as compared with the base case (0.8950p.u at node 18) and the reconfiguration using GA [24] (0.9124p.u), so the voltage profile at each node is enhanced. In addition, we shouldn't forget to point out that with this strategy we need (46.68 s) to have the solution.

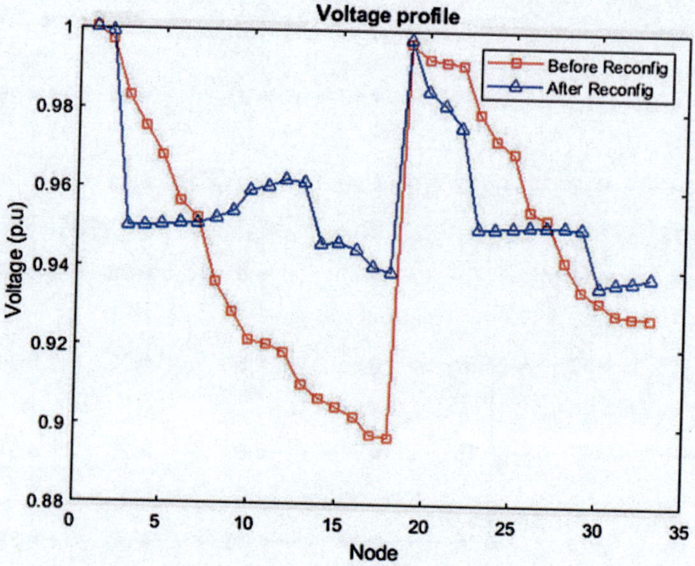

Fig. 4 Voltage profile in the case of the chaotic inertia weight

4.2 *Combination Method Results*

In this case, after reconfiguration of the network by using the combination of success rate and the chaotic inertia weight to adjust the inertia weight parameter aiming to enhance the PSO algorithm, we found the new tie line is 2–22–30–33-34 as shown in Fig. 5.

On other hand and focused on Fig. 6 that gives the value of the voltage profile at each node in two cases (before and after reconfiguration). This figure shows that the minimum voltage profile in this case equals to 0.9115p.u at node 31. Table 7 in appendices gives the value of the voltage at each node for this case.

Table 3 gives the results found by using these two strategies suggested to improve the inertia weight parameter, the other recent studies, and the base case.

Focused on the result given in the previous table, it is clear that the value of the losses in the case when the combination strategy equals to 0.115 MW, this value is lesser than the base case 0.265 MW and the [24] 0.1241 MW and 0.1331 in the case of the reconfiguration by using Prim's algorithm [26]. But compared with the case of the PSO algorithm using the chaotic inertia weight, it is noticed that this last strategy gives a better result than the combination strategies.

Concerning the voltage profile, it is clear that the minimum value of the voltage profile; in this case, it is equal to 0.9115p.u, and this value is better than the base case and is almost similar to the [24], but the other studies give more improved values than this case (0.95025p.u [26] and 0.9359p.u for the case of the PSO using the

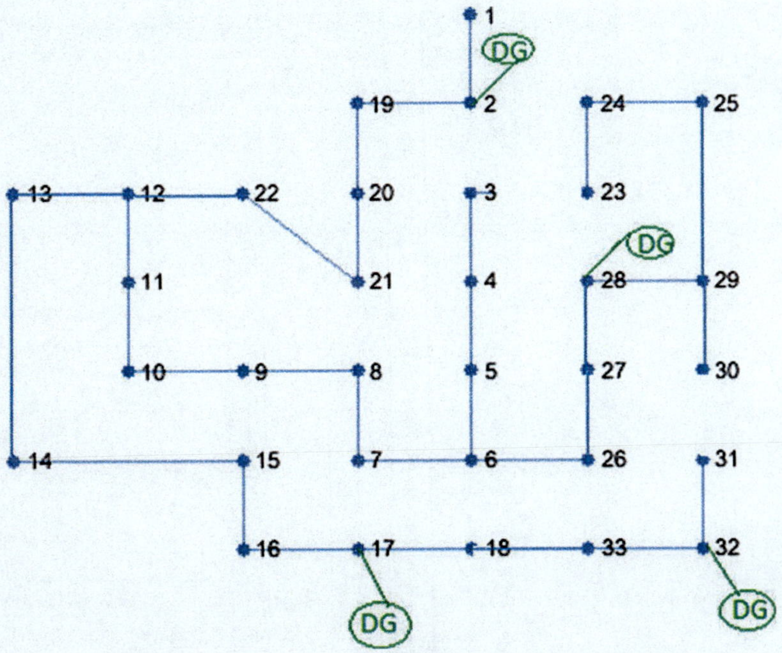

Fig. 5 Network after reconfiguration using the combination strategy

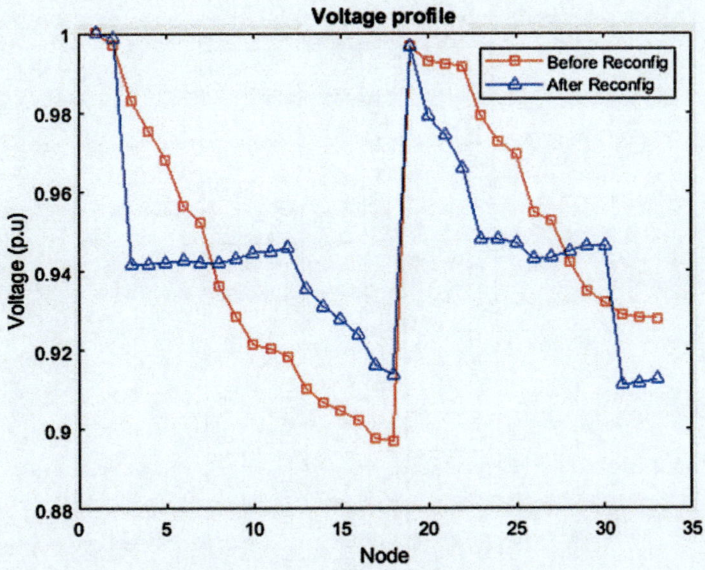

Fig. 6 Voltage profile in the case of the combination strategy

Table 3 Comparative analysis

	Base Case with DG	GA with DG [24]	Prim's algorithm with DG [26]	Proposed PSO with DG using the chaotic inertia weight	Proposed PSO with DG using the combination strategy
Tie line	33–34-35–36–37	12–15–18–21–22	12–27–33–34–35	2–13–26–29–33	2–22–30–33-34
Node of the minimum voltage	18	–	25	30	31
Minimum voltage profile (p.u)	0.8950	0.9124	0.95025	0.9359	0.9115
Total losses (MW)	0.2465	0.1241	0.1331	0.1005	0.115

chaotic inertia weight). It is interesting to point out that this strategy takes (55.178 s) to execute and give the result.

5 Conclusion

Aiming to check the condition of the generation following the consumption, several studies are interested in using the reconfiguration of the radial distribution system to reduce the losses in the electrical system. So, the main objective of this study is to find the new reconfiguration of the network by using two kinds of inertia weight (the chaotic inertia weight and the hybrid of the chaotic inertia weight and the success rate) to improve the results found by the PSO algorithm used in other recent studies.

To perform the reliability of these strategies, I select to test these strategies in the case of IEEE 33 bus with the presence of the DGs. And a comparative study is done to compare the results found by this paper with other recent studies. In the end, using these strategies helps to enhance the network; also, the reconfiguration using the chaotic inertia weight to adjust the PSO algorithm gives an improved result than the combination strategy.

In the next research, it seems interesting to encourage the authors to do the next search about using the PSO algorithm to find the optimal allocation and sizing of DGs to improve the reconfiguration and reduce losses in the network.

Appendix

See Tables 4, 5, 6, and 7

Table 4 Set of the loops for IEEE 33 bus [8]

Loops	Dimensions	Switches
1	Sd1	8–9–10–11–21–33–35
2	Sd2	2–3–4–5–6–7–18–19–20
3	Sd3	12–13–14–34
4	Sd4	15–16–17–29–30–31–36–32
5	Sd5	22–23–24–25–26–27–28–37

Table 5 Parameters of the proposed PSO

Parameter	Value
C_1	2 * rand(1)
C_2	2 * rand(1)
W_{max}	0.9
W_{min}	0.4
Population size	20
Dimension of search space	5
Maximum iteration	100

Table 6 Line and load data of IEEE 33 bus [22]

			Line data		Load data	
Branch N°	From bus	To bus	R (ohm)	X(ohm)	Pl(Kw)	Ql(kvar)
1	1	2	0.0922	0.047	100	60
2	2	3	0.493	0.2511	90	40
3	3	4	0.366	0.1864	120	80
4	4	5	0.3811	0.1941	60	30
5	5	6	0.819	0.707	60	20
6	6	7	0.1872	0.6188	200	100
7	7	8	0.7114	0.2351	200	100
8	8	9	1.03	0.74	60	20
9	9	10	1.04	0.74	60	20
10	10	11	0.1966	0.065	45	30
11	11	12	0.3744	0.1238	60	35
12	12	13	1.468	1.155	60	35
13	13	14	0.5416	0.7129	120	80
14	14	15	0.591	0.526	60	10
15	15	16	0.7463	0.545	60	20
16	16	17	1.289	1.721	60	20
17	17	18	0.732	0.574	90	40
18	2	19	0.164	0.1565	90	40
19	19	20	1.5042	1.3554	90	40
20	20	21	0.4095	0.4784	90	40
21	21	22	0.7089	0.9373	90	40
22	3	23	0.4512	0.3083	90	50
23	23	24	0.898	0.7091	420	200
24	24	25	0.896	0.7011	420	200
25	6	26	0.203	0.1034	60	25

(continued)

Table 6 (continued)

			Line data		Load data	
26	26	27	0.2842	0.1447	60	25
27	27	28	1.059	0.9337	60	20
28	28	29	0.8042	0.7006	120	70
29	29	30	0.5075	0.2585	200	600
30	30	31	0.9744	0.963	150	70
31	31	32	0.3105	0.3619	210	100
32	32	33	0.341	0.5302	60	40
Tie lines						
33	8	21	2	2		
34	9	15	2	2		
35	12	22	2	2		
36	18	33	0.5	0.5		
37	25	29	0.5	0.5		

Table 7 Voltage Profile in the case of using the chaotic inertia weights and the combination strategy

Bus	The chaotic inertia weight parameter	The combination strategy
1	1	1
2	0.999172639148097	0.998909397340444
3	0.950119615573435	0.941816046915740
4	0.950197864595776	0.941915894499951
5	0.950384769075342	0.942172719114258
6	0.950736541598505	0.942745626162071
7	0.950801915260580	0.942272190455390
8	0.952178374115133	0.942044120286597
9	0.953986728146658	0.943153382841413
10	0.959346721345702	0.944666257465305
11	0.960258850341073	0.945036272631173
12	0.961996158279058	0.945788205077869
13	0.961283469701170	0.935454716873337
14	0.945915236387610	0.931030577352812
15	0.946340607670596	0.927628810209586
16	0.944554248961896	0.923934940402118
17	0.940230512256466	0.916188864414032
18	0.939000617939255	0.913919720659541
19	0.997624030527326	0.996806758504366
20	0.984803200597631	0.979011354882791
21	0.981438432428948	0.974175549465857

(continued)

Table 7 (continued)

Bus	The chaotic inertia weight parameter	The combination strategy
22	0.975847552100016	0.965979448836100
23	0.950113458716196	0.948086801989005
24	0.950239293567017	0.947921749672686
25	0.950513319395836	0.947084596452350
26	0.950698911078949	0.942977540694717
27	0.950729767133923	0.943325296224972
28	0.950681330399383	0.945007617974172
29	0.950602669016140	0.946348853096167
30	0.935947175640918	0.946457584854114
31	0.936910083625147	0.911254236300635
32	0.937366905217155	0.911662243051541
33	0.938123317946990	0.912578877413074

References

1. Peng, F.Z.: Editorial Special Issue on Distributed Power Generation. IEEE Trans. Power Electron. **19**(5), 2 (2004)
2. Carreno, E.M., Romero, R., Padilha-Feltrin, A.: An efcient codifcation to solve distribution network reconfguration for loss reduction problem. IEEE Trans. Power Syst. **23**(4), 1542–1551 (2008)
3. Salama, M.M.A., El-Khattam, W.: Distributed generation technologies, defnitions and benefts. Electric Power Systems Research **71**(2), 119–128 (2004)
4. Multon, B.: L'énergie électrique: analyse des resources et de la production.*Journées de la section électrotechnique du club EEA* (1999)
5. Strasser, T., Andrén, F., Kathan, J., Cecati, C., Buccella, C., Siano, P., Leitão, P., Zhabelova, G., Vyatkin, V., Vrba, P., Mařík, V.: A Review of Architectures and Concepts for Intelligence in Future Electric Energy Systems. IEEE Trans. Industr. Electron. **62**(4), 2424–2438 (2014)
6. Caire, R.: Gestion de la production décentralisée dans les réseaux de distribution.Institut National Polytechnique de Grenoble, tel-00007677 (2004)
7. Xie, L., Ilic, M.D.: Model predictive dispatch in electric energy systems with intermittent resources. In IEEE International Conference on Systems, Man and Cybernetics (2008)
8. JUMA, S.A.: Optimal radial distribution network reconfiguration using modified shark smell optimization. (2018) http://hdl.handle.net/123456789/4854
9. Sivkumar, M.: A Simple Algorithm for Distribution System Load Flow with Distributed Generation, In IEEE International Conference on Recent Advances and Innovations in Engineering, Jaipur, India (2014)
10. Gallego, LA, Carreno E., Padilha-Feltrin, A.: Distributed generation modeling for unbalanced three-phase power flow calculations in smart grids. In *Transmission and Distribution Conference and Exposition: Latin America (T&D-LA)* (2010)
11. Chidanandappa, R., Ananthapadmanabha, T.:Genetic algorithm based network reconfiguration in distribution systems with Multiple DGs for time varying loads. SMART GRID Techno. **21**, 460–467 (2015)
12. Ogunjuyigbe, A., Ayodele, T., Akinola, O.: Impact of distributed generators on the power loss and voltage profile of sub-transmission network. J. Electr. Syst. Inf. Technol. **3**, 94–107 (2016)
13. Ahmad, S., Asar, A.U., Sardar, S., Noor, B.: Impact of distributed generation on the reliability of local distribution system. IJACSA **8**(6), 375–382 (2017)

14. Ma, C., Li, C., Zhang, X., Li, G., Han, Y.: Reconfiguration of distribution networks with distributed generation using a dual hybrid particle swarm optimization algorithm.Hindawi Math. Probl. Eng. **2017**, 11 (2017)
15. Sudhakara Reddy, A.V., Damodar Reddy, M.: "Optimization of network reconfiguration by using particle swarm optimization. In 1st IEEE International Conference on Power Electronics, Intelligent Control and Energy Systems (ICPEICES) (2016)
16. Tandon, A., Saxena, D.: Optimal reconfiguration of electrical distribution network using selective particle swarm optimization algorithm. In International Conference on Power, Control and Embedded Systems (2014)
17. Inji Ibrahim Atteya: Hamdy Ashour, Nagi Fahmi and Danielle Strickland, Radial distribution network reconfiguration for power losses reduction using a modified particle swarm optimisation, . Open Access Proceedings J. **2017**(1), 2505–2508 (2017)
18. Bansal, J.C., Singh, P.K., Saraswat, M., Verma, A., Jadon, S.S., Abraham, A.: Inertia weight strategies in particle swarm optimization. In Third World Congress on Nature and Biologically Inspired Computing (2011).
19. Arasomwan, A.M., ADEWUMI, A.O.: On adaptive chaotic inertia weights in particle swarm optimization. In IEEE Swarm Intelligence Symposium (2013)
20. Enacheanu, F.: outils d'aide à la conduite pour les opérateurs des réseaux de distribution (2008).https://tel.archives-ouvertes.fr/tel-00245652
21. Sharma, A., Saini, M., Ahmed, M.: Power flow analysis using NR method. In International Conference on Innovative Research in Science, Technoloy and Management, Kota, Rajasthan, India (2017)
22. Baran, M.E., Wu, F.F.: Network reconfiguration in distribution systems for loss reduction and load balancing. IEEE Trans. Power Delivery **4**(2), 1401–1407 (1989)
23. Jangjoo, M.A., Seifi, A.R.: Optimal voltage control and loss reduction in microgrid by active and reactive power generation.J. Intell. & Fuzzy Syst., **27**, 1649–1658 (2014)
24. Moarrefi, H., Namatollahi, M., Tadayon, M.: Reconfiguration and distributed Generation(DG) placement considering critical system condition. In 22nd International Conference and Exhibition on Electricity Distribution (2013)
25. Kennedy, J., Eberhart, R.: Particle swarm optimization. In International Conference on Neural Networks (1995)
26. M'dioud, M., ELkafazi, I., Bannari, R.: An improved reconfiguration of a radial distribution network by using the minimum spanning tree algorithm. Solid State Technol. **63**(6), 9178–9193 (2020)

On the Performance of 5G Narrow-Band Internet of Things for Industrial Applications

Abdellah Chehri, Hasna Chaibi, Rachid Saadane, El Mehdi Ouafiq, and Ahmed Slalmi

Abstract Manufacturing industry is continuously evolving since the very beginning of the industrial era. This modernization is undoubtedly the outcome of continuous new technology development in this field, which has kept the industries on the verge, looking for new methods for improving productivity enhancement and better operational efficiency. The advent of 5G will provide the world of industry to connect its infrastructures to digitize people and machines to optimize production flows. Narrow-band-IoT addresses "Massive IoT" type use cases, which involve deploying a large energy-efficient quantity. These low-complex objects do not need to communicate very frequently. 5G will provide the ability to develop new uses previously impossible or complex to implement. Consequently, it will complement the range of network solutions already in place in the company, giving it the keys to accelerating its transformation. This paper evaluates the 5G-NR-based IoT air interface with the FEC with industrial channel models. Low-density parity-check (LDPC), polar, turbo code, and TBCC are assumed.

1 Introduction

At this time, the 4G cellular networks have existed for several years. It is time to look forward and see what the future will bring regarding the next generation of cellular networks: the fifth generation, most often referred to as 5G [1, 2].

A. Chehri
University of Quebec in Chicoutimi, 555, Boul. de L'Université, G7H 2B1 Saguenay, QC, Canada
e-mail: achehri@uqac.ca

H. Chaibi
GENIUS Laboratory, SUP MTI, 98, Avenue Allal Ben Abdellah, Hassan-Rabat, Morocco

R. Saadane (✉) · E. M. Ouafiq
SIRC/LaGeS-EHTP, EHTP, Km 7 Route, El Jadida 20230, Morocco

A. Slalmi
Ibn Tofail University, Kenitra, Morocco

M. Ben Ahmed et al. (eds.), *Networking, Intelligent Systems and Security*,
Smart Innovation, Systems and Technologies 237,
https://doi.org/10.1007/978-981-16-3637-0_19

Today, "everyone" is already online, and with the emerging Internet of Things, everything will also be online—everywhere, and always. There is a demand for ubiquitous access to mobile services.

The "Internet of Things," which must also be counted among the most significant technological trends, is often mentioned in conjunction with M2M communication. The risk of confusion is high here: Even if the two approaches overlap, they are still two different things. What they have in common is the goal of automated data exchange between devices. IIoT, or the "Industrial Internet of Things," is mainly aimed at private users. There are various carrier networks in the area of M2M communication, i.e., cellular radio or GPRS, which are an option [3].

Furthermore, classic M2M communication involves point-to-point applications. On the other hand, in the context of the IoT, a standardized, open approach is practiced. Ultimately, however, it is already foreseeable that both technologies will converge, complement each other, and one day may no longer be separable. For example, many M2M providers have already started to integrate cloud functions into their offerings [4, 5].

3G and 4Gs important goal were to achieve constant coverage for the same services in both outdoor and indoor scenarios. According to Chen and Zhao, 5G will be a heterogeneous framework, and backward compatibility will not be mandatory indoors and outdoors [6]. The improvement of the user equipment is expected to provide the ability to support simultaneous connections, both indoors and outdoors.

The advent of 5G will provide the world of industry with the means to connect its infrastructures to digitize people and machines to optimize production flows.

5G must provide a unified vision of connectivity within the enterprise regardless of the network. The core network is designed to encompass all types of access to natively. However, the arrival of technology does not mean the "end" of other systems; on the contrary. Each has its characteristics and specific uses, such as NB-IoT, a global industry standard, open, sustainable, and scalable. It complements the other technologies already defined, such as Sigfox and LoRA; NB-IoT addresses "Massive IoT" type use cases, which involve deploying a large quantity of energy-efficient. These low-complex objects do not need to communicate very frequently. 5G will provide the ability to develop new uses previously impossible or complex to implement. Consequently, it will complement the range of network solutions already in place in the company, giving it the keys to accelerating its transformation.

IoT requires massive connectivity where several low-cost devices and sensors communicate.

The deployment of wireless technology in wider and more critical industrial applications requires deterministic behavior, reliability, and predictable latencies to integrate the industrial processes more effectively. Real-time data communication and information reliability in the wireless channels are some of the major concerns of the control society regarding NB-IoT, and hence, suitable improvements in NB-IoT are required to ensure desired reliability and time sensitivity in emergency, regulatory, and supervisory control systems.

This is being labeled as the fourth industrial revolution or Industry 4.0. There are many advantages brought by 5G cutting-edge technologies for industrial automation scenarios in the drive for industry 4.0

This paper is organized as follows. Section 2 describes the prominent 5G use cases. Section 3 introduces terminology and description of the Industrial Internet of Things, the main pillar of Industry 4.0. Section 4 describes the 5G NR (New Radio) interface. The performance of 5G narrow-band Internet of Things for industrial applications is given in Sect. 5. Finally, Sect. 6 presents our conclusions.

2 5G Use Cases

To give the reader an idea about which communication challenges 5G is expected to solve, the most important use cases will be presented. Use cases are descriptions of interactions between an actor, typically the user, and a system, to achieve a goal. They also make it clearer to understand the background for the 5G requirements, and they are likely to be the driver for the 5G technology that will be developed [7].

Some of the 5G applications will be old and familiar while introducing new and more diverse services is also expected. The existing dominating human-centric communication scenarios will be complemented by an enormous increase in communication directly between machines or fully automated devices. Hence, the use cases for 5G may be split into two following main categories: mobile broadband and the Internet of Things (IoT).

According to forecasts, the growth in data traffic volumes will be exponential for the next years. In the next decade, the total volume of mobile traffic is expected to increase a thousand times the traffic volume today, mostly due to the increasing number of connected devices.

Compared to the mobile broadband use case, the IoT category covers the use cases where devices and sensors communicate directly with each other, without a user being present at all times. This is referred to as machine-to-machine (M2M) communication [8], and this type of communication will be an essential part of the emerging IoT. Here, devices and items embedded with sensors, electronics, software, and network connectivity collect and exchange data. Sensors or devices for M2M communications will be integrated daily using objects such as cars, household appliances, textiles, and health-critical appliances [9].

Various standardization organizations have technical working groups responsible for machine-to-machine communication (M2M) and the Internet of the future. The third-generation partnership project (3GPP) deals with M2M communication under the term machine-type communication (MTC) and has started to standardize it in the 3GPP specification Release 10. In Release 11, some of the proposed suggested functions address and control the devices [10].

The Internet Engineering Task Force (IETF) oversees many of the necessary protocols in the Internet of Things. These include tried and tested protocols such as IP, TLS, and HTTP, which are widespread and ensure interoperability.

However, newer protocols have also been developed to consider the changed conditions in M2M communication, such as the constrained application protocol (CoAP) and the communication protocol IPv6 over low-power wireless personal area network (6LoWPAN). Other necessary protocols in M2M communication are message protocols such as Extensible Messaging and Presence Protocol (XMPP), MQ Telemetry Transport (MQTT), and Advanced Message Queuing Protocol (AMQP). Other protocols that enable the management of the devices, such as device management (DM) and lightweight (LW) M2M from Open Mobile Alliance (OMA) and TR-069 from Broadband Forum (BBF), were also proposed [11].

To ensure and develop M2M standards, the European Telecommunications Standards Institute (ETSI) founded a technical committee in 2009. The requirements were defined, which in addition to security and communication management, also the functional requirements a horizontal platform for M2M communication. This platform should ensure that communication with a wide variety of sensors and actuators is possible in a consistent manner for different applications.

3 Industrial Internet of Things (IIoT)

IIoT is a variant of the IoT that is used in the industrial sector. The Industrial Internet of Things can be used in many industries, in manufacturing, in agriculture, in hospitals, in institutions, in the field of health care, or the generation of energy and resources. One of the most critical aspects is improving operational effectiveness through intelligent systems and more flexible production techniques [12–15].

With IIoT, industrial plants or gateways connected to them send data to a cloud. Gateways are hardware components that connect the devices of the industrial plants and the sensors to the network. The data is processed and prepared in the cloud. This allows employees to monitor and control the machines remotely. Besides, an employee is informed if, for example, maintenance is necessary [16].

The Industrial Internet of Things also uses object-oriented systems connected to the network and can interact with each other. These devices are equipped with sensors whose role is to collect data by monitoring the production context in which they operate. The information stored in this way is then analyzed and processed, helping to optimize business processes.

From predictive maintenance to the assembly line, the Industrial Internet of Objects (IIoT) offers a significant competitive advantage regardless of the industry.

Sensors play a central role in IIoT. They collect vast amounts of data from different machines in one system. To be able to meet this challenge, the sensors in the IIoT area must be significantly more sensitive and precise than in the IoT environment. Even minor inaccuracies in the acquisition of the measurement data can have fatal consequences such as financial losses. The IIoT offers numerous advantages for industry:

1. Production processes can be automated.

2. Processes can be adapted flexibly and in real-time to changing requirements.
3. Machines recognize automatically when they need to be serviced.
4. Some of the maintenance can be carried out by the machines themselves.
5. Disturbances and production interruptions are minimized.
6. Throughput and production capacity increase.

The IIoT also brings some challenges that need to be addressed in the future intensive maintenance of the machines in terms of software and firmware. Both have to be kept up to date to close security gaps or prevent them from arising in the first place.

Secure transmission of the encrypted data to the cloud is a prerequisite. Otherwise, it will be easy for hackers to get hold of sensitive data.

So far, IIoT devices from different manufacturers are not compatible with each other. The reason is the lack of uniform standards and manufacturer-specific protocols.

High effort for the processing, protection, and storage of the data. In order to cope with the substantial amounts of data, applications, and databases from the big data area have to be used.

NB-IoT (for narrow-band IoT) is a serious asset when the goal is energy use. This infrastructure technology could provide ten years of autonomy to a 5Wh battery through energy optimization and ambient energy recovery. NB-IoT operates a 200 kHz frequency band and is used for fixed sensors that do not need a small volume of data like water or electricity meters.

LTE-M technology is widely used in mobile telephony since it is compatible with existing networks and does not require new modems. Its transfer rate is high, LTE-M will also evolve in environments such as remote monitoring or autonomous vehicles. Another advantage: LTE-M supports voice exchanges and the mobility of objects.

4 5G NR Radio Interface

With the advent of the IoT, the issues related to Industry 4.0, and experts' prediction to have more than 75 billion objects connected using a wireless network by 2025, it is necessary to create technologies adapted to these new needs. This standard allows connected objects to communicate large volumes of data over very large distances with very high latency.

NB-IoT or narrow-band IoT, or LTE-M2M is low consumption and long-range technology (LPWAN) validated in June 2016, operating differently.

Like LoRa and Sigfox, this standard allows low-power objects to communicate with external applications through the cellular network.

The communication of these objects via NB-IoT is certainly not real-time but must be reliable over time. By relying on existing and licensed networks, operators are already in charge of their quality of service. They will thus be able to guarantee a quality of service (QoS) sufficient for this operation type.

NB-IoT builds on existing 4G networks from which several features and mechanisms are inherited. It is therefore compatible with international mobility thanks to roaming, also called roaming. This also means that these networks are accessible under license and are managed by operators specialized in the field. Experts, therefore, manage the quality of the network in the area.

NB-IoT is considered 5G ready, which means that it will be compatible with this new transmission standard when it is released.

For NR, the relevant releases are Release 14, 15. In Release 14, a number of preliminary activities were done to prepare for the specification of 5G. For instance, one study was carried out to develop propagation models for spectrum above 6 GHz. Another study was done on scenarios and requirements for 5G and concluded at the end of 2016. Besides, a feasibility study was done of the NR air interface itself, generating several reports covering all aspects of the new air interface. Rel'15 will contain the specifications for the first phase of 5G.

NR DownLink (DL) and UpLink (UL) transmissions are organized into frames. Each frame lasts 10 ms and consists of 10 subframes, each of 1 ms. Since multiple OFDM numerologies are supported, each subframe can contain one or more slots.

There are two types of cyclic prefix (CP): normal CP, each slot conveys 14 OFDM symbols. Extended CP, each slot shares 12 OFDM symbols. Besides, each symbol can be assigned for DL or UL transmission, according to the slot format indicator (SFI), which allows flexible assignment for TDD or FDD operation modes. In the frequency domain, each OFDM symbol contains a fixed number of sub-carriers.

One sub-carrier allocated in one OFDM symbols is defined as one resource element (RE). A group of 12 RE is defined as one resource block (RB). The total number of RBs transmitted in one OFDM symbol depends on the system bandwidth and the numerology. NR supports scalable numerology for more flexible deployments covering a wide range of services and carrier frequencies. It defines a positive integer factor m that affects the sub-carrier spacing (SCS), the OFDM symbol, and cyclic prefix length.

A small sub-carrier spacing has the benefit of providing a relatively long cyclic prefix in absolute time at a reasonable overhead. In contrast, higher sub-carrier spacings are needed to handle, for example, the increased phase noise at higher carrier frequencies [17]. Note that the sub-carrier spacing of 15, 30, and 60 kHz wide apply to carrier frequencies of 6 GHz or lower (sub-6), while the sub-carrier spacing of 60, 120, and 240 kHz apply to above 6 GHz carrier frequencies [18].

An NB-IoT channel is only 180 kHz wide, which is very small compared to mobile broadband channel bandwidths of 20 MHz. So, an NB-IoT device only needs to support the NB-IoT part of the specification. Further information about the specification of this category can be found in the 3GPP technical report TR 45.820: Cellular system supports for ultralow complexity and low throughput Internet of Things [19] (Fig. 1).

The temporal and frequency resources which carry the information coming from the upper layers (layers above the physical layer) are called physical channels [20].

There are several physical channels to specify for the uplink and downlink:

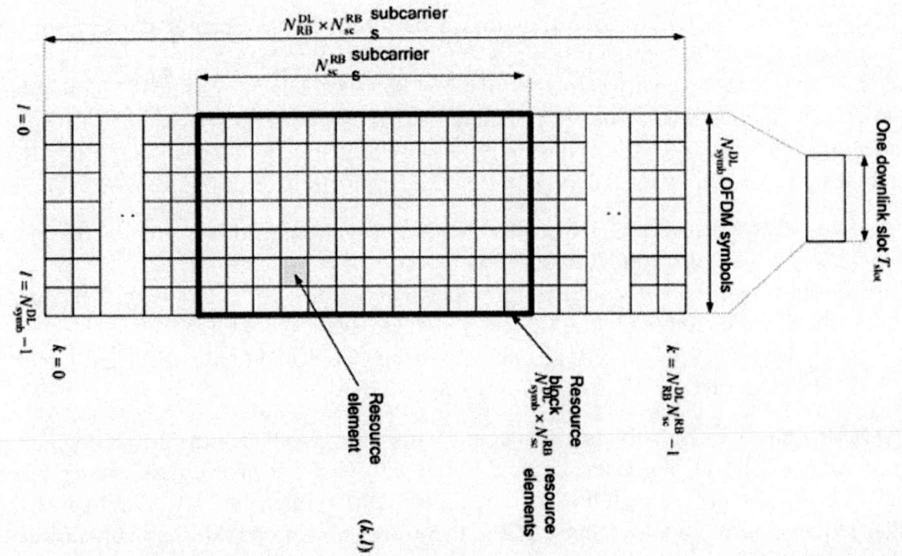

Fig. 1 NR framing structure

1. Physical downlink shared channel (PDSCH): used for downlink data transmission.
2. Physical downlink control channel (PDCCH): used as a downlink for information control, which includes the scheduling decisions required for the reception of downlink data (PDSCH) and for scheduling granting authorization to transmit data uplink (PUSCH) by a UE.
3. Physical broadcast channel (PBCH): used for broadcasting system information required by a UE to access the network;
4. Physical uplink shared channel (PUSCH): used for uplink data transmission (by a UE).
5. Physical uplink control channel (PUCCH): used for uplink control information, which includes: HARQ acknowledgment (indicating whether a downlink transmission was successful or not), schedule request (request network time-frequency resources for uplink transmissions), and downlink channel status information for link adaptation.
6. Physical random-access channel (PRACH), used by a UE to request the establishment of a connection called random access.

When a symbol is sent through a physical channel, a delay created by the propagation signal taking different paths may cause the reception of several copies of the same frame. A cyclic prefix is added to each symbol to solve this problem, consisting of samples taken from its end and tied at its beginning. The goal is to add a guard time between two successive symbols. If a CP's length is longer than the maximum channel propagation, there will be no inter-symbol interference (ISI), which means that two successive symbols will not interfere. It also avoids inter-carrier interference

(ICI), which causes the loss of orthogonality between the sub-carriers. It uses a copy of the last part of the symbol that plays a guard interval [21].

5G NR can use a spectrum from 6 GHz to 100 GHz. The 5G system's bandwidth is increased by ten times (from 100 MHz in LTE-A to 1 GHz +) compared to LTE-A technology. Bands for NR are basically classified as low, middle, and high bands, and these bands can be used depending on the application described below:

1. Low bands below 1 GHz: most extended range, e.g., mobile broadband and massive IoT, e.g., 600, 700, 850/900 MHz
2. Medium bands 1 GHz to 6 GHz: wider bandwidths, for example, eMBB and critical, for example, 3.4–3.8 GHz, 3.8–4.2 GHz, 4.4–4.9 GHz
3. High bands above 24 GHz (mm-Wave): extreme bandwidths, for example, 24.25–27.5 GHz, 27.5–29.5, 37–40, 64–71 GHz.

OFDM has 15 kHz sub-carrier spacing with a 7% (4.69 μs) cyclic prefix. The numerology for LTE has been specified after an extensive investigation in 3GPP. For NR, it was easy for 3GPP to aim for OFDM numerology similar to LTE-like frequencies and deployments. 3GPP, therefore, considered different sub-carrier spacing options near 15 kHz as basic numerology for NR. There are two important reasons to keep LTE numerology as the base numerology:

1. Narrow-band-IoT (NB-IoT) is a new radio access technology (already deployed since 2017) supporting massive machine-type communications. NB-IoT provides different deployments, including in-band deployment within an LTE operator enabled by the selected LTE numerology. NB-IoT devices are designed to operate for ten years or more on a single battery charge. Once such an NB-IoT device is deployed, the incorporating carrier will likely be rearmed to NR during the device's life.
2. NR deployments can take place in the same band as LTE. With an adjacent LTE TDD carrier, the network controller should adopt the same uplink/downlink switching model as the LTE TDD protocol. Each numerology where (an integer multiple of) a subframe is 1 ms can be aligned to regular subframes in LTE. In LTE, duplex switching occurs in special subframes. To match the direction of transmission in special subframes, the same numerology as in LTE is required. This implies the same sub-carrier spacing (15 kHz), the same OFDM symbol duration (66.67 μs), and the same cyclic prefix (4.69 μs)

5 Performance of 5G Narrow-Band Internet of Things for Industrial Applications

A particular emphasis has been done on the design of forwarding error correction (FEC) solutions to support the underlying constraints efficiently. In this regard, it has been taken into account the codes used in 5G-NR (Rel' 15) channels, low-density parity-check (LDPC) for data and polar code for control, and those code (TBCC) for control. This scenario's potential requirements are the use of lower-order modulation schemes with shorter block size information to satisfy low-power requirements.

The advanced channel coding schemes with robust error protection with low-complexity encoding and decoding is preferred. The candidate coding scheme for the next 5G based IoT system is: polar code, low-density parity-check (LDPC), turbo code, and tail-biting convolutional code (TBCC) [22].

In the time domain, physical layer transmissions are organized into radio frames. A radio frame has a duration of 10 ms. Each radio frame is divided into ten subframes of 1 ms duration. Each subframe is then divided into locations [23].

This section evaluates the 5G-NR-based IoT air interface with the FEC schemes previously and with industrial channel models.

The large-scale wireless channel characteristics were evaluated from 5 to 40 GHz frequency for industrial scenario (Fig. 2).

Since the data traffic generated by IoT applications requires a small volume, it has been considered a data bit range from 12 up to 132 with a step of 12. The segmentation (and de-segmentation) block has not been considered due to the small data packet to transmit. Furthermore, from upper layers is randomly generated and does not refer to a specific channel. LDPC, polar, turbo code, and TBCC are assumed.

3GPP agreed to adopt polar codes for the enhanced mobile broadband (eMBB) control channels for the 5G NR (new radio) interface. At the same meeting, 3GPP agreed to use LDPC for the corresponding data channel [24].

The polar code keeps performing better than other codes, achieving 3 dB, with respect to the turbo code (Fig. 3).

6 Conclusion

Modularity, flexibility, and adaptability of production tools will be the rule in Industry 4.0. 5G will allow the integration of applications that leverage automation and robotization. The optimized flow rates and the ability to integrate numerous sensors to ensure preventive and predictive maintenance of production tools constitute a prospect of increasing the reliability of Industry 4.0. The consolidation of industrial wireless communication into standards leads to an increase in deployments throughout various industries today. Despite the technology being considered mature, plant operators are reluctant to introduce mesh networks into their process, despite their very low energy profiles. While very promising, 5G will not take hold quickly, with its high costs slowing mass distribution. Between its business model, the price of the connection, and the cost of electronics, it will take a few years to see it flourish everywhere.

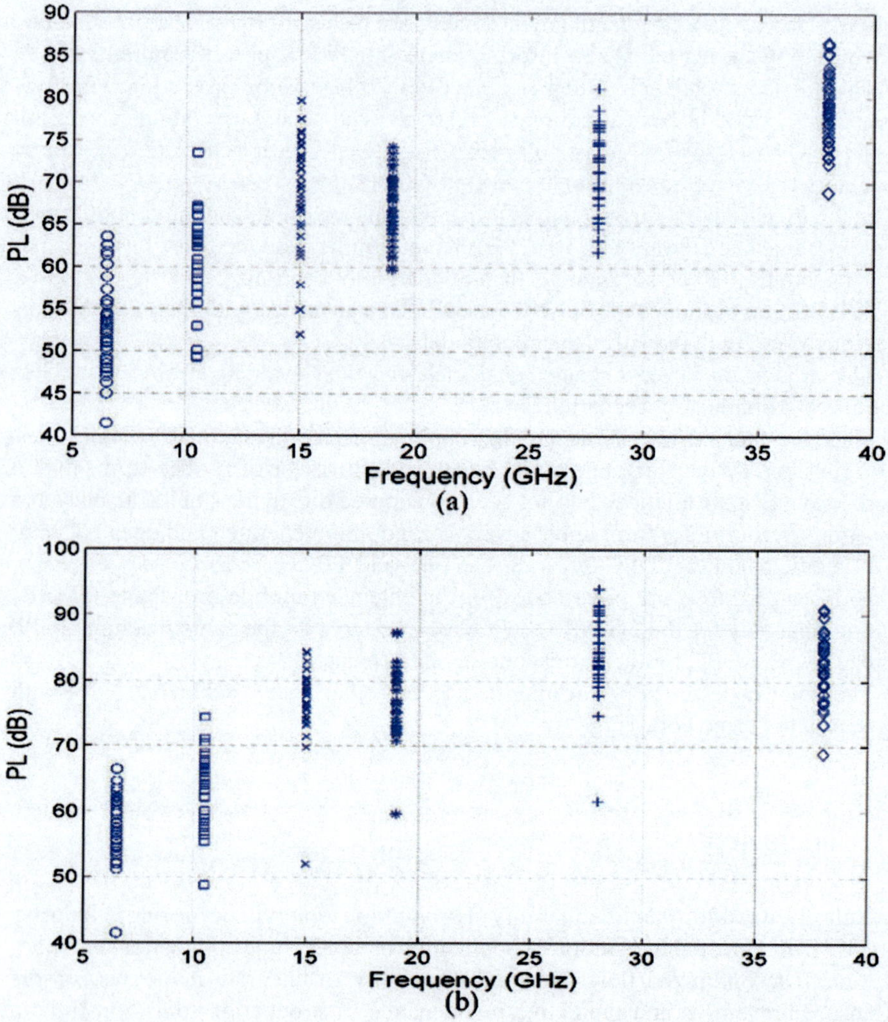

Fig. 2 Path loss versus frequency for V-V and V-H polarization for indoor channels of mm-wave bands [25]

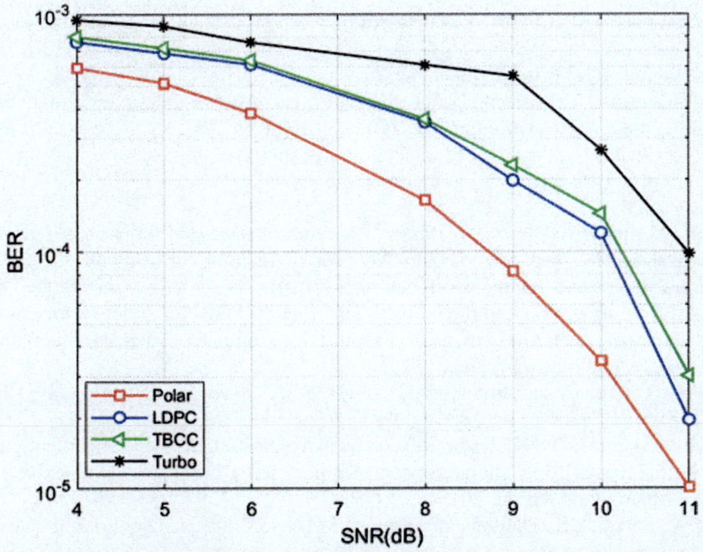

Fig. 3 BER vs. SNR for different FEC techniques

References

1. GPP: Release 15. Retrieved March 2, 2017. http://www.3gpp.org/release-15
2. Carlton, A.: 5g reality check: Where is 3gpp on standardization? Retrieved March 18th, 2017
3. Slalmi, A., Chaibi, H., Saadane, R., Chehri, A., Jeon, G.: 5G NB-IoT: Efficient Network Call Admission Control in Cellular Networks. In: concurrency and computation: practice and experience. Wiley, e6047. https://doi.org/10.1002/cpe.6047
4. Slalmi, A., Chaibi, H., Chehri, A., Saadane, R., Jeon, G., Hakem, N.: On the ultra-reliable and low-latency communications for tactile internet in 5G era. In: 24th International Conference on Knowledge-Based and Intelligent Information & Engineering Systems, Verona, Italy, 16–18 September 2020
5. Slalmi A., Saadane R., Chehri A., Kharraz H.: How Will 5G Transform Industrial IoT: Latency and Reliability Analysis. In: Zimmermann A., Howlett R., Jain L. (eds) Human Centred Intelligent Systems. Smart Innovation, Systems and Technologies, vol 189. Springer, Singapore (2020)
6. Chen, S., Zhao, J.: The requirements, challenges, and technologies for 5 g of terrestrial mobile telecommunication. IEEE Comm. Magazine **52**(5), 36–43 (2014)
7. Nokia: White paper: 5 g use cases and requirements (2014)
8. Sabella, A., Wbben, D.: Cloud technologies for flexible 5 g radio access networks. IEEE Communications Magazine **52**(5), 68–76
9. Tehrani, M.N., Uysal, M., Yanikomeroglu, H.: Device-to-device communication in5g cellular networks: Challenges, solutions and future directions. IEEE Communications Magazine **52**(5), 86–92 (2014)
10. Kunz, A., Kim, H., Kim, L., Husain, S.S.: Machine type communications in 3GPP: From release 10 to release 12. 2012 IEEE Globecom Workshops, Anaheim, CA, pp. 1747–1752 (2012). https://doi.org/10.1109/GLOCOMW.2012.6477852
11. Wahle, S., Magedanz T., Schulze, F.: The OpenMTC framework—M2M solutions for smart cities and the internet of things. IEEE International Symposium on a World of Wireless, Mobile and Multimedia Networks (WoWMoM), San Francisco, CA (2012)

12. Slalmi, A., Saadane, R., Chehri, A.: Energy Efficiency Proposal for IoT Call Admission Control in 5G Network. In: 15th International Conference on Signal Image Technology & Internet Based Systems, Sorrento (NA), Italy, November 2019
13. Chehri, A., Mouftah, H.: An empirical link-quality analysis for wireless sensor networks. Proc. Int. Conf. Comput. Netw. Commun. (ICNC), 164–169 (2012)
14. Chehri, A., Chaibi, H., Saadane, R., Hakem, N., Wahbi, M.: A framework of optimizing the deployment of IoT for precision agriculture industry, vol 176, 2414–2422 (2020). ISSN 1877-0509, KES 2020
15. Chehri, A.: The industrial internet of things: examining how the IIoT will improve the predictive maintenance. Ad Hoc Networks, Lecture Notes of the Institute for Computer Sciences, Smart Innovation Systems and Technologies, Springer (2019)
16. Chehri, A.: Routing protocol in the industrial internet of things for smart factory monitoring: Ad Hoc networks, Lecture Notes of the Institute for Computer Sciences, Smart Innovation Systems and Technologies, Springer (2019)
17. GPP, 5G NR; Overall Description; Stage-2, 3GPP TS 38.300 version 15.3.1 Release 15, October 2018
18. GPP TS 38.331 v.15.1.0: NR. Radio Resource control (RRC), Protocol Specification, 2015
19. GPP. TS 45.820 v2.1.0: Cellular System Support for Ultra Low Complexity and Low Throughput Internet of Things, 2015
20. Furuskär A., Parkvall, S., Dahlman, E., Frenne, M.: NR: The new 5G radio access technology. IEEE Communications Standards Magazine (2017)
21. Chehri, A., Mouftah, H.T.: New MMSE downlink channel estimation for Sub-6 GHz non-line-of-sight backhaul. In: 2018 IEEE Globecom Workshops (GC Workshops), Abu Dhabi, United Arab Emirates, pp. 1–7 (2018). https://doi.org/10.1109/GLOCOMW.2018.8644436
22. GPP. TS 38.213 v15.1.0: Physical Layer Procedures for Control, 2018
23. Vardy, T.: List decoding of polar codes. IEEE Trans. Inf. Theory **61**(5), 2213–2226 (2015)
24. Tahir, B., Schwarz, S., Rupp, M.: BER comparison between Convolutional, Turbo, LDPC, and Polar codes. I: 2017 24th International Conference on Telecommunications (ICT), Limassol, pp. 1–7 (2017)
25. Al-Samman, A.M., Rahman, T.A., Azmi, M.H., Hindia, M.N., Khan, I., Hanafi, E.: Statistical Modelling and Characterization of Experimental mm-Wave Indoor Channels for Future 5G Wireless Communication Networks. PLoS ONE **11**(9), (2016)

A Novel Design of Frequency Reconfigurable Antenna for 5G Mobile Phones

Sanaa Errahili, Asma Khabba, Saida Ibnyaich, and Abdelouhab Zeroual

Abstract The purpose of this paper is to design a new frequency reconfigurable patch antenna that operates in two different frequency bands. The planned antenna is designed on dielectric substrate of Rogers RT5880 with 2.2 relative permittivity. The total size of the antenna is $6 \times 5.5 \times 1.02$ mm^3. The proposed antenna is formed of a positive intrinsic negative diode (PIN diode), placed at the radiating patch to achieve frequency reconfigurability based on the switching state of the PIN diode. The simulation of the proposed antenna is implemented using CST microwave studio. The performance of antenna is analyzed from the reflection coefficient, the surface current distribution, and the radiation pattern. The antenna has two resonant frequencies for 5G applications: 26.15 GHz and 46.1 GHz.

1 Introduction

With the emergence of new standards, telecommunication systems must be able to combine several standards on the same antenna. Reconfigurable antennas [1–3] are an important part of applications in wireless communication because their operation can be modified dynamically [4], which can be very advantageous for several applications. In addition, reconfigurability allows the antenna to offer more functionality.

Reconfigurable antennas must be able to adapt to their environment by changing their operating frequency [5], and/or their polarization [6] and/or radiation pattern [7, 8].

There are several reconfigurability techniques on designing the reconfigurable antenna like employing electronic, mechanical, or optical switching [9–11]. However, the electronic switching is more frequently used compared to other approaches, that is due to her efficiency and reliability. The technics of electronic switching comprise the PIN diodes, varactor diodes, field effect transistor (FET),

S. Errahili (✉) · A. Khabba · S. Ibnyaich · A. Zeroual
I2SP Laboratory, Faculty of Science, Cadi Ayyad University Marrakech, Marrakech, Morocco
e-mail: sanaa.errahili@ced.uca.ac.ma

M. Ben Ahmed et al. (eds.), *Networking, Intelligent Systems and Security*,
Smart Innovation, Systems and Technologies 237,
https://doi.org/10.1007/978-981-16-3637-0_20

and radio-frequency microelectromechanical system(RF MEMS) switches. Other approaches based on substrate agility. Many reconfigurable antenna is proposed in the recent years.

In this paper, we propose a noval reconfigurable patch antenna that can be tuned in two frequency bands by changing the geometry of the radiation patch by using a PIN diode. The proposed reconfigurable antenna is based on the PIN diode ON and OFF status between radiating elements, and it is able to select a very separate frequency band. The proposed antenna covers two bands of the fifth generation: 26.15 and 46.1 GHz.

2 Design of the Proposed Antenna

The design of the proposed reconfigurable antenna is presented in Fig. 1, which shows the side view of the antenna.

The proposed patch [12, 13] antenna has designed on 1.6 mm thickness of Rogers RT5880 dielectric with relative permittivity 2.2, and the dimension of the substrate is $6 \times 5.5 \times 1.02$ mm^2. The metallization of the patch and the ground plane is considered as copper, which has a uniform thickness t. The excitation is launched through microstrip feed line, which has a width W_f.

Parameters value of the proposed antenna is listed in Table 1.

As shown in Fig. 1, a PIN diode is located on the radiating patch. The PIN diode

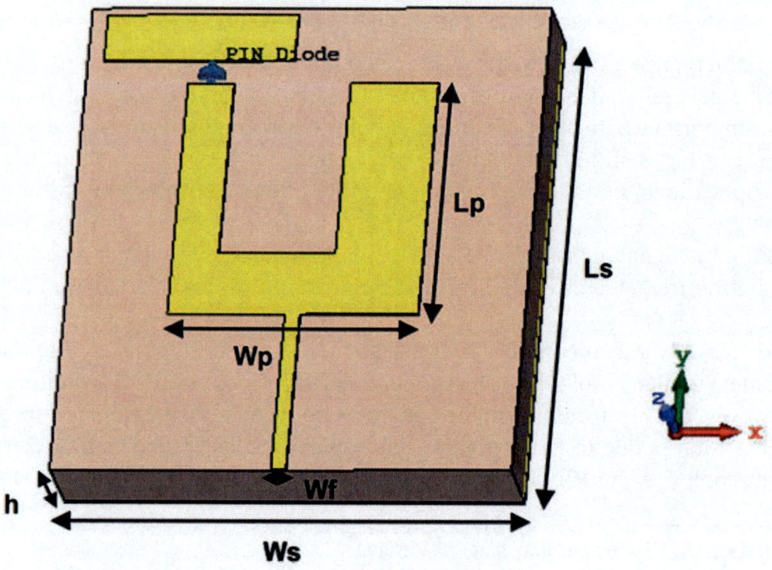

Fig. 1 Geometry of the proposed antenna

Table 1 Parameters of the proposed antenna

Parameter	Value (mm)
W	60
L	100
W_s	5.5
L_s	6
W_p	3
L_p	3
h	0.95
t	0.035
W_f	0.18
L_f	2

is used to connect the two parts of the patch. When the PIN diode is on state OFF, the antenna has only the main patch and the antenna operates at 26.15 GHz. The second configuration, when the PIN diode is on state ON, so the proposed antenna includes the two parts of patch, the proposed antenna operates at 46.1 GHz.

Figure 2 represents the equivalent circuit model of a PIN diode. The model used is that proposed in [14, 15], it is a simplified RLC equivalent circuit of the PIN diode that does not take account of the "surface mounting" effect. It consists of a parasitic inductor (*L*) in series with an intrinsic capacity (*C*) and an intrinsic resistance (*R*), which are connected in parallel (Fig. 2b). When the PIN diode is in the OFF state, the values of *R*, *L*, and *C* are, respectively, equal to R_2, L_1 and C_1. Conversely, when the PIN diode is in the ON state, the capacitance does not intervene, and the values of *R* and *L* are, respectively, equal to R1 and L1 (Fig. 2a).

In this work, the PIN diode MPP4203 is used as a switch. The circuit parameters are $L_1 = 0.45$ nH, $R_1 = 3.5\Omega$, $R_2 = 3$ Ωk, and $C_1 = 0.08$ pF.

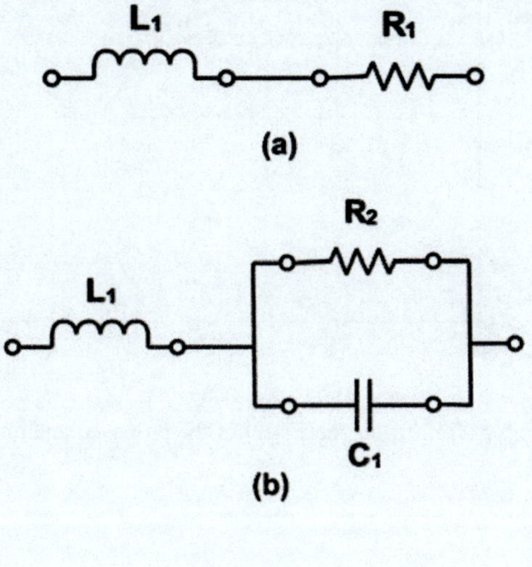

Fig. 2 PIN diode equivalent circuit [16]: **a** ON state, **b** OFF state

3 Results and Discussion

In this section, the simulated results of the proposed reconfigurable antenna: reflection coefficient S11, surface current distribution, and radiation pattern are presented.

The proposed antenna is designed, optimized, and simulated using CST studio suite.

Figures 3 and 4 show the simulated reflection coefficient of the proposed reconfigurable antenna.

There are two operating frequencies that can be varied by varying the diode PIN states (which is inserted in the antenna): OFF and ON states.

The resonant frequencies are:

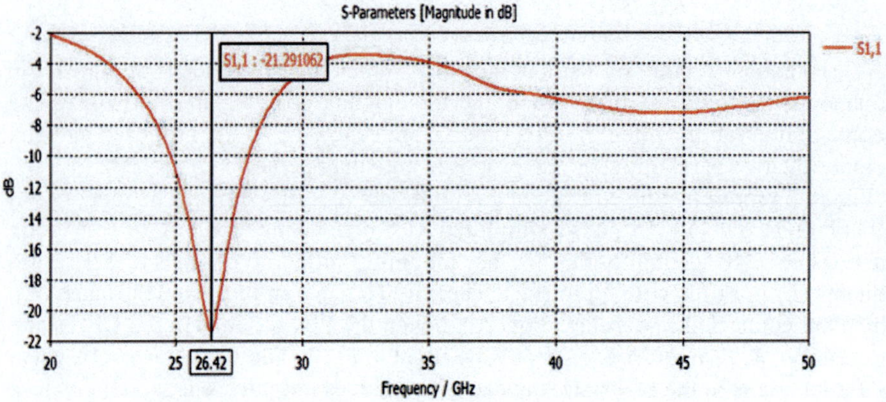

Fig. 3 Reflection coefficient of the proposed antenna when the PIN diode is in OFF state

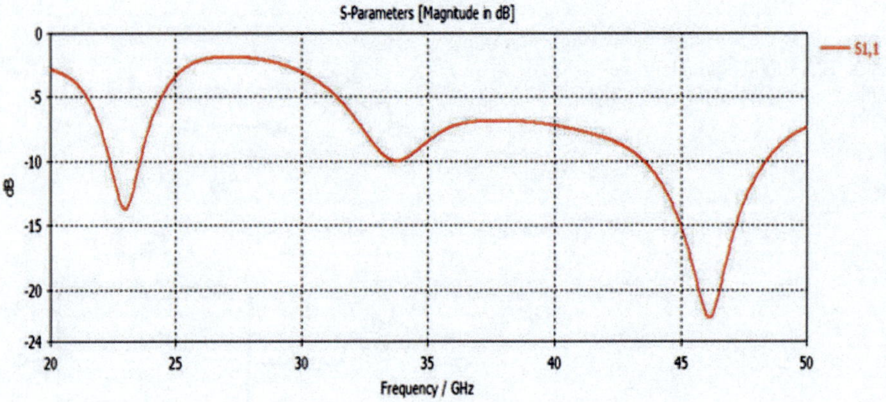

Fig. 4 Reflection coefficient of the proposed antenna when the PIN diode is in ON state

- The first resonant frequency $f_1 = 26.15$ GHz with the reflection coefficient of −21.29 dB, the value of the bandwidth at −10 dB is 24.91–27.62 GHz, if the diode PIN is OFF.
- The second resonant frequency $f_2 = 46.1$ GHz with the reflection coefficient of −22.21 dB, the value of the bandwidth at −10 dB is 43.22–47.34 GHz,if the diode PIN is ON.

Figures 5 and 6 represent the surface current distribution of the proposed reconfigurable antenna for the resonant frequencies.

In the first configuration, the PIN diode is ON, the two radiators are connected as seen in Fig. 6, the strong current distribution appears from the feed position to the

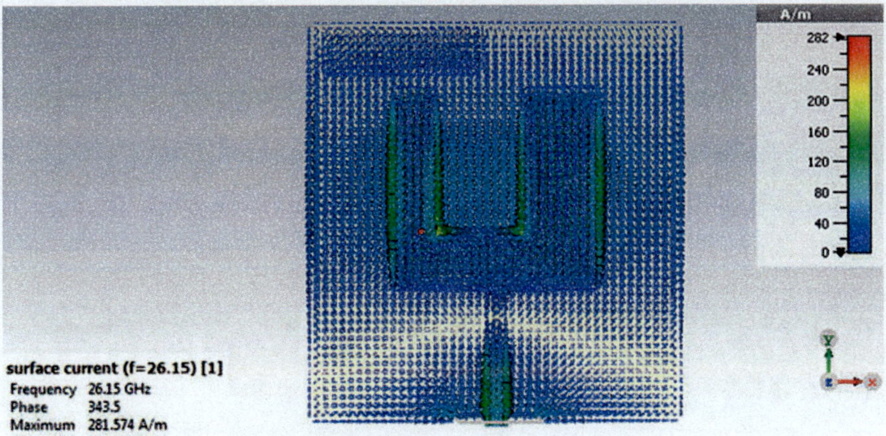

Fig. 5 Surface current distribution of the proposed antenna if the PIN diode in OFF state

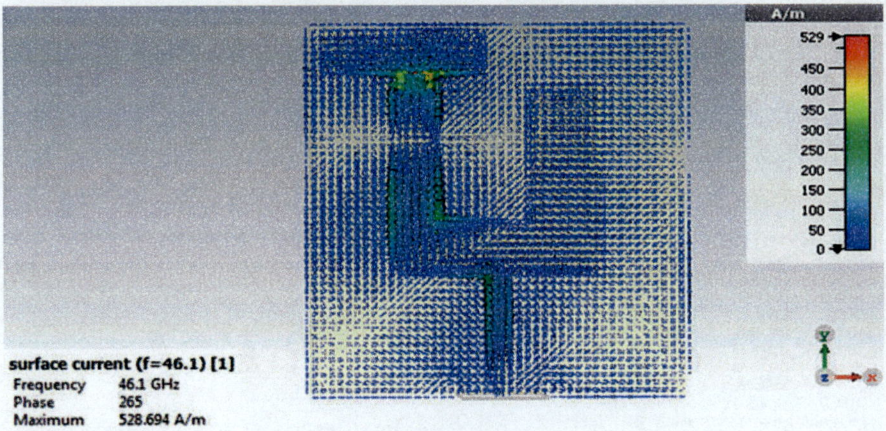

Fig. 6 Surface current distribution of the proposed antenna if the PIN diode in ON state

left side of the main radiator, and it passes from the diode to the second radiator on the left.

In the second configuration, the proposed antenna operates with only the main radiator when the PIN diode is OFF. As shown in Fig. 5, the strong current distributes from the feed position to the top sides of the main triangle radiator.

Figures 7 and 8 show the simulated radiation patterns of the proposed antenna with different switching states of the PIN diode, and it is plotted at 26.15 GHz and 46.1 GHz in 3D view. As observed, the antenna presents a good radiation performance with a max gain value of 5.34 dB for the ON state of the PIN diode and 6.08 dB for the OFF state.

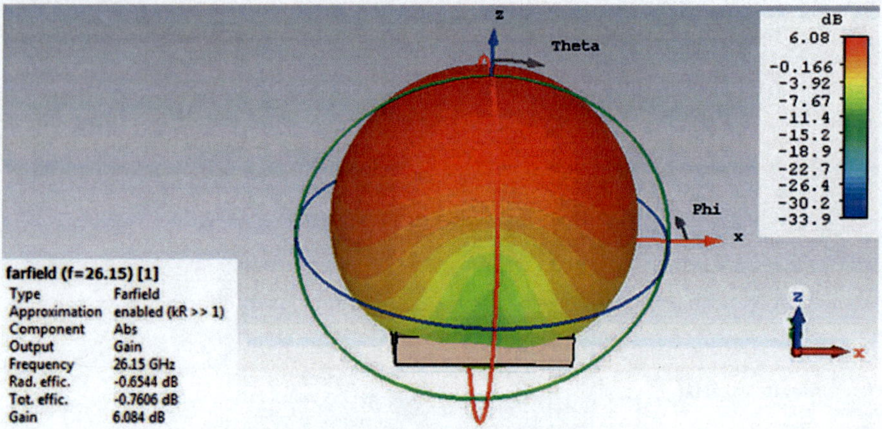

Fig. 7 Radiation pattern of the proposed antenna if the PIN diode in OFF state

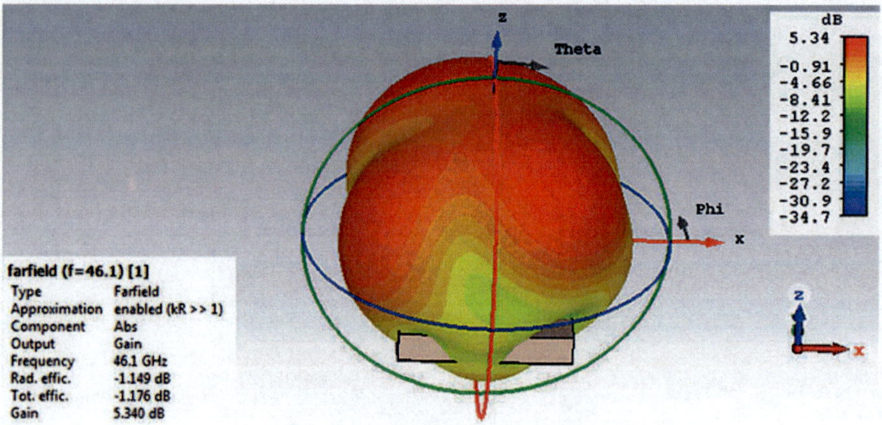

Fig. 8 Radiation pattern of the proposed antenna if the PIN diode in ON state

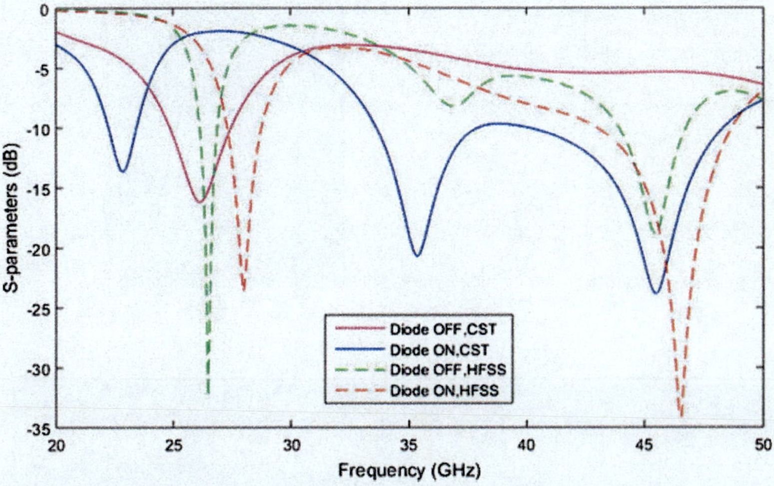

Fig. 9 Reflection coefficient of the proposed antenna simulated with CST and HFSS

4 Validation of the Results

For checking the previous results obtained by the CST microwave studio, we use another simulator program named Ansys HFSS software. The reflection coefficient of the proposed antenna for ON and OFF states of the PIN diode connected in the antenna is shown in Fig. 9. Firstly for the simulation with HFSS software, if the PIN diode in ON state, we obtained three resonant frequencies between them we have the principal frequency at 46.6 GHz.

Secondly, if the PIN diode in OFF state, we obtained two resonant frequencies so that the important frequency to us is 26.4 GHz.

So the resonant frequencies obtained by HFSS are shifted a little compared to those obtained by CST. Also, we notice that some resonant frequencies become more or Less significative. But the resonant frequencies still in the band of 24.25–27.5 GHz if the PIN diode is on OFF state and 45.5–50.2 GHz if the PIN diode is on ON state. These changes because the simulators are not same, due to different computational techniques involved, HFSS is based on finite element method (FEM) which is more accurate for designing antennas, while CST is based upon finite integration technique (FIT) [17] and is also popular among antenna designers due to ease in simulations.

5 Proposed Millimeter Wave Antenna Array for 5G

The proposed array contains eight reconfigurable antenna placed on the top of a mobile phone PCB like it shows Fig. 10 [18–20]. The overall size of the mobile phone PCB is $60 \times 100\ mm^2$. Simulations have been done using CST software to

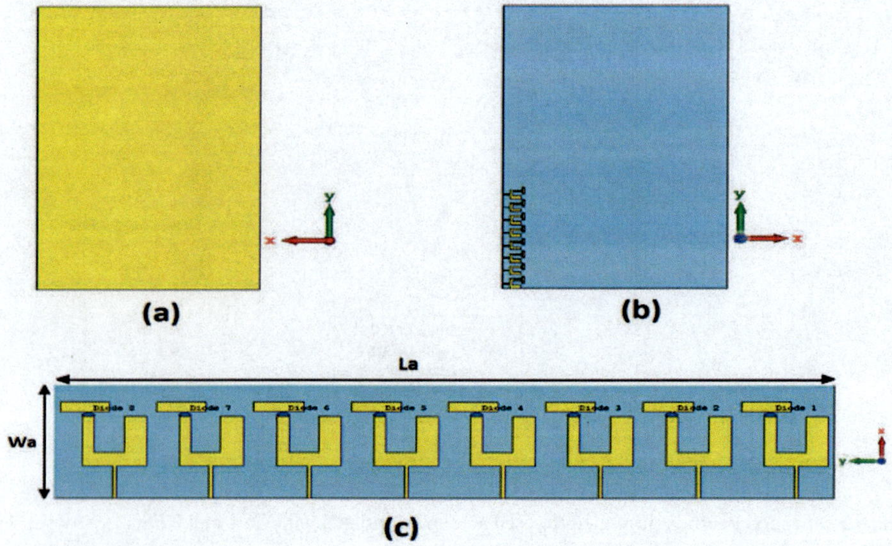

Fig. 10 Configuration of the proposed MIMO Antenna for 5G: **a** back view, **b** front view, and **c** zoom view of the antenna array

validate the feasibility of the proposed frequency reconfigurable array antenna for Millimeter Wave 5G handset applications [21, 22].

It can be seen that the proposed 5G array is compact in size with dimensions $L_a \times W_a = 25 \times 3.2\ \text{mm}^2$ (Fig. 1c). Furthermore, there is enough space in the proposed mobile phone antenna to include 3G and 4G MIMO antennas [23, 24]. The antenna is designed on a Rogers RT5880 substrate with thickness h and relative permittivity 2.2.

Figures 11 and 12 show the S-parameters (S1,1–S8,1) of the array for the two conditions of the PIN diodes (ON/OFF conditions). As illustrated, the mutual coupling between antennas elements of array is good. Furthermore, it can be seen that the array has good impedance adaptation at 26.15 GHz (all diodes in OFF state) and at 46.1 GHz (all diodes in ON state).

The 3D radiation paterns of the proposed antenna at 26.15 and 46.1 GHz are illustrated in Figs. 13 and 14 showing that the proposed reconfigurable antenna array a good beam steering property with the max gain value of 6.39 and 6.3, respectively.

6 Conclusion

In this paper, a new frequency reconfigurable patch antenna is designed, optimized, and simulated with CST studio suite. The proposed antenna can reconfigure on two different frequency bands of the fifth generation using PIN diodes, with the reflection

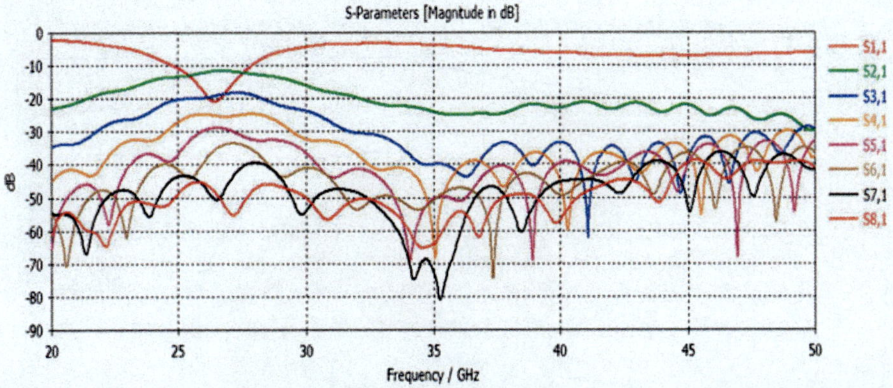

Fig. 11 Simulated S-parameters of the proposed 5G mobile phone antenna, if all diodes are in OFF state

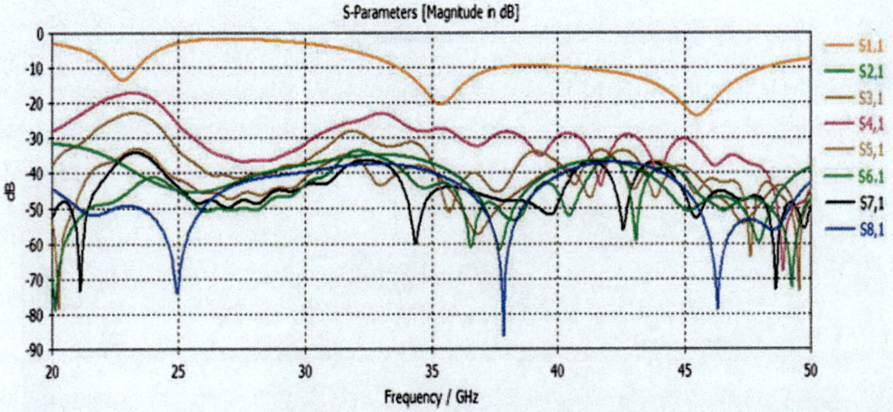

Fig. 12 Simulated S-Parameters of the proposed 5G mobile phone antenna, if all diodes are in ON state

coefficient less than −10 dB those are 24.91–27.62 GHz and 43.22–47.34 GHz. The overall structure size of the designed antenna 6 × 5.5 × 1.02 mm^3.

This antenna is useful for 5G applications [25].

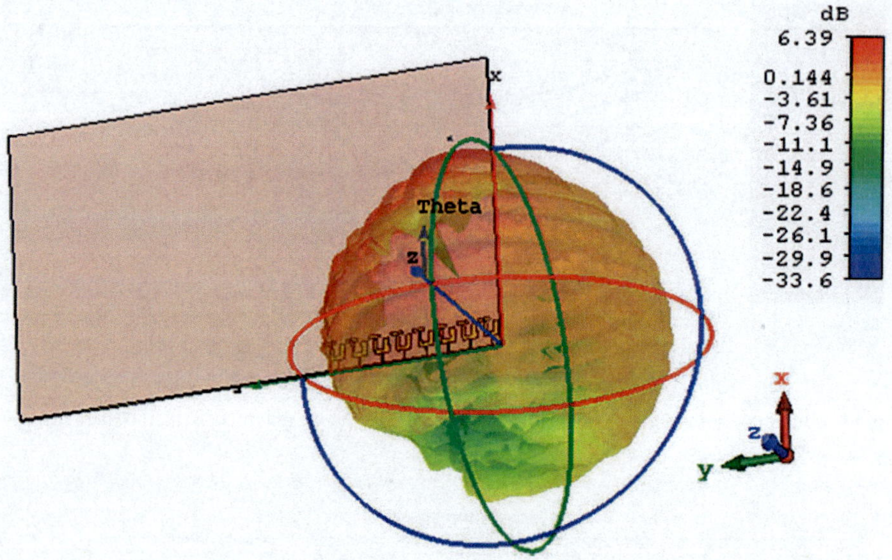

Fig. 13 Simulated of the proposed 5G mobile phone antenna, if all diodes are in OFF state

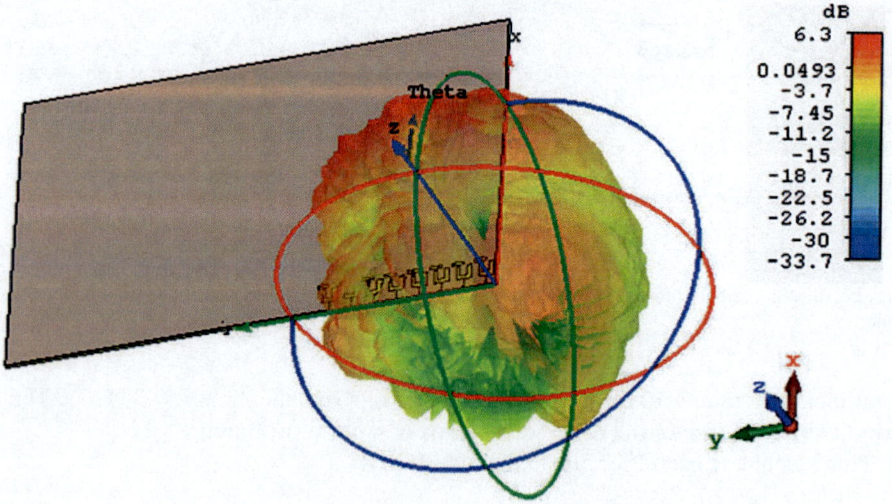

Fig. 14 Simulated of the proposed 5G mobile phone antenna, if all diodes are in ON state

References

1. Lim, E.H., Leung, K.: Reconfigurable Antennas. In: Compact multifunctional antennas for wireless systems, pp. 85–116. Wiley (2012). https://doi.org/10.1002/9781118243244.ch3

2. Loizeau, S.: Conception et optimisation d'antennes reconfigurables multifonctionnelles et ultra large bande. Ph.D. Dissertation (2009)
3. Guo, Y.J., Qin, P.Y.: Reconfigurable antennas for wireless communications. In: Chen, Z., Liu, D., Nakano, H., Qing, X., Zwick, T. (eds) Handbook of antenna technologies. Springer, Singapore (2016). https://doi.org/10.1007/978-981-4560-44-3_119
4. Bernhard, J.T.: Reconfigurable antennas. Morgan & Claypool Publishers (2007). ISNB 1598290266, 9781598290264
5. Ismail, M.F., Rahim, M.K.A., Zubir, F., Ayop, O.: Log-periodic patch antenna with tunable frequency (2011)
6. Hsu, Shih-Hsun, Chang, Kai: A novel reconfigurable microstrip antenna with switchable circular polarization. IEEE Antennas Wirel. Propag. Lett. **6**(2007), 160–162 (2007)
7. Dandekar, K.R., Daryoush, A.S., Piazza, D., Patron, D.: Design and harmonic balance analysis of a wideband planar antenna having reconfigurable omnidirectional and directional patterns. 5 (2013)
8. Nikolaou, S., Bairavasubramanian, R., Lugo, C., Car-rasquillo, I., Thompson, D.C., Ponchak, G.E., Papapolymerou, J., Tentzeris, M.: Pattern and frequency reconfigurable annular slot antenna using PIN diodes. IEEE Trans. Antennas Propag. **54**,(2), 439–448 (2006)
9. Sarah El Kadri. 2011. Contribution à l'étude d'antennes miniatures reconfigurables en fréquence par association d'éléments actifs. Ph.D. Dissertation
10. Kumar, D., Siddiqui, A.S., Singh, H.P., Tripathy, M.R., Sharma, A.: A Review: Techniques and Methodologies Adopted for Reconfigurable Antennas. In: 2018 International Conference on Sustainable Energy, Electronics, and Computing Systems (SEEMS). 1–6 (2018). https://doi.org/10.1109/SEEMS.2018.8687361
11. Salleh, S.M., Jusoh, M., Seng, L.Y., Husna, C.: A review of reconfigurable frequency switching technique on micostrip antenna. J. Phys.: Conf. Ser. **1019**, 012042 (2018). https://doi.org/10.1088/1742-6596/1019/1/012042
12. Fang, D.G.: Antenna theory and microstrip antennas. CRC Press, Taylor & Francis Group, New York (2015)
13. Zhang, Z.: Antenna design for mobile devices. Wiley (2011). Print ISBN:9780470824467, Online ISBN:9780470824481, https://doi.org/10.1002/9780470824481
14. Ismail, M.F., Rahim, M.K.A., Majid, H.A.: The Investigation of PIN diode switch on reconfigurable antenna. In: 2011 IEEE International RF & Microwave Conference. IEEE, pp. 234–237 (2011)
15. Lim, J., Back, G., Ko, Y., Song, C., Yun, T.: A reconfigurable PIFA using a switchable pin-diode and a fine-tuning varactor for USPCS/WCDMA/m-WiMAX/WLAN. IEEE Trans. Antennas Propag. **58**(7), 2404–2411 (2020). https://doi.org/10.1109/TAP.2010.2048849
16. Balanis, C.A.: Modern antenna handbook. Wiley, Hoboken (2008)
17. Balanis, C.A.: Antenna Theory: Analysis and Design, 3rd ed. John Wiley, Hoboken, NJ (2005) MLA (8th ed.)
18. Sanayei, S., Nosratinia, A.: Antenna selection in MIMO systems. IEEE Commun. Mag. **42**(10), 68–73 (2004). https://doi.org/10.1109/MCOM.2004.1341263
19. Li, Y., Sim, C., Luo, Y., Yang, G.: 12-Port 5G Massive MIMO Antenna Array in Sub-6 GHz Mobile Handset for LTE Bands 42/43/46 Applications. IEEE Acc. **6**, 344–354 (2018). https://doi.org/10.1109/ACCESS.2017.2763161
20. Sanayei S., Nosratinia, A.: University of Texas at Dallas, Antenna Selection in MIMO Systems, IEEE Communications Magazine, October (2004)
21. Rappaport, T.S., Sun, S., Mayzus, R., Zhao, H., Azar, Y., Wang, K., Wong, G. N., Schulz, J. K., Samimi, M., Gutierrez, F.: Millimeter wave mobile communications for 5G cellular: It will work! IEEE Access **1**, 335–349. [6515173] (2013). https://doi.org/10.1109/ACCESS.2013.2260813
22. Liu, D., Hong, W., Rappaport, T.S., Luxey, C., Hong, W.: What will 5G Antennas and Propagation Be? IEEE Trans. Antennas Propag. **65**(12), 6205–6212 (2017). https://doi.org/10.1109/TAP.2017.2774707

23. Sharawi, M.S.: Printed MIMO antenna engineering. Electronic version: https://books.google.co.ma/books?id=7INTBAAAQBAJ&lpg=PR1&ots=aHyFM1I5Wi&dq=mimo%20antenna&lr&hl=fr&pg=PR7#v=onepage&q=mimo%20antenna&f=false
24. Li, Y., Desmond Sim, C.-Y., Luo, Y., Yang, G.: 12-Port 5G Massive MIMO Antenna Array in Sub-6 GHz Mobile Handset for LTE Bands 42/43/46 Applications, 2169–3536 (c) (2017) IEEE, https://doi.org/10.1109/access.2017.2763161
25. Hong, W., Baek, K.-H., Lee, Y., Kim, Y., Ko, S.-T.: Study and prototyping of practically large-scale mm wave antenna systems for 5G cellular devices, IEEE Communications Magazine, September (2014)

Smart Security

A Real-Time Smart Agent for Network Traffic Profiling and Intrusion Detection Based on Combined Machine Learning Algorithms

Nadiya El Kamel, Mohamed Eddabbah, Youssef Lmoumen, and Raja Touahni

Abstract Cyber-intrusions are constantly growing due to the ineffectiveness of the traditional cyber security tools and filtering systems-based attacks detection. In the last decade, significant techniques of machine and deep learning were employed to resolve the cyber security issues. Unfortunately, the results are still imprecise with a lot of shortcomings. In this paper, we present a real-time cyber security agent based on honeypots technology for real-time data collection and a combination of machine learning algorithms for data modeling that enhances modeling accuracy.

1 Introduction

On the Internet, each connected node represents a target for a black hat, while the reasons behind cyber-attacks are various [1–4]. By exploiting vulnerabilities, intruders gain access into private networks, they may spy, steal, or sabotage the data, etc. Hence, in order to protect their sensitive data, companies deploy more and more security solutions. On the other hand, attackers develop their tools too, by adopting new techniques to avoid detection and filtering systems. In December 2020s, many leader companies compromised over the SUNBURST hack campaign including even security tools providers [5]. Intruders exploited the Solarwinds Orion platform update

N. El Kamel (✉) · R. Touahni
Laboratory of Electronic Systems, Information Processing, Mechanics and Energetics, Faculty of Sciences, Ibn Tofail University, Kenitra, Morocco
e-mail: nadiya.elkamel@uit.ac.ma

R. Touahni
e-mail: touahni.raja@uit.ac.ma

M. Eddabbah
LABTIC Laboratory ENSA, Abdelmalek Essaadi University Tangier, Tangier, Morocco

Y. Lmoumen
CIAD UMR 7533, Univ. Bourgogne Franche-Comté, UTBM, 90010 Belfort, France

M. Ben Ahmed et al. (eds.), *Networking, Intelligent Systems and Security*,
Smart Innovation, Systems and Technologies 237,
https://doi.org/10.1007/978-981-16-3637-0_21

file to add vulnerabilities and backdoors. They had employed a combination of techniques to infiltrate into 17 K of the Solarwinds customers' networks [6]. In order to detect such sophisticated attacks, we design a smart agent-based attacks detection using a combination of machine learning algorithms for flow modeling and honeypot techniques for an updatable database conception. The honeypot is a security resource implemented for being probed, attacked, or compromised [7–9]. It was proposed to automatically consider any interaction detected as a malicious activity. The generated log files data will be aggregated and modeled using a combination of machine learning classifiers, to enhance precision and future attacks detection. Next sections are devoted firstly to discussing some related works and secondly to explain the smart agent functions, advantages, and use cases.

2 Related Works

Many cyber security solutions proposed in the last decade, but results still present some limitations and shortcomings [10] while all Internet-providers seek to protect themselves against fraudulent use of their data, stealing, sabotage and against all malicious activities on computer systems. The most recent works in cyber security focus on machine and deep learning algorithms for attacks data modeling [11, 12].

Pa, Y et al. [13] suggest a honeypot-based approach, for malware detection based on corresponding signatures generated from a honeypot system. This method is still limited and unable to detect new signatures or new kinds of malwares. Moreover, a machine learning-based solution represents a promising candidate to deal with such a problem, due to its ability to learn and teach over time.

R. Vishwakarma et al. [14] present an IoT combating method against DDoS attacks, based on IoT honeypot-generated data for dynamic training of a machine learning model. The proposed method allows detecting zero-day DDoS attacks, which has emerged as an open challenge in defending IoT against DDoS attacks.

P. Owezarski et al. [10] studied an unsupervised anomalies learning-based attacks characterization, using honeypot system for data construction. This study is based on clustering techniques as subspace clustering, density-based clustering, subspace clustering, and evidence accumulation for classifying flow ensembles in traffic classes. The proposed study in this work does not require a training phase.

K. Lee et al. [15] used machine learning algorithms (SVM) to automatically classify social spam for network communities such as Facebook and MySpace based on a social honeypot information collection.

T. Chou et al. [16] suggest a three-layer hierarchy structure-based intrusion detections, consisting of an ensemble of classifiers groups, each consisting of a combination feature selecting classifiers. They applied different machine learning algorithms and feature subset to solve uncertainty problems and maximize the diversity. In terms of detection rate (DR), false-positive rate (FPR) and classification rate (CR), the results demonstrate that the hierarchy structure performs better than a single classifier-based intrusion detection.

G. Feng et al. suggest in [17] a linkage defense system for improving private network security by linking honeypots with the network security. The defense network centroid honeypot treats suspicious flows arriving from the traditional tools, while blocking network access depends on the honeypot state; if the honeypot is damaged, then the correspondent intruder will be blocked by the firewall.

3 A Real-Time Security Agent Based on Machine Learning and Honeypot Technologies

Manual cyber security tools reconfiguration represents some shortcomings in terms of time and money. While it is an illusion to think that a lock and a key represent perfect security defense, intruders develop their strategies constantly to add back-doors, to break the peer username/password, to hide their command traffic, and even to hide data in media (Steganography technique). IDS, Firewalls, and IPS protect systems from traditional hack tools and tactics but are still ineffective to detect hidden command and data traffic. There is no contact between them to block intrusion detected in an IDS by firewalls [17], and they represent a passive solution when it is about zero-day and future attacks [18]. The real cyber security challenge is accepting the probability of an imminent attack and understanding what is really going on within information systems.

The main objective of this work is to design a real-time cyber security agent, based on machine learning algorithms combination and honeypots technology. Machine learning algorithms-based intrusion modeling allows automatic attacks detection, while the honeypot system is a deceptive technology that allows intruders control of fack machines [19], traffic capturing, data collection and ensure the logging of newly coming malware features [20, 21].

Based on these technologies, the smart agent takes functions of packet interception, information collection, suspicious profile creation (database construction), comparison of suspicious profiles with attacker's database profiles for making decisions, database update, and firewall alarm if an attacker profile is detected (Fig. 1).

3.1 Profiles Creation (Phase 1)

Information security policy focuses on mechanisms that insure data integrity, availability, and confidentiality, which consists of traffic monitoring and filtering. While unknown profiles detection's time is the most critical point for intrusion detection systems. For this reason, we develop a smart agent for detecting attacks and for constructing a shared and an updatable database for protecting Internet content-providers from current and future attacks. In the first stage, the agent takes functions of packet interception and hackers profile creation based on the transport and

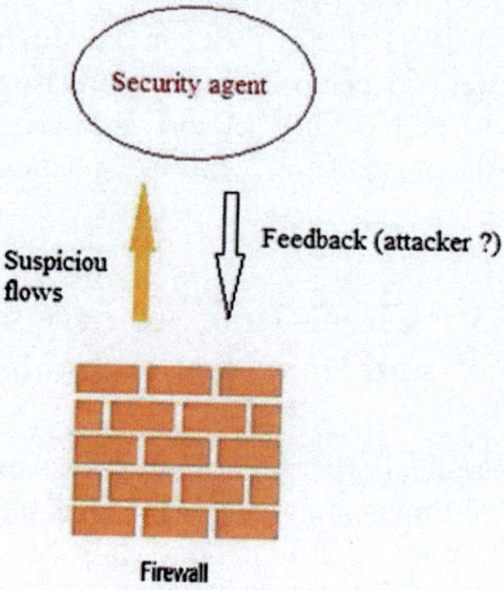

Fig. 1 Security agent-based attacks detection

application information gathered within a honeypot that emulates fake services, and machine learning for data modeling using an hierarchical structure of algorithms that maximize the detection precision.

The originality of the honeypot lies in the fact that the system is voluntarily presented as a weak source able to hold the attention of attackers [8]. The general purpose of honeypots is to make the intruder believe that he can take control of a real production machine, which will allow the smart agent to model the compromising data gathered as a profile, and making decision to send alarms when it is about an intruder profile. The honeypot classification depends on its interactions level. Low-interaction honeypot offered a limited services emulation set; for example, it cannot emulate an file transfer protocol (FTP) service on port 21, but emulate only the login command or just one other command, and it records a limited set of information, monitor only known activities. The advantage of low-interaction honeypots lies in their simplicity of implementing and managing and pose little risk since the attacker is limited. Medium-interaction honeypots give little more access than low-interaction honeypots [22] and offer better emulation services. So a medium-interaction honeypots enable logging more advanced attacks, but they require more time to implement and a certain level of expertise. For high-interaction honeypots, we provide the attacker with real operating systems and real applications [23]. This type of honeypots allows gathering a lot of information about the intruder as it interacts with a real system, examining all behaviors and techniques employed, and checking if it is about a new attack.

In this work, we employ a high-level interaction honeypot for extracting a huge amount of information about intruders, and we exploit it again after the decision phase, and if an attacker is detected, we configure the firewall in the way that it redirect him into the honeypot system one more time, for limiting his capacity of developing tools and tactics. In a network company, the productive network consists of, for example, HTTP, database, monitoring, and management servers. A network of honeypot must be deployed and configured to run the fake services of the productive

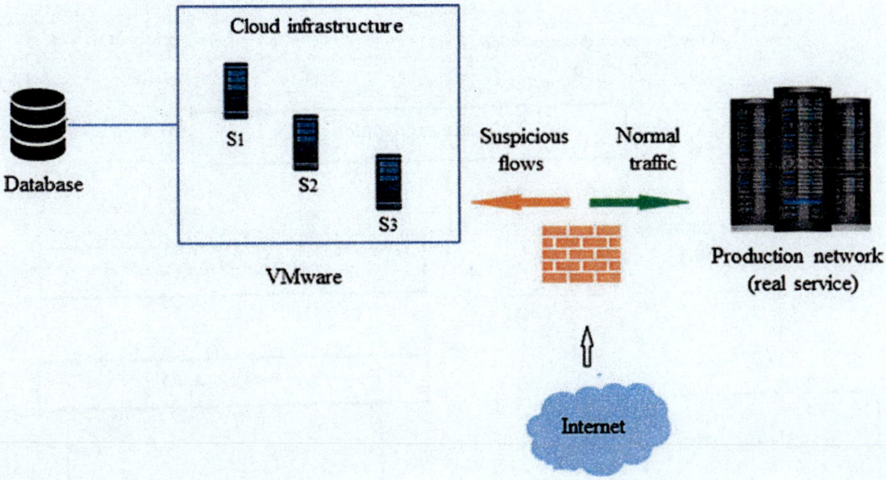

Fig. 2 Hacker profile creation architecture

network (S1, S2, S3, etc.), which suspicious flows are redirected to by the network firewall system (Fig. 2).

At the profile creation phase (Fig. 2), suspicious flows will be redirected to a network of honeypots servers, allowing intruders the control of fack servers [24]; the transport and application layer collected information will be aggregated into vectors V_{user}. Qualitative data such as IP address is stocked directly in the profile, while quantitative data is classified into homogeneous classes (time inter packet, number of packets per flow… etc.) using an hierarchical structure of machine learning algorithms combine classification algorithm and linear regression (Algorithm 1).

Machine learning techniques combination consists of mixing classification and regression algorithms, whose purpose is to maximize the precision level of the fitting models.

In this paper, we propose a combination composed of two-layers hierarchy structure, consisting of a classification algorithm, that divides flows quantitative data into homogenous subsets at the first layer, and linear regression for each subset modeling and classifying at the second layer (Fig. 3). The advantages of the hierarchical classification method lie in increasing modeling precision, reliable detection, and false alarms avoiding. For the smart agent development, this technique represents a keystone for suspicious and attacker profiles creation and update.

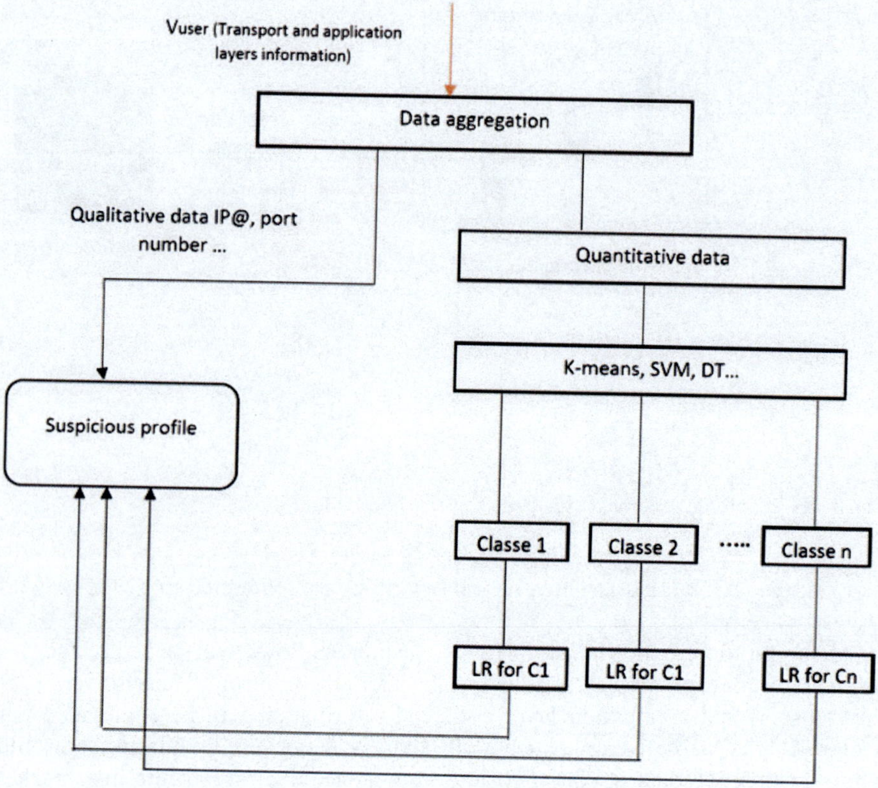

Fig. 3 Flow modeling process

Algorithm 1: Learning

INPUT

K //number of clusters

$V_{ij} = (V_1^j, V_2^j, \ldots, V_i^j)$ *//Hacker j data array (vector of vectors)*

START

$\mathbf{A_{kl}} = Q_{1A}$ ($\mathbf{V_{ij}}$) *//qualitative adaptation*

$\mathbf{A_{k'l'}} = Q_{2A}$ ($\mathbf{V_{ij}}$) *//quantitative adaptation*

f = c-1*//linear regression order = space dimension -1*

for j=1; i<c; j++ //For each row of $\mathbf{A_{k'l'}}$

(Cj [K], Rj [K]) = K-means ($\mathbf{A_{il'}}$*, K) //creation of clusters center et radius*

for j=1; i<c; j++

$CL_j[f]$ *= Linear Regression (*C_j *[K],* R_j *[K],* $\mathbf{A_{il'}}$ *//Regression of every clusters*

OUTPUT//Hacker profile

$\mathbf{A_{kl}}$*//qualitative adaptation*

CLc[f] *//linear regression coefficients*

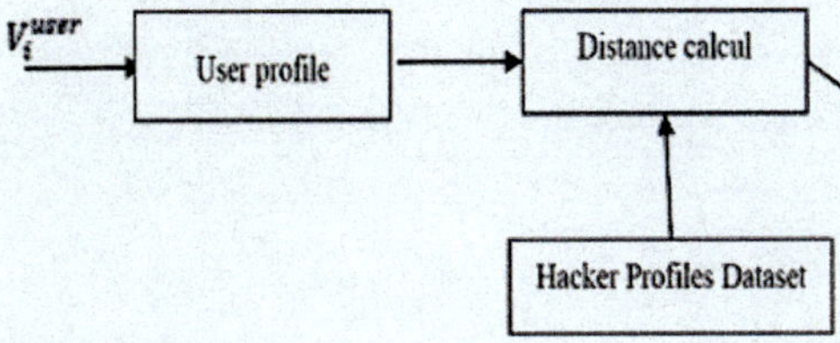

Fig. 4 Decision stage

At the profile creation phase, the initialization of these three parameters is crucial: the number of clusters, centroids initialization, and the linear function weights. The higher initialization precision of these parameters increases the precision of the fitting model.

3.2 Attacks Detection (Phase 2)

In the decision phase (Algorithm 2), the smart agent functions are packet interception, information collection, suspicious profiles creation, comparison of suspicious profiles with database profiles based on the distance metric (Fig. 4) [20], and administrator alarm if an attacker profile is detected. Hence, it represents a traffic cop that detects malicious profiles and alarms firewall systems to cut the route. The advantages of this contribution lie in the automatic detection of malicious profiles, construction of new dataset that reflect current network situations and the latest attack trends, and limiting the capacity of developing intruder's tools by redirecting him into the honeypot system one more time. Hence, by making the intruder interact again with the honeypot, the profiles will be enriched with the gathered information and then reinforcing the learning.

Algorithm 2: Decision

INPUT

K //number of clusters

$V = (V_1^j, V_2^j, \ldots, V_i^j)$ *//user j information array (vector of vectors)*

(HA$_{kl}$, HCLc[f]) *//Hackers profiles*

START

UA$_{kl}$= Q_{1A} (V$_{ij}$) *//qualitative adaptation*

A$_{k'l'}$= Q_{2A} (V$_{ij}$) *//quantitative adaptation*

f = c−1*//linear regression order = space dimension -1*

for j=1; i<c; j++ //For each row of ***A$_{k'l'}$***

*(Cj [K], Rj [K]) = K-means (****A$_{il'}$****, K) //creation of clusters center et radius*

for j=1; i<c; j++

$UCL_j[f]$ *= Linear Regression (*C_j *[K],* R_j *[K],* ***A$_{il'}$****) //Linear Regression of every clusters*

OUTPUT

Distance(HCLc[f], UCLj[f])

Isequal(HA$_{kl}$, UA$_{kl}$)

In the Euclidean sense, $HCL_c[f]$ is the closest vector to $HCL_j[f]$.

4 Conclusion

In this paper, we have presented a smart cyber security agent based on machine learning algorithms combination for automatic decision instead of manual treatment and the deceptive honeypot for data collection. The proposed agent is not just an automatic security tool for new and zero-day attacks detection, but it allows constructing a new updatable database that reflects current network situations, latest attack trends, and future attack techniques. Next works will be devoted to implementing the smart agent in a real environment (cloud infrastructure) in order to test its performances and compare it with other cyber security solutions.

References

1. Matin, I.M.M., Rahardjo, B.: The Use of Honeypot in Machine Learning Based on Malware Detection: A Review. 1–6 (2020)
2. Matin, I.M.M., Rahardjo, B.: Malware detection using honeypot and machine learning. 7th International Conference on Cyber and IT Service Management **7**, 1–4 (2020)
3. Singh, J., Singh, J.: A survey on machine learning-based malware detection in executable files. Journal of Systems Architecture 101861 (2020)
4. Iwendi, C., Jalil, Z., Javed, A.R., Reddy, T., Kaluri, R., Srivastava, G., Jo, O.: Keysplitwatermark: Zero watermarking algorithm for software protection against cyber-attacks. IEEE Access **8**, 72650–72660 (2020)
5. Bowman, J.: How the United States is Losing the Fight to Secure Cyberspace. (2021)
6. Oxford, A.: SolarWinds hack will alter US cyber strategy. Emerald Expert Briefings (2021)

7. Jiang, K., Zheng, H.: Design and Implementation of A Machine Learning Enhanced Web Honeypot System. pp. 957–961. IEEE, (Year)
8. Spitzner, L.: Honeypots: tracking hackers. Addison-Wesley Reading (2003)
9. Karthikeyan, R., Geetha, D.T., Vijayalakshmi, S., Sumitha, R.J.I.j.f.R., Technology, D.i.: Honeypots for network security. 7, 62–66 (2017)
10. Owezarski, P.: Unsupervised classification and characterization of honeypot attacks. pp. 10–18. IEEE, (2014)
11. Berman, D.S., Buczak, A.L., Chavis, J.S., Corbett, C.L.: A survey of deep learning methods for cyber security. Information **10**, 122 (2019)
12. Liu, H., Lang, B.: Machine learning and deep learning methods for intrusion detection systems: A survey. applied sciences 9, 4396 (2019)
13. Pa, Y.M.P., Suzuki, S., Yoshioka, K., Matsumoto, T., Kasama, T., Rossow, C.: IoTPOT: Analysing the rise of IoT compromises. (2015)
14. Vishwakarma, R., Jain, A.K.: A honeypot with machine learning based detection framework for defending IoT based botnet DDoS attacks. pp. 1019–1024. IEEE, (Year)
15. Lee, K., Caverlee, J., Webb, S.: Uncovering social spammers: social honeypots + machine learning. pp. 435–442. (2010)
16. Chou, T.-S., Fan, J., Fan, S., Makki, K.: Ensemble of machine learning algorithms for intrusion detection. pp. 3976–3980. IEEE, (2019)
17. Feng, G., Zhang, C., Zhang, Q.: A design of linkage security defense system based on honeypot. pp. 70–77. Springer, (2013)
18. Matin, I.M.M., Rahardjo, B.: Malware detection using honeypot and machine learning. pp. 1–4. IEEE, (2020)
19. Seungjin, L., Abdullah, A., Jhanjhi, N.Z.: A Review on Honeypot-based Botnet Detection Models for Smart Factory. International Journal of Advanced Computer Science and Applications 11, (2020)
20. El Kamel, N., Eddabbah, M., Lmoumen, Y., Touahni, R.: A Smart Agent Design for Cyber Security Based on Honeypot and Machine Learning. Security and Communication Networks 2020, (2020)
21. Ng, C.K., Pan, L., Xiang, Y.: Honeypot frameworks and their applications: a new framework. Springer (2018)
22. Negi, P.S., Garg, A., Lal, R.: Intrusion detection and prevention using honeypot network for cloud security. pp. 129–132. IEEE, (2020)
23. Wang, H., Wu, B.: SDN-based hybrid honeypot for attack capture. pp. 1602–1606. IEEE, (2019)
24. Naik, N., Jenkins, P., Savage, N., Yang, L.: A computational intelligence enabled honeypot for chasing ghosts in the wires. Complex & Intelligent Systems 1–18 (2020)

Privacy Threat Modeling in Personalized Search Systems

Anas El-Ansari, Marouane Birjali, Mustapha Hankar, and Abderrahim Beni-Hssane

Abstract Personalized search systems simplify information access on the Web by providing adapted results for each user's query according to his preferences and needs. Such systems depend mainly on collecting user data to create profiles that represent user interests. While collecting user data is essential for improving the quality of the returned results, it has also raised serious concerns regarding user privacy. Giving the user a personalized browsing experience usually comes at the cost of his privacy. Thus, most people are afraid of using such applications. Current search engines lack dedicated privacy-preserving features and do not fulfill people's expectations in terms of privacy. Researchers have been investigating solutions to overcome this issue and create privacy-aware personalized systems. In this paper, we present an overview of the current privacy threats in personalized search systems with a threat modeling approach. Furthermore, we examine techniques and solutions for user data security and privacy protection.

1 Introduction

The information amount available on the Web grows continuously preventing people from easily obtaining desired items or information [1]. This dilemma highlights a pressing demand for effective personalized search systems capable of simplifying information access and item discovery considering the user's interests and preferences. Most studies usually focus on improving personalization quality and disregarding user privacy issues. One of the challenges facing personalized systems is the privacy protection problem [2] since giving the user a personalized browsing experience comes at the cost of his privacy. Thus, people are afraid of using such applications.

As the current search engines lack dedicated privacy-preserving features and do not fulfill people's expectations in terms of privacy, alternative search engines

A. El-Ansari (✉) · M. Birjali · M. Hankar · A. Beni-Hssane
LAROSERI Laboratory, Computer Science Department, Sciences Faculty, Chouaib Doukkali University, El-Jadida, Morocco

M. Ben Ahmed et al. (eds.), *Networking, Intelligent Systems and Security*, Smart Innovation, Systems and Technologies 237,
https://doi.org/10.1007/978-981-16-3637-0_22

have emerged: metasearch engines (e.g., DuckDuckGo [3]) and search engines (e.g., Qwant [4, 5]). The former enhances existing search engines by focusing on the privacy protection of their users, while the latter develops a search engine that does not exploit users' information. Nevertheless, these alternatives do not implement any specific privacy-preserving mechanisms. Instead, they claim, in their terms of service, that they do not collect any personal information of their users. For instance, DuckDuckGo affirms that they also store searches, but not in an identifiable form, as they do not collect IP addresses or any identifiable user information.

The lack of protection for such data raises privacy issues. For instance, the AOL query logs scandal [6] when AOL research released a file on its Web site with over twenty million search queries for about 650,000 users. New York Times identified a user from the published file by cross-referencing it with some phone-book listings. AOL admitted it was an error and removed the file, yet others redistributed the file on mirror sites. This example not only raises panic among users but also dampens the data publishers' enthusiasm in offering improved personalized services.

Besides, as these systems' implementations are not publicly available and as they do not explicitly provide the data they log, users cannot be confident in the privacy protection obtained by these solutions. Users can only trust these services and hope that their data are in a safe and privacy-preserving storage space. Researchers have been investigating solutions to overcome this issue and create search engines that ensure a privacy protection by design as in [7].

For organizations that collect or manage user data, security and privacy should be mandatory as it is for individuals who own it. It is the primary concern when undertaking the process of protecting fundamentally sensitive information such as identities, finances, and health records.

Without privacy and security measures, other malicious third parties would gain access to large amounts of possibly damaging data [8]. However, the distinction between user privacy and user data security is not clear for everyone, and the terms are often misplaced or confused as the same thing.

Data security and user privacy are two fundamental components for a successful data protection strategy [9], so safeguarding data is not limited to one single concept of the two.

The difference between both topics is not in the execution, implementation, or results but the philosophy supporting them. Specifically, it is a matter of which data require protection, what protection mechanism is employed, from whom, and who is responsible for this protection.

Data security aims at preventing unapproved access to user data via leaks or breaches, despite who the unapproved party is. To achieve this, companies use technologies and tools such as user authentication, firewalls, network limitations, and even internal security measures to prevent such access. Also, security technologies like encryption and tokenization further protect user data by making it unreadable at the moment a breach occurs and can stop attackers from potentially revealing massive amounts of the user's sensitive data.

Privacy, however, focuses on ensuring that the sensitive data an organization stores, processes, or transmits are ingested compliantly and with consent from the owner of that sensitive data. It means informing users upfront of which types of data the system collects, for what purpose, and who has access to it. The individual must then agree to the terms of use, allowing the organization that ingests data to use it in line with its stated purposes.

So, privacy is more about using data responsibly and following the wishes of users to prevent using it by unauthorized parties. However, it can also include security-type measures to ensure privacy protection. For instance, efforts to prevent the linking of sensitive data to its data subject or natural person such as de-identifying personal data, obfuscating it, or storing it in different places to reduce the possibility of re-identification are other privacy provisions.

A personalized system can apply security controls without satisfying privacy requirements, yet privacy issues are hard to control without employing efficient security practices.

In this work, we study the privacy threat in personalized search systems, and we propose a threat model for privacy-preserving personalized search systems. Moreover, we discuss privacy protection and cryptographic solutions that can be used in personalized search systems.

The paper is organized as follows. The next section addresses the privacy threat modeling in personalized search. In Sect. 3, we present privacy protection solutions. While the fourth section addresses the cryptographic solutions. The paper concludes and points to future work ideas in the conclusion section.

2 Privacy in Personalized Search

To protect privacy in personalized search systems, we must consider two contradicting effects. First, such systems improve search quality by collecting more user data. Second, it must hide the sensitive data collected and stored in a user profile to place the privacy risk under control.

2.1 Personalized Search Systems

We can classify the personalized search systems into three distinct structures, based on the storage place of the user profile (server side or client side) and its use for personalization [10].

Server-side personalization: The system stores user profiles on the server side. This structure requires the user to have an identifying account. The server then creates and updates the profile either explicitly from the user's input (requesting the user to list his interests) or implicitly by collecting the user's browsing history (e.g., query and click-through history). The latter method needs no additional work from users and contains a better description of his interests. Some search engines, like Google Personalized, adopted this architecture. Most systems with such a structure ask users to provide consent before collecting and using their data. If the user accords his permission, the search system will hold all the personally identifiable data possibly available on the server. Thus, from the user's perspective, this architecture provides a low level of privacy protection.

Client-side personalization: Storing the profile on the user device, the client sends queries to a search engine and receives results, as in an ordinary Web search scenario. The client agent also performs a query expansion to generate a new personalized query before sending it to the search engine. Furthermore, as in [11], the client agent ranks the search results to match user preferences.

Client–Server collaborative personalization: This structure is a balance of the previous two. The profile is still on the client side, but the server also participates in search personalization. At query time, the client agent extracts a sub-profile from the user profile to send it to the search engine along with the query. The search engine then uses the received context to personalize the results.

2.2 *Privacy Threats*

Personalized search systems pose several risks to users' privacy. This section discusses potential privacy threats based on the location of the breach on the user's sensitive data.

Threats on the client side: Exploiting a user device to store the collected sensitive data (user profile) by personalized search systems introduces risks. Some of these threats may lead to critical problems for both users and service providers.

Cross-site scripting (XSS) is among the prevalent vulnerabilities in recent Web applications. An attacker can execute scripts within the context of the Web site under attack. Different types of XSS exist with the same result allowing for the execution of malicious codes in the browser of the user, allowing the attacker to access sensitive user data.

Client-side SQL injection (csSQLi) is a new form of well-known SQL injection attacks that have emerged recently due to the introduction of database support on clients (Google Gears and HTML 5 SQL databases). A popular mechanism used in conjunction with SQL injections is a mechanism called stacked queries (SQ), allowing an attacker to execute his query irrespective of the original one. The attacker adds the SQ to an original one through the use of a semicolon. SQL injection with SQ can be a powerful combination, as it allows executing arbitrary SQL commands on the database, especially on old browsers' versions.

Client-side data corruption or leakage is when the user, or an attacker controlling his device, changes or corrupts the stored data or retrieves sensitive information. Numerous attacks can result in a data leakage/corruption, including malware, XSS, csSQLi, and threats exploiting vulnerabilities in the user's browser or device. To ensure data security and lower the risk of exploiting client-side data storage vulnerabilities, both service providers and users need to implement preventive measures (e.g., encryption, digital signatures, and access control mechanisms). Furthermore, output encoding mechanisms and parameterized queries prevent XSS and csSQLi, respectively.

Threats on the server side: Reports of privacy breaches on the server side affecting personalized systems dominate the news with increasing frequency. Most personalization systems store the collected user data on their servers (the first target for attackers). We can classify server-side privacy threats into two categories: insider threats and outsider threats.

Insider privacy threats come from inside the service-providing organization (malicious or negligent insiders, infiltrators, or even the organization's intentions). This category involves data leakage (as in the AOL scandal), data misuse (as in the Facebook–Cambridge Analytica scandal) when using data for other purposes. Data brokerage is also an insider threat (when aggregating data and reselling the valuable categories of customers to third parties). The theft of private or commercially relevant information can come from inside the organization.

Outsider privacy threats are ever-present and pose a real danger. Any system connected to the Internet is at risk. Historically, the most famous data breaches were typically outsider type. In 2016, the email giant and search engine Yahoo had their systems compromised in a data breach, causing an information loss of more than 500 million users [12]. eBay also reported that an attacker exposed its entire list of 145 million clients' accounts in May 2014 [13]. Many more companies on the web have suffered from data breaches (Adobe, Canva, LinkedIn, Zynga, etc.). While these breaches cost millions of dollars, outsider threats are usually the ones targeted with the traditional security measures (firewall, passwords, encryption, etc.) to prevent potential attacks (malware attacks, phishing, XSS, SQL injections, password attacks, etc.). However, securing the server is a challenging and complicated task that is hard to accomplish.

Failure to implement efficient security controls like patches/updates or secure configurations, replacing default accounts, or disabling unnecessary back-end services can compromise data confidentiality and integrity. Moreover, introducing an additional measure to enhance security may increase vulnerability and expose the system to further threats. The answer to this problem is to understand system vulnerabilities and implement a risk-mitigation approach taking into consideration insider and outsider threats.

Communication channel threats: Internet serves as an electronic chain connecting a client to a server. Messages on this network travel an arbitrary path from the client device to a destination point. The message passes through a number of intermediate nodes on the network before arriving at the final destination. It is difficult to ensure that each computer on the Internet, through which messages pass, is secure and non-hostile.

Web applications often use the HTTP protocol for client–server communication, which communicates all information in plain text. Even when they provide transport-layer security through the use of the HTTPS protocol, if they ignore certificate validation errors or revert to plain text communication after a failure, they can jeopardize security by revealing data or facilitating data tampering. Compression side-channel attacks such as CRIME and BREACH presented concrete and real-world examples of HTTPS vulnerabilities in 2012 and 2013. Since then, security experts confront new attacks on TLS/SSL every year, especially with servers using version 1.2 of the TLS communication protocol, which supports encryption and compression. The combination of encryption and compression algorithms presented security flaws that allowed the attackers to open the content of the encrypted HTTP header and use the authentication token within the cookie to impersonate a user.

These attacks, among others, can lead to a privacy threat called *eavesdropping* (a sniffing or man-in-the-middle attack), which consists of collecting information as it is sent over a channel by a computer or other connected devices. The attacker takes advantage of unprotected network communications to obtain information in-transit or once received by the user. Moreover, the advancement toward the future 5G networks is rapid and expected to offer high data speed, which will eventually result in increasing data flows in the communication channels between users and servers. This fact will raise the user's concerns about privacy protection in these networks.

2.3 Privacy Threat Model

Since implementing effective privacy-preserving systems requires understanding the range of potential privacy threats in this field, it is important to study and define a privacy threat model. The threat model described in this section is based on the aforementioned privacy threats in personalized search systems.

The idea is to model how an attacker could threaten the user's privacy and conduct an attack. As described in Fig. 1, we investigate three different threat scenarios:

- The attacker controls the user's device and has access to the whole user profile.
- The communication channel is insecure, the attacker can eavesdrop and collect the user's queries with portions of his profile, and then using auxiliary information (online ontology) and social engineering skills, he can guess the user's profile.

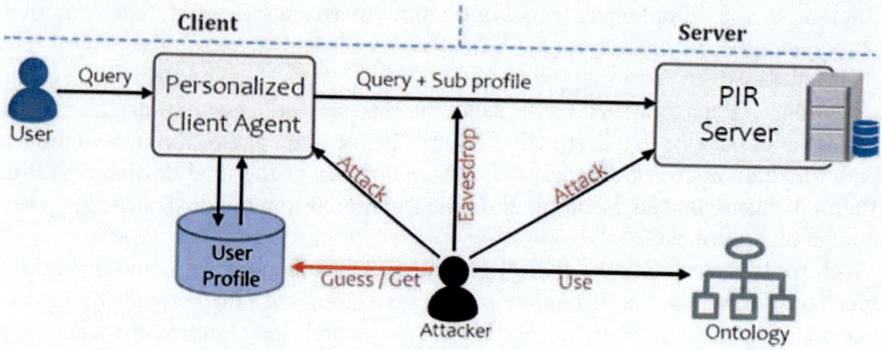

Fig. 1 Threat model for personalized search systems

- The server is compromised, and the attacker can collect data sent by the user and then guess the original user profile as in the second scenario.

In Sect. 2.2, we summarized potential methods that an attacker may use to access the client's device, the server, or the communication channel. Since the user profile is generally stored on the client side, encryption is mandatory to secure the user's data and reduce the risk in the first scenario.

During the browsing session in the personalized search system, a user sends many queries to the server, with short portions of his profile. As we mentioned in the second and third scenarios, an attacker can obtain a significant amount of the original profile, by collecting the sub-profiles and using the online ontology to figure the rest.

Considering the user profile P, each time a user enters a query q the system sends a part of P. If the attacker captures each generalized profile G_i, it is possible after n query to guess a significant portion of the profile P using the online taxonomy. And, even if the generalized profile G_i contains no private data, the attacker can still obtain the profile by comparing the G_n to the ontology.

$$\sum_{i=1}^{n} G_i = G_n \rightarrow P \tag{1}$$

where n is a number of queries (depends on the user activity and time).

To illustrate how an attacker can breach user privacy, Fig. 2 shows an example of a user profile (a) with two generalized profiles (G_a and G_b).

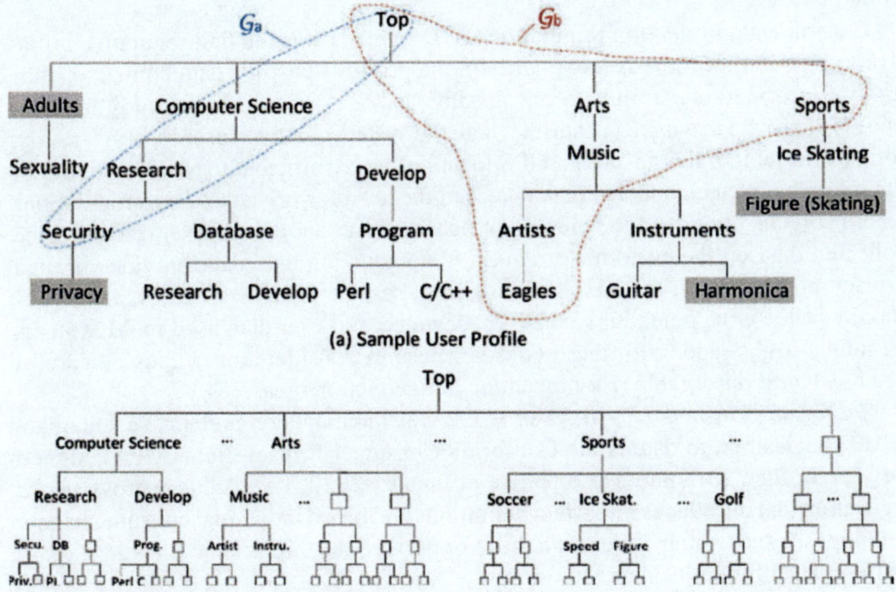

Fig. 2 Ontology-based user profile

The gray concepts in this figure reflect the user's private data. And the generalized profiles contain no sensitive data because the system stops at the parent nodes. However, in G_a for example, the attacker can retrieve the sub-tree of security relying on the taxonomy (*b*) in the same Fig. 2, where *security* is the parent of two nodes including a private one (*privacy*). Therefore, if the probability of touching both branches is equal, the attacker has 50% confidence in *privacy* leading to high privacy risk.

3 Privacy Protection Solutions

To reduce the aforementioned risks, a number of regulations and measures can be adopted in personalized search systems.

3.1 Privacy Laws and Regulations

Data are considered as the new oil, enabling new opportunities for advanced analytics (e.g., personalization) like never before, but this does not come without cost. By collecting user information, companies have to guarantee the use of data according to the latest privacy regulations. And various types of data require different measures; those generated in the healthcare sector, for example, are sensitive from a privacy point of view.

General Data Protection Regulations (*GDPR*) [14] was the first regulative initiative to mention location data explicitly in the privacy-sensitive data context, as this kind of information can infer some sensitive user information. The study in reference [15] proved that using a dataset where the system collected user location hourly, four spatio-temporal points (e.g., GPS location) were sufficient to recognize 95% of the individuals. Such conclusions increased the need for privacy-related regulations leading organizations to adopt new approaches such as anonymizing or removing the collected data on the customer's request. Anonymizing user data provides limited privacy protection. In the EU AI Guidelines, service providers and organizations should evaluate the potential adverse consequences of user data used in AI systems on human rights and, with these consequences in consideration, choose a careful strategy based on suitable risk prevention and reduction measures.

California Consumer Privacy Act (CCPA) is the most comprehensive state data privacy legislation to date is the California Consumer Privacy Act (CCPA), signed into law in June 2018 and was effective in January 2020. The CCPA is cross-sector legislation that introduces important definitions and broad individual consumer rights and imposes substantial duties on entities or persons that collect personal data about or from a California resident.

State Data Privacy Laws: In addition to the general regulations and laws, the USA has several data security and privacy laws among its states, territories, and localities. Currently, 25 state attorney generals in the USA oversee data privacy laws governing the collection, use, storage, safeguarding, and disposal of personal information collected from their residents, especially regarding the data breach notifications or the security of the social security numbers. Some apply to governmental entities only, some apply to private entities, and some apply to both.

Many organizations have already begun addressing this issue by implementing privacy-enhancing technologies (PET), which are not obligatory as per current privacy regulations yet represent the next measure toward a more ethical and secure data usage.

3.2 *Privacy-Enhancing Technologies*

Researchers try to address the privacy protection problems in personalized systems by protecting user identification using techniques such as the pseudo-identity, no identity, the group identity, and no personal information. Most efforts focus on the second level. For example, [16] provided online anonymity for users by creating a group profile of a number of users. The de-identification techniques are usually used in personalized systems that collect user's personally identifiable information (PII) and works on aggregate data. However, these techniques are vulnerable to attacks such as the unsorted matching attacks, temporal attacks, and complementary release attacks. Authors in [17] provided a detailed survey on de-identification techniques for privacy protection.

Others focus on techniques protecting the sensitive user data (user profile). In this type of privacy-preserving system, three main technique lines are available, namely differential privacy, randomized perturbation, and individual privacy.

Differential privacy (DP) is becoming widely accepted as a model for privacy protection during the past years. This solution preserves privacy by making it difficult for the attacker to assume the presence/absence of an individual in the dataset. This technique is designed to work on aggregate data and is most suitable for big data. For instance, authors in [18] proposed another DP model for neighborhood-based collaborative filtering capable of selecting neighbor privately but failed to maintain a reliable trade-off between privacy and accuracy. Authors in [19] designed a privacy-built-in client agent that perturbs user data on the client device. However, the perturbed data utility decreased due to an inhered process volatility. Another work in [20] proposes a probabilistic model for mobility dataset releasing based on differential privacy to give users control over privacy level in location-based services.

The main limitations of differential privacy are first, the noise amount added to the data (more noise for higher privacy risks), which reduces the data utility. It is also vulnerable against insider threats that come from inside the personalization server. Authors in [21] presented a survey of DP techniques for more details on the subject.

Randomized perturbation (RP) is another noise-based technique proposed in [22]. Authors claim that they can obtain accurate recommendations while adding randomness from a specific distribution to the original user data to counter information exposure. The chosen range of randomness is based only on experience, and this method does not have a provable privacy protection guarantee. Another work in [23] proposed a multi-leveled privacy-preserving approach for CF systems by perturbing ratings before submitting them to the server. Yet the results showed a decrease in utility. Authors in [24] presented a hybrid method for a privacy-aware recommender system by combining DP with RP to offer more privacy protection. However, the recommendation accuracy loss with this approach is significant. Most noise-based techniques share the problem of utility loss [25].

Both DP and RP techniques are designed for aggregate data privacy, ignoring each user's privacy requirements, and they both decrease the personalization accuracy. Lately, users' privacy concerns increased due to unethical data aggregation practices in many recommendation systems. For this reason, in our work, we focus on individual data privacy.

Individual privacy (IP): As an IP solution, authors in [26] claim that they can achieve better results with a privacy guarantee if the personalization is only performed based on less sensitive user data. The idea is to expose only the insensitive part of the profile to the search engine. Yet, even when using this approach, an attacker or the server can still collect a significant portion of the user profile. Authors in reference [27] proposed a privacy-protecting framework for book search based on the idea of constructing a group of likely fake queries associated with each user query to hide the sensitive topics in users' queries. This approach focused on limited query types with no support for general ones and no practical implementation.

4 Cryptographic Solutions

Cryptographic techniques such as homomorphic encryption (HE), searchable symmetric encryption (SSE), or garbled circuits used as a mechanism to protect private data confidentiality. Authors in [28] present a privacy-preserving solution to generate recommendations using HE and data packing. Also, authors in [29] used partially HE to design two privacy-preserving protocols for trust-oriented POI recommendation based on the off-line encryption and parallel computing.

A recent work [30] proposed a fully HE scheme for private IR in the cloud; the paper contains theoretical analysis but with no proof-of-concept implementation. A survey by [31] on keyword search with public key encryption can help understand the use of encryption in private information retrieval.

4.1 Secure k-NN Algorithm

The secure k-nearest neighbor (SkNN) algorithm [32] is used in the searchable encryption area to encrypt documents and queries presented in a vector space model of size m. SkNN algorithm allows calculating the similarity [33] (the dot product) between an encrypted document vector and an encrypted query vector without any need for decryption. The SkNN algorithm is composed of three functions: KeyGen, Enc, and Eval.

4.2 *Attribute-Based Encryption*

The attribute-based encryption (ABE) is an encryption method used to apply an access control policy to control access to a data collection [34]. It consists of adding some characteristics to the encrypted ciphertext and the user's private key. During the decryption process, ciphertext decryption is possible only if the number of matching attributes between the ciphertext and the private key exceeds a certain threshold.

Authors in [35] proposed an encryption method called key-policy attribute-based encryption (KP-ABE) that consists of storing an access structure in the user's secret key and attributes in the encrypted data. This method can achieve fine-grained access control and brings more flexibility in the management of the users than the previous technique. Nevertheless, KP-ABE has the disadvantage of not being intuitive.

Authors in [36] proposed an alternative method called CP-ABE (ciphertext-policy attribute-based encryption) that works in the same way as the KP-ABE method except that the access structure was stored in the ciphertext, while the attributes are in the user private key. During the decryption process, if some private key attributes satisfy the data access policy, then ciphertext decryption is possible. Otherwise, the user has no right to access this data, and his private key cannot decrypt the ciphertext. For example, (Pediatrics∧(Doctor1∨Doctor2)) is the access policy in the data. If the user secret key contains the attributes "Pediatrics" and "Doctor1", then the user has the right to access the data and can decrypt the ciphertext using his private key. After that, several approaches based on the CP-ABE method appeared in the literature.

4.3 *Homomorphic Encryption (HE)*

Homomorphic encryption is a cryptosystem for performing mathematical operations on encrypted data generating encrypted results which, after decryption, is the same as the result of these operations on unencrypted data [37]. Several HE systems appeared in the literature.

Paillier cryptosystem is an asymmetric encryption probabilistic algorithm. It is an additive homomorphic encryption method that can calculate two ciphertexts addition without any need for decryption. The Paillier cryptosystem applications can be in many fields that need to protect data privacy, such as the searchable encryption area.

Ideal lattice-based fully homomorphic encryption: Gentry and Boneh proposed the first fully homomorphic encryption (FHE) scheme based on ideal lattices [38].

The encryption process consists of hiding the message by adding noise. Decryption consists of using the secret key to remove the noise from the ciphertext. This scheme can perform homomorphic operations on arbitrary depth circuits. For that, the authors start by constructing a somewhat homomorphic encryption (SWHE) scheme performing limited operation number. However, the noise size grows after each arithmetic operation, mostly when it consists of multiplication until it becomes not possible to decrypt. To avoid this problem, the authors proposed a technique called "bootstrapping" that consists of refreshing the ciphertext by reducing the noise size. This technique allows us to transform a SWHE scheme into a FHE scheme. Nevertheless, this scheme is still a theoretical model because of its inefficiency.

Fully homomorphic encryption over the integers: Authors in [39] proposed a HE scheme that is similar to the one proposed by Gentry and Boneh [38] except that it is simpler and less efficient since it works with integers instead of ideals. This scheme has semantic security based on the hardness hypothesis of the approximate great common divisor (GCD) problem.

Homomorphic encryption from learning with errors: Brakerski and Vaikuntanathan proposed an asymmetric fully homomorphic encryption (FHE) scheme that operates over bits [40]. This scheme uses the ring learning with error (RLWE) assumption proposed in [41]. It manipulates polynomials using integer coefficients.

Leveled fully homomorphic encryption: The term (leveled) describes this approach because of the private key updating performed at each level of a circuit. The latter represents an arithmetic function and is composed of a set of gates. Each gate refers to the operation of addition or multiplication. The leveled homomorphic encryption [40] introduces two main techniques, key switching and modulus switching.

5 Conclusion and Perspectives

A personalized search system must ensure privacy protection to earn the user's trust. Otherwise, only a minority of users will use it, to whom the personalized experience is more important than their privacy.

In this paper, we reviewed different risks and threats on the user privacy in a threat modeling approach. Moreover, we presented several techniques to preserve the user's privacy in personalized search along with cryptographic solutions to protect the user sensitive data.

In the future, we plan to implement those solutions on a search system presented in [42] personalized using an ontology-based profiling method described in [43].

References

1. El-Ansari, A., Beni-Hssane, A., Saadi, M.: An improved modeling method for profile-based personalized search. In Proceedings of the 3rd International Conference on Networking, Information Systems & Security (pp. 1–6). (2020). https://doi.org/10.1145/3386723.3387874
2. El Makkaoui, K., Ezzati, A., Beni-Hssane, A., Motamed, C.: Cloud security and privacy model for providing secure cloud services. In 2016 2nd international conference on cloud computing technologies and applications (CloudTech) (pp. 81–86) (2016). https://doi.org/10.1109/CloudTech.2016.7847682
3. Parsania, V. S., Kalyani, F., Kamani, K.: A comparative analysis: DuckDuckGo vs. Google search engine. GRD J.S-Glob. Res. Dev. J. Eng.-Ing **2**(1), 12–17 (2016)
4. Tisserand-Barthole, C.: Qwant. com, un nouveau moteur français. Netsources **104**, 14 (2013)
5. El-Ansari, A., Beni-Hssane, A., Saadi, M.: An ontology based social search system. In: Networked Systems: 4th International Conference, NETYS 2016. https://doi.org/10.13140/RG.2.2.28161.79200
6. Roelofs, W.: The AOL scandal: an information retrieval view (2007)
7. El-Ansari, A., Beni-Hssane, A., Saadi, M.: An enhanced privacy protection scheme for Profile-based personalized search. International J. Adv. Trends Comput. Sci. Eng. **9**(3), (2020). https://doi.org/10.30534/ijatcse/2020/241932020
8. El Makkaoui, K., Ezzati, A., Hssane, A. B.: Challenges of using homomorphic encryption to secure cloud computing. In 2015 International Conference on Cloud Technologies and Applications (CloudTech) (pp. 1–7). IEEE. (2015). https://doi.org/10.1109/CloudTech.2015.7337011
9. Chen, D., Zhao, H.: Data security and privacy protection issues in cloud computing. In: 2012 International Conference on Computer Science and Electronics Engineering, Vol. 1, pp. 647–651 (2012)
10. El-Ansari, A., Beni-Hssane, A., Saadi, M., El Fissaoui, M.: PAPIR: privacy-aware personalized information retrieval. J. Ambient. Intell. Humaniz. Comput. (2021). https://doi.org/10.1007/s12652-020-02736-y
11. Hawalah, A., Fasli, M.: Dynamic user profiles for web personalisation. Expert Syst. Appl. **42**(5), 2547–2569 (2015)
12. Trautman, L.J., Ormerod, P.C.: Corporate Directors' and Officers' Cybersecurity Standard of Care: The Yahoo Data Breach. Am. UL Rev. **66**, 1231 (2016)
13. Minkus, T., Ross, K. W.: I know what you're buying: Privacy breaches on ebay. In: International Symposium on Privacy Enhancing Technologies Symposium, pp. 164–183. Springer, Cham (2014)
14. Voigt, P., Von dem Bussche, A.: The eu general data protection regulation (gdpr). A practical guide, 1st ed.. Springer International Publishing, Cham (2017)
15. De Montjoye, Y.A., Hidalgo, C.A., Verleysen, M., Blondel, V.D.: Unique in the crowd: The privacy bounds of human mobility. Sci. Rep. **3**, 1376 (2013)
16. Zhu, Y., Xiong, L., Verdery, C.: Anonymizing user profiles for personalized web search. In: Proceedings of the 19th international conference on World wide web, pp. 1225–1226 (2010)
17. Tomashchuk, O., Van Landuyt, D., Pletea, D., Wuyts, K., Joosen, W.: A Data Utility-Driven Benchmark for De-identification Methods. In: International Conference on Trust and Privacy in Digital Business, pp. 63–77. Springer, Cham (2019)
18. Zhu, T., Li, G., Ren, Y., Zhou, W., Xiong, P.: Differential privacy for neighborhood-based collaborative filtering. In: Proceedings of the 2013 IEEE/ACM International Conference on Advances in Social Networks Analysis and Mining, pp. 752–759 (2013)
19. Shen, Y., Jin, H.: Epicrec: Towards practical differentially private framework for personalized recommendation. In: Proceedings of the 2016 ACM SIGSAC conference on computer and communications security, pp. 180–191 (2016)
20. Zhang, J., Yang, Q., Shen, Y., Wang, Y., Yang, X., Wei, B.: A differential privacy based probabilistic mechanism for mobility datasets releasing. J. Ambient. Intell. Hum.Ized Comput. 1–12 (2020)

21. Desfontaines, D., Pejó, B.: Sok: Differential privacies. Proceedings on Privacy Enhancing Technologies **2020**(2), 288–313 (2020)
22. Zhu, J., He, P., Zheng, Z., Lyu, M.R.: A privacy-preserving QoS prediction framework for web service recommendation. In 2015 IEEE International Conference on Web Services, pp. 241–248. IEEE (2015)
23. Polatidis, N., Georgiadis, C.K., Pimenidis, E., Mouratidis, H.: Privacy-preserving collaborative recommendations based on random perturbations. Expert Syst. Appl. **71**, 18–25 (2017)
24. Liu, X., Liu, A., Zhang, X., Li, Z., Liu, G., Zhao, L., Zhou, X.: When differential privacy meets randomized perturbation: a hybrid approach for privacy-preserving recommender system. In International Conference on database systems for advanced applications, pp. 576–591. Springer, Cham (2017)
25. Siraj, M.M., Rahmat, N.A., Din, M.M.: A survey on privacy preserving data mining approaches and techniques. In Proceedings of the 2019 8th International Conference on Software and Computer Applications, pp. 65–69 (2019)
26. Shou, L., Bai, H., Chen, K., Chen, G.: Supporting privacy protection in personalized web search. IEEE Trans. Knowl. Data Eng. **26**(2), 453–467 (2012)
27. Wu, Z., Li, R., Zhou, Z., Guo, J., Jiang, J., Su, X.: A user sensitive subject protection approach for book search service. J. Am. Soc. Inf. Sci. **71**(2), 183–195 (2020)
28. **Erkin, Z., Veugen, T., Toft, T., Lagendijk, R.L.: Generating private recommendations efficiently** using homomorphic encryption and data packing. IEEE Trans. Inf. Forensics Secur. **7**(3), 1053–1066 (2012)
29. Liu, A., Wang, W., Li, Z., Liu, G., Li, Q., Zhou, X., Zhang, X.: A privacy-preserving framework for trust-oriented point-of-interest recommendation. IEEE Access **6**, 393–404 (2017)
30. **Wang, X., Luo, T., Li, J.: An Efficient Fully Homomorphic Encryption Scheme for Private** Information Retrieval in the Cloud. Int. J. Pattern Recognit Artif Intell. **34**(04), 2055008 (2020)
31. Zhou, Y., Li, N., Tian, Y., An, D., Wang, L.: Public Key Encryption with Keyword Search in Cloud: A Survey. Entropy **22**(4), 421 (2020)
32. Lei, X., Liu, A. X., Li, R.: Secure knn queries over encrypted data: Dimensionality is not always a curse. In 2017 IEEE 33rd International Conference on Data Engineering (ICDE), pp. 231–234. IEEE (2017)
33. Erritali, M., Beni-Hssane, A., Birjali, M., Madani, Y.: An approach of semantic similarity measure between documents based on big data. Int. J. Electr. Comput. Eng. **6**(5), 2454 (2016). https://doi.org/10.11591/ijece.v6i5.10853
34. Li, J., Zhang, Y., Ning, J., Huang, X., Poh, G. S., Wang, D.: Attribute based encryption with privacy protection and accountability for CloudIoT. IEEE Transactions on Cloud Computing (2020)
35. Li, J., Yu, Q., Zhang, Y., Shen, J.: Key-policy attribute-based encryption against continual auxiliary input leakage. Inf. Sci. **470**, 175–188 (2019)
36. Qiu, S., Liu, J., Shi, Y., Zhang, R. (2017). Hidden policy ciphertext-policy attribute-based encryption with keyword search against keyword guessing attack. Sci. China Inf. Sci. **60**(5), 052105.
37. **El Makkaoui, K., Beni-Hssane, A., Ezzati, A.: Cloud-ElGamal: An efficient homomorphic** encryption scheme. In: 2016 International Conference on Wireless Networks and Mobile Communications (WINCOM), pp. 63–66. IEEE. (2016). https://doi.org/10.1109/WINCOM.2016.7777192
38. Gentry, C., Boneh, D.: A fully homomorphic encryption scheme, vol. 20, No. 9, pp. 1–209. Stanford University, Stanford (2009)
39. Van Dijk, M., Gentry, C., Halevi, S., Vaikuntanathan, V.: Fully homomorphic encryption over the integers. In Annual International Conference on the Theory and Applications of Cryptographic Techniques, pp. 24–43. Springer, Berlin, Heidelberg (2010)
40. Brakerski, Z., Gentry, C., Vaikuntanathan, V.: (Leveled) fully homomorphic encryption without bootstrapping. ACM Transactions on Computation Theory (TOCT) **6**(3), 1–36 (2014)
41. Lyubashevsky, V., Peikert, C., Regev, O.: On ideal lattices and learning with errors over rings. Journal of the ACM (JACM) **60**(6), 1–35 (2013)

42. El-Ansari, A., Beni-Hssane, A., Saadi, M.: A multiple ontologies based system for answering natural language questions. In: Europe and MENA cooperation advances in information and communication technologies, pp. 177–186. Springer, Cham. (2017). https://doi.org/10.1007/978-3-319-46568-5_18
43. El-Ansari, A., Beni-Hssane, A., Saadi, M.: An ontology-based profiling method for accurate web personalization systems. J. Theor. Appl. Inf. Technol. **98**(14), 2817–2827 (2020)

Enhanced Intrusion Detection System Based on AutoEncoder Network and Support Vector Machine

Sihem Dadi and Mohamed Abid

Abstract In the recent years, Internet of Vehicles (IoV) became the subject of many searches. IoV relies on Vehicular Ad-hoc NETwork (VANET) which is based on Vehicle-to-Vehicle (V2V), Vehicle-to-Infrastructure (V2I) communication, and Vehicle-to-Everything (V2X). Due to heterogeneous communications in VANET, vehicles are increasingly becoming vulnerable to intruders or attackers. Several research works proposed solutions to detect intrusion, some of them used deep learning which uses a set of algorithms called neural networks in different fields, such as health, economic, transport and many others domains. In this paper, an enhanced intrusion detection system (IDS) based on AutoEncoder (AE) network and support vector machine (SVM) is proposed. Our goals are to detect five main types of attacks (DoS, DDoS, Wormhole, Black hole and Gray hole attack) that VANET may face by combining the ability of support vector machine (SVM) to exploit large amounts of data with the strength of features extraction provided by AutoEncoder (AE). The experimental results show that the enhanced intrusion detection system (IDS) is capable to reach a high level of accuracy, also we prove by security analysis that our new solution detects successfully these attacks.

1 Introduction

Smart and connected vehicles are inspiring researchers to develop new uses cases to exploit their benefits. That why, *Vehicular Ad-hoc NETwork* (VANET) is the backbone to intelligent transportation system (ITS) research. VANET is a mobile network allowing vehicles to communicate with each other, with the aim of improving road safety through the exchange of alerts between vehicles.

S. Dadi (✉) · M. Abid
Faculty of sciences of Gabes, Laboratory Hatem Bettaher Irescomtah, University of Gabes, Gabes, Tunisia
e-mail: sihem.dadi@fsg.u-gabes.tn

M. Abid
e-mail: mohamed.abid@enig.rnu.tn

M. Ben Ahmed et al. (eds.), *Networking, Intelligent Systems and Security*, Smart Innovation, Systems and Technologies 237,
https://doi.org/10.1007/978-981-16-3637-0_23

In VANET, there are many types of communications, while the most popular ones are: *Vehicle to Vehicle* V2V and *Vehicle to Infrastructure* (V2I) communication which means a communication between the *Road Side Units* (RSU), a fixed infrastructure and vehicle. Also, vehicles can communicate with any objects and this case is known as *Vehicle to Everything* (V2X).

Through this network, vehicles can exchange control, alert or "other" messages, depending on the application and the environmental context. Due to the communication of vehicles within VANET and with others entities in the Internet, they can be easily a target for many attacks. Detecting successful or failed attack attempts is important to secure networks (servers, end-hosts and other assets). Some attackers want to change the content of the information exchanged in the network, owing to dynamic topology which may result loss of connectivity as well as unreliable connections. To secure VANET, many tools are used such as ciphering exchanging data, *Intrusion Detection System* (IDS) and *Intrusion Prevention System* (IPS). IDS are defined as the last line of defense during an attack. They allow the detection of attacks targeting a vehicle or a network. However, they do not offer mechanisms in response to attacks. IDS informs administrator that an attack has been detected; however, this information is only relevant if used. On the other hand, intrusion prevention system (IPS) has the capabilities of detecting, isolating and blocking malicious attacks in real time.

In parallel, a new technology has emerged and succeeded in a short time to penetrate several domains for different purposes, which is deep learning. The latter is a subfield of machine learning: a set of algorithms that try to imitate the functioning of the human brain. Among the areas in which deep learning has been introduced, we cite: Military, health, biology, transport,... and many others. In the transport field, deep learning is used for many objectives, but the most important among them are: object detection, vehicle trajectory prediction, traffic flow prediction and intrusion detection. Also, deep learning provides intelligent solutions since it is based on neural network.

The idea of our solution is to design an enhanced IDS based on deep learning, *AutoEncoder* (*AE*) network and *Support Vector Machine* (*SVM*).

This paper is organized as follows: Sect. 2 discusses the theoretical background that mean the technologies and tools to be used in our system. Section 3 describes related works. Section 4 is devoted to the proposed solution using neural networks. We utilized the UNSW-NB15 data set to evaluate the performance metrics of the proposed enhanced IDS, and we analysed its *Security Analysis* . Finally, we conclude the paper and give some future works.

2 Theoretical Background

This section presents IDS, VANET and deep learning.

2.1 Intrusion Detection System

An intrusion detection system is a software or hardware device that automates the intrusion detection process which consists of supervising events coming from network in order to detect signs of intrusions [1]. Therefore, an IDS is a monitor-only application designed to identify and report on anomalies but after a hackers already damages network infrastructure.

Hereafter, we present the different classes of IDS and the working process of each model.

- IDS classes: Mainly as shown in Fig. 1, an IDS can be categorized into two main classes: based on its location in the network or by its detection method. The latter type is subdivided into two subclasses:
  Misuse/Signature-based detection: It is based on signatures which are stored in knowledge base; this model has a high accuracy rate of detection but only for known attacks.
  Anomaly-based detection: This model is smarter than the previous one, because it does not require signatures to detect intrusion and also it can identify unknown attacks depending on similar behavior of other intrusions. It can be accomplished via the use of artificial intelligence and deep learning algorithms. This method is prone to high false positive rate of detection.
- IDS working process: An IDS working process depends on the detection model. In the case of signature-based detection, it monitors all the network packets and based on analyzing signatures stored on the knowledge base, it detects potential malware and matches the suspicious activities in the network. In contrast, an anomaly-based detection follows these steps to successfully detect intrusions: It starts monitoring the network traffic and analyses its pattern against predefined norms or knowledge base. Then, if an abnormal traffic is identified, then it launches alerts to report unusual behavior.

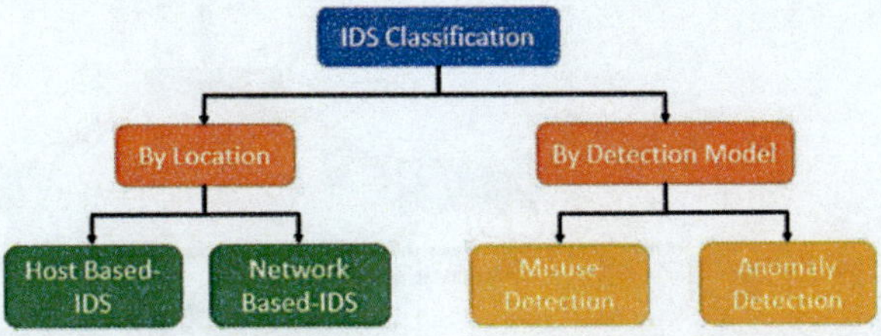

Fig. 1 Intrusion detection system classification

2.2 Vehicular Ad Hoc Network VANET

Vehicular Ad hoc NETwork called "VANET", a derivation of Mobile Ad hoc NETwork (MANET), was firstly introduced in 2001 to guarantee vehicle and driver security.

- VANET Architecture: Typically, there are four main components that make up the VANET networks: OBU, RSU, TA and AU [2] (As shown in Fig. 2).
  On Board Unit (OBU): As its name suggests, is a unit attached to each vehicle supporting *Intelligent Transport System* ITS, with the aim of exchanging information with the approximate OBUs or with RSUs.
  Road Side Unit (RSU): Standing for the Base Station and acting as the gateway between vehicles and road services provided by VANET.
  Trusted Authority (TA): Responsible for assigning digital certifications (unique identifier) to RSUs and OBUs in the network.
  Application Unit (AU): A device equipped within the vehicle that uses the applications provided by the provider using the communication capabilities of the OBU.
- Communication types in VANET: Fig. 3 shows the most popular communication types in VANET: Vehicle-to-Vehicle (V2V) and Vehicle-to-Infrastructure (V2I) [3]:
  Vehicle-to-Vehicle (V2V): Which allows vehicles to communicate with each other using IEEE WAVE [3]. **Vehicle-to-Infrastructure (V2I)**: Which allows the vehicle to communicate with the infrastructure such as traffic light or road side units (RSU), using Wi-Fi and 4G/LTE [3].
  A vehicle in VANET can communicate with other entities outside its original network. This leads to tremendous growth in the number of its communications. Hence, many new types of communication emerged such as V2N (Vehicle-to-

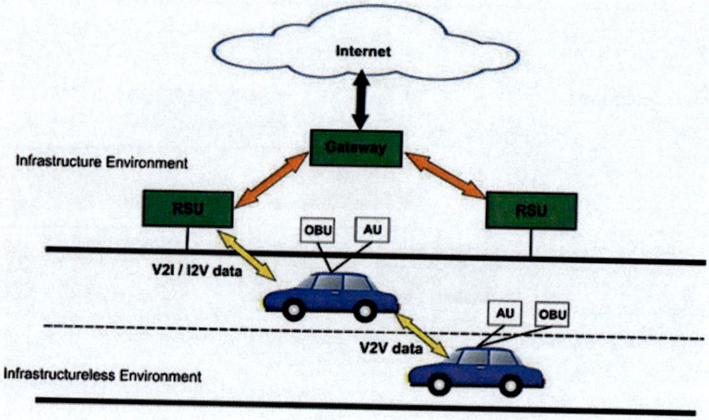

Fig. 2 VANET architecture

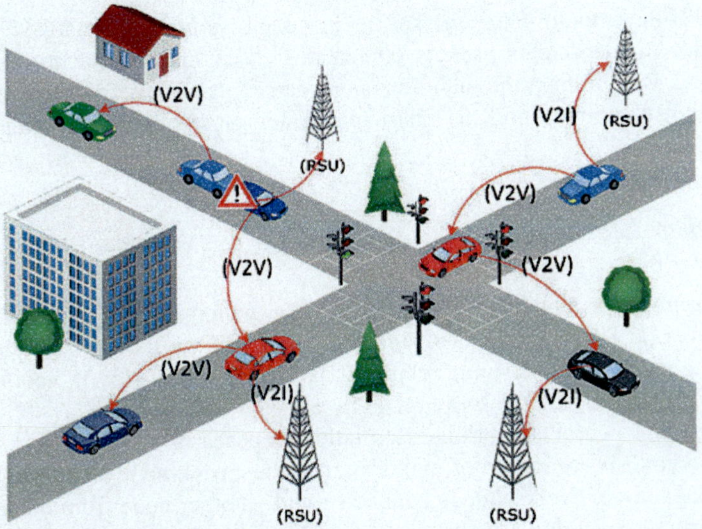

Fig. 3 Communication types in VANET

Network), V2P (Vehicle-to-Pedestrian), V2S (Vehicle-to-Sensor), V2G (Vehicle-to-Grid) and V2D (Vehicle-to-Device) and more others in the near future ....

- Attacks in VANET: The number of attacks that a car can face, in a VANET, is increasingly countless. These attacks can either shut down or degrade network performance [4]. Among these threats, we focus on five attacks that appear to be the most dangerous: Denial of Service, Distributed Denial of Service, Wormhole, Black and Gray hole attacks.
  Denial of Service Attack (DoS): is the act of blocking or preventing access to authorized users. This attack is the most dangerous one, because it threatens the network availability.
  Distributed Denial of Service Attack (DDoS): In DDOS attack, multiple malicious vehicles launch attack on a legitimate vehicle from different locations, and they may use different time slots to send those messages. So, it is very difficult to prevent or track this attack.
  Wormhole Attack: It is an attack in which two attackers locate themselves strategically in the network. Then, the attackers keep on listening to the network and record the wireless information.
  Black hole Attack: It happens when one or more nodes drop communications between other nodes. It is mainly a kind of denial of service, where the black hole is a node that always responds positively with an Route Reply (RREP) to every Route Request (RREQ) message , even if it does not have a legitimate route to the destination.

Gray Hole Attack: This attack is a category of a black hole attack in that the malicious node deletes packets but, in this case, this is done in a partial or a selective way. In fact, the malicious node can switch between two modes: Either it remains in normal mode as a harmless node or it switches to the attack mode.

2.3 Deep Learning

Deep learning is a subfield of machine learning, which is essentially a multi-layer neural network. The latter is a set of algorithms which mimic the human brain operates in order to recognize underlying relationships in a set of data [5]. A neural network has the following architecture in Fig. 4.

In general, a neural network is essentially composed of:

Input Layer: It is composed of one or more nodes; each node receives input from an external source. Every node is connected with another node from the next layer, with a particular weight.

Hidden Layers: As their name indicates, these layers are hidden from the external world, a neural network has at least 5 to 10 layers or even more. Calculations are executed in these layers, on data received from input layer in order to generate results.

Output Layer: Receives the results generated from the hidden layers.

Nowadays, there are many types of neural networks in deep learning which are used for different purposes. Among these types, we cite: *Convolutional Neural Networks* (CNN) and its derivations (DCN, DN, . . .), *Recurrent Neural Networks* (RNN) and its improvements (LSTM and GRU), also *Generative Adversarial Networks* (GAN) and its subtypes (DCGAN, WGAN, . . .) and finally *AutoEncoder network* (AE) and its subcategories (SEA, SSEA, . . .).

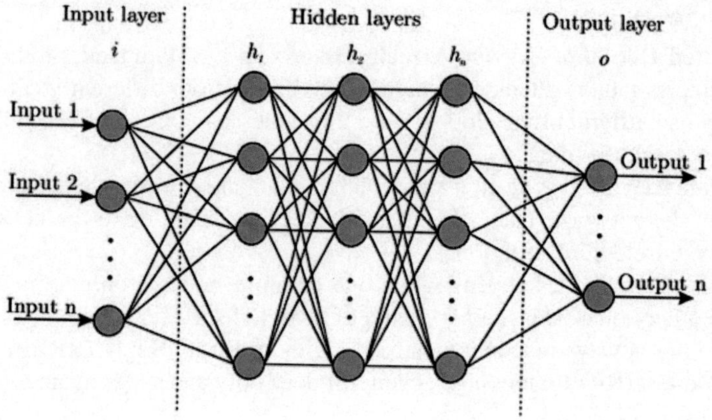

Fig. 4 Neural network structure

There are as many neural network classes as there are classifiers; among these classification algorithms, we cite: *Logistic Regression* (LR), *Naïve Bayes*, *Stochastic Gradient Descent*, *K-Nearest Neighbours* (K-NN), *Decision Tree* (DT), *Random Forest* (RF), *Support Vector Machine* (SVM), ... In this paper, we chosen to use of AE and SVM which are detailed in the following items.

- AutoEncoder (AE) network: AutoEncoder is a neural network approach that teaches a network to ignore noise, in order to learn efficient data representations (encoding) [6]. it is an unsupervised learning technique.
  AutoEncoders are widely used for: dimensionality reduction, image compression, denoising and generation and for features extraction ... and many others. As shown in Fig. 5, an AutoEncoder network is composed of two main gates which are:
  Encoder: receives the input and transforms it to a new representation, which is usually called a code or latent variable.
  Decoder: receives the generated code at the encoder and transforms it to a reconstruction of the original input.
  AE hyperparameters: There are four hyperparameters that we need to set before training an AE:
  - Code size: number of nodes in the middle layer.
  - Number of layers: the autoencoder can be as deep as we like.
  - Number of nodes per layer: the AE architecture we are working on is called a stacked AE since the layers are stacked one after another.
  - Loss function: we either use mean squared error (mse) or binary crossentropy. If the input values are in the range [0, 1], then we typically use crossentropy, otherwise we use the mean squared error.

  Similarly for the feature extraction phase, it can be achieved via many algorithms such as Kernel PCA, neural networks, ... In this paper, we select AE to realize this task because according to [7], this neural network is the best choice to extract features.
- Support vector machine (SVM): SVMs are a family of machine learning algorithms that solve problems of classification, regression or anomaly detection [8]. They are known for their solid theoretical guarantees, their great flexibility and their ease of use even without great knowledge of data mining. Figure 6 illustrates SVM principle which is simple: separate the data into classes using a border as "simple" as possible, so that the distance between the different groups of data and the border which separates them is maximum . This distance is also called "margin" and SVMs are thus referred to as "wide margin separators", the "support vectors" being the data closest to the border. There are many methods to classify a set of data such as: artificial and deep neural networks, decision tree, random forest, ... But and based on many comparative studies such as [9], it shows that support vector machine (SVM) is more powerful in the classification phase and that will be our choice to classify the selected features.

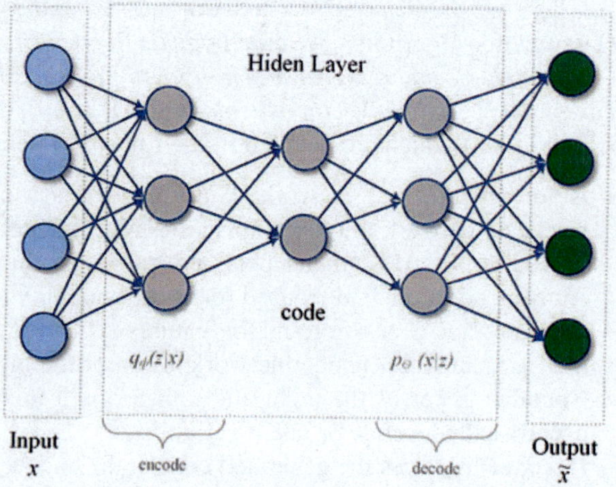

Fig. 5 AutoEncoder architecture

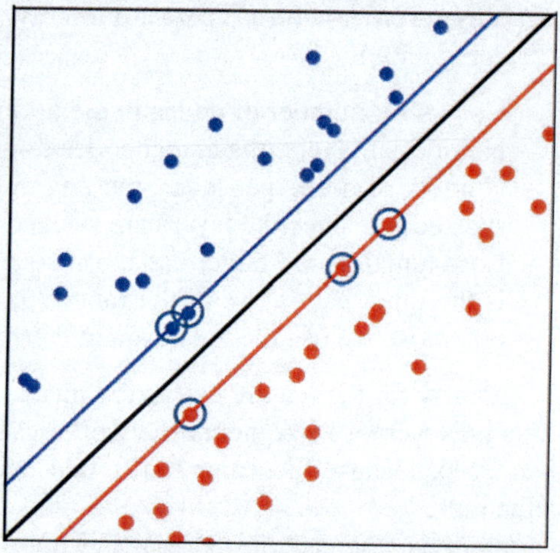

Fig. 6 SVM principle

3 Literature Survey

Thanks to its ability to detect attacks with high accuracy, intrusion detection is the most authentic technique to protect VANET. This is why it is necessary to have a strong detection mechanism such as approaches based on deep learning. In this context, several approaches have been proposed:

Multilayer Perceptron-Based Distributed Intrusion Detection System for Internet of Vehicles is an intrusion detection approach proposed by Anzer and Elhadef [10]. DoS, User to Root attack (U2R), Root to Local attack (R2L) and probe attacks are successfully identified, in this approach. Results are in form of prediction, classification and confusion matrix.

Sangve et al. suggested an algorithm for detecting rogue nodes in VANET [11]. Rogue nodes broadcast false information report that's why, in the actual approach, authors use anomaly-based detection approach to detect rogue nodes which are introduced in the system and are successfully detected.

Zhao et al. deployed an intrusion detection system for detecting intrusion in VANET [12]. In this work, they used two neural networks: *Deep Belief Network* (DBN) to extract features and *Probabilistic Neural Network* (PNN) in the classification phase.

In [13], Maglaras combined the dynamic agents and static detection to design intrusion detection system in VANET. By this approach, only DoS attack is successfully identified and detected.

Zeng et al. implemented *DeepVCM* which is a deep learning-based method for intrusion detection in VANET [14]. DeepVCM consists of two models *Convolutional Neural Network* (CNN) algorithm for features extraction and *Long Short-Term Memory* (LSTM) algorithm for classification. In this paper, authors identify the Dos, DDoS, Black hole, Wormhole and Sybil attack.

In order to detect distributed denial-of-service attack in VANET, Gao et al. developed a distributed network intrusion detection system [15]. They used *Spark-ML-RF*-based algorithm to train the system, and both *Random Forest algorithm* for classification and *T Decision Tree* to perform the classification process.

In the same context, Vatilcar et al. introduced an intrusion detection system in VANET using deep learning method [16]. This work was achieved by the use of *Deep Belief Network* (DBN) to extract the meaningful features and *Restricted Boltzmann Machine* (RBM) in classification phase. In the next section, we present the design of our proposed solution.

4 Enhanced IDS Based on AE and SVM

In this section, we present the design of an enhanced IDS which aims to detect attacks in the VANET, using AutoEncoder network and SVM algorithm. This solution is better comparing with previous ones since none of the latter can detect all the five attacks mentioned earlier, and they are cumbersome and the tools used are difficult to train.

Also, we use of a special dataset to train and test our system. We manage to study its performances. Finally, we analyse the security of our solution.

Our solution is based on the same architecture (see Fig. 7) presented in paper [17] which is the common architecture for all the previous solutions. It is composed of two modules:

Profiling module: contains the features trained off-line.

Monitoring module: detects a type of an incoming packet after feature extraction. If the monitoring module identifies a new attack type, the profiling module may update the database of the profiling module for upcoming packets.

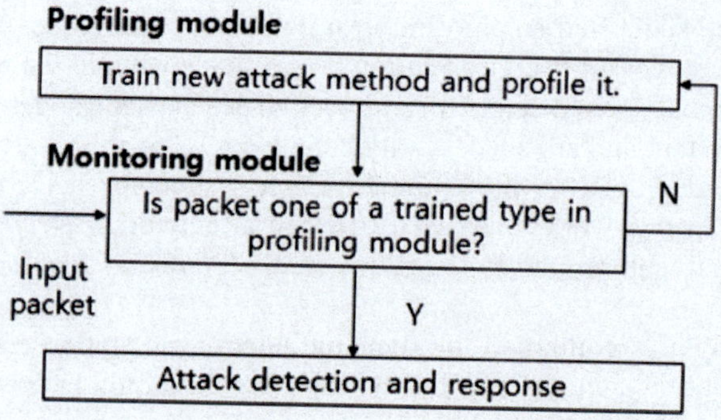

Fig. 7 IDS based on deep learning algorithms common architecture

4.1 The Proposed Architecture

The enhanced IDS aims to successfully identify: DoS, DDoS, Wormhole, Black and Gray hole attacks. Our method provides an enhanced IDS that combines the advantages of AE to extract features and SVM algorithm to classify features already extracted to solve intruders detection problem.

We implemented the new IDS and make performance analysis using UNSW NB-15 dataset which was developed in 2015 [18], and it is a dataset for network intrusion detection systems containing two million and 540,044 records. This data set involves nine attack categories and 49 features.

The partitioned dataset contains ten categories, one normal and nine attacks, namely generic, exploits, fuzzers, DoS, reconnaissance, analysis, backdoor, shell-code and worms. Figure 8 shows in detail the class distribution of the UNSW-NB15 dataset.

Category	**Training Dataset**	**Testing Dataset**
	UNSW_NB15_Training-Set	**UNSW_NB15_Testing-Set**
Normal	56,000	37,000
Generic	40,000	18,871
Exploits	33,393	11,132
Fuzzers	18,184	6062
DoS	12,264	4089
Reconnaissance	10,491	3496
Analysis	2000	677
Backdoor	1746	583
Shellcode	1133	378
Worms	130	44
Total	175,341	82,332

Fig. 8 Class distribution of the UNSW-NB15 dataset

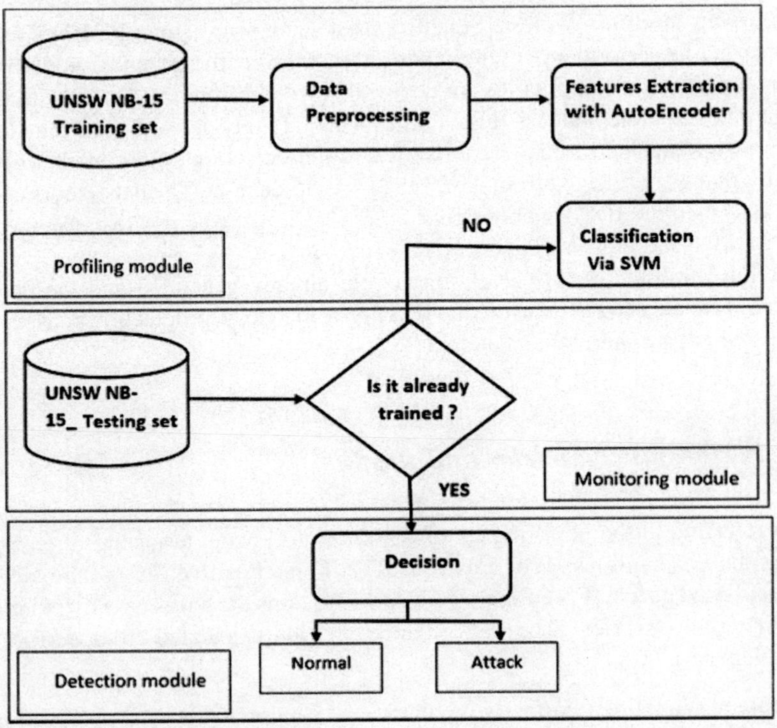

Fig. 9 Overall of the proposed system

Based on the common architecture already mentioned in the previous works, our IDS is composed of three modules, which are respectively: Profiling, Monitoring and Detection module as shown on Fig. 9.

Profiling Module: is divided into three main phases which are :

- *Data preprocessing*: In this phase, both encoding and normalization stages take place: the encoding process consists of converting symbol features to numerical values, and the normalization one refers to range these encoded values between 0 and 1.
- *Features selection/extraction*: Aims to reduce the number of features in a dataset by creating new features from the existing ones (and then discarding the original features). For our IDS, this phase is reached via the use of AutoEncoder (AE) network, the encoder receives preprocessed data from the previous phase and compresses it to a latent space representation, and then the encoder sends it to the decoder which had to reconstitute the input from the latent space representation.
- *Classification*: As its name indicates, the classification phase refers to predicting which class a feature belongs to. In our system, we use SVM to classify extracted features.

Monitoring module: We tested the proposed security system with UNSW NB-15 testing set. The performance of the enhanced IDS is directly related to anomaly detection algorithm. If the anomalies are detected correctly from the data set, it provides a high detection rate and less false alarms. The anomaly detection has the ability to detect novel attacks. In this phase, we tested the detection system with significant features that were selected from the UNSW NB-15 data set. The behaviours are analysed, and then the IDS generates four types of alarms: true positive, true negative, false positive and false negative.

Detection module: At this level, alarms are already generated and the detection accuracy rate are used to measure the IDS performance. The detection phase has two outputs which are: normal and attack.

4.2 *Performance Metrics and Results*

When referring to the performance of IDSs, the following terms are often used to discuss their capabilities: **True Positive (TP), False Positive (FP), True Negative (TN), False Negative (FN)**. Figure 10 clearly explains the terms already mentioned.

Performance metrics The performance metrics calculated from performance parameters are:

- Accuracy is a ratio of correctly predicted observation to the total observations.
  Accuracy = TP+TN/TP+FP+FN+TN.
- Precision is the ratio of correctly predicted positive observations to the total predicted positive observations.
  Precision = TP/TP+FP.
- Recall is the ratio of correctly predicted positive observations to the all observations in actual class.
  Recall = TP/TP+FN.
- F1-score is the weighted average of Precision and Recall.
  F1-score = 2*(Recall * Precision) / (Recall + Precision).

Results

After applying our approach, the results obtained are detailed in Table 1).

		Predicted class	
Actual Class		Class = Yes	Class = No
	Class = Yes	True Positive	False Negative
	Class = No	False Positive	True Negative

Fig. 10 Performance patameters

Table 1 Experimental results

Obtained results			
Parameters	Values	Parameters	Values (%)
True Positive (TP)	99.3 %	False Positive (FP)	5.78
True Negative (TN)	98.7 %	False Negative (FN)	0.85
Accuracy	96.7%		
Precision	94.4%		
Recall	99.1%		
*F*1-score	96.6%		

Table 2 Comparison between approaches

Work	DoS	DDoS	Black hole	Wormhole	Gray hole
Anzer et al.	+	−	−	−	−
Zhao et al.	+	−	−	−	−
Maglaras	+	−	−	−	−
Zeng et al.	+	+	+	+	−
Gao et al.	+	+	−	−	−
Our approach	+	+	+	+	+

We conclude that our enhanced IDS obtained good performances and reach its design goals especially a high level of accuracy.

4.3 Security Analysis

Table 2 presents a comparison between previous solutions and our proposed one. We observe that other solutions do not detect all attacks in VANET:

When using this approach, the five main attacks mentioned above are detected by our enhanced IDS.

5 Conclusion

Vehicular Ad hoc NETwork (VANET) is more and more vulnerable to attackers who try always to develop new techniques to hack it, by forging false information which may result in accidents. This evolution in attacks must leads to an evolution in protection methods such as intrusion detection system (IDS) which is defined as the last line of defense in network during an attack. At the same time, neural networks have seen a great evolution and have been able to penetrate several fields: health, transport, . . . with different utilities such as: intrusion detection, object detection, ... In this paper, we designed an enhanced IDS based on AutoEncoder (AE) network and support vector machine (SVM) in order to take advantage of their benefits and detect attacks in an effective manner. The experiment reflects the performance of the enhanced IDS in reaching high level of accuracy. Furthermore, this approach detected the various attacks that VANET may face such as DoS, DDoS, Wormhole, Black and Gray hole attack.

As future work, we could apply this solution and test its feasibility in different networks such as Flying Ad hoc NETworks (FANET) and Sea Ad hoc NETworks (SANET).

References

1. Alamiedy, T., Anbar, M., Al-Ani, A., Al-Tamimi, B., Faleh, N.: A review on feature selection algorithms for anomaly-based intrusion detection system. In: Proceedings of the 3rd International Conference of Reliable Information and Communication Technology (IRICT 2018) (2019). https://doi.org/10.1007/978-3-319-99007-1-57
2. Al-Sultan, S., Al-Doori, M.M., Al-Bayatti, A.H., Zedan, H.: A comprehensive survey on vehicular Ad Hoc network. J. Netw. Comput. Appl. (2014). https://doi.org/10.1016/j.jnca.2013.02.036
3. Kumar, G., Saha, R., Rai, M.K., Kim, T.: Multidimensional security provision for secure communication in vehicular ad hoc networks using hierarchical structure and end-to-end authentication. IEEE Access (2018). https://doi.org/10.1109/ACCESS.2018.2866759
4. Zaidi, T., Syed, F.: An overview: various attacks in VANET. In: 4th International Conference on Computing Communication and Automation (ICCCA) (2018). https://doi.org/10.1109/ccaa.2018.8777538
5. Deng, L.: A tutorial survey of architectures, algorithms and applications for deep learning. APSIPA Trans. Signal Inf. Process. (2014). https://doi.org/10.1017/atsip.2013.9
6. Mohammadi, M., Al-Fuqaha, A., Sorour, S., Guizani, M.: Deep learning for IoT big data and streaming analytics: a survey. IEEE Commun. Surv. Tutor. (2018). https://doi.org/10.1109/COMST.2018.2844341
7. Yan, B., Han, G.: Effective feature extraction via stacked sparse autoencoder to improve intrusion detection system. IEEE Access (2018). https://doi.org/10.1109/ACCESS.2018.2858277
8. Zhiquan, Q., Ying, J.T., Yong, S.: Robust twin support vector machine for pattern classification. Pattern Recogn. (2013). https://doi.org/10.1016/j.patcog.2012.06.019
9. Phan, T.N., Martin, K.: Comparison of Random Forest, k-Nearest Neighbor, and support vector machine classifiers for land cover classification using Sentinel-2 imagery. Sensors (Basel) (2018). https://doi.org/10.3390/s18010018
10. Anzer, A., Elhadef, M.: A multilayer perceptron-based distributed intrusion detection system for internet of vehicles. In: 2018 IEEE 4th International Conference on Collaboration and Internet Computing (CIC (2018). https://doi.org/10.1109/CIC.2018.00066
11. Sunil, M.S., Reena, B., Vidhya, N.G.: Intrusion detection system for detecting rogue nodes in vehicular ad-hoc network. In: International Conference on Data Management, Analytics and Innovation (ICDMAI) (2017). https://doi.org/10.1109/ICDMAI.2017.8073497
12. Zhao, G., Zhang, C., Zheng, L.: Intrusion detection using deep belief network and probabilistic neural network. In: International Conference on Computational Science and Engineering (CSE) and IEEE International Conference on Embedded and Ubiquitous Computing (EUC) (2017). https://doi.org/10.1109/CSE-EUC.2017.119

13. Maglaras, L.A.: Intrusion detection using deep belief network and probabilistic neural network. Int. J. Adv. Comput. Sci. Appl. (IJACSA) (2015). https://doi.org/10.14569/IJACSA.2015.060414
14. Zeng, Y., Qiu, M., Zhu, D., Xue, Z., Xiong, J., Liu, M.: DeepVCM: a deep learning based intrusion detection method in VANET. In: 5th International Conference on Big Data Security on Cloud (BigDataSecurity), IEEE International Conference on High Performance and Smart Computing, (HPSC) and IEEE Intl Conference on Intelligent Data and Security (IDS) (2019). https://doi.org/10.1109/BigDataSecurity-HPSC-IDS.2019.00060
15. Gao, Y., Wu, H., Song, B., Jin, Y., Luo, X., Zeng, X.: A distributed network intrusion detection system for distributed denial of service attacks in vehicular ad hoc network. IEEE Access (2019). https://doi.org/10.1109/ACCESS.2019.2948382
16. Vatilkar, R.S., Thorat, S.S.: A review on intrusion detection system in vehicular ad-hoc network using deep learning method. Int. J. Res. Appl. Sci. Eng. Technol. (IJRASET) (2020). https://doi.org/10.22214/ijraset.2020.5258
17. Kang, M., Kang, J.: Intrusion detection system using deep neural network for in-vehicle network security. PLOS ONE (2016). https://doi.org/10.1371/journal.pone.0155781
18. Moustafa, N., Slay, J.: UNSW-NB15: a comprehensive data set for network intrusion detection systems (UNSW-NB15 network data set). In: Military Communications and Information Systems Conference (MilCIS) (2015). https://doi.org/10.1109/MilCIS.2015.7348942

Comparative Study of Keccak and Blake2 Hash Functions

Hind EL Makhtoum and Youssef Bentaleb

Abstract Hashing is so crucial for integrity. Indeed, in order to respond to new technologies requirements, hash functions have always been subject to evolution and optimization. In the same context, NIST organized a competition in 2012 for a Sha-3 hashing function. BLAKE and KECCAK were two interesting functions in terms of speed and security. Keccak was the winner of the competition. It is different from the antecedent of SHA family because it combines the sponge construction to the Keccak-f-permutations. Therefore, it is optimized for security and hardware use. The Blake function is based on the same principle of the Sha-2, and it also presents many advantages of security and speed. The Blake was improvised for the BLAKE2, which is more optimized, faster, and more efficient. In this paper, we present the Keccak and the BLAKE2 algorithms; then, we will compare the performance parameters of the functions: Blake-256, Blake-512, Keccak-256, Keccak-512, Blake2b, and Blake2s.

1 Introduction

With the drastic evolution of technologies and the growth of resource-constrained smart objects,, all the related fields are led to ensure a higher level of performance and security. Data integrity is based on cryptographic hash functions, which play an essential role in digital signatures, message authentication codes, file checksums, and many other protocols in the security schemes.

In the same context, The National Institute of Standards and Technology (NIST) opened a public competition on November 2, 2007, to develop the new Sha-3 hash function, and the winner was the KECCAK[1600]. Another function of the fifth finalist that presented pretty good results is the BLAKE that was then improvised to BLAKE2 to provide better performance. Blake2 has also shown considerably less

H. EL Makhtoum (✉) · Y. Bentaleb
Engineering Sciences Laboratory, ENSA, Ibn Tofail University, Kenitra, Morocco
e-mail: hind.elmakhtoum@uit.ac.ma

Y. Bentaleb
e-mail: youssef.bentaleb@uit.ac.ma

M. Ben Ahmed et al. (eds.), *Networking, Intelligent Systems and Security*,
Smart Innovation, Systems and Technologies 237,
https://doi.org/10.1007/978-981-16-3637-0_24

energy consumption and less computational cost in resource-constrained devices of IoT domain. This paper concentrates on the finalist of SHA-3 Keccak and the improvised BLAKE2. We compared and analyzed Blake, Blake2, and Keccak. We examined differences in terms of basic parameters, speed, and storage.

2 Hash Functions Algorithms

2.1 *Keccak*

Keccak is the cryptographic hash function that won the NIST SHA-3 hash function competition. Keccak is a family of hash functions based on the sponge construction and used as a building block of a permutation from a set of seven permutations [1]. The basic component is the *Keccak-f* permutation, consisting of several simple rounds with logical operations and bit permutations.

The fundamental function of Keccak is chosen from a set of seven Keccak-f permutations, denoted by Keccak-f[b], with b ∈ {25, 50, 100, 200, 400, 800, 1600} is the width of the permutation. "b" is also the state's width in the sponge construction. The state is an array of 5 × 5 [2].

Algorithm The internal state size is $1600 = r + c$ bits, and it starts with the value 0 for SHA-3, and $r =$ 576bits; $c =$ 1024bits in case of Keccak-512 because $c =$ 512(digest size) × 2

In the absorbing phase, the input message is first padded and divided into small parts of r bits: the first r bits are XORed with the internal state, then the first r bits of the resulted in the internal state (padding) XOR with the first r bits of the precedent message, as shown in Fig. 1.

The squeezing phase is a reverse operation to get the digest. It consists of applying only permutations on the state and finally get parts of the digest. It is repeated up to generate the necessary length of the output (Fig. 2).

The f function here represents permutation is the Keccak-f(1600). It consists of 5rounds which repeat 24 times.

Θ: Calculate the parity of the 5 × w columns (of 5 bits) of the state, then calculate the or exclusive between two neighboring columns (Fig. 3).

ρ: Circularly shift the 25 words of a triangular number (Fig. 4).

Π: Permutation of the 25 words with a fixed pattern (Fig. 5).

χ: Combine bit by bit lines (Fig. 6).

ι: Calculate the XOR of a constant with a state word, which depends on the iteration number (n): This last step aims to break the last symmetries left by the previous ones.

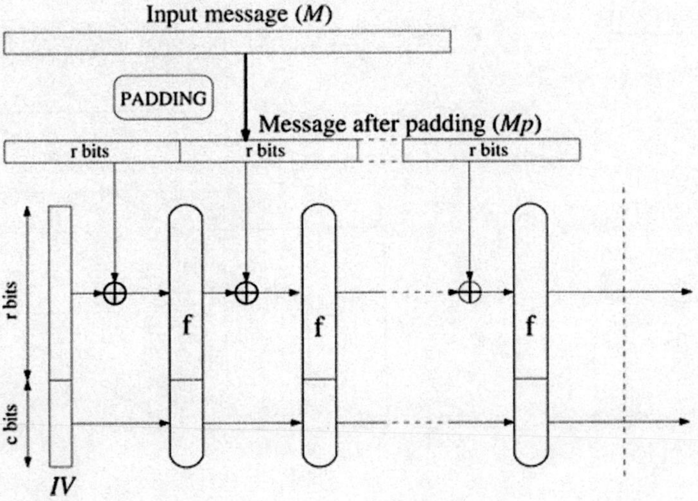

Fig. 1 Absorbing phase [1]

Fig. 2 Squeezing phase [1]

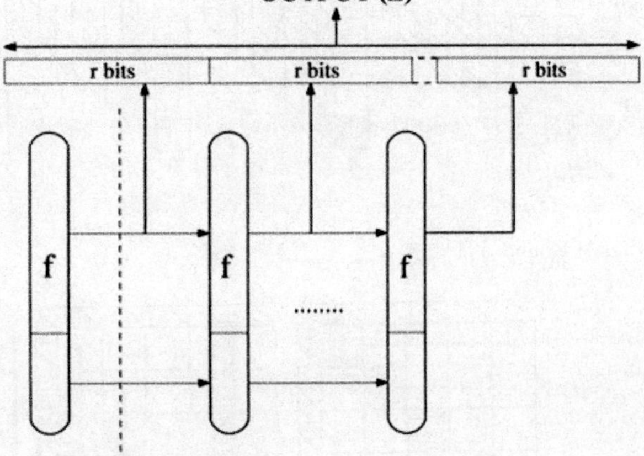

2.2 *Blake2*

BLAKE2 [3] is an ARX cryptographic hash function, a successor of the Blake family. It shares many similarities with the original design. However, differences occur at every level: internal permutation, compression function, and hash function construction. It is designed to have the best performances on software implementations, and it is claimed to be faster than sha-2 and sha-3.

Blake2 consists of two main algorithms: BLAKE2b that is optimized for 64-bit platforms and produces digests of a size up to 64 bytes, and BLAKE2s is optimized for 8- to 32-bit platforms and produces digests of a size up to32 bytes. [4].

Fig. 3 Round Θ [1]

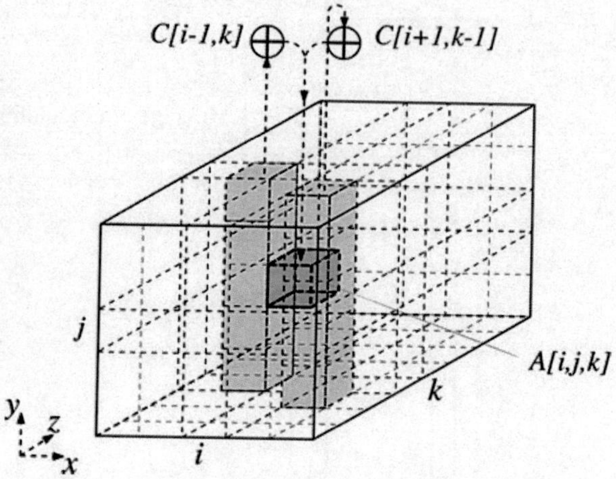

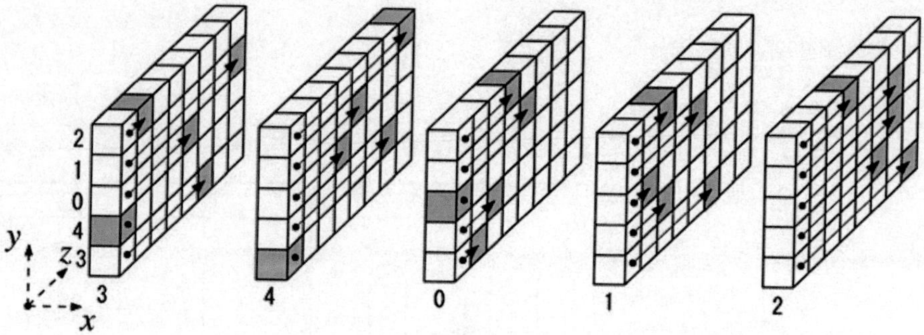

Fig. 4 Round ρ [1]

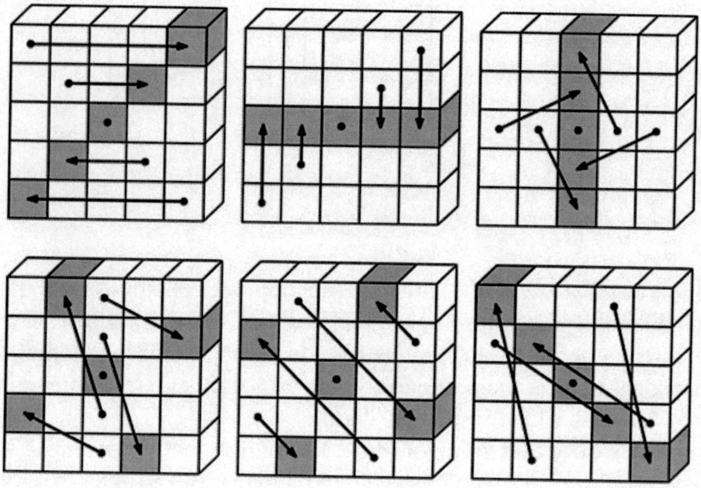

Fig. 5 Round Π [1]

Fig. 6 Round χ [1]

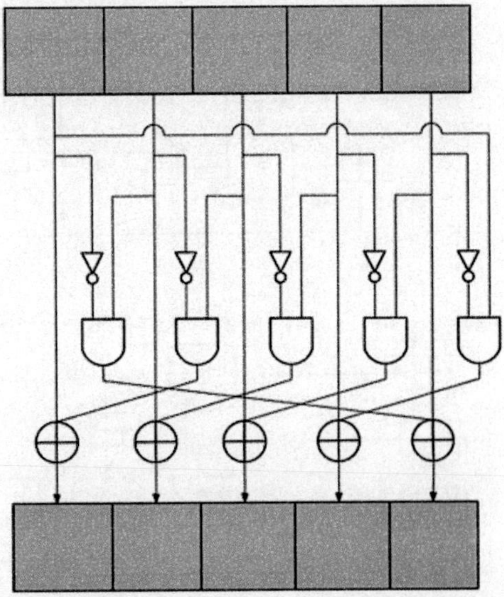

They are portable to any CPU, but they are twice fast when used on the CPU size for which it is optimized.

Algorithm
The message is divided into blocks(d) of 512bits. The internal state v is initialized by affecting the matrice IV (values below) XORed with the digest length to the h values, which is an 8words block.

h0 = IV ⊕ P,
for i = 0 to N−1,
hi + 1 = compress(h i,l i,m i),
return hN,

This way, the state matrice V is initialized with the first eight items of h, and the remaining items are initialized from the initialization vector iv. For 10 (Blake2s) or 12 (Blake2b) rounds, the G function is applied to columns and diagonals of the v matrice in combination with the message bloc (d0), so we get the h values, which will be compressed once again with the next block of the message d1 (Fig. 7).

3 Performance

The main objective of BLAKE2 is to provide several parameters to use in applications without the need for additional constructions and modes and also to speed-up the hash function to a level of compression rate close to MD5. Added changes are the

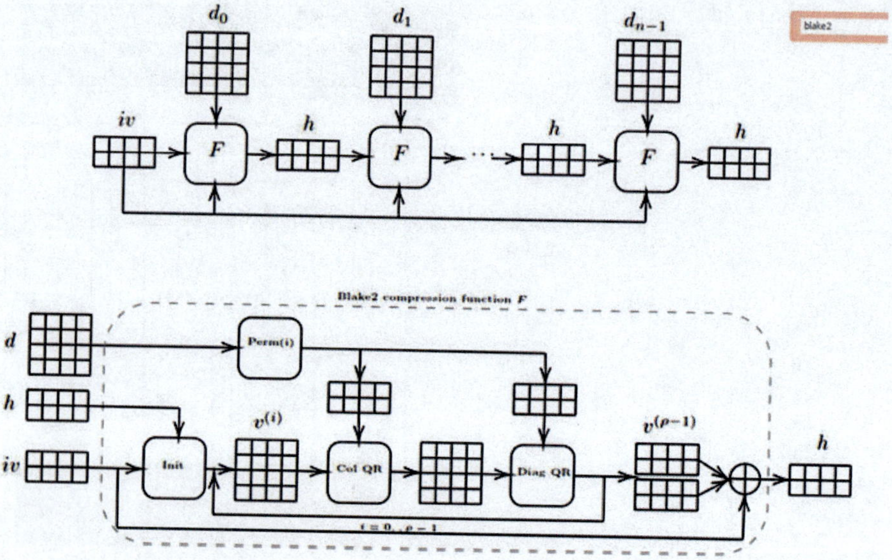

Fig. 7 Blake2 Algorithm and the compress function of BLAKE2 [3]

following: 16-word constant was removed, simplified padding was used, and rotation was reduce from 16 to 12 from BLAKE2b and from 14 to 10 in BLAKE2s. Also, BLAKE2 supported additional functionalities such as keyed salt, personalization block, and tree hashing models. The blake2 is an optimization of the Blake; it is faster, and it consumes less memory than the Blake (32%).

Hardware vs Software While Blake offers excellent software performance, Keccak uses less hardware area than BLAKE and less energy per hashed bit. Keccak also has a "permutation" structure that allows the same hardware to be efficiently reused for applications beyond hashing. Therefore, Keccak consumes fewer physical resources than Blake.

Storage BLAKE used eight-word constants as IV plus 16-word constants [5] for use in the compression function; BLAKE2 uses a total of 8-word constants instead of 24 [5]. It saves 128 ROM bytes and 128 RAM bytes in BLAKE2b implementations and 64 ROM bytes and 64 RAM bytes in BLAKE2s implementations. It means that Blake2 consumes 32% less memory than Blake [6].

In 64 platforms, Blake2s needs 168bytes, Blake2b 336 bytes [6], less space than the Keccak, that needs 1600Bits RAM for operations, which means that the Keccak consumes less storage than the BLAKE2b and more storage than blake2s.

Both functions are considered as lightweight protocols. So, the choice between them is basically related to the application requirements.

Speed (fast) Many works compared BLAKE2 variants to Keccak variants in terms of speed, and it proves that the Blake2 is faster. It presents better results in terms of speed. [5].

Table 1 Performance of 512-bit versions of hash algorithms, speed(Intel Core i3-2310 M-Sandy bridge)

Algorithm	Keccak 512	Blake 512	BLAKE2b 512
Round of compression	24	16	12
Major operations in a round	AND, OR, XOR, ROT, SHR	Modular addition, XOR, and rotate operations	Modular addition, XOR, and rotate operations
Speed cycle/byte [7]	20.46 cycle/byte	14.69 cycle/byte	9cycle/byte
Hash applications	Lightweight	Website links of Perl, PHP, Java script	Argon2, WinRAR, OpenSSL

Table 2 Performance of 256-bit versions of hash algorithms, Speed(Intel Core i3-2310 M-Sandy bridge)

Algorithm	Keccak 256	Blake 256	BLAKE2s
Round of compression	7	14	10
Major operations in a round	AND, OR, XOR, rotate operations, shr	Modular addition, XOR, and rotate operations	Modular addition, XOR, and rotate operations
Speed cycle/byte [7]	10.87	16.38	5.50
hash applications	Ethereum	Blakecoin	WireGuard, checksum, 8th, peerio

Tables 1 and 2 compare the 512 versions of the hash functions Blake and Keccak, and it shows with various parameters.

4 Conclusion

In this paper, we selected two hashing functions of The NIST competition. Both functions are suitable for constrained objects, but each presents different parameters. The comparison is necessary to make the right choice depending on the requirement of the application. Regarding critical applications, such as authentication of the IOT constrained objects, that must consider the limited capacities in storage and processing of objects. These operations need a less consuming, fast, and light hashing function to ensure a high-level security and avoid infiltration attacks.

References

1. Dat, T.N., Iwai, K., Matsubara, T., Kurokawa, T.: Implementation of high-speed hash function Keccak on GPU. IJNC **9**(2), 370–389 (2019)
2. Kavun, E.B., Yalcin, T.: A lightweight implementation of Keccak hash function for radio-frequency identification applications. In: Ors Yalcin, S.B. (ed.) Radio frequency identification: Security and privacy issues, vol. 6370, pp. 258–269. Springer, Berlin Heidelberg (2010)
3. Ramos-Calderer, S., Bellini, E., Latorre, J.I., Manzano, M., Mateu, V.: Quantum search for scaled hash function preimages. , arXiv:2009.00621[quant-ph], September 2020, Consulté le: Décember 28, 2020
4. Sugier,J.: Implementation efficiency of BLAKE2 cryptographic algorithm in contemporary popular-grade FPGA devices. In: Kabashkin, I., Yatskiv, I., Prentkovskis, O. (eds.) Reliability and statistics in transportation and communication, vol. 36, pp. 456–465. Springer International Publishing, Cham (2018)
5. Rao, V., Prema, K.V.: Comparative study of lightweight hashing functions for resource constrained devices of IoT. In 2019 4th International Conference on Computational Systems and Information Technology for Sustainable Solution (CSITSS), Bengaluru, India, pp. 1–5 (2019)
6. Aumasson, J.-P., Neves, S., Wilcox-O'Hearn, Z., Winnerlein, C.: BLAKE2: simpler, smaller, fast as MD5. In: Jacobson, M., Locasto, M., Mohassel, P., Safavi-Naini, R. (eds) Applied cryptography and network security, pp. 119–135, vol. 7954. Springer, Berlin, Heidelberg (2013)
7. Aumasson, J.-P., Meier, W., Phan, R.C.-W., Henzen, L: The hash function BLAKE. Springer, Berlin Heidelberg (2014)

Cryptography Over the Twisted Hessian Curve $H^3_{a,d}$

Abdelâli Grini, Abdelhakim Chillali, and Hakima Mouanis

Abstract In this paper, we will give some properties of the twisted Hessian curve over the ring $\mathbb{F}_q[\epsilon]$ denoted by $H^3_{a,d}$, with $\mathbb{F}_q$ is a finite field of order $q = p^b$, where p is a prime number ≥ 5 and $b \in \mathbb{N}^*$, and we prove that when p doesn't divide $\#(H_{a_0,d_0})$, then $H^3_{a,d}$ is a direct sum of H_{a_0,d_0} and $\mathbb{F}^2_q$, where H_{a_0,d_0} is the twisted Hessian curve over $\mathbb{F}_q$. Other results are deduced from, we cite the equivalence of the discrete logarithm problem on the twisted Hessian curves $H^3_{a,d}$ and H_{a_0,d_0}, which is beneficial for cryptography and cryptanalysis as well, and we give an application in cryptography.

1 Introduction

In [1], Bernstein et al. introduced the twisted Hessian curves over a field. In [6, 7], we defined this curve over the ring $\mathbb{F}_q[\epsilon]$, $\epsilon^2 = 0$, and in [8] we studied the coding over twisted Hessian curves over the same ring. In this article, our objective is to study the twisted Hessian curve defined over the ring $\mathbb{F}_q[\epsilon]$, $\epsilon^3 = 0$. The goal of this work is the search for new groups of points of a twisted Hessian curve over a finite ring, where the complexity of the discrete logarithm calculation is good for use in cryptography.

We started this article by studying the arithmetic of the ring $\mathbb{F}_q[\epsilon]$, $\epsilon^3 = 0$, where we establish some useful results which are necessary for the rest of this work. In the third section, we will define the twisted Hessian curves over $\mathbb{F}_q[\epsilon]$ and we will classify the elements of the twisted Hessian curve $H^3_{a,d}$. Afterwards, we will define

A. Grini (✉) · H. Mouanis
Faculty of Science Dhar El Mahraz-Fez, Sidi Mohamed Ben Abdellah University,
P.O. Box 1796, Atlas-Fez, Morocco
e-mail: abdelali.grini@usmba.ac.ma

A. Chillali
Sidi Mohamed Ben Abdellah University, FP, LSI, Taza, Morocco
e-mail: abdelhakim.chillali@usmba.ac.ma

M. Ben Ahmed et al. (eds.), *Networking, Intelligent Systems and Security*,
Smart Innovation, Systems and Technologies 237,
https://doi.org/10.1007/978-981-16-3637-0_25

the group law of $H^3_{a,d}$ and we will show that $H^3_{a,d}$ is a direct sum of H_{a_0,d_0} and $\mathbb{F}_q^2$, when p doesn't divide $\#(H_{a_0,d_0})$.

Other purpose of this paper is the application of $H^3_{a,d}$ in cryptography. Thereby, from Corollary 3 we deduce that the discrete logarithm problem in $H^3_{a,d}$ is equivalent to that in H_{a_0,d_0} and $\#(H^3_{a,d}) = p^{2b}\#(H_{a_0,d_0})$.

Other cryptographic applications are given in Section 4 which we will establish the coding over twisted Hessian curve $H^3_{a,d}$.

2 Arithmetic Over the Ring $\mathbb{F}_q[\epsilon]$, $\epsilon^3 = 0$

Let p be a prime number ≥ 5 such that -3 is not a square in $\mathbb{F}_p$. We consider the quotient ring $R_3 = \mathbb{F}_q[X]/(X^3)$, where $\mathbb{F}_q$ is the finite field of characteristic p and q elements. Then the ring R_3 can be identified with the ring $\mathbb{F}_q[\epsilon]$, $\epsilon^3 = 0$. In other words,

$$R_3 = \{a + b\epsilon + c\epsilon^2 / a, b, c \in \mathbb{F}_q\}.$$

Now, we will give some results concerning the ring R_3, which are useful for the rest of this paper.

Let two elements in R_3 represented by $X = x_0 + x_1\epsilon + x_2\epsilon^2$ and $Y = y_0 + y_1\epsilon + y_2\epsilon^2$ with coefficients x_i and y_i are in the field $\mathbb{F}_q$ for $i = 0, 1, 2$.

The arithmetic operations in R_3 can be decomposed into operations in $\mathbb{F}_q$, and they are computed as follows:

$$X + Y = (x_0 + y_0) + (x_1 + y_1)\epsilon + (x_2 + y_2)\epsilon^2$$

$$X.Y = (x_0y_0) + (x_1y_0 + x_0y_1)\epsilon + (x_2y_0 + x_1y_1 + x_0y_2)\epsilon^2$$

Similar as in [2, 9, 15], we have the following results:

- $(R_3, +, .)$ is a finite unitary commutative ring.
- R_3 is a vector space over $\mathbb{F}_q$ and has $(1, \epsilon, \epsilon^2)$ as basis.
- R_3 is a local ring. Its maximal ideal is $M = (\epsilon) = \epsilon\mathbb{F}_q$.
- An element $X = x_0 + x_1\epsilon + x_2\epsilon^2$ is invertible in the ring R_3 if and only if $x_0 \neq 0 \ mod\ p$. In this case we have:

$$X^{-1} = x_0^{-1} + (-x_0^{-2}x_1)\epsilon + (x_0^{-3}x_1^2 - x_0^{-2}x_2)\epsilon^2$$

We denote by π the canonical projection defined by

$$\begin{aligned} R_3 &\to \mathbb{F}_q \\ a + b\epsilon + c\epsilon^2 &\mapsto a \end{aligned}$$

3 Twisted Hessian Curves Over the Ring R_3

Definition 1 A twisted Hessian curve over the ring R_3 is a curve in the projective space $P_2(R_3)$, which is given by the equation: $aX^3 + Y^3 + Z^3 = dXYZ$, where $a, d \in R_3$ and $a(27a - d^3)$ is invertible in R_3, and denoted by $H^3_{a,d}$. So we have:

$$H^3_{a,d} = \{[X : Y : Z] \in P_2(R_3) \backslash \ aX^3 + Y^3 + Z^3 = dXYZ\}.$$

3.1 *Classification of Elements of $H^3_{a,d}$*

To have a clear idea on the twisted Hessian curves over the ring R_3, we can classify its elements according to their projective coordinate. This is the subject of the following proposition.

Proposition 1 *Every element in $H^3_{a,d}$ is of the form $[1 : Y : Z]$ (where Y or $Z \in R_3 \setminus M$) or $[X : Y : 1]$ (where $X \in M$), and we write:*
$H^3_{a,d} = \{[1 : Y : Z] \in P_2(R_3) \backslash a + Y^3 + Z^3 = dYZ$, *and* Y *or* $Z \in R_3 \setminus M\} \cup \{[X : Y : 1] \backslash aX^3 + Y^3 + 1 = dXY$, *and* $X \in M\}$.

Proof Let $[X : Y : Z] \in H^3_{a,d}$, where X, Y and $Z \in R_3$.

- If X is invertible, $[X : Y : Z] = [1 : X^{-1}Y : X^{-1}Z] \sim [1 : Y : Z]$. Suppose that Y and $Z \in M$; since $a + Y^3 + Z^3 = dYZ$ then $a \in M$, which is absurd. So, Y or $Z \in R_3 \setminus M$.
- If X is non invertible, then $X \in M$, so $X = x_1\epsilon + x_2\epsilon^2$, where x_1 and $x_2 \in \mathbb{F}_q$. So we have two cases for Z:
  - Z invertible :
    $[X : Y : Z] = [XZ^{-1} : YZ^{-1} : 1] \sim [X : Y : 1]$.
  - Z non invertible:
    We have X and $Z \in M$, since $aX^3 + Y^3 + Z^3 = dXYZ$, then $Y^3 \in M$ and so $Y \in M$. We deduce that $[X : Y : Z]$ isn't a projective point since (X, Y, Z) isn't a primitive triple [[11], pp. 104–105]

Lemma 1 *Let $[X : Y : 1] \in H^3_{a,d}$, where $X \in M$.*
If $X = x_1\epsilon + x_2\epsilon^2$, then

$$[X : Y : 1] = [x_1\epsilon + x_2\epsilon^2 : -1 - \frac{1}{3}d_0x_1\epsilon - \frac{1}{3}(d_1x_1 + d_0x_2)\epsilon^2 : 1].$$

Proof Let $X = x_1\epsilon + x_2\epsilon^2$, $Y = y_0 + y_1\epsilon + y_2\epsilon^2$, $d = d_0 + d_1\epsilon + d_2\epsilon^2$ and $a = a_0 + a_1\epsilon + a_2\epsilon^2$.

Since $[X : Y : 1] \in H_{a,d}^3$, then $[X : Y : 1]$ verifies the equation:

$$aX^3 + Y^3 + Z^3 = dXYZ \tag{1}$$

So:

$$aX^3 + Y^3 + 1 = dXY \tag{2}$$

implies that:

$$\begin{aligned}&(a_0 + a_1\epsilon + a_2\epsilon^2)(x_1\epsilon + x_2\epsilon^2)^3 + (y_0 + y_1\epsilon + y_2\epsilon^2)^3 + 1\\&= (d_0 + d_1\epsilon + d_2\epsilon^2)(x_1\epsilon + x_2\epsilon^2)(y_0 + y_1\epsilon + y_2\epsilon^2)\end{aligned}$$

This means that:

$$\begin{cases} y_0 = -1 \\ y_1 = -\frac{1}{3}d_0x_1 \\ y_2 = -\frac{1}{3}(d_1x_1 + d_0x_2). \end{cases}$$

Thus $[X : Y : Z] = [x_1\epsilon + x_2\epsilon^2 : -1 - \frac{1}{3}d_0x_1\epsilon - \frac{1}{3}(d_1x_1 + d_0x_2)\epsilon^2 : 1]$.

3.2 The Group Law Over $H_{a,d}^3$

After classifying the elements of twisted Hessian curve $H_{a,d}^3$, we will define the group law on it.

We first consider the mapping defined by:

$$\begin{aligned}\tilde{\pi} : \quad H_{a,d}^3 &\rightarrow H_{a_0,d_0}\\ [X : Y : Z] &\mapsto [\pi(X) : \pi(Y) : \pi(Z)]\end{aligned}$$

where H_{a_0,d_0} is the twisted Hessian curve over $\mathbb{F}_q$.

Then, we are ready to define the group law on $H_{a,d}^3$.

Theorem 1 *Let $P = [X_1 : Y_1 : Z_1]$ and $Q = [X_2 : Y_2 : Z_2]$ two points in $H_{a,d}^3$.*

1. Define:

$$X_3 = X_1^2Y_2Z_2 - X_2^2Y_1Z_1,$$

$$Y_3 = Z_1^2X_2Y_2 - Z_2^2X_1Y_1,$$

$$Z_3 = Y_1^2X_2Z_2 - Y_2^2X_1Z_1.$$

If $\tilde{\pi}([X_3 : Y_3 : Z_3]) \neq [0 : 0 : 0]$ then $P + Q = [X_3 : Y_3 : Z_3]$.

2. *Define:*

$$X_3' = Z_2^2 X_1 Z_1 - Y_1^2 X_2 Y_2,$$

$$Y_3' = Y_2^2 Y_1 Z_1 - a X_1^2 X_2 Z_2,$$

$$Z_3' = a X_2^2 X_1 Y_1 - Z_1^2 Y_2 Z_2.$$

If $\tilde{\pi}([X_3' : Y_3' : Z_3']) \neq [0 : 0 : 0]$ *then* $P + Q = [X_3' : Y_3' : Z_3']$.

Proof By using [1, Theorems 3.2 and 4.2] we prove the theorem.

Corollary 1 $(H^3_{a,d}, +)$ *is an abelian group with* $[0 : -1 : 1]$ *as identity element.*

The group law is now defined on $H^3_{a,d}$, we will give some of its properties and homomorphisms defined on it.

3.3 The $\tilde{\pi}$ Homomorphism and Results

Theorem 2 *Let* $a = \tilde{a} + a_2\epsilon^2$, $d = \tilde{d} + d_2\epsilon^2$, $X = \tilde{X} + X_2\epsilon^2$, $Y = \tilde{Y} + Y_2\epsilon^2$ *and* $Z = \tilde{Z} + Z_2\epsilon^2$ *be elements of* R_3, *which verified the equation of twisted Hessian:*

$$aX^3 + Y^3 + Z^3 = dXYZ.$$

Then

$$\tilde{a}\tilde{X}^3 + \tilde{Y}^3 + \tilde{Z}^3 = \tilde{d}\tilde{X}\tilde{Y}\tilde{Z} + (D + AX_2 + BY_2 + CZ_2)\epsilon^2,$$

where

$$\begin{aligned} D &= d_2 X_0 Y_0 Z_0 - a_2 X_0^3, \\ A &= d_0 Y_0 Z_0 - 3a_0 X_0^2, \\ B &= d_0 X_0 Z_0 - 3Y_0^2, \\ C &= d_0 Y_0 X_0 - 3Z_0^2. \end{aligned}$$

Proof Let $a = \tilde{a} + a_2\epsilon^2$, $d = \tilde{d} + d_2\epsilon^2$, $X = \tilde{X} + X_2\epsilon^2$, $Y = \tilde{Y} + Y_2\epsilon^2$ and $Z = \tilde{Z} + Z_2\epsilon^2$ be elements of R_3.
Then

$$\begin{aligned} Y^3 &= \tilde{Y}^3 + 3\tilde{Y}^2 Y_2\epsilon^2 \\ Z^3 &= \tilde{Z}^3 + 3\tilde{Z}^2 Z_2\epsilon^2 \\ aX^3 &= \tilde{a}\tilde{X}^3 + 3\tilde{a}\tilde{X}^2 X_2\epsilon^2 + a_2\tilde{X}^3\epsilon^2 \\ dXYZ &= \tilde{d}\tilde{X}\tilde{Y}\tilde{Z} + (d_2\tilde{X}\tilde{Y}\tilde{Z} + \tilde{d}\tilde{X}\tilde{Y}Z_2 + \tilde{d}\tilde{X}Y_2\tilde{Z} + \tilde{d}\tilde{Y}\tilde{Z}X_2)\epsilon^2. \end{aligned}$$

Since $[X : Y : Z] \in H^3_{a,d}$, then:

$$aX^3 + Y^3 + Z^3 = dXYZ,$$

so

$$\tilde{a}\tilde{X}^3 + \tilde{Y}^3 + \tilde{Z}^3 = \tilde{d}\tilde{X}\tilde{Y}\tilde{Z} + (d_2X_0Y_0Z_0 - a_2X_0^3 + (d_0Y_0Z_0 - 3a_0X_0^2)X_2$$
$$(d_0X_0Z_0 - 3Y_0^2)Y_2 + (d_0Y_0X_0 - 3Z_0^2)Z_2)\epsilon^2,$$

thus:

$$\tilde{a}\tilde{X}^3 + \tilde{Y}^3 + \tilde{Z}^3 = \tilde{d}\tilde{X}\tilde{Y}\tilde{Z} + (D + AX_2 + BY_2 + CZ_2)\epsilon^2,$$

where,

$$\begin{aligned} D &= d_2X_0Y_0Z_0 - a_2X_0^3, \\ A &= d_0Y_0Z_0 - 3a_0X_0^2, \\ B &= d_0X_0Z_0 - 3Y_0^2, \\ C &= d_0Y_0X_0 - 3Z_0^2. \end{aligned}$$

Lemma 2 *The mapping*

$$\begin{aligned} \tilde{\pi} : \quad & H^3_{a,d} \quad \to H_{a_0,d_0} \\ & [X : Y : Z] \mapsto [\pi(X) : \pi(Y) : \pi(Z)] \end{aligned}$$

is a surjective homomorphism of groups.

Proof From Theorem 2; $\tilde{\pi}$ is well defined, and from Theorem 1 we prove that $\tilde{\pi}$ is a homomorphism.

Let $[X_0 : Y_0 : Z_0] \in H_{a_0,d_0}$, then there exists $[X : Y : Z] \in H^3_{a,d}$ such that $\tilde{\pi}([X : Y : Z]) = [X_0 : Y_0 : Z_0]$.

Indeed, by Theorem 2, we have

$$D = -(AX_2 + BY_2 + CZ_2)$$

Coefficients $-A$, $-B$ and $-C$ are partial derivative of a function

$$F(X, Y, Z) = aX^3 + Y^3 + Z^3 - dXYZ$$

at the point $[X_0 : Y_0 : Z_0]$ cannot be all three null. We can then, at last, conclude that $[X_2 : Y_2 : Z_2]$.

Finally, $\tilde{\pi}$ is a surjective.

We define on the set $\mathbb{F}_q^2$ the law $*$ by:

$$(x_1, x_2) * (x_1', x_2') = (x_1 + x_1', x_2 + x_2' + \frac{1}{3}d_0 x_1 x_1')$$

Lemma 3 *$(\mathbb{F}_q^2, *)$ c $(0, 0)$ as unity.*

Lemma 4 *Let $[X : Y : 1]$ and $[X' : Y' : 1] \in H^3_{a,d}$, where X, Y, X', Y' are as in Lemma 1, we have:*

$$[X : Y : 1] + [X' : Y' : 1] = [(x_1 + x_1')\epsilon + (x_2 + x_2')\epsilon^2 + \tfrac{1}{3}d_0 x_1 x_1'\epsilon^2, -1 - \tfrac{1}{3}d_0(x_1 + x_1')\epsilon - (\tfrac{1}{3}d_1(x_1 + x_1') + \tfrac{1}{3}d_0(x_2 + x_2') + \tfrac{1}{9}d_0^2 x_1 x_1')\epsilon^2, 1] \quad (3)$$

Proof By using the Theorem 1, we prove the lemma.

Lemma 5 *The subset $G = \{[X : Y : 1], X \in M\}$ is a subgroup of $H^3_{a,d}$ and every element in G, not unity, is of order p.*

Proof Let $P = [x_1\epsilon + x_2\epsilon^2, -1 - \frac{1}{3}d_0 x_1\epsilon - \frac{1}{3}(d_1 x_1 + d_0 x_2)\epsilon^2, 1] \in G$, we denote $2P = P + P$ and $(n + 1)P = nP + P$ for all $n \geq 2$. We have from Lemma 4:

$$\begin{aligned} pP &= [px_1\epsilon + (px_2 + \tfrac{p}{3}d_0 x_1^2)\epsilon^2, -1 - \tfrac{p}{3}d_0 x_1\epsilon - \tfrac{1}{3}(pd_1 x_1 + pd_0 x_2 + \tfrac{p}{3}d_0^2 x_1^2)\epsilon^2, 1] \\ &= [0 : -1 : 1] \; mod \; p \end{aligned}$$

So, we claim that $pP = [0 : -1 : 1]$.

Lemma 6 *The mapping*

$$\begin{aligned} \phi : (\mathbb{F}_q^2, *) &\rightarrow (H^3_{a,d}, +) \\ (x_1, x_2) &\mapsto [x_1\epsilon + x_2\epsilon^2, -1 - \tfrac{1}{3}d_0 x_1\epsilon - \tfrac{1}{3}(d_1 x_1 + d_0 x_2)\epsilon^2, 1] \end{aligned}$$

is an injective homomorphism of groups.

Proof From Lemma 1, we deduce that ϕ is well defined.
We have $\phi(0, 0) = [0 : -1 : 1]$ and for all (x_1, x_2) and $(x_1', x_2') \in \mathbb{F}_q^2$ we have by Lemma 4:

$$\begin{aligned} \phi((x_1, x_2) * (x_1', x_2')) &= [(x_1 + x_1')\epsilon + (x_2 + x_2')\epsilon^2 + \tfrac{1}{3}d_0 x_1 x_1'\epsilon^2, -1 - \tfrac{1}{3}d_0(x_1 + x_1')\epsilon \\ &\quad -(\tfrac{1}{3}d_1(x_1 + x_1') + \tfrac{1}{3}d_0(x_2 + x_2') + \tfrac{1}{9}d_0^2 x_1 x_1')\epsilon^2, 1] \\ &= [x_1\epsilon + x_2\epsilon^2, -1 - \tfrac{1}{3}d_0 x_1\epsilon - \tfrac{1}{3}(d_1 x_1 + d_0 x_2)\epsilon^2, 1] + [x_1'\epsilon + x_2'\epsilon^2, \\ &\quad -1 - \tfrac{1}{3}d_0 x_1'\epsilon - \tfrac{1}{3}(d_1 x_1' + d_0 x_2')\epsilon^2, 1] \\ &= \phi(x_1, x_2) + \phi(x_1', x_2'), \end{aligned}$$

then ϕ is an homomorphism of groups.

It remains to prove that ϕ is injective. Let $(x_1, x_2) \in \mathbb{F}_q^2$ such that

$$\phi(x_1, x_2) = [0 : -1 : 1].$$

Then,

$$[x_1\epsilon + x_2\epsilon^2, -1 - \frac{1}{3}d_0x_1\epsilon - \frac{1}{3}(d_1x_1 + d_0x_2)\epsilon^2, 1] = [0 : -1 : 1];$$

therefore $x_1 = x_2 = 0$. This prove that ϕ is injective.

Lemma 7 *Ker* $\tilde{\pi} = Im\ \phi$.

Proof Let $[x_1\epsilon + x_2\epsilon^2, -1 - \frac{1}{3}d_0x_1\epsilon - \frac{1}{3}(d_1x_1 + d_0x_2)\epsilon^2, 1] \in Im\phi$, then

$$\tilde{\pi}([x_1\epsilon + x_2\epsilon^2, -1 - \frac{1}{3}d_0x_1\epsilon - \frac{1}{3}(d_1x_1 + d_0x_2)\epsilon^2, 1]) = [0 : -1 : 1]$$

and so, $Im\phi \subset Ker\ \tilde{\pi}$.

Conversely, let $[X : Y : Z] \in Ker\ \tilde{\pi}$, then

$$[x_0; y_0; z_0] = [0 : -1 : 1],$$

so Z is invertible, and from Proposition 1: $X \in M$ so, $[X : Y : Z] \sim [X : Y : 1]$; and from Lemma 1

$$[X : Y : Z] \sim [x_1\epsilon + x_2\epsilon^2, -1 - \frac{1}{3}d_0x_1\epsilon - \frac{1}{3}(d_1x_1 + d_0x_2)\epsilon^2, 1] \in Im\ \ \phi.$$

So $Ker\ \tilde{\pi} \subset Im\ \phi$.

Finally: $Ker\ \tilde{\pi} = Im\ \phi$.

From Lemmas 2, 6 and 7, we deduce the following corollary.

Corollary 2 *The sequence*

$$O \longrightarrow Ker\tilde{\pi} \xrightarrow{i} H^3_{a,d} \xrightarrow{\tilde{\pi}} H_{a_0,d_0} \longrightarrow 0$$

is a short exact sequence which defines the group extension $H^3_{a,d}$ *of* H_{a_0,d_0} *by* $Ker\tilde{\pi}$, *where* i *is the canonical injection.*

Theorem 3 *Let* $n = \#(H_{a_0,d_0})$ *the cardinality of* H_{a_0,d_0}. *If* p *doesn't divide* n, *then the short exact sequence:*

$$O \longrightarrow Ker\tilde{\pi} \xrightarrow{i} H^3_{a,d} \xrightarrow{\tilde{\pi}} H_{a_0,d_0} \longrightarrow 0$$

is split.

Proof of Theorem 3. p doesn't divide n, then exists an integer f such that $nf = 1$ $mod\ p$. So, there is an integer m such that $1 - nf = p\,m$.
Let $[1 - nf]$ the homomorphism defined by:

$$\begin{array}{rcl} [1-nf] : H_{a,d}^3 & \to & H_{a,d}^3 \\ P & \mapsto & (1-nf)P \end{array}$$

There exists an unique morphism φ, such that the following diagram commutes:

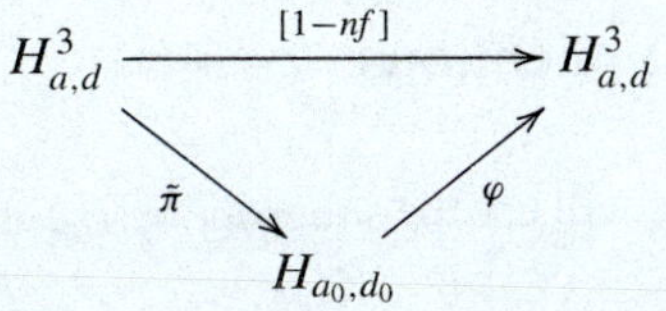

Indeed, let $P \in ker(\tilde{\pi}) = Im\,\phi$, then: $\exists (x_1, x_2) \in \mathbb{F}_q^2$ such that:

$$P = [x_1\epsilon + x_2\epsilon^2, -1 - \frac{1}{3}d_0x_1\epsilon - \frac{1}{3}(d_1x_1 + d_0x_2)\epsilon^2, 1].$$

We have from Lemma 5:

$$(1 - nf)P = pmP = [0 : -1 : 1],$$

then $P \in ker([1 - nf])$. It follows that $ker(\tilde{\pi}) \subseteq ker([1 - nf])$, this proves the above assertion.

Now we prove that $\tilde{\pi} \circ \varphi = id_{H_{a_0,d_0}}$. Let $P' \in H_{a_0,d_0}$, since $\tilde{\pi}$ is surjective, then there exists a $P \in H_{a,d}^3$ such that $\tilde{\pi}(P) = P'$. We have $nP' = [0 : -1 : 1]$, then

$$n\tilde{\pi}(P) = [0 : -1 : 1]\ and\ \tilde{\pi}(nP) = [0 : -1 : 1]$$

implies that $nP \in ker(\tilde{\pi})$ and so, $nfP \in ker(\tilde{\pi})$; therefore, $\tilde{\pi}(nfP) = [0 : -1 : 1]$. Moreover,

$$\varphi(P') = (1 - nf)P = P - nfP,$$

then

$$\tilde{\pi} \circ \varphi(P') = \tilde{\pi}(P) - [0 : -1 : 1] = P'$$

and so, $\tilde{\pi} \circ \varphi = id_{H_{a_0,d_0}}$.

Finally, the sequence is split.

Corollary 3 *If p doesn't divide $\#(H_{a_0,d_0})$ then, $H_{a,d}^3$ is isomorphic to $H_{a_0,d_0} \times \mathbb{F}_q^2$.*

Proof From the Theorem 3, the sequence

$$O \longrightarrow Ker\tilde{\pi} \xrightarrow{i} H^3_{a,d} \xrightarrow{\tilde{\pi}} H_{a_0,d_0} \longrightarrow 0$$

is split then, $H^3_{a,d} \cong H_{a_0,d_0} \times ker(\tilde{\pi})$, and since $ker(\tilde{\pi}) \cong Im\,\phi \cong \mathbb{F}_q^2$, then the corollary is proved.

4 Cryptographic Applications

Let $P \in H^3_{a,d}$ of order k, we will use the subgroup $< P >$ of $H^3_{a,d}$ to encrypt message, and we denote $E = <P>$.

4.1 Coding of Elements of E

We will give a code to each element $Q = rP \in H^3_{a,d}$, where $r \in 1, 2, \ldots, k$.

Let $Q = [X : y_0 + y_1\epsilon + y_2\epsilon^2 : z_0 + z_1\epsilon + z_2\epsilon^2]$, where $y_i, z_i \in \mathbb{F}_{p^b}$, for $i = 0, 1, 2$. We set:

$$y_i = c_{0,i} + c_{1,i}\alpha + c_{2,i}\alpha^2 + \ldots + c_{b-1,i}\alpha^{b-1} = c_{0,i}c_{1,i}c_{2,i}\ldots c_{(b-1),i}$$

where α is a primitive root of an irreducible polynomial of degree b over $\mathbb{F}_p$, then, we code Q as it follows:

- If $Q = [1 : y_0 + y_1\epsilon + y_2\epsilon^2 : z_0 + z_1\epsilon + z_2\epsilon^2]$, then $Q = \underbrace{10\ldots0y_0y_1y_2z_0z_1z_2}_{3\times b\times 3}$
- If $Q = [x_1\epsilon + x_2\epsilon^2 : y_0 + y_1\epsilon + y_2\epsilon^2 : 1]$, then $Q = \underbrace{0\ldots0x_1x_2y_0y_1y_2 1\ldots0}_{3\times b\times 3}$

4.2 Exchange of Secret Key

Ali and Badr want exchange the secret key, for this they start publicly with integer b, a twisted Hessian curve $H^3_{a,d}$, a point $P \in H^3_{a,d}$ of order k and the coding method over $E = <P>$:

- Ali chooses a secret random integer $s_A \in [O, k-1]$, computes $K_A = s_A\,P$ and sends K_A to Badr.

- Badr chooses a secret random integer $s_B \in [O, k-1]$, computes $K_B = s_B P$ and sends K_B to Ali.
- Ali computes $S = s_A K_B$.
- Badr computes $S = s_B K_A$.
- Their secret common key is then $S = s_B\ s_A\ P$. Choose the p code of point S as a private key, which transformed on the decimal code S_D.

Ali and Badr can encrypt and decrypt the message (m), with the secret key S_D such as the decimal code of point S.

4.3 Twisted Hessian Curve Key Generation Block Diagram

The procedure to generate a public key in $H^3_{a,d}$ is outlined as follows:

- The users Ali and Badr compute their secret common key $S = s_B\ s_A\ P$ which transformed on the decimal code S_D.
- Encode the message (m) on point $P_m \in H^3_{a,d}$.
- Choose a random integer $t \in [0, k-1]$ and compute $Q = tP_m$.
- Compute $R = S_D Q$

Then, the public key is $\{a,\ d,\ P,\ Q,\ R\}$, and the private key is $\{s_A,\ s_B,\ t,\ S_D\}$. This operation is shown in Fig. 1.

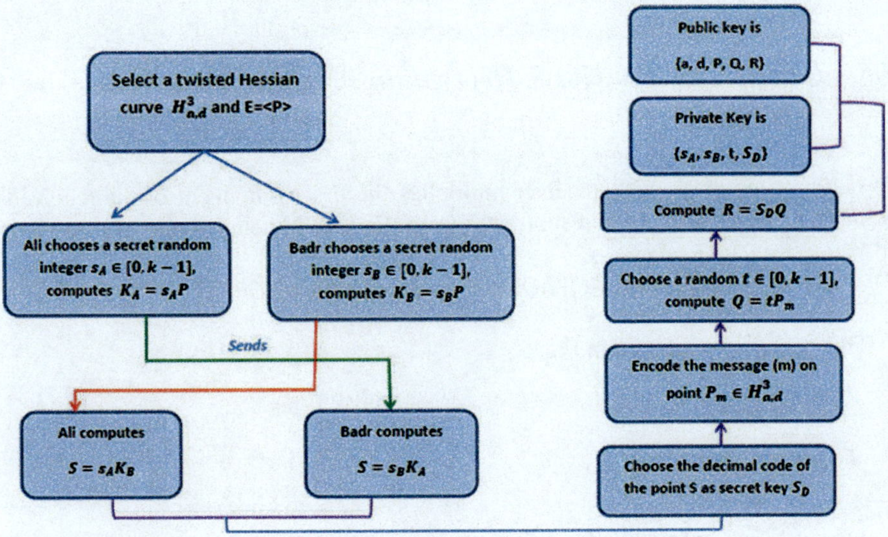

Fig. 1 Twisted Hessian curve Key generation

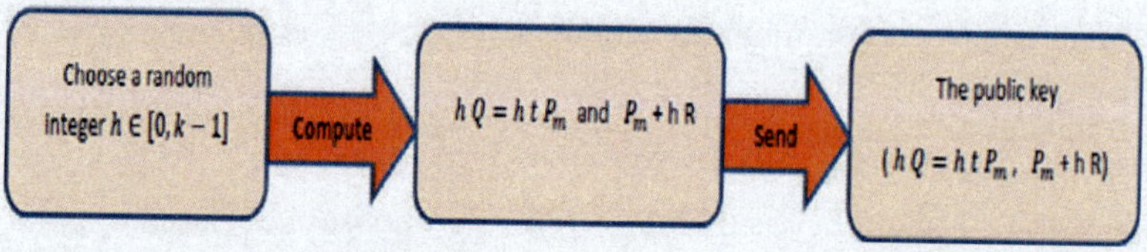

Fig. 2 Twisted Hessian curve encryption

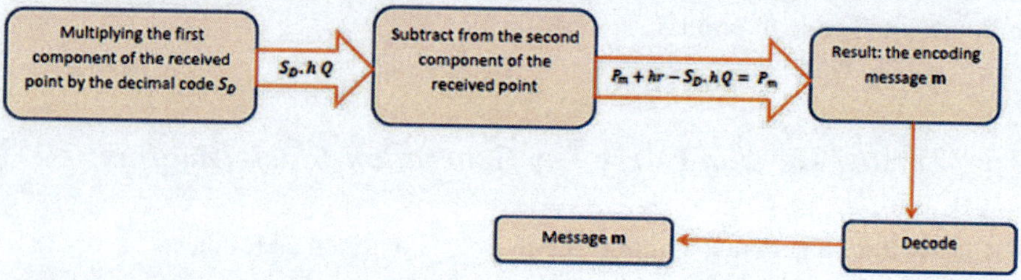

Fig. 3 Twisted Hessian curve decryption

4.4 *Twisted Hessian Curve Encryption Process Block Diagram*

As shown in Fig. 2, to encrypt P_m, a sender chooses an integer h at random and sends the point $(hQ, P_m + hR)$.

4.5 *Twisted Hessian Curve Decryption Process Block Diagram*

To decrypt this message, a receiver multiplies the first component of the received point by the secret key S_D and subtract it from the second component:

$$(P_m + hR) - S_D.hQ = P_m + hS_D.Q - S_D.hQ = P_m. \qquad (4)$$

This operation is shown in Fig. 3.

5 Conclusion

In this work, we have extended the results of twisted Hessian curves over R_3 and we have proved the bijection between $H^3_{a,d}$ and $H_{a_0,d_0} \times \mathbb{F}^2_q$. In cryptography applications, we have established the coding over twisted Hessian curves $H^3_{a,d}$; furthermore, we deduce that the discrete logarithm problem in $H_{a,d}^3$ is equivalent to that in $H_{a_0,d_0} \times \mathbb{F}_q^2$ and $\#(H_{a,d}^3) = p^{2b}\#(H_{a_0,d_0})$. Our future work will focus on the generalist; these studies for all integers $n > 3$, $\epsilon^n = 0$, which are beneficial and interesting in cryptography.

References

1. Bernstein, D.J., Chuengsatiansup C., Kohel D., Lange T.: Twisted Hessian curves. In: Lauter, K., Rodrguez-Henrquez, F. (eds.) Progress in Cryptology—LATINCRYPT 2015. Lecture Notes in Computer Science, vol. 9230, pp. 269–294. Springer, Cham (2015). https://doi.org/10.1007/978-3-319-22174-8_15
2. Chillali, A.: Elliptic curves of the ring $F_q[\epsilon]$, $\epsilon^n = 0$. Int. Math. Forum (2011)
3. Chuengsatiansup, C., Martindale, C: Pairing-friendly twisted Hessian curves. In: Chakraborty, D., Iwata, T. (eds.) Progress in Cryptology INDOCRYPT 2018. INDOCRYPT 2018. Lecture Notes in Computer Science, vol. 11356. Springer, Cham. https://doi.org/10.1007/978-3-030-05378-9_13
4. Diffie, W., Hellman, M.: New directions in cryptography. IEEE Trans. Inf. Theory **22**(6), 644. https://doi.org/10.1109/TIT.1976.1055638
5. ElGamal, T.: A public key cryptosystem and a signature scheme based on discrete logarithms. In: Blakley, G.R., Chaum, D. (eds.) Advances in Cryptology. CRYPTO 1984. Lecture Notes in Computer Science, vol .196. Springer, Berlin, Heidelberg. https://doi.org/10.1007/3-540-39568-7_2
6. Grini, A., Chillali, A., Mouanis, H.: The binary operations calculus in $H_{a,d}^2$. Boletim da Sociedade Paranaense de Matematica. http://www.spm.uem.br/bspm/pdf/next/next.html (2020)
7. Grini, A., Chillali, A., ElFadil, L., Mouanis, H.: Twisted Hessian curves over the ring $F_q[e]$, $e^2 = 0$. Int. J. Comput. Aided Eng. Technol. https://www.inderscience.com/info/ingeneral/forthcoming.php?jcode=ijcaet (2020)
8. Grini, A., Chillali, A., Mouanis, H.: Cryptography over twisted Hessian curves of the ring $F_q[\epsilon]$, $\epsilon^2 = 0$. Adv. Math.: Sci. J. **10**(1), 235–243 (2021). https://doi.org/10.37418/amsj.10.1.24
9. Hassib, M.H., Chillali, A., Elomary, M.: Abdou: elliptic curve over a chain ring of characteristic 3. In: International Workshop of Algebra and Applications, 2014, FST Fez, Morocco. Journal of Taibah University for Science (2015). https://doi.org/10.1016/j.jtusci.2015.02.001
10. Joye, M., Quisquater, J.: Hessian elliptic curves and sidechannel attacks. In: Cryptographic Hardware and Embedded Systems—CHES 2001, Lecture Notes in Computer Science, vol. 2162. Springer, pp. 402–410 (2001) https://doi.org/10.1007/3-540-44709-1_33
11. Lenstra, H.W: Eliptic curves and number-theoretic algorithms. In: Processing of the International Congress of Mathematicians, Berkely, California, USA, (1986)
12. Sahmoudi, M., Chillali, A.: Key Exchange over Particular Algebraic Closure Ring. Tatra Mountains Math. Publ. **70**, 151–162 (2017). https://doi.org/10.1515/tmmp-2017-0024
13. Silverman, J.H.: The Arithmetic of Elliptic Curves. GTM, vol. 106. Springer, New York (2009). https://doi.org/10.1007/978-0-387-09494-6
14. Smart, N.: The Hessian form of an elliptic curve. Cryptographic hardware and embedded systems-CHES: (Paris), 118–125. Lecture Notes in Computer Science, vol. 2162. Springer, Berlin (2001)
15. Tadmori, A., Chillali, A., Ziane, M.: Coding over elliptic curves in the ring of characteristic two. Int. J. Appl. Math. Inf. **8** (2014)
16. Zeriouh, M., Chillali, A., Boua, A.: Cryptography based on the matrices. Boletim da Sociedade Paranaense de Matematica **37**(3), 75–83 (2019). https://doi.org/10.5269/bspm.v37i3.34542

Method for Designing Countermeasures for Crypto-Ransomware Based on the NIST CSF

Hector Torres-Calderon, Marco Velasquez, and David Mauricio

Abstract Crypto-ransomware are malicious programs that encrypt the data of an infected machine, making it a hostage until the owner of the device decides to pay the fee to recover their information. This has become a complex cybersecurity problem causing more and more economic damage. Crypto-ransomware has rendered cybersecurity models not adequate since they do not establish specific guidelines for the design of countermeasures. This paper proposes a method for the design of countermeasures related to crypto-ransomware attacks based on the NIST 800–53 revision 4 standard and the Information Security Maturity Model published by ISACA in the COBIT Focus magazine. The model consists of five phases: identify vulnerabilities, assess vulnerabilities, propose countermeasures, implement countermeasures, and evaluate countermeasures. This allows an organization to measure its current cybersecurity state, know cybersecurity measures oriented to crypto-ransomware and its prioritization through criticality indexes in a simple, adaptive and easy to implement way. A case study in a Peruvian company shows the simplicity and ease of use of the method, which allows the design of countermeasures with which the level of cybersecurity can be improved by 55.6%.

1 Introduction

The inclusion of sensitive data in digital platforms within companies has prompted crime around the world to migrate from a face-to-face to virtual context. Since the 90s, malicious programs have been developed in order to perform some kind of damage to organizations in order to benefit the criminals [1]. One of the most popular malicious programs is crypto-ransomware or encryption ransomware, a malicious program that encrypts all personal data on the infected machine, holding it hostage until the owner of the device decides to pay the fee and obtain the means to recover their information [2].

H. Torres-Calderon (✉) · M. Velasquez · D. Mauricio
Universidad Peruana de Ciencias Aplicadas, Prolongación Primavera 2390, Lima, Perú
e-mail: u201510124@upc.edu.pe

M. Ben Ahmed et al. (eds.), *Networking, Intelligent Systems and Security*,
Smart Innovation, Systems and Technologies 237,
https://doi.org/10.1007/978-981-16-3637-0_26

A well-known crypto-ransomware is SamSam, a highly complex program which deactivates the protection of computers, allowing information to be encrypted quickly and subsequently requesting a reward of up to fifty thousand dollars per attack, achieving an estimated total of 6.5 million dollars between 2016 and 2018 [3]. For this reason, over time, multiple companies have been forced to invest more and more in resources for the defense of their information, operational continuity and reputation [4]. However, 87% of organizations do not have the budget to offer levels of cybersecurity and resilience that satisfy optimal protection against possible attacks [4]. In Latin America, the periodic training rates in companies on social engineering incidents, malicious code incidents, access control to sensitive information reaches 18%, 28%, 20% and 27%, respectively [5].

Various companies and organizations have shown their interest in the common good in terms of cybersecurity by providing alternative frameworks and standards in order to mitigate the risk of cyber-attacks. One of them is the National Institute of Standards and Technology (NIST) which created the Cybersecurity Framework (CSF), one of the most used frameworks worldwide, created to improve the critical infrastructure of companies based on a series of recommendations and references according to the categorization of different information security functions and cybersecurity outcomes that allow raising the protection index of companies [6].

However, the CSF does not show the specific guidelines to design countermeasures against a specific type of malware, specifically crypto-ransomware, making the design arduous, requiring more time, money and specialists. This article proposes a method for the design of countermeasures against crypto-ransomware attacks, which consists of five steps: test, analysis, diagnosis, recommendations and continuous improvement. The proposal is based on the CSF, an analysis of intrusion methods, characteristics of a crypto ransomware, and the COBIT maturity model [7].

The article is organized into 5 sections. In Sect. 2, a literature review on crypto ransomware and its detection is made. The proposed method is described in Sect. 3, and its validation through a case study is presented in Sect. 4. Finally, the conclusions follow in Sect. 5.

2 Literature Review

2.1 Methods for Designing Cybersecurity Countermeasures

The methods for designing cybersecurity countermeasures are composed of a series of phases that allow users to identify vulnerabilities and security risks in order to establish countermeasures. In general, these include the following phases: identify vulnerabilities, evaluate vulnerabilities, propose countermeasures, implement countermeasures and evaluate countermeasures, as described in Table 1.

An inventory of some methods for designing cybersecurity countermeasures and their phases are presented in Table 2, where it is observed that most of them contem-

Table 1 Description of the identified phases

Phase	Description	References
Identify vulnerabilities	In this phase, it is explored which elements, mainly vulnerabilities, related to cybersecurity the organization has	[8–13]
Asses vulnerabilities	In this phase, the elements identified later are analyzed and evaluated to determine the impact they have on the organization	[1, 8–13]
Propose countermeasures	After evaluating, countermeasures should be considered to control and mitigate the impact defined in the previous phase	[1, 8–14]
Implement countermeasures	The implementation of the proposed countermeasures should be planned to depend on the needs and resources of the organization	[1, 11, 14]
Evaluate countermeasures	After a period, the effectiveness, efficiency and efficacy of the implemented countermeasures must be evaluated in order to identify opportunities for improvement	[14]

Table 2 Phases of the methods studied

Author	Identify vulnerabilities	Asses vulnerabilities	Propose countermeasures	Implement countermeasures	Evaluate countermeasures
[1]		Understand risks	Develop adequate policies	Establish best practices for users	
[14]			Implementation of the awareness program	Training focused on countermeasures	Evaluation of effectiveness
[8]	Identify vulnerabilities	Monitor risks	Deploy countermeasures		
[9]	Identify enablers of change	Identify the threat of crypto-ransomware	Define response tools		
[10]	Identify the risks of a cyber attack	Identify the impact of the cyber attack	Define a response strategy		
[11]	Identify new trends	Explore identified trends	Exploit trends explored	Adapt trends to business	
[12]	Identify vulnerabilities	Measure risks	Implement countermeasures		
[13]	Identify processes	Assess security maturity	Plan an improvement plan		

plate only three out of the five phases, and only the work of Torten et al. [14] considers the countermeasures evaluation phase.

2.2 *Mechanisms and Cases of Infection*

Infection mechanisms are those vectors that are taken advantage of or exploited by threats to infect a victim's devices. These seek to violate the confidentiality, integrity and/or availability of the different information and technology assets. A group of these intrusion vectors is related to social engineering since they depend on misinformation and ignorance about information security on the part of their victims. Some of them are presented in Table 3.

Table 3 Intrusion vectors found

Intrusion vectors	Description	Articles
Phishing	A set of techniques that seeks to deceive a victim through identity theft to manipulate him/her and make them disclose confidential and sensitive information	[15–21]
Drive-by download	Mode related to the involuntary download of malicious components that endanger the integrity, confidentiality or availability of a device	[15, 16, 22]
Malvertising	Modality that consists of the use of online advertising for the distribution of malware through failures or security breaches	[15, 16, 18–20, 22]
Social networks	Modality related to the access to unvalidated malicious links shared through publications on social networks	[15, 18, 20]
Cloud applications	Phishing-related modality that consists of the use of cloud storage applications to share malicious files	[15]
Critical vulnerabilities	Set of functionalities or characteristics of a software that have errors or failures not foreseen by the developer that cause security breaches waiting to be discovered and exploited	[15, 20–23]
Brute force attacks	Trial-and-error-based mode that seeks to find keys or access vectors by testing all possible related combinations	[15, 22]
Macros	A mode that takes advantage of algorithms and autorun programs to infiltrate malicious programs	[21–23]

3 Method

To solve the problem presented above, we developed a method for the design of countermeasures against computer attacks of the crypto-ransomware type in order to allow companies to have a tool to increase the effectiveness of their protection schemes. The method is based on the NIST CSF to know the cybersecurity outcomes associated with crypto-ransomware, the NIST standard 800-53 revision 4 [24] to know how to obtain these outcomes and the published "Information Security Maturity Model (ISMM)" maturity model by ISACA [7] which is used to measure the level of implementation of controls in an organization.

Figure 1 shows the five phases that make up the flow of activities related to the application of the method based on what has been analyzed in the literature review. These consist of answering the questions that make up the questionnaire, which are designed considering the controls. Then, the current state must be constructed based on the responses to the questionnaire, which is stipulated in the controls and the reference processes. Next, the current state must be contrasted with the desired state to identify the gap that exists between the maturity levels and the opportunities for improvement. Later, these are analyzed to define the recommendations to increase the maturity level of the current state. Next, an implementation plan for the recommendations should be defined and its viability should be analyzed. Depending on the feasibility, adjustments will have to be made or the implementation should be started. After a period stipulated by the user, they will have to evaluate the performance of the countermeasures implemented to determine their effectiveness and efficiency.

3.1 Fundamentals

The method is based on the NIST CSF to know the cybersecurity outcomes associated with crypto ransomware, the NIST standard 800–53 [6] to know how to obtain these outcomes and the published "Information Security Maturity Model (ISMM)" maturity model. by ISACA [7] which is used to measure the level of implementation of controls in an organization.

We use the NIST CSF because, among cybersecurity frameworks, it best fits the objectives we seek to achieve with the method:

- Easy to use
- Specialization in cybersecurity
- Flexibility

To verify this, we performed benchmarking (Table 4) where we compared four widely used cybersecurity frameworks, considering three criteria that cover these objectives and two predefined criteria associated with their age and maturity. The criteria associated with the objectives had the same weighting among themselves, equivalent to the other two criteria. Next, an assessment is used using the Likert scale

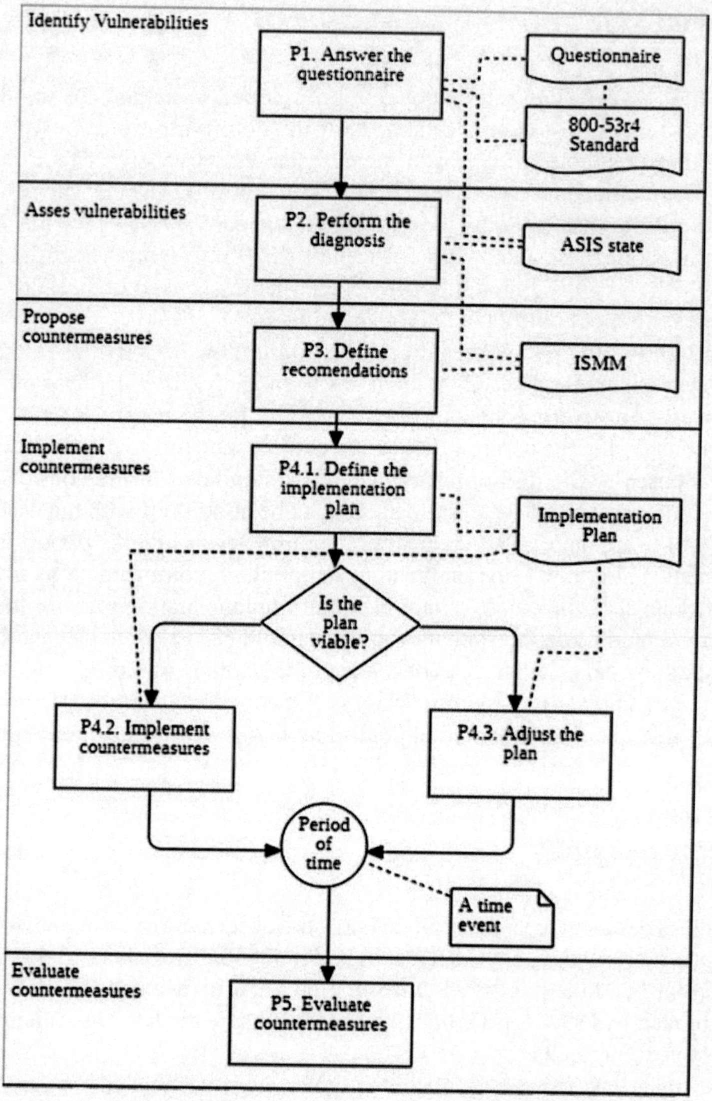

Fig. 1 Flow chart of method application

(1: Totally disagree; 2: Disagree, 3: Neither agree nor disagree, 4: Agree, 5: Totally agree). For example, in usability, the NIST CSF, ISO 27,000 and COBIT 5 are used and recognized internationally by different organizations and government agencies; however, they present weights of 5, 5 and 3 respectively because the first two present ease of use, and the last one because it is more complex despite its details; in IASME governance, the valuation was 3 because it is little known, and the studies are limited to the UK.

Table 4 Benchmarking results

Criteria	Weight (%)	NIST CSF	COBIT 5 for SI	IASME governance	ISO 27,000 family
Usability	25	5	3	3	5
Cybersecurity Specialty	25	5	4	5	5
Antiquity	12.5	2	5	3	5
Flexibility	25	5	4	3	1
Maturity	12.5	2	5	5	4
Total		4.3	4	3.8	3.9

For the standard used, we decided to use one of the tools provided by the CSF. This tool is known as the core and provides different bibliographic references that act as guides on how to achieve the outcomes. In this way, we selected the NIST 800–53 revision 4 standard because it is open access and is published by the same body that published the framework.

For the maturity levels, we decided to use those proposed by the information security group or ISG of the largest bank in India as of today [7]. These maturity levels have been published and approved by ISACA, so we consider that they are the indicated tools to allow us to evaluate the maturity of the controls published in the selected standard.

3.2 *Cybersecurity Controls*

The NIST 800–53 standard initially has 18 families that group approximately 180 controls in total. A first analysis of different case studies was carried out to identify the main intrusion methods and vulnerabilities related to crypto-ransomware attacks. With this analysis, we were able to put together a list of 25 vulnerabilities present in these types of attacks, and we were able to identify the frequency of each of these. With this information, we were able to select the outcomes of the core of the NIST framework related to crypto-ransomware attacks, which allowed us to select the standard controls related to vulnerabilities, reducing the number to 144 controls.

Subsequently, a second analysis was carried out that consisted of identifying the relationship between the controls and the vulnerabilities in order to know how important the controls are depending on how many vulnerabilities they cover. In this way, we were able to assign an expected maturity level to each control depending on its importance, and it allowed us to reduce the number of controls to the 20 most important in relation to crypto-ransomware, which are found in Table 5.

To calculate the criticality index shown in Table 5, a study was carried out where the weight of each of the 25 vulnerabilities mentioned above was determined based on their frequency. A matrix was used to define which vulnerabilities each control

Table 5. 20 most important controls.

Control	Name	Criticality Index (%)
IR-6	Incident reporting	100
PL-2	System security plan	100
PM-7	Enterprise architecture	100
RA-3	Risks evaluation	100
IR-5	Incident monitoring	93
PL-8	System security architecture	93
SA-11	Developer security testing and evaluation	91
RA-5	Vulnerability Scanning	90
IR-4	Incident handling	90
PM-8	Critical infrastructure plan	88
CA-7	Continuous monitoring	82
AU-6	Report, analysis and review of audits	79
AU-13	Information disclosure monitoring	79
PM-5	Inventory of information systems	79
SI-4	Information system monitoring	79
SI-5	Security alerts, warnings and policies	79
SI-7	Software, firmware and information integrity	79
CA-8	Penetration testing	70
RA-2	Categorization of security	64
SI-2	Failure remediation	64

covers, where a control can cover several vulnerabilities and a vulnerability can be covered by several controls. Once we defined the relationship between controls and vulnerabilities, we proceeded to calculate the criticality index as shown in Eq. (1). In this, iC is the criticality index, x takes the value of 1 if the control covers the vulnerability and 0 if not and pV is the weight of the vulnerability.

$$iC = \sum x \times pV \tag{1}$$

Using the criticality index, the expected maturity levels were determined by control. For this, we divide 100% by five equal parts and distribute them between the criticality indexes. So, the expected maturity level is five for controls with a criticality greater than 80%.

3.3 P1: Answering the Questionnaire

The questionnaire is a document that contains between three to five questions for each control to determine the current state of the organization and is at https://bit.ly/3fyJCDa. The questions for control are of the form: Is there control? Is there a process or tool for reporting? Who is responsible? The questionnaire is applied to the head of cybersecurity.

3.4 P2: Perform Diagnosis

The objective of the diagnosis is to show the contrast between the current level of maturity of each control and the desired level of maturity. For this, the degree of compliance of each control in the organization is analyzed according to the domains and levels of the maturity model.

For the analysis of the fulfillment of each cybersecurity function, a table is considered that relates the controls with the functions of the CSF where the frequency with which the former are cited as a reference is identified and a weight is calculated based on this. An example is seen in Table 6 for the "Protect" function.

Also, by presenting the diagnosis grouped by the functions of the framework used, we allow the organization to carry out its own analysis to determine which aspects are most important to its business and decide which are more prudent to improve first.

For the protection indexes, five risk scenarios were proposed based on the five most frequent vulnerabilities identified in the initial analysis of the research. These scenarios include a description, the information asset violated, the compromised

Table 6 Control weights according to the protect function of the NIST CSF

NIST function	Control code	Frequency	Weight	Current state of maturity	Target maturity state
Protect	CA-7	2	0.18		5
	PL-2	1	0.09		5
	PL-8	1	0.09		5
	RA-3	1	0.09		5
	RA-5	1	0.09		5
	SA-11	1	0.09		5
	SI-2	1	0.09		4
	SI-4	1	0.09		4
	SI-7	2	0.18		4

aspect of security, the associated vulnerability, the threat that exploits the vulnerability, the dimension where the scenario impacts and the controls related to managing the scenario.

With this last variable, it is possible to calculate the protection index of the organization related to the said scenario according to the level of maturity that the organization possesses with respect to the related controls. This index is calculated through the average of the current maturity of the controls related to each scenario contrasted with the maximum maturity level.

3.5 *P3: Define Recommendations*

The recommendations provided by the method are based on the maturity levels, and their objective is to improve the current levels of the organization using the tool.

It is convenient for the organization to review the content of the standard since it provides the basic guidelines related to each control and a list of improvement proposals for these. In addition, it is important to note that the maturity levels are effective in identifying the lack of basic functionalities stipulated by the standard in relation to the controls; however, it does not take into consideration the proposals for improvement in the analysis, so these are optional.

3.6 *P4: Implement Countermeasures*

The implementation is a phase designed to help organizations adopt the recommendations so that they can improve the level of maturity of their controls. It is made up of three sub-phases.

3.6.1 P4.1 Define the Implementation Plan

The plan consists of 3 deliverables: a work-breakdown structure (WBS) to define the scope of the implementation, a Gantt chart to define the time, budget and resources allocated to WBS activities, and a matrix of risks associated with the implementation that allows the organization to identify risk exposure and a response plan. Subsequently, this plan must be reviewed and evaluated to obtain the agreement of the sponsors.

3.6.2 P4.2 Adjust the Plan

If there is an observation that results in the modification of the previously stipulated plan, this must be managed using formal documentation and the respective communication channels.

3.6.3 P4.3 Implement Countermeasures

After obtaining the final version of the implementation plan, it must be implemented. Once completed, a period must be stipulated that will define the performance evaluation of the controls and recommendations implemented.

3.7 P5: Evaluate Countermeasures

After the period defined in phase 4, a work team assigned to evaluate the performance of the improved controls must be defined. This must determine if the efficiency and effectiveness of the controls were improved through the implementation of the recommendations and if this improvement is reflected in the organization's security indicators.

4 Validation

The validation of the proposed method was carried out through a case study in a company in Peru.

4.1 Case Study

For the application of the method, a Latin American organization that specializes in the trade of construction machinery and mining equipment was selected. This organization has different companies located in various Latin American countries. Its headquarters are located in Peru, which has a security area where a security officer, specialists and experts in information security operate.

The security team that participated in the validation was made up of three specialists with an average of 10 years of experience each in auditing and information security. One of them has been president of a chapter of ISACA and another is a SAP security specialist.

4.2 Method Application

Regarding the application of the model in the organization presented, we followed the flow shown in Fig. 1. For the first phase, it began with the request for approval for the use of company information for the validation of the method. Once obtained, we proceeded to coordinate meetings by video calls with the work team. During these, questions from the questionnaire were asked to obtain the different points of view and to be able to build a current state more faithful to reality.

In the second phase, we used the responses obtained in the meetings to analyze each control and determine the maturity levels of the associated domains. We then averaged the maturity levels of each domain to determine the maturity level of the control. We locate this in Table 6, in the column of current maturity level. Likewise, the calculation of Table 8, Fig. 2 and the protection indices was carried out. The third phase consisted of consolidating a report requested by the organization made up of three main sections:

- Executive summary: Summary of the results obtained and how to interpret the recommendations.
- Results of the analysis: Report by control on the maturity of the related domains that allows the analysis on where to begin the improvement of the controls.
- Recommendations: Consolidated of the recommendations provided by the maturity model and the standard improvement proposals.

Due to the nature of the situation generated in the country and the organization's policies, in the fourth phase, it was decided to present recommendations for implementation in the form of an EDT and a matrix of possible risks.

Fig. 2 Current maturity level vs. target maturity level

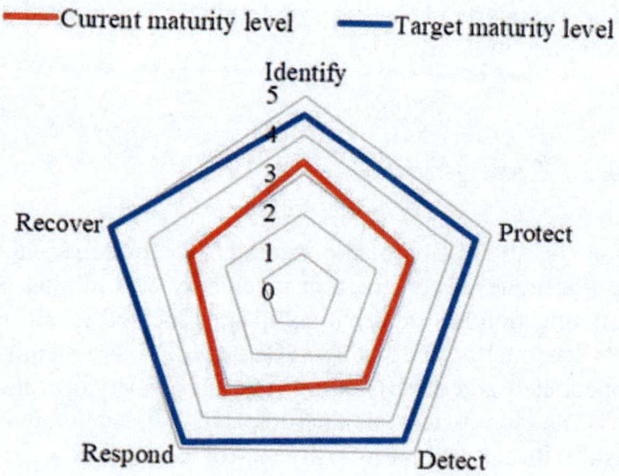

Table 7 Expected maturity level vs. current maturity level per control

Control	Current maturity level	Expected maturity level
AU-13	2	4
AU-6	5	4
CA-7	1	5
CA-8	4	4
IR-4	3	5
IR-5	3	5
IR-6	3	5
PL-2	4	5
PL-8	4	5
PM-5	2	4
PM-7	4	5
PM-8	4	5
RA-2	3	4
RA-3	4	5
RA-5	4	5
SA-11	1	5
SI-2	4	4
SI-4	3	4
SI-5	4	4
SI-7	3	4

4.3 Results

The results shown below are the product of processing the information obtained from the questionnaire. In Table 7, we determined that the organization does not meet the expected level of maturity in 16 of the 20 controls that were studied. Also, it was observed that the organization did not meet the expected maturity levels by function of the NIST CSF as shown in Table 8.

Table 8 Expected maturity level vs. current maturity level by NIST CSF role

NIST CSF Role	Current maturity level	Expected maturity level
Identify	3.3	4.5
To protect	2.9	4.6
Detect	2.8	4.6
Reply	3.2	4.8
Recover	2.9	5.0

$$\text{Function MA} = \sum \text{control MA} \times \text{Control weight per function} \quad (2)$$

The calculation shown in Table 8 was given through Eq. (2) using the values shown in Tables 6 and 7 where "MA" refers to the current maturity level. Additionally, Fig. 2 was constructed to visually contrast the gap between the current maturity level and the target maturity level and understand approximately how much improvement could be made in the organization with respect to cybersecurity crypto-ransomware scenarios.

After obtaining and analyzing the diagnosis, a report was prepared and divided into 3 main segments, an executive summary, the results of the analysis and recommendations based on the results. This report was sent to the employees of the organization who worked together with us, and an attempt was made to schedule a final meeting to present the results and obtain feedback.

Likewise, a work-breakdown structure (WBS) was built with a series of activities and recommended work packages for a subsequent implementation plan. Additionally, a matrix of risks associated with the implementation of the recommendations was developed.

5 Conclusions

In the present work, a method has been proposed for the design of countermeasures related to crypto ransomware attacks based on the NIST 800–53 standard and the Information Security Maturity Model published by COBIT and consists of 5 phases: identify vulnerabilities, evaluate vulnerabilities, pose countermeasures, implement countermeasures and evaluate countermeasures. The method allows an organization to measure its current cybersecurity status, learn about cybersecurity measures in the form of the 20 most important controls and prioritize these through criticality indexes in a simple, adaptable and easy way.

The implementation of the proposed method in a Peruvian capital goods company oriented to the trade of machinery for the agricultural and construction sectors shows that the proposal is easy to apply. It is agile because it required 15 days of work. In addition, it identified an intermediate level to more (3.02 out of 5) of maturity in cybersecurity and that this can be improved up to a 55.6% by implementing countermeasures.

Among the challenges to be developed, the method must be extended to consider other malicious programs and cybersecurity standards; likewise, a tool must be developed that supports the application of the method.

Acknowledgements The authors thank the organization of the case study and the UPC for the partial funding of the following research.

References

1. Richardson, R., North, M.: Ransomware: Evolution, mitigation and prevention. Int. Manag. Rev. **13** (2017)
2. Hassan, N.A.: Ransomware Revealed, 1st ed. Springer Science, New York (2019)
3. Toapanta Toapanta, S.M., Mafla Gallegos, L.E., Benavides Quimis, B.S., Huilcapi Subia, D.F.: Approach to mitigate the cyber-environment risks of a technology platform. Proc 3rd Int. Conf. Inf. Comput. Technol. ICICT 2020, 390–396 (2020). https://doi.org/10.1109/ICICT50521.2020.00069
4. EY: ¿La ciberseguridad es algo más que protección? Quito (2018)
5. Jurado Pruna, F.X., Yarad Jeada, P.V., Carrión Jumbo, J.L.: Análisis de las características del sector microempresarial en latinoamérica y sus limitantes en la adopción de tecnologías para la seguridad de la información. Rev. Científica. Ecociencia. **7**,1–26 (2020). https://doi.org/10.21855/ecociencia.71.303
6. NIST: Marco para la mejora de la seguridad cibernética en infraestructuras críticas (2018). https://doi.org/10.6028/NIST.CSWP.04162018
7. Salvi, V., Kadam, A.W.: Information Security Management at HDFC Bank: Contribution of Seven Enablers. COBIT Focus **1**, 8 (2014)
8. Tolubko, V., Vyshnivskyi, V., Mukhin, V., et al.: Method for determination of cyber threats based on machine learning for real-time information system. Int. J. Intell. Syst. Appl. **10**, 11–18 (2018). https://doi.org/10.5815/ijisa.2018.08.02
9. Connolly, L., Wall, D.S.: The rise of crypto-ransomware in a changing cybercrime landscape: Taxonomising countermeasures. Comput. Secur. **87**, 101568 (2019). https://doi.org/10.1016/j.cose.2019.101568
10. Mañas-Viniegra, L., Niño González, J.I., Martínez Martínez, L.: Transparency as a reputational variable of the crisis communication in the media context of wannacry cyberattack. Rev. Comun la SEECI., 149–171
11. Carayannis, E.G., Grigoroudis, E., Rehman, S.S., Samarakoon, N.: Ambidextrous cybersecurity: The seven pillars (7Ps) of cyber resilience. IEEE Trans. Eng. Manag., 1–12 (2019). https://doi.org/10.1109/TEM.2019.2909909
12. Rea-Guaman, A.M., Mejía, J., San Feliu, T., Calvo-Manzano, J.A.: AVARCIBER: a framework for assessing cybersecurity risks. Cluster. Comput. **23**, 1827–1843 (2020). https://doi.org/10.1007/s10586-019-03034-9
13. Al-Matari, O.M.M., Helal, I.M.A., Mazen, S.A., Elhennawy, S.: Adopting security maturity model to the organizations' capability model. Egypt. Informatics. J. (2020). https://doi.org/10.1016/j.eij.2020.08.001
14. Torten, R., Reaiche, C., Boyle, S.: The impact of security awarness on information technology professionals' behavior. Comput. Secur. **79**, 68–79 (2018). https://doi.org/10.1016/j.cose.2018.08.007
15. Sood, A.K., Bajpai, P., Enbody, R.: Evidential Study of Ransomware. **5**, 1–10 (2018)
16. Ali A (2017) Ransomware: A Research and a Personal Case Study of Dealing with this Nasty Malware. Issues Informing Sci Inf Technol 14:087–099. https://doi.org/10.28945/3707
17. Sipior, J.C., Bierstaker, J., Borchardt, P., Ward, B.T.: A ransomware case for use in the classroom. Commun. Assoc. Inf. Syst. **43**, 598–614 (2018). https://doi.org/10.17705/1CAIS.04332
18. Thomas, J.E.: Individual Cyber Security: Empowering Employees to Resist Spear Phishing to Prevent Identity Theft and Ransomware Attackss. Int. J. Bus. Manag. **13**, 1 (2018). https://doi.org/10.5539/ijbm.v13n6p1
19. Gupta, B.B., Tewari, A., Jain, A.K., Agrawal, D.P.: Fighting against phishing attacks: state of the art and future challenges. Neural. Comput. Appl. **28**, 3629–3654 (2017)
20. Hull, G., John, H., Arief, B.: Ransomware deployment methods and analysis: views from a predictive model and human responses. Crime. Sci. **8**, 2 (2019). https://doi.org/10.1186/s40163-019-0097-9

21. Patyal, M., Sampalli, S., Ye, Q., Rahman, M.: Multi-layered defense architecture against ransomware. Int. J. Bus. Cyber. Secur. **1**, 52–64 (2017)
22. Wilner, A., Jeffery, A., Lalor, J., et al.: On the social science of ransomware: Technology, security, and society. Comp. Strateg. **38**, 347–370 (2019). https://doi.org/10.1080/01495933.2019.1633187
23. Watson, F.C.: Petya/NotPetya Why It Is Nastier Than WannaCry and Why We Should Care. ISACA **6**, 1–6 (2017)
24. NIST: NIST Special Publication 800–53: Security and Privacy Controls for Federal Information Systems and Organizations. NIST SP-800–53 Ar4 400+ (2013). https://doi.org/10.6028/NIST.SP.800-53Ar4

Comparative Study Between Network Layer Attacks in Mobile Ad Hoc Networks

Oussama Sbai and Mohamed Elboukhari

Abstract Amid the most recent decade, a few research endeavors have explored developing Internet of Things (IoT) and Mobile Ad hoc Networks (MANETs) application situations in a new concept called IoT-MANETs. One of the constraints of these applications is the security of the communication between the nodes. In this article, we analyze and compare a simulation result of the impact of DATA Flooding, Link-spoofing and Replay attacks with Optimized Link State Routing Protocol (OLSR) routing protocol (RFC 3626) and DATA Flooding, RREQ Flooding and HELLO Flooding attacks with Ad hoc On-Demand Distance Vector (AODV) routing protocol (RFC 3561) on using ns-3 simulator. In this comparison, we took into consideration the density of the network by the number of nodes included in the network, the speed of the nodes, the mobility model, and even, we chose the IEEE 802.11ac wireless local-area network (WLAN) standard for the MAC layer, in order to have a simulation, which deals with more general and more real scenarios.

1 Introduction

Mobile Ad Hoc Networks (MANETs) alludes to a class of wireless networks that can be shaped dynamically and randomly without the need for infrastructural setups. Such networks are capable of adapt and reconfigure themselves at the fly in keeping with node mobility and changing network topologies [1]. Each wireless node can be a simple device or even a router. Once the node is a simple device it could send messages to any specified destination node through some route or perhaps receive messages from other nodes. When the node functions as a router, it can relay the packet to the destination or next router in the route. Whenever important, each node can buffer packets awaiting transmission [2]. Present and future MANET applications cover a variety of areas [1]:

O. Sbai (✉) · M. Elboukhari
Department of Applied Engineering, ESTO (Higher School of Technology), Mohammed 1st University, Oujda, Morocco
e-mail: o.sbai@ump.ac.ma

M. Ben Ahmed et al. (eds.), *Networking, Intelligent Systems and Security*, Smart Innovation, Systems and Technologies 237,
https://doi.org/10.1007/978-981-16-3637-0_27

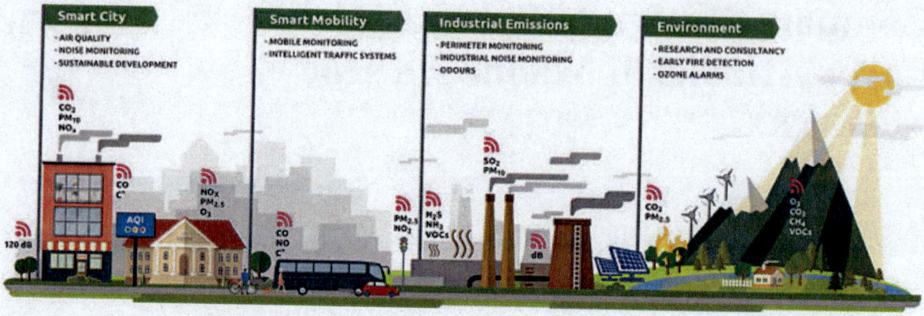

Fig. 1 The fields of application of MANETs

- Home applications: smart wireless sensor nodes and actuators embedded in consumer electronics.
- Environmental applications including tracking movements of animals, chemical/biological detection, precision agriculture, tracking of weather and earth activities.
- Body area networks (BAN).
- Tactical networks (military communications and automated battlefields) (Fig. 1).

This article is a continuation of the work started in the following articles [3–6]. We will compare and analyze the impact of attacks targeting the OLSR protocol [7] of one party: Link-spoofing and Replay and DATA Flooding attacks, and the AODV protocol [8] of another party: DATA Flooding, RREQ Flooding, and HELLO Flooding attacks. In the scenarios simulated for each protocol, we tried to set the simulation parameters most appropriate to the use case of each protocol, and close to one of the real cases.

The organization of the paper is as follows: Sect. 2 presents and defines the different attacks studied and their implementation protocols; also, this section cites some related work. The following section presents the experimental environment set-up (Simulation time, IEEE 802.11ac, mobility model, etc.). Performance metrics are presented in Sect. 4. Sections 5 and 6 present simulation comparison of Link-spoofing, Replay, DATA Flooding, RREQ Flooding, and HELLO Flooding attacks in ns-3 simulator. Finally, a conclusion is given in the last section.

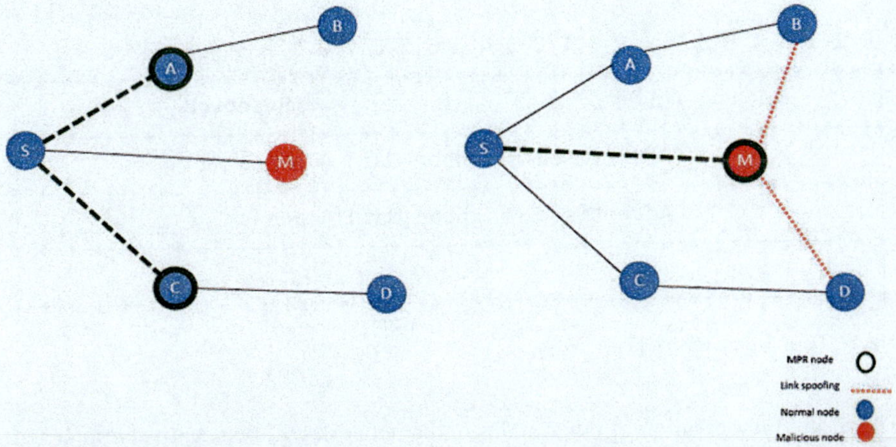

Fig. 2 Link-spoofing attack [3]

2 Background

2.1 *Link-Spoofing Attack in MANETs*

This attack is implemented in the network using the OLSR protocol. The link-spoofing malicious nodes claim that they are its one-hop neighbor node, by the injection of wrong information about non-existent nodes or (and) insert false information about neighbors using Hello messages [9, 10] (Fig. 2).

2.2 *Replay Attack in MANETs*

When the networks using OLSR protocol, infected by the replay attack, the attackers replicate expired TC messages and forward them to the network's nodes afterwards.

With the nature of MANETs: network topology on permanent changing, this control packet inject by malicious nodes will be expired. Therefore, the routing tables of nodes will be updated on using wrong information, so the routing operation is disrupted (Fig. 3).

2.3 *HELLO Flooding Attack in MANETs*

In AODV protocol, the HELLO-Flooding malicious nodes modify the value of HELLO_INTERVAL on reducing the interval time in HELLO-message sanded to their nodes neighbors so that they can inject a grand number of hello packets.

```
 0                   1                   2                   3
 0 1 2 3 4 5 6 7 8 9 0 1 2 3 4 5 6 7 8 9 0 1 2 3 4 5 6 7 8 9 0 1
+-+-+-+-+-+-+-+-+-+-+-+-+-+-+-+-+-+-+-+-+-+-+-+-+-+-+-+-+-+-+-+-+
|              ANSN             |           Reserved            |
+-+-+-+-+-+-+-+-+-+-+-+-+-+-+-+-+-+-+-+-+-+-+-+-+-+-+-+-+-+-+-+-+
|               Advertised Neighbor Main Address                |
+-+-+-+-+-+-+-+-+-+-+-+-+-+-+-+-+-+-+-+-+-+-+-+-+-+-+-+-+-+-+-+-+
|               Advertised Neighbor Main Address                |
+-+-+-+-+-+-+-+-+-+-+-+-+-+-+-+-+-+-+-+-+-+-+-+-+-+-+-+-+-+-+-+-+
|                              ...                              |
+-+-+-+-+-+-+-+-+-+-+-+-+-+-+-+-+-+-+-+-+-+-+-+-+-+-+-+-+-+-+-+-+
```

Fig. 3 TC message format [7]

Therefore, the operation of other nodes in the network will be negatively impacted. [11].

2.4 *RREQ Flooding Attack in MANETs*

The RREQ Flooding Attack is dedicated to the reactive routing protocols in MANETs. In this attack, a malicious node injects a huge number of RREQ packets to an unknown destination (non-existent IP address in the network) (Fig. 4).

Therefore, no node in the network will respond with a RREP packet, and the DATA packets cannot be transmitted, because the network gets saturated by the flooded RREQs packets.

```
 0                   1                   2                   3
 0 1 2 3 4 5 6 7 8 9 0 1 2 3 4 5 6 7 8 9 0 1 2 3 4 5 6 7 8 9 0 1
+-+-+-+-+-+-+-+-+-+-+-+-+-+-+-+-+-+-+-+-+-+-+-+-+-+-+-+-+-+-+-+-+
|     Type      |J|R|G|D|U|   Reserved          |   Hop Count   |
+-+-+-+-+-+-+-+-+-+-+-+-+-+-+-+-+-+-+-+-+-+-+-+-+-+-+-+-+-+-+-+-+
|                            RREQ ID                            |
+-+-+-+-+-+-+-+-+-+-+-+-+-+-+-+-+-+-+-+-+-+-+-+-+-+-+-+-+-+-+-+-+
|                    Destination IP Address                     |
+-+-+-+-+-+-+-+-+-+-+-+-+-+-+-+-+-+-+-+-+-+-+-+-+-+-+-+-+-+-+-+-+
|                  Destination Sequence Number                  |
+-+-+-+-+-+-+-+-+-+-+-+-+-+-+-+-+-+-+-+-+-+-+-+-+-+-+-+-+-+-+-+-+
|                    Originator IP Address                      |
+-+-+-+-+-+-+-+-+-+-+-+-+-+-+-+-+-+-+-+-+-+-+-+-+-+-+-+-+-+-+-+-+
|                  Originator Sequence Number                   |
+-+-+-+-+-+-+-+-+-+-+-+-+-+-+-+-+-+-+-+-+-+-+-+-+-+-+-+-+-+-+-+-+
```

Fig. 4 Route Request (RREQ) message format [8]

In addition, the route table of the intermediate nodes will overflow, so that no new RREQ packet can be received, and in result, the network services are denied (Denial-of-Service attack).

Besides, unnecessarily forwarding these fake RREQs packets causes a serious loss of node resources like energy and bandwidth.

2.5 DATA Flooding Attack in MANETs

The malicious DATA Flooding node(s) send or inject in network a great volume of futile DATA packets. The immoderate injected DATA packets engorge the network, what makes the communication between the nodes of the network cannot be completed, because the available network bandwidth will be exhausted.

The destination node gets submerged and cannot work normally, because of the excessive packets generated by the malicious node(s).

2.6 Related Works

In our previous works [3, 5, 6], we studied and analyzed the performance of AODV and OLSR protocols in MANET, the malicious nodes are introduced in the network to analyses the impact of attacks on network's performances. We used ns-3 simulation tool, by taking into consideration the use cases of each protocol by a specific configuration of the simulation scenarios. The presence of malicious nodes in the network decreases the performance of this last, and we compare the packet delivery ratio, routing overhead, normalized routing load, and end-to-end delay with and without the presence of the attacker node in the network.

In this research [12], the authors had analyzed three types of routing protocols in MANET witch are AODV, ZRP [13] and LAR [13], by implementing a DDoS using CBR traffic flooding. The simulator tool using id ns-2. Additionally, the efficiency of routing protocols was analyzed using performance parameters of the throughput and end-to-end delay.

In this paper [14], the authors had focused their study on Blackhole, Grayhole, and Wormehole attacks in OLSR protocol. And in [15], the authors had analyzed the impact of Selfishness attack on AODV and DSDV protocols, on using ns-2 simulator, The network's performance metric used are throughput, packet delivery ratio, and average delay.

Table 1 Simulation parameters

Parameter	Value	
Dimension	1000 × 1000 m	
Simulation time	60 s	
Number of packet	60	
Packet size	56 bytes	
Mobility model	Random Way point Mobility Model	
MAC Layer	IEEE 802.11ac	
Mobility speed	Uniform Random Variable [Min = 0|Max = 30]	
Protocol	AODV	OLSR
Attack	Link-spoofing Replay DATA Flooding	DATA Flooding RREQ Flooding HELLO Flooding
Pause	Constant Random Variable [Constant = 0]	Constant Random Variable [Constant = 30]

3 Experimental Environment Set Up

In this section, we discuss the experimental environment setup and the values taken for different parameters in ns-3 simulator [16]. The network area consists of 1000 m × 1000 m where nodes are randomly distributed and mobile on used the Random Waypoint Mobility Model to have more general node mobility [17].

The simulation is started during 60 s, on sending one packet per second. We use the protocol IEEE 802.11ac in MAC layer (recent version of 802.11n) thanks to the advantage (higher data throughput, high capacity, low latency, efficient use of power, etc.) defined in [18].

The main experimental parameters used are presented in Table 1 and all experimental value refers to average values of experiments.

4 Performance Metrics

4.1 Packet Delivery Rate

The Packet delivery rate (PDR) is the division of the total number of packets received by the destination (PR), by the total number of packets sent by the source (PS), multiply 100%.

$$\text{PDR} = \frac{\sum_{j=1}^{n} PRj}{\sum_{i=1}^{n} PSi} \times 100 \tag{1}$$

4.2 *Normalized Routing Load*

Normalized Routing Load (NRL) represents the total number of control packets (C), divided by the total number of DATA packets received by the destination (PR).

$$\text{NRL} = \frac{\sum_{j=1}^{n} Cj}{\sum_{j=1}^{n} PRj} \tag{2}$$

5 Comparison Between Types of Flooding Attack

5.1 *Packet Delivery Rate*

Figures 5 and 6 presents a comparison's PDR between the implementation of single and multiple of three types of Flooding attack, the results show that the impact

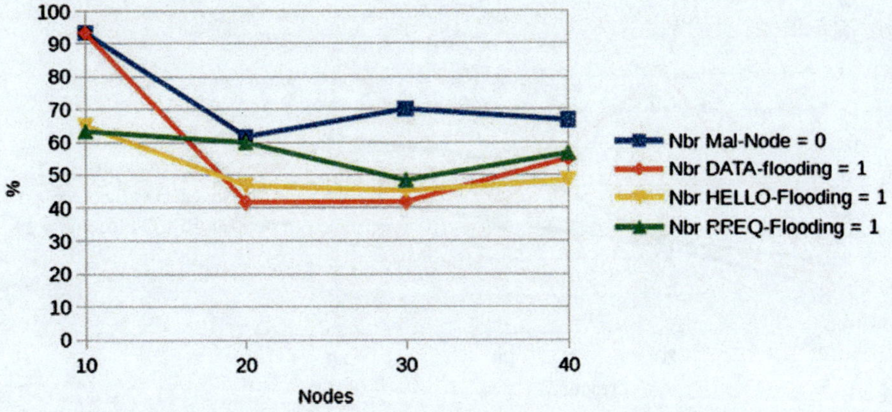

Fig. 5 Single malicious node's PDR simulation result

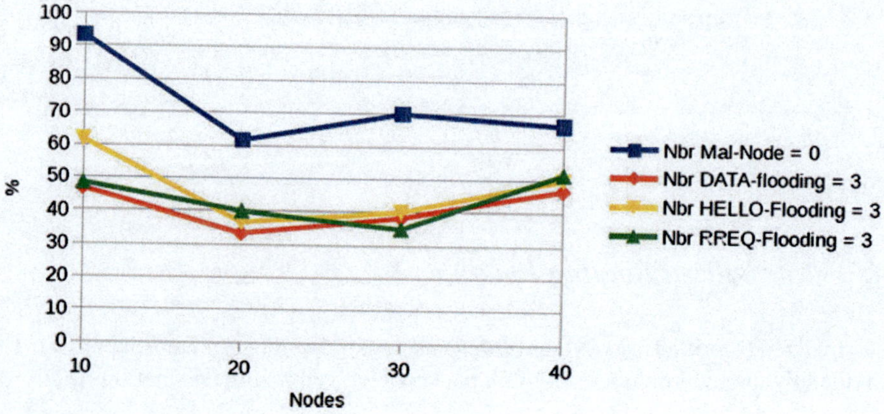

Fig. 6 Multiple malicious nodes PDR simulation result

of DATA Flooding attack that is more negative than HELLO Flooding and RREQ Flooding attacks for the two cases.

5.2 *Normalized Routing Load*

Figures 7 and 8 presents a comparison's NRL between the implementation of three types of Flooding attack in of single and multiple attacker, these results show that, the NRL of DATA Flooding attack is superior than normal network and HELLO Flooding and RREQ Flooding attacks for the two cases (single and multiple attack). Therefore, the DATA flooding attack is more dangerous than other attacks.

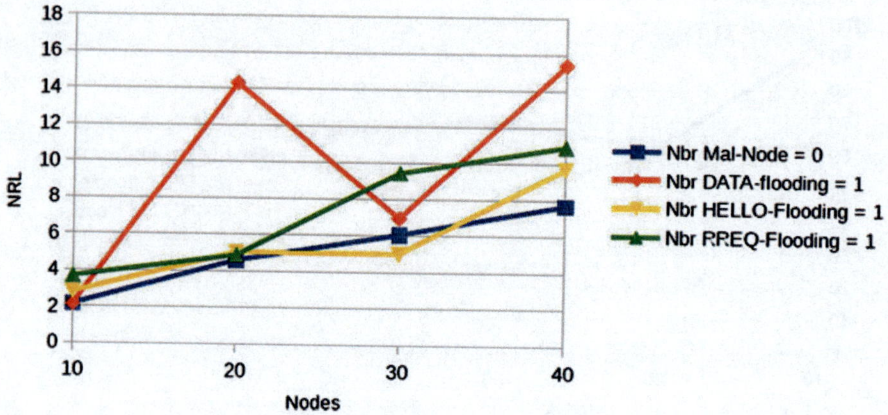

Fig. 7 Single malicious node's NRL simulation result

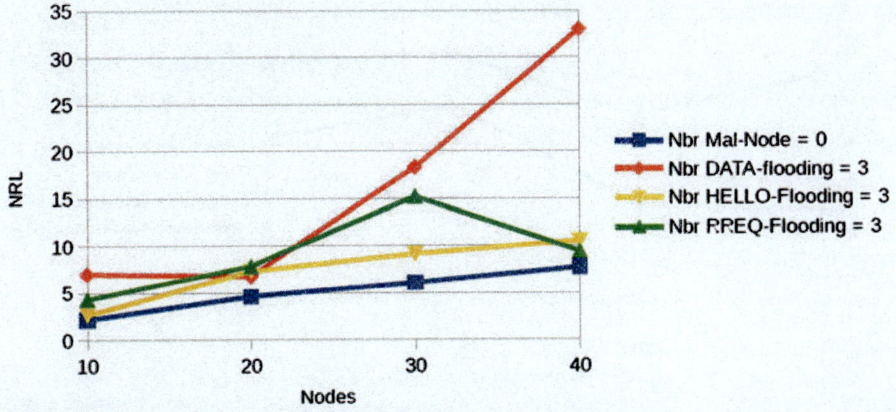

Fig. 8 Multiple malicious nodes NRL simulation result

6 Comparison Between Link-Spoofing, DATA Flooding and Replay Attacks

6.1 Packet Delivery Rate

Figures 9 and 10 presents a comparison's PDR between the implementation of single and multiple Link-spoofing, DATA Flooding and Replay attacks, the results show that the three attacks have the same negative impact on network with single malicious node, but for multiple malicious nodes, the simulation results show that the DATA Flooding attack is more disruptive than others.

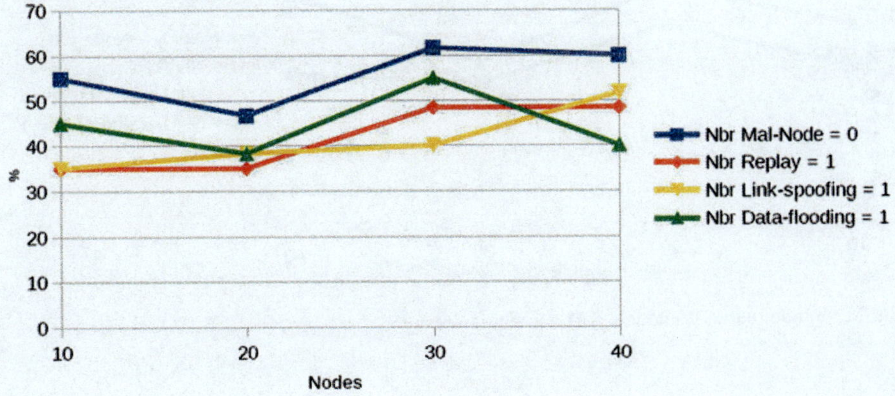

Fig. 9 Single malicious node's PDR simulation result

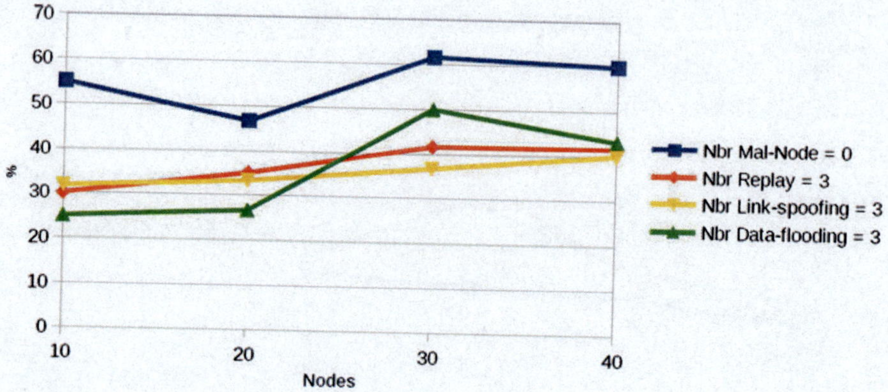

Fig. 10 Multiple malicious nodes PDR simulation result

6.2 Normalized Routing Load

Figures 11 and 12 presents a comparison's NRL between the implementation of single and multiple Link-spoofing, DATA Flooding and Replay attacks, the results show that, the NRL of both Replay and Link-spoofing attacks in single malicious node has approximately the same impact on network, and the DATA flooding attack is the least vulnerable. In case of multiple attack, the results indicate that the DATA Flooding attack is the most defective of the network then others attack.

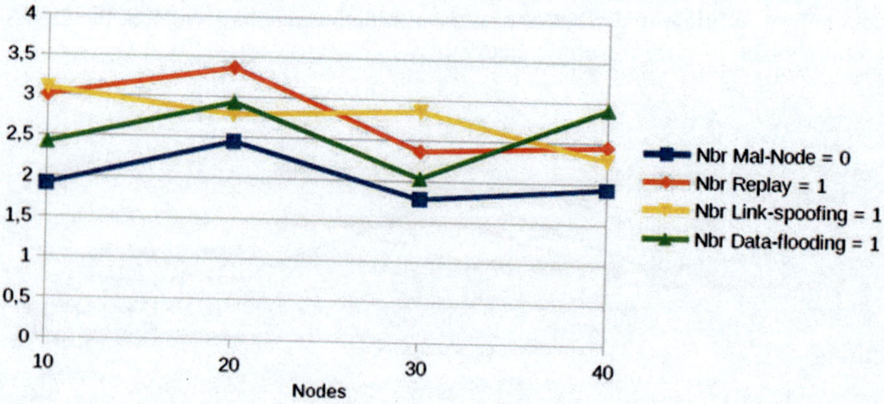

Fig. 11 Single malicious node's NRL simulation result

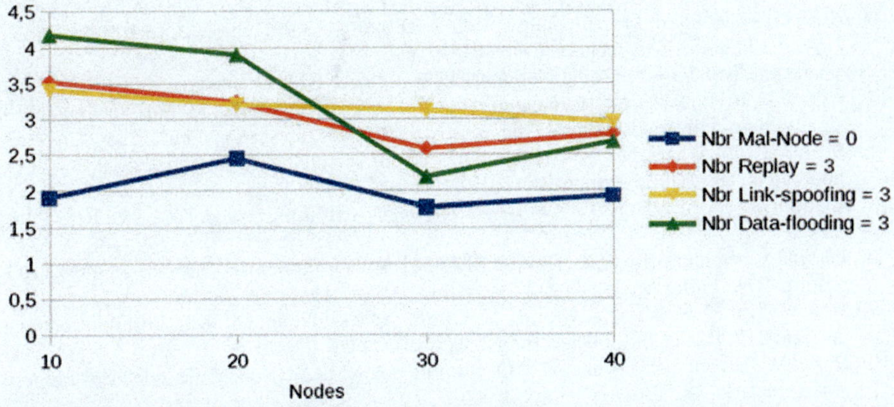

Fig. 12 Multiple malicious nodes NRL simulation result

7 Conclusion

In this paper, we have compared the impact of RREQ Flooding, HELLO Flooding, and DATA Flooding attacks in AODV routing protocols, and Link-spoofing, DATA Flooding, and Replay attacks OLSR routing protocols against MANETs. These attacks have been implemented in ns-3 simulator.

After comparison between the three derivatives of Flooding attack, DATA Flooding attack is more vulnerable than others attacks: low PDR and high NRL; and RREQ Flooding attack has less negative impact than others. For the difference between the Link-spoofing, DATA Flooding, and Replay attacks, we conclude that the three attacks have approximately the same negative impact on network's performances in single malicious node case, but multiple malicious node case, the DATA Flooding attack is the most defective and disruptive of the network.

Consequently, the RREQ Flooding, HELLO Flooding, DATA Flooding, Link-spoofing, and Replay attacks have a higher significant effect on the network performance.

References

1. Datta, R., Marchang, N.: Security for mobile Ad Hoc Networks. In: Handbook on securing cyber-physical critical infrastructure. Elsevier, pp. 147–190 (2012)
2. Fazeldehkordi, O.A., Elahe, A.,, Sadegh, I.,: Effect Of Black Hole Attack On AODV Routing Protocol In MANET. A study of black hole attack solutions: On aodv routing protocol in manet. Syngress, vol. 4333, March (2015)
3. Sbai, O., Elboukhari, M.: A simulation analysis of MANET's link-spoofing and replay attacks with ns-3. In Proceedings of the 4th International Conference on Smart City Applications—SCA '19, pp. 1–5 (2019)

4. Sbai, O., Elboukhari, M.: Simulation of MANET's Single and Multiple Blackhole Attack with NS-3. Colloq. Inf. Sci. Technol. Cist. **2018**, 612–617 (2018)
5. Sbai, O., Elboukhari, M.: A Simulation Analyses of MANET's Attacks Against OLSR Protocol with ns-3. In: Ben Ahmed, M., Boudhir, A.A., Santos, D., El Aroussi, M. (eds.) Innovations in smart cities applications edition 3, pp. 605–618. Springer International Publishing, Cham
6. Sbai, O., Elboukhari, M.: A simulation analyse of MANET's RREQ flooding and HELLO flooding attacks with ns-3, pp. 1–5 (2019)
7. Clausen, T., Jacquet, P.: RFC3626: Optimized link state routing protocol (OLSR). RFC Editor (2003)
8. Perkins, C., Belding-Royer, E., Das, S.: RFC3561: Ad hoc on-demand distance vector (AODV) routing. RFC Editor (2003)
9. Jeon, Y., Kim, T.H., Kim, Y., Kim, J.: LT-OLSR: Attack-tolerant OLSR against link spoofing, pp. 216–219. Proc. Conf. Local Comput. Networks, LCN (2012)
10. Desai, V.: Performance evaluation of OLSR protocol in MANET under the influence of routing attack, pp. 138–143 (2014)
11. Madhavi, S., Duraiswamy, K.: Flooding attack aware secure AODV. J. Comput. Sci. **9**(1), 105–113 (2013)
12. Abdelhaq, M., et al.: The resistance of routing protocols against DDOS attack in MANET. Int. J. Electr. Comput. Eng. **10**(5), 4844–4852 (2020)
13. Vinet, L., Zhedanov, A.: A 'missing' family of classical orthogonal polynomials. Antimicrob. Agents Chemother. **58**(12), 7250–7257 (2010)
14. Bhuvaneswari, R., Ramachandran, R.: Denial of service attack solution in OLSR based manet by varying number of fictitious nodes. Cluster Comput. **22**(S5), 12689–12699 (2019)
15. Abdelhaq, M., et al.: The impact of selfishness attack on mobile ad hoc network. Int. J. Commun. Networks Inf. Secur. **12**(1), 42–46 (2020)
16. Kristiansen, S.: Ns-3 Tutorial (2010)
17. Nisar, M.A., Mehmood, A., Nadeem, A., Ahsan, K., Sarim, M.: A two dimensional performance analysis of mobility models for MANETs and VANETs, **3**(5), 94–103 (2014)
18. Paper, T.W., 802.11ac: The Fifth Generation of Wi-Fi, March, pp. 1–25 (2014)

Security of Deep Learning Models in 5G Networks: Proposition of Security Assessment Process

Asmaa Ftaimi and Tomader Mazri

Abstract 5G networks bring a new design paradigm that will revolutionize telecommunications and other sectors, such as industry 4.0, smart cities, and autonomous vehicles. However, with the inherent advantages, many challenges can emerge. Today, the community is increasingly motivated to address these challenges by leveraging deep learning models to improve the 5G end-user quality of experience (QoE). However, this alternative approach would exhibit network assets to a series of security threats that could compromise the availability, integrity, and privacy of 5G architectures. This paper will extensively examine the vulnerabilities in the 5G models to draw the community's attention to the threats they may involve when integrated without sufficient prevention. Its main contribution lies in the comprehensive and adaptive approach it proposes to appraise deep learning models' security by assessing and managing the vulnerabilities they may present before their implementation in 5G network architectures.

1 Introduction

With the arrival of smart cities, connected objects, and augmented reality, several constraints have emerged. Indeed, the 4G network can no longer meet today's needs in throughput, latency, communication reliability, and connected objects' density. Hence, the arrival of 5G networks has taken over to offer a distributed and flexible architecture providing high performances.

5G is a new generation of mobile networks offering performance far beyond the one provided by the LTE network [1]. Nevertheless, several techniques are needed to deal with the complicated challenges associated with 5G services, e.g., to cope with the massive amount of data traffic carried in the 5G network, the classical traffic management and resource allocation features must be deployed automatically [2]. Thus, 5G network components require the involvement of high analysis and

A. Ftaimi (✉) · T. Mazri
Laboratory of Advanced Systems Engineering, Ibn Tofail Science University, Kenitra, Morocco

M. Ben Ahmed et al. (eds.), *Networking, Intelligent Systems and Security*,
Smart Innovation, Systems and Technologies 237,
https://doi.org/10.1007/978-981-16-3637-0_28

prediction capabilities and must be equipped with intelligence to ensure the necessary performance, hence the new trend of deploying deep learning models in 5G networks.

Deep learning models have received increasing interest in the research community. Their ability to predict, personalize, and detect behavior has prompted several researchers to integrate them into various applications. Their deployment in 5G networks is no exception, and the results obtained to date are promising, whether in the optimization of resource allocation, orchestration between different NFVs and SDNs, prediction of traffic and user mobility according to user behavior and historical data, and automation of complex and laborious tasks. Deep learning has enabled the transformation of 5G into cognitive networks with high adaptability to end-user demand to ensure a better quality of user experience [3].

Nevertheless, several studies have been conducted to examine the potential vulnerabilities that may emerge from the expressiveness and the flexibility aspect of deep learning models [4]. It has been shown that deep learning models enfold flaws that can be harnessed by attackers to carry out illicit activities. Indeed, Papernot et al. [5] have carried out an experiment to test machine learning models' resilience to adversarial attacks. They have successfully performed malicious attacks toward deep learning models without needing specific access to the model's features or dataset. The obtained unexpected results have motivated the research community to focus on the field of adversarial learning to develop secure and robust deep learning models [6].

Although, the research community has been interested in deep learning models' security without examining this aspect in a specific domain of application. However, this approach can be misleading when assessing the impact of attacks and their robustness [7, 8]. For instance, the analysis of the impact of attacks targeting deep learning models cannot be dissociated from the system's criticality built on this model. Indeed, a less robust attack targeting a model exploited in e-health or 5G network is extremely critical and has a huge impact compared to a highly robust attack aiming models used in video games. Hence, the importance of this work focuses on examining the security of deep learning models in their application in the 5G network.

In this paper, we will first present an overview of 5G networks and the innovative features they import. Then we will discuss the challenges associated with these functionalities in terms of management and computational and storage resources and the contribution of deep learning models in this area as well as the efficient and powerful solutions they provide. Afterward, we will study the challenges in terms of security resulting from artificial intelligence integration in 5G networks. We will scrutinize the vulnerabilities that are enfolded in deep learning models. Finally, we will propose a method to evaluate the models' security before incorporating them in the 5G network components.

2 Overview of 5G Architecture

Today, 5G offers a variety of services designed to meet the needs of several sectors. Indeed, besides the classic telephony and file transfer service, 5G provides the connectivity of smart cities and autonomous vehicles, connected objects, augmented reality, and the 4.0 industry. All these market segments require different qualities of service. Hence, experts have defined key performance indicators to make a taxonomy of the envisaged 5G services. As shown in Fig. 1, in addition to reliability, eight other key performance indicators have been used to highlight the 5G market's different needs clearly. However, meeting all these criteria would be unfeasible, which explains the approach followed by the 5G network that considers a polymorphic system to define several declinations; each can fulfill a certain number of constraints while using a well-determined configuration of the 5G network. The three declinations of the 5G network are as follow.

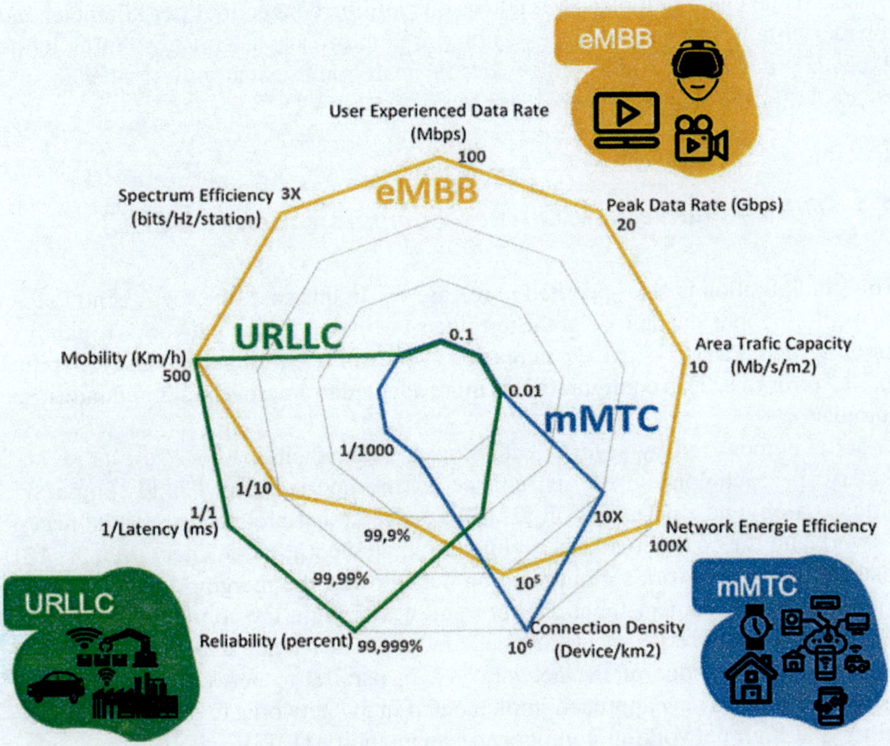

Fig. 1. 5G KPI's and declinations

2.1 Enhanced Mobile BroadBand (eMBB)

This configuration was specified in release 15, which was produced in 2018 during the first phase of the 3GPP standardization process [9]. In this case, user throughput is privileged over latency, the density of terminals, and reliability. This declination allows peak data rate up to 20 Gbps, and user experienced data rate up to 100 Mbit/s. It was designed to satisfy the need for high throughput required in Mobile Cloud Computing, smart devices, and UHD streaming applications.

2.2 Massive Machine Type Communication (mMTC)

It was specified during 2020 in release 16. This configuration is not as concerned with peak data rate and traffic capacity but rather with the density of terminals to be managed and energy efficiency. It allows one million connections per kilometer and a battery lifetime of ten years. It was proposed to offer an adequate solution to the expanding number of connected devices in smart homes, smart cities, e-health, and wearables [10].

2.3 Ultra-Reliable Low Latency Communication (URLLC)

This configuration is also specified in release 16. In this case, the focus is no longer on the throughput and density of the terminals but mainly on the network's reliability, mobility, and latency. URLLC can reach a latency of one millisecond and reliability of 10^{-9} error rate. This configuration is more adapted to smart vehicles and industrial automation.

5G is empowered by several technologies. Nevertheless, above all, its revolutionary approach comes from its reliance on an important trio: Cloud computing, virtualization, and softwarization [2]. The conventional architecture of the access network and the core network is no longer maintained in the 5G network [11, 12]. Instead, the 5G network's cloudification has allowed the merging of the system's control and management functionalities and has brought the computing capacities closer to the end-user to meet his needs in terms of latency and throughput, and availability at the edge of the network [1]. In parallel to network cloudification, softwarization is a new approach implemented in 5G networks to deliver great flexibility and high networking components' adaptability [13]. It involves employing software programs to provide the functionalities offered by network equipment and services. This approach requires a complete redesign of the networks [14]. However, it provides many advantages for simplifying complex management operations, such as coordination and load balancing between network components [2].

Virtualization has contributed significantly to the transformation of physical nodes performing complex operations into software blocks that can be easily clustered and combined to create sophisticated network operations [15]. In this way, Network Function Virtualization (NFV) technology alleviates the dependency that has always existed on hardware components in the network to ensure improved flexibility and high network functionality and efficiency [16]. However, these technologies are highly correlated in the 5G network. They require powerful computing resources to orchestrate between NFVs, optimize resources and automate complex tasks efficiently, hence the crucial necessity to harness deep learning networks' potential to fulfill such tasks.

3 Deep Learning Models Applications in 5G

Deep learning models significantly exceed many of the traditional methods of automation and prediction. They are characterized by their great potential for generalization and their high expressivity, enabling them to substitute conventional approaches in several disciplines such as telecoms and networking. In the following sub-sections, we highlight several potentials of deep learning models and the significant profits they can deliver to 5G networks, prominently in terms of predicting behaviors, helping in optimizing resources, and detecting events in the network [17] as illustrated in Fig. 2.

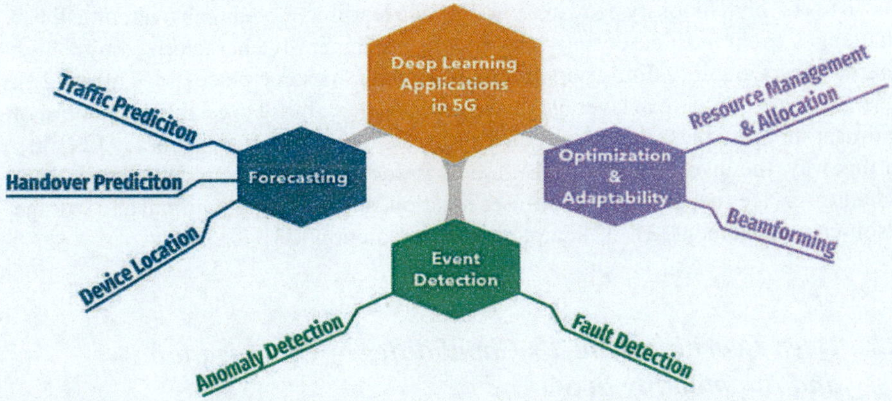

Fig. 2 Deep learning applications in 5G

3.1 Deep Learning model's Capabilities of Forecasting in 5G

In the next paragraphs, we will explore the predictive potentials and capabilities offered by deep learning models in 5G networks.

Traffic Prediction. With the growing volume of managed traffic in the 5G network, traffic prediction becomes crucial to optimizing resource allocation. The role of DL in such a context is primordial. Several research projects have focused on traffic prediction through the deployment of deep learning models for network slicing [18, 19] and congestion prevention [20]. In [21–23], the authors adopted an interesting approach by examining the temporal and spatial properties of end-users to predict the traffic they generate in the network.

Handover Prediction. In telecom networks, the handover is an essential process to guarantee service quality while ensuring mobility for network users. The handover process consists in swiping the user from one base station to another based on predefined values of the Reference Signal Receiver Power (RSRP), the Reference Signal Receiver Quality (RSRQ). However, this process is not as simple as that. Errors can occur during handovers, and communications can sometimes be interrupted, hence using deep learning to facilitate this operation [24, 25]. According to the work of Khunteta et al. [26], a well-trained deep learning model can predict the success or failure rate of the next handover based on previous experiences and several data collected in the network. Ozturk et al. [27] used a deep learning model to predict the likelihood of a handover occurrence to prepare the network in advance.

Device Location Prediction. Predicting the location of users is an important feature in telecom networks. This functionality appears to be complex in 5G networks due to users' high mobility. However, using deep learning models can overcome these challenges as, in most cases, user movements are generally characterized by a high degree of repeatability. End-users frequently attend the same places very often [28]. Deep learning models can leverage this aspect of repeatability to establish a pattern of end-user mobility behavior and predict future locations with high accuracy [29, 30]. In this way, the involvement of DL models in location prediction can significantly enhance services depending on end-user location, such as computational and storage resource management [31, 32] and handover prediction [33].

3.2 Deep Learning Model's Capabilities of Optimization and Adaptability in 5G

In the following paragraphs, we will outline the power of deep learning models in helping to resolve optimization problems in 5G network components by adapting their dimensioning according to the end-user's requirements.

Optimization of Resource Management and Allocation. 5G is deeply based on network slicing technology. The latter leverages the network infrastructure to satisfy the various needs and requirements of multiple market segments. Gutterman et al. [34] have examined the utilization of DL models to forecast resource allocation measures to align each network slice with the intended usage category. Yan et al. [35] have demonstrated the relevance of DL in historical data processing to help in decision making and therefore optimizing resource allocation planning for a network slice following Service Level Agreement (SLA) requirements. In [36], a deep learning model was proposed to minimize energy consumption during resource allocation for network slices. In this study, the authors approached the subject of energy consumption by employing a DL model and considering several parameters such as channel characteristics, end-user requests, duration of communication, etc. Ahmed et al. [37] have studied the integration of a DL model in a multi-cell network to achieve maximum throughput while helping in optimizing the allocation of resources in the network regarding the CSI channel and the user's location.

Optimization of Beamforming. Maksymyuk et al. [38] studied the use of a reinforcement learning model in the beamforming technology that marks the 5G network. The model evaluates each antenna's required phases and amplitudes in the MIMO technology to estimate the signal coverage according to the users' location. When users are scattered in separated areas, strong signal coverage is needed to reach the maximum of the end devices. When they are concentrated in the same location, the signal needs to be focused and aimed to reach them all.

3.3 Deep Learning model's Capabilities of Event Detection in 5G

This section highlights the key applications of deep learning in event detection, especially regarding anomalies and failures that may occur in 5G networks.

Anomaly Detection. Anomaly detection is an important aspect that must be optimized in a 5G network to ensure its service continuity. It consists in detecting malicious activities occurring in the network and eventually identifying where the problem is originating. However, the network's heterogeneity and the large volume of traffic it handles contribute considerably to the complexity of detecting anomalies in the 5G network [39, 40]. Several works have proposed the introduction of deep learning models for the detection of cyber-attacks. Many researchers have designed models to extract normal traffic properties to detect malicious threats [41]. Others have rather focused on learning normal network behavior to alert in case of antenna failure or network congestion [41, 42].

Fault Detection. Fault detection is important to increase network performance and reduce latency. Ensuring fault detection is critical for the URLLC component of the 5G network. This mission is both vital and complex because when a failure occurs in an antenna, for example, the fault must first be detected, and then the source of the problem needs to be located. This task seems simple; however, it hides a great complexity, especially in 5G networks because of the large number of antennas used and the employed equipment's heterogeneity [43, 44]. Thus, Chen et al. [45] proposed a method based on a neural network model to detect and localize antenna faults in mmWave systems in 5G networks.

After delving into the potentials delivered by deep learning models when employed in 5G networks, we will present in the next section the challenges that may be met when incorporating Artificial intelligence in 5G. We will mainly focus on the security aspects and the vulnerabilities that can reside in deep learning models with an extensive study of the impact of adversarial attacks targeting deep learning-built systems in 5G networks.

4 Vulnerabilities in Deep Learning Models Employed in 5G Networks

Today deep learning networks have attracted growing interest from the research community. Several implementations of DL networks have been developed. While some have already been implemented, others are being proposed. However, various works have examined these models' security aspects and have shown that they are vulnerable [46]. Indeed, several studies have demonstrated that such models could be targeted by adversarial attacks using carefully and precisely designed perturbation injected into the dataset to mislead the model [47].

Several attack strategies have been developed to assess machine learning models' robustness toward adversarial manipulations in this context. Papernot et al. [48] have conducted adversarial attacks on machine learning models requiring zero accessibility to the model's dataset or features following a Jacobian saliency map-based attack (JSMA). In 2013, Szegedy et al. [46] have examined adversarial attacks in deep neural networks by introducing noise into the input dataset and have succeeded in reducing the model's classification meticulousness. Goodfellow et al. [49] have proposed a new approach known as Fast Gradient Sign Method to generate adversarial examples, Carlini and Wagner [50] and Liu et al. [51] have suggested optimization-based methods to craft noisy perturbations that could be utilized in attacks targeting machine learning models.

In [52], Usama et al. have carried out a white-box Carlini and Wagner attack [50] against a CNN model used in an end-to-end Modulation classification system. They used the L2 norm constraint as a metric of perturbations introduced into the input dataset. They easily succeeded in diminishing the classifier accuracy dramatically. Their findings have revealed the severity of the deep learning models' vulnerability in 5G networks. Usama et al. [52] have also examined the fragility of the Channel Autoencoder unsupervised model deployed in 5G architectures. They introduced an additive white Gaussian noise (AWGN) as an adversarial disturbance and have witnessed a high drop in the model accuracy indicated by a large expansion of block error rate (BLER). The conducted study has highlighted the flaws embedded in the unsupervised autoencoders model, which have drastically compromised its integrity.

Besides, Goutay et al. [53] have developed an attack against autoencoders using deep reinforcement learning models with noisy feedback. In an attempt to approximate the real scenario, the researchers have performed a black-box attack where no knowledge of the targeted system is assumed, using a substitution model instead and taking advantage of the adversarial examples' transferability property. This approach is founded on multiple studies that have highlighted the similarity in several models' behavior against additive perturbations to the input data set, even if they appear to be different. Usama et al. [52] have succeeded in dropping the model's accuracy rate from 95 to 80%, thereby reducing its confidence in its results.

Suomalainen et al. [54] have also scrutinized the criticality of harnessing inherent vulnerabilities in deep learning models to cause large-scale damage to 5G networks. Indeed, deep learning models can be leveraged in the load management of network resources. The importance of such functionality is high since it accommodates the end-users' demand without compromising the optimal and efficient use of network resources [55]. Load balancing combines criticality and complexity as it requires resource orchestration and planning, traffic prioritization, classification, and prediction [56]. However, Jani Suomalainen et al. have suggested that attacker risks carrying out a DOS attack toward load balancing models and influences them to redirect traffic to certain resources, causing an overload of some components while others remain unused.

In addition to the aforementioned use cases, we have encountered several other scenarios describing the harness of inherent vulnerabilities and flows in deep learning models incorporated in 5G network. Therefore, we propose in the section below to develop a taxonomy of groups and categories of different threats identified in these models. Indeed, system security's classical approach inspired us to design classification of threats that may reside in deep learning models applied in 5G. This classification is founded on three key categories that are confidentiality, integrity, and availability as illustrated in Fig. 3.

1. Confidentiality threats: Regularly, this class of threat involves obtaining unauthorized access to information transmitted through the network by an adversary. This menace can lead to harmful effects, including leakage of sensitive information such as revealing information about end-user behavior or divulging critical data. An opponent can potentially corrupt the model and even escalate privileges to gain unauthorized access to network resources.
2. Availability threats: Attackers can jeopardize network availability by performing denial of service attacks, either by causing network congestion or by overloading network infrastructure components. The opponent can also conduct denial of detection to prevent the network from detecting failures, allowing the attacker to interrupt its normal operation.
3. Integrity threats: This threat is essentially related to traffic interception and the modification of data transmitted in 5G networks. Indeed, the attacker could inject carefully crafted infinitesimal perturbations in the traffic to mislead the model. Other opponents possessing strong capabilities can completely alter the model's behavior by influencing its decisions.

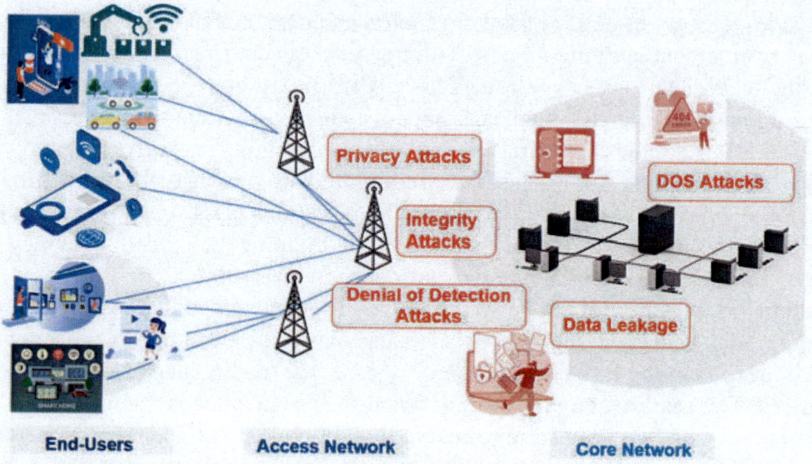

Fig. 3 Threats in deep learning model in 5G architectures

Therefore, we can assume that several scenarios exist. An attacker can leverage the vulnerabilities inherent in deep learning models to accomplish a malicious intent threatening the availability, integrity, or confidentiality of 5G network services. Consequently, it has become apparent that deep learning models'.

5 Proposition of Security Assessment Process for Deep Learning Models

The integration of deep learning models in 5G networks provides optimal solutions to the challenges confronted in such innovative generation of telecom networks. However, it also imposes additional challenges related to their security and resilience to adversarial attacks. Indeed, there is no doubt about the automation and forecasting benefits that the application of deep learning in the 5G network can bring. Nevertheless, it is noteworthy that the 5G network would be exploited in a wide range of application areas, including e-health, autonomous vehicle, smart homes, and many other extremely critical structures. Hence, the necessity to examine their security very carefully to avoid unexpected security incidents. So far, there are no reliable method to evaluate the security of machine learning models. For this reason, we have proposed in this section a new approach to assess deep learning models' security and their robustness toward adversarial attacks before they can be applied in 5G architectures. Our approach, built on our previous works in [57], proposes a process of vulnerabilities assessment and management articulated around three essential steps to obtain a comprehensive evaluation of the model from a security perspective as shown in Fig. 4.

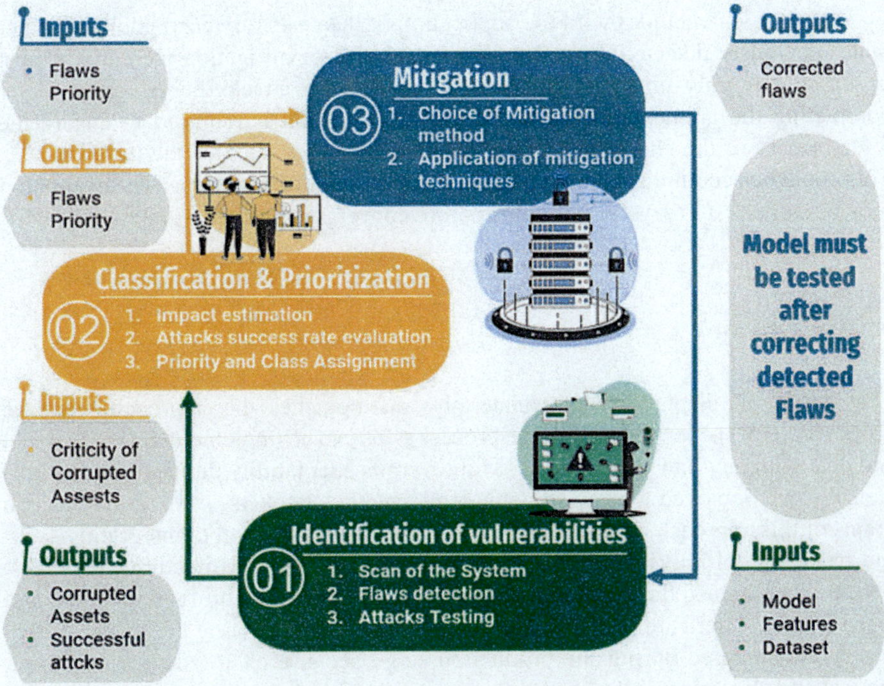

Fig. 4 The process of Assessment and management of deep learning model vulnerabilities

This process is initiated with the identification of existing vulnerabilities in the model under consideration. In this framework, a security test of vulnerabilities is highly recommended. According to Tian-yang et al. [58], security vulnerabilities testing involves a real simulation of attack scenarios that the attacker may follow. In this case, acting as an attacker will permit us to observe the system through the adversary's eyes and thus will help to examine the system effectively and identify the different vulnerabilities that could potentially compromise the model and succeed the attack. In this regard, we strongly recommend testing several types of attacks among those existing in the literature, with a particular focus on the most robust and adaptative ones. This step leads to the identification of corrupted assets and successful attacks that can exploit them.

Following the identification of vulnerabilities and the attacks that could potentially harness them, a work of expurgation needs to be performed. Indeed, an estimation of the impact of successful attacks and their success rate is necessary and can contribute considerably to the prioritization of vulnerabilities. In this step, the criticality of the assets containing the vulnerabilities can significantly influence the decision to be taken regarding the attack's impact. The assessment must consider three important criteria: the success rate of the attack, its impact, and the criticality of the assets affected by the attack. The job is rounded off by drawing up a final prioritized list of detected breaches.

The process concludes with the application of the necessary corrections to remediate previously detected critical flaws. The choice of mitigation techniques must be based primarily on their effectiveness against the attacks being tested. After completing the application of fixes, a testing operation is required to ensure the effectiveness of the chosen mitigation methods in the context of the tested model. This could be accomplished by repeating the first step of the process. The evaluation may be subjected to several iterations before converging to a secure model.

6 Conclusion

The aspect of adaptability and generality distinguishes the approach we have proposed in this paper. Indeed, all the process steps can be implemented in any model and deal with any flaws and attacks. Moreover, its adaptability derives from its efficiency when deployed for any strategies followed by attackers. Among the positive grants of this approach, it is essential to mention its contribution to integrating security in the model implementation cycle in the 5G network architectures resources. The results obtained during the implementation of the vulnerability assessment and management pattern are realistic. They are extremely approaching the scenarios that can be encountered during the production stage because all the data used during the test and the vulnerability scan are consistent with the one transmitted over the network. The attacks carried out during the identification of vulnerabilities simulate the attacker's behavior and approximate the real world's security incidents. Finally, in addition to its apparent flexibility, this process designed as an iterative wheel adds a great advantage to this approach since it allows continuous improvement of the models to enhance their security and reduce their flaws.

References

1. Chang, C.-Y., Nikaein, N.: Cloudification and slicing in 5G radio access network, http://www.theses.fr/2018SORUS293/document (2018)
2. Barakabitze, A.A., Ahmad, A., Mijumbi, R., Hines, A.: 5G network slicing using SDN and NFV: A survey of taxonomy, architectures and future challenges. Comput. Netw. **167**, 106984 (2020). https://doi.org/10.1016/j.comnet.2019.106984
3. Santos, G.L., Endo, P.T., Sadok, D., Kelner, J.: When 5G meets deep learning: A systematic review. Algorithms. **13**, 208 (2020). https://doi.org/10.3390/a13090208
4. Barreno, M., Nelson, B., Sears, R., Joseph, A.D., Tygar, J.D.: Can machine learning be secure? In: Proceedings of the 2006 ACM Symposium on Information, computer and communications security - ASIACCS '06. p. 16. ACM Press, Taipei, Taiwan (2006). https://doi.org/10.1145/1128817.1128824.
5. Papernot, N., McDaniel, P., Goodfellow, I., Jha, S., Celik, Z.B., Swami, A.: Practical Black-Box Attacks against Machine Learning. arXiv:1602.02697 [cs] (2017)
6. Huang, L., Joseph, A.D., Nelson, B., Rubinstein, B.I.P., Tygar, J.D.: Adversarial machine learning. In: Proceedings of the ACM Conference on Computer and Communications Security.

pp. 43–57. ACM Press, New York, New York, USA (2011). https://doi.org/10.1145/2046684.2046692
7. Chakraborty, A., Alam, M., Dey, V., Chattopadhyay, A., Mukhopadhyay, D.: Adversarial Attacks and Defences: A Survey. arXiv:1810.00069 [cs, stat] (2018)
8. Mello, F.L. de: A survey on machine learning adversarial attacks. J. Inf. Secur. Cryptogr. **7**, 1–7 (2020). https://doi.org/10.17648/jisc.v7i1.76
9. Tani, N.: IoT-driven evolution and business innovation. NTT DOCOMO Technical J. **19**, 82 (2018)
10. Sabella, D., Vaillant, A., Kuure, P., Rauschenbach, U., Giust, F.: Mobile-Edge Computing Architecture: The role of MEC in the Internet of Things. IEEE Consumer Electron. Mag. **5**, 84–91 (2016). https://doi.org/10.1109/MCE.2016.2590118
11. Kekki, S., Featherstone, W., Fang, Y., Kuure, P., Li, A., Ranjan, A., Purkayastha, D., Jiangping, F., Frydman, D., Verin, G., Wen, K.-W., Kim, K., Arora, R., Odgers, A., Contreras, L.M., Scarpina, S.: ETSI White Paper No. 28 MEC in 5G networks (2018)
12. Hassan, N., Yau, K.-L.A., Wu, C.: Edge computing in 5G: A review. IEEE Access. **7**, 127276–127289 (2019). https://doi.org/10.1109/ACCESS.2019.2938534
13. Sayadi, B., Gramaglia, M., Friderikos, V., von Hugo, D., Arnold, P., Alberi-Morel, M.-L., Puente, M.A., Sciancalepore, V., Digon, I., Crippa, M.R.: SDN for 5G Mobile networks: NORMA perspective. In: Noguet, D., Moessner, K., and Palicot, J. (eds.) Cognitive radio oriented wireless networks. pp. 741–753. Springer International Publishing, Cham (2016). https://doi.org/10.1007/978-3-319-40352-6_61.
14. Trivisonno, R., Guerzoni, R., Vaishnavi, I., Soldani, D.: SDN-based 5G mobile networks: architecture, functions, procedures and backward compatibility: SDN-based 5G mobile networks: architecture, functions, procedures and backward compatibility. Trans. Emerging Tel. Tech. **26**, 82–92 (2015). https://doi.org/10.1002/ett.2915
15. Giannoulakis, I., Kafetzakis, E., Xylouris, G., Gardikis, G., Kourtis, A.: On the Applications of Efficient NFV Management Towards 5G Networking. In: Proceedings of the 1st International Conference on 5G for Ubiquitous Connectivity. ICST, Levi, Finland (2014). https://doi.org/10.4108/icst.5gu.2014.258133.
16. Siddiqui, M.S., Escalona, E., Trouva, E., Kourtis, M.A., Kritharidis, D., Katsaros, K., Spirou, S., Canales, C., Lorenzo, M.: Policy based virtualised security architecture for SDN/NFV enabled 5G access networks. In: 2016 IEEE Conference on Network Function Virtualization and Software Defined Networks (NFV-SDN). pp. 44–49. IEEE, Palo Alto, CA (2016). https://doi.org/10.1109/NFV-SDN.2016.7919474
17. McClellan, M., Cervelló-Pastor, C., Sallent, S.: Deep Learning at the Mobile Edge: Opportunities for 5G Networks. Appl. Sci. **10**, 4735 (2020). https://doi.org/10.3390/app10144735
18. Bega, D., Gramaglia, M., Fiore, M., Banchs, A., Costa-Perez, X.: DeepCog: Cognitive Network Management in Sliced 5G Networks with Deep Learning. In: IEEE INFOCOM 2019 - IEEE Conference on Computer Communications. pp. 280–288. IEEE, Paris, France (2019). https://doi.org/10.1109/INFOCOM.2019.8737488
19. Guo, Q., Gu, R., Wang, Z., Zhao, T., Ji, Y., Kong, J., Gour, R., Jue, J.P.: Proactive Dynamic Network Slicing with Deep Learning Based Short-Term Traffic Prediction for 5G Transport Network. In: Optical Fiber Communication Conference (OFC) 2019. p. W3J.3. OSA, San Diego, CA (2019). https://doi.org/10.1364/OFC.2019.W3J.3
20. Zhou, Y., Fadlullah, ZMd., Mao, B., Kato, N.: A Deep-Learning-Based Radio Resource Assignment Technique for 5G Ultra Dense Networks. IEEE Network **32**, 28–34 (2018). https://doi.org/10.1109/MNET.2018.1800085
21. Chen, L., Yang, D., Zhang, D., Wang, C., Li, J., Nguyen, T.-M.-T.: Deep mobile traffic forecast and complementary base station clustering for C-RAN optimization. J. Netw. Comput. Appl. **121**, 59–69 (2018). https://doi.org/10.1016/j.jnca.2018.07.015
22. Huang, C.-W., Chiang, C.-T., Li, Q.: A study of deep learning networks on mobile traffic forecasting. In: 2017 IEEE 28th Annual International Symposium on Personal, Indoor, and Mobile Radio Communications (PIMRC). pp. 1–6. IEEE, Montreal, QC (2017). https://doi.org/10.1109/PIMRC.2017.8292737

23. Zhang, C., Zhang, H., Yuan, D., Zhang, M.: Citywide Cellular Traffic Prediction Based on Densely Connected Convolutional Neural Networks. IEEE Commun. Lett. **22**, 1656–1659 (2018). https://doi.org/10.1109/LCOMM.2018.2841832
24. Hosny, K.M., Khashaba, M.M., Khedr, W.I., Amer, F.A.: New vertical handover prediction schemes for LTE-WLAN heterogeneous networks. PLoS ONE **14**, e0215334 (2019). https://doi.org/10.1371/journal.pone.0215334
25. Svahn, C., Sysoev, O., Cirkic, M., Gunnarsson, F., Berglund, J.: Inter-Frequency Radio Signal Quality Prediction for Handover, Evaluated in 3GPP LTE. In: 2019 IEEE 89th Vehicular Technology Conference (VTC2019-Spring). pp. 1–5. IEEE, Kuala Lumpur, Malaysia (2019). https://doi.org/10.1109/VTCSpring.2019.8746369
26. Khunteta, S., Chavva, A.K.R.: Deep Learning Based Link Failure Mitigation. In: 2017 16th IEEE International Conference on Machine Learning and Applications (ICMLA). pp. 806–811. IEEE, Cancun, Mexico (2017). https://doi.org/10.1109/ICMLA.2017.00-58
27. Ozturk, M., Gogate, M., Onireti, O., Adeel, A., Hussain, A., Imran, M.A.: A novel deep learning driven, low-cost mobility prediction approach for 5G cellular networks: The case of the Control/Data Separation Architecture (CDSA). Neurocomputing **358**, 479–489 (2019). https://doi.org/10.1016/j.neucom.2019.01.031
28. Xiong, H., Zhang, D., Zhang, D., Gauthier, V., Yang, K., Becker, M.: MPaaS: Mobility prediction as a service in telecom cloud. Inf Syst Front. **16**, 59–75 (2014). https://doi.org/10.1007/s10796-013-9476-z
29. Cheng, Y., Qiao, Y., Yang, J.: An improved Markov method for prediction of user mobility. In: 2016 12th International Conference on Network and Service Management (CNSM). pp. 394–399. IEEE, Montreal, QC, Canada (2016). https://doi.org/10.1109/CNSM.2016.7818454
30. Qiao, Y., Yang, J., He, H., Cheng, Y., Ma, Z.: User location prediction with energy efficiency model in the Long Term-Evolution network: User location prediction with energy efficiency model. Int. J. Commun. Syst. **29**, 2169–2187 (2016). https://doi.org/10.1002/dac.2909
31. Gante, J., Falcao, G., Sousa, L.: Beamformed Fingerprint Learning for Accurate Millimeter Wave Positioning. In: 2018 IEEE 88th Vehicular Technology Conference (VTC-Fall). pp. 1–5. IEEE, Chicago, IL, USA (2018). https://doi.org/10.1109/VTCFall.2018.8690987
32. Gante, J., Falcão, G., Sousa, L.: Deep Learning Architectures for Accurate Millimeter Wave Positioning in 5G. Neural Process Lett. **51**, 487–514 (2020). https://doi.org/10.1007/s11063-019-10073-1
33. Wang, C., Zhao, Z., Sun, Q., Zhang, H.: Deep Learning-Based Intelligent Dual Connectivity for Mobility Management in Dense Network. In: 2018 IEEE 88th Vehicular Technology Conference (VTC-Fall). pp. 1–5. IEEE, Chicago, IL, USA (2018). https://doi.org/10.1109/VTCFall.2018.8690554
34. Gutterman, C., Grinshpun, E., Sharma, S., Zussman, G.: RAn resource usage prediction for a 5G slice broker. In: Proceedings of the International Symposium on Mobile Ad Hoc Networking and Computing (MobiHoc). pp. 231–240. Association for Computing Machinery, New York, NY, USA (2019). https://doi.org/10.1145/3323679.3326521
35. Yan, M., Feng, G., Zhou, J., Sun, Y., Liang, Y.-C.: Intelligent Resource Scheduling for 5G Radio Access Network Slicing. IEEE Trans. Veh. Technol. **68**, 7691–7703 (2019). https://doi.org/10.1109/TVT.2019.2922668
36. Luo, J., Tang, J., So, D.K.C., Chen, G., Cumanan, K., Chambers, J.A.: A Deep Learning-Based Approach to Power Minimization in Multi-Carrier NOMA With SWIPT. IEEE Access. **7**, 17450–17460 (2019). https://doi.org/10.1109/ACCESS.2019.2895201
37. Ahmed, K.I., Tabassum, H., Hossain, E.: Deep Learning for Radio Resource Allocation in Multi-Cell Networks. IEEE Network **33**, 188–195 (2019). https://doi.org/10.1109/MNET.2019.1900029
38. Maksymyuk, T., Gazda, J., Yaremko, O., Nevinskiy, D.: Deep Learning Based Massive MIMO Beamforming for 5G Mobile Network. In: 2018 IEEE 4th International Symposium on Wireless Systems within the International Conferences on Intelligent Data Acquisition and Advanced Computing Systems (IDAACS-SWS). pp. 241–244. IEEE, Lviv (2018). https://doi.org/10.1109/IDAACS-SWS.2018.8525802

39. Fernandez Maimo, L., Perales Gomez, A.L., Garcia Clemente, F.J., Gil Perez, M., Martinez Perez, G.: A Self-Adaptive Deep Learning-Based System for Anomaly Detection in 5G Networks. IEEE Access. **6**, 7700–7712 (2018). https://doi.org/10.1109/ACCESS.2018.2803446
40. Parwez, M.S., Rawat, D.B., Garuba, M.: Big Data Analytics for User-Activity Analysis and User-Anomaly Detection in Mobile Wireless Network. IEEE Trans. Ind. Inf. **13**, 2058–2065 (2017). https://doi.org/10.1109/TII.2017.2650206
41. Fernández Maimó, L., Huertas Celdrán, A., Gil Pérez, M., García Clemente, F.J., Martínez Pérez, G.: Dynamic management of a deep learning-based anomaly detection system for 5G networks. J. Ambient. Intell. Human Comput. **10**, 3083–3097 (2019). https://doi.org/10.1007/s12652-018-0813-4
42. Hussain, B., Du, Q., Zhang, S., Imran, A., Imran, M.A.: Mobile Edge Computing-Based Data-Driven Deep Learning Framework for Anomaly Detection. IEEE Access. **7**, 137656–137667 (2019). https://doi.org/10.1109/ACCESS.2019.2942485
43. Hu, P., Zhang, J.: 5G-Enabled Fault Detection and Diagnostics: How Do We Achieve Efficiency? IEEE Internet Things J. **7**, 3267–3281 (2020). https://doi.org/10.1109/JIOT.2020.2965034
44. Yu, A., Yang, H., Yao, Q., Li, Y., Guo, H., Peng, T., Li, H., Zhang, J.: Accurate Fault Location Using Deep Belief Network for Optical Fronthaul Networks in 5G and Beyond. IEEE Access. **7**, 77932–77943 (2019). https://doi.org/10.1109/ACCESS.2019.2921329
45. Chen, K., Wang, W., Chen, X., Yin, H.: Deep Learning Based Antenna Array Fault Detection. In: 2019 IEEE 89th Vehicular Technology Conference (VTC2019-Spring). pp. 1–5. IEEE, Kuala Lumpur, Malaysia (2019). https://doi.org/10.1109/VTCSpring.2019.8746510
46. Szegedy, C., Zaremba, W., Sutskever, I., Bruna, J., Erhan, D., Goodfellow, I., Fergus, R.: Intriguing properties of neural networks. arXiv:1312.6199 [cs]. (2014)
47. Ibitoye, O., Abou-Khamis, R., Matrawy, A., Shafiq, M.O.: The Threat of Adversarial Attacks on Machine Learning in Network Security—A Survey. arXiv:1911.02621 [cs]. (2020)
48. Papernot, N., McDaniel, P., Sinha, A., Wellman, M.: Towards the Science of Security and Privacy in Machine Learning. arXiv:1611.03814 [cs]. (2016)
49. Goodfellow, I.J., Shlens, J., Szegedy, C.: Explaining and Harnessing Adversarial Examples. arXiv:1412.6572 [cs, stat]. (2015)
50. Carlini, N., Wagner, D.: Towards Evaluating the Robustness of Neural Networks. In: 2017 IEEE Symposium on Security and Privacy (SP). pp. 39–57. IEEE, San Jose, CA, USA (2017). https://doi.org/10.1109/SP.2017.49
51. Liu, Y., Ma, S., Aafer, Y., Lee, W.-C., Zhai, J., Wang, W., Zhang, X.: Trojaning Attack on Neural Networks. Department of Computer Science Technical Reports. (2017).
52. Usama, M., Mitra, R.N., Ilahi, I., Qadir, J., Marina, M.K.: Examining Machine Learning for 5G and Beyond through an Adversarial Lens. arXiv:2009.02473 [cs]. (2020)
53. Goutay, M., Aoudia, F.A., Hoydis, J.: Deep Reinforcement Learning Autoencoder with Noisy Feedback. arXiv:1810.05419 [cs, math]. (2019).
54. Suomalainen, J., Juhola, A., Shahabuddin, S., Mammela, A., Ahmad, I.: Machine Learning Threatens 5G Security. IEEE Access. **8**, 190822–190842 (2020). https://doi.org/10.1109/ACCESS.2020.3031966
55. Le, L.-V., Sinh, D., Lin, B.-S.P., Tung, L.-P.: Applying Big Data, Machine Learning, and SDN/NFV to 5G Traffic Clustering, Forecasting, and Management. In: 2018 4th IEEE Conference on Network Softwarization and Workshops (NetSoft). pp. 168–176. IEEE, Montreal, QC (2018). https://doi.org/10.1109/NETSOFT.2018.8460129.
56. Zhang, S., Zhang, N., Zhou, S., Gong, J., Niu, Z., Xuemin, Shen: Energy-Sustainable Traffic Steering for 5G Mobile Networks. arXiv:1705.06663 [cs, math]. (2017).
57. Ftaimi, A., Mazri, T.: Analysis of Security of Machine Learning and a proposition of assessment pattern to deal with adversarial attacks. E3S Web Conf. 229, 1004 (2021). https://doi.org/10.1051/e3sconf/202122901004.
58. Tian-yang, G., Yin-sheng, S., You-yuan, F.: Research on Software Security Testing. International Journal of Computer and Information Engineering. 4, 9 (2010). https://doi.org/10.5281/zenodo.1081389.

Effects of Jamming Attack on the Internet of Things

Imane Kerrakchou, Sara Chadli, Mohammed Saber, and Mohammed Ghaouth Belkasmi

Abstract The capacity of the Internet to be connected to everything, everywhere, and at every time, thus offering great ease to users, has gained popularity over the years. The automation of daily tasks is no longer a dream. The Internet of things has reached a peak, where a wealth of gadgets and applications are successfully connecting to sensors and the Internet, offering a new age of relief and communications. The heterogeneous combination of sensors, things, and the Internet together has resulted in a variety of protection and privacy problems, making security difficult. Since IoT is used for commercial purposes, it has paved the way for attackers and hackers to enter the private lives of the public at large, as well as to organizing crimes and attacks. In this paper, we focus on wireless sensor networks (WSN) because malicious nodes easily attack these networks. We perform an implementation of the Jamming attack and a detailed analysis of the results obtained. The protocol chosen to evaluate the performance of the WSN is the S-MAC protocol. Different scenarios are proposed in order to evaluate the performance of an attacked network and the severity of the Jamming attack.

1 Introduction

Today, we live in a rapidly changing world. Rightly so, because technological advances made what might once have been a fairy tale a reality. Internet of things

I. Kerrakchou (✉) · S. Chadli · M. Saber · M. G. Belkasmi
Mohammed First University Oujda, ENSA Oujda, SmartICT Lab, Oujda, Morocco
e-mail: i.kerrakchou@ump.ac.ma

S. Chadli
e-mail: s.chadli@ump.ac.ma

M. Saber
e-mail: m.saber@ump.ac.ma

M. G. Belkasmi
e-mail: m.belkasmi@ump.ac.ma

M. Ben Ahmed et al. (eds.), *Networking, Intelligent Systems and Security*,
Smart Innovation, Systems and Technologies 237,
https://doi.org/10.1007/978-981-16-3637-0_29

(IoT) is a new technology that will make human life smarter and much simpler. Utilizing IoT, a user can monitor everything and everywhere to accommodate his or her convenience. Smart objects can connect through many heterogeneous network technologies like wireless local area networks (WLAN), radio frequency identification (RFID), cellular services (3G, 4G, LTE, and 5G), and wireless sensor networks (WSN) [1]. The WSN name is a three-word combination. These sensors imitate human sensors. In other words, they are able to collect information on sound, sight, smell, and temperature. Wireless sensor networks are composed of sensor nodes with a low cost, limited capacity, and communicate with other nodes over short distances utilizing a significantly low rate of power. WSNs are usually randomly scattered in the destination area, and they execute a combined strategy to send predefined parameters from the environment to the target field. Due to their large variety of uses, they could be utilized in many different applications varying from the military fields to the health system. In the majority of cases, sensor nodes operate in fairly difficult environments and thus present a higher risk of physical harm than conventional networks. The vulnerabilities of sensor networks in terms of security can be further exploited to generate many different forms of threats [2]. For this reason, the security of the WSN must achieve several objectives such as confidentiality, availability, authenticity, integrity, etc., in order to protect the network against all types of attacks.

The attacks known in sensor networks can be divided into various categories based on various taxonomic models. In this article, the wireless sensor network attacks are classified according to the several layers of OSI whose function and operations are affected, damaged, or destroyed. Attacks at the MAC layer have generated a great deal of attention, and researches exist in this regard because the MAC protocol has a significant impact on network performance, such as throughput, energy, etc. Attacks on the MAC layer are primarily aimed to achieve denial of service (DoS). Attackers typically aim to limit the access of authorized users to the wireless canal by affecting the operation of the system thus affecting the availability and connectivity of the network. They can also aim to unfairly consume the canal resources, causing severe damages in the real world, where the attacker follows the media access control (MAC) protocol and sends over the common canal, periodically or continuously, to respectively target all the communications. The following is a brief description of certain types of attacks for the three layers (application layer, network layer, and perception layer). Then, we will implement a Jamming type attack that targets the perception layer, more precisely the MAC layer using the S-MAC protocol, in order to visualize the effects of this attack on the network performance [3].

The remainder of this paper is presented as follows. A detailed description of the protocol used (S-MAC) is provided in Sect. 2. Attacks on IoT and their classifications are presented in Sect. 3. In Sect. 4, we simulate and analyze the effects of the attack on network performance. Finally, the conclusion will be presented in Sect. 5.

2 MAC Protocols for WSN

2.1 MAC Protocols

MAC protocols enter into play once a node is authorized to send its packet in order to avoid collisions on the receiver end and to control any physical layer access consistently. When a physical canal is used by all the nodes in a network, the MAC protocols are decisive to ensure that access to the canal is controlled and coordinated between the different nodes so that the information could be transmitted between nodes. Different MAC protocols with various goals are suggested to support wireless sensor networks. The primary objective of the MAC protocol on the focus of the authors is energy efficiency. Other major attributes are extensibility and the ability to adapt to changes such as node density, network size, and topology. The second objective is latency and throughput. The MAC layer provides two primary criteria for the design of an effective MAC protocol. The first is to identify the causes of energy loss and the second is the communications model that the network uses to decide the traffic behavior that the MAC protocol will process [4]. There are many MAC protocols, including S-MAC, T-MAC, D-MAC, X-MAC, B-MAC, and others.

2.2 S-MAC Protocol

Sensor-MAC is the first protocol in the MAC layer for sensor networks that takes into account the energy limits of battery-powered sensor nodes. The principal idea to conserve energy is to switch off the radio transmitters when no pertinent communication is happening. In periodic listening/sleeping cycle T_c (see Fig. 1), the nodes exchanging synchronization packets (SYNC) in the period of synchronization T_{sync},

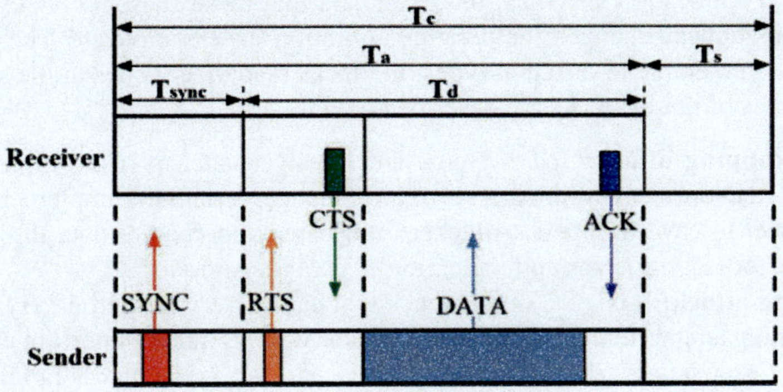

Fig. 1 Frame for S-MAC protocol

sending data packets during the data period T_d and turning off the radio in the sleeping period T_s, i.e., $T_c = T_a + T_s$ with the period of activity $T_a = T_{sync} + T_d$.

If a node decides to transmit data, it must first transmit a request (RTS) while the receivers are listening. Once the receiver responds with a CTS packet, the node sends the data. In the end, an ACK will be sent by the receiver to indicate that reception is established with success. In determining an energy-efficient listening/sleeping cycle when entering a network zone, the node will first listen to its neighbors for a while. If a SYNC packet is received, the node will follow the listen/sleep cycle defined by the SYNC packet. This node is named follower. If the node does not receive a SYNC packet, it selects its proper listen/sleep cycle and begins to broadcast SYNC packets. This node is named synchronizer (see Fig. 2). Each node establishes a schedule table that contains the schedules of all neighboring nodes [5].

3 Attacks on IoT

In this part, we will describe the basic architecture of an IoT system that typically includes three layers: perception layer, network layer, and application layer. Then, we are going to address some of the common issues and threats related to all of those layers (see Fig. 3).

3.1 Attacks Perception Layer

This layer is often identified as the layer of a sensor. With the support of different technologies such as wireless sensor network (WSN), RFID sensor network (RSN), and radio frequency identification (RFID), this layer is responsible for identifying things, capturing, and collecting data from the sensors [6]. These sensors are selected according to application specifications. The data that these sensors collect can be about the position, air changes, temperature, vibration, acceleration, etc. The perception layer is vulnerable to different types of attacks that affect nodes in the sensors. Popular forms of these attacks are described below:

- **Eavesdropping attack:** IoT systems are mostly, made up of multiple nodes distributed in open environments. As a consequence, certain IoT implementations are subject to eavesdroppers. Attackers can listen and capture data throughout various processes, such as authentication or data transmission.
- **Jamming attack:** It is a DoS type attack that can be very dangerous. By simply transmitting an interference signal, the attacker will essentially interrupt communication on a wireless canal, stop normal service, create performance failures, and even destroy the control system.
- **Booting attack:** All services of security are activated while the system is in operating mode. But, at the moment of start-up, there's a window for the attackers

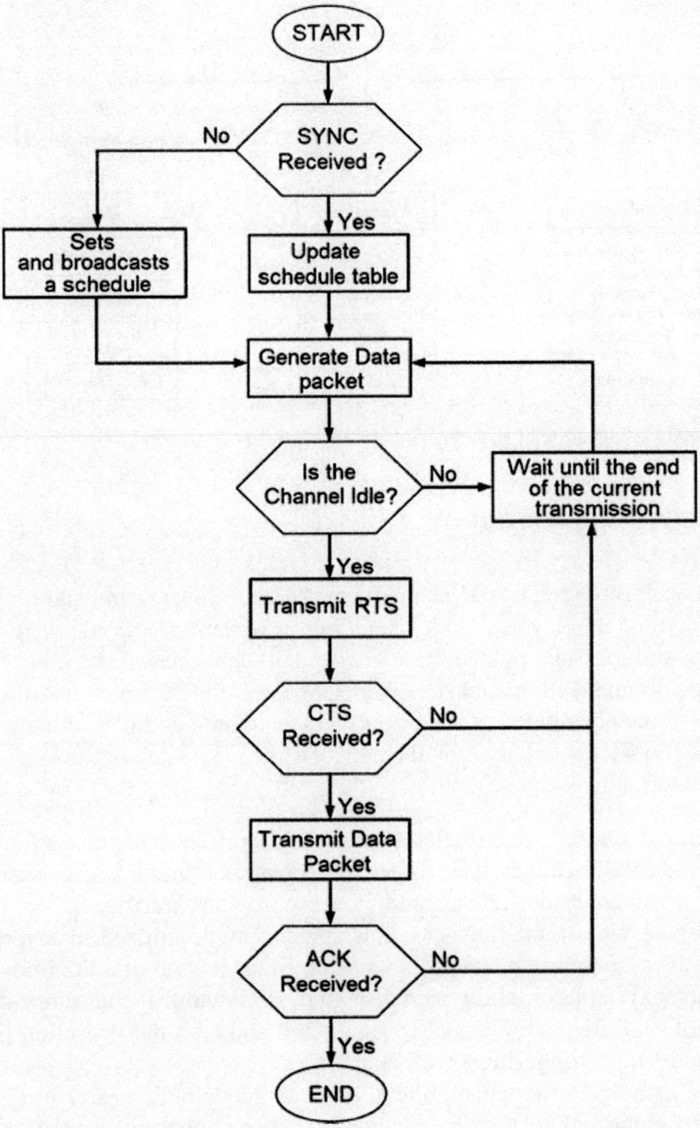

Fig. 2 Diagram of S-MAC protocol

to target the nodes of the system. Devices that are low energy consumption have consistent sleep–wake cycles making them more vulnerable to this attack [7].

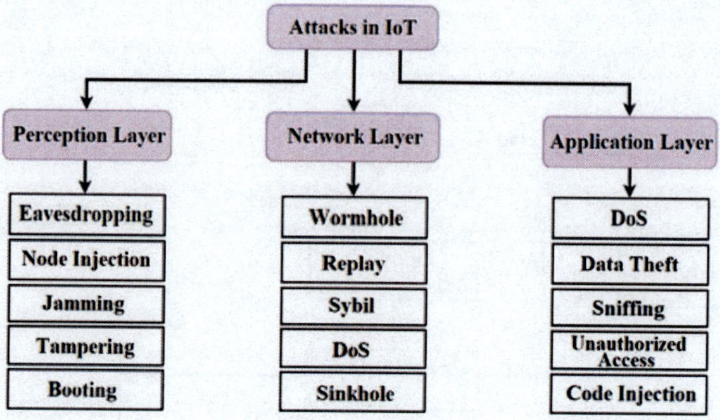

Fig. 3 Layered classification of IoT attacks

3.2 *Attacks Network Layer*

Network layer is often referred to as the access gateway layer or transport layer. The main objective of this layer includes handling information with message routing, subscription and message publishing management, data transmission, etc. By way of differing communication canals such as GSM, Wi-Fi, Ethernet, etc., the data is obtained from the layer below it. Current access technologies and protocols, such as IPV6/ IPV4/ LoWPAN, are used by the network layer [8]. Among the most popular types of network attacks, we can cite:

- **Sybil attack:** On this, one malicious node takes different identities (defined as Sybil node) and localizes itself at various positions within the network. This conducts to an enormous redistribution of resources arbitrarily.
- **Denial of Services (DoS) attack:** This attack may be utilized in several ways to make the system ineffective. It is not utilized to steal or alter the information, but to target system availability and disable it. A system that can transmit a huge amount of radiofrequency signals may disrupt and stop the operation of every sensor node, thus, triggering a DoS attack.
- **Sinkhole attack:** In this attack, the attacker targets a node nearer to the SINK (identified as the sinkhole node), and makes it appear attractive for the rest of the nodes on the system, and therefore, draws network traffic to it [9].

3.3 *Attacks Application Layer*

- It is the upper layer representing the applicable elements of the system. Its key role is to give the service needed by IoT-specific users. This layer uses several protocols that usually include constrained protocol for application (CoAP) and message queue telemetry transport (MQTT). These protocols help to easily supply the user with the desired service [10]. As a result, it presents specific security problems that are not presented in the rest of the layers, like privacy issues and data theft. The main security problems faced by the application layer are examined below.

- **Data theft:** The information or data obtained by sensors from the devices of IoT is most sensitive while in transit. Attackers with the intentions of utilizing credentials for private use or reselling it to the highest buyer will steal the data quite easily if appropriate security procedures are not respected.
- **Sniffing attack**: Data packets could be captured and sensible data could be removed by sniffing if the data packets have little or no encryption during transmission [11].
- **Unauthorized access:** Access control involves providing access to legitimate users and refusing access to non-authorized users. With non-authorized access, attackers can steal data or get access to confidential data.

4 Jamming Attack Implementation on S-MAC Protocol

4.1 Proposed System

In this part, we will implement the Jamming attack in a WSN at the MAC layer, in order to visualize the effect and severity of this type of attack on network performance. The system we proposed in our work consists of the SINK, i.e., the base station and a group of nodes. The role of the SINK is to collect and process the packets sent by the nodes in a centralized model.

In this article, we will implement two types of Jamming attacks that target the base station. In the first type, the attacker will only use request to send (RTS) control packets to trigger the attack in the network. While in the second type, the attacker will use data packets. Once the malicious node is implemented in the system, it analyzes the traffic to detect the protocol used, which in our case is S-MAC, so that it can synchronize with the network. Then, it continuously transmits a large number of packets in the transmission channel to the SINK.

The objective of the simulation of the control packet Jamming attack (RTS) and data packet Jamming attack (DATA) is to visualize and analyze the impact of each of these two attacks and compare their gravity on the network in order to develop a precise mechanism that protects against the most effective attack. Fig. 4 shows the activity modeling of Jamming attack.

Fig. 4 Jamming attack flowchart

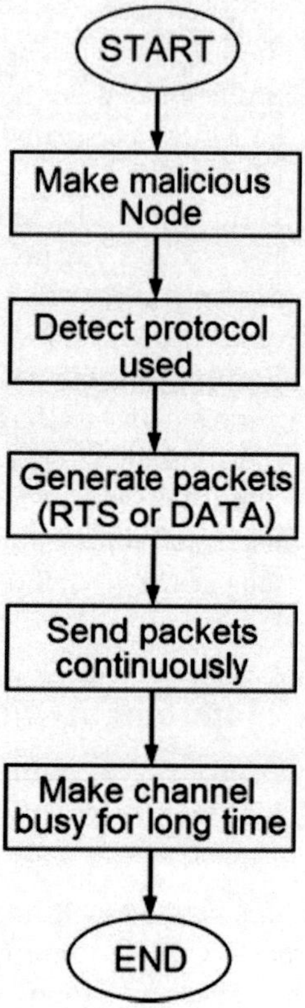

4.2 Simulation Experiment

The simulation is realized in the OMNeT++ environment under Linux. We used a set of twenty-five nodes, where node 0 is the SINK. The channel utilized in simulation is a wireless channel. The MAC protocol used is S-MAC. Therefore, the nodes update their sleep schedule using this protocol. The simulation period was set to 200 s. The initial energy of each node is 18720 J. We utilized a star architecture in which all the nodes send their packets to the SINK. The number of packets sent equals five packets per second with a transmission power set at 36.3 mW. Table 1 illustrates the simulation parameters.

In our simulation, two scenarios are proposed. The first scenario (see Fig. 5) represents the normal case of the system, i.e., no malicious node is implemented in the network. The nodes synchronize with each other and then transfer control and data packets to the base station.

The second scenario (see Fig. 6) represents the Jamming attack by adding a malicious node, which is node 25, in the network. In this scenario, we will simulate the two types of Jamming attacks we explained above. Both have the same network architecture and the same simulation parameter values. The only difference between

Table 1 Simulation parameters

Parameters	Values
Simulation time (s)	200
Simulation area (m)	60 × 60
Number of nodes Mobility model Topology	25 No Mobility Star
Transmit power (mW)	36.3
Packet rate (pps) Data packet size (bytes) End time Protocol	5 100 End of simulation S-MAC
SYNC packets size (bytes)	11
RTS packets size (bytes)	13
Contention period (ms)	10
Frame time (ms)	610

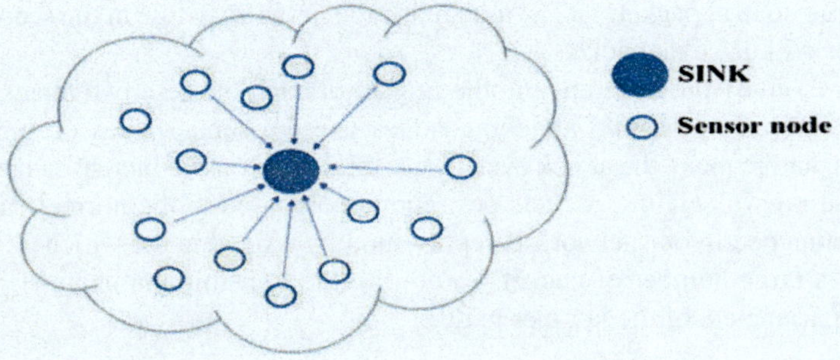

Fig. 5 Scenario 1—Normal case

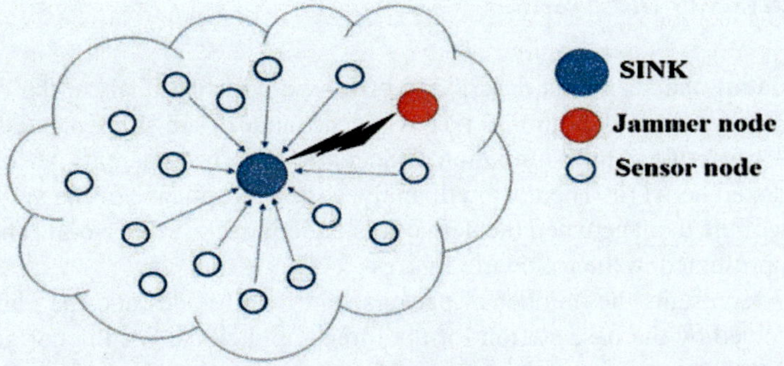

Fig. 6 Scenario 2—Network with jamming attack

Table 2 Simulation parameters for Jammer node

Parameters	Values
Number of jammers	1
Trajectory	Fixed
Transmit power (mW)	57.42
Packet rate (pps)	50
Protocol	S-MAC
SYNC packets size (bytes)	11
RTS packets size (bytes)	200
Data packets size (bytes)	200
Frame time (ms)	610
Contention period (ms)	10
End time (s)	End of simulation

these two types of attack is the attacker node. In the first case, the malicious node will use large control packets (RTS) to damage the network, while in the second case the attacker will use data packets.

In order to study the difference in the impact of each of these two attacks under the same conditions, the two attacking nodes use the same values of simulation parameters, and to make the attack even more efficient, we have increased the packet size and the rate of sending packets per second compared to the normal case. The Jamming launched in our network is represented by a fixed node, which allows the sending of a large number of packets saturating the transmission channel. Table 2 shows the parameters of the jammer node.

4.3 Analysis of the Results

After simulating the scenarios described above, we will first analyze the network behavior under normal conditions, i.e., when no attacks are implemented in the system. The objective of this simulation of Scenario 1 is to compare its performance with an attacked network. Then, we will analyze the performance of the system and the severity of the damage when the network is under attack (Scenario 2). The result obtained is presented in the following figures.

Figure 7 represents the number of packets sent by all nodes and the number of packets received by the base station for the three simulations. For the normal case, we notice that the number of packets sent is almost the same as the number of packets received. So the number of lost packets is very low, which explains that the network is working correctly. When the attack is implemented, either for the RTS control packet attack or the DATA packet attack, we notice that the number of packets sent by the nodes is decreased compared to the normal case. This decrease in the number of packets sent is due to the high-speed broadcast of false packets generated by the

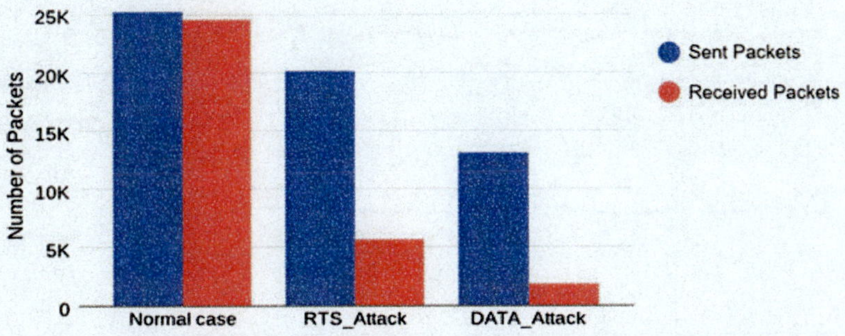

Fig. 7 Number of packets sent and received

attacker in the network, which increased the traffic, made the transmission channel busy, and ultimately prevent other nodes in the network from sending their packets correctly.

Concerning the reception of packets, for the attacked network, we can see that a very limited number of packets was received compared to the number sent. Figure 8 shows more precisely the packet loss rate in the network. As we noticed at the beginning, the packet loss rate for the normal case is too low, but it is very high in the case of an attack. Most of the packets sent to the SINK were not received. The reason for the decrease in the number of packets received is the high traffic generated by the malicious node and the number of false packets sent to the SINK, which made the SINK busy by the reception of these packets and unable to receive all the packets intended for it by the legitimate nodes.

Figure 9 shows the energy consumption by each node of the network for the three simulations. For Scenario 1, when no attacks are implemented, the network functions correctly, and the S-MAC protocol has a better energy efficiency thanks to the implementation of SYNC packets. Once the attack is implemented (Scenario 2), we notice that the power consumption has doubled compared to the normal case.

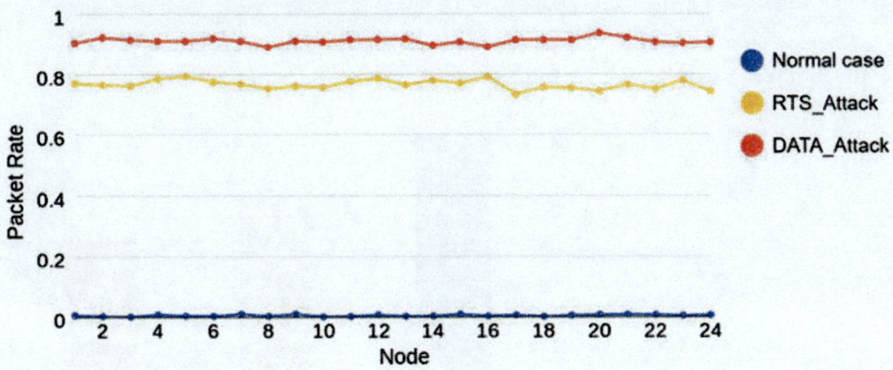

Fig. 8 Packet loss rate per node

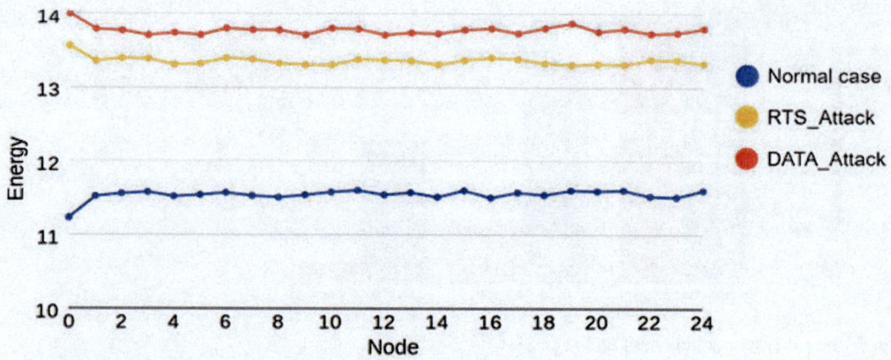

Fig. 9 Energy consumption per node (J)

The reason for the high consumption at the base station level for both simulations (attack with RTS packets and attack with DATA packets) is the reception and processing of a large number of false packets sent by the attacker. This repeated sending of the false packets resulted in high traffic in the transmission channel, which caused the legitimate nodes to consume more power in unnecessary retransmissions of their packets to the SINK.

Energy is a term that is often used synchronously with the lifetime of the network. This means that maximum energy consumption leads to a minimum network life. Figure 10 shows the estimated network lifetime for the three simulations. It can be seen that the battery of the attacked networks is quickly depleted compared to the normal case because of the Jamming attack which resulted in high energy waste and thus a short lifetime.

As mentioned above, the goal of simulating the two types of attack (RTS and DATA) is to analyze the effect of each of these two attacks and to compare their severities on the network. According to the results obtained, we notice that the Jamming attack by data packets (DATA) is more efficient compared to the attack by control packets (RTS).

Fig. 10 Estimated network lifetime (days)

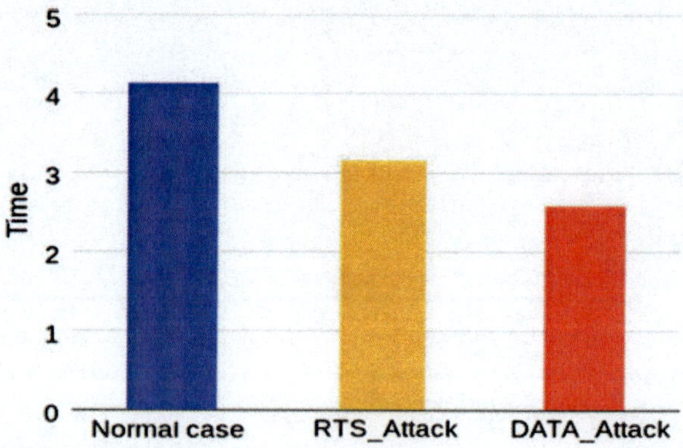

The attack by data packets (DATA) affects the entire network and degrades its performance in a very significant way, which subsequently leads to the destruction of the system.

5 Conclusion

As the emergence of IoT has occurred, several vulnerabilities, varying from attacks on the devices to attacks on the data in transits, have attracted the attention of the research community. Besides, the inexpensive design of sensor nodes and the facility to reprogram them will make sensor networks highly vulnerable to intentional attacks. In this paper, we have given an overview of the use of the internet of things, the objectives of security, the MAC protocols as well as the classification of attacks according to the different layers of an IoT application. Then, we analyzed the effects of the Jamming attack on the network using the S-MAC protocol. Two types of Jamming attacks were analyzed: The first is the attack by control packets (RTS) and the second is the attack by data packets (DATA). The parameters used to determine system efficiency are the number of packets delivered, packet loss rate, power consumption, and network lifetime. In the end, the results showed that the Jamming attack, specifically by DATA packets, is a very dangerous type of attack that can be effectively used to deteriorate network performance and quality of service and then damage the device. In the future work, we will try to design and implement a mechanism to protect networks from Jamming attacks by data packets.

References

1. Hasan, Ali , Khattak: Munam Ali Shah, Sangeen Khan, Ihsan Ali, Muhammad Imran, Perception layer security in Internet of Things. Futur. Gener. Comput. Syst. **100**, 144–164 (2019)
2. Mohanta, B.K., Jena, D., Satapathy, U., Patnaik, S.: Survey on IoT security: Challenges and solution using machine learning, artificial intelligence and blockchain technology. Internet of Things **11,** 100227 (2020).
3. Jagriti, D.K.: Energy consumption reduction in S-MAC protocol for wireless sensor network. Procedia Comp. Sci., **143**, 757–764 (2018).
4. Ouaissa, M., Ouaissa, M.: Rhattoy, Enhanced and Efficient Multilayer MAC Protocol for M2M Communications. Adv. Intell. Syst. Comput. **1165**, 539–547 (2021)
5. Sakya G., Singh P.K.: Medium access control protocols for mission critical wireless sensor networks. In: Singh P., Bhargava B., Paprzycki M., Kaushal N., Hong WC. (eds.) Handbook of wireless sensor networks: issues and challenges in current scenario's. Advances in Intelligent Systems and Computing, vol 1132. Springer, Cham (2020).
6. Aarika, K., Bouhlal, M., Ait Abdelouahid, R., Elfilali, S., Benlahmar, E.: Perception layer security in the internet of things. Procedia Comp. Sci. **175**, 591–596 (2020)
7. Hassija, V., Chamola, V., Saxena, V., Jain, D., Goyal, P., Sikdar, B.: A Survey on IoT Security: Application Areas, Security Threats, and Solution Architectures. IEEE Access **7**, 82721–82743 (2019)

8. Jha, R.K., Puja, H.K., Kumar, M., Jain, S.: Layer based security in Narrow Band Internet of Things (NB-IoT), Comp. Netw. 107592 (2020).
9. Sengupta, J., Ruj, S., Das Bit, S.: A comprehensive survey on attacks, security issues and blockchain solutions for IoT and IIoT, J. Net. Comp. Appl., **149**, 102481 (2020).
10. da Cruz, M.A.A., Joel, J.P.C., Lorenz, P., Solic, P., Al-Muhtadi, J., Albuquerque, V.H.C.: A proposal for bridging application layer protocols to HTTP on IoT solutions. Future Generation Comp. Sys., **97**, 145–152 (2019)
11. Anand, S., Sharma, A.: Assessment of security threats on IoT based applications. Materials Today: Proceedings (2020)

H-RCBAC: Hadoop Access Control Based on Roles and Content

Sarah Nait Bahloul, Karim Bessaoud, and Meriem Abid

Abstract Social networks, smartphones, mobile applications... produce an avalanche of data on a large-scale and in an unstructured way. The phenomenon *Big Data* was born in order to address the different challenges including data storing, data analysis, data querying and so on. Technological advances always carry new security vulnerabilities that are not taken into consideration at the beginning. Security aspects usually require time to be addressed. The information system security is the set of measures to prevent any failure or threat including unauthorized accesses. Perfect protection must contain the four basic building blocks that are: authentication, access control, auditing, and encryption. In our work, we are specifically interested in access control. We first analyze the well-known access control models that were applied to Big Data. We then investigate the most important security projects. Most of these approaches and projects rely mainly on coarse-grained access control policies. In this work, we propose a novel approach called H-RCBAC that relies on two known models: the role-based access control (RBAC) and the content-based access control (CBAC). H-RCBAC is a new architecture that refines the access control process by considering a set of taboo words to guarantee fine-grained access control.

1 Introduction

Digital data explosion has forced researchers to find new ways to analyze and exploit the world, to manage new scalability issues about the capture, the storage, the analysis, and the representation of data. Big Data has quickly become an unavoidable

S. Nait Bahloul (✉)
Computer Science Department, LSSD Laboratory, University of Science and Technology of Oran Mohamed Boudiaf, Oran, Algeria
e-mail: sarah.naitbahloul@univ-usto.dz

K. Bessaoud · M. Abid
Computer Science Department, University of Mostaganem, Mostaganem, Algeria
e-mail: karim.bessaoud@univ-mosta.dz

M. Abid
e-mail: meriem.abid@univ-mosta.dz

M. Ben Ahmed et al. (eds.), *Networking, Intelligent Systems and Security*,
Smart Innovation, Systems and Technologies 237,
https://doi.org/10.1007/978-981-16-3637-0_30

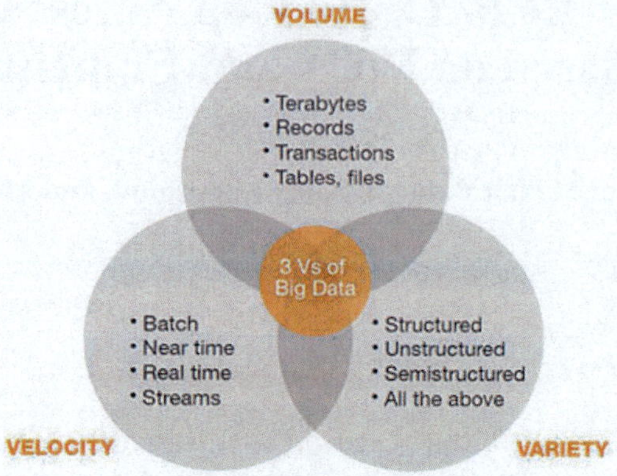

Fig. 1 Big data 3Vs

trend for many industrial players because of the contribution it offers in terms of storage, processing, data analysis, and decision support tools, by its computing power which is distributed on a cluster that groups together a set of machines. It provides a response to very complex queries and supports a variety of data sources (Fig. 1).

The research community agrees that the Big Data is characterized by the 3Vs problem [1]. (1) Volume refers to the amount of data, (2) Variety refers to several types of data, and (3) Velocity refers to the speed of data processing. According to the 3Vs model, the challenges of big data management result from the expansion of all three properties, and not just the volume.

Big Data projects require the choice of a storage method, operating technology, and data analysis tools to optimize processing time on large data. In this context, one solution has emerged among the different works, namely, Hadoop. Apache Hadoop is an open-source software framework used for distributed storage and processing of very large datasets. Various other frameworks have emerged since. The most known and used is Apache Spark. It is significantly faster than Hadoop since it uses memory to execute the treatments.

As a first step, we focus in this work on the Hadoop Framework. Despite its performance compared to Spark, Hadoop is largely used as a solution for storage and processing large distributed data.

Hadoop, as many new incoming technologies, was not fitted to address security threats at the beginning. Hadoop was first utilized to manage a large amount of non-sensitive data. The success of Hadoop and its adoption in various utilization contexts raised new data security issues. One question then emerges: How to process and exploit such a large volume of data while ensuring their *security*?

Like any information system, for enhanced protection, Hadoop must contain the four basic bricks of security [2], namely:

- **Authentication**: To protect an information system, it is essential to integrate a tool allowing a person (or a process) identification before allowing it to connect. There are several authentication techniques that can be classified into two categories:

  - Basic authentication: Who am I (login), prove it (authenticator (password)).
  - Strong authentication: simple authentication reinforced by the addition of "locks": biometric fingerprint, smart card, retinal recognition, ...

- **Authorization**: Once connected, the person is not allowed to do everything or to reach all the resources. It is, therefore, necessary to control its access. Several techniques and approaches are proposed to meet the different needs of companies and/or applications. In our work, we will focus more specifically on this security brick. We will see later the different approaches used to ensure access control in the Big Data.
- **Audit**: Although the person is identified, authenticated, and authorized to reach certain resources, a well-secured system must track all the operations. This is tractability, an essential process to recover problems.
- **Encryption**: In order to effectively combat ever-evolving threats, a data-centric approach is also needed to protect sensitive information. To do this, we use encryption for certain data, stored safely, and made unintelligible via an encryption algorithm preserving their confidentiality. The encryption key must be strictly protected to make this process effective.

We focused on our work on access controls on Hadoop. Although many approaches were proposed to enable Hadoop's access control policy, most of them rely mainly on coarse-grained access control policies.

In this paper, we propose a novel approach to define and manage fine-grained access control rules on Hadoop.

The rest of the paper is organized as follows: Sect. 2 presents the framework Hadoop. Section 3 presents the proposed access control models on Big Data. In Sect. 4, we present the different security projects to enhance security on Hadoop and the Big Data ecosystem. We present and detail our approach in Sect. 5. We conclude in Sect. 6.

2 Hadoop

Hadoop is an open-source framework providing tools to use a cluster to store and manipulate large volumes of data quickly and optimally. It exploits the computing power of the cluster machines. It is managed by the Apache Foundation.

The two main components of Hadoop are the HDFS and the MapReduce programming model [3].

2.1 HDFS: Hadoop Distributed File System

HDFS is a file system allowing to store a large volume of data on a cluster by disregarding the capacities of the machines or their states. The system is fault-tolerant.

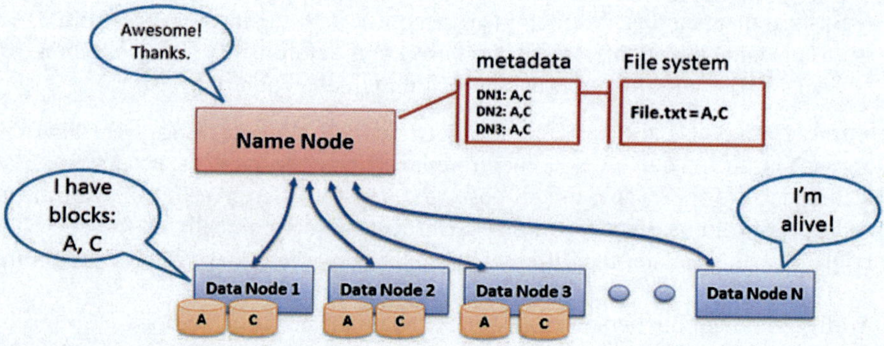

Fig. 2 HDFS architecture

Its architecture is based and thought according to this. The system uses data blocks of a larger size than the known file systems. This value is 64 MB (but can be changed). Each file is thus divided into blocks, and these blocks are distributed on the cluster. There is a replication rate of the default blocks set to 3 (but there again modifiable). Each block is present on different nodes at a time. The loss of a node is not a problem since the lost blocks are replaced by other nodes.

HDFS consists of a main node, called the *NameNode*. This node is very important, and it manages the data location. It matches a file with its associated blocks (the metadata of a file) and knows also on which nodes each block is located. The second type of node is *DataNodes*. A DataNode handles the data blocks. The DataNode very often keeps the NameNode aware of the blocks it manages. So, the NameNode can detect problems and ask block replication. DataNodes do not support files, but only blocks. The NameNode manages the file. It will be able to open, close, drop files, and send these changes to the relevant DataNodes. It will ask DataNodes to create blocks, drop, read, or write them. Figure 2 summarizes the HDFS architecture.

2.2 *MapReduce*

MapReduce [3] is a programming model introduced by Google for the processing of a large volume of data. MapReduce divides a query into tasks and assigns them to many computers. Later, the results are collected in one place and integrated to form the final result. The MapReduce algorithm has two important tasks, namely Map and Reduce.

- The Map task takes a set of data and converts it into another set of data, where each element is transformed into key–value pairs.
- The Reduce task takes the Map output as an input and combines those key–value pairs into a smaller set of tuples. The Reduce task is always performed after the Map task.

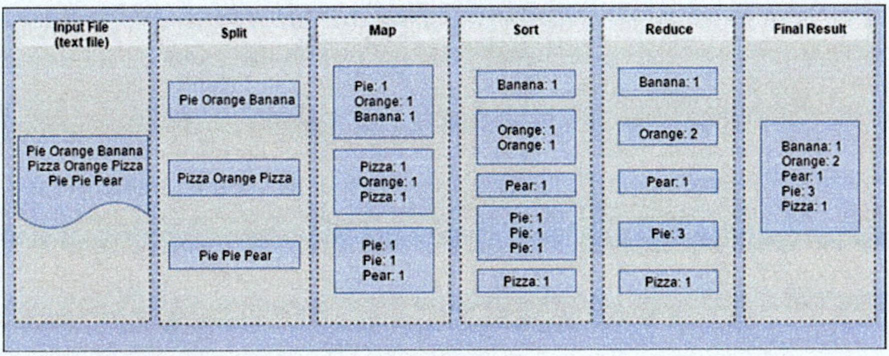

Fig. 3 MapReduce step

Figure 3 details a MapReduce processing that calculate the occurrences of the words in a file.

- **Split**: The file is divided into blocks.
- **Map**: It is a user-defined function, which calculates the occurrences of words in the bloc. Keys are created with an associated value. In this example, a key is a word and the value is a number n which signifies that the word is present n time.
- **Shuffle**: All identical keys are grouped.
- **Reduce**: A processing is carried out on all the values of the same key. In this example, the values are summed, which makes possible to get the number of occurrences of the words in the file.

The final result is stored in a file and returned to the requester.

3 Access Control Models on Big Data

Access control is one of the most commonly used methods for managing authorizations. Data security is likely to be compromised due to a poor access control strategy.

Added to this, the data source multiplicity can generate problems in choosing a good strategy of access control. We will show in this section a survey on the adaptation over time of access control approaches in Big Data.

3.1 Access Control Lists (ACLs)

Before the ACL implementation, the HDFS authorization model was equal to traditional UNIX permissions (Permission Bits). In this template, permissions for each file or directory are managed by a set of three distinct user classes: Owner, Group, and Other. There are three permissions for each user class: Read, Write, and Execute. Thus, for any file system object, the permissions can be encoded in 3 * 3 = 9 bits. When a user tries to get access to a file, HDFS applies permissions based on the most specific class of users applicable to that user. If the user is the owner, HDFS checks the rights of the owner class. If the user is a member of the file group, HDFS checks the permissions group class. Otherwise, HDFS checks for other class permissions. This model can adequately handle a large number of security requirements.

In Hadoop, HDFS supports POSIX ACLs [4]. By default, ACL support is disabled, and the NameNode prohibits ACLs creation. It must be activated by reconfiguring the NameNode.

POSIX ACLs allows to assign different permissions to different users or groups, even if they are not the original owner. For example, user John creates a file. It does not allow anyone in the group to get access to the file it owns, except for another user, Antony (even if other users belong to the John group). This means that the owners or groups of the file can agree to other users and groups to manage access to that file through POSIX ACLs.

3.2 *Role-Based Access Control (RBAC)*

The role-based access control model (RBAC) uses the role as an intermediary between users and permissions [5]. It simplifies administration tasks by reducing the number of assignments to be handled. This model is widely adopted by companies and manufacturers. There is no need to change the permission each time a person joins or leaves an organization. By this characteristic, RBAC is considered as an "ideal" model for companies with high turnover rates. It minimizes the risk of errors and unintended permissions by reducing the workload of administrators. The RBAC standard was proposed by the National Institute of Standard and Technology (NIST) in 2001 and formally adopted by the ANSI standard in 2004.

The RBAC model provides the necessary security functionality for the Big Data environment. The most well-known security project using RBAC is Sentry (we will detail this project below) which allows fine granularity access control for Impala, Apache Hive, MapReduce, Apache Pig…

Some works have been based on the concept of role to propose more advanced solutions. We can cite the work of Gupta et al. [6] who proposed a formal multi-layer access control model (called HeAC) for Hadoop ecosystem. They extend HeAC base model to provide a cohesive object-tagged role-based access control (OT-RBAC) model, consistent with generally accepted academic concepts of RBAC. Besides inheriting advantages of RBAC, OT-RBAC offers a novel method for combining RBAC with attributes.

3.3 Content-Based Access Control (CBAC)

To satisfy the new access control needs in Big Data without explicitly identifying each subject and object, a new approach has emerged, named the content-based access control (CBAC) [7]. The approach is still in the proposal phase.

The CBAC is added as a second layer of access control and added to the base layer of Hadoop. The CBAC model offers finer access granularity. For a user query, the CBAC function evaluates $f(u, d)$ to true or false. u represents the subject and d the data object. The function is evaluated during query processing. If the function returns True, access to the file will be granted. Otherwise, access will be denied.

In the model, a subject u is represented by a set of objects it possesses. The objects are represented by a bag of words. The decision function will receive the maximum similarity between the object of the candidate and all the files of the owner and will compare it with a planned threshold:

$$f(u, d_i) = \max_j(\mathrm{SIM}_d(d_u, j, d_i)) \geq T$$

This threshold will determine the number of files that will be accessed by the candidate. So, a poor choice of the threshold can be fatal. The approach Top-K similarity is proposed for the choice of a perfect and dynamic threshold.

In [8], the authors propose an access control framework, which enforces access control policies dynamically based on the sensitivity of the data. The framework enforces access control policies by harnessing the data context, usage patterns and information sensitivity.

3.4 Attribute-Based Access Control (ABAC)

In the ABAC approach [9], access to protected resources is based on the user attributes (name, date of birth, ...). This approach makes possible to associate the user attributes with other attributes (IP address, time of day, ...) to make an access decision. Rather than using the role of a user to decide whether to grant access to a resource, ABAC can combine several attributes to make a contextual decision.

ABAC uses attributes as building blocks to define access rules. This is done through a structured language called eXtensible Access Control Markup Language (XACML). For example, a rule could declare: Allow managers to get access to finance-type data only if they come from the Department of Finance. This would allow users with role attributes = "Manager" and Department = "Finance" to access data with the category attribute = "Finance".

Other works are based on the ABAC approach. We can cite among others, the work of Gupta et al. [10]. They proposed a fine-grained attribute-based access control model, referred to as HeABAC, catering to the security and privacy needs of the multi-tenant Hadoop ecosystem.

4 Security Projects on Big Data

Several projects have contributed to the improvement and strengthening of Hadoop security. We detail in the following the most important projects: Sentry, Apache Ranger, and Rhino. We first introduce the basic security in Hadoop. We will focus on the access control aspect of these projects.

4.1 Basic Hadoop Security

At the beginning of Hadoop, developers did not focus on data security because of the specific use of Hadoop. Having become democratized, security has become one of the main goals of the various actors. Various tools within the platform, such as Hive, Impala, Pig . . . have drawn up their own security needs.

Security mechanisms have emerged, including the authentication to verify the identities of users. The method chosen for Hadoop was Kerberos, a well-established protocol and is common in business systems such as Microsoft Active Directory. After authentication came the authorization system. Hadoop uses (as seen above) ACLs, coarse access permission to HDFS files. In other words, a user has the right to get access to either the complete document or nothing. Hadoop added the encryption of the data transmitted between the nodes, as well as the data stored on the disk [11].

4.2 Apache Sentry

Apache Sentry [12] is one of the first security projects that offer fine-grained authorization. Sentry is integrated with SQL query framework, Apache Hive, and Impala de Cloudera.

Sentry provides the ability to control and enforce specific levels of privileges to users or applications that authenticate to a Hadoop cluster. It is designed to be an authorization engine for Hadoop. It allows fine-grained authorization rules to be defined and allows a user or application to get access to Hadoop's requested resources. Sentry can authorize permissions for various varieties of data models.

Sentry aims to make the authorization for components of the Hadoop ecosystem in a harmonized way, so that, security administrators can easily control what users and groups have access without needing to know the ins and outs of every single component in Hadoop.

A data processing tool (example, Hive) identifies the user who requests access to a data item (such as reading a row from a table). The tool instructs Sentry to process the user's query to enable or deny access. Sentry uses rules to define permissions and roles to combine or consolidate rules, making it easy and flexible to administer group permissions for different objects.

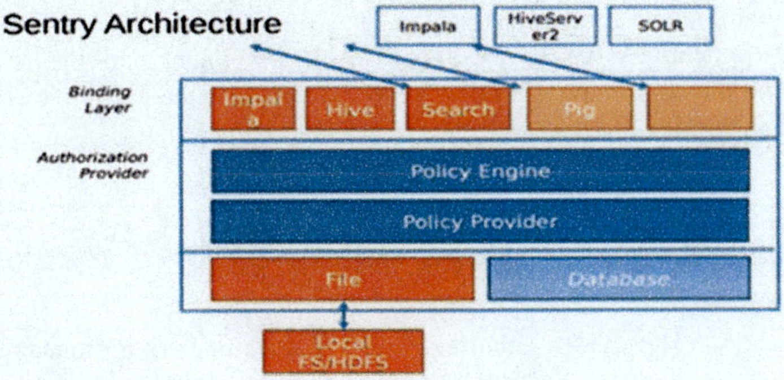

Fig. 4 Sentry architecture

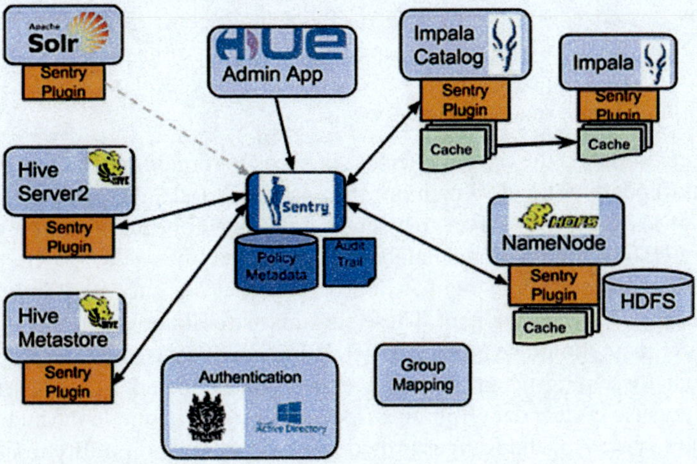

Fig. 5 Integration of sentry into the hadoop ecosystem

As shown in Fig. 4, the Sentry architecture consists of an authorization provider and a link layer (binding layer) which is the bridge between the tools (Hive, Impala, ...) and Sentry.

The authorization provider consists of two components: The policy engine is responsible for validating and evaluating the policy and then determines whether the requested action is allowed by checking the provider (Policy Provider). The latter is the storage mechanism for policies.

The Sentry architecture is extensible, and any developer can develop a link for different components (e.g., Pig...). Each mandatory component in Sentry implements a privilege model for specific binding. For example, Hive binding implements a privilege model to perform fine-grained permissions.

As shown in Fig. 5, Apache Sentry works with multiple components of the Hadoop environment. At the core, there is Server Sentry that stores authorization metadata and provides tools to retrieve and modify these metadata.

The actual authorization decision is made by a policy engine that runs in data processing applications like Hive or Impala. Each component loads the Sentry plug-in. This component contains the client service to process the permissions by using Sentry and the policy provider.

4.3 Apache Ranger

Apache Ranger [13] provides a centralized security framework for managing fined-granularity access control. Security administrators can easily manage policies for accessing files, folders, databases, tables, or columns. These policies can be defined for individual users or groups, and then applied inside Hadoop.

Formerly known as Apache Argus, Apache Ranger concurrence Apache Sentry since it also handles permissions. It adds a layer of permissions in Hive, HBase, and Knox. It has the advantage over Sentry in defining column-level permissions in Hive.

The architecture of Apache Ranger consists of:

- The Ranger portal: is the central interface of security for administration. Users can create and update policies, which are then stored in a database. The portal also consists of a verification server that sends audit data collected from plug-ins for storage in HDFS or a relational database.
- Ranger Plug-ins: are lightweight Java programs that are integrated into the processes of each cluster component. These plug-ins use policies to determine whether an access request should be granted or not. When a request is delivered through the component, these plug-ins intercept the request and evaluate it against the security policy; they also collect data from the user's request that is sent to the audit server.
- User group sync: Apache Ranger provides a synchronization utility to users and groups from Unix or LDAP or Active Directory. The user or group information is stored in Ranger portal and used to define the policy.

In the Apache Ranger version 0.5.0, the community made the first step toward a real Attribute-Based Access Control (ABAC), providing an access control framework based on dynamic rules.

4.4 Rhino Project

Rhino [14] is an open-source project developed and maintained by Intel, which aims to improve the data protection mechanism in Hadoop at different levels. The aim is to fill the gaps representing insecurity in the Hadoop environment and provide several safety components. The various improvements consist of:

- Encryption: It provides the ability to encrypt or decrypt data stored in HDFS.
- Authentication: It supports multiple authentication mechanisms, such as public key cryptography. In the current Hadoop security system, users and services are authenticated with Kerberos. The Rhino project enables authentication of users in a centralized service by developing Hadoop with single sign-on.
- Access Control: It provides better access granularity (at the cell level) in HBase. It is more comfortable with column-oriented databases. Each cell is associated with a visibility label. When the user queries the data, it provides the group to which it belongs and the role it has. This information will be verified with the visibility label by concluding whether or not the user can access this cell.

After presenting the different access control models applied in the Big Data and the different implemented projects, we present in the following table a summary by highlighting the relationship between the approaches and projects.

	ACL	RBAC	ABAC	CBAC
Access control granularity	Files, Tables	Columns	Cells	Cells
Projects	\	Sentry, Rhino	Ranger	\
Tools	HDFS, HBase, Pig, Zookeper, Hue...	Hive, Impala, HBase, Sqoop, ...	HDFS, Hive, HBase...	\

5 H-RCBAC: Hadoop Access Control Based on Roles and Content

The evolution of the security needs and the technological environments to be controlled means that the authorization systems must be adapted constantly. We defined in the previous sections the different access control models implemented in the Big Data (ACLs, RBAC, CABC, ABAC) and the projects that participate in the evolution and reinforcement of security in Hadoop (Sentry, Rhino, Apache Ranger). Nevertheless, these projects don't answer all the access control issues in Big Data. For example, the access control model implemented in HDFS provides coarse-grained access control. Indeed, we recall that the mechanism set up for HDFS (ACLs) allows only to define whether the user accesses to the full document or not. This coarse-grained access control particularly attracted our attention. Such strict access control (full access or no access) can affect the rights of users. In fact, in some cases, a user can access a part of the document without threatening the data confidentiality.

To address this problem, we propose an approach combining the RBAC and CBAC models with an algorithm that filters the taboo words to offer a fine-grained access control.

It is worth noticing that there are two ways to control access: (1) Assume that the user has the default right to access everything and set (via rules) interdictions or (2) assume that user has no right to access and set (via rules) access permissions. In our approach, we opted for the first assumption.

The RBAC model in our approach simplifies administrative tasks by reducing the assignments' management. Each user is assigned to a role that allows access to certain documents. However, this is not sufficient in the context of HDFS since the permissions will always be based on the ACLs. The use of this model does not offer enough flexibility. Let us take an example of a hospital database. We would like to give physicians access to a certain document but which contains some sensitive data that must be anonymous. By using ACLs, we have two choices: either take the risk, make the document accessible, or prohibit all physicians to access this document. To handle this rough decision, we propose to combine the RBAC with the CBAC model by integrating an algorithm that filters the taboo words. This solution applied to our example allows physicians to access the document parts that do not contain sensitive data (taboo words).

We based our approach on the Discretionary Access Control (DAC) model. This model is based on the "owner" concept. Each user is responsible for its documents. In other words, it is up to him to define the access rules of his data. He can give access to the complete document to a role X. Prohibit access to a role Y. It can define finer accesses by filtering certain parts of the document through a set of taboo words and the model CBAC.

We recall here that the CBAC rules are based on a decision function which takes as input a document and a list of words and compares the words' occurrences in the document with a threshold.

The idea of our approach is as follows: Each user is allowed (through his role) to access a certain number of objects. These accesses are defined via rules based on the CBAC model. We have added a layer of restrictions to the CBAC that allows the user to access only to a "subset" of the document. For a user u with a role r that sends a query q for access to a document d. It will be necessary to calculate, via the decision function, the taboo words' occurrences of the role r in the document d. If it is greater than the defined threshold, the user u will not access the document d. If the threshold is not reached, the user u will access a filtered document d' (without the taboo words).

5.1 H-RCBAC Implementation

The implementation of H-RCBAC is divided into two parts (Fig. 6):

- The first step is to give the possibility to the document owner (when he puts the document in HDFS) to define his prohibitions to a document d (define the taboo words). At this step, it is necessary to pre-calculate (with the known word count MapReduce) the frequency of each word in the document. This will prevent us from calculating, at each query, the occurrence of taboo words throughout the

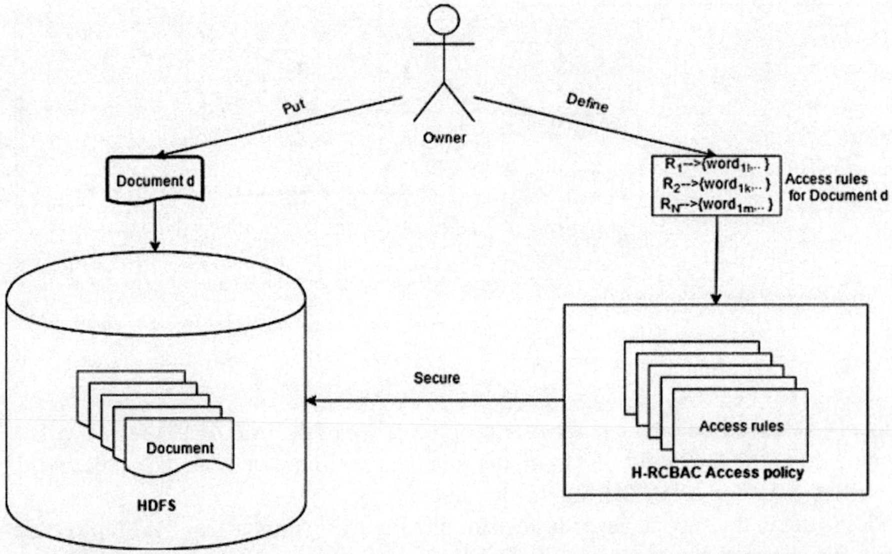

Fig. 6 Access rules definition

document. Indeed, by pre-calculating the frequency of each word at the document input, we will execute only one time the word count algorithm.
- The second part deals with the execution or not of a query user.

Access Rules Definition
Our approach facilitates decision making. When a document has loaded, the interdictions that are specific to it are loaded into the access policies. Each policy must provide all necessary information to enable decision making. To achieve this, we have to integrate a plug-in in HDFS that allows the user to manage their access policies without going through an administrator.

The plug-in offers the possibility to the owner to:

- Choose roles: These roles are defined upstream by the administrator and assigned to the different users of the system. When creating a document, the owner retrieves the list of roles and assigns a set of prohibitions for each role.
- Define taboo words: Lists taboo words (forbidden access) for one or more roles.
- Execute the word count algorithm on the document.

Queries Answering
After defining the access rules, it is important to intercept the queries and filter the results. For this, we have implemented our approach in the NameNode of HDFS. When a user requests access to an HDFS file, the NameNode will use our plugin. The latter will check if the user (who sent the request) has the right to access this file. Figure 7 details how and when our algorithm interacts to answer to queries.

We suppose that a user u sends a query. For our example, the query is a simple word count algorithm on a document d. The first step is to verify if the query can be executed or not. For that, our plug-in will calculate the decision function of the

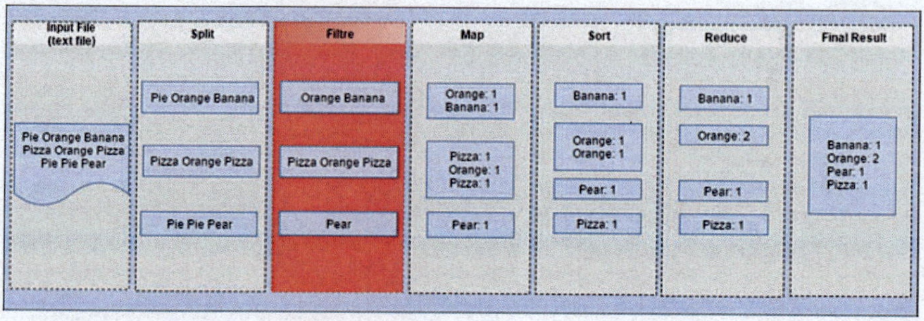

Fig. 7 MapReduce step with taboo words filter

model CBAC, based on the taboo words of the role r of the user u. The result will be compared to the threshold. If it is higher, the query will not be executed. Otherwise, it will be executed after filtering the document.

Let's detail the second case via an example. Figure 7 summarizes the MapReduce step. We suppose that the user has no right to access to the taboo word "Pie".

- In the mapping phase: The query and the list of taboo words will be sent to the various nodes containing the file partitions. Before executing the query, each partition (a copy) is filtered (deletion of taboo words). In our example, the word "Pie". The result of the map phase will be stored on each node.
- In the Reduce phase: The query will continue and the final result will be stored in the HDFS file system.

In this case, the query will be executed on a filtered document. This offers fine-grained access control and better data confidentiality.

6 Conclusion and Perspectives

The Big Data offers computing power due to its speed of processing queries. It supports a wide variety of data from heterogeneous environments while delivering high performance for loading and analysis. However, all this accumulated data from various sources requires protection. By default, HDFS works with ACLs, but this is not enough to effectively protect data. Various security solutions are offered to Hadoop and its ecosystem thanks to the various existing projects. That said, there are limits to each approach.

We have proposed in our work a new approach that allows reinforcing the access control in Hadoop. Our approach is based on the combination of two existing models: RBAC and CBAC. This solution offers a fined-grained access control mechanism by taking into consideration a set of taboo words before the querying process. H-RCBAC offers a supplementary access control compared to CBAC. Indeed, when CBAC offers access to the whole document, our solution excludes the taboo words before authorizing the access. Our solution gives good results when the query is independent of the taboo words list. If the taboo words are used in the calculation of the query result, the result can be inconsistent in addition to being incomplete.

As perspectives, we aim to design a module that identifies the query feasibility and the results consistency, based on the nature of the query and the taboo words list. We can easily distinguish between three categories of results: correct and complete results, correct but incomplete results, and the impossibility to execute the query.

This work can also be improved by integrating it into existing projects. Indeed, the implemented projects (Sentry, Rhino …) offer an already effective framework and present in several tools of the Hadoop ecosystem.

References

1. Chen, M., Mao, S., Liu, Y.: Big data: a survey. Mobile Netw. Appl. **19**(2), 171–209 (2014)
2. Sharma, P.P., Navdeti, C.P.: Securing big data hadoop: a review of security issues, threats and solution. Int. J. Comput. Sci. Inf. Technol. **5** (2014)
3. Guo, C., Wu, H., Tan, K., Shiy, L., Zhang, Y., Luz, S., Mapreduce: Simplified data processing on large clusters. In: Proceedings of OSDI, San Francisco, CA, USA (2004)
4. Apache Hadoop. https://hadoop.apache.org/docs/current/hadoop-project-dist/hadoop-hdfs/HdfsPermissionsGuide.html. Last accessed on 22 Feb 2021 (2017)
5. Ferraiolo, D.F., Sandhu, R., Gavrila, S., Kuhn, D.R., Chandramouli, R.: Proposed NIST standard for role-based access control. ACM Trans. Inf. Syst. Secur. (TISSEC) **4**(3), 224–274 (2001)
6. Gupta, M., Patwa, F., Sandhu, R.: Object-tagged RBAC model for the hadoop ecosystem. In: IFIP Annual Conference on Data and Applications Security and Privacy, pp. 63–81. Springer (2017)
7. Zeng, W., Yang, Y., Luo, B.: Access control for big data using data content. In: 2013 IEEE International Conference on Big Data, pp. 45–47. IEEE (2013)
8. Ashwin Kumar, T.K., Liu, H., Thomas, J.P., Hou, X.: Content sensitivity based access control framework for hadoop. Digital Commun, Netw. **3**(4), 213–225 (2017)
9. Cavoukian, A., Chibba, M., Williamson, G., Ferguson, A.: The importance of ABAC: Attribute-based access control to big data: privacy and context (2015)
10. Gupta, M., Patwa, F., Sandhu, R.: An attribute-based access control model for secure big data processing in hadoop ecosystem. In: Proceedings of the Third ACM Workshop on Attribute-Based Access Control, pp. 13–24 (2018)
11. Das, D., O'Malley, O., Radia, S., Zhang, K.: Adding security to apache hadoop. Hortonworks, IBM (2011)
12. Apache Sentry: https://sentry.apache.org/. Last accessed on 22 Feb 2021 (2016)
13. Apache Ranger: http://ranger.apache.org. Last accessed on 22 Feb 2021 (2014)
14. Rhino Project: https://github.com/intel-hadoop/project-rhino/ Last accessed on 22 Feb 2021 (2015)

Toward a Safe Pedestrian Walkability: A Real-Time Reactive Microservice Oriented Ecosystem

Ghyzlane Cherradi, Azedine Boulmakoul, Lamia Karim, and Meriem Mandar

Abstract Mobility is one of the key factors to consider in order to make cities more efficient, a necessity taking into account the millions of citizens travel on a daily basis to places known or unknown. Road safety, in particular, is of tremendous importance. Pedestrian accidents that cause too much injury or even death are serious problems in cities. In this work, we present a real-time reactive system, the aim of which is to provide the safest route among all possible routes for given source and destination at a particular time period, based on a modeling of the network by a fuzzy graph. Its main advantages over the proposed solutions lie in the robustness to incomplete data modeled as fuzzy information. The system involves the pgRouting opensource library that extends the PostGIS/PostgreSQL geospatial database, to provide geospatial routing functionality. So, we offer a web-location based service allowing pedestrians to enter their destination, then select a route uses an intelligent algorithm, providing them with the safest route possible instead of the fastest route. This service will certainly help save lives and, to a certain extent, reduce pedestrian accidents.

1 Introduction

The advent of smart mobility has enabled the creation and development of applications for mobility support addressed to different categories of road users, including pedestrians. Indeed, each of us has different preferences when it comes to transportation, but at one time or another everyone is a pedestrian. Unfortunately, pedestrian accidents are on a rise all over the world. Statistics of accidents in the world indicate

G. Cherradi · A. Boulmakoul (✉)
LIM/IOS FSTM, Hassan II University, Casablanca, Morocco

L. Karim
LISA Laboratory ENSA Berrechid, Hassan 1st University, Settat, Morocco

M. Mandar
ENS, Hassan II University, Casablanca, Morocco

M. Ben Ahmed et al. (eds.), *Networking, Intelligent Systems and Security*,
Smart Innovation, Systems and Technologies 237,
https://doi.org/10.1007/978-981-16-3637-0_31

that despite the efforts made, pedestrians remain the most vulnerable road users, with road accidents taking place through the fault of both drivers and pedestrians. Since a significant part of pedestrian crossings is non-signalized, the number of accidents on them, including accidents where people died, is much higher. At the same time, people die as a result of road traffic crashes outside the city almost 20 times less than in the cities themselves. This is due to the ratio of the intensity of pedestrian and traffic flows. As pedestrians, children are at even greater risk of injury or death from traffic crashes due to their small size, inability to judge distances and speeds, and lack of experience with traffic rules.

The capacity to respond to pedestrian safety is an important component of efforts to prevent road traffic injuries. With advancements in technology and the availability of geolocation solutions in every field of mobility is fostering the creation of novel approaches, such as the so-called Mobility-as-a-Service (MaaS) paradigm [1]. The MaaS concepts aim to offer users an integrated, comprehensive and simple service. The concept of walkability is gradually acquiring worldwide a key position in mobility planning. Pedestrian dynamics are difficult to characterize as they have many effects from different sources. Walking, unlike other models of travel, is not tied to a vehicle on a lane, and the underlying infrastructure is very heterogeneous. In addition, environmental factors (traffic lights, public furniture, advertising, etc.), as well as the total waiting time of pedestrians, the average distance between the vehicles following one after another, and the atmospheric conditions (wind, rain, etc.) directly affect walking. In this context, modeling pedestrian safety with respect to the uncertain nature of these factors is an important research objective.

This article presents an integrated platform to collect, identify, and provide a variety of pedestrian-related communication and control functions to demonstrate how it is possible to create a scalable and integrated mobility service for safe pedestrian walking, through the coordination of reusable components on a microservice architecture. In addition, the article proposes a risk evaluation model using fuzzy set theory [2, 3] to consider the incomplete and uncertain nature of the risk parameters. The system involves the open-source pgRouting library which extends the PostGIS/PostgreSQL geospatial database to provide geospatial routing functionality. Thus, it offers a web location-based service allowing pedestrians to enter their destination; then select a route using an intelligent algorithm; provide them with the safe and short paths to take when they navigate the city. The remainder of this paper is organized as follows. After discussing some related works in Sect. 2, we introduce our microservice development approach in Sect. 3, focusing in particular on the advantages that this paradigm introduces on the basis of a real, currently developing infrastructure. Then, Sect. 4 details the proposed model of pedestrian safety for a given urban area. Finally, Sect. 5 concludes the paper with some final remarks and directions for future work.

2 Related Works

Within the field of urban computing, our work is related mainly to research on pedestrians' risk modeling, on pedestrian safety systems, and on urban navigation. Here, we review these lines of research and their connections to our study. Pedestrians' risk modeling: Traditionally, pedestrian risk is generally assessed using crash frequency models, based on historical data [4–7]. Crash frequency models have been developed using spatial zones or intersection-level data. However, crash frequencies can be affected by different factors including traffic, speeds, geometry, and built environment, etc. Authors in [8] proposed a composite indicator of pedestrian's exposure, taking into account pedestrian characteristics, road and traffic conditions, as well as pedestrian compliance with traffic rules. The traffic conflicts technique has also been used for measuring the exposure of pedestrians at specific crossing locations [9]. Lassare et al. [10] developed an approach of accident risk based on the concept of risk exposure used in environmental epidemiology, such as in the case of exposure to pollutants. Saravanan et al. [11] present an accident pre-diction approach based on fuzzy logic. Their study evaluates the accident risk for pedestrians as well as vehicles and meets the accident zones on a given road network. Mandar et al. [12] present a new measurement of virtual pedestrians and vehicles mutual accidents risk indicator. Pedestrians' dynamics are modeled using the basic fuzzy ant model [13], to which they have integrated artificial potential fields. Another factor of risk is the conflict between the vehicular and pedestrian flows at left-hand corners [14]. Different models have been proposed for risk assessing and working out of measures to improve the pedestrian safety. Thus, the model proposed in the research [15] allows to measure the impact of potential risk factors on pedestrians' intended waiting times. In research [16], the author proposes a multivariant method of risk analysis consisting of two hierarchically generalized linear models, characterizing two different facets of unsafe crossing behavior, and uses a Bayesian approach with the data augmentation method to draw statistical inference for the parameters associated with risk exposure.

To date, several pedestrian safety systems have been proposed; David et al. [17] discuss the response time for smartphone applications that warn users of collision risks, analyzing approaches based on a centralized server and on ad-hoc communication connections between cars and pedestrians. The use of mobile phones by pedestrians (for talking, texting, reading) affects their awareness of the surrounding environment, hence augmenting the risk of incidents. Authors in [18] proposed an android smartphone application (WalkSafe) that aids people that walk and talk, improving the safety of pedestrian mobile phone users. It uses the back camera of the mobile phone to detect vehicles approaching the user, alerting the user of a potentially unsafe situation. In research [19], authors assessed two systems using "standard tests" (vehicle driving toward a pedestrian dummy positioned on the course). The results attest that the functionality of the systems depends on the vehicle speed to avoid collisions, but it is limited at a certain speed (40 km/h). Although these systems are intended to effectively reduce pedestrian injury outcomes, the collision avoidance performance of these systems remains limited.

Urban navigation: To support pedestrian route choices to minimize potential dangers, route qualities (including safe crossing facilities, motor traffic volume and speeds) need to be considered in the design of pedestrian navigation systems [20]. Systems with new navigation objectives that go beyond the shortest path have recently appeared in the literature, mainly tailored to the personal interests and spatiotemporal constraints of users [16, 21–23], or routes that take into account traffic information gathered from historical speed patterns and GPS devices [21–24]. Indeed, pedestrian navigation and routing systems need to be developed in a user-friendly manner that enhances road safety and provides an optimal user experience.

3 System Model

We consider the system model as an integrated solution offering various services: geolocation, real-time traceability, and pedestrians routing. By leaning toward a microservice architecture [25], the intention is to develop and execute a plurality of microservices, which are developed, tested, deployed, and run isolated from the rest. Moreover, the implementation reveals benefits of microservice architecture regarding scalability, extensibility, flexibility, and fault tolerance. Figure 1 concisely shows the main components and the architecture of the proposed system and their connections.

Data collection and pre-processing: To improve the quality of analysis and decision making, a powerful, scalable, and interoperable data collection service is required. Therefore, a smart data collection service has been developed in our system. Based on concepts of IoT allied to wireless sensor networks, the proposed service uses sensors to obtain information about environment, vehicle, and pedestrians in which they are inserted and have great importance to analyze several factors that influence pedestrian's walkability.

Pedestrian risk calculator: This microservice devoted to calculate in real time the risk related to a road. Risk indicators are valid and depend on how they are read. There is no absolute risk indicator. Nevertheless, the indicator we propose estimates the journey time. The residence time of a pedestrian on a section of road increases the exposure of risk. We postulate that the value of risk must be a function of the total time it takes to cross the segment. Therefore, our exposure model suggests that risk poses appositive correlation with two factors: probability of an accident and duration of risk exposure.

Pedestrians routing: The routing of the pedestrians is the service responsible for finding point-to-point safest path in a constructed time-dependent network with fuzzy risk. The implemented routing algorithm is designed to compute an optimized shortest path that surpasses the performance of classical algorithms in terms of computational time and space. In general, two kinds of data are needed as input for the proposed service, namely: (a) spatial data, which represents the transportation network as a collection of arcs with nodes loaded from OSM (OpenStreetMap)

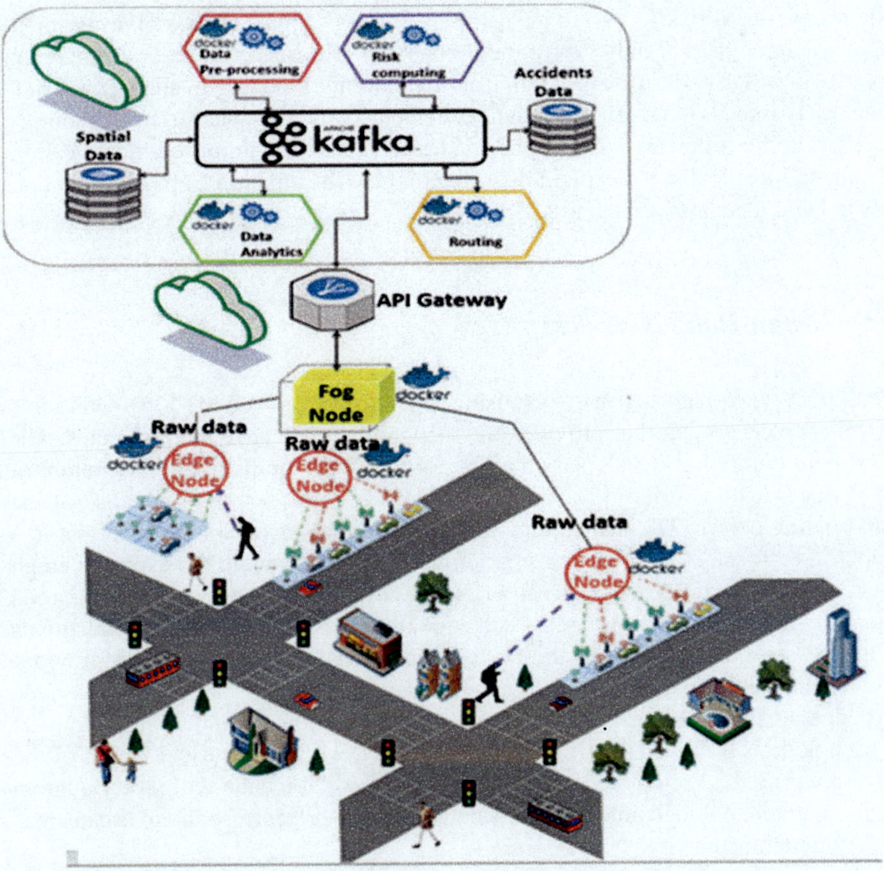

Fig. 1 The system architecture

into the PostgreSQL/PostGIS database. (b) Non-spatial attribute information for the attribute data for road such as name, category, length, risk cost.

4 Modeling Pedestrian Risk

Throughout the paper, we will represent the urban network using an undirected graph $G(V, E)$. The nodes of the graph, V, represent intersections and the edges, E, represent the road segments that connect intersections. Each road segment $s \in E$ is associated with two (unrelated) weights: its length, denoted by l_s, and its risk, denoted by r_s. The length of segments s corresponds to the actual distance between the two intersections connected by the associated street segment, while the risk of segment s can be defined as the multiplicative product of the probability of an accident and the gravity (the pedestrian at risk factor) to pedestrian if it does occur. In fact, risk assessment of pedestrian must not only derive the probability of a crash, but must estimate the severity to which pedestrian is at risk from such events. One way to analyze a pedestrian safety issue is to identify the significant factors affecting pedestrian crash injury severity. In the following subsections, we describe how the proposed model can be obtained. Specifically, we explain how we combine information from different data sets to build a pedestrian risk model.

4.1 Urban Road Networks

We extract the urban road network using OpenStreetMap (OSM) [26]. OSM is a crowd-sourced project that provides free geographic data such as road maps. The OSM community has developed a number of tools that facilitate the integration of OSM maps with a variety of applications. For our purposes, we make use of the osm4routing parser [27], that exports the map of interest from an OSM format to a graph-based format (appropriate for graph-based algorithms). In the exported graph $G(V, E)$, each vertex from V corresponds to a designated pedestrian crossing, road intersection, or dead-end. Whereas each edge from E is undirected and corresponds to a road segment connecting the corresponding vertices. Osm4routing provides supplemental information about the intersections and the street segments as node and edge attributes. Especially, the geographical location (latitude/longitude pair) of each node is provided. Each edge e is annotated with the physical length of the corresponding road segment. It is the length value we associate with every edge in the input graph. All road networks considered in this work correspond to urban areas in the Mohammedia city.

4.2 Pedestrians Risk

Preliminaries. In this section, we recall some definitions and the main results about the fuzzy set theory which will be used later. The first person who introduced fuzzy Set theory is L. Zadeh. He introduced his article regarding "Fuzzy Sets" [11].

Definition 1 (*Fuzzy set*) [11] is defined in terms of a membership function, which is a mapping from the universal set U to the interval [0, 1]. Then, a fuzzy set $\tilde{A}$ in U is a set of ordered pairs:

$$\tilde{A}(x) = \{x, \mu_{\tilde{A}}(x) | x \in U, \mu_{\tilde{A}}(x) \in [0, 1]\} \tag{1}$$

A membership function is a function from a universal set U to the interval [0, 1]. A fuzzy set $\tilde{A}$ is defined by its membership function $\mu_{\tilde{A}}(x)$ over U. When fuzzy set theory was presented, researches considered decision making as one of the most attractive application fields of that theory [2, 3].

Definition 2 (*Fuzzy number*) [2, 3] a fuzzy subset $\tilde{A}$ of the real line R with membership function $\mu_{\tilde{A}}: R \to [0, 1]$ is a fuzzy number if its support is an interval $[a, b]$ and there exist real numbers s, t with $a \le s \le t \le b$ fulfilling:

i. $\mu_{\tilde{A}}(x) = 1$ for $s \le x \le t$
ii. $\mu_{\tilde{A}}(x) \le \mu_{\tilde{A}}(y)$ for $s \le x \le y \le t$
iii. $\mu_{\tilde{A}}(x) \ge \mu_{\tilde{A}}(y)$ for $t \le x \le y \le b$
iv. $\mu_{\tilde{A}}(x)$ is upper semicontinuous

We will denote the set of fuzzy numbers by FN and a fuzzy number by $\tilde{N}$. We observe that every real number N is a fuzzy number whose membership function is the characteristic function:

$$\mu_N(x) = \begin{cases} 1 \text{ if } x = N \\ 0 \text{ if } x \ne N \end{cases} \tag{2}$$

Definition 3 (*Triangular fuzzy number (TFN)*), a triangular fuzzy number $\tilde{N}$ can be defined by a triplet (a, b, and c). The triangular fuzzy number is used to represent uncertainty. The membership function is [2].

$$\mu_N(x) = \begin{cases} 0 & x < a, c < x \\ \frac{x-a}{b-a} & a < x < b \\ \frac{x-c}{b-c} & b < x < c \end{cases} \tag{3}$$

where $0 \le a \le b \le c \le 1$, a and c stand for the lower and upper values of the support of N, respectively, and b stands for the modal values. The graphical representation of a triangular fuzzy number is shown in Fig. 2.

Definition 4 (*Gaussian fuzzy number (GFN)*), compared to traditional triangular fuzzy number (TFN), GFN is more realistic to represent uncertainty because the

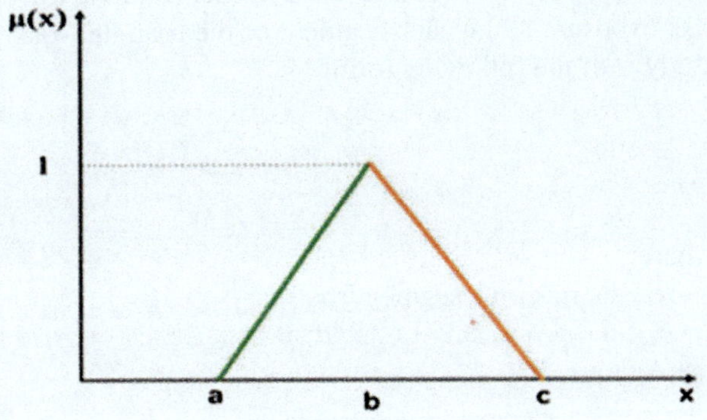

Fig. 2 The graphical representation of a triangular fuzzy number (TFN)

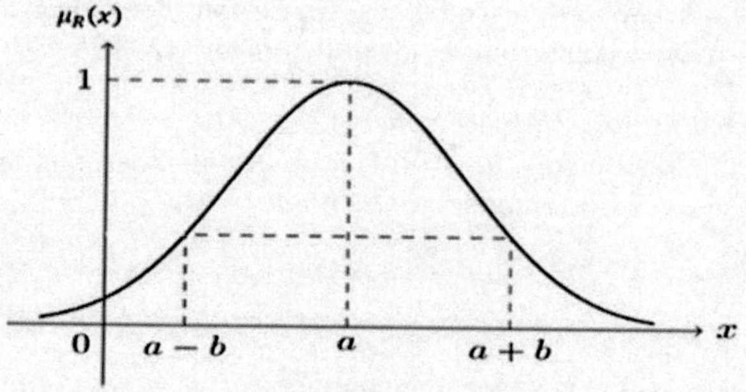

Fig. 3 Representation of a Gaussian fuzzy number (GFN)

distribution is assumed to be Gaussian. The Gaussian fuzzy function transforms the original values into a normal distribution. The fuzzy Gaussian function is given below:

$$\mu_N(x) = \exp\left[-\frac{1}{2}\left(\frac{x-a}{b}\right)^2\right], x \in \mathbb{R}, b > 0 \tag{4}$$

A Gaussian Membership function is specified by two parameters: a Gaussian membership function is determined complete by a and b; a is the expectation and b is the standard deviation. Figure 3 shows a graphical representation of a Gaussian fuzzy number (GFN).

Pedestrians risk Model. A risk model for the urban road network is essentially an assignment of a risk score r to each segment s that is proportional to the probability of an accident happening on the corresponding road segment and the pedestrians-at-risk factors. In the proposed model, the duration of risk exposure is estimated by travel time of the road segment traveled. During this time, the pedestrian is exposed to risk depending on risk factors. This travel time is one of several sources of uncertainty in pedestrian's walkability, which cannot be known or expressed exactly. This uncertainty is represented by Gaussian fuzzy number Ts~GFN (t_s, γt_s) where t_s, γt_s are the mean and the standard deviation, respectively, and γ is a proportionality parameter. The pedestrian crash injury severity associated with each factor is multiplied by the probability of accident occurrence, and the travel time to give an estimate of risk exposure along each segment of the road network. The expression used in this study is of the following form:

$$R_s = \sum_j P_{sj} {}^* F_j {}^* T_s \tag{5}$$

where:

R_s: the risk on a segment road s;

P_{sj}: the probability of accident occurrence on road segment s with respect of a risk factor j;

F_j: the severity of pedestrian crash injury on a segment s with respect to factor j;
T_s: The travel time associated with segment s.

In expression (5), some inputs parameters are GFNs, and some are constant therefore, the resulting risk is also obtained in the form of a GFN whose membership function normalized is expressed as:

$$\mu_R(x) = \exp\left[-\frac{1}{2}\left(\frac{x-a}{b}\right)^2\right], x \in \mathbb{R}, b > 0 \quad (6)$$

The total risk of a path, denoted by $R(x)$, is the summation of the risk weight of segment s_k. Where s_k is the kth segment for a path x.

$$\begin{aligned} R(x) &= \sum_k^n \sum_j P_{s_k j} {}^* F_j {}^* T_{s_k} \\ &= N\left(\sum_k^n \sum_j P_{s_k j} {}^* F_j {}^* T_{s_k}, \ \gamma \sum_k^n \sum_j P_{s_k j} {}^* F_j {}^* T_{s_k}\right) \end{aligned} \quad (7)$$

The risk R is calculated for each arc in the network N using (7). $R(R \sim N(0, 1))$ is the normalized value of the expected fuzzy risk of a segment s as shown in Fig. 4.

For each segment s, there is a variety of risk values according to an α (predetermined confidence level) to be specified according to the risk threshold not to be exceeded $\mu_R(x < \lambda_i) = \alpha_i$. (See Fig. 5).

We need to search values of λ_i the evaluation of edge according to an α_i risk level and a departure time t. Finally, the problem becomes to define the safest path in a time-dependent network as shown in Fig. 6.

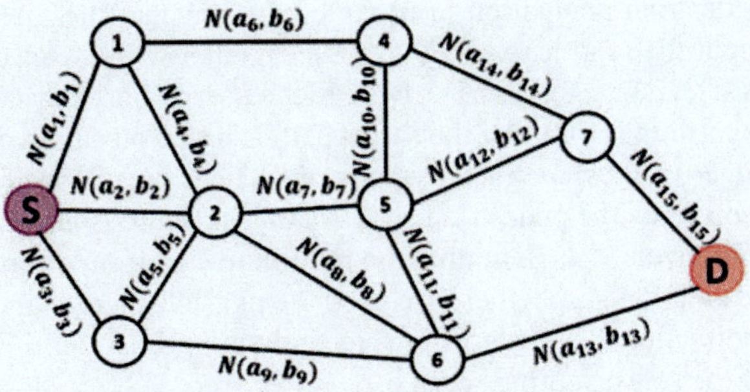

Fig. 4 Fuzzy risk graph

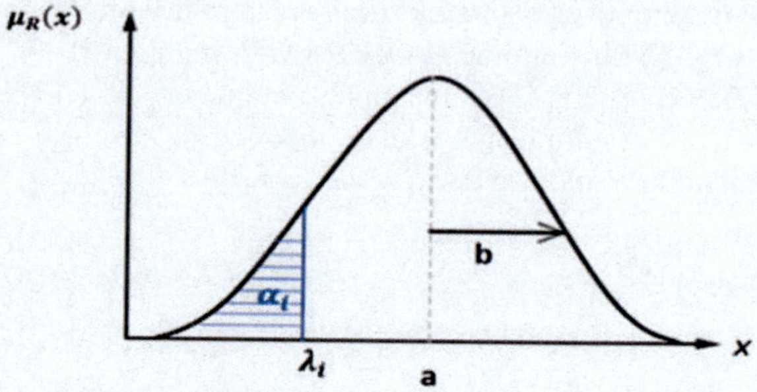

Fig. 5 Acceptable risk value λ_i according to α_i level

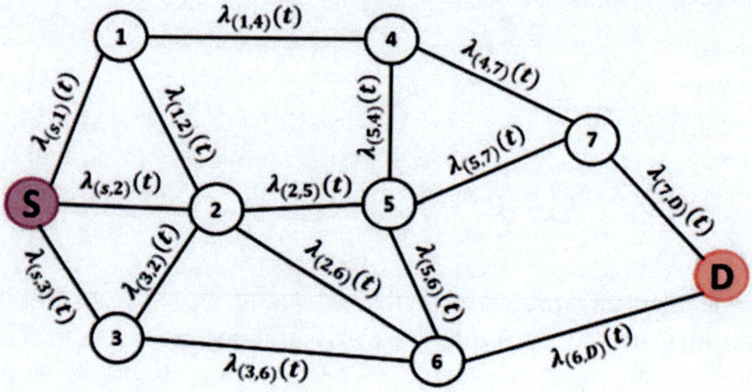

Fig. 6 Discrete time-dependent risk graph

4.3 Safest Route Determination

We now turn to an application exploiting the model presented above to provide safe pedestrian navigation in urban environments. The safest path problem takes as an input the road network G (V, E); together with a pair of source–destination nodes, (s, d), and its goal is to provide to the user a short and safe path between s and d. The algorithm for finding the safest path is based on a network composed of vertices and edges (routes), defined by pairs of vertices. Each edge has a cost which, in our case, represents the pedestrian risk. As this attribute is known, the problem of pedestrian routing consists in finding the minimum cost from a vertex of source A to a vertex of destination B specifies. We used the pgr_bdAstar () function (which implements the bidirectional A * algorithm) to find the optimal path. Figure 7 illustrates the result of the safest routing service.

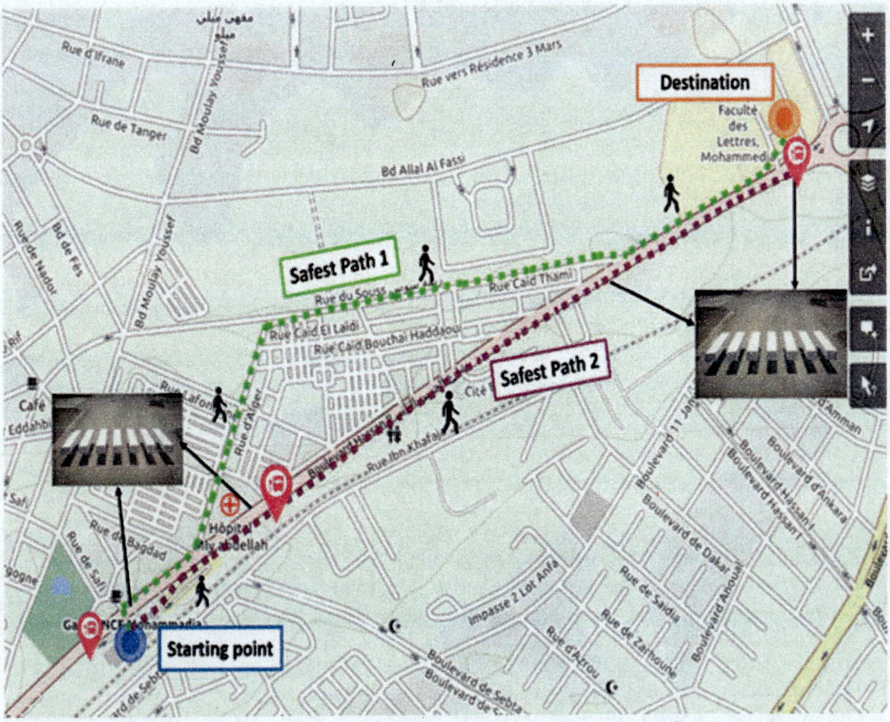

Fig. 7 An illustrative example of safest paths. The routes depicted as Paths 1–2 offer various alternative for traveling between a starting point and a destination

5 Conclusion

Smart walkability is a key element to support pedestrians in their daily activities and to offer them a livable smart city. Information about urban transportation, pedestrian crossings, and safest paths would be of great benefit in this context. In order to provide, an ecosystem to manage such services and features, we designed and prototyped a distributed information system and proposed a fuzzy pedestrian risk model under the consideration of travel time. As for the pedestrian routing problem in practice, data are very important to evaluate risk as well as cost. The accuracy and quality of the data could have significant impact on the result. Currently, data for pedestrian are insufficient and incomplete. Future research directions include how to establish sufficient databases and how to validate the proposed model.

Acknowledgments This work was partially funded by Ministry of Equipment, Transport, Logistics, and Water–Kingdom of Morocco, The National Road Safety Agency (NARSA) and National Center for Scientific and Technical Research (CNRST). Road Safety Research Program# an intelligent reactive abductive system and intuitionist fuzzy logical reasoning for dangerousness of driver-pedestrians interactions analysis.

References

1. MaaS Global. http://whimapp.com/. Last accessed 28 Nov 2020
2. Zadeh, L.A.: Fuzzy sets as a basis for a theory of possibility. Fuzzy Sets Syst. **1**(1), 3–28 (1978)
3. Zadeh, L.A.: Probability measures of fuzzy events. J. Math. Anal. Appl. **23**(2), 421–427 (1968)
4. Brüde, U., Larsson, J.: Models for predicting accidents at junctions where pedestrians and cyclists are involved. How well do they fit? Accid. Anal. Prev. **25**, 499–509 (1993).
5. Cameron, M.: A method of measuring exposure to pedestrian accident risk. Accid. Anal. Prev. **14**, 397–405 (1982)
6. Lee, J., Abdel-Aty, M., Xu, P., Gong, Y.: Is the safety-in-numbers effect still observed in areas with low pedestrian activities? A case study of a suburban area in the United States. Accid. Anal. Prev. **125**, 116–123 (2019)
7. Ni, Y., Wang, M., Sun, J., Li, K.: Evaluation of pedestrian safety at intersections: a theoretical framework based on pedestrian-vehicle interaction patterns. Accid. Anal. Prev. 118–129 (2016)
8. Van der Molen, H.H.: Child pedestrian's exposure, accidents and behavior. Accid. Anal. Prev. **13**(3), 193–224 (1981)
9. Gårder, P.: Pedestrian safety at traffic signals: a study carried out with the help of a 601 traffic conflicts technique. Accid. Anal. Prev. **21**(5), (1989)
10. Lassarre, S., Papadimitriou, E., Yannis, G., Golias, J.: Measuring accident risk exposure for pedestrians in different micro-environments. Accid. Anal. Prev. **39**(6), 1226–1238 (2007)
11. Saravanan, S., Sabari, A., Geetha M.: Fuzzy-based approach to predict accident risk on road network. Int. J. Emerg. Technol. Adv. Eng. 4(5), (2014). ISSN 2250-2459, ISO 9001:2008 Certified Journal
12. Mandar, M., Boulmakoul, A., Lbath, A.: Pedestrian fuzzy risk exposure indicator. Transp. Res. Procedia **22**, 124–133 (2017)
13. Boulmakoul, A., Mandar, M.: Fuzzy ant colony paradigm for virtual pedestrian simulation. Open Oper. Res. J. **2011**, 19–29 (2011). ISSN: 18742432. https://doi.org/10.2174/1874243201105010019
14. Di Stasi, L.L., Megías, A., Cándido, A., Maldonado, A., Catena, A.: The influence of traffic signal solutions on self-reported road-crossing behavior. Span. J. Psychol. **17**(103), 1–7 (2014)
15. Quistberg, D.A., Howard, E.J., Ebel, B.E., Moudon, A.V., Saelens, B.E., Hurvitz, P.M., Rivara, F.P.: Multilevel models for evaluating the risk of pedestrian–motor vehicle collisions at intersections and mid-blocks. Accid. Anal. Prev. **84**, 99–111 (2015)
16. Ayala, I., Mandow, L., Amor, M., Fuentes, L.: An evaluation of multi-objective urban tourist route planning with mobile devices. LNCS Ubiquitous Comput. Ambient Intell. **7656**(2012), 387–394 (2012)
17. David, K., Flach, A.: Car-2-x and pedestrian safety. Vehic. Technol. Mag. **5**(1), 70–76 (2010)
18. Wang, T., Cardone, G., Corradi, A., Torresani, L., Campbell, A.T.: WalkSafe: a pedestrian safety app for mobile phone users who walk and talk while crossing roads. In: Proceedings of the Twelfth Workshop on Mobile Computing Systems & Applications, pp. 1–6 (2012)
19. Ando, K., Tanaka, N.: An evaluation protocol for collision avoidance and mitigation systems and its application to safety estimation. In: Proceedings of the 23rd International Technical Conference on the Enhanced Safety of Vehicles. Seoul, Republic of Korea (2013)
20. Czogalla, O., Herrmann, A.: Parameters determining route choice in pedestrian networks. In: TRB 90th Annual Meeting Compendium of Papers DVD, pp. 23–27. Washington, DC (2011)
21. Gonzalez, H., Han, J., Li, X., Myslinska, M., Sondag, J.: Adaptive fastest path computation on a road network: a traffic mining approach. In: 33rd International Conference on Very Large Data Bases (2007)
22. Joo, J.Y., Kim, S.H.: A new route guidance method considering pedestrian level of service using multi-criteria decision-making technique. J. Korea Spat. Inf. Soc. 1983–1991 (2011)
23. Kanoulas, E., Du, Y., Xia, T., Zhang, D.: Finding fastest paths on a road network with speed patterns. In: IEEE International Conference on Data Engineering (2006)
24. Yuan, J., Zheng, J., Xie, X., Sun, G.: T-drive: enhancing driving directions with taxi drivers' intelligence, In: IEEE Transactions on Knowledge and Data Engineering (2012)

25. Newman, S.: Building Microservices: Designing Fine-Grained Systems. O'Reilly Media, Inc. (2015)
26. Openstreetmap Homepage. http://www.openstreetmap.org. Last accessed 28 Nov 2020
27. Osm4routing: https://github.com/tristramg/osm4routing. Last accessed 28 Nov 2020
28. Baibing, L.: A model of pedestrians' intended waiting times for street crossings at signalized intersections. Transp. Res. Part B. **51**, 17–28 (2013)
29. Routledge, D., Repetto-Wright, R., Howarth, I.: Four techniques for measuring the 605 exposure of young children to accident risk as pedestrians. In: Proceedings of the 606 International Conference on Pedestrian Safety, Haifa, Israel (1976)
30. Routledge, D., Repetto-Wright, R., Howarth, I.: The exposure of young children to 603 accident risk as pedestrians. Ergonomics **17**(4), 457–480 (1974)

Image-Based Malware Classification Using Multi-layer Perceptron

Ikram Ben Abdel Ouahab, Lotfi Elaachak, and Mohammed Bouhorma

Abstract Classification of malware variants is the most challenging task in the cybersecurity landscape. Malware developers keep one step ahead of defenders for the sake of using advanced artificial intelligence techniques. That is why we are in extreme need of an efficient malware classifier. In this paper, we proposed and experiment a malware classifier able to affect each inputted malware into its corresponding family. To do so, we use the multi-layer perceptron algorithm with malware visualization technique. It refers to converting a malware binary into grayscale images. Besides, to reach a great accuracy, we experiment different architectures by varying hidden layers, neurons and activation functions. Then, we obtained an accuracy of 97.6%. At the end, we compare the obtained results with literature, and we conclude that the multi-layer perceptron algorithm is a good malware classifier with specified hyperparameters that were used.

1 Introduction

Malware is malicious software like adware, ransomware, virus, worm, etc. There are a variety of malware types, and each one of them have a special goal and target. Common targets are industrial networks, personal computers, mobile phones and IoT devices.

The damage caused by a malware attack varies and depends on the hacker's attention. Malware attacks could bypass access controls, gather sensitive information, disturb operations, block access and loss information. Sometimes the damage is very costly.

I. Ben Abdel Ouahab (✉) · L. Elaachak · M. Bouhorma
Computer Science, Systems and Telecommunication Laboratory (LIST), Faculty of Sciences and Techniques, University Abdelmalek Essaadi, Tangier, Morocco
e-mail: ibenabdelouahab@uae.ac.ma

M. Bouhorma
e-mail: mbouhorma@uae.ac.ma

M. Ben Ahmed et al. (eds.), *Networking, Intelligent Systems and Security*,
Smart Innovation, Systems and Technologies 237,
https://doi.org/10.1007/978-981-16-3637-0_32

Most recently, dealing with malwares has been in the top of interest of every cybersecurity researcher [14]. In fact, we found various works related to malware analysis, detection and classification.

Otherwise, with the widespread use of artificial intelligence (AI) and its subfields as machine learning (ML) and deep learning (DL) in several domains [13], malware classification has also been taken advantage of intelligent algorithms [12]. For malware classification task, many ML and DL algorithms are used, giving impressive results.

To classify malware samples, features are required to characterize each family. For this purpose, we have a use of different features as API calls sequences [5], behavioral reports based on dynamic analysis [4] and others extracted from static analysis. A new malware feature inspired from computer vision converts malware binaries into images and extract images features. The malware visualization technique shows similarities in samples belonging to the same family which can be used in the classification task.

In this paper, we propose a malware classifier based on the multilayer perceptron algorithm and using the visualization technique. We present an experimental study by varying different hyperparameters of the artificial neural network algorithm.

The remainder of this paper is organized as follows. Section 2 presents related works of malware classification task. Then, in Sect. 3, we present MLP algorithm that we will use after. Section 4 focused on the dataset and data processing. Afterward, we give experimentation use cases in Sect. 5. Then, the results are discussed in Sect. 6. Finally, we conclude and give some of our future works.

2 Related Works

In literature, we found a use of many algorithms for malware classification task. For instance, K-nearest neighbor (KNN), random forest (RF), support vector machine (SVM), multi-layer perceptron (MLP), convolutional neural networks (CNN), recurrent neural networks (RNN), etc.

In [7], authors proposed a deep learning-based malware classification model by using an 18-layers deep residual network. Experiment result reaches an average accuracy of 86%. In [6], researchers performed CNN classifier with Malimg database using visualization technique. Their proposed model achieved an accuracy of 98.52%. In addition, using the same CNN model architecture with another dataset named Microsoft malware dataset gives highest accuracy of 99.97%.

Otherwise, researchers in [9] combined two feature named Gabor wavelet transform (GWT) and GIST, then build a malware classifier using feed forward artificial neural network (ANN). Experiment results give an accuracy of 96.35%.

In a previous work [1], we performed a KNN malware classifier based on the visualization technique. By varying many parameters, we reached an accuracy of 97.92%. In addition, a similar classifier was made [2] using KNN and the visualization technique, but this time, we have the use of only 50-GIST features instead of 320. The given result is very interesting. The accuracy reached using 50 features is 97.67% in less duration.

Another deep architecture was adopted in [8], where authors proposed CNN malware images classifier. They classify malware images into 32 different families, by extracting local binary pattern (LBP) feature. Finally, they got an accuracy of 93.92%.

3 Multi-layer Perceptron

Multi-layer perceptron (MLP) algorithm is a part of artificial neural networks (ANN) [3]. The basic unit of an ANN is a network with only one node called a perceptron. The perceptron is made of inputs, weights, activation function and output. The activation function takes each input and its weight, multiplies them and gives the result as output.

While the MLP is a class of ANN, it should have at least three nodes. Nodes of MLP are arranged in layers: the input layer, the hidden layers and the output layer. In general, there must be one or more hidden layers. For the training, MLP uses backpropagation. The problem with MLP is to find a suitable number of hidden layers and choose an appropriate number of neurons in each layer. So using evaluation metrics, the MLP architecture is evaluated each time we change parameters.

Returning to the activation functions, which are very important in ANN, because they act directly in the learning phase by mapping inputs to outputs using nonlinear complex functions. In practice, these are the common activation functions used with MLP hidden layers: identity, logistic sigmoid function, hyperbolic tan function (tanh) and the rectified linear unit function (relu). Corresponding equation used with each activation function is given in Table 1.

Hence, in this paper, we use two evaluation metrics: accuracy and confusion matrix. Accuracy is obtained by dividing correct classifications by the total number of samples. Also, the confusion matrix gives a clearer view of predicted values and real ones. At the end, we added more evaluation metrics as precision, recall, f1-score and hamming loss for comparison.

Table 1 Activation functions definitions

Activation function	Mathematical formula
(1) identity	$f(x) = x$
(2) logistic	$f(x) = 1/(1 + \exp(-x))$
(3) tanh	$f(x) = \tanh(x)$
(4) relu	$f(x) = \max(0, x)$

4 Data Processing

Malimg dataset is one of the most known malware datasets released by the vision research lab of UCSB. The Malimg dataset contains 9339 malware images belonging to 25 different families (Fig. 1). The resulting malware images used visualization technique that was firstly proposed in [10] in 2011. Visualization technique refers to the process of converting a binary into grayscale image; an extract of the used database is presented in Fig. 2.

In order to extract features from images, there are several descriptors that have been used in many researches. For instance, the local binary pattern (LBP), discrete wavelet transformation (DWT), Gabor wavelet transform (GWT) and GIST are the most used in malware image processing.

Our proposed classifier uses GIST descriptor as malware image features. Global texture GIST descriptor feature is a 320-dimensional vector that was firstly proposed by AUDE OLIVA and their colleagues in [11]. Then it is been used in several applications.

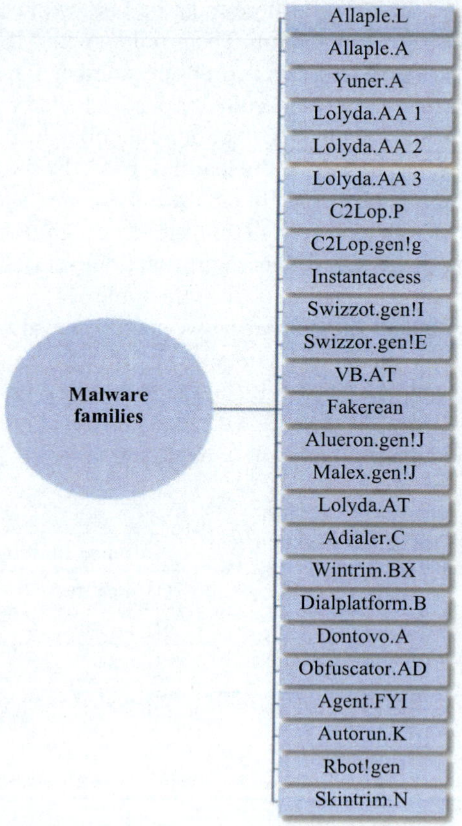

Fig. 1 Malware families present in the Malimg database

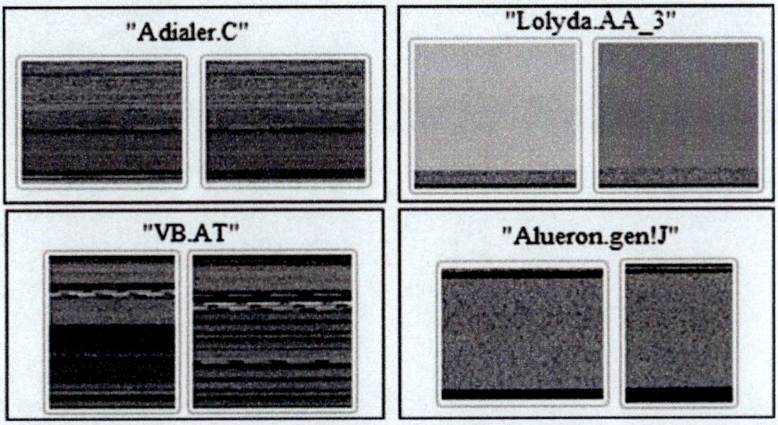

Fig. 2 Extract from the malware images database

5 Experimentation

The main goal in this experimentation is to have an efficient malware classifier using advanced techniques. To do so, we chose to use the multi-layer perceptron algorithm due to its significant performances in many applications. In addition, the visualization technique allows us to visualize a malware binary, as a grayscale image. We adopted this method because it is an easy, efficient and rapid manner to deal with malwares.

Furthermore, we propose two MLP architectures presented in two use cases:

- CASE 1: MLP architecture with one hidden layer is used for classifier malware. Then, we variate number of units in this unique hidden layer.
- CASE 2: Another MLP architecture using 2 hidden layer is performed to classify malwares into their corresponding families. Then, we evaluate the model by varying the number of units in each hidden layer.

The proposed MLP architecture of the case 1 is illustrated in the scheme of Fig. 3, where input layer presents malwares images features. In other words, the input layer is the GIST descriptor vector of malware images. Then, we add one hidden layer. And we have the result as output layer.

In the same manner, we perform the case 2 which concerns the MLP classifier architecture using 2 hidden layers (Fig. 4). As mentioned before, the input layer regards malware features, and the output layer presents the classification result.

For each case, we change the number of units in every hidden layer, and we vary the activation function with all instances. Activation functions used are: identity, logistic, relu and tanh.

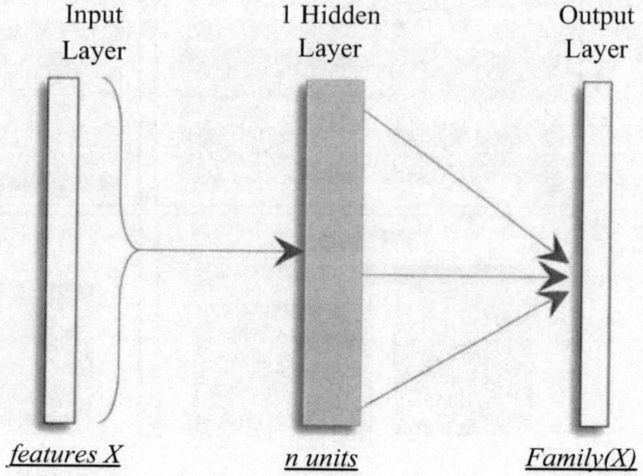

Fig. 3 MLP model architecture with one hidden layer (Case 1)

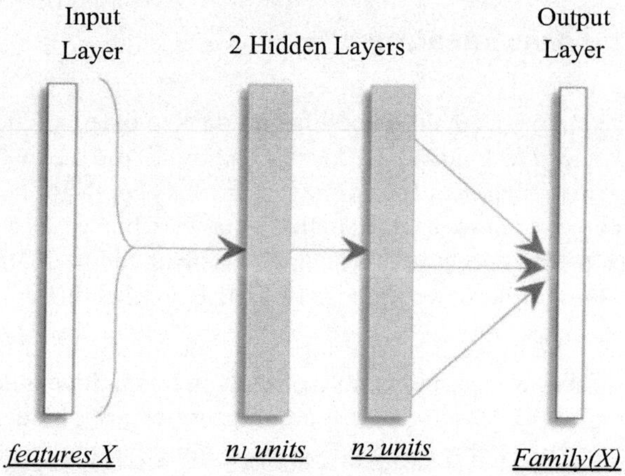

Fig. 4 MLP model architecture with two hidden layers (Case 2)

6 Results and Discussion

In this section, we present and discuss the obtained results. As the experimentation, the results are given for each case: case 1 using MLP with 1 hidden layer and case 2 using MLP classifier with 2 hidden layers.

The neurons for our architecture were set in range of 1 to 101 units in a one-hidden layered architecture. Table 2 presents the results that we obtain. It gives accuracy variation each time we change the units in the hidden layer, with different activation functions. Likewise, Fig. 5 shows more clearly the ups and downs of the accuracy while increasing units in the hidden layer using the four mentioned activation functions.

Results of case 1 show that the MLP malware classifier using 1 hidden layer reached a highest accuracy of 0.9745. This highest value is obtained when we use

Table 2 Accuracy by units' variation using different activation function for one hidden layer architecture

Number of neurons	Accuracy			
	Identity	Logistic	Relu	tanh
1	0.4980	0.3983	0.3222	0.4133
11	0.9623	0.9010	0.9595	0.9584
21	0.9684	0.9555	0.9638	0.9656
31	0.9699	0.9591	0.9656	0.9738
41	0.9702	0.9591	0.9673	0.9702
51	0.9720	0.9612	0.9656	0.9720
61	0.9709	0.9666	0.9677	0.9724
71	0.9727	0.9648	0.9691	0.9738
81	0.9720	0.9663	0.9706	0.9738
91	0.9699	0.9677	0.9720	0.9717
101	0.9731	0.9670	0.9702	0.9745

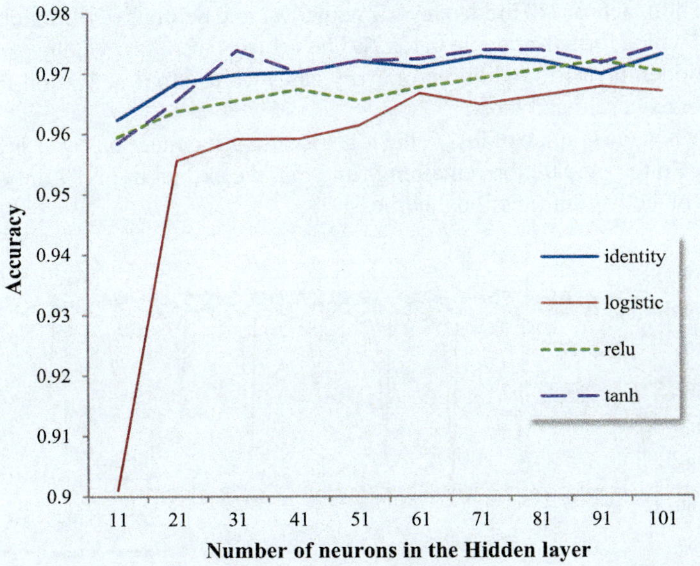

Fig. 5 Representation of accuracy during units' variation in case 1

101 neurons in the single hidden layer architecture with the hyperbolic activation function.

Moreover, we can see that the classifier had a smaller accuracy when using few units in the hidden layer. For instance, when using only 11 neurons, the accuracy

Table 3 Important results in case 1

Accuracy		Activation function	Neurons
Best	0.974524578	tanh	101
Worse	0.900968784	logistic	11

Table 4 Important results in case 2: MLP with two hidden layers

Accuracy		Activation function	Neurons
Best	0.976059813	tanh	[30, 100]
Worse	0.821672049	logistic	[10, 10]

obtained is 0.9009 with logistic sigmoid function. A summary of highest and lowest accuracy obtained is given in Table 3.

Overall, the variation of activation function shows significant gap. In our case, the malware classification using visualization technique, and using these combinations of parameters, we can say that the most suitable activation function is tanh.

Then again, for the second experimental case, we use MLP malware classifier using two hidden layers. Results are presented in Table 4, Figs. 6 and 7. In this case, we variate both neurons of hidden layer 1 named $n1$ and neurons of the hidden layer 2 named $n2$, which form the couple $[n1, n2]$. The neurons were set in range of 10 to 100 for both hidden layers. In addition, in every case, we use four activation functions: identity, logistic, relu and tanh.

As shown in the graphic of Fig. 6, the logistic activation function curve is below all the others. So to compare the remaining function, we extract in Fig. 7 only identity, relu and tanh activation functions' curves.

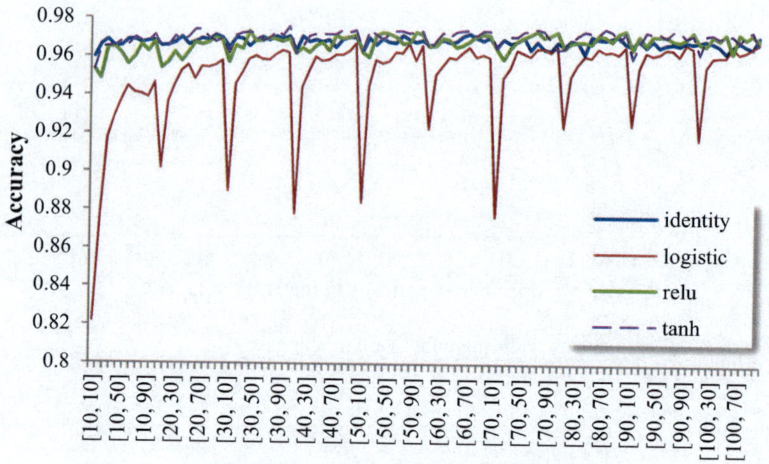

Fig. 6 Representation of accuracy during units' variation in case 2 with all activation functions

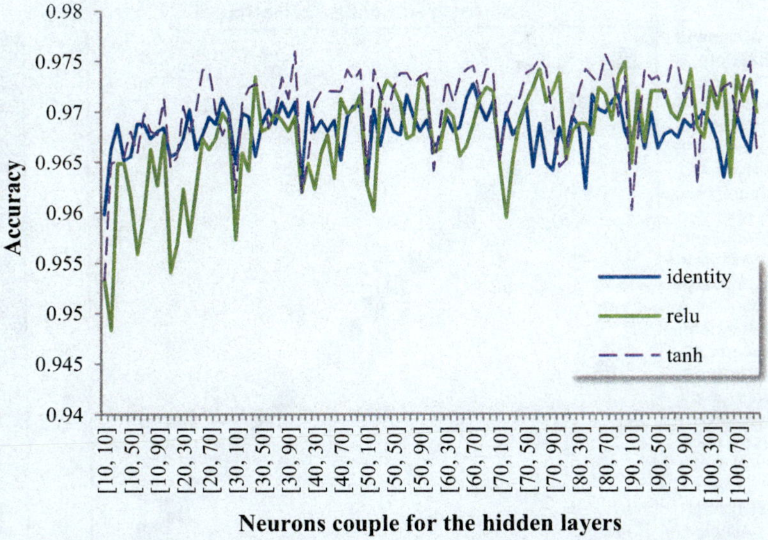

Fig. 7 Representation of accuracy during units' variation in case 2, with the three activation functions

It is obvious that activation functions presented in Fig. 6 give close accuracy values. On top, we find the tanh curve in most cases. The highest accuracy reached is 0.9760 when we use: 30 neurons in hidden layer 1, 100 neurons in hidden layer 2 and the hyperbolic tan as activation function. The lowest accuracy is obtained when we use relu activation function. A summary of highest and lowest accuracy obtained in case 2 is given in Table 4.

To sum up, the proposed malware classifier based on MLP network and visualization technique gives great accuracy when we use architecture of two hidden layers and the hyperbolic tan activation function. The classifier reached 0.9760 accuracy value with a network composed of 30 neurons in the first hidden layer, and 100 neurons in the second hidden layer.

Confusion matrix in Fig. 8 presents more clearly the result of this specific case. The classifier failed totally in the family 'Autorun.K,' based on test data. Generally, the MLP malware classifier performs correctly.

To summarize briefly the above-obtained results, we present in Table 5 the detailed MLP architecture and hyperparameters as well as other's evaluation metrics.

Let's compare results we got with literature review. We present in Table 6 a summary of related works showing the used classifier and the resulting accuracy sorted.

Obtained results depend on several factors such as the nature and structure of the database, features, classifier algorithm and others hyper parameters. That is why you can see very different results in [6] and [8] while using CNN. In addition, KNN classifier in [1] outperforms the current MLP classifier with small difference.

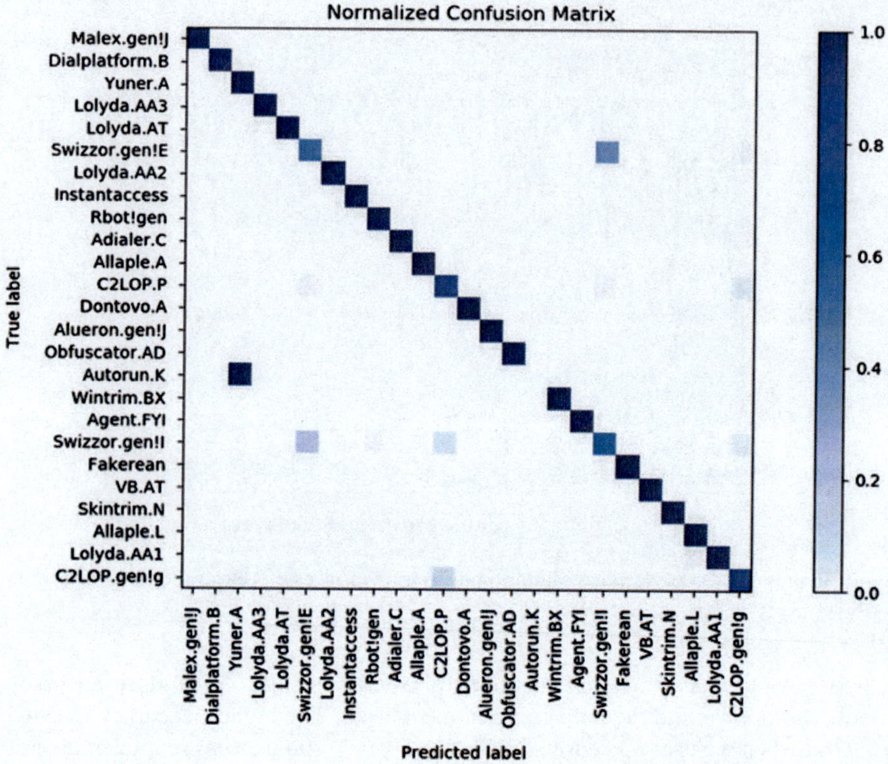

Fig. 8 Confusion matrix of the best case

Table 5 MLP architecture, hyper parameters and evaluation metrics

Model architecture and hyper parameters	
Number of hidden layers	2
Neurons per hidden layer	[30, 100]
Activation function	tanh
Solver	adam
Learning rate	0.001 (constant)
Other evaluation metrics	
Accuracy	0.976089343
F1-score macro	0.912807811
F1-score weighted	0.967928184
Hamming loss	0.026910657
Recall macro	0.91549644
Precision macro	0.910930803
Precision weighted	0.961172179

Table 6 Summary of related works

Reference	Method	Accuracy (%)
[6]	CNN	98.52
[1]	KNN + visualization	97.92
ours	*MLP + visualization*	*97.60*
[9]	Feed forward ANN	96.35
[8]	CNN	93.92
[7]	Deep residual network	86

7 Conclusion and Perspective

In conclusion, this paper proposed a malware classifier, and its main goal is to classify malware samples into their families effectively in short timing. For this purpose, we performed a multi-layer perceptron classifier using malware visualization technique based on grayscale images. In order to have the highest accuracy, we made several experimentations include variations of: hidden layers, neurons in each hidden layer and activation functions. At the end, the highest accuracy was reached when using two hidden layers with, respectively, 30 and 100 neurons each, and by using the hyperbolic tan as activation function. The performance of our classifier was evaluated using accuracy which gives 0.9760 and also confusion matrix.

As future work, we intend to use the proposed classifier to defend against malware attacks in real environments following a special process. Also, to improve the classifier accuracy, we could use a hybrid solution. In other words, we suppose that a combination of machine learning and deep learning algorithms could give better results.

Acknowledgements We acknowledge financial support for this research from the "Centre National pour la Recherche Scientifique et Technique", CNRST, Morocco.

References

1. Ben Abdel Ouahab, I., et al.: Classification of grayscale malware images using the K-nearest neighbor algorithm. In: Ben Ahmed, M., et al. (eds.) Innovations in Smart Cities Applications, 3rd edn., pp. 1038–1050. Springer International Publishing, Cham (2020). https://doi.org/10.1007/978-3-030-37629-1_75
2. Ben Abdel Ouahab, I. et al.: Speedy and efficient malwares images classifier using reduced GIST features for a new defense guide. In: Proceedings of the 3rd International Conference on Networking, Information Systems & Security. Association for Computing Machinery, Marrakech, Morocco (2020). https://doi.org/10.1145/3386723.3387839
3. Bishop, C.M., Bishop, P., of N.C.C.M.: Neural Networks for Pattern Recognition. Clarendon Press (1995)
4. Galal, H.S., et al.: Behavior-based features model for malware detection. J. Comput. Virol. Hack. Tech. **12**(2), 59–67 (2016). https://doi.org/10.1007/s11416-015-0244-0

5. Jerlin, M.A., Marimuthu, K.: A new malware detection system using machine learning techniques for API call sequences. J. Appl. Secur. Res. **13**(1), 45–62 (2018). https://doi.org/10.1080/19361610.2018.1387734
6. Kalash, M., et al.: Malware classification with deep convolutional neural networks. In: 2018 9th IFIP International Conference on New Technologies, Mobility and Security (NTMS), pp. 1–5 (2018). https://doi.org/10.1109/NTMS.2018.8328749
7. Lu, Y., et al.: Deep Learning Based Malware Classification using Deep Residual Network, p. 7 (2019)
8. Luo, J., Lo, D.C.: Malware image classification using machine learning with local binary pattern. In: 2017 IEEE International Conference on Big Data (Big Data), pp. 4664–4667 (2017). https://doi.org/10.1109/BigData.2017.8258512
9. Makandar, A., Patrot, A.: Malware analysis and classification using artificial neural network. In: 2015 International Conference on Trends in Automation, Communications and Computing Technology (I-TACT-15), pp. 1–6 (2015). https://doi.org/10.1109/ITACT.2015.7492653
10. Nataraj, L., et al.: Malware images: visualization and automatic classification. In: Proceedings of the 8th International Symposium on Visualization for Cyber Security—VizSec '11, pp. 1–7. ACM Press, Pittsburgh, Pennsylvania (2011). https://doi.org/10.1145/2016904.2016908
11. Oliva, A., Torralba, A.: Modeling the shape of the scene: a holistic representation of the spatial envelope. Int. J. Comput. Vision **42**(3), 145–175 (2001). https://doi.org/10.1023/A:1011139631724
12. Sikos, L.F.: AI in Cybersecurity. Springer (2018)
13. Soufyane, A., et al.: An intelligent chatbot using NLP and TF-IDF algorithm for text understanding applied to the medical field. In: Ben Ahmed, M., et al. (eds.) Emerging Trends in ICT for Sustainable Development, pp. 3–10. Springer International Publishing, Cham (2021). https://doi.org/10.1007/978-3-030-53440-0_1
14. Souri, A., Hosseini, R.: A state-of-the-art survey of malware detection approaches using data mining techniques. Hum. Cent. Comput. Inf. Sci. **8**(1), 3 (2018). https://doi.org/10.1186/s13673-018-0125-x

Preserving Privacy in a Smart Healthcare System Based on IoT

Rabie Barhoun and Maryam Ed-daibouni

Abstract Internet of things (IoT) refers to a new technology made up of many heterogeneous resources that are interconnected, controlled, monitored, and analysed through the Internet. These connected devices and resources are part of the smart environment. Recently, many IT organizations and companies adopted this technology. Today, modern societies are highly dependent on technology information and communications, especially, based on IoT which becomes an emerging IT technology being used increasingly in industry, government and healthcare. In a smart healthcare system based on IoT, there are usually many distributed care units with shared resources. One of the most challenging tasks in a smart healthcare system based on IoT is the preservation of privacy. In fact, protecting personal information from unauthorized persons is considered an essential requirement for many smart systems. The traditional access control models do not support privacy requirements and inflexible. In this paper, we propose a new access control model, called activity and attribute-based access control (ABAC), which can efficiently preserving privacy in a smart healthcare system based on IoT.

1 Introduction

Today's smart environments are highly dependent on information and communication technologies based on the Internet of Things, which is becoming an emerging technology and increasingly used in smart industry, smart government, smart education and smart healthcare. In a recent report [1], considering that "things" in connected smart environments are growing enormously, it is predicted that there will be up to billions of devices by 2025.

In a smart healthcare environment, IoT "things" are computationally constrained devices, such as sensors, that can detect, measure and extend connectivity between systems and medical actors via the Internet in an omnipresent manner. Figure 1

R. Barhoun (✉) · M. Ed-daibouni
Faculty of Science Ben M'sik, University Hassan II, Casablanca, Morocco

M. Ben Ahmed et al. (eds.), *Networking, Intelligent Systems and Security*,
Smart Innovation, Systems and Technologies 237,
https://doi.org/10.1007/978-981-16-3637-0_33

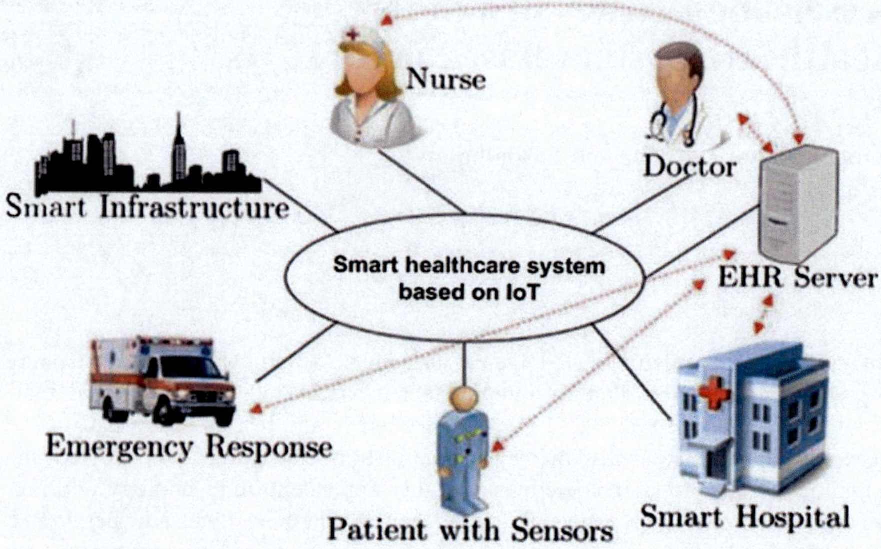

Fig. 1 Smart healthcare system based on IoT

presents a typical smart healthcare environment, where several heterogeneous objects/nodes are installed.

In a smart healthcare system based on IoT, there are usually many distributed care units with shared resources. These generate many problems such as degradation or loss of control over access to information and privacy issue. One of the most challenging tasks in a smart healthcare system based on IoT is privacy concerned with protection of the personal information (PI) against unauthorized, which is considered a critical requirement for many smart environments. The traditional access control models do not support privacy requirements and inflexible. In this paper, we first present an overview of access control models in a distributed smart environment to retrieve a suitable model for such an environment based on IoT. Secondly, we propose a new access control model, called activity-attribute-based access control (AABAC), which can effectively enhance privacy in the smart healthcare system and produce a more perfect and flexible mechanism of access control. In this model, the concept of activity is crucial for the preservation of privacy. The principle of our approach is that the tasks performed, and actions are determined by the activity purpose, and therefore, access outside the purpose will be considered a violation of privacy. For example, access to a patient's medical record during a consultation activity, each access to this record outside the purpose activity will be considered a violation of the patient's privacy.

The rest of the paper is organized as follows: Sect. 2 introduced an overview of the access control models. In Sect. 3, we have a comparative analysis of access control models for distributed smart environments. In Sect. 4, we propose an improvement of ABAC model for a smart healthcare system based on IoT by integrating activity concepts into the basic ABAC model. Section 5 presents the design and implementation of our AABAC model in smart healthcare systems based on IoT, and Sect. 6 concludes this paper.

2 Access Control

In computer security, access control is the enforcement of a predefined access policy on objects to define the actions of an entity to only those requests and object to which it is entitled. A fine-grained access control system gives control over methods, objects and collections of resources. The dissertation thesis [2] presents an access control policy for the IoT environment that applies a fine-grained access control model at both the function and object-level to ensure that users with permission to call a function also have appropriate access to target objects that are required to successfully perform a particular task.

The access control is the mechanism by which services decide whether to accept or deny requests according to the permissions affected to users. There are four parts to the problem:

A. **Identification**: Specifying a responsible party for actions. A responsible party may be a person or a non-person entity (NPE), such as a computer or a router. We will use the term user to cover both cases.
B. **Authentication**: Used to prove the right to use an identity, take on a role or prove possession of one or more attributes.
C. **Authorization**: Paper [3] express the access policy by explicitly granting a right.
D. **Access Decision**: Applying some combination of the other three to decide whether a request should be honored.

Many access control models are used, including mandatory access control (MAC), discretionary access control (DAC), identity-based access control (IBAC), role-based access control (RBAC) and attribute-based access control (ABAC).

2.1 Dictionary Access Control (DAC)

DAC is a method of restricting access to an object based on the entity identity (e.g., user, process and group). This type of access control is called discretionary because the entity with a certain privilege can pass its privileges or security attributes to another entity. This results in, however, the problem of loss of confidentiality of information. DAC is widely used in many networks and operating systems.

2.2 Mandatory Access Control (DAC)

MAC is a method of constraining an entity from accessing or performing operations on a resource, based on predefined security attributes or labels assigned to the entity and the target object.

2.3 Identity-Based Access Control (IBAC)

Access control based on the identity (IBAC—identity-based access control) is among the first access control models. This model introduces the basic concepts of subject, action and object. The objective of this model IBAC is to control direct access of subjects to objects using actions. This control is based on the identity of the subject and the object identifier.

2.4 Role-Based Access Control (RBAC)

The papers [4–7] proved that the RBAC model is versatile and conforms closely to the organizational model used in firms. RBAC meets this requirement by separating users from roles. Work [8] discusses that access rights are given to roles, and roles are assigned to users. Here, the role combines users and privileges. Roles are created for various job functions, and users are assigned roles based on their qualifications and responsibilities. RBAC is thus more scalable than user-based security specifications and greatly reduces the cost and administrative overhead.

This model is very simple and easy to use. And, it is considered to be the best model of access control for the local domain. The roles are assigned to the user statically by the security administrator that is not preferable in a dynamic environment. Also, it is difficult to change the privilege of the user without changing the role of the user. Furthermore, RBAC becomes problematic in a distributed and dynamic environment and has no delegation models which are required in such an environment. Also, work [4] discusses that problem of "Role Explosion" that can arise when used to support dynamic attributes in large organizations, in thousands of separate roles for different collections of permissions.

2.5 Attribute-Based Access Control (ABAC)

In traditional policies, the users can obtain their privileges through roles or directly, but it is possible that users can give certain privileges that they do not really need. This is contradictory to the least privilege principle which requires a subject that must be able to access only the information and resources that are necessary for its purpose. Currently, how to assign privileges to subjects so as to achieve this principle is still not solved. Papers [9, 10] present that ABAC is a recent policy that has drawn particular attention; its principle of decision is based on taking into account the attributes of different actors (subject, object and environmental conditions) before giving access to resources.

The overview of AABAC model is shown in Fig. 2, the permissions containing the combination of an object and operations, where the subject access to the object according to certain conditions. The operation describes the instructions to execute on the objects. Access rights can be defined in a subject attribute and permission. Paper [9] explains that ABAC model can dynamically assign permissions to subjects and objects. ABAC uses subject, object and their environmental attributes. Papers [11, 12] show that in model ABAC before using the attributes to make an access control decision, the integrity and validation of these attribute resource is verified.

The ABAC model is very flexible and supportive in an environment which is large, open, distributed, sharable and collaborative and where the numbers of users are very high and most of the users are unknown before and the roles of the users are statically or not defined in advance. Furthermore, it supports the global agreement features such as the user attributes that are provided in one domain can be forwarded to the other domain at the point of domain-to-domain interaction.

This model has the high complexity due to the specification and maintenance of the policies. In addition, there is a problem of mismatching and confusing attributes especially when those attributes provided by the user do not necessarily match to those used by the service provider of a web-based system or service. Furthermore, it increases the privacy, flexibility, sharing and global agreement and provides the interoperability among several service providers which can use these attributes data dynamically and can decide upon user rights.

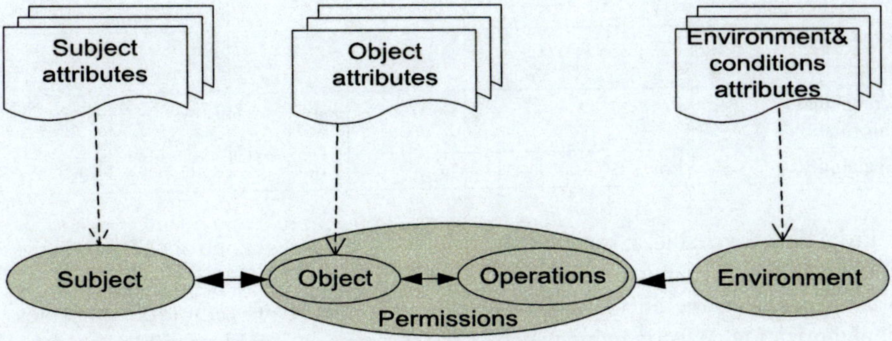

Fig. 2 Attribute-based access control

3 Comparative Analysis of Access Control on Distributed Smart Environment

In this section, we present a comparative analysis of MAC, DAC, IBAC, RBAC, ABAC and AABAC with some point of issues. This comparison is based on different factors, which will help us to choose the most efficient access control model suitable for a smart environment based on IoT. Those factors are dynamicity, distributed systems, global agreement, flexibility, simplicity, authorization decision, granularity, manageability, trust, changing privileges, policies specification and maintenance, role explosion, contextual information and scalability.

The following table summarizes this comparison:

	MAC	DAC	RBAC	IBAC	ABAC
Dynamicity	No	No	No	No	Yes
Distributed systems	No exactly	No exactly	No exactly	No exactly	Yes
Global agreement	No	No	No	No	Yes
Flexibility	No	No	No	No	Yes
Simplicity	Yes	Yes	Yes	Yes	No
Authorization decision	Locally	Locally	Locally	Locally	Globally
Granularity	Low	Low	Low	Low	High
Manageability	Simple	Simple	Simple	Complex	Complex
Trust	Locally	Locally	Locally	Locally	Globally
Changing privileges	Simple individual user cannot change access rules	Simple	Complex	Simple	Simple
Policies specification and maintenance	Low	Low	Simple	Low	Complex
Role explosion problem	No	No	Yes	No	No
Contextual information	Low	Low	Low	Low	High
Scalability	Low	Low	Low	Low	Yes

From the above table, it is clear that attribute-based access control (ABAC) is the most suitable for dynamic distributed smart environment. This model granted access to users based on the attributes of the requesting user. It uses multiple attributes for authorization decision, which enables the system to be highly flexible, scalable, interoperable, and multifunctional access control that may deal with diverse security requirements in distributed environment based on IoT.

4 Our Proposed AABAC Model Based on IoT

The ABAC model has drawbacks with reference to privacy concerns. Indeed, because of the descriptive nature of subject attributes, implementation of attributes by sharing capabilities causes a problem of increasing the risk of privacy violation of personally identifiable information by dint of involuntary exposure of attribute data to untrusted third parties or aggregation of sensitive information in environments less protected. A second consideration is that releasing attributes to the policy evaluating engine is a sensitive activity as the third party may not be trusted.

According to security constraints, the principle of least privilege as an access control policy allows the assignment of least rights to different actors. This principle is important in preserving privacy and sensitive resources in the design of an access control policy in a smart environment. In our previous work [13], we proposed a model called medical activity-attribute-based access control model (MA-ABAC) which extends the functionality of ABAC by introducing the concept of medical activity, defined by the purposes of treatment, to meet privacy concerns in a collaborative healthcare environment. In MA-ABAC model, if a privilege is used outside the purpose of the activity, then it is considered a violation of the principle of least privilege and then a violation of privacy. To take into account the smart environment based on IoT, we propose the activity-attribute-based access control (AABAC) model shown in Fig. 3.

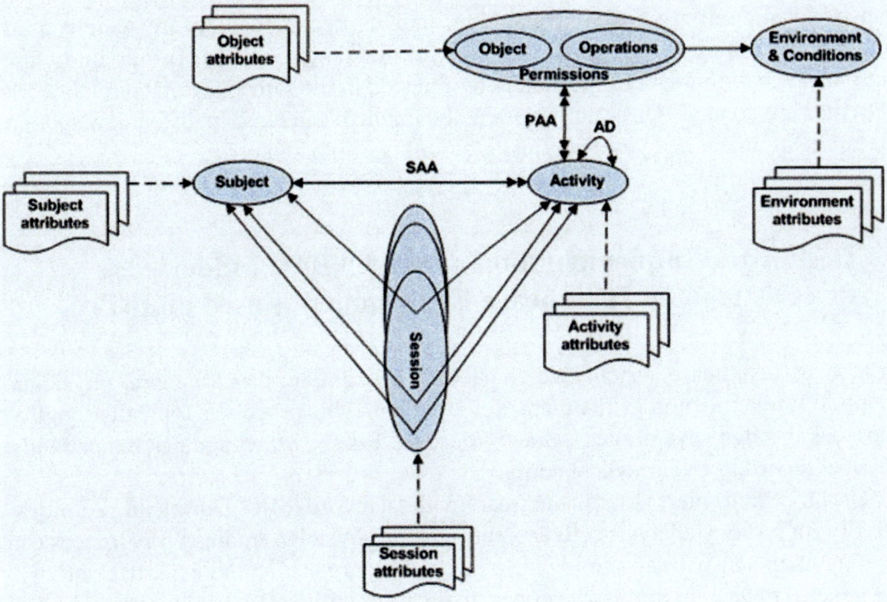

Fig. 3 Activity-attribute-based access control (AABAC) model for smart healthcare system based on IoT

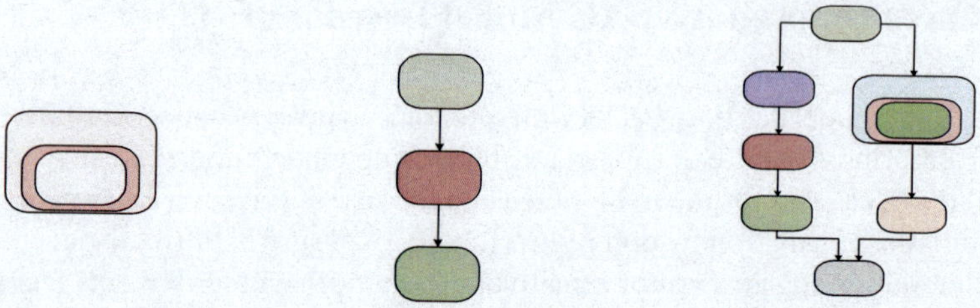

(b) Hierarchical structure of activities (a) Linear structure of activities (c) Hybrid structure of activities

Fig. 4 Different structures of activities based on IoT

In this design, activity (A) is an abstraction of the collaborative works, which provides a scope for resources, privileges and roles defined by a medical activity purpose (P) in a medical unit, in which the participants (collaborators) realize their tasks in order to achieve the medical activity purposes. Indeed, according to our approach, the smart healthcare system allows a smart group of medical actors to work together in order to treat a patient admitted to a hospital.

In a smart collaborative healthcare system, many medical activities exist, and new activities are created. An activity can also span across multiple environments based on IoT. In most cases, the distributed collaborative environments need to be structured hierarchically. A medical collaboration specification needs to support the decomposition of tasks into activities/sub-activities and planning how such activities can perform. Activity concept, planning, and its decomposition are widely used in collaboration systems, specifically in workflow environments. In our study, the activities/sub-activities can be classified or nested in the same way that human work activities are created. These activities can be mainly structured in different manners: hierarchical, linear or hybrid as illustrated in Fig. 4a, b and c.

5 Design and Implementation of the AABAC Model in a Distributed Healthcare Environment Based on IoT

In a smart healthcare system, we consider, for example, that a general physician trying to find a diagnosis (treatment) for a patient. This process is typically lengthy and goes through several medical activities, see Fig. 5, which updates the patient's medical record in each medical activity.

The physician enters notes in the smart medical record (MR). During the radiology activity (or biology analysis activity), the general physician studies x-ray images (or biology analysis) with a radiologist (or with a biologist). At the expertise activity, the general physician discusses proper medication with colleagues while browsing medicine catalogs. Otherwise, it is necessary to start another medical activity, as shown in Fig. 6.

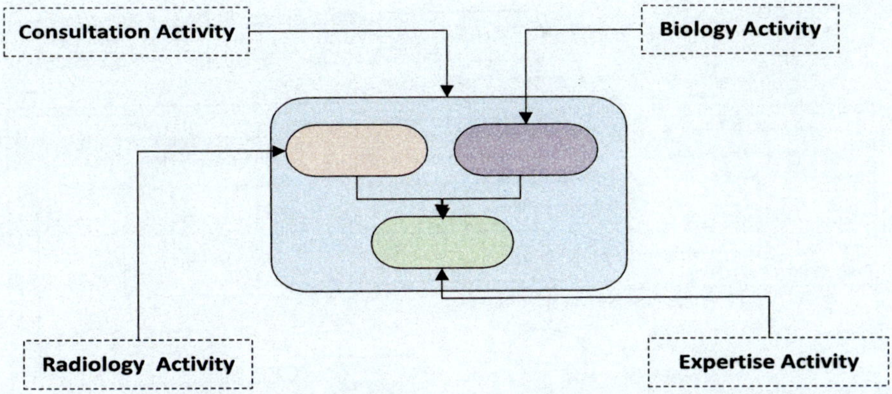

Fig. 5 Hybrid medical activities of general diagnosis in smart healthcare based on IoT

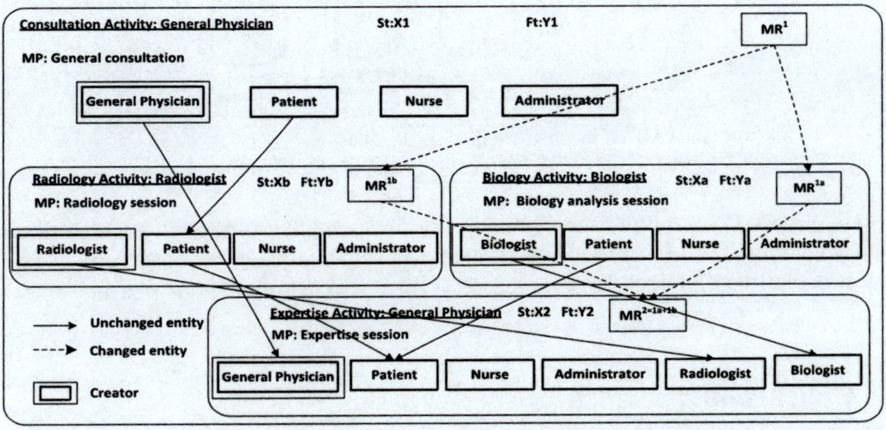

Fig. 6 The activities of general diagnosis process of a patient in smart healthcare

The proposed architecture of our mechanism for preserving privacy and protected resources in smart healthcare system is illustrated in Fig. 7.

When initiating an activity, the activity decision module (ADM) retrieves the attributes of the triggered activity from policy information point (PIP) entity. The policy activity server will generate an XML file containing all the permissions for the activity created. The ADM manager will decide whether an action (executed in the medical activity) is allowed or not by consulting the XML file relating to the active medical activity on the one hand, and on the other hand based, on other attributes relating to subject, environments and object retrieved from PIP entity, respectively, by the managers: subject manager module (SMM), environment manager module (EMM) and object manager module (OMM). The administration activity policy (AAP) is used to define medical activities and their policies, see Fig. 8.

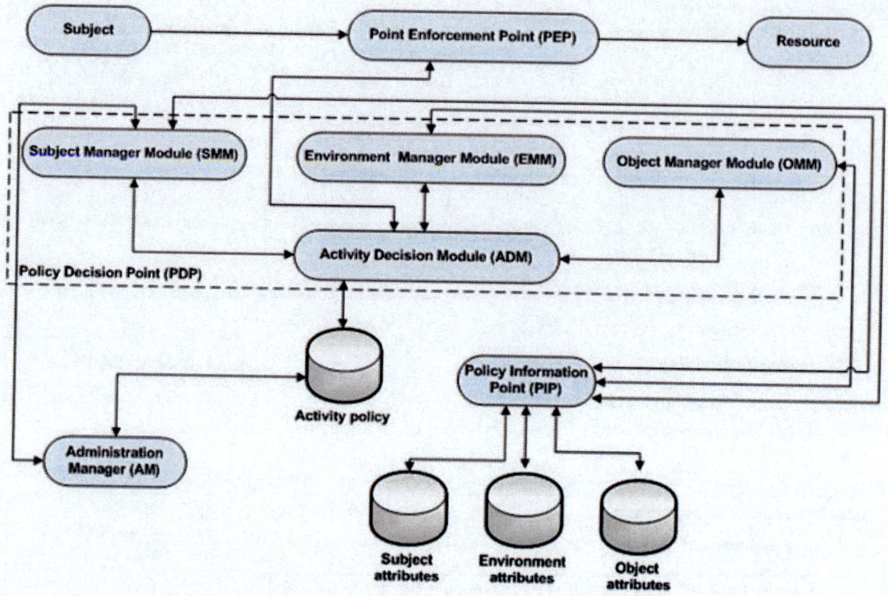

Fig. 7 Architecture of AABAC mechanism in smart healthcare system

The prototype developed was based on the Java language and API framework on Linux Ubuntu 16.04 (32 bits). The choice of the Java language is based on the fact that it is platform-independent; therefore, it runs on a heterogeneous platform.

6 Conclusion

In this paper, we have presented a critical review of the main access control models. We have examined the advantages and disadvantages of each model. This investigation led us to choose the ABAC model as the most suitable model for all distributed smart environments, although this model presents some drawbacks such as privacy violation. We then improved this weakness by proposing a new model called AABAC based on activity concept. In this model, if a privilege is used outside the activity purpose, then this access is considered as a violation of privacy. Our access control model not only guarantees the properties of access control but also preserving privacy in IoT environment.

We presented the design of our implementation in the smart healthcare system based on IoT. In the current implementation, we focused on the medical activity concept; this concept encompasses healthcare requirements as well as the implementation of the principle of least privileges, and then the preservation of privacy. We believe that our approach can be adapted to support any distributed smart environment.

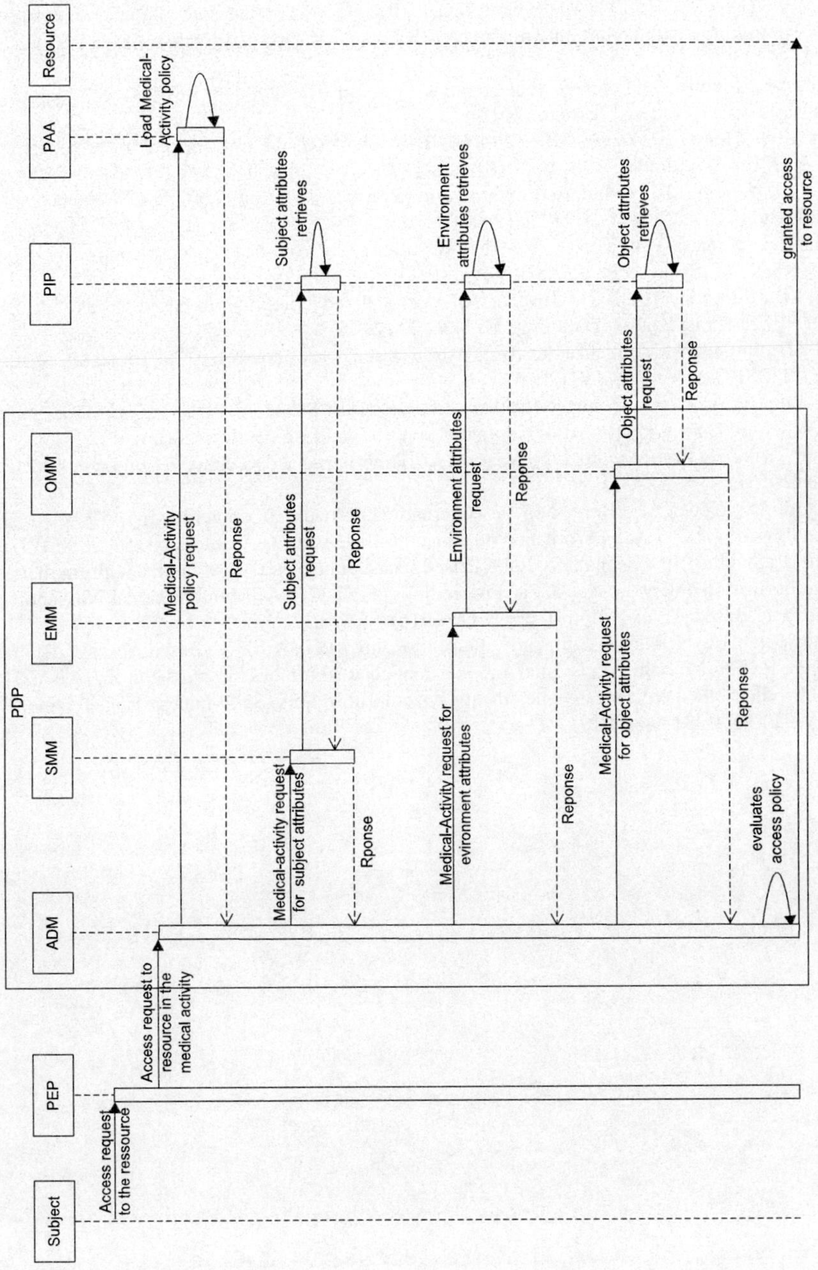

Fig. 8 The AABAC authorization process—diagram sequence

References

1. Lee, Y.T., Hsiao, W.H., Lin, Y.S., Chou, S.C.T.: Privacy-preserving data analytics in cloud-based smart home with community hierarchy. IEEE Trans. Consum. Electron. **63**(2), 200–207 (2017)
2. Faraji, M.: Identity and access management in multi-tier cloud infrastructure. Doctoral Dissertation, University of Toronto (2013)
3. Karp, A.H., Haury, H., Davis, M.H.: From ABAC to ZBAC: the evolution of access control models. J. Inf. Warfare **9**(2), 38–46 (2010)
4. Ahn, G.J., Sandhu, R.: Role-based authorization constraints specification. ACM Trans. Inf. Syst. Secur. (TISSEC) **3**(4), 207–226 (2000)
5. Bertino, E., Bonatti, P.A., Ferrari, E.: TRBAC: a temporal role-based access control model. ACM Trans. Inf. Syst. Secur. (TISSEC) **4**(3), 191–233 (2001)
6. Joshi, J.B., Bertino, E., Latif, U., Ghafoor, A.: A generalized temporal role-based access control model. IEEE Trans. Knowl. Data Eng. **17**(1), 4–23 (2005)
7. Li, N., Tripunitara, M.V.: Security analysis in role-based access control. ACM Trans. Inf. Syst. Secur. (TISSEC) **9**(4), 391–420 (2006)
8. Kalajainen, T.: An Access Control Model in a Semantic Data Structure: Case Process Modelling of a Bleaching Line. Department of Computer Science and Engineering (2007)
9. Hu, V.C., Kuhn, D.R., Ferraiolo, D.F., Voas, J.: Attribute-based access control. Computer **48**(2), 85–88 (2015)
10. Brossard, D., Gebel, G., Berg, M.: A systematic approach to implementing ABAC. In: Proceedings of the 2nd ACM Workshop on Attribute-Based Access Control, pp. 53–59 (2017)
11. Biswas, P., Sandhu, R., Krishnan, R.: Label-based access control, an ABAC model with enumerated authorization policy. In: ABAC '16 Proceedings of the 2016 ACM International Workshop on Attribute Based Access Control, pp. 1–12 (2016)
12. Mukherjee, S., Ray, I., Ray, I., Shirazi, H., Ong, T., Kahn, M.G.: Attribute Based Access (2017)
13. Barhoun, R., Ed-daibouni, M., Namir, A.: An extended attribute-based access control (ABAC) model for distributed collaborative healthcare system. Int. J. Serv. Sci. Manage. Eng. Technol. (IJSSMET) **10**(4), 81–94 (2019)

Smart Digital Learning

Extracting Learner's Model Variables for Dynamic Grouping System

Noureddine Gouasmi, Mahnane Lamia, and Yassine Lafifi

Abstract Collaborative learning is of great help in improving the training of learners in e-learning systems. But the efficiency of collaborative learning depends on the quality of the learning groups formed. The question is therefore: Which variables should be used to group learners in order to reach the most effective collaboration between them? In this paper, we present a literature review of variables describing learner's model, which can be used for grouping learners in collaborative e-learning systems. Variables are ranged in 11 classes, and possible associations between variables are considered, after extracting frequent patterns from variables using FP-growth algorithm, and generating association rules. The results show that four variables from learner's are mostly used with association with other variables. These variables are: "Number of accesses to the platform", "Number of complete activities", "Platform time" and "Final Grades".

1 Introduction

Collaboration is the act of working with one or more other people. It involves a cyclical process of renegotiation. Collaboration evolves as parties interact over time [12].

In the learning domain, "collaborative learning" is any learning activity carried out by members of a group of learners having a common objective, in order to succeed in their learning. Each learner being a source of information, motivation, interaction,

N. Gouasmi (✉) · Y. Lafifi
Labstic, 08 Mai 45 University, P.O. Box 401, 24000 Guelma, Algeria
e-mail: gouasmi.noureddine@univ-guelma.dz

Y. Lafifi
e-mail: lafifi.yacine@univ-guelma.dz

M. Lamia
LRS Laboratory, University of Badji Mokhtar, P.O. Box 12, 23000 Annaba, Algeria
e-mail: mahnane_lamia@yahoo.fr

M. Ben Ahmed et al. (eds.), *Networking, Intelligent Systems and Security*,
Smart Innovation, Systems and Technologies 237,
https://doi.org/10.1007/978-981-16-3637-0_34

mutual assistance and each benefiting from the contributions of others and the help of a trainer to facilitate individual and collective learning [14].

Collaborative learning systems help learners to work in groups. It allows document sharing between members and provides tools to upload, download and archive several types of documents [11]. It also provides means for exchanging information between learners of the same group [7], enables learners to develop cognitive capacity and knowledge necessary for the development of collaborative skills [3], gives learner the possibility of managing his time and providing a wider choice of activities more suited to their needs and interests, increases reflection time during collaborative online learning and allows to combine learner's contributions [14].

For a collaboration to be effective, it is important to choose the "best" members of a collaborative group. According to Bekele [2], the process of grouping learners is focused on three key points: how to assign a learner to a group, size of groups and whether the group is heterogeneous or homogeneous [15].

The paper is organized in two main parts. The first one shows variables used to describe learner's model, which can be used for grouping learners. The second part presents the analyses of learner's model variable, by describing the methodology used in this work, and after, by showing results of data analysis.

2 Variables Used to Describe Learner's Model

This research analysed 63 primary studies published between 2001 and 2019. After analysis of the 63 articles approved, it was possible to find 76 variables. The variables were categorized into seven layers for a better understanding, and each layer represents only the subject that variable approaches. A variable can be categorized in more than one layer. For example, the variable "area of employment" could fit in layers "User's Information" and "Psychological"; however, in this work each variable will be presented in only one category. The layers are detailed in Table 1.

Table 1 Variables' layers

Layer	Number of variables
Communication	14
Access	8
Activities	8
Actions	16
Time	10
User's information	8
Psychological	12

Table 2 Variables found that are referring to communication

Sublayer	Variables
General	Which communication resource was used
	Communication level
Forum	Participation
	Total amount of posts
	Posts per forum
	Total amount of replies
	Number of access
	Access frequency
	Added files
	Content of the message
Chat	Number of messages
	People the student interacted with
Comments	Number of comments made
	Number of comments read

- **Communication**

The first layer, which addresses communication, represents all the data found that refers to some conversation, whether by posts in forums or by direct messages, this communication can also be considered between students or between the student and the teacher.

Communication is an essential element that must be analysed. According to Takaffoli, 2012 [9], in order to fully appreciate the participation of students, we need to understand their patterns of interactions and answer questions like who is involved in each discussion, who is the active/peripheral participant in a discussion thread.

The number of interactions between students or teachers can represent how engaged the student is. According to Zhang [16], "for using the discussion board, the total number of postings (all were qualified ones) by each student on the discussion board was used as the measure for participation or communication in class. This measure reflects how involved a student was in his or her learning."

During the analysis, 14 variables emerged, referring to communication. These variables divided into four sublayers relating to the origin of this communication, and these variables are presented in Table 2.

- **Access**

The second layer is referring to access. Social learning systems provide data about students' access, and it means that it is possible to know when students were online and which pages they were accessing. It is important to detect what are the primary students' access pages to understand what type of content they are accessing.

Table 3 Variables found that are referring to access

Layer	Variables
Access	Number of accesses per day to the platform
	Number of accesses to the platform
	Access platforms in specific sessions
	First and last file accessed
	Single views to specific pages
	Number of pages visited
	Number of views per page
	Time viewing specific pages

Table 4 Variables found related to the activities

Layer	Variables
Activities	Number of complete activities
	Answers given in the activity
	Tests done
	Reading of activities
	Grades in activities
	Time spent on activities
	Activities done correctly
	Number of attempts on the same activity

> The number of visualizations is one order of magnitude above creations and updates, which in turn are on order above deletions. […]. This situation clearly derives from two facts: the first is that for accessing every creation, update or delete page, we need first to visualize it; the second, relates to the natural curiosity of people, associated with the fear to 'act' in spite of just 'observe'. (Figueira 2017) [4]

Eight variables were found, referring to access. These variables are presented in the following Table 3.

- **Activities**

As in the face-to-face learning mode, the activities done by the student are a way to measure his interest, performance and learning. Social learning systems provide information to the teacher about the proposed activities. It is possible to analyse not only activities done by students but also the period in which the student has delivered it and whether it is delayed. This layer presents eight variables that can be found in Table 4.

All the record of actions made by the user within the platform is detected through logs, clicks or events. This information allows identifying how the student behaves inside the platform.

Figueira [4] explains "We have counted the number of distinct students accesses, of different resources being used; of events logged by the system. Then, we were

Table 5 Variables found that are referring to the actions of users

Sub-layer	Variables
Events	Affected user
	Which event
	Context of the event
	Complete events
	Time the event occurred
	Number of events
Logs	General logs
	Keywords
	Number of logs
Clicks	General clicks
	Clicks by activity
	Hyperlinks
	Number of clicks
	Number of clicks per session
Downloads	Downloaded files
	Size of downloaded files

able to differentiate between four types of actions performed by students: creations, visualizations, updates, and deletions."

- **Actions**

This layer can be divided into four sublayers, information about logs, clicks, downloads and events made by the user (Table 5).

- **Time**

Information about the student's study time is essential to detect how engaged that student is in his or her studies. While, in the face-to-face modality, it is not possible to identify how much time the user is studying at home, social learning system provides data reporting about how long the student is online.

A student who devotes his time to studies may present better results, and this relationship is confirmed by Romero and Barbera [10] "We observed a slight correlation between the time-on-task devoted by students on a weekly basis and their academic performance."

Table 6 presents the variables of this layer divided into two categories, the one in which it considers the duration and the one that considers the schedule of the action.

- **User's Information**

Personal information about the student is often used to better identify each student's profile. Primary studies currently point some data like gender, where the students live. The eight variables found in this layer are presented in Table 7.

Table 6 Variables found related to time

Sub-layer	Variables
Duration	Platform time
	Time between activities
	Time between the original post and the activity done
	Time between posting activity and deadline
	Time per session
	Average idle time
	Average time per week
Schedule	Time of day
	Weeks without posting
	Date of access

Table 7 Variables found related to the user's information

Layer	Variables
User's information	ID
	Full name
	Age
	Gender
	IP address used
	Appearance
	Culture
	Student profile

- **Psychological Layer**

The psychological layer uses data related to the psychological profile of the learner (Table 8).

However, it is clear that none of these variables can be considered singly and separately. All studies approved for this research used the variables together to achieve the desired result and a better understanding of the data.

3 Data Analysis

After establishing a list of variables used to describe the learner's model, we analyse these variables to determine if it is possible to extract any relationships or associations between them. For this, we will start by extracting frequent patterns combining the variables, and then from these patterns, we will generate association rules. The extraction of frequent patterns is carried out with the FP-growth algorithm, and then from the result obtained, we generate association rules.

Table 8 Variables found related to psychological layer

Layer	Variables
Psychological	Group work attitude
	Sociability
	Openness
	Extroversion
	Agreeableness
	Neuroticism
	Engagement
	Self-confidence
	Conscientiousness
	Area of employment
	Intent
	Learner's preferences

3.1 Frequent Patterns Mining

Research on frequent pattern algorithms started in the 1990s. The goal was to detect sets of data that appears recurrently on a database, not to classify instances, but to determine which items emerge frequently and which items are associate [5].

Frequent pattern mining is mostly used in customer transaction analysis. It attempts to identify associations, or patterns, between various items that have been bought by a particular costumer, which tends to costumer behaviour analysis, where we seek to extract associations such as: "a customer who buys item *x*, also buys item *y*".

Extraction of frequent patterns consists in searching in a dataset (itemset) for groups of items which appear *s* times, at least. *s* is the minimum support, corresponding to the smallest value from which a set of items is considered to be frequent [1].

FP-Growth Algorithm

FP-growth instead of *Apriori* algorithm is not based on candidate generation. It stands on two paradigms [3]:

- A compact representation of the database, with a structure called FP-tree.
- A divide-and-conquer strategy for exploring the data.

The FP-growth algorithm can be described as follows [1]:

1. Constructing FP-tree by compressing the DB representing frequent items,
2. Extracting conditional FP-tree for a selected item,
3. From the conditional FP-tree, generate frequent itemsets.

The main advantage of FP-growth algorithm is that it reduces the scan time compared to *Apriori* algorithm [6], while its disadvantage is that it requires large memory space for conditional FP-tree in the worst case [8].

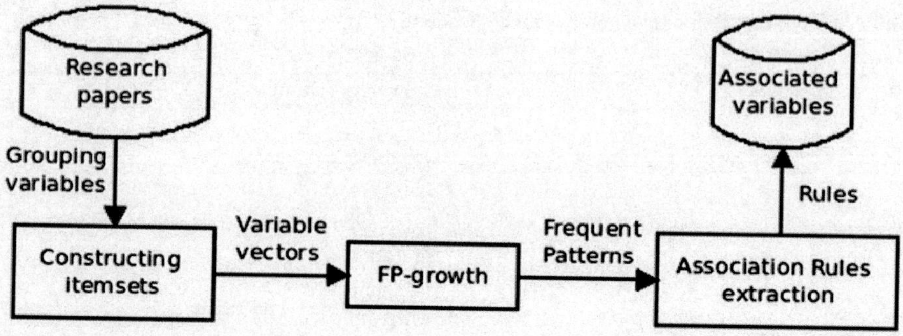

Fig. 1 The analysis process

Association Rules
Frequent patterns are often used to generate association rules. They describe correlations among items in a pattern.

An association rule is an implication *X* => *Y*, where *X* and *Y* are sets of items. The confidence value is the ratio of the support of *X* U *Y* to that of the support of *X*. It means that if the antecedent *X* is satisfied, then it is probable that the consequent *Y* is satisfied too [13].

Association rules are generated in three steps [1]:

1. Generate all the frequent patterns in the itemset at a minimum support level,
2. Extract all the rules from the frequent patterns,
3. Keep only rules whose confidence is greater than a minimum level of confidence.

3.2 Analysis and Results

For our experimentation, we want to find possible association rules between 76 learner's model variables. A PHP program, implementing FP-growth algorithm and association rules generation, was executed on the variable. Figure 3 shows interfaces of the application. FP-growth algorithm and rules association were applied with a min support equal to 2, which will allow us to take into account only the variables appearing in at least two papers. The confidence value is set to 0.75.

For implementing FP-growth algorithm, we first transform the values to constitute a binary vector corresponding to a transaction of the itemset.

The analysis process steps are (see Fig. 1):

1. Construct a table which columns are student model variables.
2. Each row of the table matches a referenced paper, so if a variable is cited in the paper, its corresponding column receives the value 1, 0 otherwise. Figure 2 shows a representation the variables vectors, where the first paper cites the first variable, but not the second one, and so on.

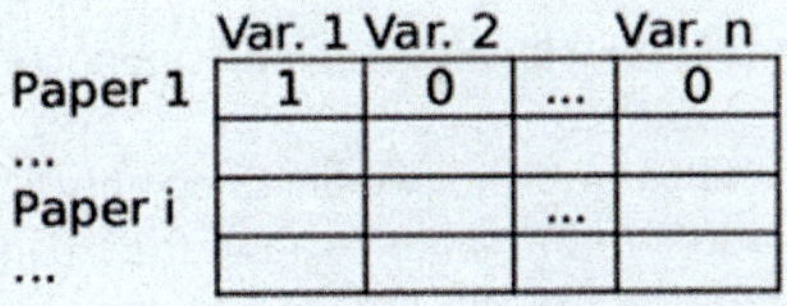

Fig. 2 Variables' vectors

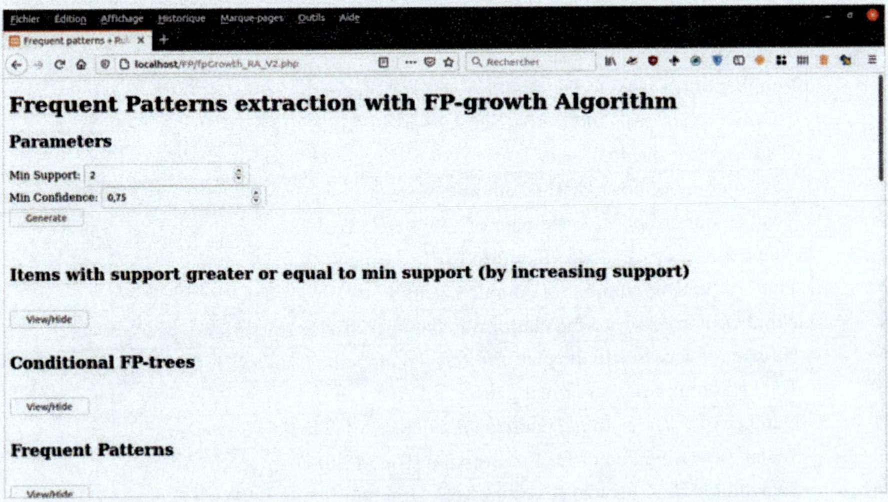

Fig. 3 Interface of FP-growth and AR generation program

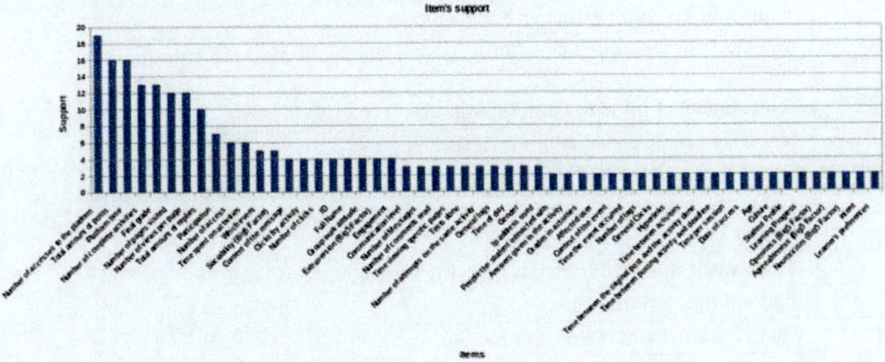

Fig. 4 Extracted frequent patterns histogram

3. After having created all vectors of the student model variables, they are merged to constitute a transaction. The set of transactions represents the itemset to be analysed with the FP-growth algorithm.
4. Frequent patterns in the itemset are generated using the FP-growth algorithm.
5. Finally, the association rules are extracted from frequent patterns.

Variables, with their relative supports for 63 papers (for min support = 2), are shown in Fig. 4.

After applying FP-growth algorithm, we obtained only 52 frequent patterns (with a support value great or equal to 2). Table 9 shows the frequent patterns extracted.

From these frequent patterns, 36 rules were generated (Table 10).

Table 9 Extracted frequent patterns

No.	Frequent pattern	Support
1	Number of accesses to the platform & Total amount of posts	8
2	Number of accesses to the platform & Platform time	7
3	Platform time & Final grade	6
4	Total amount of posts & Total amount of replies	5
5	Total amount of posts & Number of views per page	5
6	Number of accesses to the platform & Final grade	5
7	Platform time & Number of complete activities	5
8	Number of accesses to the platform & Number of views per page	4
9	Number of accesses to the platform & Total amount of posts & Final grade	4
10	Total amount of posts & Final grade	4
11	Number of accesses to the platform & Number of complete activities	4
12	Sociability (Big5 Factor) & Extraversion (Big5 Factor)	3
13	Sociability (Big5 Factor) & Group work attitude	3
14	Number of accesses to the platform & Number of pages visited	3
15	Platform time & Number of complete activities & Final grade	3
16	Number of complete activities & Final grade	3
17	Number of accesses to the platform & Total amount of posts & Platform time	3
18	Total amount of posts & Platform time	3
19	Number of complete activities & Tests done	2
20	Number of accesses to the platform & Number of comments read	2
21	Sociability (Big5 Factor) & Group work attitude & Extraversion (Big5 Factor)	2
22	Group work attitude & Extraversion (Big5 Factor)	2
23	Platform time & Full Name	2
24	Clicks by activity & Number of clicks	2

(continued)

Table 9 (continued)

No.	Frequent pattern	Support
25	Total amount of posts & Total amount of replies & Content of the message	2
26	Total amount of posts & Content of the message	2
27	Total amount of replies & Content of the message	2
28	Platform time & Number of complete activities & Final grade & Time spent on activities	2
29	Platform time & Number of complete activities & Time spent on activities	2
30	Platform time & Final grade & Time spent on activities	2
31	Platform time & Time spent on activities	2
32	Number of complete activities & Final grade & Time spent on activities	2
33	Number of complete activities & Time spent on activities	2
34	Final grade & Time spent on activities	2
35	Total amount of posts & Participation	2
36	Number of accesses to the platform & Total amount of posts & Platform time & Total amount of replies	2
37	Number of accesses to the platform & Total amount of posts & Total amount of replies	2
38	Number of accesses to the platform & Platform time & Total amount of replies	2
39	Number of accesses to the platform & Total amount of replies	2
40	Total amount of posts & Platform time & Total amount of replies	2
41	Platform time & Total amount of replies	2
42	Number of accesses to the platform & Total amount of posts & Number of views per page	2
43	Number of accesses to the platform & Number of pages visited & Number of views per page	2
44	Number of pages visited & Number of views per page	2
45	Platform time & Number of complete activities & Number of pages visited	2
46	Platform time & Number of pages visited	2
47	Number of complete activities & Number of pages visited	2
48	Number of accesses to the platform & Total amount of posts & Platform time & Final grade	2
49	Number of accesses to the platform & Platform time & Final grade	2
50	Total amount of posts & Platform time & Final grade	2
51	Number of accesses to the platform & Total amount of posts & Number of complete activities	2
52	Total amount of posts & Number of complete activities	2

Table 10 Extracted frequent patterns

No.	Association rule	Confidence
1	Group work attitude & Extraversion (Big5 Factor) ⇒ Sociability (Big5 Factor)	1
2	Extraversion (Big5 Factor) ⇒ Sociability (Big5 Factor)	0.75
3	Group work attitude ⇒ Sociability (Big5 Factor)	0.75
4	Total amount of replies & Content of the message ⇒ Total amount of posts	1
5	Total amount of posts & Content of the message ⇒ Total amount of replies	1
6	Final grade & Time spent on activities ⇒ Platform time & Number of complete activities	1
7	Number of complete activities & Time spent on activities ⇒ Platform time & Final grade	1
8	Number of complete activities & Final grade & Time spent on activities ⇒ Platform time	1
9	Platform time & Time spent on activities ⇒ Number of complete activities & Final grade	1
10	Platform time & Final grade & Time spent on activities ⇒ Number of complete activities	1
11	Platform time & Number of complete activities & Time spent on activities ⇒ Final grade	1
12	Number of complete activities & Time spent on activities ⇒ Platform time	1
13	Platform time & Time spent on activities ⇒ Number of complete activities	1
14	Final grade & Time spent on activities ⇒ Platform time	1
15	Platform time & Time spent on activities ⇒ Final grade	1
16	Final grade & Time spent on activities ⇒ Number of complete activities	1
17	Number of complete activities & Time spent on activities ⇒ Final grade	1
18	Platform time & Total amount of replies ⇒ Number of accesses to the platform & Total amount of posts	1
19	Total amount of posts & Platform time & Total amount of replies ⇒ Number of accesses to the platform	1
20	Number of accesses to the platform & Total amount of replies ⇒ Total amount of posts & Platform time	1
21	Number of accesses to the platform & Platform time & Total amount of replies ⇒ Total amount of posts	1
22	Number of accesses to the platform & Total amount of posts & Total amount of replies ⇒ Platform time	1

(continued)

Table 10 (continued)

No.	Association rule	Confidence
23	Number of accesses to the platform & Total amount of replies ⇒ Total amount of	1
24	Platform time & Total amount of replies ⇒ Number of accesses to the platform	1
25	Number of accesses to the platform & Total amount of replies ⇒ Platform time	1
26	Platform time & Total amount of replies ⇒ Total amount of posts	1
27	Number of pages visited & Number of views per page ⇒ Number of accesses to the platform	1
28	Number of complete activities & Number of pages visited ⇒ Platform time	1
29	Platform time & Number of pages visited ⇒ Number of complete activities	1
30	Number of complete activities & Final grade ⇒ Platform time	1
31	Total amount of posts & Platform time & Final grade ⇒ Number of accesses to the platform	1
32	Number of accesses to the platform & Platform time & Final grade ⇒ Total amount of posts	1
33	Total amount of posts & Final grade ⇒ Number of accesses to the platform	1
34	Number of accesses to the platform & Final grade ⇒ Total amount of posts	0.8
35	Total amount of posts & Number of complete activities ⇒ Number of accesses to the platform	1
36	Total amount of posts & Platform time ⇒ Number of accesses to the platform	1

4 Discussion and Proposed Model

The first observation to be made about variables is that the most frequent variables only belong to six layers: grades layer, time layer, activities layer, access layer, communication layer (forum sublayer). This leads to the conclusion that in the papers studied, the profile of the learner is built mainly from his grade and his acts on the e-learning systems (activities, access and communication). Naturally, studying learner activities and communication leads to study his accesses to the learning platform, which is confirmed by rules (19, 24, 27, 31, 33, 35 and 36).

Other observation, papers using "**Total amount of posts**" as learner profile variable (rules 21, 23, 26, 32 and 34) uses also access variable ("**Number of accesses to the platform**") coupled with other layer variable (time layer and communication layer). For papers using "Platform time", they mainly use "**Number of complete activities**" or "**Number of accesses to the platform**" coupled with other variable (rules 8, 12, 14, 22, 25, 28 and 30). For the variable "**Number of complete activities**", the link can be explained by the fact that completing activities for a learner is closely linked to deadlines, and therefore to platform time. We can also observe that the counterpart is also true, while "**Number of complete activities**" is associated with "**Platform time**" coupled with other variables (rules 10, 13, and 29).

"**Sociability**" variable is associated with two other variables ("Group work attitude" and "**Extroversion**"—rules 1, 2 and 3). It is clear that sociabilty and extroversion are very related while both are defined by social engagement, and thus attitudes in group (the "**Group work attitude**" variable).

Finally, "**Final Grades**" variable is associated with combinations of three variables (**Platform time, Number of complete activities** and **Time spent on activities**) in rule 11 (and also rules 15 and 17).

From this analysis, we can draw as a conclusion that to have learners profile which can be used for grouping learners, we can build this model from the variables: "**Number of accesses to the platform**", "**Number of complete activities**", "**Platform time**" and "**Final Grades**". These variables appear mainly in the most association rules with other variables. Moreover, only a few papers adopt multiple intelligences theories in a dynamic grouping learner [17].

5 Conclusion

This paper presents a review of learner's model variables which could be used for dynamic group formation for collaborative learning. These variables were analysed with frequent pattern extraction method (FP-growth) and association rule generation concerning the variables. The results show that, from the analysed papers, only four variables from learner's are mostly used with association with other variables. These variables are: "**Number of accesses to the platform**", "**Number of complete activities**", "**Platform time**" and "**Final Grades**". They can be good candidates as criteria for grouping learners. The next step is to use them with multiple intelligences theories in a dynamic grouping system.

References

1. Aggarwal, C.C., Bhuiyan, M.A., Al Hasan, M.: Frequent pattern mining algorithms: a survey. In: Frequent Pattern Mining, pp. 19–64. Springer (2014)
2. Bekele, R.: Computer-assisted learner group formation based on personality traits. Ph.D. thesis. Staats-und Universitätsbibliothek Hamburg Carl von Ossietzky (2005)
3. Benraouane, S.A.: Guide pratique du e-learning: stratégie, pédagogie et conception avec le logiciel Moodle. Dunod (2011)
4. Figueira, Á.: Mining moodle logs for grade prediction: a methodology walk-through. In: Proceedings of the 5th International Conference on Technological Ecosystems for Enhancing Multiculturality, pp. 1–8 (2017)

5. Fournier-Viger, P., Lin, J.C.W., Nkambou, R., Vo, B., Tseng, V.S.: High-Utility Pattern Mining. Springer (2019)
6. Gu, X.F., Hou, X.J., Ma, C.X., Wang, A.G., Zhang, H.B., Wu, X.H., Wang, X.M.: Comparison and improvement of association rule mining algorithm. In: 2015 12th International Computer Conference on Wavelet Active Media Technology and Information Processing (ICCWAMTIP), pp. 383–386. IEEE (2015)
7. Gweon, G., Jun, S., Lee, J., Finger, S., Rosé, C.P.: A framework for assessment of student project groups on-line and off-line. In: Analyzing Interactions in CSCL, pp. 293–317. Springer (2011)
8. Kavitha, M., Selvi, S.: Comparative study on apriori algorithm and FP growth algorithm with pros and cons. In. J. Comput. Sci. Trends Technol. (IJCST) **4** (2016)
9. Rabbany K.R., Takaffoli, M., Zaiane, O.R.: Social network analysis and mining to support the assessment of on-line student participation. ACM SIGKDD Explor. Newslett. **13**(2), 20–29 (2012)
10. Romero, M., Barbera, E.: Quality of learners' time and learning performance beyond quantitative time-on-task. Int. Rev. Res. Open Distrib. Learn. **12**(5), 125–137 (2011)
11. Stahi, G., Koschmann, T., Suthers, D.D.: Computer-supported collaborative learning. The Cambridge Handbook of the Learning Science, pp. 409–425 (2006)
12. Thomson, A.M., Perry, J.L.: Collaboration processes: inside the black box. Public Admin. Rev. **66**, 20–32 (2006)
13. Ventura, S., Luna, J.M.: Supervised Descriptive Pattern Mining. Springer (2018)
14. Walckiers, M., De Praetere, T.: L'apprentissage collaboratif en ligne, huit avantages qui en font un must. Distances et savoirs **2**(1), 53–75 (2004)
15. Zamani, M.: Cooperative learning: homogeneous and heterogeneous grouping of iranian efl learners in a writing context. Cogent Educ. **3**(1), 1149,959 (2016)
16. Zhang, X.: An analysis of online students' behaviors on course sites and the effect on learning performance: a case study of four LIS online classes. J. Educ. Library Inf. Sci. **57**(4), 255–270 (2016)
17. Zheng, Y., Subramaniyan, A.: Personality-aware collaborative learning: Models and explanations. In: International Conference on Advanced Information Networking and Applications, pp. 631–642. Springer (2019)

E-learning and the New Pedagogical Practices of Moroccan Teachers

Nadia El Ouesdadi and Sara Rochdi

Abstract The strategic vision 2015–2030 is the most recent school reform in Morocco. It promotes the use of information and communication technologies in the professionalization of teaching in all schools and in all disciplines. Teachers are therefore called upon to develop professionally throughout their careers by innovating their teaching practices. This article aims to answer the question: How could an online device meet the needs of teachers in terms of ICT use in the classroom? As well as to define distance learning education, its characteristics, namely the added value of using ICT in the classroom to enhance learning and to stay in touch with the teacher-learner through the Collab platform for teachers of all disciplines in Morocco. The experimentation we have conducted, presents to teachers-learners an online device, involving the consultation of content and optimization of interactive activities to be developed through software. This is one of the innovative models developed with this in mind to reveal the benefits on the implementation of a learning device and apply it in the classroom.

1 Introduction

During the last few years, information and communication technologies (ICT) have experienced a remarkable boom in all areas including that of teaching and learning. The integration of ICT in education is one of the main concerns of the Ministry of National Education in Morocco. So that they can improve the quality of teaching and learning. The strategic vision 2015–2030 is the most recent reform in Morocco's education system. It promotes the use of technology. Cited in Lever 12 [1]. Development of an open, diversified, efficient, and innovative pedagogical model: "Strengthen the integration of educational technologies to improve the quality of learning, through the implementation of a new national strategy, able to accompany and support innovations likely to promote the development of institutions." This will be achieved

N. El Ouesdadi · S. Rochdi (✉)
CLHEN, Linguistics, Didactics and Communication, EST, University Mohammed First, Oujda, Morocco

M. Ben Ahmed et al. (eds.), *Networking, Intelligent Systems and Security*, Smart Innovation, Systems and Technologies 237,
https://doi.org/10.1007/978-981-16-3637-0_35

"by integrating digital media and interactive tools in teaching and learning activities, research, and innovation."

The development, new specific pedagogy in teaching/learning has appeared thanks to technology and the explosion of the Internet, a new mode of distance learning has emerged whether it is school, university or professional. Online education allows a multitude of new possibilities such as learning, exchange, collaboration, etc.

In other words, it opens new avenues for learning in general and it corresponds to the use of computer technologies that help learners to improve their performance and knowledge through the exchange of necessary information which allows for total autonomy of users. Some authors have proposed a typology of the pedagogical uses of ICT in five categories, namely ICT to exchange, communicate, collaborate, and cooperate, ICT to produce, create and publish, ICT to research and document, ICT to train and self-train, ICT to animate and organize [2].

Teachers are invited to innovate and evolve their pedagogical concepts and practices. Indeed, "ICTs are powerful cognitive tools. But, if they offer multiple solutions to many of today's educational problems, they will only be truly useful if the trainer agrees to transform, or even change his or her conceptions and practices" [3]. It is therefore useful for teachers to be receptive and trained in the proper use of technological tools in classroom practices. They should also be trained in scriptwriting and the creation of pedagogical content so that they can design content that meets their pedagogical objectives and the needs of their learners. This will allow interactivity and accessibility for the learner and the effective use of educational technologies.

Mangenot [4] asserts that we can only speak of an integration of ICTE when "the computer tool is used effectively to support learning". For all these reasons, the Ministry of National Education has programmed a series of face-to-face training sessions for teachers in information and communication technologies. And other online training courses are set up for teachers to familiarize them with the use of distance learning platforms, the use of the digital environment, the pedagogical scripting, and to introduce them to the mediatization of online courses.

It is in this context that this article is written, with the aim of answering our fundamental question: How could an online device meet the needs of teachers in terms of using ICT in the classroom?

To answer our question, we have made available a set of questions:

1. What is distance education?
2. What is collaborative learning?
3. How can we learn online?
4. How can pedagogical scenarios be developed and used in the classroom?
5. What is the role of a tutor?

Based on a pre-investigation on the platform in which we have been beneficiaries and subsequently tutors, we have drawn up the following hypotheses:

- An online device could help teachers to design their courses online and then apply them in the classroom through educational software.
- An online device promotes collaboration and exchange between learners.

In our research, we will initially define the key concepts of our research, namely distance learning, educational technologies according to the strategic vision 2015–2030, collaboration in e-learning, etc. Then we will discover the universe of the Collab platform, its aims, its content, etc. Then we will discuss the results of our experience with the Collab platform, and we will conclude with the contributions of this eminent experience.

2 Methodology

The objective of this research is to identify the use of ICT use in the professional process of primary and secondary school teachers in Morocco in order to answer our main question: How could an online device meet the needs of teachers in terms of ICT use in a classroom?

In order to do so, we opted for a quantitative approach to collect the data that we will present and which are the result of a survey and research in the Collab platform in the form of description, tutoring, and evaluation of the homework given by the beneficiaries in the second part of the training which contains the development of interactive exercises with the Rubis software. This method was chosen because it will allow us to answer our research question.

3 Theoretical Framework

As mentioned above, the world of education today is witnessing transformations that affect teaching and learning methods and modes with the emergence of the:

Educational technologies. These are defined by the CSEFRS [5] as follows:

Educational technologies are information and communication technologies adapted to education. They intervene in learning, training and supervision, in planning, organization and management, and in evaluation.

Educational technologies include interactive programs and software, digital resources, various electronic tools and equipment, as well as networks and information systems and the services they offer such as distance learning, video conferencing, digital libraries, etc.

The integration of educational technologies aims:

- To raise the quality of education and training by facilitating the acquisition of knowledge, increasing learner motivation, and enhancing the attractiveness of the school;
- To qualify the learner to access the knowledge society, to master the strategies of distance learning, and to build personal projects in research and innovation;
- To rationalize educational governance by relying on integrated information systems;

- To create networks for exchange, sharing, and development of collective intelligence and support for teamwork in education;
- To involve and help educational stakeholders in isolated areas.

Distance Learning

The notion of distance in training is opposed to presence, in geographical, psychic, and relational terms in which the need for contact and trust comes into play [6]; thus, "distance training is defined as opposed to face-to-face training, by breaking the spatial co-presence between the teacher and his learners" [7]. It "brings together students, teachers, one or more knowledge objects, and technical supports such as Moodle-type platforms, the Internet, electronic files, not forgetting traditional printed courses" ([6]: 397).

Among the practices in use, Papi et al. [8] list four intentions of the devices:

- Reduce distance: counter geographical distance (virtual classroom);
- Enrich the experience: diversify learning experiences (wiki, webinar);
- Support interaction: connect to counter isolation (videoconferencing);
- Develop skills: achieve learning objectives through the uses of the social web (forum).

Virtual learning spaces and third places

A distance course contains several virtual learning spaces: information, collaboration, production, assistance, self-management, and evaluation spaces [9]. Some spaces may be formal (the institution's digital learning environment) or informal (social networks, personal learning spaces). In organizing virtual spaces, one should not ignore what happens outside the classroom; one can even seek to enhance these informal spaces [10].

Distance learning is supervised and tutored. It is in fact a framework that seeks to guide the learners in a device and offers them flexibility in teaching. Indeed, tutoring is based on a follow-up of the learners to accompany and guide them with the help of a tutor.

Tutoring

Distance tutoring is a pedagogical accompaniment in a device. According to Bourdet [11], tutoring is seen as a "formative function which is characterized by the exercise of varied and sometimes contradictory roles (joint empathy/validation of stages, general guidance/adaptation to specificities)."

Today we talk about the active approach in the teaching–learning process. This insists on the participation of the learner in the construction of his learning. The teacher is no longer the person who holds the knowledge; he has become a mediator, a coach, and a facilitator of learning.

The coaching scenario is developed in relation to the learning scenario and takes into account several aspects [12]:

- Tutorial objective;

- Actors involved;
- Types of tutoring: individual, collective, peer tutoring, etc.;
- Learning support plans: cognitive, socio-affective, motivational, and metacognitive;
- Tutoring methods: synchronous, asynchronous, proactive, and reactive;
- Frequency and positioning in the learning scenario;
- Tutor support and resources to be produced;—Reusability of resources and approaches.

The role of the tutor

The role of the tutor is the person who "accompanies a learner or a group of learners at a distance by the means of communication and training that computer, multimedia, and the Internet allow today" Lisowski [13]. It has several functions within a computer-mediated device. It has a social role because it ensures the organization of the learners' work and the constitution of groups to interact and exchange.

The tutor is the one who constitutes the groups, distributes the work tasks for the learners. He is, according to Legendre, "a guide, an instructor who teaches one person or a small group of students at a time: he is a student advisor" [14]. In addition to this, the tutor animates the work of the group and encourages the learners to make a collective production and work toward the same goal while working in a digital environment and in professional learning conditions.

Collaborative distance learning

Thanks to the digital evolution in recent years, the avenues of virtual collaboration have developed considerably. This new modality has undergone important theorizations, such as the definitions of cyberculture [15] and the new learning theory, connectivity [16] (Downes, 2012, May 21).

Henri and Lundgren-Cayrol [17] proposed a definition of collaborative learning: "Collaborative learning is an active process in which the learner works on building knowledge…The learner commits to working with group members toward a common goal while reconciling personal interests and objectives…Exchanges with the group and the regulation of a collective task allow the learner to share his or her discoveries, negotiate the meaning to be given to his or her work, and validate newly constructed knowledge".

Among the tools for collaborative work in an online device, we have.

The Wiki

The Wiki is a collaborative distance working tool; it allows learners to work together on the same document, and they can edit or comment on it.

Its advantages:

- The teacher will be able to stimulate learners, correct, and orient their work.
- Allows learners to work together on course topics.
- Learners and teachers are notified of any new modifications or notifications.

The forum

The forum is also a tool for remote collaborative work. It is composed of a set of online discussions on different topics, and it allows to generate debates in the form of questions and answers between learners, or between learners and teachers.

Its advantages:

- Allows the teacher to structure and direct the exchange.
- Promotes communication between learners.
- Discusses course topics.
- Motivates learners.

Thanks to these tools that are easy to use, the shared knowledge allows the construction; the collective co-construction of knowledge facilitates the interaction between the teacher and the learners. We move from individual learning to collaborative learning.

We are going to discover the environment of the Collab system.

4 Presentation of the Collab [18]

COLLAB is an e-learning device designed and produced by the National Center for Pedagogical Innovations and Experimentation—Distance Learning Division—of the MENFP. This platform offers e-learning on the use of free and open-source software (Chaînes éditoriales scenari) to the entire Moroccan educational community (see Fig. 1).

Its objectives:

- To help beneficiaries navigate the platform to take advantage of its training content.
- Navigate the platform.

Fig. 1 Collab platform web page

- Access the different courses.
- Locate course components: Blocks, Resource Activities…
- Use the activity modules presented in the course.
- Interact with tutors and other beneficiaries.

The trainings.

The Collab platform offers three types of training courses:

- Development of educational multimedia content
- Distance learning in team sports
- Tutor training.

The first training consists of two modules (see Fig. 2). In our research, we will focus on the second module "Elaboration of interactive exercises."

First module: development of digital resources.

SCENARI In this section, the learner provides an overview of the course environment. By clicking on the development of educational multimedia content (DCME) course, learners can access the first module of the course, which is the SCENARI software.

Beneficiaries are called to install the Opale Software: Opale advanced is a complete version of Opale intended for the personnel of the ICTE units, Web professionals, and people already autonomous on Opale starter (see Fig. 3).

When learners have completed the first module, they will find a link to the second module.

Module 2: Developing Interactive Exercises.

RUBIS which is an editorial chain for the creation of exercisers and which allows to create exercise activities in two modes: self-learning and evaluative. The RUBIS course is subdivided into three weeks; each week contains courses to view or files to download and an activity to complete (see Fig. 4).

Fig. 2 Development of educational multimedia content

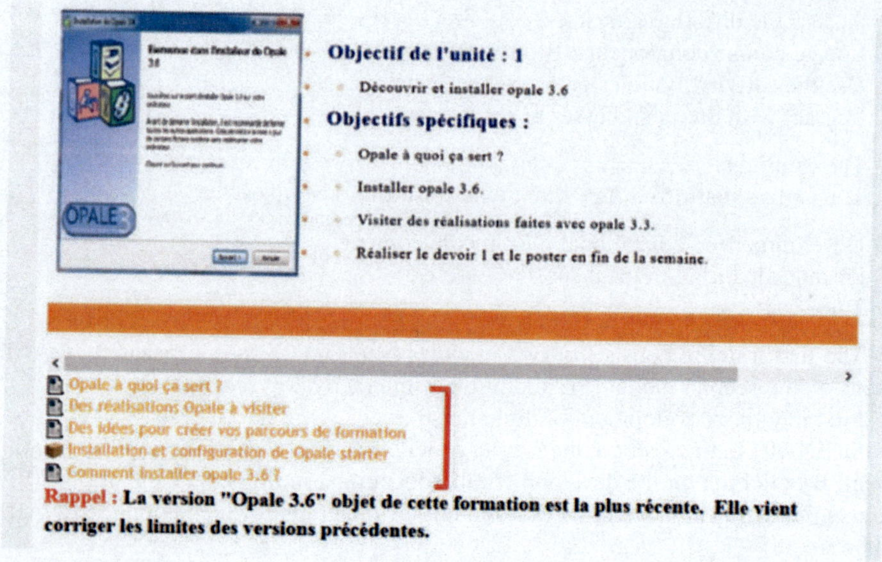

Fig. 3 Opale advanced software

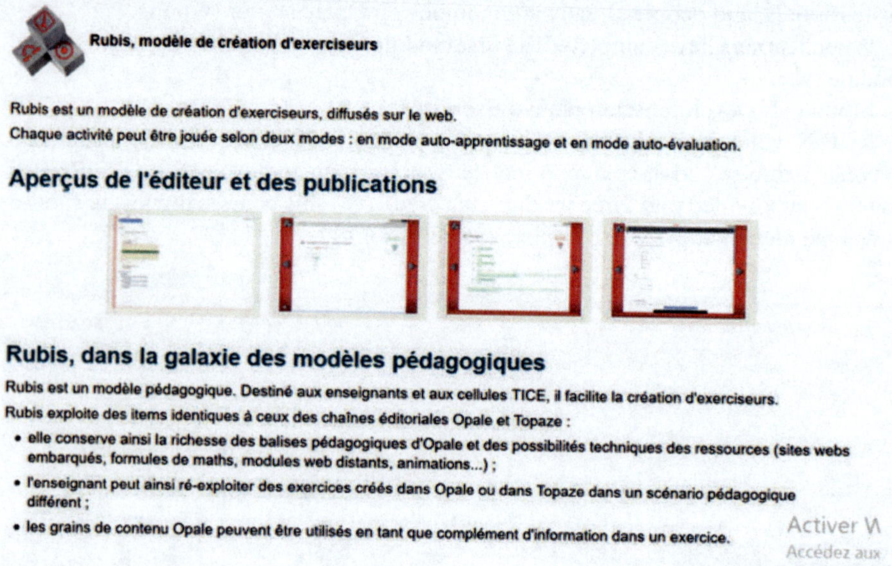

Fig. 4 Ruby software

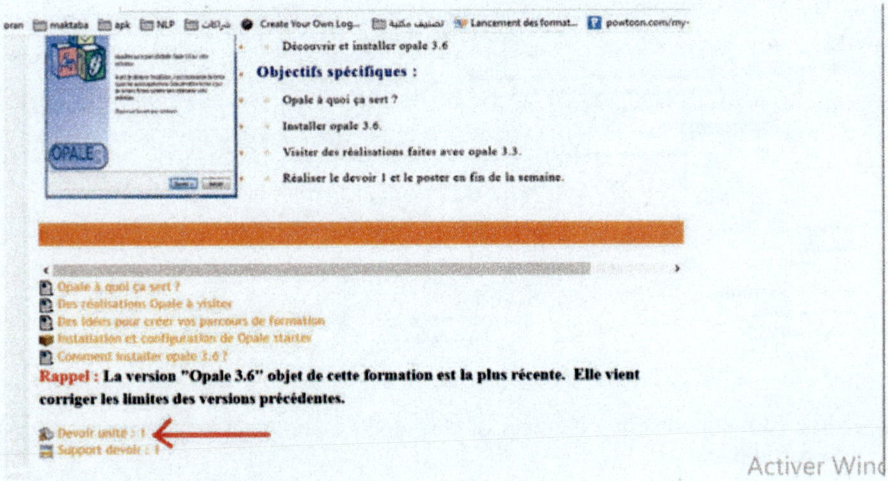

Fig. 5 Homework assignment

The Forum

Each course has its own forum. The course forum is usually located at the beginning of each course. The forums in this course are considered complementary collaborative learning tools.

The DCME course is supervised under the heading "Homework," with learners receiving support and follow-up from online trainers called "tutors." The trainers schedule activities in the form of assignments at the end of each week of training (see Fig. 5). These assignments are limited in time. This is indicated in the header of each week, and the tutors indicate the final deadline for completing the assignment.

Each week of training contains one activity to be submitted by the end of that week. By clicking on "Activity1" for example; a window will open showing the place to click to hand in the assignment and by clicking on add (in the same way as posting a topic on the forum).

5 Results of the Experiment

We monitored and tutored the learners and evaluated their performance through the homework assignments handed in each weekend. The number of participants registered in the training is 399.

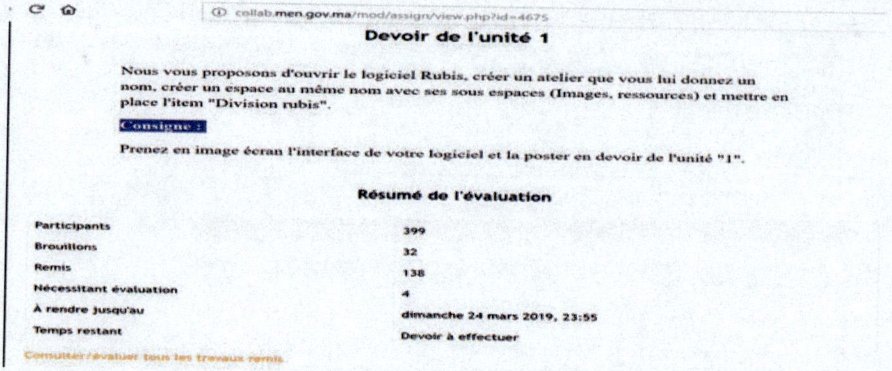

collab.men.gov.ma/mod/assign/view.php?id=4675

Devoir de l'unité 1

Nous vous proposons d'ouvrir le logiciel Rubis, créer un atelier que vous lui donnez un nom, créer un espace au même nom avec ses sous espaces (Images, ressources) et mettre en place l'item "Division rubis".

Consigne :

Prenez en image écran l'interface de votre logiciel et la poster en devoir de l'unité "1".

Résumé de l'évaluation

Participants	399
Brouillons	32
Remis	138
Nécessitant évaluation	4
À rendre jusqu'au	dimanche 24 mars 2019, 23:55
Temps restant	Devoir à effectuer

Consulter/évaluer tous les travaux remis

– Homework assignment 1 Number of participants who handed in the assignment: 142 participants

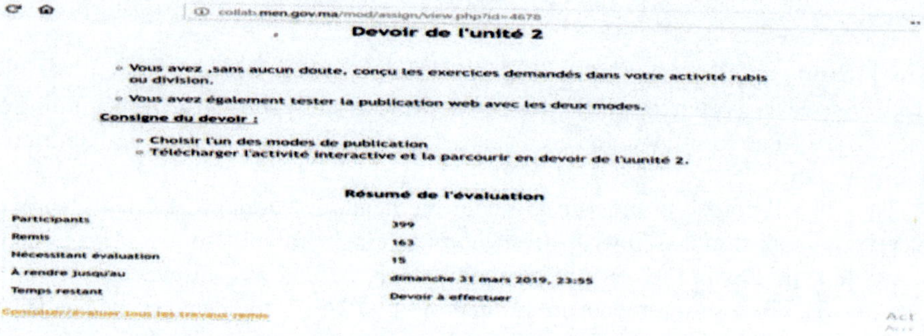

collab.men.gov.ma/mod/assign/view.php?id=4678

Devoir de l'unité 2

- Vous avez ,sans aucun doute, conçu les exercices demandés dans votre activité rubis ou division.
- Vous avez également tester la publication web avec les deux modes.

Consigne du devoir :

- Choisir l'un des modes de publication
- Télécharger l'activité interactive et la parcourir en devoir de l'uunité 2.

Résumé de l'évaluation

Participants	399
Remis	163
Nécessitant évaluation	15
À rendre jusqu'au	dimanche 31 mars 2019, 23:55
Temps restant	Devoir à effectuer

Consulter/évaluer tous les travaux remis

Act

Homework assignment 2 Number of participants who handed in the assignment: 178 participants

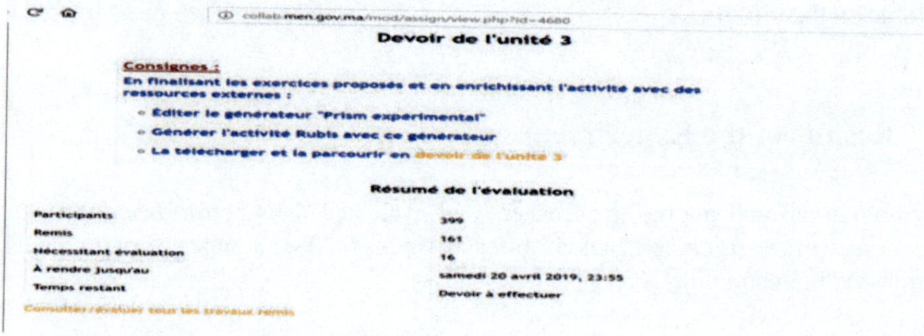

collab.men.gov.ma/mod/assign/view.php?id=4680

Devoir de l'unité 3

Consignes :

En finalisant les exercices proposés et en enrichissant l'activité avec des ressources externes :

- Éditer le générateur "Prism expérimental"
- Générer l'activité Rubis avec ce générateur
- La télécharger et la parcourir en devoir de l'unité 3

Résumé de l'évaluation

Participants	399
Remis	161
Nécessitant évaluation	16
À rendre jusqu'au	samedi 20 avril 2019, 23:55
Temps restant	Devoir à effectuer

Consulter/évaluer tous les travaux remis

– Homework assignment #3 Number of participants who handed in the assignment: 177 participants

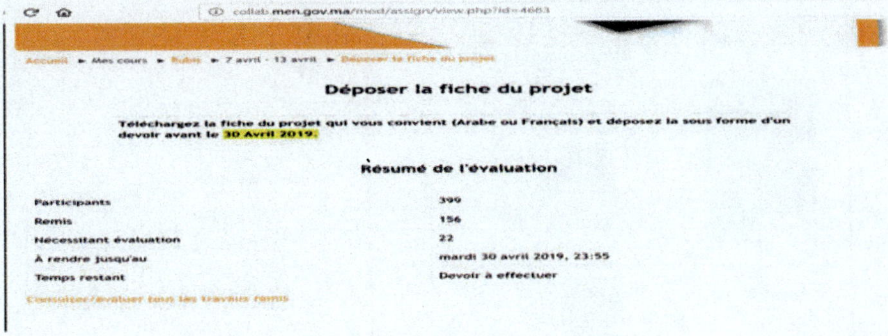

The submission of the project

– Number of participants who submitted the final project is: 178 participants

The discussion forum

We observed learners' interactions in the forum and found that learners interact with other beneficiaries and with the tutors on each week's courses.

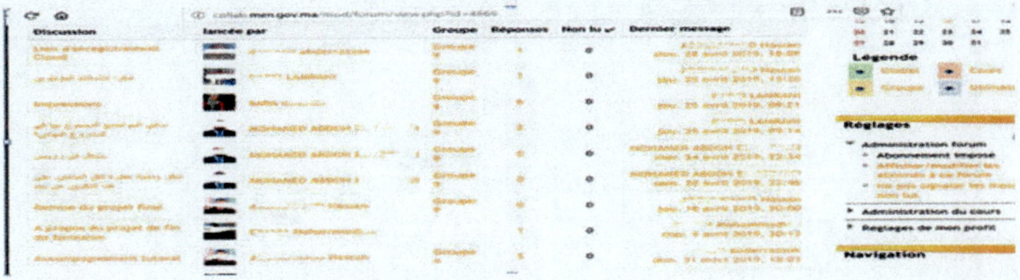

Example of a submitted job

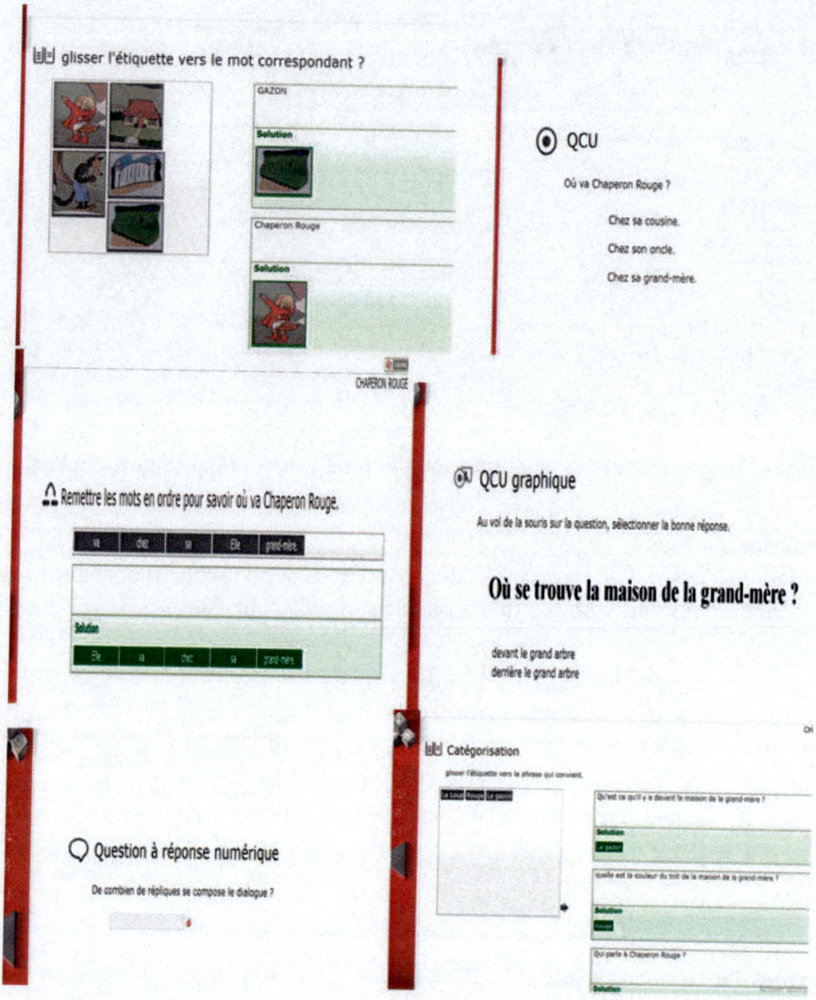

6 Commentary

Conducting this experiment has shown us that the place of information and communication technologies has a great place in teacher training. The Ministry of National Education in Morocco relies heavily on pedagogical innovation; but according to our experimentation, this remains insufficient because we believe that the actors of the educational system still show resistance to the use of technological tools. The results above show a reticence and abandonment of the beneficiaries (399 registered and 178 learners who were able to follow and submit their final project). Regarding the interactions in the forum, we observed that the discussion space is a beneficial and effective way to exchange, share, and collaborate between learners and learners with their tutors.

The projects delivered at the end of the training show the involvement of the teachers who benefit from distance learning and their desire to use ICT tools in their classroom practices in order to innovate. The software proposed in the training (Opal and Ruby) is tools that allow the designing teachers to realize a digital pedagogical scenario as well as to evaluate the learners' achievements with various types of exercises. This has been approved through the feedback from the beneficiary teachers.

7 Conclusion

Today, digital technology has become an essential tool for all teacher training. This is why the COLLAB e-learning system offers tools and learning methods free of charge, so as to innovate, break with the traditional method and arouse the curiosity of teachers. It is a method that enables a better understanding of how technology can have a real impact on teaching and learning. In this way, it helps the teacher understand that integrating ICT has a positive impact on student learning. This allows us to confirm our hypotheses put forward above as the Collab platform has allowed the beneficiary teachers to design their courses online, and then they can easily apply them in the classroom through the educational software offered as well as the online device has fostered collaboration and exchange between learners.

We believe then that the integration of technology in teaching and learning requires a considerable effort on the part of all components of the educational system to promote a better integration of the technological tool within the Moroccan school.

References

1. Levier 12, la vision stratégique 2015–2030
2. Basque, J., Karin, L.-C.: A typology of ICT applications in education. Sciences et techniqueséducatives **9**(3–4), 263–298 (2002)
3. Karsenti, Peraya, Viens: Review and prospects for research on teacher training for the pedagogical integration of ICTs. Revue des sciences de l'éducation **XVIII**(2), 459.470 (2002). www.erudit.org/revue/rse/2002/v28/n2/007363ar.pdf
4. Mangenot, F.: Classification des apports d'Internet à l'apprentissage des langues. p. 145sur: www.alsic.univ-fcomte.fr (1998)
5. Conseil supérieur de l'éducation, de la formation et de la recherche scientifique, Pour une école de l'équité de la qualité et de la promotion; la vision stratégique 2015–2030 p. 95
6. De la Berchoud, M.J.: distance - prendre en compte des publics loinignés et décentrer la réflexion méthodologique: Sociodidactique ou sémiodidactique ? Ela. Etudes de linguistique appliquée **4**(168), 395–405 (2012)
7. Peraya, D.: Distances, absence, proximities and presences: concepts in motion. Distances Knowl. Mediation [Online], **8** (2014)

8. Papi, C., Brassard, C., Bédard, J.L., Medoza, G.A., Sarpentier, C.: L'interaction en formation à distance: entre théories et pratiques. TransFormations – Recherches en éducation et formation des adultes (17) (2017)
9. Paquette, G.: L'ingénierie pédagogique: Pour construire l'apprentissage en réseaux. Sainte-Foy, QC, PUQ (2002)
10. Stockless, A, Villeneuve, S.: Les compétences numériques chez les enseignants: doiton devenir un expert? Dans Romero, M. et al (dir.), Usages créatifs du numérique pour l'apprentissage au XXIe siècle, pp. 141–148. Québec, QC, PUQ (2017)
11. Bourdet, J.-F.: Tutorat en ligne et création d'un espace formatif. Alsic [Online] **1002**(30), 23–32 (2007)
12. Rodet, J.: Propositions pour l'ingénierie tutorale. Tutorales, Revue de la communauté de pratiques des tuteurs à distance **7**, 6–28 (2010)
13. Lisowski, M.: L'e-tutorat. http://www.centreinffo.fr/IMG/pdf/AFP2204357.pdf (2010)
14. Legendre, R Dictionnaire actuel de l'éducation. p. 1378 (2005)
15. Lévy, P.: L'intelligence collective. Pour une anthropologie du cyberespace. La Découverte, Paris (1994).
16. Siemens, G.: Connectivism. A Learning Theory for the Digital Age. Retrieved from http://www.elearnspace.org/Articles/connectivism.htm (2004)
17. Henri, F., Lundgren-Cayrol, K. Collaborative distance learning: understanding and designing virtual learning environments. Sainte-Foy: Presses universitaires du Québec., p. 42 (2003)
18. http://collab.men.gov.ma/pluginfile.php/12817/mod_resource/content/8/guide_papier.pdf

A Sentiment Analysis Based Approach to Fight MOOCs' Drop Out

Soukaina Sraidi, El Miloud Smaili, Salma Azzouzi, and My El Hassan Charaf

Abstract Over the past few decades, a new model of education has emerged in the world of education under the acronym MOOC (Massive Open Online Courses). This model has made it possible to create transitional educational solutions by attracting large populations and ensuring a worldwide online presence. However, there is still a dark and fuzzy interface regarding the high dropout rate of learners at the end of the courses. This issue leads us to ask: How can we make online mega-courses more reliable and attractive? In this article, we explore new tools and methods (quality and machine learning) to analyze the different limitations and difficulties learners face in MOOCs and then build a deeper and better understanding of the main causes of the high dropout rate.

1 Introduction

The rapid rhythm of educational development as well as the emergence of new learning styles have led human innovation to develop new tools and methods to exchange knowledge. The aim is to provide an interactive and productive space for learners in order to make the learning process more flexible and accessible to all. In this context, a new style of distance learning cited as global, free, large-scale and open—known as MOOC (Massive Open Online Courses)—has emerged in recent years, taking advantage of new technologies.

Although the concept is attracting more and more learners thanks to the new flexible features it offers to learners. However, there is still a dark and fuzzy interface related to the dropout rate at the end of such courses. On the other hand, the proliferation of different means of communication and the advent of social networks (FACEBOOK, Twitter; ….) as well as the forums included in the MOOCs has increasingly

S. Sraidi (✉) · E. M. Smaili · S. Azzouzi · M. E. H. Charaf
Faculty of Sciences, Ibn Tofail University, Kenitra, Morocco
e-mail: sraidi.soukaina@uit.ac.ma

E. M. Smaili
e-mail: smaili.miloud@uit.ac.ma

M. Ben Ahmed et al. (eds.), *Networking, Intelligent Systems and Security*,
Smart Innovation, Systems and Technologies 237,
https://doi.org/10.1007/978-981-16-3637-0_36

facilitated the social interaction of different users who can express themselves, ask questions or even share their experiences. Therefore, the huge data obtained through the exchanges of learners on MOOCs' forums and social networks (Facebook, or twitter for example) present important information that can help to identify the main causes of MOOCs dropping out by analyzing the difficulties the learners' encounter as well as their motivation.

To this end, we suggest in this paper to gather data from MOOC forums and associated social media groups. Then, we use sentiment analysis to extract and analyze useful data such as participants' feedback, learning methodology, programs, and so on. More precisely, a combination of a good number of quality tools (Pareto, cause effects, etc.) and machine learning approaches (Naives bayes, *k*-medoid) can be used on the basis of the comments of the forums, to determine the learners' motivation rate but also to identify and group the causes of failures. This information will help to understand learners' needs and the relationship between a problem and all possible root causes. The idea is to anticipate the learners' performance and then make the appropriate changes to ensure continuous improvement.

The document is organized as follows: The second section describes some basic concepts. Then, we describe materials and methods to implement our model in Sect. 3, Afterwards, we introduce the problematic statement and some related works respectively in Sects. 4 and 5. Then, we explain in Sects. 6 and 7 our MOOC analysis model as well as our quality approach. In the last section, we give some conclusions and identify future works.

2 Preliminaries

2.1 Massive Open Online Course (MOOC)

Recently, there has been a new style of distance learning, known as MOOC (Massive Open Online Courses), emerging in the world of education [1]. It provides a space for collaborative sharing of techno-pedagogical practices to promote both the individual's involvement as well as the common construction of knowledge [1]. The " massive " character of the MOOC refers to the large number of learners who participate and enroll in these programs. These massive courses are made up of different resources and activities that allow both communicating knowledge and learning as well as monitoring and supervising learners during their training paths.

2.2 *Quality Approach*

In a complex world, competitive and especially exigent, the quality remains a way to distinguish competitive organizations. Indeed, the quality is defined as the set of activities that guide and control an organization in terms of quality. The principle is:

- Organizations understand present and future needs.
- The leaders establish the purpose and the orientations of the organization.
- People at all levels are the essence of an organization and full involvement on their part allows their abilities to be used for the benefit of the organization.
- A desired result is achieved more efficiently when resources and related activities are managed as a process
- Identify, understand and manage interrelated processes as a system contributes to the effectiveness and efficiency of the organization to achieve its objectives.
- Continuous improvement of the overall performance of an organization should be a permanent objective of the organization and effective decisions are based on the analysis of data and information.
- An organization and its suppliers are interdependent and mutually beneficial relationships increase the capacity of both organizations to create value.

3 Materials and Methods

3.1 *Naïve Bayesienne Algorithm*

Naive Bayesian is a very simple and powerful algorithm for supervised machine learning, and it is one of the most popular classification algorithms, it allows to predict the probability of an event based on the conditions we know for the events in question.

The algorithm is easy to understand and guarantees good performance. It's also fast and easy to train even with a small data set.

3.2 K *Medoids Algorithm*

Description. *k*-medoids is a machine learning clustering algorithm that identifies groups of observations with similar characteristics, as well as individuals in different groups that stand out as much as possible.

Principe. The unsupervised *k*-medoids algorithm starts with a set of medoids, then iteratively replace one by another if it helps to reduce the overall distance. It is more efficient for small data. *K*-medoid is a robust alternative to *k*-means clustering (Fig. 1).

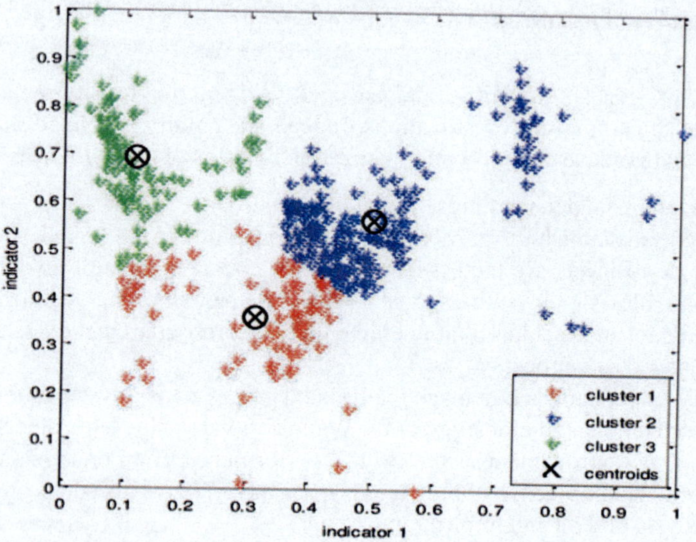

Fig. 1 *k*-medoid algorithm

3.3 *Pareto*

Description. The Pareto chart is a tool for showing graphically the problems affecting a given situation, listed in descending order. It is used to prioritize issues based on their frequency of occurrence (number of impressions). Pareto is useful to identify the reasons on which to significantly improve the situation. This will avoid wasting energy on things that have little impact.

Principe. In general, the method aims to sort any aggregate into two parts: the vital problems and the more secondary problems. This tool highlights the 80/20 rule. In other words, acting on 20% of the causes makes it possible to treat 80% of the effects. Below is a representation that illustrates this principle (Fig. 2).

4 Problematic Statement

Today, the MOOCs offer a space for exchange that allows a large number of learners around the world to easily access courses. They represent one of the best-known means of transmitting and disseminating knowledge. However, such motivation should be linked to the quality and performance of the training.

In fact, even of the MOOCs' popularity, the platforms they proposed don't face many challenges such as the high dropout rate. Our objective is to suggest a quality approach to assess the quality and performance of a MOOC. Therefore, the goal is to:

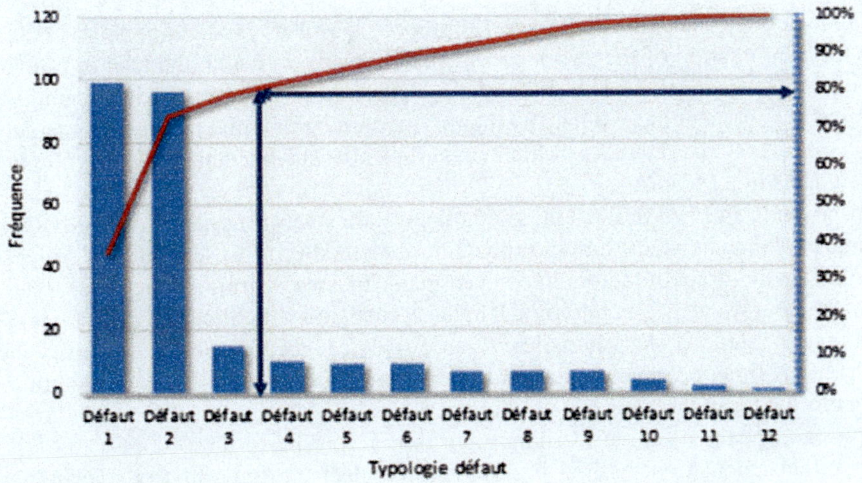

Fig. 2 Pareto diagram

- Collect and analyze the information provided by the learners in order to classify them according to three types (motivated, demotivated and neutral)
- Extract and group problems linked to a MOOC at all levels that could affect the high rate of dropout. (Platform defaults, E-Course Quality default, (student) Instruction Support defaults, Evaluation quality defaults, Infrastructure quality defaults)
- Identify all the causes that have a more or less direct influence on an observed problem in order to prioritize the efforts to be made for the problem resolution
- Highlight the most important causes on the total number of effects and thus take targeted measures to improve the course.

To this end, we propose tousesomemethods to extract, analyze, and control different constraints encountered by each learner in a MOOC in order to enhance the quality of courses.

5 Literature Review

Currently, there is little recent research devoted to the correction and prevention of the problems described in the previous section. Indeed, the authors in [2] present a visual analysis system for the purpose of exploring anomalous learning patterns and aggregating them into the data. The system integrates an anomaly detection algorithm that allows the interactive detection of anomalies between and within groups on the basis of semantic and interpretable data summaries by group.

On the other hand, the authors in [3] propose a MOOC prediction method based on the combination of the two tensors (global, local), to tackle the task of prediction of abandonment. In [4], the authors contribute to a deeper understanding of learner engagement in a MOOC by identifying three influential parameters, namely visits, attempts and feedback, which are sufficiently independent to allow grouping of students in a MOOC.

The study in [5] describes the methodology for analyzing behavior in MOOCs using the *k*-means algorithm and the "elbow method". In [6], the authors suggest to classify the dropout factors into seven major themes: learning experience, interactivity, course design, technology, language, time and situation. The authors in [7] examine the general characteristics of large-scale MOOC courses and quantify the influences of these characteristics on student performance. Furthermore, the authors propose in [8] an analysis of the characteristics of MOOCs users using the unsupervised *k*-means machine learning algorithm according to three stages: the first, the weight calculation method is designed to select the important characteristics according to the weight, the second is to optimize the algorithm for the initial cluster center, and the third is to determine the optimal number of clusters.

The article [9] provides an analysis of the expected dropout rate for MOOC students. This analysis automatically extracts functionality from click data and filters functionality using clustering tools and weighted MaxDiff to improve the accuracy of prediction. Moreover, the authors in [10] present an analysis with focus on three dimensions of learner behavior: Course Activity Profiles; Test activity profiles and the most relevant forum peers or best friends. The article [11] makes three major contributions to the literature related to the design and evaluation of open online courses: (1) an expanded assessment tool for MOOC teaching methods to be used by learned designers and researchers in their own contexts, (2) an illustration of how to use group analysis Close neighbors to identify educationally similar MOOCs, and (3) a preliminary analysis of clusters to take into account characteristics and factors contributing to educational similarity between massive, intra-clustered online courses.

The work [12] presents a deeper and better understanding of the behavior of MOOC actors by bringing together and analyzing the different objectives of these actors. The main finding was a set of eight clusters, namely blended learning, flexibility, high quality content, instructional design and learning methodologies, lifelong learning, learning by network, openness and student-centered learning.

Another study [13] apply models of unsupervised students initially developed for synchronous didactic dialogue to MOOC forums. They use a clustering approach to group similar articles, compare clusters with manual annotations by MOOC researchers. Besides, the authors in [14] review the factors that lead to a high number of dropouts in order to predict, explain and solve the problem related to both students and MOOCs. To this end, they suggest to use machine learning tools as well as artificial intelligence. Furthermore, the paper [15] suggests a CNN-LSTMATT method based on MOOC dropout prediction time series. The aim is to improve student course completion rate and to obtain a better prediction results using LSTM to extract temporal characteristics and CNN to extract local abstract characteristics.

Another model for MOOC dropout prediction based on extracting learning behaviors features is introduced in [16]. Their work is based on click path data and optimizing SVRmodel parameters using IQPSO. The approach presented in [17] gives an analysis and interpretation of the dropout phenomenon. Their study is performed on a dataset with four MOOC courses and 49,551 enrolled learners. They use also feature selection methods and machine learning algorithms in order to ensure the prediction and classification of learners in a MOOC.

Finally, we suggest in [18, 22] a new system to analyze the learner traces in the forum posts in order to understand and predict the causes of failure in MOOC learning environments. Moreover, we propose in [19] a system to analyze the performance of a MOOC by determining the relation between the learners' satisfaction rate and the main drop out causes using the combination of both ISHIKAWA quality method and a machine learning algorithms.

This paper can be considered as a continuity of previous works [18–21] where we propose to increase the courses completion rate and fight against the dropping out. The purpose is to make the MOOC adapted to each learner based on the knowledge and preferences of each one.

6 MOOC Analysis Model

In this section, we describe our approach to analyze the learners' engagement on MOOCs. We aim to provide an appropriate system to identify the root causes of dropouts. More specifically, we aim to analyze the comments left on social networks concerning MOOCs in order to deal with the various dropout causes.

Therefore, we highlight below the basic concepts of our analysis model. As shown in Fig. 3, the model is implemented in five steps:

6.1 Data Collection Approach

Data collection. We are moving from information societies to recommendation and sharing societies and social networks offer a perfect space for sharing information. In this context, we approach our subject by defining, collecting and storing MOOC forum data (forum posts) from Facebook and Twitter groups in Coursera sessions.

Therefore, the dataset is obtained via application programming interfaces (APIs) and also automatically via web scraping.

Preprocessing. As social media data is not standardized, we will need to preprocess such collected information. In fact, we pre-process the textual data for further analysis by replacing and removing all symbols with their emoticons (hashtags and URLs, stop words, non-English words, punctuation marks, symbols and numbers).

Generally, the data obtained remains incomplete, noisy and inconsistent due to the lack of attribute or errors and outliers for example.

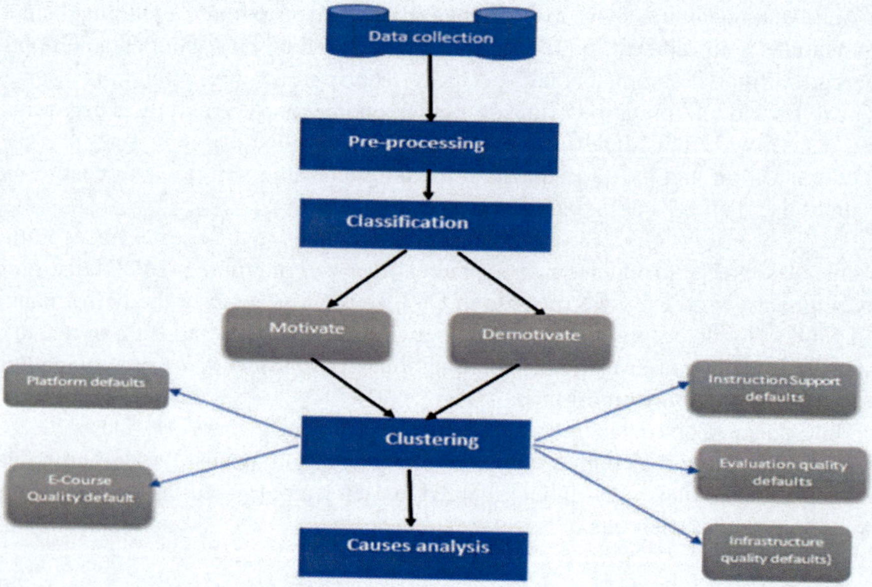

Fig. 3 MOOC analysis model

6.2 Classification System

We aim through this step to detect the sentimental polarity (positive, negative, neutral) of a given text using the supervised Naive Bayes algorithm. The name comes from Bayes' theorem, which can be written mathematically as follows:

$$P(A|B) = \frac{P(B|\mathrm{A}).P(A)}{P(B)} \tag{1}$$

With:

- $P(A)$ and $P(B)$ are respectively the probability of events: A, B and
- $P(B)$ is greater than 0.

The algorithm remains predictive despite the fact that the hypothesis of independence of the explanatory variables conditioned by the classes is difficult to justify. The extraction of opinion targets and the sentiment expressed towards these targets will help us to extract the causes of success (positive ranking) and failure (negative ranking) of the MOOC.

6.3 Clustering

The classification of the feelings expressed in the previous step will help us to distribute the data concerning the demotivation of learners in clusters using the unsupervised algorithm k-medoids with $k = 5$: (Platform defaults, E-Course Quality default, (student) Instruction Support defaults, Evaluation quality defaults and Infrastructure quality defaults). In order to extract the causes of failure in the MOOC, the algorithm works as follows:

- Choose k data points in the cloud as the starting points for the cluster centers.
- Calculate their distance from all points of the point cloud.
- Classify each point of the cluster where it is closest to the center.
- Select a new point in each cluster that minimizes the sum of the distances of all the points in this cluster from itself.
- And at the end Repeat step 2 until the centers stop changing in a data set.

At this stage, we specify also the indicators needed to assess the obtained results. After having the problemsgrouped, we move to the analysis of each problem using an adequate quality tool. The aim is to identify the causes of a problem and to understand the relationship between a problem and all the possible causes.

6.4 Causes Analysis

After collecting, classifying and grouping the causes into categories, we proceed to the presentation of the most important causes of failure of a MOOC using the PARETO quality control method.

The objective is to highlight the important causes regarding the total number of effects which allow to take corrective and preventive actions and by the way to improve the quality of the MOOC.

7 Proposed Quality Approach

7.1 Prototype

Figure 4 gives an overview of our prototype. It describes the proposed approach starting with the data collection from the social networks until the detection of the various causes of failure of the MOOCs.

The approach can serve as a system for detecting the root causes of failure but also to establish corrective and preventive actions for help decision-makers to improve the quality of MOOCs.

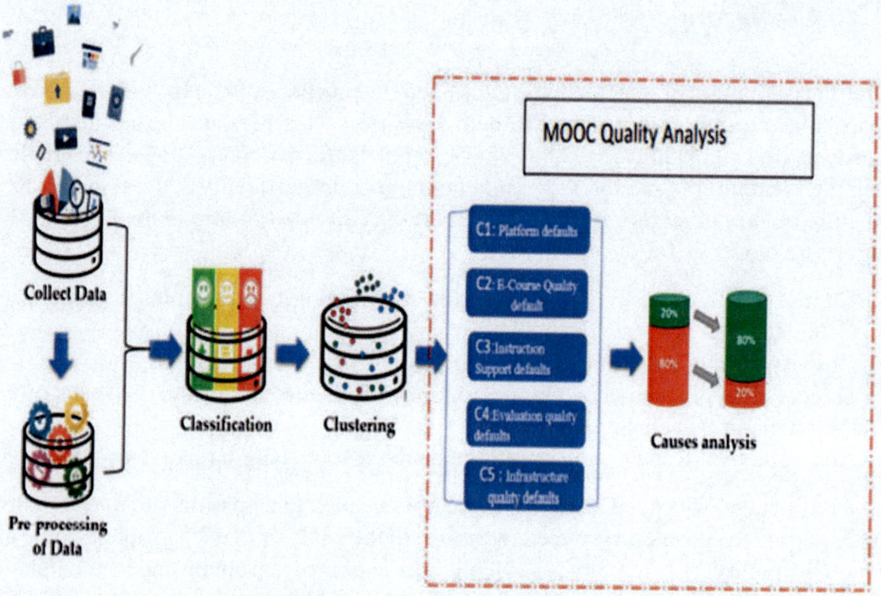

Fig. 4 MOOC quality analysis approach

7.2 *Model Description*

The aim of our approach is to extract the root causes of failures and provide an overview of the most important causes in order to overcome the low completion rate and improve the quality of MOOCs.

As shown in Fig. 4: All learners' interactions (forum post) are defined, collected and stored through application programming interfaces (APIs) or obtained automatically through web scraping. Afterwards, the data is preprocessed and classified in order to extract various characteristics and to determine polarities (motivate, demotivate, or neutral) using the machine's naive bayes algorithm. After classifying the learners, we group the causes of success and failure into five groups ((Platform defaults, E-Course Quality default, (student) Instruction Support defaults, Evaluation quality defaults, Infrastructure quality defaults).

Finally, we calculate the degree of impact of each given phenomenon in relation to other phenomena using the law of 20/80. The objective is to detect the 20% of the causes that generates 80% of the consequences in order to enhance the performance and the quality of the MOOCs.

8 Conclusion

As conclusion, the work presented in this article deals with the dropping out of learners during their MOOC training. Therefore, the main objective is to analyze the obstacles and problems encountered by learners during their online courses. To this end, we follow in the footsteps of the learners during a course, then we analyze their interactions on MOOCs forums as well as social media platforms using machine learning algorithms. The basic idea is to classify learners' data according to their polarities (negative, positive or neutral) then grouping the difficulties encountered by the learners (with negative polarity) into five clusters (Platform defaults, E-Course Quality default, Instruction Support defaults, Evaluation quality defaults and Infrastructure quality defaults) which allow us to extract the causes of failure in MOOCs.

Therefore, our approach aims to provide the necessary support to the decision makers to define the corrective and preventive actions to eliminate the causes which generate a higher dropout rate. Finally, our approach is limited to the use of data according to Latin language as our experience regarding the use of the Arabic language presents many difficulties. As prospects, we plan to propose new quality methods with other different strategies to evaluate the learning process within MOOCs to further improve our approach.

References

1. Tahiri, J.S., et al.: MOOC…Un espace de travail collaboratif mature: Enjeux du taux de réussite. La 2éme conférence francophone sur les systemes collaboratifs (SysCo'14). SEPTEMBER, pp. 131–144 (2014)
2. Mu, X., et al.: MOOCad: visual analysis of anomalous learning activities in massive open online courses. EuroVis (Short Papers). **2019**, 91–95 (2019). https://doi.org/10.2312/evs.20191176
3. Liao, J., et al.: Course drop-out prediction on MOOC platform via clustering and tensor completion. Tsinghua Sci. Technol. 24(4), 412–422 (2019). https://doi.org/10.26599/TST.2018.9010110..
4. Shi, L., et al.: Revealing the hidden patterns: a comparative study on profiling subpopulations of MOOC students. In: Proceedings of the 28th International Conference on Information Systems Development: Information Systems Beyond 2020, ISD 2019 (2019)
5. Shi, L., et al.: Social interactions clustering MOOC students: an exploratory study. arXiv. (2020).
6. Goopio, J., Cheung, C.: The MOOC dropout phenomenon and retention strategies. J. Teach. Travel Tourism 00, 00, 1–21 (2020). https://doi.org/10.1080/15313220.2020.1809050.
7. Xing, W.: Exploring the influences of MOOC design features on student performance and persistence. Distance Educ. **40**(1), 98–113 (2019). https://doi.org/10.1080/01587919.2018.1553560
8. Xiao, L.: Clustering research based on feature selection in the behavior analysis of MOOC users. J. Inf. Hiding Multim. Signal Process. **10**(1), 147–155 (2019)
9. Ai, D., et al.: A dropout prediction framework combined with ensemble feature selection. In: ACM International Conference Proceeding Series (New York, NY, USA, Mar. 2020), 179–185 (2020)

10. Liu, Z., et al.: MOOC learner behaviors by country and culture; an exploratory analysis. In: Proceedings of the 9th International Conference on Educational Data Mining, EDM 2016, pp. 127–134 (2016)
11. Quintana, R.M., Tan, Y.: Characterizing MOOC pedagogies: exploring tools and methods for learning designers and researchers. Online Learn. J. **23**(4), 62–84 (2019). DOI:https://doi.org/10.24059/olj.v23i4.2084.
12. Yousef, A.M.F., et al.: A cluster analysis of MOOC stakeholder perspectives. RUSC. Univ. Knowl. Soc. J. **12**(1), 74 (2015). https://doi.org/10.7238/rusc.v12i1.2253
13. Ezen-Can, A., et al.: Unsupervised modeling for understanding MOOC discussion forums: a learning analytics approach. In: ACM International Conference Proceeding Series. 16–20-Mar (2015), pp. 146–150. https://doi.org/10.1145/2723576.2723589.
14. Fisnik, D., et al.: MOOC Dropout Prediction Using Machine Learning Techniques: Review and Research Challenges (2018). https://doi.org/10.1109/EDUCON.2018.8363340
15. Min. C., et al.: A dropout prediction method based on time series model in MOOCs. J. Phys.: Conf. Ser. **1774**(2021) 012065 IOP Publishing (2021). https://doi.org/10.1088/1742-6596/1774/1/012065.
16. Cong, J.: (2020). MOOC student dropout prediction model based on learning behavior features and parameter optimization. Interact. Learn. Environ. https://doi.org/10.1080/10494820.2020.1802300
17. Youssef, M., et al.: (2019) A machine learning based approach to enhance MOOC users' classification. Turk. Online J. Distance Educ.-TOJDE April 2020 ISSN 1302-6488
18. Soukaina, S., et al.: Quality approach to analyze the causes of failures in MOOC. Proceedings of the 5th International Conference on Cloud Computing and Artificial Intelligence: Technologies and Applications November 24–26, pp. 1–5 (2020). https://doi.org/10.1109/CloudTech49835.2020.9365904
19. Soukaina, S., et al.: MOOCs performance analysis based on quality and machine learning approaches. In: Proceedings of the 2nd IEEE International Conference on Electronics, Control and Computer Science, 2–3 Dec 2020, Kenitra Morocco (2020). https://doi.org/10.1109/ICECOCS50124.2020.9314606
20. Miloud, S., et al.: An adaptive learning approach for better retention of learners in MOOCs. (2020). In: Proceedings of the 3rd International Conference on Networking, Information Systems & Security (NISS2020), Article 26, pp. 1–5 (2020). https://doi.org/10.1145/3386723.3387845
21. Smaili, E.M., et al.: An optimized method for adaptive learning based on PSO algorithm. In: Proceedings of the 2nd IEEE International Conference on Electronics, Control and Computer Science, 2–3 Dec 2020, Kenitra Morocco (2020). https://doi.org/10.1109/ICECOCS50124.2020.9314617
22. Smaili E., et al.: (2021) Towards Sustainable e-Learning Systems Using an Adaptive Learning Approach. In: Emerging Trends in ICT for Sustainable Devel Science, Technology & Innovation (IEREK Interdisciplinary Series for Sustainable Development). Springer, Cham. https://doi.org/10.1007/978-3-030-53440-0_38

The Personalization of Learners' Educational Paths E-learning

Ilham Dhaiouir, Mostafa Ezziyyani, and Mohamed Khaldi

Abstract The personalization of learners' educational paths in MOOC (Massive Open Online Courses) is one of the major questions that help teachers in their functions of teaching and monitoring of learners. Being intended for a huge and multiple number of learners, MOOCs are problematic for teachers in monitoring and understanding each learner's profile. It is for these reasons that we propose a semantic web (SW) approach to build a recommendation system in a graphical interface that we created using the NetBeans IDE. The goal is to enable learners who follow the MOOC to directly head for modules that meet the characteristics and requirements of their profiles. Moreover, the proposed approach is based on some essential technologies such as resource description framework (RDF), ontology, web ontology language (OWL) and RDF query language (SPARQL). Thus, the main objective of this paper is to present another beneficial technique aimed at improving distance courses by allowing teachers and learners to work in better conditions. The primary results showed that the performance of the (SW) based recommendation system is effective.

1 Introduction

Online classes or E-learning represent a set of educational resources (courses, videos, activities, projects, quizzes, etc.) that are accessible for everyone and free via Internet. Indeed, it is the current most useful way of learning in universities and institutions all over the world because of the current circumstances of the COVID-19 pandemic we are experiencing so far.

In an online course, the learner is at the center of the learning process because he must be autonomous in the organization of the learning process, as for the teacher who

I. Dhaiouir (✉) · M. Khaldi
Laboratory of Mathematics and Application FST, Tangier, Morocco

M. Ezziyyani
Laboratory of S2IPU ENS, UAE, Tetouan, Morocco

M. Ben Ahmed et al. (eds.), *Networking, Intelligent Systems and Security*, Smart Innovation, Systems and Technologies 237,
https://doi.org/10.1007/978-981-16-3637-0_37

provides supervision in a distance training, it plays a very important and primordial role in the preparation of educational resources (courses, activities, exercises, videos) and also in the simplification of interactions between them and the learners. But since MOOCs are accessible to everyone and free, then this presents a real problem in terms of the massive number of registered learners with different backgrounds, levels and learning interests.

Indeed, according to the collected statistics during the ABC MOOC of Project Management, we have noticed this variety [1] because each learner has different expectations, basic comprehension, priorities or even ways of learning that varies from one learner to another. But presently, we find that in most MOOCs, there is only one path offered to learners, this latter does not necessarily meet the expectations and needs of all the learner profiles. So, the apprentices who participate in the training and pursue it to the end have a low rate of 10%. This issue is at the center of the current research on MOOCs through the analysis of learner behavior [2–4] to improve this retention rate [4–6]. So, as a solution to this problem, we need the personalization of educational content to adapt to the characteristics of the learner.

The personalization of learning in a MOOC is a learning process that seeks to improve the quality of online education and training. The latter has become very important in the educational field because a good pedagogy requires a good design of courses that adapt to the profiles of the learners.

The personalization operation obliges having an idea about the knowledge of the learners [7] based on their experience, their level of knowledge, their format and resource preferences [8] as well as the monitoring of learners in their learning process to have any information that may be useful in defining the progress situation of learners.

To guarantee the personalization of a MOOC, the teacher must present content adjusted to the profile of each learner taking into account their characteristics and to meet their needs as regards the learner, he must play an important role in the lead and control of learning [9].

This is why our objective in this article is to create a graphical interface intended for the use of learners who follow distance training by recommending them personalized educational content appropriate to their profiles. First, we will propose the work related to the question of the customization of courses in MOOCs. Secondly, in our approach we will collect all the beneficial traces left by the learners during the registration in the formation also during the passage of the diagnostic test. Thirdly, on the basis of these traces we will determine the information on the basis of which we will create ontologies thanks to the "Protégé" software. Fourth, on the basis of these ontologies we create a recommendation system in a graphical interface that we create using NetBeans IDE to be able to recommend personalized educational content that is appropriate to the learning levels and preferences of each learner.

2 Related Work

Many approaches have been appointed by numerous author that allows personalization either by focusing on the observation of the learner himself, or by recommending or adapting the content to the profiles of the learners.

Nonetheless, most of the approaches have limits and, in most cases, they offer learners internal resources for the followed MOOC that we find reasonably limited. Because in case the MOOC does not manage to respond to a difficulty confronted by the learner, then it is useful to offer them information on the web that can be useful for their noted difficulty or to widen their knowledge.

The approach of [10], proposed a model named PERSUA2MOOC which allows teachers and MOOC designers to personalize learners' trails by offering them content that fulfills their educational objectives, taking into consideration analyzing the traces left by learners during their participation in the platform and carrying out activities. As in the work of [11], which is formulated to guide and advise participants with low knowledge and low know-how in the MOOC. My learning mentor intends to increase the level of this section of learners by nudging them to be independent in their learning and by empowering them with planning and cues to help them get the most out of MOOCs by supporting self-learning. Regardless, this work is still at an initial stage and needs to be implemented and evaluated with real MOOCs. Moving on to another work, which is that of [12], who considers that the strategy of personalization comprises in establishing a process of transformation of learning scenarios, to resources for the semantic personalization of learning experiences MOOC, via the plan of a method for estimating the relationships between abilities based on a structured competency classification, whose purpose is to match the skills of learners with those who are involved in other elements of the learning plan. This comparison method has been executed within a professional platform called TELOS. The example illustrated here is quite general, and thus, it is not that manageable to implement.

The work of [13], has made it feasible to find the MOOCs which suit a learner, by asking them what are their learning objectives, expressed throughout a taxonomy of the studied field. This approach [13] is established on an internal study, by directly asking questions to the learners to specify their necessities. But the difficulty that occurs here is the massive number of learners, which makes it impossible to analyze all the responses of the learners to find them MOOCs suitable, for their learning objectives. Another work that was able to formulate the first adaptive MOOC platform (MOOC) is that of [14]. The platform, offers a solid academic framework and a personalized learning experience in a MOOC learning environment. [14] Expanded the first MOOC in the field of computational molecular dynamics (CMD). In this work, [14] we describe the design, improvement and deployment of this MOOC which has managed to handle the heavy loads of the massive open online course and the stress of the users. The convincingness of this approach [14] has not yet been approved, because the data on learner behaviors obtained by the authors are under study and analysis and have not been presented definitely. Another cited project is that of [5], which recommended the examination of the requested activities to learners on the MOOC platform, to detect the problems overlooked by learners which demotivate them and which push them to with-draw the MOOC. This approach [5], consists of estimating the basic deviation between the objectives set by the learners and the current accomplished level, to offer corrective classes for learners, who have not achieved their goals. This project had thresholds concerning the taking into account of all the learners are given their massive number.

3 Materials and Methods

Our work is in the field of computer learning environments (EIAH) rather than in the field of information and communication technologies applied to education (NTICE) and pedagogical engineering which is defined in [15], as: "A method that is supporting the analysis, design, control and planning of the dissemination of education systems, combining the concepts, processes and precepts of educational design, software engineering and cognitive engineering. The pedagogical design according to [15] which is a form of engineering attempted to improve educational practices." The submitted protocol is an approach aimed at the creation of a graphical interface facilitating the orientation of learners by recommending personalized content. On the basis of a set of criteria: mark, level, sector, language, age … which must first meet their needs and secondly facilitate teachers' monitoring and supervision of learners. Our interface also allows learners who need support the possibility of reuse of free educational resources (OER) delivered on the web by other teachers and public institutions which, according to the report from the 2002 UNESCO forum, stand out as follows: 'The recommended definition of open educational resources is: The open provision of educational resources, made possible by information technologies and of communication, for consultation, use and adaptation of the user community for non-commercial purposes'. It has a heterogeneous identity, which strives to help learners in their learning, as well as to help teachers in the operation schooling. They are privileged, which means no one can edit them, but they can be reused and shared. A quality OER is very expensive in terms of preparation time and effort.

Thus, to regulate the diversity of OER, we will use the ontologies of the semantic web and the LOM metadata standard to use the external educational resources of the web when needed by the learner. Therefore, in our work, we will use the information entered by the learners when registering for the training, then these learners will be directed to take a diagnostic test to determine the prerequisites of each of them. On the basis of these traces, we will create an ontology on "Protégé" which will allow us subsequently by using the RDF / XML and OWL language in NetBeans IDE to create a graphical interface which introduces a system of recommendation that recommends to each learner, the modules followed in this training that fits with his specialty, his level and his skills.

3.1 Research Framework and Context

The study involves a sample of 120 students with a bachelor's degree and who have graduated from the following three majors: Mathematical and Computer Sciences, Physics and Biology.

This is a supplement concerning various modules appropriate to the three courses mentioned above which was started in the 1st of December 2020. The content of the modules was structured in six main sessions, corresponding to the six weeks of lessons.

Each of these weeks was itself structured into short sub-sequences each comprising a short video, a manuscript and a self-assessment quiz of the concepts covered in the video and the manuscript.

In addition, each set of sequences was accompanied by an introduction that determines the objectives of the week, the necessary prerequisites that the learner must have as well as the necessary time that must be devoted to each part.

The learners are led to read the manuscripts shared by the teacher, to see videos which facilitate the comprehension of the lessons as well as to carry out individual assignments in the form of exercises that are requested at the end of each week.

3.2 The Traces

The reuse of traces, within the framework of computerized human learning environments (EIAH), has emerged in a movement of increasing complexity of technologies supporting ILE and their uses. The question of collecting traces, analyzing them and using them is far from new. The problem is not only how to analyze the traces but also how to really complete and exploit the traces resulting from the observations and left by the learners in the platform to improve the learning of the learners.

The digital traces left by the students in the registration form were of a heterogeneous and massive character but useful for us, because thanks to them we can have an impression on the profiles of the learners registered in the MOOC.

In our case, we will use all the traces found during the registration of learners in the training by filling out the registration form. Once learners have completed their registration they will automatically be taken for a diagnostic test, and on this basis the teacher can get a general idea of each learner's level, prerequisites, knowledge and gaps in order to address them. Then, guide them to courses that meet their expectations, preferences and in general their profile. These traces can be used to create ontologies in the "Protégé" software (see Fig. 1).

According to the figure above, the processing of all the traces will be carried out based on two types of data: The first type concerns the registration data retained when filling out the registration form, the second type concerns the prerequisite data retained during the diagnostic test. From this group of data, we will create ontologies on the "Protégé" software in order to attribute to each student the modules and the

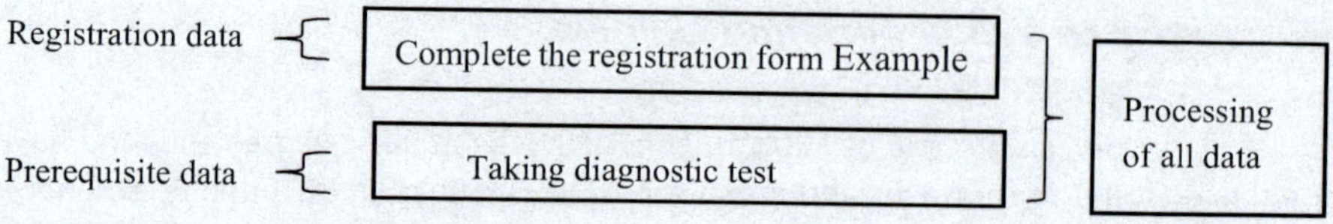

Fig. 1 All the traces to be processed

course that are appropriate to their characteristics. The figure above shows all the traces that seem useful to us in our study and which will help us subsequently to create ontologies.

3.3 Semantic Web

The semantic web represents an evolution of the World Wide Web. This term of (SW) involves specific techniques recommended by the World Wide Web Consortium (W3C) that improves the current web and its use. These techniques provide a meaning of the data without worrying about their representation [16] and both humans and machines can understand it. Additionally, the objective of (SW) is to create a global data base of connected data on the web [17]. Thus, the semantic web technologies are ontology, resource description framework (RDF), web ontology language (OWL), and SPARQL query language.

3.4 Web Ontology Language (OWL)

Web ontology language (OWL) allows users to represent and write their ontologies in a specific domain, it is built on the top of RDF [18]. For creating these ontologies, we can use a free open-source "Protégé" ontology editor, this platform is popular in the semantic web field and it was developed in JAVA. With this editor, the user could create and manipulate the ontologies in various representation formats.

The use of these traces allowed us to get an idea about each learner profile. We will use the secure software for the creation of OWL2 (Web Language Ontologies) which are defined by an IRI. Ontology is a central element of the semantic web which seeks, on one hand, to rely on the modeling of web resources from conceptual representations of the concerned fields and on the other hand, enable to make assignments [19]. An ontology is not only the identification and the classification of theories but they are also elements which are attached to them and which one calls here properties that can be evaluated. "Protégé" is a graphical ontology development environment developed by Stanford SMI.

In "PROTÉGÉ" knowledge model, ontologies are made up of a skeleton of classes that have attributes (slots), which can themselves have certain properties (facets) [20].

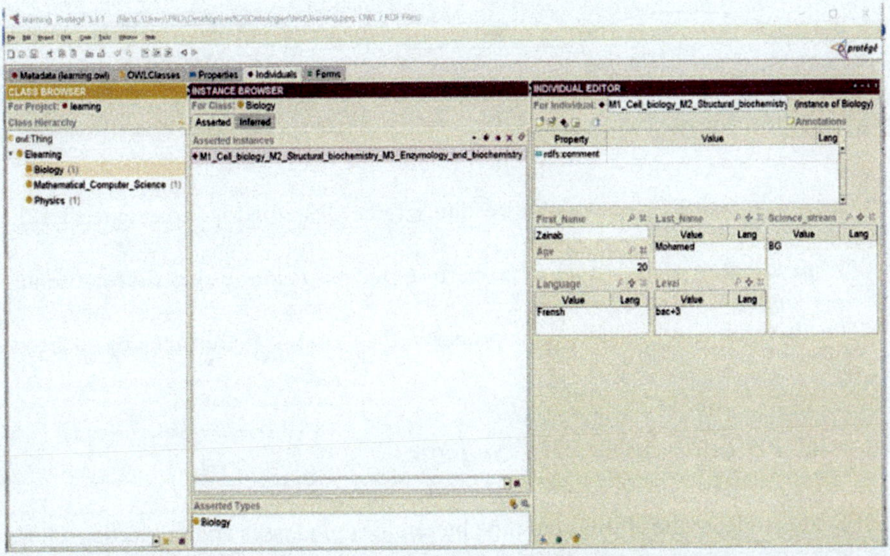

Fig. 2 Example of our ontology created by "Protégé"

The development of the list is done via the graphical interface without needing to produce a formal language. In our case, we chose on three types of route (see Fig. 2).

In the figure above, we have created an ontology on "Protégé" for the three specialties: Biology, Mathematical Computer Science and Physics.

For each specialty, we have proposed three modules. Learners are automatically classified according to their specialty based on the information they entered, while registering for the training.

3.5 *Resource Description Framework (RDF)*

Resource description framework (RDF) is a model to describe the resource on the web and their metadata or semantics in order to interchange the data on the web. The RDF consists of three concepts [21] such as: Resource: this term refers to a web page, image, person, video, fragment page, which means anything that has a URI can be considered as a resource. Description: for describing the resource in order to be more understandable, also the description has attributes, features, and specific relation between resources. Framework: for providing the language syntaxes of these descriptions. The information on the web is structured in the form of the triple by RDF [22] (see Fig. 3).

The RDF language which represents a formal framework for describing resources according to a graph model, by expressing the relations between these resources in the form of triples: Subject-Predicate-Object.

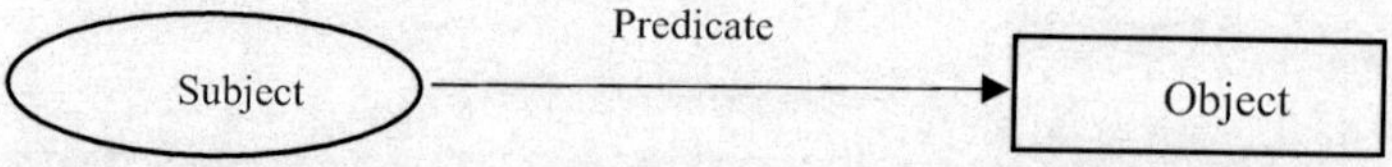

Fig. 3 The RDF triples

The subject that represents the resource to be described is represented by the ellipse.

The predicate means the type of property applicable to this resource is represented by a rectangle.

The object that is the value of the property (the literal) is represented by an arrow.

3.6 RDFS Language (RDF Scheme)

RDFS adds to RDF the ability to define hierarchies of classes and properties whose applicability and range of values can be constrained using the rdfs: domain and rdfs: range properties. Each application domain can thus be associated with a scheme identified by a particular prefix and corresponding to a URI.

The ***rdfs*:*Class***: allows a resource to be declared as a class for other resources.

For example, we can define in RDFS the ***Biology*** class which describes one of the streams of our training.

The ***subclassOf*** property is used to define class hierarchies for our example the Biology sector is an existing course in our E-learning course.

```
<rdfs:Class rdf:ID="Biology">
  <rdfs:subClassOf rdf:resource="#Elearning"/>
  <rdfs:Class>
```

- **RDFS** clarifies the notion of property defined by RDF by allowing to give a type or a class to the subject and to the object of triples. For this, RDFS defines:
- **The rdfs property domain**, which allows you to define the class of subjects linked to a property.
- **The rdfs property**, which allows you to define the class or the data type of the values of a property.

3.7 SPARQL Protocol and RDF Query Language (SPARQL)

SPARQL protocol and RDF query language are a query language used for querying and updating the RDF documents [23]. The purpose of using the SPARQL query language is that the RDF is based on the XML language, and for making the communication and exchanging the information, we need the SPARQL query. The query

```
PREFIX ns: <http://www.owl-ontologies.com/learning.owl#>
PREFIX rdf: <http://www.w3.org/1999/02/22-rdf-syntax-ns#>
SELECT ?elearning
WHERE{
    ?elearning ns:Module ?Module.
```

Fig. 4 Example of SPARQL query language

language must understand the syntax of RDF and also must understand the data model and semantic vocabulary of RDF [24] (see Fig. 4).

Where, PREFIX ns: http://www.owl-ontologies.com/learning.owl# is the namespace of ontology, PREFIX rdf: http://www.w3.org/1999/02/22-rdf-syntax-ns# is the namespace of RDF, rdf: type is the property, and ns: E-learning is the object. So, when execute this query, will return all E-learning. Therefore, all the steps taken to develop our recommendation system proposed, in our graphical interface based on the semantic web approach, in order to recommend the appropriate modules for each learner, during the training are illustrated in the flowchart below (see Fig. 5).

4 Result

In our work, we have been able to obtain satisfying results by creating a graphical interface, incorporating a recommendation system that will facilitate the task of monitoring and supervising the teachers who provide this training. Our web interface, is based on the SW approach, developing HTML5, JavaScript, CSS and a technique for dynamically create HTML code (JSP) "Java Server Pages," were all used in order to implement the web page, in addition the Bootstrap framework was used for the design of our web interface.

And for the development of our recommendation system based on the semantic web, eclipse IDE for Enterprise Java developers and the Apache Jena framework (for creation of the semantic web application) were used for the purpose of implementing all classes and methods by working with the ontology file created in the "Protégé "software. This ontology is in RDF format and having semantic data.

To create our ontology, we based ourselves on traces left by learners when filling out the registration form to participate in the training. These data consists of specific parameters to recommend the modules that each learner must follow during the training and which meet the characteristics of his learning profile. In Fig. 6, we find the source code that allowed us to create our graphical interface in NetBeans IDE (see Fig. 6).

Once our ontology had been created in the "Protégé" software, we will then create a graphical interface by NetBeans IDE in order to recommend to the learners the modules that they will follow during the training and which are suitable for their profiles.

Fig. 5 Course recommendation system based on the semantic web approach design flowchart

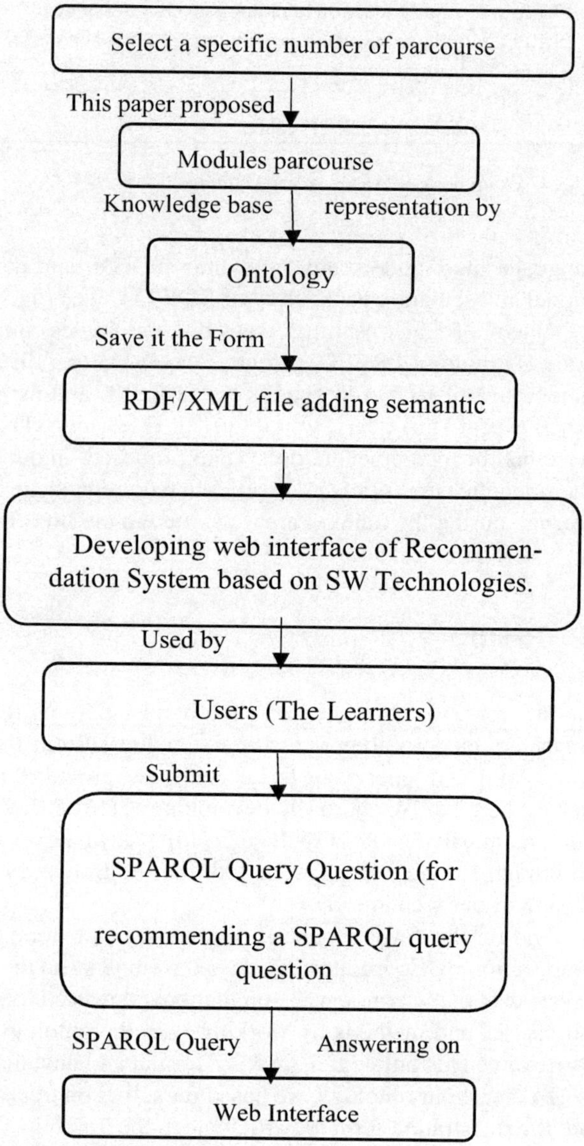

To demonstrate our ontology based recommendation system approach we will develop a computer application on Java. The software will be able to consult and analyze our ontological model created in order to reveal the necessary data to formulate the recommendation data to learners. For this, we use the following elements.

The ***Sparkle Query*** language for the selection and the ***JenaOnthology*** package to create the ontological model where we will load the data as well as a file for the

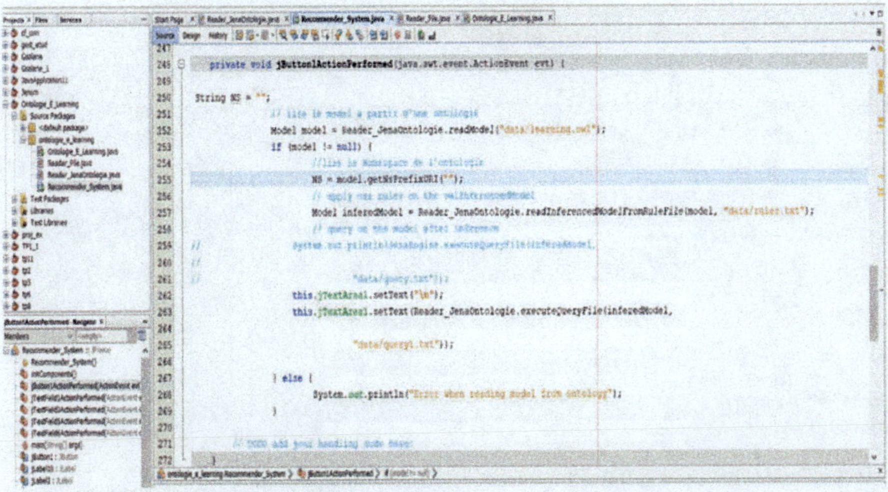

Fig. 6 The source code of our graphical interface created on NetBeans

rules. The data revealed by the recommendation will be interpreted in the form of a text on a graphical interface see Fig. 7.

For example, according to the figure above, the 20-year-old student Karim Zerhouni with a bachelor's degree in MCS, was recommended to his three modules in English to follow during the training:

- Module 1: Mathematical programming.
- Module 2: Commutative algebra.
- Module 3: Java Programming.

When we click on Submit SPARQL Query, the query uses all the parameters inserted by the teacher (see Fig. 8).

When we click on submit response to the SPARQL request of the recommended module according to these settings, that is to say: First Name = Karim, Last Name = Zerhouni, Age = 20, Level = Bac + 3, Science Stream = MSC, Language = English.

Then, according to this person's settings, modules that meet the characteristics of his profile will be recommended to him to follow during the online training.

5 Conclusion

In this article, we proposed a recommendation system based on the semantic web approach in the field of E-learning to better help teachers to monitor and supervise learners as well as to help them follow the MOOC until the end. In our graphical interface, our recommendation system has the function of recommending the appropriate

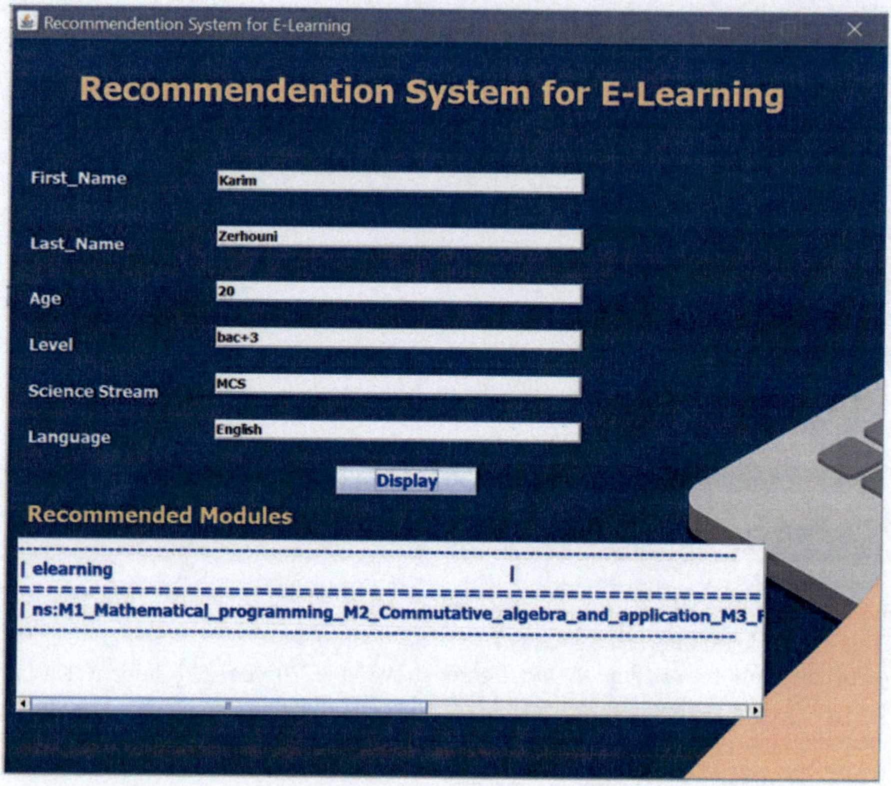

Fig. 7 Web interface of recommendation system

```
PREFIX ns: <http://www.owl-ontologies.com/learning.owl#>
PREFIX rdf: <http://www.w3.org/1999/02/22-rdf-syntax-ns#>
SELECT ?elearning
WHERE{
        ?elearning ns:Age ?Age.
        FILTER(?Age = 20)
        ?elearning ns:First_Name "Karim".
        ?elearning ns:Last_Name "Zerhouni".
        ?elearning ns:Level "bac+3".
        ?elearning ns:Language "English".
        ?elearning ns:Science_stream "MCS".
         ?elearning rdf:type ns:Elearning.
```

Fig. 8 SPARQL Query based on parameters provided by the teacher modules for each learner, according to the specific criteria that have been dedicated when registering learners in the training. Our objective was to help teachers as well as learners by improving web research, improving efficiency and give the answer quickly thanks to the technologies of the semantic web. Additionally, this system facilitates the exchange of both human and mechanical information is why we can say that this is the smart web. The results showed that the recommendation system based on the SW approach is effective. Our perspective is to improve this system in future work.

References

1. Cisel, M.: Qui étaient les participants du MOOC Gestion de Projet ? Blog La révolution MOOC. http://blog.educpros.fr/matthieu-cisel/2013/08/16/qui-etaient-lesparticipants-du-mooc-gestion-de-projet
2. Baker, R., Evans, B., Dee, T.: Understanding persistence in MOOCs: descriptive & experimental evidence. EMOOCs **2014**, 5–10 (2014)
3. Willems, C., Renz, J., Staubiz, T., Meinel, C.: Reflections on enrollment numbers and success rates at the openhpi MOOC platform. EMOOCs **2014**, 101–106 (2014)
4. Halawa, S., Mitchell, J.: Dropout prediction in MOOCs using learner activity features. EMOOCs **2014**, 58–65 (2014)
5. Miranda, S., Mangioni, G., Orciuoli, F., Loia, V., Salerno, S.: The SIRET training platform: facing the dropout phenomenon of MOOC environments. EMOOCs **2014**, 107–113 (2014)
6. Liyanagunawardena, T.R., Parslow, P., Williams, S.A.: Dropout: MOOC participants' perspective. EMOOCs **2014**, 95–100 (2014)
7. Brusilovsky, P.: Adaptive and intelligent technologies for web-based education. http://www.kuenstliche-Intelligenz.de/archive/ (2001)
8. Höök, K.: Steps to Take Before IUI Becomes Real. The Reality of Intelligent Interface Technology, Edinburgh (1997)
9. Lebow, D.: Constructivist values for instructional systems design: five principles toward a new midset. Educ. Tech. Res. Dev. **41**(3), 3–16 (1993)
10. Lefevre, M., Guin, N., Jean-Daubias, S.: Personnaliser des activités pédagogiques de ma-nière unifiée: une solution à la diversité des dispositifs. STICEF 19, (2012)
11. Gutiérrez-Rojas, I., Alario-Hoyos, C., Pérez-Sanagustin, M., Leony, D., Delgado-Kloos, C.: Scaffolding self-learning in MOOC. EMOOCs **2014**, 43–49 (2014)
12. Gilbert Paquette, O.M.: Competency-based personalization for massive online learning. Smart Learn. Environ., pp. 1–19 (2015)
13. Gutiérrez-Rojas, I., Leony, D., Alario-Hoyos, C., Pérez-Sanagustin, M., Delgado-Kloos, C.: Towards an outcome-based discovery and filtering of moocs using moocrank. EMOOCs **2014**, 50–57 (2014)
14. Sonwalkar, N.: The first adaptive MOOC: a case study on pedagogy framework and scalable cloud architecture—part I. MOOCs Forum, pp. 22–29 (2013)
15. Paquette, G.: L'ingénierie pédagogique: pour construire l'apprentissage en réseau. Presses de l'Université du Québec, Québec (2002)
16. Robu, I., Robu, V., Thirion, B.: An introduction to the semantic web for health sciences librarians. J. Med. Libr. Assoc. **94**(2), 198–205 (2006)
17. Laufer, C.: Guia_Web_Semantica, p. 133 (2015)
18. McGuinness D.L., Van Harmelen, F.: OWL web ontology language overview. W3C Recomm. **2004**, 10 (2004).
19. Charlet, J., Bachimont, B., Troncy, R.: Ontologies pour le Web sémantique, s. d., 31.2004.

20. https://protege.stanford.edu/index.shtml. Author, F.: Article title. J. **2**(5), 99–110 (2016)
21. Gandon, F., Krummenacher, R., Han, S.-K., Toma, I.: The Resource Description. Framework and its Schema. Handbook of Semantic Web Technologies (Issue January). http://www.springer.com/us/book/9783540929123 (2011)
22. Cyganiak, R., Wood, D., Lanthaler, M.: RDF 1.1 concepts and abstract syntax. W3c Recomm. 2014; 25
23. Group WSW: SPARQL 1.1 Overview. W3C Recomm. W3C (2013)
24. Che,H.: A semantic web primer. J. Am. Soc. Inf. Sci. Technol. **57**(8) (2006). https://doi.org/10.1002/asi.20368

Formulating Quizzes Questions Using Artificial Intelligent Techniques

Abdelali El Gourari, Mustapha Raoufi, and Mohammed Skouri

Abstract Formulating the question is one of the most important things for evidence-based practice, regardless of the discipline in question. Formulating questions, and even answering questions, is a powerful key to our profession, and our practice can be seen through research in information science and by significant developments in evidence-based practice. We should therefore develop a comprehensive classification of the types of frequently asked questions and prioritize initial research. In this paper, we will discuss how to create and make tests questions based on artificial intelligence techniques and algorithms and make a simple comparison between the most recent methods used to make questions, and finally, we will do an analytical study at Cadi Ayyad University Marrakech Morocco by one of the ways based on artificial intelligence (AI) algorithms.

1 Introduction

The process of making questions is to focus on using computer technology to create questions from one or more sources of data. In other words, we are creating an artificial intelligence-based system consisting of inputs and outputs that take existing sources from data sources as inputs, as these sources can be organized by (datasets, excel, tables, and texts…) or unorganized sources (such as texts: books, science articles, email messages we send and receive…). In addition, other sources may have expressed questions that are made by the teacher in order to obtain questions similar to those he has entered, and when the system handling this data, it will provide us with an output, which can be one type or divided into several types. So what do

A. El Gourari (✉) · M. Raoufi · M. Skouri
Laboratoire de Didactique Et de Pédagogie Universitaire, Centre d'Etudes, d'Evaluation Et de Recherches Pédagogiques, Université Cadi Ayyad Marrakech, Marrakech, Morocco
e-mail: abdelali.elgorari@uca.ac.ma

M. Raoufi
Laboratory of Materials Energy and Environment, Faculty of Sciences Semlalia, Cadi Ayyad University, Marrakech, Morocco

M. Ben Ahmed et al. (eds.), *Networking, Intelligent Systems and Security*,
Smart Innovation, Systems and Technologies 237,
https://doi.org/10.1007/978-981-16-3637-0_38

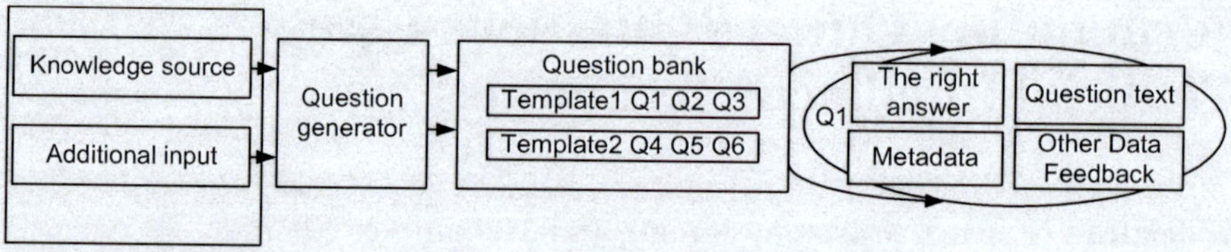

Fig. 1 Process based on the use of computer technologies (AI) to create questions

we mean by making the questions? The complex process is based primarily on the use of computer technologies (AI) to create questions from one or more sources of available data.

Through Fig. 1, we note that the makers of these questions are taking data sources such as the reference book for the material, and other forms of data sources, which can also be observed that the outputs are expressed from the back of these questions [1]. If we take, for example, question one where it contains the text of the question, the correct answer, and other data, such as the revised feed and the description data, which determine how difficult the question or topic is.

2 Quizzes Questions Vocabulary Types and Standards for Their Construction

Questions have an important role to play in assessing learning outcomes and assessing the level of achievement of learning goals; the skill of question formulation is, therefore, one of the most important criteria to be included in the quality of the calendar, and the test vocabulary is classified into two types [2].

2.1 *Response Production Questions*

This type requires the learner to write his answer to the problem posed to him and was divided into (perfect questions—knowledge of terminology—pictures and drawings—article questions (short—long)).

2.2 *Response Selection Questions [3]*

In this type of question, the learner is given several answers to the question or solutions to the problem, and he has to choose (identify) the correct answer or solution among them or the best ones. These questions are called objective questions because they are objective, that is, they do not differ from one individual to another. It is divided into (choice of multiple—right and wrong—the mood—rearranging).

Here, we show each of these species, in terms of the educational output is measured by each of these kinds of questions and the rules for preparing each type of question. This is a frame of reference against which to measure the quality of the question in achieving the educational output that it should measure.

Article Questions [3]
The vocabulary of the article is of the kind that allows the learner to answer it in his own words. The learner asks to summon from memory the information relevant to the question, and in this kind of question, he is required to organize and present facts, terms, concepts, or ideas, that is, to engage in creative activity, and these questions usually start with words like (discuss—explain—compare—write what you know about—remember…). So, what characterizes a vocabulary is the freedom of response it provides to the examiner, as this type of test presents a particular question that requires an answer. The examiner is free to decide how to interpret the problem and the information he uses, how to organize the answer, and how to structure it. The questions of the article thus help to measure specific objectives, such as the ability to innovate and organize and to integrate and express ideas, using the specific scanned language, which other types of test singles cannot achieve to the same degree.

Educational output for which the vocabulary of the article measures [4]

- The ability to express in writing, where the primary importance lies in this type of capacity to produce, integrates, and expresses ideas.
- The ability to select, organize, and link information as the learner summons and reorganizes the answer linking different elements of the decision and composes it in the way he sees it.
- Makes the learner active in his choice of information on the problem asked by the question and then organizes, links, and brings it out in an integrated subject.
- It is useful to verify higher mental processes because it requires conclusions, comparisons, analyses, and judgments on knowledge of different types.
- If the article's questions are well crafted, they lead learners to become accustomed to good school habits that enable them to learn the important facts, to understand the relationships between them, and to understand and absorb the subject.

Defects of article questions [3]
The formulation of the vocabulary may often lead students to differ in their understanding of the meaning of the article. This leads to a lack of compromise in arriving at the required answer not because of their low level in the article, but because of the lack of clarity in the question of the article. Some students may have a linguistic skill in written expression, and in the way information is presented, linked, and organized. This affects the correction, giving a very high degree to a subject that may not contain substantive ideas of value, in the sense that the framework within which the

answer is placed may affect the quality of the degree regardless of the integrity and accuracy of the subjects. It takes a long time to correct the questions of the article over and above the teacher's stress because each student tries to write as many pages as possible because he believes that quantity has a significant impact on the degree to which it is obtained, even if it is unrelated to the substance of the subject.

Rules for preparing article questions

The question must be clear and specific so that the problem it poses is the same in the minds of learners. This can be achieved by testing the exact terminology and reviewing the question several times to make sure that it is clear. Also to measure the output of education, the question must avoid words such as "who and what, when, I remember and identify" and other words associated with remembering facts and information. Other phrases that would measure the higher levels of goals, such as "why, explain, compare, link, explain, analyze and criticize," can be used.

Judging tests on article questions

- The question provides a great deal of freedom for the learner to express his views, thinking, and conscience.
- The question presents the problem in a way that enables the learner to organize and link information related to the problem to produce an integrated theme.
- The question achieves a focus on different mental processes, requiring conclusions, comparisons, analyses, and judgments.
- The question encourages good school habits that enable students to understand the important facts and the relationships between them.
- The question measures a learner's creativity.
- The question is formulated in a clear and unambiguous manner.

Substantive questions [5]

Substantive questions mean questions that can be objectively valued, meaning that there is an agreement in the provisions, if the substantive question is corrected, by a set of corrections, where all arrive at the same scores without room for the intervention of subjective judgments, and of the types of objective questions (complete—multiple choice, right and wrong, identical).

Completion questions

Completion questions are words written by a teacher, from which one or more words have been deleted, each deleted word has been placed one or several points, and the learner is asked to place deleted words that make the meaning of the phrase complete and clear. Merits of completion questions (ease of placement and correction—relatively comprehensive for the scientific material to be tested—the field of speculation is relatively weak).

Education output measured by completion questions

- Test of vocabulary and terminology.

- Test output on simple facts such as names, dates, events, places, and descriptions.
- Test output on principles.
- Test output on method and method knowledge.
- Test output for measuring simple interpretation of information.

Rules for preparing completion questions

- The missing phrase must be formulated in as brief a form as possible so that the question is clear and unambiguous so that the answer is fully determined, and the vacuum is filled only by the required answer.
- The phrase should not contain a large number of spaces because the presence of many spaces leads to the ambiguity of the question and thus to the wide variety of answers. Only keywords should be deleted from the phrase.
- As a general rule, the phrase should contain only one or two incomplete words so that learners can understand and answer the question.

Questions of right and wrong
Questions of right and wrong are the type of language to which the learner responds by one of the following means: (Right/wrong) (Yes/no) (True/wrong).

Merits of right and wrong questions

- Objective in its rectification and it takes no effort to correct it.
- Easy to develop, compared to other objective tests.
- It consumes a large area of the paper.

Educational output measured by right and wrong questions

- Measuring knowledge of simple facts and concepts and meaning of terminology.
- Measure the ability of the learner to detect common concepts that are incompatible with scientific realities.
- Measure the ability of the learner to distinguish truth from opinion.
- Measuring the ability of learner's to think critically. In this case, data on a particular subject may be presented to the learner in the form of an experiment, map, graphs, or tables.
- Measuring the ability of know the malaise relations. To measure this goal, we put two issues into one phrase and the learner rules if the relationship between them is true or wrong.

Rules for Building Right and Wrong Questions

- The phrase or question must contain only one idea, be key and important, and be in a prominent place of the phrase.
- The phrase must be formulated tightly so that it is completely correct or false and has no argument about its correctness or error.

- The phrases in the test must not be consistently longer than the wrong phrases, but the correct phrases must be approximately equal in length.

Multiple selection questions [4]
In its simplest form, multiple-choice questions consist of a problem and several alternative solutions. The problem arises either in an inquisitive form or in the form of an imperfect phrase. The problem is called the origin of the question. Alternatives include one correct answer, several false answers called distractions or camouflage, and the function of distractions distracting learners who do not know the correct answer.

The educational output measures

- Measure of remembrance: Such as the recollection of information, facts, terminology, principles, procedures, methods, generalizations, theories, and other first-level outputs of knowledge education objectives.
- Measuring understanding and assimilation: In this case, the position contained in the question must be new to the learner, but if the question contains a position identical to that experienced by the learners, the question will only measure the level of recollection of information and facts.
- Measuring analytical capacity: Examples include the ability to infer function, the implicit structure of a written text, the ideas used in a document, the ability to analyze the elements of a situation new to the learner, or the use of certain criteria to analyze a situation.
- Measurement of calendar capacity: Examples include the ability to judge a particular act or idea, to identify the values and views used in an act, to judge an action by comparing it to another act, or to use a set of criteria for making a judgment.

Rules for Building Multiple-Choice Question

- The origin of the question must contain a well-defined problem so that the learner understands the meaning of the problem without having to use alternatives to clarify it. The main criterion for containing the origin of the question on an entire problem is the ability to respond to the origin of the question without considering alternatives.
- The problem must be carefully formulated: So that the learner does not have to speculate on what he means by asking, so the question must contain all the information needed for the solution, even if it seems self-evident.
- Negation formula must be avoided whenever possible, and a line must be placed under the negation device so that the learner can pay attention to it and take it into account when answering the question.
- Each alternative must be linguistically appropriate to the origin of the question. If the first word of the correct alternative is an actor, all other alternatives must start with an actor and so on. Some studies have indicated that respondents choose long alternatives more than short ones.

- The connections between the origin of the question and the correct answer must be avoided. A word in the correct answer is often marked because it is similar to a word in the origin of the question. But it is okay to put such words in the wrong answers.
- The use of this alternative has proved to be very easy if it is the correct alternative. On the other hand, if it is wrong, it is sufficient to exclude it simply by suspecting that one of the other alternatives is wrong.

3 The Required Background to Complete and Develop This Field

We are dealing with sources of the content data organized and unorganized, so we must have a background in a treatment of natural languages, texts, images, and for the manufacturer of questions; we are in the process of using educational and evaluation theories, so we must have a background in education and learning. We do not limit ourselves to the direct application of these technologies while developing new ways to benefit the educational and the evaluation communities, also, when we get products we try to make as close as possible to the questions that people make. That means, it must be free from the spelling mistakes, so we should have this background in the treatment of natural language generation. Figure 2 illustrates in detail what we have explained.

All of these areas that have been mentioned depend on machine learning, and deep learning is employed very much. For this reason, the background required from them must be present strongly. Also in terms of the answer-making must be objective, credible, and must contain a goal of transparency and stability. But the question is how to achieve credible and objective results? That is what we will talk about in financial terms.

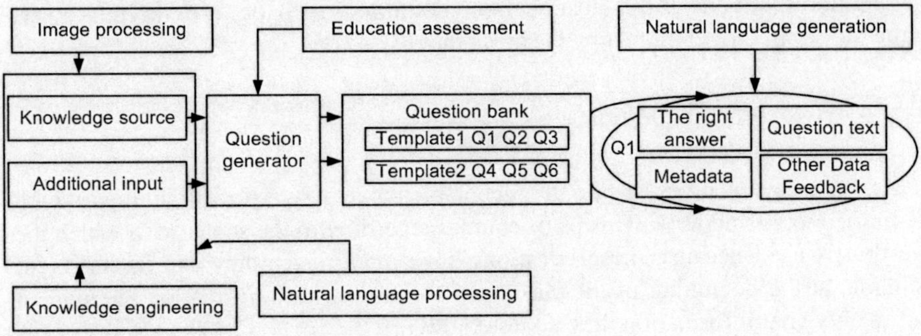

Fig. 2 Sources of the content data organized and unorganized

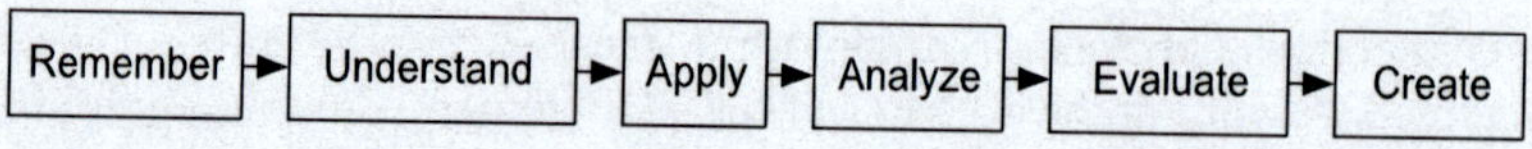

Fig. 3 Tools used to determine the cognitive level

4 Some of the Ingredients that Will Help Us Verify the Credibility and Stability of the Question

There are many ingredients, but we have just focused on three ones because over the past years, there has been researching interest in this area, we mean, most researchers focus on them to make questions, which are as follows:

4.1 *Cognitive Level*

These are all the cognitive processes [6] that the answers to the question include, for example, when a student answers the question, what are the processes that are activated in his mind, is he trying to remind his or her courses? Or is trying to use in-depth analysis, or what?

Figure 3 divides cognitive levels into six phases, beginning with a capacity to remember the teachers that the student has studied and ended with the capacity to invent and create new things.

4.2 *Difficulties*

There are many definitions of how to measure difficulty that has been addressed in the literatures, but one of the most common definitions is that of the correct answers to the question; the relationship (1) shows that [7]:

$$\text{Difficult question} = \frac{\text{number of correct answers}}{\text{total number of answers}} \times 100 \tag{1}$$

There are several proposals for the acceptable range of question difficulties, which are limited to 20 and 80%. This is, of course, according to the scenario in which the question is used because these questions have many uses, they can be used, for example, at the beginning of the classroom to know what information the student knows and what information they do not know.

Example: We have a question for 60 students, among them 40 ones answered correctly. Difficult question = 40/60 * 100 = 66.67%.

4.3 Discrimination

The question is how to distinguish between different-level tests; of course, there are many equations to calculate the coefficient of differentiation [8], but there is a well-known equation that depends on the order of students to step up by grade, where artificial intelligence algorithms can be used, especially unsupervised learning algorithms that classify students by each student's level. We can calculate this discrimination rate (DR) with the following question (2), which (NCAT) is Number of Correct Answers in the Top group and the (NCAL) is Number of Correct Answers in the Lower group:

$$\text{DR} = \frac{\text{NCAT} - \text{NCAL}}{\text{individuals for one group}}\ \text{DR} = \frac{\text{NCAT} - \text{NCAL}}{\text{individuals for one group}} \quad (2)$$

The acceptable range is as follows: The discriminating factor is positive, more than 19.

Example: We have three questions for 100 students in the following distribution of answers to the upper and lower groups (Table 1).

If we take the example of question three, 20% of students in the upper group versus 20% of students in the lower group have answered the question, which makes us unable to know and distinguish between high-level and low-level students.

5 Uses of Artificial Intelligence in the Industry of Questions

Artificial intelligence is used in many areas to solve some complex problems. Education is one of the areas that it care about using artificial intelligence to solve its problem. For example, it is used to guide and assist students [9]. Also, linking it to the concepts of e-learning and distance learning to measure, evaluate and track students during the educational process [10]. It can also be used to create questions. These are techniques and algorithms of artificial intelligence to make high-quality questions.

In Fig. 4, it can be seen that the use of artificial intelligence, especially deep learning, in the creation of questions can give high accuracy and credibility to the question, classifying good questions and false questions.

Table 1 Coefficient of differentiation

Question number	The upper groups (%)	The lower groups (%)	The coefficient of differentiation
Question 1	14	4	+0.6
Question 2	4	14	−0.6
Question 3	20	20	0

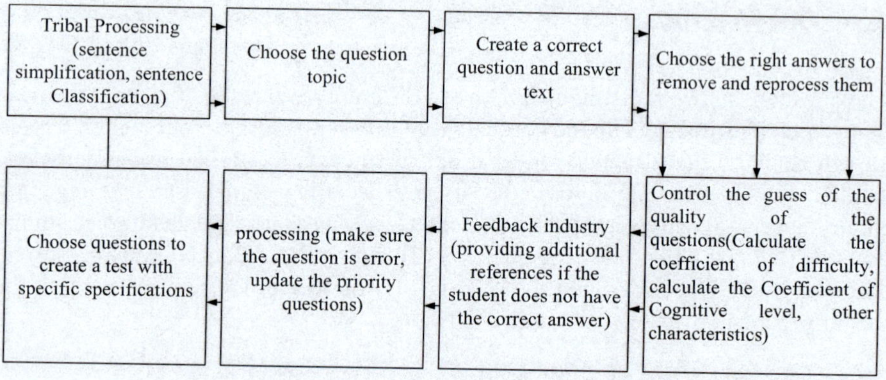

Fig. 4 Uses of artificial intelligence in the industry of questions

5.1 Methods for Formulating and Making Questions

Template-based methods [11]
The blocks identify the surface building for the questions also; templates contain static text and variable parts. These templates describe the specification of terms that can replace changing parts where they can be based on linguistic or moral approaches.

Rules-based methods [12]
Rules often accompany text-based methods as inputs rely on append descriptive data to the input; this metadata is used to match the entries to the pattern in the rules. The rule is implemented to turn matching entries into questions.

Example: identify the appropriate questioning tool (where from, why?).

Methods based on statistical approaches [8]
Learning how to make questions from training data (pairs of questions and answers).

Example: Guess the location of the word or words to be replaced by blank (e.g., in questions to fill in the blank).

6 Results

All of these results were obtained using the following AI algorithms (artificial neural network (ANN), convolutional neural network (CNN), and recurrent neural network (RNN)) (Table 2).

In Fig. 5, we observe the quality of the accuracy for statistical method with a scale ranging from 0.05 to 0.1% in each episode that we did compare to the other two methods.

Table 2 AI algorithms used to solve our problem

Model	ANN [13]	CNN [14, 15]	RNN [14]
Use it for	Tabular datasets Classification prediction Regression prediction	Image data Classification prediction Regression prediction	Text data Classification prediction Regression prediction

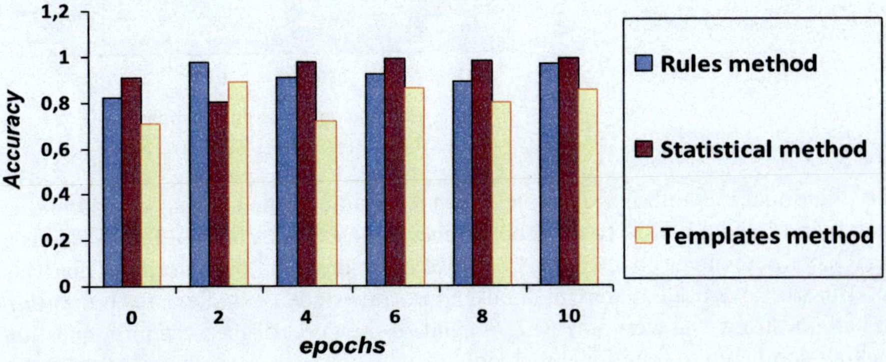

Fig. 5 Accuracy-based comparison of the three methods

Illustrations in Figs. 6 and 7 show how important it is to use advanced computer technology instead of old traditional techniques to develop tests questions through an analytical study we conducted at Cadi Ayyad University Marrakech, Morocco in 2020 when most questions were expressed about the University's multiple-choice questions. The results were analyzed after several criteria were taken into account, including the average level of difficulty, and the quality of the questions and the discrimination factor must be positive. Every time the rate of bad questions is measured, the method based on artificial intelligence algorithms is much better than

Fig. 6 Ratio of bad questions using the traditional method

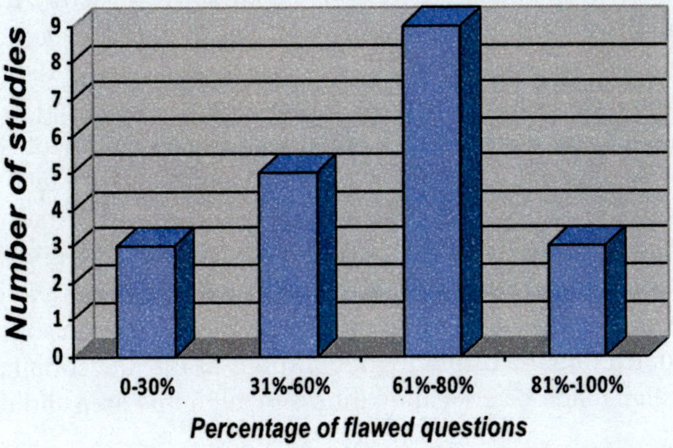

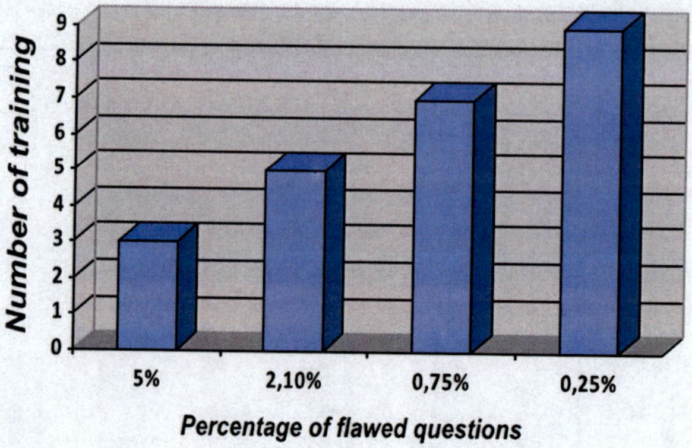

Fig. 7 Ratio of bad questions using AI algorithms

the traditional method. For example, when we studied nine studies, we found that they represented 61–80% of the total number of questions analyzed there. This is a very large percentage that was bad because it violated one of the criteria that was set. But when we used modern methods in the same study (9 studies), the proportion of bad questions was very low (0.25%) and so the good of the system in question making compared to traditional methods.

Finally, we can conclude that there are many bad questions that are used to assess students. What is the reason? Maybe one of the residents did not have the background that we mentioned earlier, so to address these forms, we have decided to use and develop a smart system based on AI (deep learning) techniques. This type of issue is designed to improve a quality and avoid errors in the use of traditional techniques.

7 Conclusion

There are many areas in which we can make significant progress, especially in the formulating quizzes questions, for example, enriching shapes and organizing questions mean designing a good and attractive structure to stimulate student skills and improve the formulation of questions that can be demonstrated in a similar way to human making, as well as creating and enriching data sources used. In this paper, we study how to create good tests questions based on artificial intelligence techniques and algorithms. Doing that is based on one of the most important criteria to be included in the quality of the calendar and the test vocabulary. The comparison between the most recent methods used to make traditional questions, and one of the ways based on AI algorithms at Cadi Ayyad University Marrakech Morocco we confirm that using the AI improve a quality and avoid errors in the use of traditional techniques. In particular, also organized data sources; explore other aspects of controlling the difficulty of questions as one question has many characteristics such as language or ethical qualities. Finally, this area should be developed so that only a one-touch teacher can make a complete test, and this is the purpose we seek to achieve with these modern technologies.

References

1. Bala, K., Kumar, M., Hulawale, S., Pandita, S.: Chat-Bot for college management system using A.I. Int. Res. J. Eng. Technol. 2–6 (2017)
2. http://www.khayma.com/mohgan73/101msdcf/21.htm
3. Caldwell, D.J., Pate, A.N.: Effects of question formats on student and item performance. Am. J. Pharm. Educ. **77**, 1–5 (2013). https://doi.org/10.5688/ajpe77471
4. Bonham, S.W., Deardorff, D.L., Beichner, R.J.: Comparison of student performance using web and paper-based homework in college-level physics. J. Res. Sci. Teach. **40**, 1050–1071 (2003). https://doi.org/10.1002/tea.10120
5. Jia, J., Chen, Y., Ding, Z., Ruan, M.: Effects of a vocabulary acquisition and assessment system on students' performance in a blended learning class for English subject. Comput. Educ. **58**, 63–76 (2012). https://doi.org/10.1016/j.compedu.2011.08.002
6. Tobin, K.: The role of wait time in higher cognitive level learning. Rev. Educ. Res. **57**, 69–95 (1987). https://doi.org/10.3102/00346543057001069
7. Mieloo, C., Raat, H., van Oort, F., et al.: Validity and reliability of the strengths and difficulties Questionnaire in 5–6 year olds: differences by gender or by parental education? PLoS One 7 (2012). https://doi.org/10.1371/journal.pone.0036805
8. Tuwor, T., Sossou, M.A.: Gender discrimination and education in West Africa: strategies for maintaining girls in school. Int. J. Incl. Educ. **12**, 363–379 (2008). https://doi.org/10.1080/13603110601183115
9. El Gourari, A., Raoufi, M., Skouri, M., & Ouatik, F. (2021). The Implementation of deep reinforcement learning in e-Learning and distance learning: Remote practical work. *Mobile Information Systems*. https://doi.org/10.1155/2021/9959954.
10. El Gourari, A. Skouri, M., Raoufi, M., & Ouatik, F. (2020) The future of the transition to e-learning and distance learning using artificial intelligence. In: 2020 Sixth International Conference on e-Learning (econf) pp. 279–284. https://doi.org/10.1109/econf51404.2020.9385464.
11. Wu, S., Zhang, Y.: A comprehensive assessment of sequence-based and template-based methods for protein contact prediction. Bioinformatics **24**, 924–931 (2008). https://doi.org/10.1093/bioinformatics/btn069
12. Douglas, K.M., Mislevy, R.J.: Estimating classification accuracy for complex decision rules based on multiple scores. J. Educ. Behav. Stat. **35**, 280–306 (2010). https://doi.org/10.3102/1076998609346969
13. Agatonovic-Kustrin, S., Beresford, R.: Basic concepts of artificial neural network (ANN) modeling and its application in pharmaceutical research. J. Pharm. Biomed. Anal. **22**, 717–727 (2000). https://doi.org/10.1016/S0731-7085(99)00272-1
14. Retrieval, D.: Natural language processing for prolog programmers. Data Knowl. Eng. **12**, 246–247 (1994). https://doi.org/10.1016/0169-023x(94)90017-5
15. Banerjee, I., Ling, Y., Chen, M.C., et al.: Comparative effectiveness of convolutional neural network (CNN) and recurrent neural network (RNN) architectures for radiology text report classification. Artif. Intell. Med. **97**, 79–88 (2019). https://doi.org/10.1016/j.artmed.2018.11.004

Smart Campus Ibn Tofail Approaches and Implementation

Srhir Ahmed and Tomader Mazri

Abstract This paper describes the concept and studies of "smart campus" using different methodologies and introduces a global strategy and use case for each environment in the University Park. Our main goal is to define, present the general smart campus principles and objectives which revolve around the different IOT and cellular infrastructures in the university, introducing a general ICT architecture for their coordination and also detailing their direct use in the management of the university and applications for learning and research, reduce the cost and take one step forward a university smart campus.

1 Introduction

The smart campus concept has been the main focus of many researchers recently due to the valuable insights gained toward developing smart campus. The university campus apparently is a small city where it delivers to all variant user's a variant service. There are several factors that attract the researchers to study the smart campus including protecting the environment delivering high-quality services, protecting and saving the cost. Smart campus is a related ecosystem, applications, services and use cases, among the main standards that are directly related with focus on:

- Urban planning
- Transport
- Energy
- Education
- Health care
- Mobility
- Logistic

S. Ahmed (✉)
Ibn Tofail University, Kénitra, Morocco

T. Mazri
National School of Applied Sciences of Kenitra, P.B: 242, 14000 Kénitra, Morocco

M. Ben Ahmed et al. (eds.), *Networking, Intelligent Systems and Security*,
Smart Innovation, Systems and Technologies 237,
https://doi.org/10.1007/978-981-16-3637-0_39

- E-government.

The Internet of Things is a fundamental part of the smart campus, and it is inescapable getting it invoked. The Internet of Things is a communication model that gained its tenacity from its capability of connecting variety of everyday life objects to the Internet [1]. These objects include alarms, security locks, sensors, drones, appliances, robots, office equipment and so on. Even though IOT is in its early stages, there are many applications and standardization that has been developed in many domains including home automation, smart grids, water and waste management, traffic control, smart vehicles, healthcare assistance and industrial automation.

The IOT is involved everywhere in smart campus especially in:

- Facilities
- Electrical systems
- Safety
- Classroom technologies
- Tutoring spaces
- Residential services
- Physical and mental health.

This paper provides an overview of all the use case that we can implemented in our smart campus.

2 Smart Campus Architecture

Smart campus is one of the innovations that will be developed at Ibn Tofail University, one of the main aspects that being considered in developing smart campus is the infrastructure [2] that is why we introduce a technical architecture for smart campus and proposes a local operational model. It also presents a case study of the digitalization of the university.

The architecture of smart campus which described in (Fig. 1) defines three basic aspects for smart campus:

1. Smart education that consists on: e-learning, RFID tags, virtual classroom
2. Smart parking, a parking system that provides a vehicle location tracking [21], reporting information and also provides information when the parking is full and specifying the number of availability places
3. Smart room: System that provides information related to the building temperature control systems, electrical systems and lighting systems.

As described in [26, 27], there are several concepts to be implemented to achieve smart campus among them we find:

- Internet of Things (IoT)—The Internet of Things integrates sensors, controllers, machines, people and things in a new way to realize intelligent identification, location, tracking and monitoring

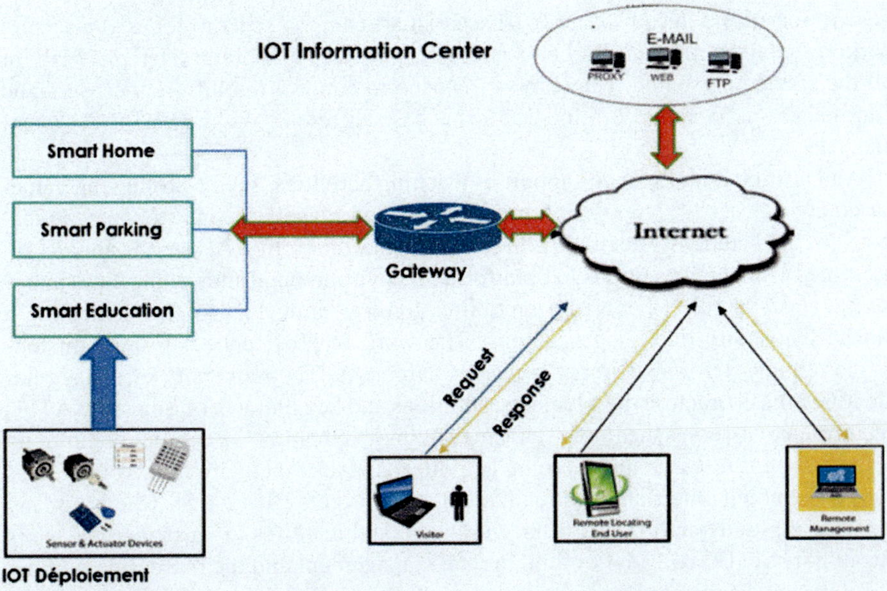

Fig. 1 System design of smart campus

- Big data technologies include mass data acquisition, mining, storage and processing. Big data technology in the wisdom of the campus [28, 29] in all aspects of the application will make its management services to a higher level
- Cloud computing: Cloud computing requires a combination of grid computing, parallel computing, powerful integrated computing, distributed computing only the use of open, integrated, high scalability, on-demand service cloud computing model can provide good infrastructure support, collaborative information architecture and dynamically configurable resources
- Business intelligence: Business intelligence utilizes data warehousing and data mining techniques to systematically store and manage user data, provide analysis reports, provide decision-making information for a variety of university activities [25], basis and analyze user data through various statistical analysis tools
- Social networking: The social network covers all forms of network services [14] with the core of human society. It is a social or social characteristics of the network services.

3 Related Work

In terms of applying the smart campus concept to learning activities, many research projects have been proposed to develop many implementation of high-level use case for smart capabilities that could improve education; the work in [3] presents a brief description of the design of an IoT-based smart campus scheme focused on smart parking, smart room and smart education. Also studies of an integrated platform for all the service presented, where Wi-Fi is used to connect the different sensors and end devices associated with the platform. Also different works aim for the smart analysis.

And artificial intelligence support of teaching activities, such as following game-based approaches [4], supporting multimedia conferences [5], smartphone applications [6], and students' health [7]. In terms of smart mobility, Alvarez-Campana [8] presented a university campus IoT platform for environmental and people flows monitoring, and Toutouh [9] describe a mobility prediction mechanism, implementing it on the University of Málaga campus. The work in [10] focuses instead on IoT-supported disaster management for smart campuses. The university of malaga has identified the characteristics elements, solutions and key features of SmartUMA [11] and present it assesses the impact and level of involvement of the campus community in the different research and learning activities, and also there are numerous studies conducted about smart campus realizations or concepts [9, 12–15], whereof many are focusing on technological frameworks and architectures in particular use cases the authors of [12] consider optimizing the management and the increasing number of connected devices and propose a very precise model of dividing campus networks according to roles; using authentication controllers based on Open Flow and virtualization technologies throughout most of these work, particularly focused on IoT and monitoring systems, the reader is confronted with a situation in which many do not focus on security-related issues of these remote technologies, which will be a significant concern for industries looking to utilize and exploit these technologies.

4 Proposed Method

Our work consists to present paper focuses on the description as a relevant smart campus example case, of the characteristics, elements, solutions and key features of outsmart campus. Secondly, it defines the impact and level of involvement of the campus community in the different research and learning activities, providing recommendations for all environment that we define in our university park.

Our approach consists in defining four functional definition parts (Fig. 2) that will describe our fields of view, and then, we will define all the description relevant smart campus use case with each part according to well-defined and detailed axes for each field.

After the functional decomposition, we will associate each functional definition among the four functions defined in (Fig. 2) by all the involved use cases.

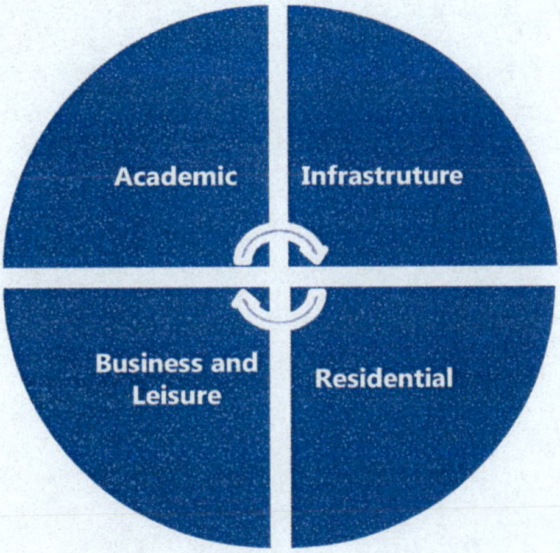

Fig. 2 **Functional definition** for smart campus

4.1 Academic Function

As defined in the previous section, which defines our smart campus architecture separate into three parts.

- Smart home
- Smart education
- Smart parking.

For the part of functions academic, we break it down into different use case as mentioned below to cover our university environment:

- Conference room
- Laboratories
- Study place for individual use
- Library
- Academic hospital
- Special conference seating.

4.2 Infrastructure Function

A key benefit of smart campus infrastructure is that it helps universities which offer incoming students the resources they want while keeping costs down, and it involves all the logistic and transport parts as seen below:

- Parking spaces
- Means of transportation
- Accessibility (car, bus)

- Reserved space for bicycles.

4.3 *Residential Function*

This part will define everything that is related to the accommodation and housing of students, both external and internal, as well as guests at university events

- Student accommodation (national)
- Student accommodation (short-term)
- Faculty housing
- Short stay apartment.

4.4 *Business and Leisure*

The goal in this part is to involve activities that will help the students to integrate among themselves either through sports competitions or through paid activities offered by companies in collaboration with the university.

- Eternal conference spaces
- Cultural centers associating
- Sports and technical competitions
- Activities that combine work and learning.

5 Use Case for Smart Campus

The subject and its exploitation possibilities being vast, we have chosen to focus on four proposals as seen in (Fig. 3) where the installation of sensors seems interesting as well as the corresponding scenarios. Also the use of the Internet of Things (IoT) on the intelligent campus provides the ability to monitor every aspect of the work environment such as the temperature, overcrowding, the availability of equipment

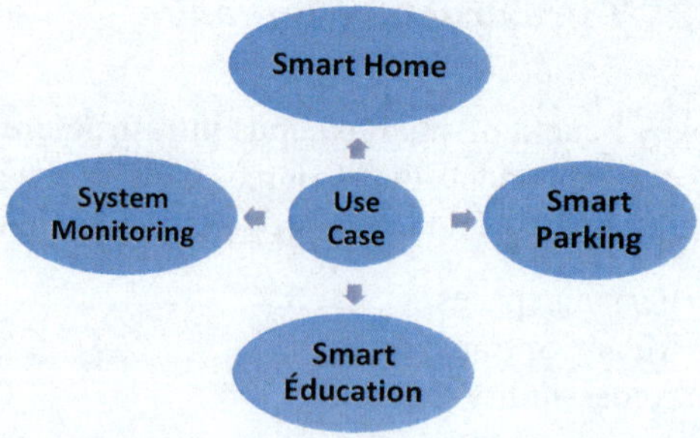

Fig. 3 Smart campus Ibn Tofail use case

and catering facilities. This data can be combined with the onboarding process that collects feedback data from staff to match with and improve the work experience, and we have defined these scenarios for the use of our smart campus Ibn Tofail.

Below, we will define each use case separately and define the proposals and studies for each case.

5.1 Smart Home

The trend in recent years has been toward "Smart Building" and "Smart Home," but the notion of a smart building is still rather vague, and there are many definitions [12]. An overall definition could be as follows: automation implemented to make the management and operation of the building more efficient. This automation can take several forms as well as the final goal. Smart buildings are often associated with buildings that are energy autonomous and regulate their consumption themselves [13]. Thus, smart campus projects already existing in France such as the one in Versailles are mainly concerned with this energy aspect. In the context of our project, the meaning we are interested in is to make the Campus "Smart" by connecting it and offering its users information on its current use, the conditions that prevail there [19]. But to offer this type of service, like smart buildings and smart home, it is necessary to set up a network of sensors that allow the desired information to be transmitted and processed [20, 22]; for this purpose, we have defined our work area for smart home on the following points:

- Temperature control systems
- Electrical systems
- Lighting systems
- Water sensors
- Fire alarm systems
- Security alarms
- Fire detection
- Temperature monitoring
- Visual management.

5.2 Smart Parking

A new need was identified, with suggesting the idea of being able to find out the number of spaces available in the parking lot.

In this section, we will install presence sensors in the parking lot and then register them from a sensor API. In this way, the newly collected data will be stored and can be evaluated. Following this, a user will be able to create a service that allows users to view the number of spaces remaining in the parking lot [15]. For example, a student driving to class in the morning would like to know whether there are any free parking spaces left in the parking lot before he or she goes to class. If there are none left, it is more interesting for him or her to go directly to another parking lot. He therefore accesses a web application of his choice [17] (website, smartphone application) connected to our platform and can visualize the number of remaining spaces in the parking.

A little later, a study group wants to know the average occupancy of the parking lot during the day in order to know if it is interesting to activate the barriers [18]. To do this, this group logs on to the platform and retrieves information on the parking lot occupancy over the last few months and can perform calculations using this data.

The work area for smart parking was defined on the following points:

- Number of cars
- Control departure/arrival
- Traffic control
- Monitoring
- Parking control.

Considering the increase of urban and traffic congestion in our smart campus, smart parking is the best strategic issue to work on, not only in the research field but also in economic interests.

5.3 Smart Education

Access to education is vital for the economic, social and cultural development of all societies. In this section, we have set ourselves many goals to help to connect all the students in the digital age. First of all, we want to provide higher education with an online learning platform or e-learning. It should equip academic institutions and open digital spaces with an autonomous virtual desktop infrastructure (VDI) with its own processing, hosting and storage capacity. The smart education aims on the one hand to modernize the digital communication infrastructure data processing and storage, and on the other hand, to deploy technological platforms to improve teaching and learning in universities, elementary schools, colleges and high schools, it will enable as follows:

- The online learning of students through the "e-learning platform"
- The production and recording of courses and didactic and pedagogical content through the "Virtual Classroom"
- The modernization of networks and strengthen the security of the connectivity of the universities
- The deployment of a multimedia room in universities. To provide students with connectivity and access to personal data processing and storage space through VDI rooms.
- Laboratory access through campus cards only
- Smart printers in campus, access using campus cards

- Biometric/card-based attendance system for students
- Online booking facility for hall and classrooms
- Online leave application and tracking
- Online book recommendation from library for the faculty. e-Notice facilities over Intranet.

5.4 System Monitoring

This section interacts in all the defined domains and functionalities that have been defined in the previous sections (academic, infrastructure and residential) [16]. All the equipment on our intelligent campus must be supervised in real time, hence the need to set up certain points:

- IP video surveillance
- Fire alarm systems
- Access to electronic doors
- IP-based police and security teams
- Police vehicles over IP
- Smart monitoring of gas leakage in residences, flats.

Among the main advantages that we will propose for our smart campus defined in the following:

- RFID tags
- Notification, centralization, alerts
- Physical safety
- Digitalization.

Smart campuses are safe and secure; it allows to ensure optimal student attendance through the use of RFID tags also for each activity; a smart school gains a competitive advantage through the use of SMS and email communication integrated into the software [21, 22], and the management can send instant notifications and alerts, and the installation of CCTV cameras and other surveillance systems on the premises ensures total security for students, teachers, staff and school equipment. Also a variety of sensors are used at the street layer for a variety of smart campus use cases. Here is a short representative list [23]:

- Magnetic sensor
- Lighting controller
- Video cameras combined with video analytics
- Air quality sensor
- Device counters.

The magnetic sensor detects a parking event by analyzing changes, such as a car or a truck, [24] comes close to it for the lighting controller which can dim and brighten a light according to a combination of temporal and ambient light conditions to save more energy, and for cameras combined with video analytics, video can detect faces, vehicles and traffic conditions for a variety of traffic and security use cases for our campus.

For the air quality sensor, it allows to detect and measure the concentrations of gases and particles to give a perspective of the pollution in a given area and avoid the risk of fire either on our parking lot or on our smart home, and for the device counters, he gives an estimate of the number of devices in the area.

6 Conclusions

Certain requirements for the establishment of a smart campus pose several technological challenges, including the following:

- How to collect data?
- What are the different data sources, including hardware terminals and software?
- What type of network connectivity is best suited to each type of data to be collected?
- What type of power availability and other infrastructure, such as storage, is required?
- How can data from different sources be combined to create a unified view?
- How can the final analysis be made available to specialized intelligent campus staff, such as traffic operators, parking control officers, lighting operators?

Each intelligent campus needs a suitable and structured computer model that enables distributed data processing with the level of resilience, scale, speed and mobility required to efficiently and effectively deliver the value that the generated data can create when properly processed on the network. For this purpose, the principles and driving characteristics of the University of Ibn Tofail Smart Campus approaches learning and research activities are detailed. A general system design that describes the main technological infrastructures of a smart campus is presented associated with functional definition and will define a scheme for the implementation of smart campus based on four functional definition parts that will describe our fields view and define all the description relevant smart campus use case with each part according to well-defined and detailed axes for each field:

1. Smart education
2. Smart parking
3. Smart home.

And describe their implementation all this field can be developed on IOT Technology which is the main key for the deployment of smart campus.

References

1. Alghamdi, A., Thanoon, M., Alsulami A.: Toward a Smart Campus Using IoT: Framework for Safety and Security System on a University Campus
2. Jurva, R., Matinmikko-Blue, M., Niemelä, V., Nenonen, S.: Architecture and Operational Model for Smart Campus Digital Infrastructure
3. Sari, M.W., Ciptadi, P.W., Hardyanto, R.H.: Study of Smart Campus Development Using Internet of Things
4. Zhai, X., Dong, Y., Yuan, J.: Investigating learners' technology engagement—a perspective from ubiquitous game-based learning in smart campus. IEEE Access **6**, 10279–10287 (2018)
5. Zhang, W., Zhang, X., Shi, H.: MMCSACC: a multi-source multimedia conference system assisted by cloud computing for smart campus. IEEE Access **6**, 35879–35889 (2018)
6. Kim, T., Ramos, C., Mohammed, S.: Smart city and IoT. Futur. Gener. Comput. Syst. **78**, 160–162 (2017)
7. Gao, X., Sun, Y., Hao, L., Yang, H., Chen, Y., Xiang, C.: A new soft pneumatic elbow pad for joint assistance with application to smart campus. IEEE Access **6**, 38967–38976 (2018)
8. Lin, Y.B., Chen, L.K., Shieh, M.Z., Lin, Y.W., Yen, T.H.: CampusTalk: IoT devices and their interesting features on campus applications. IEEE Access **6**, 26036–26046 (2018)
9. Alvarez-Campana, M., López, G., Vázquez, E.V., Villagrá, V., Berrocal, J.: Smart CEI Moncloa: **an IoT-based platform for people flow and environmental monitoring on a smart university** campus. Sensors **17**, 2856 (2017)
10. Van Merode, D., Tabunshchyk, G., Patrakhalko, K., Yuriy, G.: Flexible technologies for smart campus. In: 13th International Conference on Remote Engineering and Virtual Instrumentation (REV), 2016.
11. Toutouh, J., Arellano, J., Alba, E.: BiPred: a bilevel evolutionary algorithm for prediction in smart mobility. Sensors **18**, 4123 (2018)
12. Hannan, A., Arshad, S., Azam, M., Loo, J., Ahmed, S., Majeed, M., Shah, S.: Disaster management system aided by named data network of things: architecture, design, and analysis. Sensors **18**, 2431 (2018)
13. Universidad de Málaga: Smart-campus, Vicerrectorado de Smart-campus. Available online: https://www.uma.es/smart-campus
14. Chen, C., Chen, C., Lu, S.-H., Tseng, C.-C.: Role-based campus network slicing. In IEEE 24th International Conference on Network Protocols (ICNP) Workshop on Control, Operation and Application in SDN Protocols, 2016
15. Qian, Lv.: Establishment of smart campus based on cloud computing and Internet of Things. Comput. Sci. **38**(10), 18–21 (2011)
16. Nie, X.: Constructing smart campus based on the cloud computing platform and the internet of things. In: Proceedings of the 2nd International Conference on Computer Science and Electronics Engineering (ICCSEE), 2013
17. Bahl, P., Padmanabhan, V.N.: Radar: an in-building rf-based user location and tracking system. In: IEEE INFOCOM 2000
18. Kaur, V., Tyagi, A., Kritika, M., Kumari, P., Salvi, S.: Crowdsourcing based android application for structural health monitoring and data analytics of roads using cloud computing. In: 2017 International Conference on Innovative Mechanisms for Industry Applications (ICIMIA, 2017) pp. 350–360
19. Sharma, K., Suryakanthi, T.: Smart system: IoT for university. In: International Conference on Green Computing and Internet of Things (ICGCloT), pp. 1586–1593 (2015)
20. Wang, C., Vo, H.T., Ni, P.: An IoT application for fault diagnosis and prediction. In: IEEE International Conference on Data Science and Data Intensive Systems, pp. 726–731 (2015)
21. Lee, N.K., Lee, H.K., Lee, H.W., Ryu, W.: Smart home web of object architecture. In: International Conference on Information and Communication Technology Convergence, pp. 1200–1216 (2015).

22. Hager, M., Schellenberg, S., Seitz, J., Mann, S., Schorcht, G.: Secure and QoS-aware communications for smart home services, in 2012 35th International Conference Telecommunications and Signal Processing (TSP), 2012, pp. 10–19.
23. Luo, L.: Data acquisition and analysis of smart campus based on wireless sensor. Wirel. Pers. Commun. **102**, 2897–2911 (2018)
24. Prandi, C., Monti, L., Ceccarini, C., Salomoni, P.: Smart campus: fostering the community awareness through an intelligent environment. Mob. Netw. Appl. 2019.
25. CiscoIOT https://idoc.pub/documents/ciscopressiotfundamentals-6nq80ejgd9nw,last accessed 2020/08/10
26. SmartBuilt: http://www.greenbang.com/from-inspired-to-awful-8-definitions-of-smart-buildings_18078.html. Last accessed 2020/07/20
27. Tien, J.M.: Big Data: unleashing information. J. Syst. Sci. Syst. Eng. **2013**(02) (2013)
28. Guo, H., Wang, L., Chen, F., Liang, D.: Scientific big data and digital Earth. Chin. Sci. Bull. **2014**(35) (2014)
29. Chen, J., Xiang, L.G., Gong, J.Y.: Virtual globe-based integration and sharing service method of GeoSpatial Information. Sci. China (Earth Sci.) **2013**(10) (2013)

Boosting Students Motivation Through Gamified Hybrid Learning Environments Bleurabbit Case Study

Mohammed Berehil

Abstract Gamification refers to the use of game elements in non-game context, it has been highly used in different fields like business, where stakeholders try to create enjoyable experiences and motivate the users. Moreover, gamification also have been frequently applied in education, where many teachers applied game mechanics to design learning activities that can increase the students' engagement and motivation in the learning environment. In our research, we tried to assess the impact of hybrid gamified leaning environment on student's motivation. We have used the MDA (Mechanics, Dynamics, Aesthetics) and the Hexad (Player's type model) models to create different enjoyable activities inside a gamified learning environment called *Bleurabbit*. Furthermore, to respond to our research questions, we have designed a case study based on applied research methodology and used interviews and observation to collect data. The collected results were very satisfactory and have demonstrated the positive effect of gamification on boosting students' motivation.

1 Introduction

The pandemic spread has been a big accelerator for online learning, many universities have moved online, and in fact, it was considered as magical solution in difficult times. Nevertheless, one of the main problems facing online or hybrid learning environments is the lack of motivation, many students feel more frustrated and alone, as the physical presence of the tutor and the classmates is partly or totally suppressed. So, the learning environment plays a major in learning process, and gamifying it can solve the problem of motivation, especially that many kinds of the literature [16] suggested that gamified learning environment can enhance students' motivation [2, 20].

In our research, we assumed that that changing the learning environment could ameliorate the learning process. Consequently, we tried to combine two gamification frameworks to create more enjoyable environments and bring knowledge to the

M. Berehil (✉)
Faculty of Letters, Mohammed First University, Oujda, Morocco
e-mail: m.berehil@ump.ac.ma

M. Ben Ahmed et al. (eds.), *Networking, Intelligent Systems and Security*,
Smart Innovation, Systems and Technologies 237,
https://doi.org/10.1007/978-981-16-3637-0_40

student's favourite environment (which is related to games) [25]. In addition, we tried to **assess the effectiveness of gamified virtual learning environment on motivating students**using a case study approach, so that we can observe closely the interaction between students and different aspects of the gamified learning environment.

We divided our paper into nine parts organising it as follows: Sect. 2 was devoted to related works, in Sect. 3, we provided working definition of gamification, in Sect. 4, we discussed the used gamification frameworks, in Sect. 5, we gave an overview of Bleurabbit, in Sect. 6, we highlighted the adopted methodological approach, and in Sect. 7, we described our case study. While Sect. 8 was devoted to the results and discussion, in the ninth section, we provided conclusion and perspectives.

2 Related Works

Many kind of the literature related to the gamified learning environments used quantitative research approach [20]; in a published article *Gamification and student motivation*, researchers proposed a study where they assess the effect of the gamification learning environment on students motivation. The researchers [2] used a pre- and post-experimentation survey to test the knowledge acquisition in gamified learning environment. The results have shown positive attitude from students toward the gamified learning environment. Nonetheless, we consider that qualitative methods can be more effective in dealing with gamification, especially that we consider it as a human phenomenon that is related to a uniquely personal experience which differs from one person to another.

The same vision is advanced in a study [3] where a qualitative study was led, based on usage as cogent (College Enterprise), a gamified environment to simulate a real enterprise environment using virtual money was developed. Researchers [3] used focus group and interviews to collect data, and the results show that well-implemented gamified learning environment has very positive effects on student's motivation and increases the engagement among them. Nonetheless, the proposed study was conducted in game-based environment which differ from gamified learning environment as it needs more financial and human resources, and additionally, it cannot fit different pedagogical models.

In another paper [12], the researcher used the MDA framework to develop a design model, for eLearning gamified environment. The researcher emphasises the importance of using gamification framework as it can bridge the gap between the game elements and the design rules. As they used this model to develop the *BlackSlash.com*, which is a website for children to learn HTML, researchers have proposed promising results. This also can be observed in the work developed by researchers in Singapore University, where they use octalysis model to develop a mobile application for increasing students' activity, outside the classroom. The results were so promising, and student have shown a big motivation to such application [21]. Eventless, none of the above-mentioned works have address the issue of player type, which can be very important to create a better experience.

3 Gamification Working Definition

Gamification is a very hard term to be defined. While games are not an odd element in education, they were highly adopted in the past and adopted in by many teachers [15]; nevertheless, gamification represents much more complex concept than a simple educational game [19], as it was used in other fields rather than education like healthcare and business [7] …

There is a big misunderstanding between different concepts, including games as "*a system in which players engage in an artificial conflict, defined by rules, that results in a quantifiable outcome*" [18], serious games "*serious games describe the use of complete games for non-Entertainment purposes*" [14]**Game-based learning (GBL)**called. GBL is a pedagogical approach to learning which consists of developing games to attend pedagogical goals; the developed games are part of a learning process and are meant for developing new learning skills. [15].

The term gamification [4] was firstly used in 2008 in the digital media industry, but was not popularised until 2010 in the field of interfaces design [5]. Dren defines it as "*the use of game mechanics in non-game context*" [5, 24]. To apply gamification, the designer works carefully on deciding how to bring game elements in a serious context in a swift way and create a game-like experience[4, 22]. In education, we can identify two kinds of gamification design [10], **content gamification** "*the application of game elements, game mechanics and game thinking to alter content to make it more game-like*" [9].

This kind of gameful design acts on the content and makes it more game-like, so, for example, you can add a narrative element or make the course as story…The second type is **structural** gamification which is "*the application of game elements to propel a learner through content with no alteration or changes to the content*" [9]. This design is about making learning environment gamified without interfering with the content, for instance, you can add a leaderboard or badges or experience points…in your classroom.

In our research, as shown in Fig. 1, we consider gamification as the intersection between playful design, game elements and motivational design, each of these elements can be part of the gamification design process

4 Gamification Design Frameworks (Two Models Approach)

The gamification frameworks are the result of transdisciplinary research in different fields including psychology, neurosciences and computer science …[11, 17], in fact, many researchers and game designers have tried to regroup the different game elements into one design framework (*framework for success, gamification design process...*) [11] that as a toolkit for games like experience designers. We can identify two types of gamification frameworks:

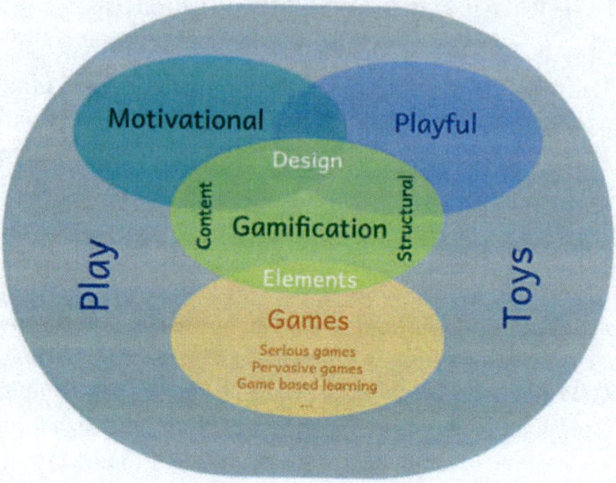

Fig. 1 Gamification

- **Game-based frameworks**: refer to gamification frameworks that are mainly concerned with the game structure and elements, including mechanics, dynamics …[10] the main concern of the designer is providing good game structure, which facilitate player journey inside the game.
- **Player-based gamification frameworks**: a framework that is player oriented, in other words, the development of the game is related to the player actions and reactions inside the game.

4.1 M.D.A

Developed by robin Hunicke et al. [8], the MDA framework, (Mechanics, Dynamics and Aesthetics), represents one of the earliest gamification frameworks used to develop, understand and depict game elements in a context that is not a game. It brokes games into three major elements

- **Mechanics** represent the different elements provided by the game designer that includes the settings, the goals, the rules... Games mechanics are constant and do not change [8].
- **Dynamics** Unlike mechanics, dynamics are produced by the player, more precisely dynamics are the results of the player behaviour that occurs on the mechanics, and in other words, dynamic are not fix matter but rather a variant that depends on the player's strategies of playing and the way he interacts with the game mechanics. [8].
- **Aesthetics** The last important component of games according to MDA creatorsis what it called aesthetics which refers to the emotional response of the user to the system; this could be experienced as passion, joy, eustress…Aesthetics are the results of the interaction between the player and the game or between mechanics and dynamics. Most of the time the wished result is FUN [8].

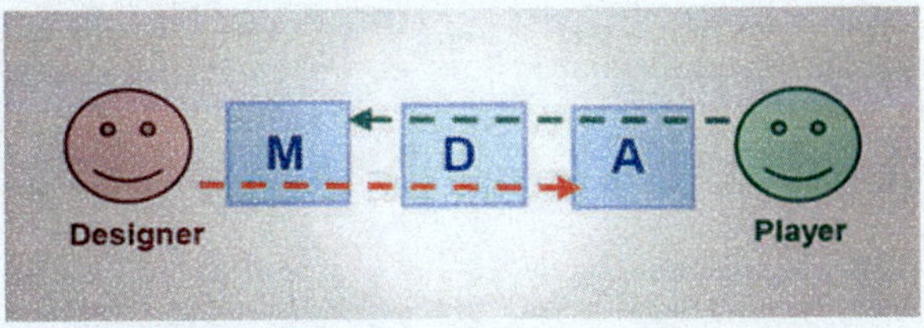

Fig. 2 MDA perspectives

4.1.1 Game Designer Versus Player Perspective

MDA represents a group of "*connected lenses*" [8] that work complementary; it helps to analyse the game from both player and designer perspectives (Fig. 2).

First, we have the designer whose main interest is the mechanics, which in his perspective gives rise to dynamics and aesthetics [8]. The second actor is the player, and the actions of the player are driven by aesthetics which are primordial to the success of gamification.

4.2 Player-Based Gamification Hexad Model

The Hexad model, which was developed by Andrzej Marczewsk [6], and based on BRTEL model [1], was empirically proven and tested [13, 23], through a questionnaire. Marczewski offers a more detailed model and identified eight player types, depending on the player's behaviour and intentions inside the game [13]:

- **Free Spirits** are motivated by autonomy and self-expression. They want to create and explore.
- **Achievers** are motivated by mastery. They are looking to gain knowledge, learn new skills and improve themselves. They want challenges to overcome.
- **Philanthropists** are motivated by purpose and meaning. This group are altruistic, wanting to give to other people and enrich the lives of others in some way, with no expectation of reward.
- **Disruptors** are motivated by change. In general, they want to disrupt the established system, either directly or through other users to force positive or negative change.
- **Players** are motivated by extrinsic rewards. They will do what is needed to collect rewards from a system and not much more. They are in it for themselves.

These types of players also fall in three big categories as shown in Fig. 3. **Willing to play**, **less willing to play** and **not willing to play**.

In our research, we tried to combine both framework (MDA and Hexad design frameworks) to create an adequate learning environment, which allows a fluid players–mechanics interaction and takes into consideration subtleties between different types of players, aiming to generate the targeted dynamics and aesthetics.

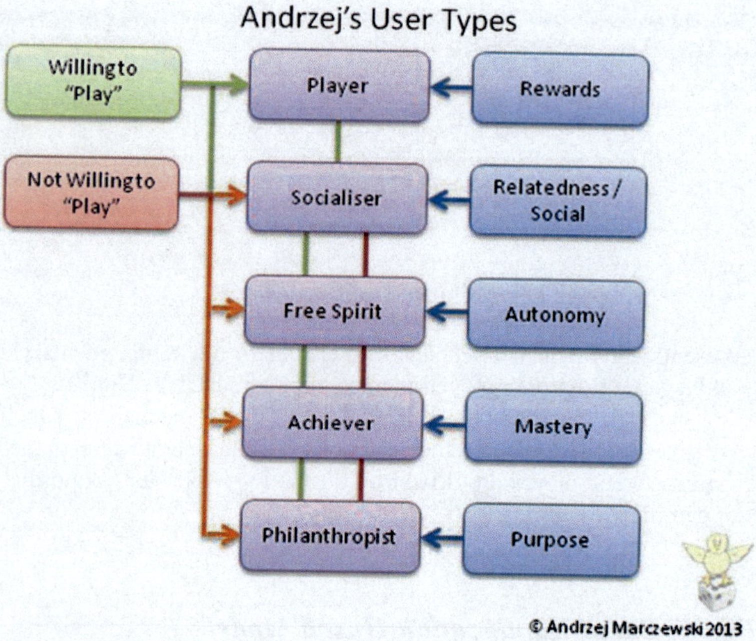

Fig. 3 HEXAD player type

5 Bluerabbit

BlueRabbit is a free Web-based gamified virtual learning environment upon which we have based our case study, developed by a Mexican computer scientist, *BlueRabbit* is built on the principles of MDA gamification design framework. The platform's first aim is to move beyond marks and grades, by establishing a system offers an enjoyable learning environment. Additionally, it allows the use of Hexad player type questionnaire for each student, which make it the propitious platform for our experience.

The platform is equipped with a very colourful interface as shown in Fig. 4 and does not need great technical skills, ergonomically speaking, it is very easy and enjoyable, and it is also responsive which mean it offers the possibility for both mobiles and computers usage. Furthermore, it is both responsive and user friendly for course designer/tutor and the student, which means it allows the use of it in both computers and mobile devices (tablet, smartphones ...). We can divide the platform componenets into three main elements:

Fig. 4 Overview of the platform interface

5.1 *Narrative Elements*

Narrative elements refer to the proposed way by the platform to exhibit the content as a story, and in other words, these elements allow the course designer to transform the content into a plot where students play the heroes, this includes:

- **Quests** are the most basic element of *BlueRabbit*, and they are the mandatory activities of the learning process.
- **Sidequest**: The sidequest represents secondary activities, nonetheless the sidequests represent another opportunity for the players to get money, badges and XP.
- **Characters** are the learners inside the platform.
- **Missions** are special activities that are limited in time and have more constraints.
- **Project** is a space for cooperative activities, which allows the interaction between different team members.
- **Challenges** are a rubric that allows you to propose activities which are challenging to the students.

5.2 *Feedback and Rewards*

We can divide the reward system provided by the *Bluerabbit* platform into two types. Extrinsic motivation includes the XP rewards and badges gained by students through realising quests or sidequests or any other activities intrinsic motivation component mainly related to the student's progress, including levels and leaderboar.

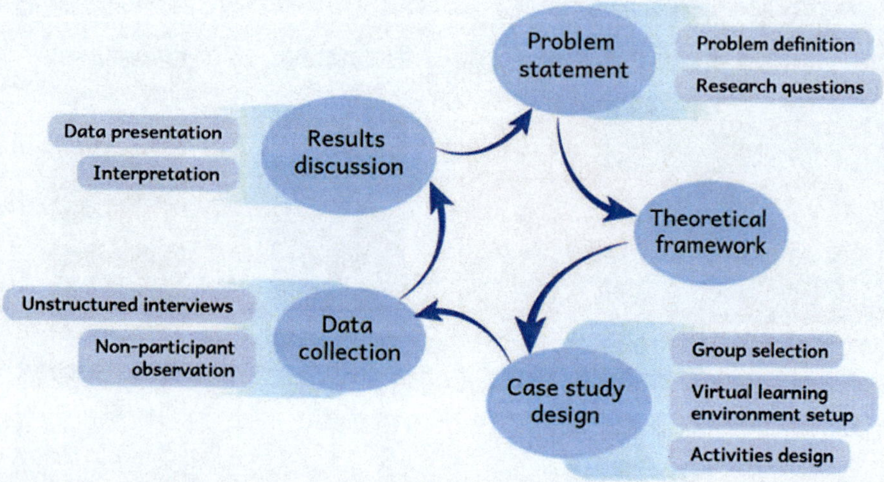

Fig. 5 Research design cycle

6 Methdological Approach

We have opted for applied research paradigm, as the nature of our research problem, and our main goal was testing the use of gamification in particular learning environment and assessing its efficiency on motivating students, as shown in Fig. 5 which we have designed our research iteratively.

We have adopted a qualitative research. So, we have chosen to design a case study and collect data through interviews and observation

7 Case Study Design

We have opted for a case study as a research method, which allows "*an in-depth exploration of a single case*" [26]. In other words, it helps studying and analysing the behaviours and reactions of a student or a group of students inside a learning environment. We constructed our case study as follows; first choosing the sample, setting up the environment, designing the activities, implementing the environment and finally observing the students.

7.1 Sampling

For our research sample, we have had sixteen students (Master1 students, *Ingénierie de formation et Technologies éducatives* Mohammed first university) to whom we introduced the *Bluerabbit* platform in one of the modules they studied in the second semester, entitled *Motivation, Objectifs et compétences*. These students were familiar with platform usage, and nonetheless, it was the first time they use a gamified learning environment.

7.2 *The Onboarding Process*

The onboarding process is the process through which we have introduced students to the learning environment and its functions. First, we started with face to face session then a series of synchronous and asynchronous communication sessions were kept to help students with any kind of issues.

To introduce the platform to the students, we have devoted a half-hour face to face session. In the first 15 min, we have introduced the main functions and mechanisms of the platform through a small practical presentation and simulation.

7.2.1 Synchronous Session

In addition to the face-to-face sessions, and as the platform did not offer the possibility of synchronous communication, a Facebook special group also was created to ensure fluid communication with the students and continuous feedback delivery, especially during the first days.

7.2.2 Welcome Badges

The first offered badges were what we called welcome badges, and these badges were offered to each student who successfully subscribes to the platform (Each student can display the badge in his profile and can be seen by other students.). To push the onboarding process, familiarise the students with the reward system.

7.2.3 Quest Design

The designed activities were part of a studied module where students were obliged to have exams, grades and final mark to pass the module. Additionally, we have been obliged to redesign the last two mandatory activities of the module in collaboration with the tutor, so we can respect the module's objectives, hierarchy and pedagogical design. The first activity was programmed a week before the second, and both activities were accounted for the final mark of the module.

We have first changed the nature of the activities by moving from the traditional notion of pedagogical activities into the realm of narrative as we identified them as quests. Additionally, we have programmed the system, so that it offered 1500 Bloo and 400 XP for submitting the task on time, and it was opened directly without the need of an extra sidequest to be done. Students were asked to write a synthesis about motivation based on the multiple intelligence theory of Gardner. A leader board was always kept available to show students who finished first the quest.

Unlike the first quest, the second quest did not open until you finish the first one and reach the second level, to reach the second level inside the game you need to acquire a given amount of XP points which can be collected through completing the sidequests.

7.3 Sidequest

Simultaneously, we have also designed four *sidequests* which appeared successively in the learning environment. The sidequests did not have any influence on the marks, and nonetheless, they help to gather virtual money (which helps to buy deadlines), open mandatory quests and reaching higher levels.

We have proposed four sidequests, each sidequest was rewarded differently depending on the difficulty of it, the asked tasks were simple and do not need a lot of time to be done, and no deadline was imposed. The students can directly respond to the sidequests on the platform and get an immediate reward.

7.4 Deadlines

Even though we have put strict deadlines, we have offered the students the possibility to buy new deadlines, in case they missed the first established deadline, which gave them more possibilities for failing and redoing the activities. Besides, the deadline buying would also push students into doing more activities (sidequest) for gathering the *money* to buy the deadlines. The flexibility of deadlines was set up to offer student several lives inside the learning environment.

7.5 Reward and Rating Mechanism

A special rating system was set up, making both the tutor and the students more comfortable and raise the competitiveness among the students. The proposed system offers different kinds of tokens including badges, marks, experience points and rating leaderboard. Furthermore, some rewards were automatically given after submitting the work, while other rewards were given by the tutor. This system was specially and carefully designed to keep students motivated using the different gamification features of the learning environment while designing it. We have tried to abide by certain pedagogical constraint imposed by the tutor like the final mark and the final exam.

7.5.1 Badges

in addition to the subscription badges, we have also provided other kinds of badges calling them *badges d'excellence*, those badges were given to students who have presented the best works, respecting the deadlines. Badges represent immediate reward as they were distributed shortly after the submission of the work. The badges were also accompanied by Bloo money and XP.

7.5.2 Experience Points

Experience points were given to students after realising the asked tasks, and the XP amount depends on the length and difficulty of each task. These points allow the student to transcend in the levels and assess his level of mastery. Each quest, sidequest or badge is rewarded with money, as shown in fig, the money allows students to buy deadline or get unblocked. We have established this type of reward to push students to do their best and maximum of the activities and collect the maximum of the money as some students can be motivated with the collection of items.

7.5.3 Leader Boards

The second kind of proposed reward was leaderboards, so we had proposed two kinds of leaderboards. The first leaderboard called performance leaderboard was a personal board kept on the left of the screen of each student, the role of this board is to remind him of his achievements, tickets, done quest, sidequests, and also it plays the role of a to-do list by alerting students about what is left to be done (Fig. 6).

The second leaderboard was a general ranking board that exhibits student's achievements in the learning environment. This board includes the finished quests and the sidequests. Additionally, the board shows the experience points gained by each student and the position of the student among the classroom, which kept student updated about their position among their classmates.

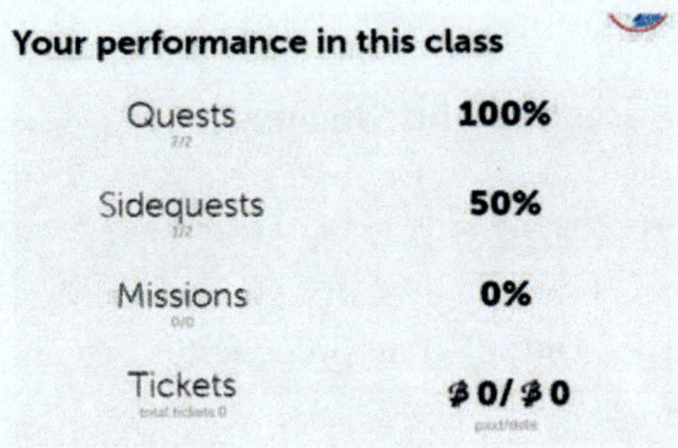

Fig. 6 Performance leaderboard

Fig. 7 Ranking leaderboard

As shown in Fig. 7 no GPA was calculated, there were no accumulated points to be counted, and we opted for XP and levels to quantify the progress of the student in the learning environment which was more practical.

8 Results and Discussion

8.1 Results

8.1.1 Quest Versus Sidequest

The comparison between the mandatory quests and optional sidequest completion would allow us to assess the motivated/non-motivated students, 4 students have completed both the proposed quests and optional sidequests, 12 students only mandatory quests.

8.1.2 Player Type Identification

The second criterion is the player type identification. We consider students who identify their play as more involved with the game and believe in the game mechanics. Four students identify their player item, two were explorers and two were free spirit. Item 12 did not respond to the questionnaire.

8.1.3 Deadlines

The third criterion was the deadline respect, and this item will influence the other mechanic as first we consider the student who respects deadlines more motivated, nevertheless who grilled the deadline would use other game mechanics like the quests or the missions which make him enjoys more the game. No deadline was grilled, all students submitted their work on time, and they were very punctual.

8.1.4 Badges

Reward distribution represents one of the major criteria that we have considered for assessing the students' motivation and interaction with the game mechanics as badges reflect the achievements of students and parallel the involvement with the game mechanics which 16 first subscribers were distributed (for all students), 4 first realisation, 4 hard workers and 0 philanthropist.

8.1.5 Communication

Communication emerges as a criterion in the middle of the experience, as we have noticed that some students communicate constantly and were very interested in the experience. So we considered it as influential in the motivation of students. Four students out of 16 have contacted us. The five students who have communicated did it through Facebook group rather than the official platform in both the synchronous and asynchronous ways. Most of the discussion was around technical issues.

8.1.6 Interviews

Even though we inform the students beforehand about the interview's time, we had interviews with only six students, the only available students, as they were full-time students and do not work. Our questions aimed at collection the student's impressions, who were relaxed as we keep the informal statue.

8.2 *Discussion*

While conducting our interviews, we have noticed that students have shown more interest in the platform visual elements, which reflects the importance of the visual elements in the learning process and especially the motivational aspect.

The results show that the mandatory quests which were directly involved in the final mark were all submitted while the optional side quests were done only by four students, and this reflects the nature of the items that motivated the student. Besides, it opens the debate about the perception of the reward and the culture of the final test, having a good final mark can be influential and hinder doing other activities. In other words, changing the nature of the reward can affect the student's activity and the learning process.

We have also noticed that only four players identified their player types by responding to the questionnaire, this was optional and not mandatory, and the same four players are those who completed all the sidequests and win the biggest number of badges, XP and Bloo. We can conclude that those four players were the most enthusiast about the platform and enjoy the different proposed game mechanics, as it was confirmed during the short interviews. Those four players all had the same player types: explorer and free spirits, which pushes us into thinking that the activities were more oriented toward this kind of players.

The same four players were also the ones who frequently communicate and ask more questions. Consequently, they have succeeded in the onboarding process and have grasped the different components of the learning environment. We can consider that communication is very important in any learning process, and ensuring a fluid dialogue with students can guaranty their success. Most of the communication process took place in *Facebook*, which is frequently used by student.

Deadlines and rewards distribution plays a major role in students' motivation. As we have seen, no deadline was grilled, which reflect the motivation of the student to the activities. Nevertheless, the amount of gained rewards, which does not affect the final mark, was poor; an only minority has done the sidequests, which means the values of reward in the real world are very important to push the student's motivation.

9 Conclusion and Perspectives

To sum up, the results extracted from this experience do not give obvious conclusions; nevertheless, we can say that the gameful design have a positive impact on the students who expressed their interest in the visual design of the platform and keep constant communication with the environment. Additionally, we can say that the visual elements play a major role in students' motivation for learning, especially when students start exploring the learning environment. Moreover, opting only for structural gamification and not changing the nature of the content can still hinder the gamification process. We can develop a more enjoyable environment applying both structural and content gamification design effectively. We also can have a larger sample and longer study, which will help us to have more exact results about gamified learning environment.

References

1. Bartle, R.: HEARTS , Clubs, Diamonds, Spades: (1996) (2014)
2. Buckley, P., Doyle, E.: Gamification and student motivation. Interactive Learning Environments **24**(6), 1162–1175 (2016). https://doi.org/10.1080/10494820.2014.964263
3. Chen, Y., Burton, T., Mihaela, V., Whittinghill, D.M.: Cogent: A case study of meaningful gamification in education with virtual currency. International Journal of Emerging Technologies in Learning **10**(1), 39–45 (2015). https://doi.org/10.3991/ijet.v10i1.4247
4. Deterding, S.: Gamification: Designing for Motivation. Interactions **19**(4), 14 (2012). https://doi.org/10.1145/2212877.2212883
5. Deterding, S., Khaled, R., Nacke, L.E., Dixon, D., et al.: Gamification: toward a definition. In: CHI 2011 Gamification Workshop Proceedings, vol. 12. Vancouver BC, Canada (2011)
6. Dixon, D.: Player Types and Gamification. CHI 2011 Workshop Gamification Using Game Design Elements in NonGame Contexts, pp. 12–15 (2011). DOIACM 978-4503-0268-5/11/05. http://gamification-research.org/wp-content/uploads/2011/04/11-Dixon.pdf
7. Hochleitner, W., Lankes, M., Nacke, L.E., Tscheligi, M., Busch, M., Mattheiss, E., Orji, R., Marczewski, A.: Personalization in serious and persuasive games and gamified interactions. In: CHI PLAY 2015—Proceedings of the 2015 Annual Symposium on Computer-Human Interaction in Play (May 2016), 811–816 (2015). https://doi.org/10.1145/2793107.2810260
8. Hunicke, R., Leblanc, M., Zubek, R.: MDA: A formal approach to game design and game research. In: AAAI Workshop—Technical Report **WS-04-04**, pp .1–5 (2004)
9. Kapp, K.M.: Choose Your Level: Using Games and Gamification to Create Personalized Instruction. Handbook on Personalized Learning for States, Districts, and Schools, pp. 131–143 (2016). http://www.centeril.org/2016handbook/resources/Cover_Kapp_web.pdf
10. Lamprinou, D., Paraskeva, F.: Gamification design framework based on SDT for student motivation. In: Proceedings of 2015 International Conference on Interactive Mobile Communication Technologies and Learning, IMCL 2015 (November), pp. 406–410 (2015). https://doi.org/10.1109/IMCTL.2015.7359631
11. Maican, C., Lixandroiu, R., Constantin, C.: Interactivia.ro—a study of a gamification framework using zero-cost tools. Comput. Hum. Behav. **61**, 186–197 (2016). https://doi.org/10.1016/j.chb.2016.03.023. http://dx.doi.org/10.1016/j.chb.2016.03.023
12. Malas, R.I., Hamtini, T.M.: A gamified e-learning design model to promote and improve learning. International Review on Computers and Software **11**(1), 8–19 (2016)
13. Marczewski, A., Holdings, M.: User Types HEXAD Gamification. Game Thinking and Motivational Design, User Type HEXAD (2016)
14. Marsh, T.: Serious games continuum: Between games for purpose and experiential environments for purpose. Entertainment Computing **2**(2), 61–68 (2011)
15. Muntean, C.I.: Raising engagement in e-learning through gamification. In: Proceedings of 6th International Conference on Virtual Learning ICVL, vol. 1, pp. 323–329 (2011)

16. Nah, F.: Gamification of Education: A Review of Literature—HCI in Business vol. 8527, pp. 401–409. Springer (2014). https://doi.org/10.1007/978-3-319-07293-7. http://link.springer.com/10.1007/978-3-319-07293-7%5Cnlink.springer.com/content/pdf/10.1007/978-3-319-07293-7.pdf
17. Novak, D., Nagle, A., Riener, R.: Linking recognition accuracy and user experience in an affective feedback loop. IEEE Transactions on Affective Computing **5**(2), 168–172 (2014). https://doi.org/10.1109/TAFFC.2014.2326870
18. Salen, K., Tekinbaş, K.S., Zimmerman, E.: Rules of Play: Game Design Fundamentals. MIT Press (2004)
19. Stott, A., Neustaedter, C.: Analysis of Gamification in Education, vol. 8, p. 36. Surrey, BC, Canada (2013)
20. Subhash, S., Cudney, E.A.: Gamified learning in higher education: A systematic review of the literature. Comput. Hum. Behav. **87**, 192–206 (2018). https://doi.org/10.1016/j.chb.2018.05.028
21. Tan, J., Sockalingam, N.: Gamification to Engage Students in Higher Education, pp. 1–14. Research Collection Lee Kong Chian School of Business (2015)
22. Tondello, G.F., Nacke, L.E.: Applying gameful design heuristics. In: Proceedings of the 2017 CHI Conference Extended Abstracts on Human Factors in Computing Systems, CHI EA '17, p. 1209–1212. Association for Computing Machinery, New York, NY, USA (2017). https://doi.org/10.1145/3027063.3027116. https://doi.org/10.1145/3027063.3027116
23. Tondello, G.F., Wehbe, R.R., Diamond, L., Busch, M., Marczewski, A., Nacke, L.E.: The gamification user types Hexad scale. In: CHI PLAY 2016—Proceedings of the 2016 Annual Symposium on Computer-Human Interaction in Play, pp. 229–243 (2016). https://doi.org/10.1145/2967934.2968082
24. Topics, T.: Using Game Elements in Non-game Contexts: Towards Determining the Effects and Applicability of Gamification, pp. 3–4 (2011)
25. Yildirim, I.: Students' perceptions about gamification of education: a Q-method analysis. Egitim ve Bilim **42**(191), 235–246 (2017). https://doi.org/10.15390/EB.2017.6970
26. Yin, R.K.: Case study Research and Applications: Design and Methods. Sage Publications (2017)

An Analysis of ResNet50 Model and RMSprop Optimizer for Education Platform Using an Intelligent Chatbot System

Youness Saadna, Anouar Abdelhakim Boudhir, and Mohamed Ben Ahmed

Abstract A chatbot is a software (or machine) that has the ability to talks with a user: it is a virtual assistant that can answer a number of user questions, and providing the correct responses. In the last few years, the use of chatbots is very popular in various fields, such as health care, marketing, educational, supporting systems, cultural heritage, entertainment, and many others. This paper proposes an intelligent chatbot system that can give a response in the form of natural language or audio to a natural language question or image input in different domains of education and will support multiple languages (English, French, and Arabic). To realize this System, we used different deep learning architectures (CNN, LSTM, Transformers), computer vision, transfer learning to extract image features vector, and natural language processing techniques. In the end, after the implementation of the proposed model, a comparative study was conducted in order to prove the performance of this system using image-response model and question-response model using accuracy and BLEU score metrics.

1 Introduction

The advance in artificial intelligence technology has many benefits in various domains, industry, economy, agriculture, education, and more.

Many researchers concluded the importance of artificial intelligence to improve human life and particularly in the education domain. Thus, chatbots in industry and education have increased considerably in the last few years. Many of them are used as a tutor for students or as customer support. In each case, the chatbot is trained to

Y. Saadna (✉) · A. A. Boudhir · M. Ben Ahmed
List Laboratory, Faculty of Sciences and Techniques, UAE University, Tangier, Morocco

A. A. Boudhir
e-mail: aboudhir@uae.ac.ma

M. Ben Ahmed
e-mail: mbenahmed@uae.ac.ma

M. Ben Ahmed et al. (eds.), *Networking, Intelligent Systems and Security*,
Smart Innovation, Systems and Technologies 237,
https://doi.org/10.1007/978-981-16-3637-0_41

carry out a question-response task. Otherwise, chatbots are known to respond to user questions on a given topic or level and do not support all types of questions (natural language question, image). The purpose of this paper is to design and implement a chatbot that has the ability to support all types of questions (natural language question, image), support multi-language questions (English, French, Arabic), and cover multiple levels of education. The system acts like an intelligent robot that explains, in different languages, the given image or text in the input by giving the response as text or audio in the output.

2 Related Works

In the literature, there are many approaches associated with chatbots, especially on e-learning systems. From the start of the last decade, the usage of AI as e-learning support has captured the interest of many researchers for its many implementations. One among these research works is [1], during which Nenkov N. et al. have looked over the realization of intelligent agents on platform IBM Bluemix with IBM Watson technology. These agents in the form of chatbots need to automate the interaction between the student and therefore the teacher within the frames of Moodle learning management system. Watson is a cognitive system that merges capabilities in analytics, NLP, and machine learning techniques. In this case, Facebook Messenger Bot GUI Builder realizes a chatbot through Facebook Messenger to simplify communication between teachers and students: it might be arranged by acquiring Moodle test basis. In Nordhaug et al. [2] proposed a game based e-learning tool named TFC (The Forensic Challenger), used to teach digital forensic investigation. A chatbot inside the learning platform helps students. A multiple-choice question based quiz is realized for kinesthetic learners, and there is a pedagogical chatbot agent that helps users. It provides easy navigation and interaction within the content. The chatbot is implemented to be a pedagogical agent for the users, which is meant for discussions and help with the topics. It also acts as a navigation tool and can play video or use the advanced wiki if there are something to ask. In Satu et al. [3] many chatbot applications based on AIML are analyzed: especially, an integrated platform that consists of basic AIML knowledge is presented. During this project, chatbot is named Tutor-bot because its functionality backing of didactics done in e-learning environments. It contains some features as natural language management, presentation of contents, and interaction with a search engine. Besides, e-learning platforms work is linked to indispensable services to web service. A continuous monitoring service has been created on e-learning platform servers, which is another controlling machine: Daemon. In Niranjan et al. [4] discussed an interesting approach using Bayesian theory to match the request of students and provide the right response. In particular, chatbot agent accepts to student's answers and extracts the keywords from the question with the use of a lexical parser, and then, the keywords are compared with the category list database. The Bayesian probabilities are obtained for all categories within the list. Once the category is chosen, keywords are compared with the questions under the category using Bayesian probability theory. The answer to the question, which has the biggest posterior probability, is then fed into the text to speech conversion module and thus the student receives the answer to his question as a voice response. In Farhan et al. [5] using a WebBot in an e-learning platform, to deal with the lack of real-time responses for the students. In fact, when a student asks a question on the e-learning platform the teacher could answer later. If there are more students and more questions, this delay is going to be increased. WebBot is a web-based chatbot that predicts future events based on keywords entered on the web. During this work, Pandora is employed, a bot that saves the questions, and answers them on XML style language,i.e., artificial intelligence markup language (AIML). This bot is trained with a sequence of questions and answers: When it cannot furnish a response to a question, a human user is in charge of responding.

3 System Architecture

The architecture of our system is composed of:

- Front-End
- Back-End
- Model
- Database (Fig. 1).

The first module is the presentation layer (front-end) where we provide a friendly interface for the user; it consists of different kinds of devices like phones, tablets, PCs, and so on.

The back-end used to handle operations that are not seen by end-user like:

- Users registration and authentication.
- Registration and authentication of users.
- Prediction of responses for question or image input.
- Conversion from text to speech to make an audio response.
- Handling multiple languages (English, French, Arabic) using translation API.

This module works in the background to better assure the demands of the user: It manages business logic and data storage, working in collaboration with the model part.

The model is where we have our models of deep learning.

The database used to store dataset after cleaning and user data.

4 Proposed Architecture

In this section, we discuss each tool that we use in the model and how they work together for solving this heterogeneous problem.

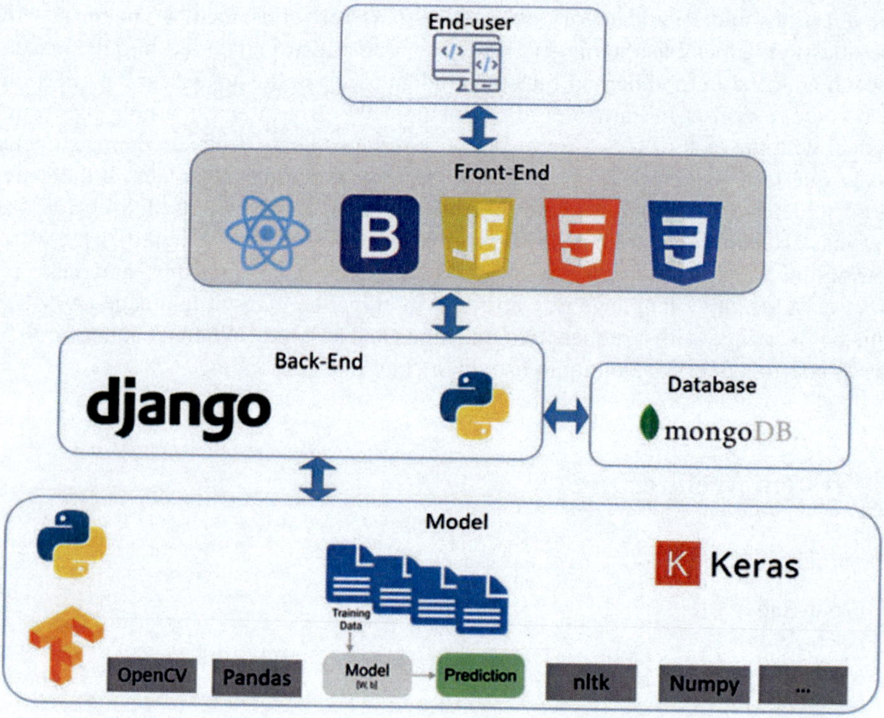

Fig. 1 System architecture

4.1 Image-Response Model

This model combines between two families of artificial intelligence, natural language processing, and computer vision. In this model, we have merged the results of two different models:

- The responses are going to be pre-processed before being indexed and encoded using our vocabulary built from the pre-processed response tokens of the entire corpus.
- The images are going to be passed to a pre-trained convolutional neural network (CNN), so as to extract the features vector of image using the transfer learning fixed feature extracted method.
- Then, we will pass the vector of response and the image features vector to our model encoder.
- The image feature vector will pass through a dropout layer to avoid model overfitting and a dense layer (fully connected layer) to obtain a 256-dimension vector output.

- The response vector will go through an embedding layer to make the correlation between the words, then a dropout layer to avoid model overfitting and an LSTM layer to obtain an output vector of dimension 256.
- As the two outputs of the last two layers have the same 256 dimension, we will merge them with an add layer, and the output is the output of our encoder that we will pass to the decoder.
- The output of the encoder will be passed to the decoder of which we have two dense layers (fully connected layer). The last dense layer contains a Softmax activation function to generate the probability distribution for the 2124 words in the vocabulary we have.

The main goal of this approach is to repeat the vector of image n times, where n is the length of the response that is fixed for the entire responses corpus; then these resulting vectors are going to be passed to an encoder and a decoder are going to generate the response at the end. An encoder is generally used to encode a sequence, in our case the sequence is the two vectors, first sequence of the response vector and the image vector, to merge them and pass them to a decoder, in order to generate a probability distribution. To obtain the next word, we choose the word with a maximum probability at each time step using greedy search algorithm (Fig. 2).

4.2 *Question-Response Model*

Here, we tend to use the transformers architecture [6] without making any change in the global architecture. We make changes only in the hyperparameters until we get high results and adapt the model to our problematic and dataset (Fig. 3).

5 Implementation

5.1 *Dataset*

- SciTail [7]: The SciTail dataset is an entailment dataset created from multiple-choice science exams and web sentences. Each question and the correct answer choice are converted into an assertive statement to form the hypothesis.
- ARC [8]: A new dataset of 7787 genuine grade-school level, multiple-choice science questions, assembled to encourage research in advanced question-answering. The dataset is partitioned into a challenge set and an easy set.
- SciQ [9]: The SciQ dataset contains 13,679 crowd-sourced science exam questions about Physics, Chemistry, and Biology, among others. The questions are in multiple-choice format with four answer options each.

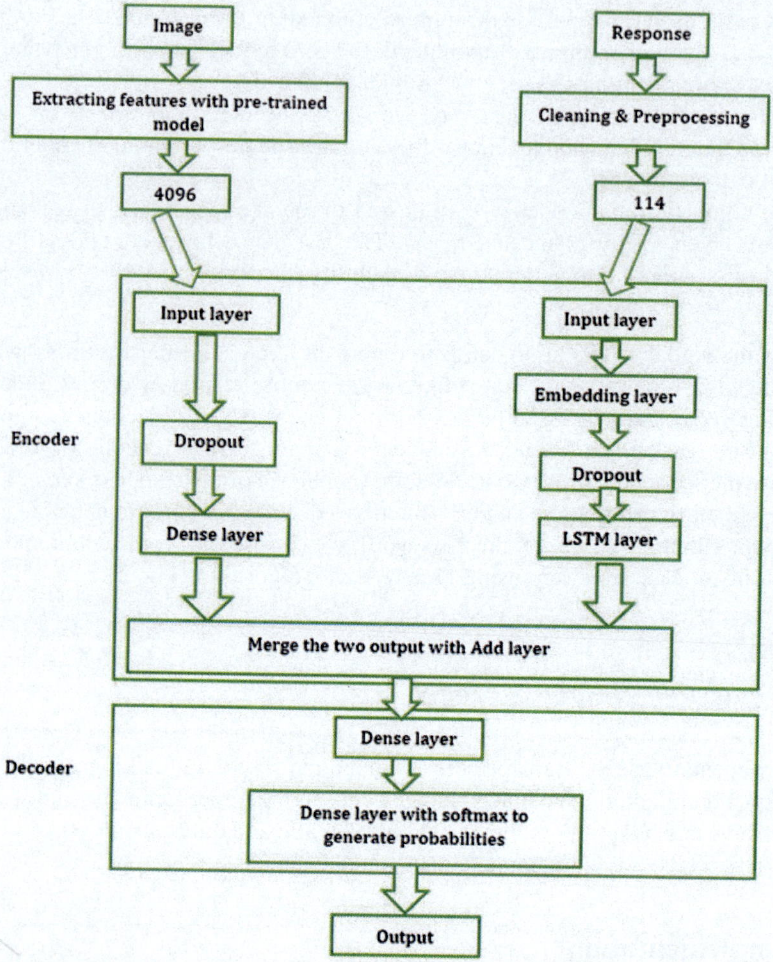

Fig. 2 Image-response model

- Question Answer Dataset [10]: There are three question files, one for each year of students: S08, S09, and S10, as well as 690,000 words worth of cleaned text from Wikipedia that was used to generate the questions.
- Physical IQA [11]: Physical Interaction QA, a new commonsense QA benchmark for naive physics reasoning focusing on how we interact with everyday objects in everyday situations.
- AI2 science [12]: The AI2 Science Questions dataset consists of questions used in student assessments in the United States across elementary and middle school grade levels. Each question is 4-way multiple-choice format and may or may not include a diagram element.

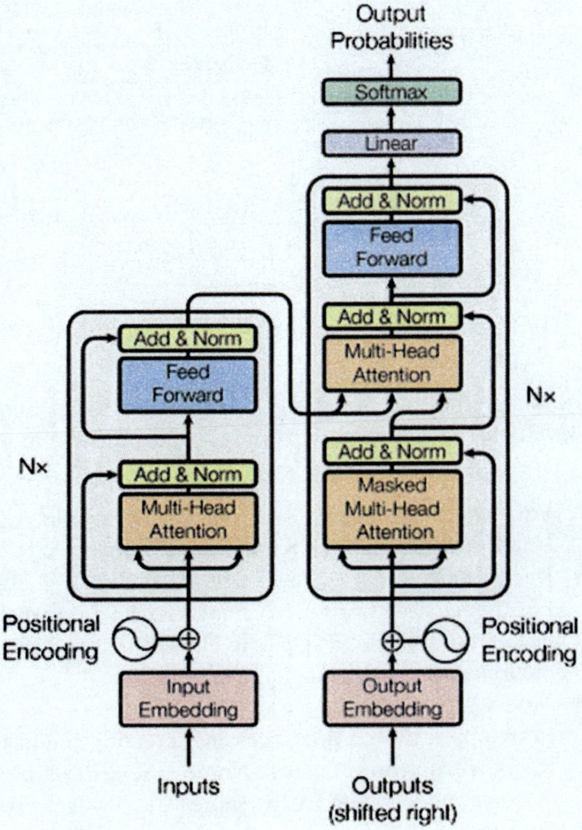

Fig. 3 Question-response model (transformers architecture)

- Image Answer Dataset: this is a dataset that we collect using google forms and it contains about 1200 image answer in different domain (Physics, Biology, Computer science, …) and level (primary, school, high school and university) of education.

5.2 *Hardware*

To train the two models, question-response and image-response models, we used a machine with the following specification (Table 1).

Table 1 Hardware specification

Item	Value
Processor	i7-8550U
RAM	24Go
Storage	1To HDD + 256Go SSD
GPU	NVidia GeForce MX130
VRAM	2Go
Operating system	Windows 10 Pro 64bit

5.3 Languages and Libraries Used

In this section, we are going to list the languages and libraries we used during the development of the system. We will decompose it into four parts:

- Front-End
  For the front-end, we used ReactJS framework with other front-end tools like HTML5, Bootstrap, CSS3, and JavaScript.
- Back-End
  For the back-end, we used Django REST Framework and Python language. We used other libraries as gTTs to convert text to speech and translate-api to handle the translation from language to other.
- Model
  In this part, during preprocessing, creating models and training them, we used Keras, TensorFlow, Pandas, NumPy, Scikit-Learn and we used NLTK library to evaluate models with BLEU score.
- Database
  To store cleaned dataset and data of users, we used MongoDB database.

5.4 Pre-processing

- Image-Response model

Images considered as entries (X) to the model. As we know, any entry to a model must be in the form of a vector. Thus, we need to convert each image into a vector with a fixed size, which can then be fed as an input to the neural network. To achieve this, we choose transfer learning by using pre-trained models like VGG16 (Convolutional Neural Network) to extract a vector of characteristics for each input image. For feature extraction, we used the pre-trained model up to part $7 \times 7 \times 512$. If we want to do a classification task, we will need to use the entire model (Fig. 4).

The model accepts as input an image of size $224 \times 224 \times 3$ and returns as output a feature vector with a dimension of $7 \times 7 \times 512$. Noting that the responses are what we want to predict. Thus, during the learning period, the responses will be the output variables (Y) that the model learns to predict. Nevertheless, the prediction of

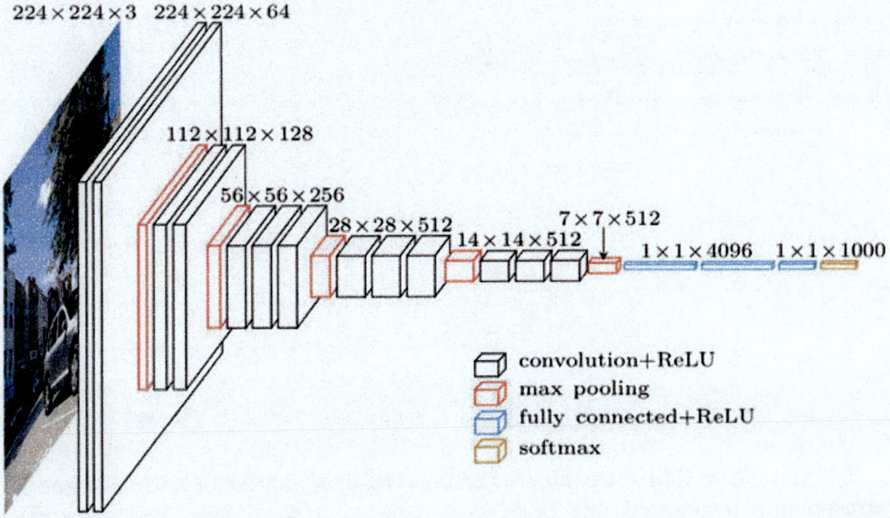

Fig. 4 Architecture of the VGG16

the entire response is not done at once. The prediction of the response is done word-by-word. Therefore, we need to code each word in a fixed size vector. Quite simply, we will represent each unique word in the vocabulary by an index (integer). We have a vocabulary with 2124 unique words in the entire corpus and thus each words are going to be represented by an integer between 1 and 2124. We will take the following example of the response: "angular is typescript based open source web application framework." We will build our vocabulary by adding the two words "startseq" and "endseq" to determine the start and end of the sequence: (Suppose we have already done the cleaning steps).

vocab = {angular, is, endseq, typescript, based, opensource, web, application, framework, startseq}

Let us give an index to each word in the vocabulary we get:

angular-1, is-4, endseq-3, typescript-9, based-7, opensource-8, web-10, application-2, framework-6, startseq-5

Let us take an example where the first characteristics vector of the image Image_1 has a logo of the angular framework and the corresponding response is "startseq angular is typescript based open source web application framework endseq."

Keep in mind that the characteristics vector of the image is the input, and the response is what we need to predict.

Although, we predict the response in the following way:

First, we provide the characteristics vector of the image and the first word as input and we attempt to predict the second word, i.e.

Input = Image_1 + 'startseq'; Output = 'angular'.

We then provide the characteristics vector of the image and the first two words as input and let us attempt to predict the third word, that is:

Table 2 Predicting a response

i	Image feature	First part of the response	Target word
1	Image_1	startseq	angular
2	Image_1	startseq, angular	is
3	Image_1	startseq, angular, is	typescript
…	…	…	…
N-1	Image_1	startseq angular is typescript based opensource web application	framework
N	Image_1	startseq angular is typescript based opensource web application framework	endseq

Input = *Image_1* + "*startseq angular*"; *Output* = "*is*".

And so on…

To sum up, the following table represents the data matrix for an image and its corresponding response (Table 2).

The model will not accept the English text of the response, but rather the sequence of indexes (integers) where each index corresponds to a unique word.

First, we will need to make sure that all the sequences are equal in terms of length. That is why, we need to add zeros at the end of each sequence.

We will add zeros that will make each sequence will have the same length, which is the maximum length of the largest response in our case 114 words (Table 3).

- Question-Response model

Since each model has an encoder-decoder architecture, they have the same basis for predicting responses. The only difference is that they have a very different architecture of the encoder and decoder, and rather than giving an image feature vector in this case, we have a vector that represents the asked question.

Table 3 Predicting a response using word indexes and sequence padding

i	Image feature	First part of the response	Target word
1	Image_1	[5, 0, 0, … , 0]	1
2	Image_1	[5, 1, 0, 0, … , 0]	4
3	Image_1	[5, 1, 4, 0, 0, … , 0]	9
…	…	…	…
N-1	Image_1	[5, 1, 4, 9, 7, 8, 10, 2, 0, 0, … , 0]	6
N	Image_1	[5, 1, 4, 9, 7, 8, 10, 2, 6, 0, 0, … , 0]	3

5.5 *Evaluation Metrics*

We used two types of evaluation metrics: accuracy and BLEU score.

Accuracy is calculated by the following equation:

$$\text{Accuracy} = \frac{\text{Item predicted correctly}}{\text{All items predicted}} \tag{1}$$

BLEU (bilingual evaluation understudy) is an algorithm for evaluating the quality of text that has been machine-translated from one natural language to another. Quality is considered the correspondence between a machine's output and that of a human. BLEU's output is always a number between 0 and 1. This value indicates how similar the candidate text is to the reference texts, with values closer to 1 representing more texts that are similar.

BLEU score is calculated by the following equation:

$$\text{BLEU} = \text{BP.exp}\left(\sum_{n=1}^{N} w_n \log(p_n)\right) \tag{2}$$

6 Results and Comparative Study

In this section, the results of our models are tested by using the test data of the datasets we used during training. It represents a comparative study with different pre-trained model and hyperparameters and with respect to accuracy and BLEU score.

- Image-Response model

For this model, we use an equivalent hyperparameters, but we modify the pre-trained model used in each training (Tables 4 and 5).

Table 4 Used hyperparameters

Hyperparameter	Value
Dropout	0.5
Optimizer	Adam
Learning rate	0.0001
Split data	0.4
Batch size	128
Epochs	30
Loss function	Categorical_crossentropy

Table 5 Results with different pre-trained model

Pre-trained model	Accuracy	BLEU score
VGG16	99.03	86.67
VGG19	99.82	85.24
Xception	99.70	86.99
ResNet50	99.79	91.88
InceptionV3	99.58	85.22

Table 6 Fixed hyperparameters

Hyperparameter	Value
Dropout	0.5
Learning rate	0.0001
Split data	0.2
Batch size	64
Epochs	150
Loss function	SparseCategoricalCrossentropy

As shown in Table 2, the pre-trained model ResNet50 provides the best BLUE score and the second best accuracy.

While the BLUE score is more valid as an evaluation metric than the accuracy within the case of text generation, we decided to use the model with ResNet50 for the deployment.

Question-Response model

For this model, we fixed some hyperparameters and changed the others (Tables 6, 7, and 8).

Table 7 Results with RMSprop optimizer and changing the other model hyperparameters

Number of heads	Number of layer	Accuracy	BLEU score
16	1	37.89	42.88
	2	37.01	27.44
8	1	36.67	41.63
	2	36.36	32.13

Table 8 Results with Adam optimizer and changing the other model hyperparameters

Number of heads	Number of layer	Accuracy	BLEU score
16	1	35.18	40.55
	2	31.47	26.72
8	1	34.32	35.50
	2	33.53	26.53

From Tables 4 and 5, we can see that the model with good results is with the Optimizer RMSprop, 16 heads and 1 layer. Thus, this is the best one for deployment.

7 Conclusion

In this paper, we proposed a chatbot system for education applications, which can support multiple languages (English, French, and Arabic) and different levels of education. The concluded results show the best performance of the pre-trained model ResNet50 compared to other models in the case of Image-Response. Otherwise, and according to the question-response model, the results conducted to the relevance of the RMSprop Optimizer compared to Adam Optimizer for well deployment interest.

The chatbot uses an API for translation to handle multiple languages. This is not favorable, because there is a loss of technical words of a specific domain, to improve this it is better to use a dataset, for example in Arabic, to train the model with it. There is many things that we can add to improve this work. In this context, we have plan in future work to improve the performance of the model in terms of accuracy and response time and add the functionality of recording a question in audio format.

References

1. Nenkov, N., Dimitrov, G., Dyachenko, Y., Koeva, K.: Artificial intelligence technologies for personnel learning management systems. In: Eighth International Conference on Intelligent Systems, 2015
2. Nordhaug, Ø., Imran, A.S., Alawawdeh, Al., Kowalski, S.J.: The forensic challenger. In: International Conference on Web and Open Access to Learning (ICWOAL), 2015
3. Satu, S., Parvez, H., AI-Mamun, S.: Review of integrated applications with AIML based chatbot. In: First International Conference on Computer and Information Engineering (ICCIE), 2015
4. Niranjan, M., Saipreethy, M.S., Kumar G.T.: An intelligent question answering conversational agent using naïve Bayesian classifier. In: International Conference on Technology Enhanced Education (ICTEE), 2012
5. Farhan, M., Munwar, I.M., Aslam, M., Martinez Enriquez, A.M., Farooq, A., Tanveer, S., Mejia P.A.: Automated reply to students' queries in e-learning environment using Web-BOT. In: Eleventh Mexican International Conference on Artificial Intelligence: Advances in Artifical Intelligence and Applications, Special Session—Revised Paper, 2012.
6. Vaswani, A., Shazeer, N., Parmar, N., Uszkoreit, J., Jones, L., Gomez, A.N., Kaiser, Ł., Polosukhin, I.: Attention is all you need. In Advances in Neural Information Processing Systems, pp. 5998–6008 (2017)
7. Khot, T., Sabharwal, A., Clark, P.: SciTaiL: A textual entailment dataset from science question answering. In: AAAI (2018)
8. Clark, P., Cowhey, I., Etzioni, O., Khot, T., Sabharwal, A., Schoenick, C., Tafjord, O.: Think you have solved question answering? Try ARC, the AI2 reasoning challenge. arxiv:1803.05457 (2018)
9. Welbl, J., Liu, N.F., Gardner, M.: Crowdsourcing multiple choice science questions. arxiv: 1707.06209 (2017)

10. Smith, N.A., Heilman, M., Hwa, R.: Question generation as a competitive undergraduate course project. In: Proceedings of the NSF Workshop on the Question Generation Shared Task and Evaluation Challenge (2008)
11. Bisk, Y., Zellers, R., Le Bras, R., Gao, J., Choi, Y.; PIQA: reasoning about physical commonsense in natural language. arXiv:1911.11641 (2020)
12. Clark, P.: Elementary school science and math tests as a driver for AI: take the aristo challenge! In: AAAI (2015)

Smart Information Systems

BPMN to UML Class Diagram Using QVT

Mohamed Achraf Habri, Redouane Esbai, and Yasser Lamlili El Mazoui Nadori

Abstract The business process model and notation (BPMN) standard provides notations in the form of diagrams, which are clearly legible for the needs of internal organizations and facilitate collaboration between enterprise components. The problem is how to find a transformation solution between the BPMN and the UML (Unified modeling language) to benefit from the simplicity of the BPMN on the one hand, the stability and the widespread UML on the other hand. We work to have a high-performance solution within the framework of model-driven architecture (MDA) that saves time, cost and quality of the software. This article represents a transformation method that allows us to go through the BPMN business process diagram to arrive at the UML class diagram, finally, to generate the code automatically by using the query views transformations (QVT) transformation language, and this transformation is a fruitful combination between the trades side and the computer side.

1 Introduction

The computer modeling language at the base is not understandable by the majority of the staff of all the specialties in the company that prevents a perfect collaboration with the information system.

In this context, object management group (OMG) has adopted the business process model and notation (BPMN), a standard for writing business processes by models, and it offers a simple and clear graphical notation for the entire body of the company. From business analysts through IT developers to simple users, this process modeling standard incorporates new symbols for business process diagrams [1].

The OMG has previously created the model-driven architecture (MDA) standard [2], and the goal is to migrate legacy information systems that exist in enterprises to new platforms, adapt them with new IT components, to protect the investment, maximize flexibility and save time and have a strong, efficient and up-to-date system. The MDA standard is based on three models, namely the computing-independent

M. A. Habri (✉) · R. Esbai · Y. Lamlili El Mazoui Nadori
Mohammed First University, Oujda, Morocco

M. Ben Ahmed et al. (eds.), *Networking, Intelligent Systems and Security*,
Smart Innovation, Systems and Technologies 237,
https://doi.org/10.1007/978-981-16-3637-0_42

model (CIM), the platform-independent model (PIM) and the platform-specific model (PSM).

This convergence between these two OMG standards prompted us to think how to move from the CIM model of the BPMN to a PIM model of another modeling language, namely UML, for example, from a process diagram to a class diagram through the use case diagram and also to consolidate the modernization already carried out by MDA, our ultimate goal is to automatically generate the code from the class diagram, so our operation is to start with a BPMN business process diagram and end up with the code. This transformation requires the use of a transformation language such as query view transformation (QVT), a language that belongs to the OMG family.

Our paper rethinks the method of transformation presented in Addamssiri et al. [3] "Generating the PIM Behavioral Model from the CIM using QVT," Rhazali et al. [4] "A New Methodology CIM to PIM Transformation Resulting from an Analytical Survey," Bảo et al. [5] "A proposal for a method to translate BPMN model into UML activity diagram," [6]. Ondrej Macek et al. "The BPM to UML activity diagram transformation using XSLT" and Oualid et al. [7] "Transforming CIM to PIM Using UML."

This article is organized as following the introduction in the first place, related work is presented in the second section, and the third contains the background of the basics used in the work by offering explanations on BPMN, MDA and QVT.

The fourth section is the part that reveals the performed task, exactly starting with the tracing of the BPMN diagram according to the case study passing through the transformation rules and the use of the QVT arriving at the desired result.

Finally, we made a conclusion, and at the same time, a vision of the tasks that we intend to accomplish in the future, which will enhance our work and even take it as a reference.

2 Related Works

In this section, we present the related works of transformation between BPMN and UML according MDA.

In [3], the authors present a method founded on a transformation and a modelization of the model CIM to the model PIM; practically, in this case, they propose the transformation of the BPMN (business process model and notation) to the UML, of the diagram of business processes to the diagram of use case and later arriving at the sequence diagram based on sequence business vocabulary rules (SBVR), and the transformation language adopted is the query views transformation (QVT), in which case the transformation rules respect the transformation rules of QVT.

The authors propose in [4] a methodology that makes it possible to transform the CIM models into the PIM models into a model-driven architecture (MDA); the idea is to first create transformable models at the CIM to facilitate the transformation using the language ATLAS TRANSFORMATION LANGUAGE (ATL); in this case, the task is to go from the BPMN at the level of the CIM arriving at the UML to model the PIM, and the example chosen is the booking service.

The authors present in [5] an idea that focuses on the conversion of BPMN into UML, precisely to go from the business process diagram to the activity diagram by concocting the context as it is, by specifying rules of direct, complex transformations are specified according to the need and the case studied.

In [6], the authors propose a new type of diagram transformation in business process modeling (BPMN) notation into a unified modeling language (UML) activity diagram. Using the XSLT, a transformational language for transforming XML documents into other documents, includes a vocabulary, an XML document style and specifying transformation rules that match the description of the BPMN in XMI and the description of the UML DA in XMI.

The authors present propose in [8] an idea that is based on the creation of the components of the platform trades Java ee 6 from the business processes modeled by the business process model notation (BPMN) 2.0; this creation consists to carry out three types of transformation, first transform the BPMN diagram into an UML class diagram, then transform it into a UML model with Java Platform profiles. To finally get a meta-object facility script (MOFScript) into Java EE components, the transformation operations are done using query view transformations (QVT) and MOF script.

3 Background Knowledge

3.1 The Business Processes Models: BPMN

A business process model and notation (BPMN) will provide companies with the ability to understand their internal procedures via graphical notation and give organizations the ability to communicate these procedures in a standard way; BPMN was originally developed by the business process management initiative (BPMI) and has been maintained by the object management group (OMG) since the merger of these two Consortia in June 2005 [1, 9].

The current version of BPMN is 2.0.2 and is from 2013. Since July 2013, it has been an international standard ISO/IEC 195106. In this context, BPMN has been used because of its clarity and simplicity, especially in the business world.

3.2 The Model-Driven Architecture MDA

The model-driven architecture is a software development approach, proposed and supported by the OMG.

MDA provides a strong framework that enables system infrastructures to evolve in response to an endless parade of platforms, while preserving and leveraging existing technology investments. It enables more efficient, faster and less costly system integration strategies [10].

MDA models:

- CIM (independent model computation) independent calculation model: describes the flows and actions on the system.
- PIM (platform-independent model) platform-independent model: describes business-oriented processes.
- PSM (platform-specific model) platform-dependent model: describes the technical details related to the implementation for a platform.

The MDA standard ensures the passage between the three types of models so it helps us a lot for the realization of our work.

3.3 The Query View Transformation: QVT

As part of saving effort and reducing errors, OMG adopted the model transformation language called QVT in 2005 [1]. Model transformation into MDA is an automated way of modifying and creating models. A model transformation usually specifies the acceptable input models and the models that it can output by specifying the meta-model that a model must conform to. The QVT standard introduces the means to query, display and transform models based on meta-object facility (MOF) 2.0 [11]. The QVT language is supported by Eclipse, so on the technical side, there will be no obstacle to ensure the passage between BPMN and UML that are already supported by Eclipse.

4 Our Proposal Approach

In our paper, we present an approach in which we realize a transformation of the independent model of computation CIM to a model independent of platforms PIM; the two models are represented by two different languages of modeling; we begin to treat the study of cases (online purchase) by a process diagram in the first place using BPMN; subsequently transforming it into a UML class diagram using the QVT transformation language, the tool used for the representation of diagrams and the development of the transformation is the ECLIPSE software, so our task has allowed us to have a complete modeling cycle starting with the CIM represented by a process diagram arriving at the code step or even the desired software the PSM.

We focused on an important and relevant transformation; from the BPMN diagram as an input, to the UML class diagram as an output, we rethink the transformation methods in the following papers:

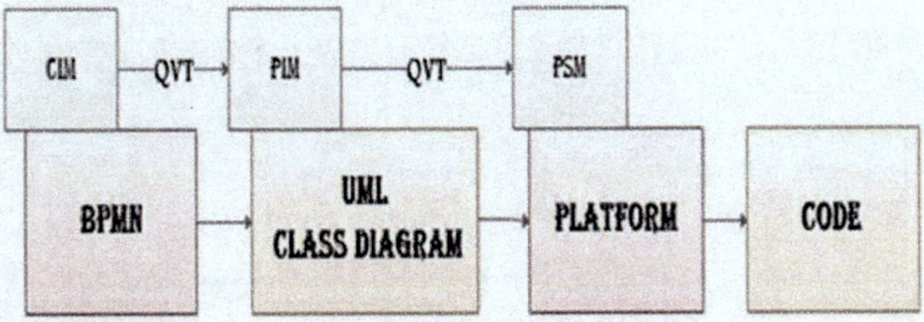

Fig. 1 Transformations of the models in MDA by projecting our study of cases

In [3, 4], the authors propose a transformation of BPMN to the use case of UML arriving at the UML sequence diagram; the authors in [5, 6] describes a transformation going from BPMN diagram arriving at the UML activity diagram; in [7], the authors made a CIM to PIM transformation in a single modeling language, namely UML.

We opted for a direct transformation between two different modeling languages; using the BPMN model as a use case and since there is a very large intersection between the BPMN model and the UML activity diagram to get directly to the class diagram from the BPMN instead of the use case diagram of UML or the UML activity diagram, then we choose the platform and we build the code (Fig. 1).

4.1 *BPMN to Class Diagram Transformation QVT Rules*

To have a transformation done correctly by the QVT, we have based on transformation rules that express each element in one language to what concerns it in the other language. Practically, the class diagram is more sophisticated, especially to design the case study unlike the BPMN which is very simple and clear so to properly determine each transformation rule, we have used conditions in the transformation rules that we will find the example in "4.3 extracted code by QVT" to distinguish each case to ensure the passage.

4.2 *Case Study: BPMN Diagram by Eclipse*

We have chosen a simple and clear case study that is the online purchase; a customer placing orders online on a website, as shown in the BPMN diagram, uses elements represented by simple symbols like "Start" that means the start of the process, tasks like simple, sent, received, for example, "the task choose product." In the middle of the process, we observe the collaboration between the client and the site that we call them participants and this is why we call our example a diagram of collaboration. In the future as a next step, we can choose a more complicated case study to use more elements thus more BPMN symbols (Fig. 2).

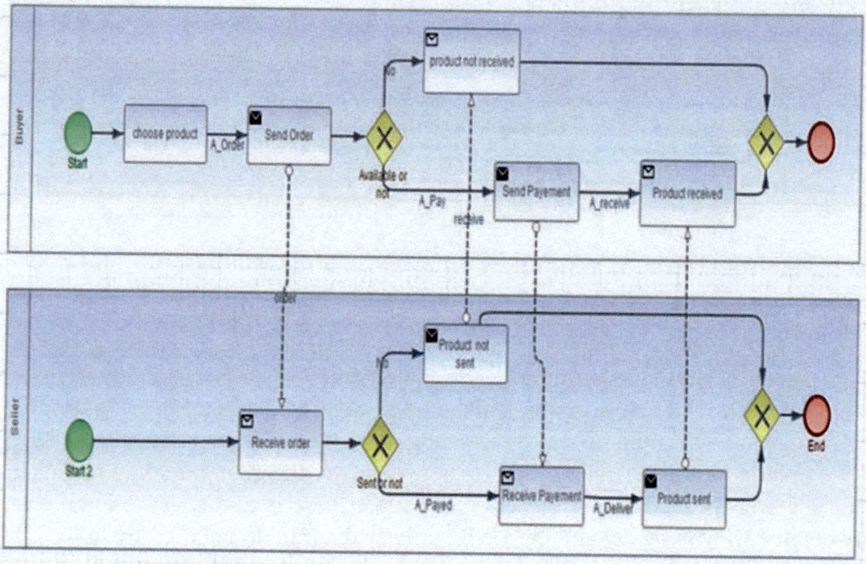

Fig. 2 BPMN diagram for our case study

4.3 *Extracted Code by QVT Language Used by Eclipse*

In this piece of the QVT code (view transformation language), we first import the sources of UML and BPMN and how to write the transformation rules in QVT, for example, the transformation between the task and the class or the operation, and to put the difference, we used the condition in the ID of the task following the conception of course [12, 13].

Figure 3 presents the principle part of the M2M transformation with QVT language.

As we showed in Fig. 3, we distinguished the transformed elements by a modification of the id of the task; we notice in Fig. 4 "properties of the 'receive order' task," for example, that each element of the BPMN model has an id, a source and a destination according to its position in the diagram.

4.4 *The Result of the Transformation*

The execution of the previous QVT code allows a transformation of the source components of the BPMN toward the destination components of the UML while respecting the transformation rules mentioned in Table 1 (Fig. 5).

```
//modeltype UML uses 'http://www.eclipse.org/qvt/1.0.0/Operational/examples/simpleuml';
modeltype UML uses 'http://www.eclipse.org/uml2/5.0.0/UML';

modeltype bpmn uses 'http://www.omg.org/spec/BPMN/20100524/MODEL';

transformation BPMN2UML(in bp : bpmn, out uml : UML);

main() {
bp.objects()[Participant]->map Partic2Class();
bp.objects()[Lane]->map Lane2Class();
bp.objects()[Task]->forEach(t){
    if(t.id.endsWith("o")){
        bp.objects()->resolve(c:Class|c.name=t.id.substringBefore("_").replace("_",""))->first().ownedOperation+= object Operation {
                name := t.name;
            };
    }else
       if (not t.id.endsWith("_n")){
        t.map Task2Class();
        }
};
bp.objects()[SequenceFlow]->forEach(t){
    if (t.name.startsWith("A_"))

        t.map SequenceFlow2Association();
};
bp.objects()[MessageFlow]->forEach(t){
    if (t.name.startsWith("A_"))

        t.map MessageFlow2Association();
        };
bp.objects()[ExclusiveGateway]->forEach(e){
    if (e.name.startsWith("A_"))
        e.map ExGate2Association();
```

Fig. 3 Result of the transformation by eclipse

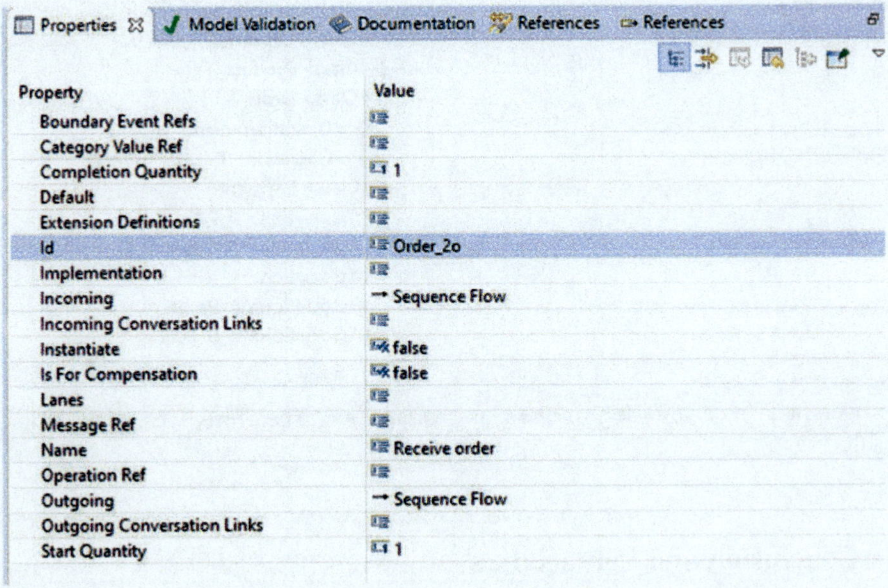

Fig. 4 Properties of the "receive order" task

Our result is the class diagram produced by the eclipse papyrus tool based on the structure of the class diagram of Fig. 4. This is the last step of our work of which we can observe the transformations mentioned in the table of transformation rules of Table 1. For example, the participant "buyer" in BPMN is transformed into a class "buyer" on the UML class diagram, the sequence flow "send_paiment" is transformed into an association "A_pay", and the received task "receive_order" into

Table 1 BPMN to class diagram transformation QVT rules

Transformation rules	Source model element	Target model element
Participant (pool) to class	Participant (pool)	Class
Lane to class	Lane	Class
Task "Activity" to class or method "with condition"|	(Task)	Class or method condition
	Receive task	
	Send task	
Sequence flow/message flow to association	Sequence flow	Association (condition)
Exclusive gateway (OR XOR …) to association	Exclusive gateway	Association or (generalization)

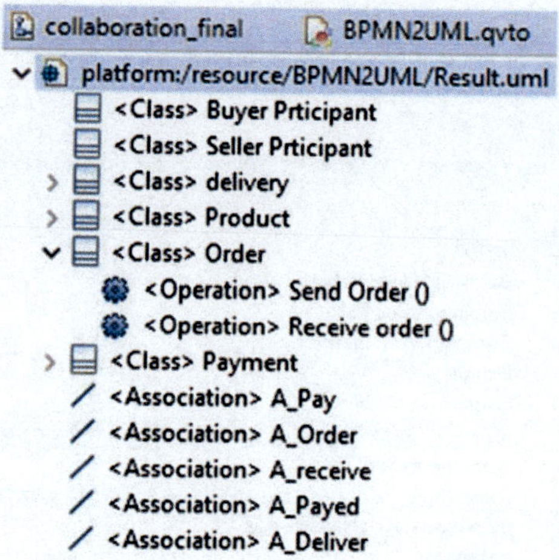

Fig. 5 Result of the transformation by eclipse

an operation "receive order" in the class "order," etc. Therefore, we arrive at our result (Fig. 6).

5 Conclusion and Perspectives

The transformation between two models is an essential step in a modeling project, and the MDA approach helps to have a perfect conception especially to properly represent each model by other modeling languages.

In our article, we have, based on the transformation between the BPMN diagram and the class diagram of the UML, given the importance of the latter for the highlighting of the software, and this transformation can be seen as a bridge between the simple users and the IT developers.

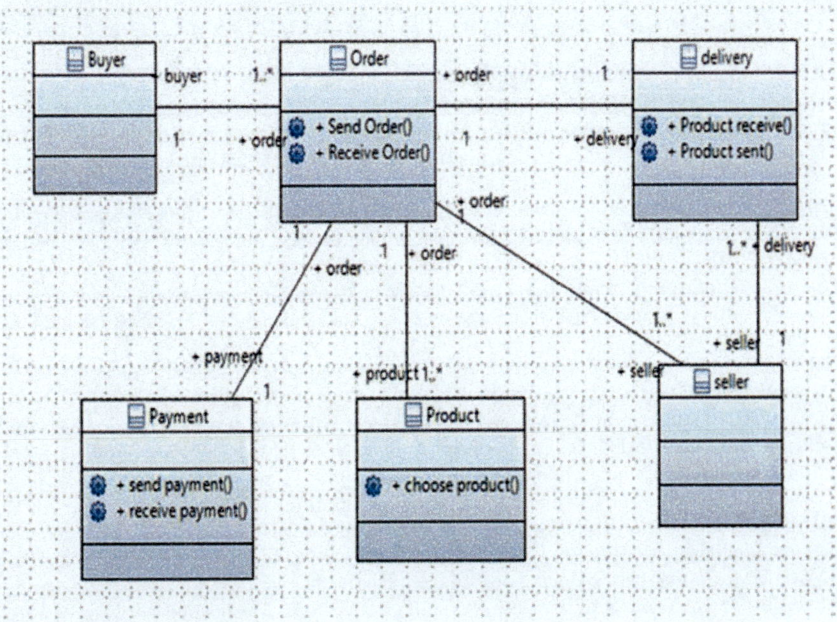

Fig. 6 Result in class diagram

This result led us to think about finding other results with other transformations using other modeling languages, (diagrams), other languages of transformations or to use the ADM architecture-driven modernization standard, practically to carry out an inverse transformation with which we use in our paper, i.e., to transform the PIM in our case, it is the class diagram of UML into a business process model, namely the BPMN which represents the CIM.

References

1. Object Management Group (OMG): https://www.omg.org/bpmn/index.htm
2. MDA en action Ingénierie logicielle guidée par les modèles, Xavier Blanc, édition Eyrolles, 2005
3. Addamssiri, N., Kriouile, A., Balouki Y., Taoufiq, G.: J. Comput. Sci. Inf. Technol. **2**(3 & 4) (2014)
4. Rhazali, Y., Hadi, Y., Mouloudi, A.: Third World Conference on Complex Systems (WCCS), 2015
5. Bảo, N.Q.: Vietnamese-German University—BIS (2010)
6. Macek, O., Richta, K.: Proceedings of the Dateso 2009 Annual International Workshop on DAtabases, TExts, Specifications and Objects, Spindleruv Mlyn, Czech Republic, April 15–17, 2009
7. Oualid, B., Saida, F., Amine, A., Mohamed, B.: Applying a model driven architecture approach: transforming CIM to PIM using UML. Int. J. Online Biomed. Eng. (iJOE) **14**(9), 170–181 (2018)

8. Debnath, N., Martinez, C.A., Zorzan, F., Riesco, D.: IEEE Trans. Ind. Informat., December 2013
9. BPM Modeling FAQ: http://www.bpmodeling.com/faq/.
10. Gotti, S., Mbarki, S.: IFVM bridge: a model driven IFML execution. Int. J. Online Biomed. Eng. (iJOE) **15**(4), 111–126 (2019)
11. Sbai, R., Arrassen, I., Meziane, A., Erramdani, M.: QVT transformation by modeling. (IJACSA) Int. J. Adv. Comput. Sci. Appl. **2**(5), (2011)
12. Argañaraz, M., Funes, A.: An MDA Approach to Business Process Model Trans-formations, université national de SAN LUIS, SADIO electronic journal of informatics operations research, janvier, 2010.
13. Esbai, R., Elotmani, F., Belkadi, F.Z.: Model-driven transformations: toward automatic generation of column-oriented NoSQL databases in Big Data context. Int. J. Online Biomed. Eng. (iJOE). **15**(09), 2019 (2019)
14. Blanc, X.: MDA en action, Ingénierie logicielle guidée par les modèles, EYROLLES. Paris, 1st edition (2005)
15. Abdelhedi, F., Ait Brahim, A., Atigui, F., Zurfluh, G.: Processus de transformation MDA d'un schéma conceptuel de données en un schéma logique NOSQL. In: Congrès INFORSID, 34ème édition, Grenoble, 31 mai–3 juin 2016
16. Braun, R., Esswein, W.: Classification of Domain-Specific BPMN Extensions. In: 7th IFIP Working Conference on The Practice of Enterprise Modeling (PoEM), Nov 2014, Manchester, United Kingdom
17. OMG: UML Infrastructure Final Adopted Specification, version 2.0, September 2003
18. Radoslava, S.K., Velin, S.K., Nina, S., Petia, K., Nadejda, B.: Design and Analysis of a Relational Database for Behavioral Experiments Data Processing. International Journal of Online and Biomedical Engineering (iJOE) **14**(02) (2018), 117–132 (2019)

Endorsing Energy Efficiency Through Accurate Appliance-Level Power Monitoring, Automation and Data Visualization

Aya Sayed, Abdullah Alsalemi, Yassine Himeur, Faycal Bensaali, and Abbes Amira

Abstract Accrediting the fast economic growth and the enhancement of people's live standards, the overall household's energy consumption is becoming more and more substantial. Thus, the need of conserving energy is becoming a critical task to help preserve energy resources and slow down climate change, which in turn, protects the environment. The development of an Internet of Things (IoT) system to monitor the consumer's power consumption behavior and provides energy saving recommendation at a timely manner can be advantageous to shape the user's energy saving habits. In this paper, we integrate the $(EM)^3$ framework into a local IoT platform named Home-Assistant to help centralize all the connected sensors. Additionally, two smart plug systems are proposed to be part of the $(EM)^3$ ecosystem. The plugs are employed to collect appliances energy consumption data as well as having home automation capabilities. Through Home-Assistant User Interface (UI), end-users can visualize their consumption trends together with ambient environmental data. The comparative analysis performed demonstrates great potential and highlights areas of future work focusing on integrating more sensing systems into the developed platform for the sake of enriching the existing database.

A. Sayed (✉) · A. Alsalemi · Y. Himeur · F. Bensaali
Department of Electrical Engineering, Qatar University, Doha, Qatar
e-mail: as1516645@qu.edu.qa

A. Alsalemi
e-mail: a.alsalemi@qu.edu.qa

Y. Himeur
e-mail: yassine.himeur@qu.edu.qa

F. Bensaali
e-mail: f.bensaali@qu.edu.qa

A. Amira
Institute of Artificial Intelligence, De Montfort University, Leicester, UK
e-mail: abbes.amira@dmu.ac.uk

M. Ben Ahmed et al. (eds.), *Networking, Intelligent Systems and Security*,
Smart Innovation, Systems and Technologies 237,
https://doi.org/10.1007/978-981-16-3637-0_43

1 Introduction

The consumer's electricity demand is projected to rise to 30% by 2040 (compared to 2017) [1, 2]. This is due to a multitude of factors including the increase of population, growth in devices and urbanization, thus, creating a strain over current electricity infrastructure [3, 4]. Moreover, in order to satisfy this amount of demand, the environment is drastically and perpetually affected, causing pollution and global warming [5, 6]. Not to mention, the energy crisis caused by the overwhelming production of CO2 which has been cited as a serious issue in the energy market [7, 8].

Further, across the world, COVID-19 social distancing measures have resulted in a significant segment of the population having to stay at home [9]. In certain regions, this move has resulted in a temporary transfer of energy demand away from business districts and urban centers to residential areas, resulting in increased utilization and higher bills, creating an economic effect on customers. Data from the US, Europe, and India indicates that during the most restrictive containment stage (covering early-mid-April), energy sector activity decreased by 15% on average, while residential electricity usage increased by 5% [9].

In this regard, a range of solutions are presented to converse energy. Internet of energy (IoE) is being one of the top energy conserving solutions[10–12]. Following smart grids, IoE has emerged as a renowned technology in the energy sector. This technology employs the Internet to collect, arrange, optimize, and control the networks energy data obtained from different edge devices to develop a distributed smart energy infrastructure [13–15]. By using sensor and communication technologies, data is collected and used to predict consumer's demand and supply to further polish their energy consumption behavior [12, 16].

Since the consumption is a human product, and through taking advantage of the aforementioned technology, collected domestic energy consumption data can be used to dissect the factors responsible of shaping behavior and then construct technology-based techniques to motive a more energy efficient habits [6, 17, 18]. In this context, because of the widespread use of the IoT devices, it becomes at hand to collect appliance-level energy consumption data, which helps in accurately analyzing power consumption traces and effectively detecting abnormal consumption behaviors [19]. This is possible either by installing individual IoT smart sensors/meters [16, 20] or through deploying the energy disaggregation technology that allows the extraction of the appliances' signatures from the main supply only by using artificial intelligence tools [21–23]. By consequence, sub-metering plays an essential role in order produce more accurate and personalized recommendations, which enable end-users in performing the best energy-saving decision-making at the right time [24].

1.1 Our Contributions

In this paper, we investigate a new approach of integrating the consumer engagement toward energy saving behavior by means of exploiting micro-moments and mobile recommendation systems $(EM)^3$ ecosystem into an IoT platform called Home-Assistant. Then, we examine commercial smart plug systems. After, we include a comparative analysis between the $(EM)^3$ smart plugs. Finally, we discuss the faced challenges while integrating $(EM)^3$ framework with the Home-Assistant platform.

1.2 Paper Organization

The rest of this article is organized as follows. Section 2 highlights related work on energy efficiency systems . Section 3 gives an overview and the characteristics of the $(EM)^3$ framework. Section 4 describes the system design along with declaring the different components it comprises. Section 5 provides an overview of Home-Assistant platform and different applications. Section 6 goes in details over the $(EM)^3$ smart plugs. Section 7 presents the evaluation of results in real-time and discuss further limitations. Finally, the paper is concluded with future work in Sect. 8.

2 Related Work

Adopting a more accountable, efficient, and environmentally-aware power consumption is becoming essential to maintain the longevity of the modern electrical grid [25, 26]. According to the authors of [27], smart meters are being implemented in millions of houses around the world, allowing a bi-directional communication between the utility companies and the households, and as a result, generating a sheer high volume of data. By using big data technologies capable of taking this hefty amount of data and extracting consumer energy usage patterns, predicting demand, and optimizing usage, the utility providers' capabilities can be revolutionized. Primarily, the paper presents an intelligent data mining model to investigate, forecast, and visualize energy time series to reveal numerous energy usage habits.

Owing to the accessibility of electricity and its integral role, it can be quite the challenge to model the consumer demeanor toward electricity saving [28]. Utilizing smart home technologies can help bridge the gap between understanding the consumer's behavior and integrating that understanding into advancing smart technology. Smart homes technologies are not only intended to improve the energy usage, but also in fact provides other benefits such as enhanced lifestyle, security, and safety. Some of the many technologies used in smart homes to change the consumer's power consumption behavior includes smart meters, smart plugs, and home automation devices. In [29], a practical case study of IoT dedicated to home automation is illustrated. The paper discusses the implementation of three hardware modules, involving a smart home, smart building, and smart city. Similar to this work, an IoT platform named Home-Assistant was used with the smart home implementation.

The authors in [30] highlight the significance of collecting and analyzing power consumption data in buildings, in order to recognize the moments when the occupants could adjust their energy patterns. Additionally, it has been described that the change in the residence's behavior can aid reaching the net-zero energy goal for a building, creating an opportunity for overall energy saving. This occurs once the total power consumption of the building is equal to the power generated from the renewable energy systems operating in that property.

Moreover, in [31], the authors present a consumer-oriented energy monitor built on the Raspberry Pi that allow smart energy monitoring and services in residences. Named YoMoPie, the system monitors both active and apparent power, saves recorded data on the board as well as enabling the energy sensor access via a Python API. The API allows executing user-designed services to increase energy quality in buildings and households.

Also, an energy management system (EMS) that regulates HVAC appliances is developed for fostering energy conservation in houses [32]. Based on the plan-do-check-act (PDCA) cycle, the EMS has to manage data collection, data processing, and execution. In this article, we are introducing a real-world EMS deployed in a real-world building. Powered by a Home-Assistant-based platform, micro-controller enabled sensing units integrate those units with actuators and the database. The units are inter-connected via a mesh network.

Another Home-Assistant contribution is the VOLTTRON-enabled EMS for homes [33]. The intention of this initiative is to develop open-source, easy-to-use EMS accessible on the market so that anyone can profit from it. The system is a mixture between a home automation system and a distributed control and sensing software (i.e., VOLTTRON). As a result, the EMS is to comply with government requirements such as system monitoring and regulation, smooth connectivity between devices, DR, intelligence, data processing, and protection.

3 Overview of the (EM)^3 Framework

The (EM)^3 framework has been developed to promote behavioral improvement for customers through improving understanding of energy use in domestic households and buildings. It involves four key steps described as: collection of data (i.e., consumption footprints and atmospheric conditions) from various appliances in institutional buildings [34, 35], processing of consumption footprints in order to abstract energy micro-moments that help identify anomalies [36, 37], implementation of consumer preference details to detect correlations between them, and generation of customized feedback to minimize abnormalities [38, 39] and visualize consumption footprints [40].

Sensing components play a crucial role in collecting and securely preserving data in the platform store. By using micro-controller unit (MCU), data from different sensors is extracted and wirelessly transfered from various cubicles to the (EM)^3 storage server housed in the Qatar University (QU) lab. Figure 1 illustrates the overall design of the (EM)^3 energy efficiency framework.

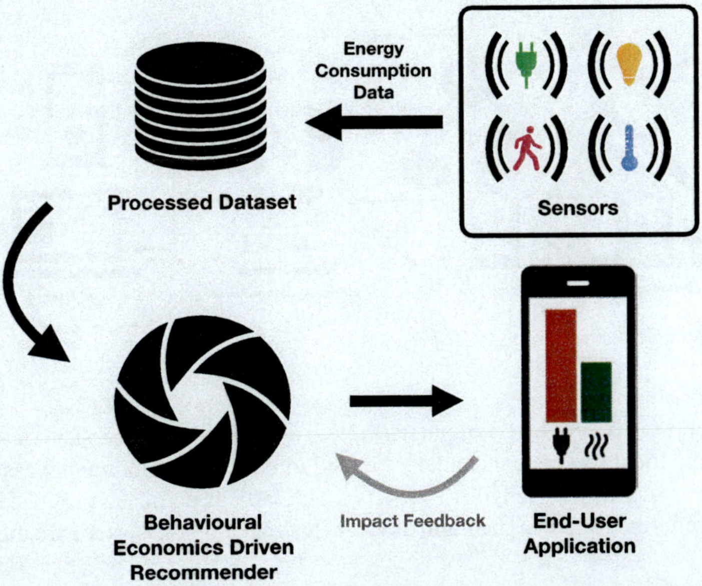

Fig. 1 Overview of (EM)^3 energy efficiency framework

In addition, a recommendation engine focused on an algorithm that considers consumer needs, energy priorities, and availability in order to optimize approval of the suggested action and improve the performance of the [41] recommended method. The algorithm is based on the user's extracted patterns involving the repetitive usage of devices at some times during the day [20]. It is derived from data on energy usage and room occupancy reported in the recent background of activities of the customer (or office) [42].

In summary, the framework is comprised of the following functions:

1. Data Collection: collect data for power usage and environmental monitoring [43].
2. Micro-Moment Classification: analyzes abnormal energy patterns [44].
3. Suggestions and Automations: provides personalized guidance to end-users to endorse energy-saving management activities coupled with recommended actions for environmental change [45] .
4. Visualization: upload results, observations, and feedback in an accessible and engaging manner via a mobile application.

4 Proposed Platform

The system encompasses several sensing components to capture different types of data such as power consumption, occupancy, light level, and air information (i.e., temperature and humidity). The collected data is sent over Wi-Fi to the local server

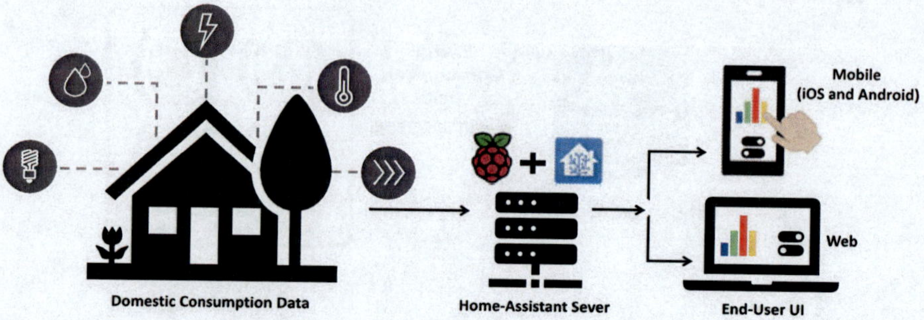

Fig. 2 System block diagram

to be viewed via a personal computer (PC) or any mobile device. The data is also stored on the local server so it can later be feed into the micro-moment classifier and the recommender system. Figure 2 depicts a block diagram of the proposed system with the predefined stages. The main devices that make up the system are mentioned next.

4.1 Home-Assistant Platform

A Raspberry Pi 4 B+ runs the latest version of Home-Assistant platform with ESPHome integration. ESPHome is used to establish a connection between ESP32 and the platform. The library also allows full control over the ESP32 via the Home-Assistant UI.

4.2 Sensing Components

All the selected sensors with their operating ranges are shown in Table 1 including:

1. The DHT22 is a digital sensor used to measure relative humidity in percent and temperature in Celsius;
2. The TSL2591 measures the intensity of light in Lux units;
3. The AM312 is a passive infra-red (PIR) motion sensor;
4. The HLW8012 is used to measure the real power of a connected appliance in Watt;
5. A 5V relay is used as a switch to control the appliances remotely; and
6. The Sonoff POW is a wirelessly connected device used to monitor electricity usage and can also be utilized as a smart switch.

Table 1 List of components

Component name	Operating range
DHT22	Temperature: −40°C−480°C and Humidity: 0–100%
TSL2591	188μLux-88,000Lux
AM312	3–5 m
HLW8012	Voltage: 90-240VAC, max current: 15A and max power: 3500W
5V Relay	10A and 250VAC
Sonoff POW	Voltage: 90-250VAC, max current: 16A and max power: 3500W

5 Overview of Home-Assistant

When investing in a home automation ecosystem, there are a lot of factors to consider. Currently, most of the devices available for purchase are linked to some kind of cloud service, offering to the end-user a level of convenience [46]. Simultaneously, numerous problems arise when using such cloud services, an obvious one being, providing the host company a full access to the user's personal data. There are several alternatives to explore when planning to regain control over your smart devices and achieving local control [47]. Home-Assistant being one of these options. Home-Assistant is a platform that helps to centralize all the sensors and gadgets available at home. The platform utilizes message queuing telemetry transport (MQTT) to communicate with other devices [48]. MQTT is a lightweight messaging protocol utilized to exchange messages between devices.

5.1 Connected Nodes

Through ESPHome, the connection between Home-Assistant and the different hardware modules is made possible. ESPHome is an add-on available through Home-Assistant to allow control over a verity of MCU, in this case ESP32. This is feasible by simply editing a configuration file, and then the nodes can be controlled remotely through ESPHome dashboard.

Two nodes are created into ESPHome to represent the two physically available units, the environmental module, and the power module, in the lab as displayed in Fig. 3. The former setup gathers contextual information on temperature, humidity, occupancy, and level of light. The latter setup provides data on power usage and also involves a relay used as a switch over, however, many distance to control any appliance.

Furthermore, in Fig. 4, the nodes included in Home-Assistant through ESPHome are indicated. The "powertest" is connected with the power unit, and the "testnode"

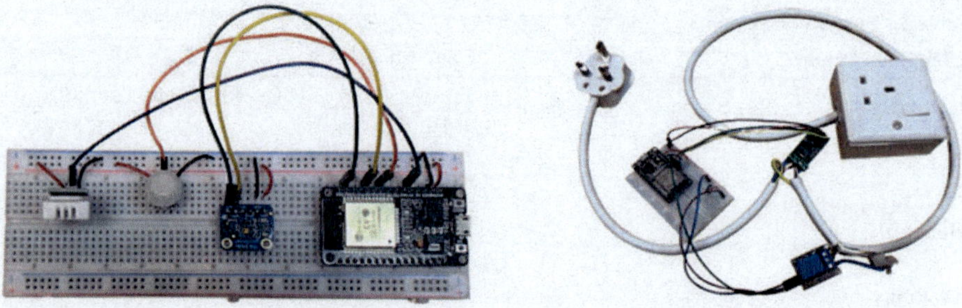

(a) Environmental module (b) Power module

Fig. 3 Hardware configuration

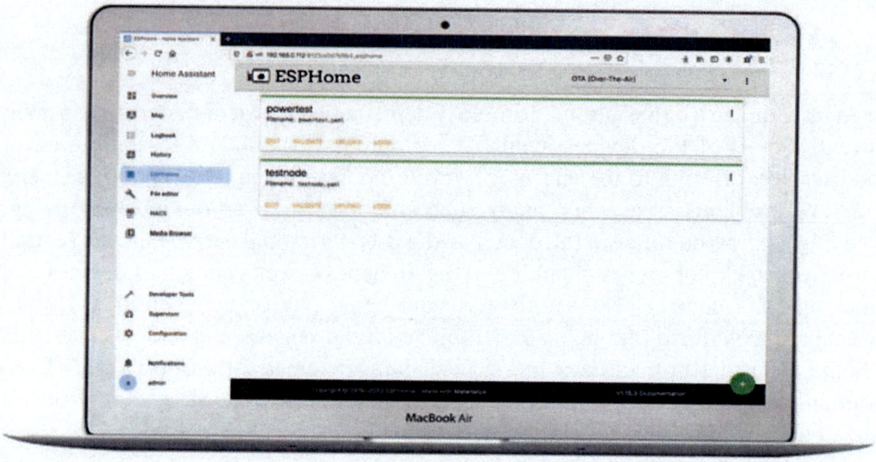

Fig. 4 The nodes in ESPHome dashboard interface

is linked with the environmental unit. In the configuration file corresponding to each node, the sensors comprising each setup are included.

5.2 Data Visualization Options

Through Home-Assistant history component, data is tracked and saved as an SQLite database. The end-user has access to all stored information. The end-user additionally can select a specific time to view data at. Power consumption data is presented in Figs. 5 and 6. The figures display the power consumption patterns of a 60W lamp.

This is also true for other sensors connected with Home-Assistant, such as data collected from the environmental unit can be viewed in the same manner as Fig. 5 and Fig. 6. Data can be also viewed instantaneously through the overview similar to what illustrated in Fig. 7.

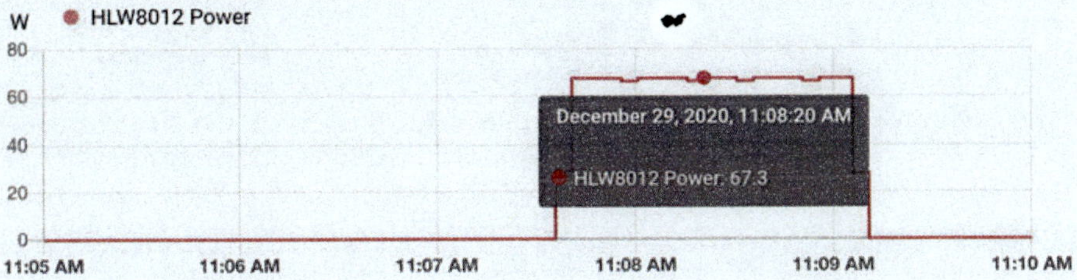

Fig. 5 Power consumption of a 60W lamp measured by HLW8012 sensor

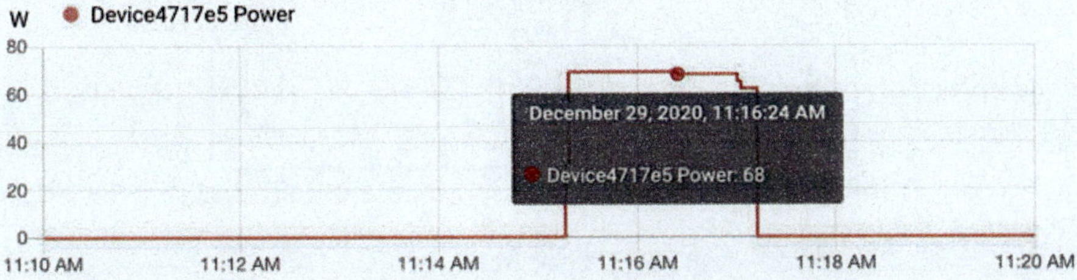

Fig. 6 Power consumption of a 60W lamp measured by the Sonoff POW device

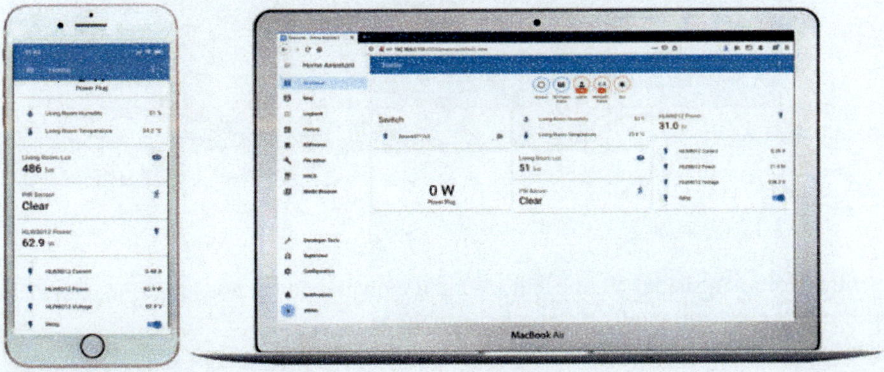

(a) Mobile UI (b) Web UI

Fig. 7 End-user interface

6 (EM)^3 Smart Plug

To determine the power consumption for a given appliance, a proper sensing device is required. To perform that task, two smart plug alternatives are proposed, which are the HLW8012 an power sensor and the Sonoff POW device.

Figure 8 provides illustration on how the connection is done between the Sonoff POW smart plug, the power source, and the load. The power plug should be connected to the grid to provide the needed power to any appliance connected to the socket.

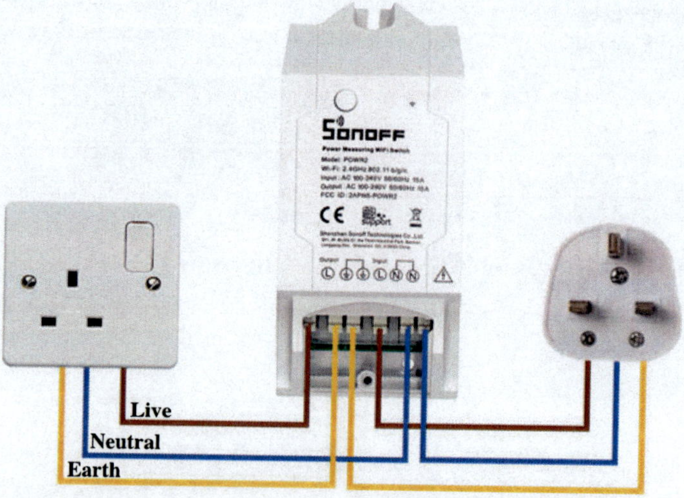

Fig. 8 Overview of the Sonoff POW smart plug connection

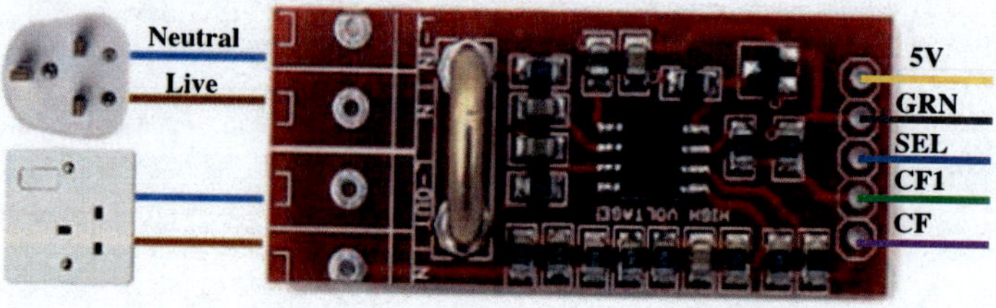

Fig. 9 Overview of the HLW8012 connection

Over the Home-Assistant UI, the end-user can view the power consumption data and additionally can turn on/off the coupled appliance.

As for the HLW8012 power sensor, the connection with the source and load was performed similar to the Sonoff POW device; however, in the case of this sensor, additional pins must be connected with an ESP32 in order to power the sensor and also acquire the power consumption reading. The sensor is powered using 5V and Ground, and the other pins are connected with digital pins on the ESP32. The connection of the HLW8012 sensor is demonstrated in Fig. 9.

7 Results and Discussion

To evaluate the new integration and the proposed smart plugs, the performance of the smart plugs is reported along with a discussion of the current limitations. The test-bed overview is demonstrated next.

Fig. 10 PX110 watt-meter

7.1 Test-Bed Overview

Figure 2 depicts the test-bed configuration. It includes either the web or the mobile application UI, local server with Home-Assistant, the various sensors to provide contextual, and power consumption information. The test-bed was designed mainly to measure the communication latency and concurrently evaluate the performance of the smart plug alternatives against a reliable watt-meter. The remaining sensors were evaluated versus a reference measurement apparatus in [49].

7.2 Comparative Analysis Between Smart Plugs

The reference measurement tool used is the PX110 watt-meter shown in Fig. 10. The meter has an accuracy of $\pm 1.5\%$ for real power [50]. Table 2 depicts the reference used and accuracy of the smart plug alternatives connected with different home appliances. The test was conceded with a duration of 45–60 s depending on the appliance. For example, the kettle operating time (i.e., time needed to boil water to 100°C) was 45 s; therefore, the readings were collected for that time period.

When assessing the efficiency of the smart plugs, it is clear from Table 2, the overall superiority of the $(EM)^3$ smart plug against the Sonoff POW. This due to the fact that the $(EM)^3$ smart plug was exactly calibrated to match the reference, while Sonoff POW unit was not.

Table 2 Smart plugs accuracy results

Appliance	$(EM)^3$ Smart plug accuracy (%)	Sonoff POW accuracy (%)
60W lamp	99.28	97.58
Air cooler	98.87	96.51
Monitor	96.55	95.95
Kettle	99.12	96.93
Laptop	96.96	97.44
Overall	98.16	96.88

7.3 Discussion

The deployed Home-Assistant platform displays the environmental data with an update interval of 1 sec, as for the power plugs the interval is set to be 5 sec. Hence, all the data can be viewed on the UI in real-time, either from the web interface or from the open-source smartphone app (on iOS and Android). The performance of the smart plugs is satisfactory, with the $(EM)^3$ in the lead with an accuracy of 98.16% and Sonoff POW finishes last with accuracy of 96.88%.

With that being said, the integration is still limited, since the micro-moment classifier and the recommender system are still not implemented into the platform. Another constraint is, data collected is solely stored on the Raspberry Pi with no backup, preventing the risk of losing valuable data. Moreover, the current setup is only implemented for one room (i.e., research cubicle) and not been verified for several end-users. However, since the current setup connection is displaying good stability with a high measured accuracy, it can be extended to support more end-users and appliances.

8 Conclusion

In this paper, the integration of $(EM)^3$ framework with Home-Assistant is presented. The framework aims to develop more environmentally conscious habits through providing energy saving recommendation at the right micro-moment when the end-user is more receptive to change. The framework is supported by a system of sensors to collect rich habitual and environmental data. The additional smart plugs proposed to be added to the system are tested against a trustworthy reference to measure their performance. High stability and accuracy are reached with future work, focusing on expanding the size of the sensing system linked with Home-Assistant with the intention of increasing the database scope and diversity.

Acknowledgements This paper was made possible by National Priorities Research Program (NPRP) grant No. 10-0130-170288 from the Qatar National Research Fund (a member of Qatar Foundation). The statements made herein are solely the responsibility of the authors.

References

1. Miglani, A., Kumar, N., Chamola, V., Zeadally, S.: Blockchain for internet of energy management: review, solutions, and challenges. Comput. Commun. **151**, 395–418 (2020)
2. Himeur, Y., Alsalemi, A., Bensaali, F., Amira, A.: A novel approach for detecting anomalous energy consumption based on micro-moments and deep neural networks. Cogn. Comput. **12**(6), 1381–1401 (2020)
3. Cao, X., Dai, X., Liu, J.: Building energy-consumption status worldwide and the state-of-the-art technologies for zero-energy buildings during the past decade. Energy Build. **128**, 198–213 (2016)
4. Alsalemi, A., Himeur, Y., Bensaali, F., Amira, A., Sardianos, C., Varlamis, I., Dimitrakopoulos, G.: Achieving domestic energy efficiency using micro-moments and intelligent recommendations. IEEE Access **8**, 15047–15055 (2020)
5. Keho, Y.: What drives energy consumption in developing countries? The experience of selected African countries. Energy Policy **91**, 233–246 (2016)
6. Himeur, Y., Alsalemi, A., Al-Kababji, A., Bensaali, F., Amira, A.: Data fusion strategies for energy efficiency in buildings: overview, challenges and novel orientations. Inf. Fusion **64**, 99–120 (2020)
7. Sardianos, C., Varlamis, I., Chronis, C., Dimitrakopoulos, G., Alsalemi, A., Himeur, Y., Bensaali, F., Amira, A.: The emergence of explainability of intelligent systems: Delivering explainable and personalized recommendations for energy efficiency. Int. J. Intell. Syst. **36**(2), 656–680 (2021)
8. Sardianos, C., Varlamis, I., Chronis, C., Dimitrakopoulos, G., Himeur, Y., Alsalemi, A., Bensaali, F., Amira, A.: A model for predicting room occupancy based on motion sensor data. In: 2020 IEEE International Conference on Informatics, IoT, and Enabling Technologies (ICIoT), IEEE, pp. 394–399 (2020)
9. Snow, S., Bean, R., Glencross, M., Horrocks, N.: Drivers behind residential electricity demand fluctuations due to covid-19 restrictions. Energies **13**(21), 5738 (2020)
10. Alsalemi, A., Himeur, Y., Bensaali, F., Amira, A.: An innovative edge-based internet of energy solution for promoting energy saving in buildings. Sustain. Cities Soc. 1–20 (2021)
11. Al-Ali, A.-R., Zualkernan, I.A., Rashid, M., Gupta, R., Alikarar, M.: A smart home energy management system using IoT and big data analytics approach. IEEE Trans. Consum. Electron. **63**(4), 426–434 (2017)
12. Shahzad, Y., Javed, H., Farman, H., Ahmad, J., Jan, B., Zubair, M.: Internet of energy: opportunities, applications, architectures and challenges in smart industries. Comput. Electr. Eng. **86**, 106739 (2020)
13. Sardianos, C., Varlamis, I., Chronis, C., Dimitrakopoulos, G., Himeur, Y., Alsalemi, A., Bensaali, F., Amira, A.: Data analytics, automations, and micro-moment based recommendations for energy efficiency. In: 2020 IEEE Sixth International Conference on Big Data Computing Service and Applications (BigDataService), IEEE, pp. 96–103 (2020)
14. Kabalci, E., Kabalci, Y.: From Smart Grid to Internet of Energy. Academic Press (2020)
15. Alsalemi, A., Himeur, Y., Bensaali, F., Amira, A., Sardianos, C., Chronis, C., Varlamis, I., Dimitrakopoulos, G.: A micro-moment system for domestic energy efficiency analysis. IEEE Syst. J. 1–8 (2020)
16. Himeur, Y., Alsalemi, A., Bensaali, F., Amira, A.: An intelligent non-intrusive load monitoring scheme based on 2d phase encoding of power signals. Int. J. Intell. Syst. **36**(1), 72–93 (2021)

17. Zhang, C.-Y., Yu, B., Wang, J.-W., Wei, Y.-M.: Impact factors of household energy-saving behavior: an empirical study of Shandong Province in China. J. Cleaner Prod. **185**, 285–298 (2018)
18. Azizi, Z.M., Azizi, N.S.M., Abidin, N.Z., Mannakkara, S.: Making sense of energy-saving behaviour: a theoretical framework on strategies for behaviour change intervention. Procedia Comput. Sci. **158**, 725–734 (2019)
19. Himeur, Y., Elsalemi, A., Bensaali, F., Amira, A.: Smart power consumption abnormality detection in buildings using micro-moments and improved k-nearest neighbors. Int. J. Intell. Syst. 1–25 (2021)
20. Elsalemi, A., Himeur, Y., Bensaali, F., Amira, A.: Appliance-level monitoring with micro-moment smart plugs. In: The Fifth International Conference on Smart City Applications (SCA), pp. 1–5 (2020)
21. Himeur, Y., Alsalemi, A., Bensaali, F., Amira, A.: Efficient multi-descriptor fusion for non-intrusive appliance recognition. In: IEEE International Symposium on Circuits and Systems (ISCAS). IEEE, vol. 2020, 1–5 (2020)
22. Himeur, Y., Alsalemi, A., Bensaali, F., Amira, A.: Improving in-home appliance identification using fuzzy-neighbors-preserving analysis based qr-decomposition. In: International Congress on Information and Communication Technology. Springer, Berlin, pp. 303–311 (2020)
23. Himeur, Y., Alsalemi, A., F.ensaali, Amira, A., Sardianos, C., Varlamis, I., Dimitrakopoulos, G.: On the applicability of 2d local binary patterns for identifying electrical appliances in non-intrusive load monitoring. In: Proceedings of SAI Intelligent Systems Conference. Springer, Berlin, pp. 188–205 (2020)
24. Himeur, Y., Alsalemi, A., Bensaali, F., Amira, A.: Building power consumption datasets: survey, taxonomy and future directions. Energy Build. **227**, 110404 (2020)
25. Al-Kababji, A., Alsalemi, A., Himeur, Y., Bensaali, F., Amira, A., Fernandez, R., Fetais, N.: Energy data visualizations on smartphones for triggering behavioral change: Novel vs. conventional. In : 2nd Global Power, Energy and Communication Conference (GPECOM). IEEE, vol. 2020, pp. 312–317 (2020)
26. Sardianos, C., Chronis, C., Varlamis, I., Dimitrakopoulos, G., Himeur, Y., Alsalemi, A., Bensaali, F., Amira, A.: Real-time personalised energy saving recommendations. In: The 16th IEEE International Conference on Green Computing and Communications (GreenCom), pp. 1–6 (2020)
27. Singh, S., Yassine, A.: Big data mining of energy time series for behavioral analytics and energy consumption forecasting. Energies **11**(2), 452 (2018)
28. Bhati, A., Hansen, M., Chan, C.M.: Energy conservation through smart homes in a smart city: a lesson for Singapore households. Energy Policy **104**, 230–239 (2017)
29. Debauche, O., Mahmoudi, S., Moussaoui, Y.: Internet of things learning: a practical case for smart building automation. In: 2020 5th International Conference on Cloud Computing and Artificial Intelligence: Technologies and Applications (CloudTech). IEEE, pp. 1–8 (2020)
30. Chou, C.-C., Chiang, C.-T., Wu, P.-Y., Chu, C.-P., Lin, C.-Y.: Spatiotemporal analysis and visualization of power consumption data integrated with building information models for energy savings. Resour. Conserv. Recycl. **123**, 219–229 (2017)
31. Klemenjak, C., Jost, S., Elmenreich, W., Yomopie: a user-oriented energy monitor to enhance energy efficiency in households. In: 2018 IEEE Conference on Technologies for Sustainability (SusTech), IEEE, pp. 1–7 (2018)
32. Najem, N., Haddou, D.B., Abid, M.R., Darhmaoui, H., Krami, N., Zytoune, O.: Context-aware wireless sensors for IoT-centeric energy-efficient campuses. In: 2017 IEEE International Conference on Smart Computing (SMARTCOMP), IEEE, pp. 1–6 (2017)
33. Zandi, H., Kuruganti, T., Fugate, D., Vineyard, E.A.: Volttron-enabled home energy management system, Tech. rep., Oak Ridge National Lab.(ORNL), Oak Ridge, TN (United States) (2019)
34. Himeur, Y., Alsalemi, A., Bensaali, F., Amira, A.: Effective non-intrusive load monitoring of buildings based on a novel multi-descriptor fusion with dimensionality reduction. Appl. Energy **279**, 115872 (2020)

35. Himeur, Y., Elsalemi, A., Bensaali, F., Amira, A.: Recent trends of smart non-intrusive load monitoring in buildings: a review, open challenges and future directions. Int. J. Intell. Syst. 1–28 (2020)
36. Sardianos, C., Chronis, C., Varlamis, I., Dimitrakopoulos, G., Himeur, Y., Alsalemi, A., Bensaali, F., Amira, A.: Smart fusion of sensor data and human feedback for personalised energy-saving recommendations. Int. J. Intell. Syst. 1–20 (2021)
37. Himeur, Y., Alsalemi, A., Bensaali, F., Amira, A., Varlamis, I., Bravos, G., Sardianos, C.: Dimitrakopoulos, Techno-economic analysis of building energy efficiency systems based on behavioral change: a case study of a novel micro-moments based solution. Appl. Energy 1–25 (2021)
38. Himeur, Y., Ghanem, K., Alsalemi, A., Bensaali, F., Amira, A.: Artificial intelligence based anomaly detection of energy consumption in buildings: a review, current trends and new perspectives. Appl. Energy **287**, 116601 (2021)
39. Himeur, Y., Elsalemi, A., Bensaali, F., Amira, A.: The emergence of hybrid edge-cloud computing for energy efficiency in buildings. In: Proceedings of SAI Intelligent Systems Conference, pp. 1–12 (2021)
40. Al-Kababji, A., Alsalemi, A., Himeur, Y., Bensaali, F., Amira, A., Fernandez, R., Fetais, N.: Interactive visual analytics for residential energy big data. Inf. Vis. 1–20 (2021)
41. Himeur, Y., Elsalemi, A., Bensaali, F., Amira, A: Appliance identification using a histogram post-processing of 2d local binary patterns for smart grid applications. In: Proceedings of 25th International Conference on Pattern Recognition (ICPR), pp. 1–8 (2020)
42. Varlamis, I., Sardianos, C., Dimitrakopoulos, G., Alsalemi, A., Himeur, Y., Bensaali, F., Amira, A.: Reshaping consumption habits by exploiting energy-related micro-moment recommendations: a case study. In: Communications in Computer and Information Science, Springer International Publishing, Cham, pp. 1–22 (2020)
43. Alsalemi, A., Ramadan, M., Bensaali, F., Amira, A., Sardianos, C., Varlamis, I., Dimitrakopoulos, G.: Endorsing domestic energy saving behavior using micro-moment classification. Appl. Energy **250**, 1302–1311 (2019). https://doi.org/10.1016/j.apenergy.2019.05.089 https://doi.org/10.1016/j.apenergy.2019.05.089
44. Himeur, Y., Alsalemi, A., Bensaali, F., Amira, A.: Robust event-based non-intrusive appliance recognition using multi-scale wavelet packet tree and ensemble bagging tree. Appl. Energy **267**, 114887 (2020)
45. Sardianos, C., Varlamis, I., Dimitrakopoulos, G., Anagnostopoulos, D., Alsalemi, A., Bensaali, F., Himeur, Y., Amira, A.: Rehab-c: recommendations for energy habits change. Future Gener. Comput. Syst. **112**, 394–407 (2020)
46. Himeur, Y., Alsalemi, A., Bensaali, F., Amira, , A., Sardianos, C., Dimitrakopoulos, G., Varlamis, I.: A survey of recommender systems for energy efficiency in buildings: Principles, challenges and prospects. Inf. Fusion 1–33 (2020)
47. Alsalemi, A., Al-Kababji, A., Himeur, Y., Bensaali, F., Amira, A.: Cloud energy micro-moment data classification: a platform study. In: 2020 IEEE/ACM 13th International Conference on Utility and Cloud Computing (UCC), IEEE, pp. 420–425 (2020)
48. Home Assistant. Available online https://www.home-assistant.io/. Accessed 30-12-2020
49. Alsalemi, A., Ramadan, M., Bensaali, F., Amira, A., Sardianos, C., Varlamis, I., Dimitrakopoulos, G.: Boosting domestic energy efficiency through accurate consumption data collection. In: IEEE SmartWorld, Ubiquitous Intelligence & Computing, Advanced & Trusted Computing, Scalable Computing & Communications, Cloud & Big Data Computing, Internet of People and Smart City Innovation (SmartWorld/SCALCOM/UIC/ATC/CBDCom/IOP/SCI). IEEE, pp. 1468–1472 (2019)
50. TRMS three- and single phase digital wattmeters. Available online: http://www.farnell.com/datasheets/3649.pdf. Accessed 30-12-2020

Towards a Smart City Approach: A Comparative Study

Zineb Korachi and Bouchaib Bounabat

Abstract There are conflicts and ambiguity regarding smart city strategies. Some works present smart city processes with vague and inconsistent steps. While others propose smart city elements and dimensions, rather than providing a clear and holistic approach. The most of smart city strategy works overlap one another, which creates ambiguity for smart city leaders. To fill this gap and reduce this ambiguity, the current paper presents and describes the main components of a smart city strategy framework, which are: strategic vision, action plan, and management strategy. To evaluate the relevance of these elements, the present paper conducts a comparative study.

1 Introduction

To be a smart city, it is first necessary to have specific goals and strategies and to be committed to fulfilling those goals [1, 2]. Despite the importance of smart cities, still few studies investigate how developing a clear and consistent smart city approach to help cities become smarter [3]. There is no agreement about the smart city definition, domains, and indicators [4]. In addition to the challenges surrounding the smart city definition and indicators, there is also an ambiguity regarding the definition of the smart city strategy [5].

Many of the smart city strategy efforts are fragmented, stressing only some aspects of the smart city, rather than approaching them in an integrated way [6–8]. Some of these works treat city objectives and indicators, whereas others emphasize solution architectures and technical details [9]. This enhances the misunderstanding and ambiguity regarding the smart city strategy, rather than resolving it and enabling actionable smart city planning [8]. Evidence of this ambiguity is presented by recent works

Z. Korachi (✉) · B. Bounabat
ADMIR Laboratory, Rabat Information Technology Center, ALQUALSADI Team, ENSIAS/Mohamed V University in Rabat, Rabat, Morocco
e-mail: zineb.korachi@um5s.net.ma

B. Bounabat
e-mail: b.bounabat@um5s.net.ma

M. Ben Ahmed et al. (eds.), *Networking, Intelligent Systems and Security*, Smart Innovation, Systems and Technologies 237,
https://doi.org/10.1007/978-981-16-3637-0_44

illustrating the existence of overlapping smart city development processes, which generate conflict and ambiguity regarding smart city development paths [7, 10–13].

Even though there is no 'one-best-way' of designing a smart city strategy, there is a need for an integrated and comprehensive framework that highlights the different components of a smart city and which provides guidelines for smart city strategy design [5]. The current work fills this gap by presenting a comparative study illustrating and describing the main smart city strategy components which are: smart city strategic vision, action plan, and management strategy.

The paper is structured as follows: the next section presents literature review, followed by methodology, problem of statement, results, discussion, and conclusion.

2 Literature Review

This section analyzes existing smart city frameworks and strategies. It shows the relevant blocks, strengths, and weaknesses of these frameworks. This analysis helps to identify and factorize the relevant components of these frameworks and models in a unified solution.

- Agbali et al. (2019) present a comparative analysis between three cities: Boston, Manchester, and San Diego. The comparison is conducted using a smart city ecosystem composed of the following domains: Smart infrastructure, smart institution, smart people [3].
- Oliveira et al. (2020) highlight the concepts surrounding cities as follow: mobility, health care, governance industry, and services [14]. These concepts are important elements that should be incorporated into the smart city strategy. Oliveira et al. (2020) cite that the citizen is the core center of the smart city system.
- Darmawan et al. (2019) identify crucial factors in the process of readiness and application of the smart city concept to regional governments in Indonesia [15]. These factors are presented as follow: Perceived of use, service quality, system quality, information quality, and intention use [15]. These factors provide city leaders with characteristics that can help to identify a successful strategy.
- Dabeedooal et al. (2019) propose the smart tourism dimensions as follow: smart infrastructure, smart business, governance, and urban metabolism. It presents a framework for smart tourism composed of the following components: Technology applications, leadership, human capital, entrepreneurs, innovation, social capital, tourism experience, tourism competitiveness [16]. The smart tourism framework should achieve the following characteristics: Attractions, accessibility, amenities, available packages, activities, ancillary services [16]. This framework concentrates only on some elements that should be highlighted by the smart tourism strategy, rather than providing a clear and integrative smart tourism implementation process.
- Gokozan et al. (2017) present the following smart city components and their definitions: smart care, smart energy, smart society, smart office, smart mobility,

and smart space [2]. It defines the smart city management center concept, which works like the brain and central nervous system of smart cities, which connect and integrate information and processes [2].

- Rotuna et al. (2019) show that the blockchain is a relevant solution for a wide range of challenges faced by the smart city, but the implementation depends on the city characteristics and context [17]. A blockchain-based smart city infrastructure has the advantages of increased efficiency due to the automated interactions with its citizens, optimized distribution of resources, and fraud reduction [17].
- Saba et al. (2019) provide smart city definitions, characteristics (Sustainability, smartness, life quality, and urbanization), trends and needs, the architecture of the smart city (sensing layer, transmission layer, data management layer, application layer), smart city main components (territory, infrastructure, people, government), and smart city pillars (sustainability, technology, flexibility, citizen involvement). This paper presents that open data is a crucial element in the development of smart cities [18].
- Einola et al. (2019) present the advantages of the open strategy in a smart city [19]. An open strategy that includes the participation of external and internal stakeholders has many undeniable benefits: increasing collective commitment and, through commitment, enabling more effective strategic actions and joint sensemaking [19]. Open strategizing can improve creativity by capturing more diverse views [19]. Einola et al. (2019) provide a smart city process that involves citizens in the definition of the strategy through crowdsourcing [19].
- Afonso et al. (2015) propose a Brazilian smart city maturity model composed of five levels namely: Simplified, managed, applied, measured, and turned [4]. The model is based on the following domains: Water, education, energy, governance, housing, environment, health, security, technology, transport [4].
- Aljowder et al. (2019) provide a systematic literature review on maturity models that assess the level of maturity for smart city projects [20]. It provides an analysis and classification of these models based on their components and perspectives. This study can help to identify the list of elements that should be highlighted through the smart city strategy (ex: education, health, energy…).
- Komninos et al. (2015) present an overall ontology for the smart city and defines its building blocks [21]. They defined a set of smart city indicators, namely: application ontology size, the maximal length of nodes, the number of the object, data properties and super classes, the position of the ontology within the overall smart city ontology, the type of the digital space, the knowledge generation processes, the highest level of innovation to be achieved [21]. The definition of the smart city ontology simplifies the smart city strategy definition and implementation.
- Kuyper (2016) defines the concept of the smart city strategy [5]. This theoretical debate was then applied to two practical examples of smart cities: Barcelona and Amsterdam, to present how have they approached the smart city implementation [5]. The comparison between Barcelona and Amsterdam is done based on the following characteristics: Direction of strategy, main focus, planning horizon, strategic choices, SMART framework, smart city reference model, citizen empowerment and inclusion, smart city pilot projects & upscaling [5].

Smart city strategy components provided by the above studies overlap one another, this creates ambiguity and misunderstanding regarding how to implement smart cities. Hence the need for a clear and consistent smart city approach. Few frameworks are addressing clear guidelines for smart city strategy development. The majority focus on providing smart city components, and dimensions, rather than integrating them in a clear and coherent smart city strategy to frame and facilitate smart city implementation.

3 Methodology

This work aims to identify a clear and consistent smart city approach, by selecting recent smart city frameworks and presents them comparatively using the smart city approach proposed by Korachi and Bounabat in [1] and [22]. This approach consists of three main processes: strategic vision definition, action plan elaboration, and management strategy definition [1]. Figure 1 illustrates the process of how developing a smart city vision. Figure 2 presents the process for the action plan definition. Figure 3 illustrates the management strategy definition process. The components of these processes are presented in Table 1. This current paper compares these processes with recent smart city frameworks to evaluate their originality and completeness.

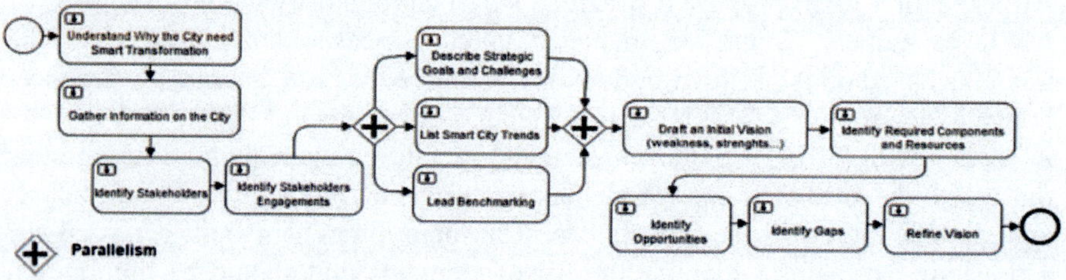

Fig. 1 Strategic vision process [22]

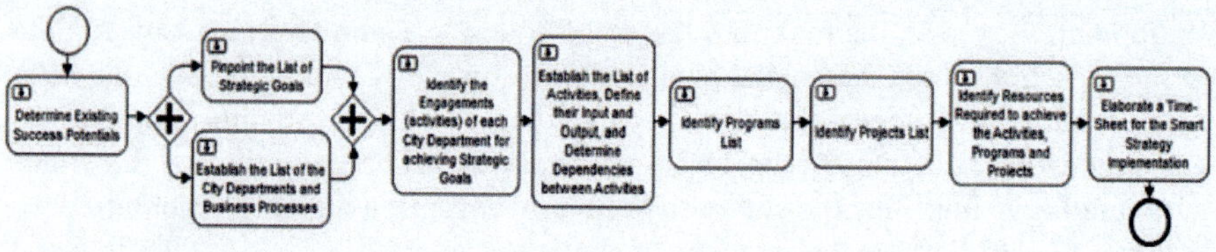

Fig. 2 Action plan process [22]

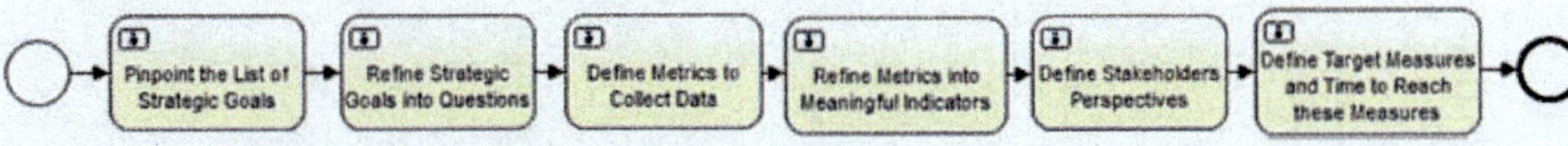

Fig. 3 Management strategy process [22]

Table 1 Korachi and Bounabat smart city approach components [1, 22, 23]

Processes	Activities [1, 22]
Strategic vision	(1) Why the city need smart transformation (2) Gather information on the internal and external environment of the city (3) Identify stakeholders and their engagements (4) Identify and describe strategic goals (5) Identify challenges (6) List smart city trends (7) Lead benchmarking (8) Determine city strengths and weaknesses (9) Identify the main components that will be highlighted through the Transformation Strategy (10) Define desired outcomes, changes, and impact of the smart transformation (SV10) (11) Identify the required components and resources for achieving desired goals and outcomes (12) Identify gaps (13) Identify opportunities
Action plan	(1) Determine existing success potentials (2) Establish a list of city departments and business processes (3) Identify the engagements (activities) of each city department for achieving strategic goals (4) Establish the list of activities, define their input and output, and determine dependencies between activities (5) Identify the programs list (6) Identify the projects list (7) Identify required resources to achieve smart city projects (8) Elaborate a timesheet for the smart strategy implementation
Management strategy	(1) Define appropriate KPIs (2) Evaluate the digital transformation maturity level (3) Smart city dashboard (4) Control the smart city evolution and rank

4 Problem of Statement

This section summarizes the above studies by comparing them according to the components illustrated in Table 1. The comparison analysis is illustrated in Table 2. It illustrates that the cited works focus mainly on describing the requirements, components, technologies, and dimensions of smart cities (Strategic Vision 9), without proposing a comprehensive approach providing mechanisms to simplify their development.

Table 2 Comparative analysis of smart city works

Strategy Source	Smart City												
	Strategic Vision 1	Strategic Vision 2	Strategic Vision 3	Strategic Vision 4	Strategic Vision 5	Strategic Vision 6	Strategic Vision 7	Strategic Vision 8	Strategic Vision 9	Strategic Vision 10	Strategic Vision 11	Strategic Vision 12	Strategic Vision 13
[3]			◆						◆				
[14]			◆						◆				
[15]									◆				
[16]									◆				
[2]									◆				
[17]									◆		◆		
[18]	◆			◆		◆			◆				
[19]			◆						◆				
[4]									◆				
[20]									◆				
[21]				◆					◆		◆		
[5]	◆		◆	◆					◆		◆		

(continued)

Table 2 (continued)

Strategy Source	Smart City Action Plan1	Action Plan 2	Action Plan 3	Action Plan 4	Action Plan 5	Action Plan 6	Action Plan 7	Action Plan 8	Management Strategy 1	Management Strategy 2	Management Strategy 3	Management Strategy 4
[3]												
[14]												
[15]	◆											
[16]												
[2]												
[17]												
[18]												
[19]									◆			
[4]										◆		
[20]										◆		
[21]									◆	◆		
[5]												

The lack of a comprehensive and practical smart city approach, integrating business and technological processes and ensuring their strategic alignment, confirms the need for a new framework, which organizes all the smart city aspects and concerns in a unified solution.

5 Results

The smart city strategy is not a global approach, which must be implemented in the same way and with the same processes all over the world [24]. However, there are some standard aspects that are common to smart cities all over the world, and this section aims to explore them.

5.1 *Smart City Objectives*

Several works describe the smart city strategic objectives and this section factors them into the list below [6, 24–26]:

- Natural resources management and protection
- Creating a competitive economy
- Networking development
- Digitization of public and private services
- The central role of technology
- Advancing human and social capital
- Reducing CO2 emissions
- Creation of an urban infrastructure that meets the expectations and needs of current and future generations
- Improving the quality of life
- Development of a clean and sustainable environment.

5.2 *Smart City Challenges*

Cloud computing, the Internet of Things, open data, semantic web, and future Internet technologies are the leading technologies for the development of the smart city. All these technologies have their challenges and limitations and they together form a complex system that increases the challenges [25] that can be grouped into the following two categories:

- **Managerial challenges** [26]: managerial attitudes and behavior, resistance to change, a blurred vision of IT management.

- **Technical challenges** [27, 28]: interoperability, processing of huge amounts of data in real-time, mashup of heterogeneous resources, high-energy consumption, privacy, and cyber-attacks.

5.3 *Smart City Main Components*

The current section identifies the main components contributing to the design and implementation of the smart city strategy. Previous studies deal with these components in an unstructured and unorganized way which makes it difficult to identify and understand them. This section aims to structure them according to the following categories: requirements, dimensions, risks, policy, recommendations, and architecture.

5.3.1 Strategic requirements of the smart city

The smart cities requirements can be divided into the following two categories:

- **Managerial requirements** [26, 29]: vision, strategy, leadership, collaboration, management, organization, governance, political context, people and communities, and culture.
- **Technical requirements** [24–30]: technology, sustainable infrastructure, environment, integration of services and applications, broadband, communication channels, and sensors.

5.3.2 Smart city dimensions

Different overlapping smart city dimensions are presented in the literature. This section aims to simplify their understanding by grouping them in Table 3.

According to Table 3, each study defines the concept of the "smart city dimension" in its own way. For example, the dimensions proposed by [31] and [27] represent the areas of the city in need of digital transformation, such as the economy, governance, citizens, life, transport, and the environment. While [37–39] suggest a set of dimensions which include city domains and the techniques required for their development such as technology, infrastructure, and innovation.

5.3.3 Smart City Risks

Smart city projects have many advantages, but the security risks for data and services cannot be avoided. To address this, Guntur & Ibrahim (2019) propose a cybersecurity strategy based on three dimensions: citizens, technologies, and institutions [40].

Table 3 Smart city dimensions

Source	Economy	Governance	People	Life	Mobility	Environment	Stakeholders	Technologies	Infrastructure	Strategy	legal	Data	Sustainability	Innovation	Context	Security
[31]	◆	◆	◆	◆	◆	◆										
[32]		◆	◆			◆			◆							
[33]		◆					◆	◆		◆						
[27]	◆	◆	◆	◆	◆	◆										
[34]	◆	◆	◆					◆			◆		◆			
[35]		◆					◆	◆		◆						
[36]		◆							◆	◆					◆	
[37]		◆						◆	◆			◆		◆		
[30]	◆			◆												◆

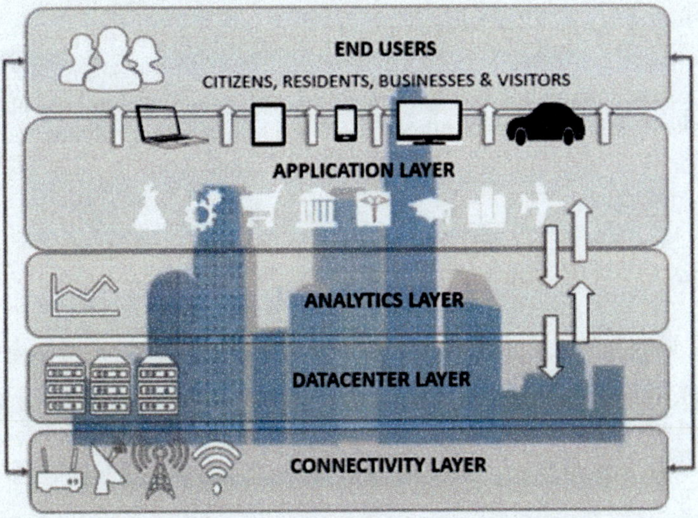

Fig. 4 Smart City layers [39]

5.3.4 Smart City Implementation Policy

The various existing smart city implementation policies are [6]:

- National versus local strategies
- Stage of development: new projects or continuation of existing projects
- Hardware-oriented strategies or software-oriented strategies
- Economically or geographically oriented strategies.

5.3.5 Smart city architecture

Smart city architecture layers [33, 39] are data collection, data processing, data analysis and integration, production, and use of information. They interact with each other as illustrated in Fig. 4.

5.3.6 Recommendations

Among the recommendations that can improve the development of smart cities are the following [6]:

- The study of what is already in place and how it can be improved
- Prioritization of areas that need to be improved urgently
- Selectivity, synergies, and prioritization are three standard core processes in the planning of a smart city.
- Stakeholders engagement.

5.4 Smart City Stakeholders

Smart city stakeholders involve those responsible for the development of the smart city and those affected by its outcome [27]. They include citizens, educational institutions, health care, public safety providers, and government organizations.

5.5 Smart City Resources

Smart city development requires a set of resources such as [25, 27]:

- **Network infrastructure**: the smart city needs a highly interconnected, cost-effective, energy-efficient, and reliable infrastructure.
- **Technology services**: consisting of the acquisition of applications, services, data management systems, IoT platforms, and data sources.
- **Devices**: sensors, smartphones, RFID, and NFC.
- **Financial resources**: the smart city project requires the acquisition of a powerful technological infrastructure, millions of sensors, thousands of network equipment, and computing devices are needed. This means the need for a huge budget.

6 Conclusion

There is still a conflict among the smart city definitions and frameworks. Various studies in the literature addressing smart city frameworks and strategies. The literature analysis shows that these studies overlap one another and create ambiguity around smart city strategies. To fill this gap, the present paper conducts a comparative analysis that allows to identify the smart city main components and processes. This study aims to reduce the ambiguity and misunderstanding that surrounds smart city approaches and frameworks, by presenting a relevant smart city approach composed of three processes namely: strategic vision definition, action plan development, management strategy identification. The approach was evaluated comparatively with recent smart city approaches. The result of the comparative analysis shows that the approach is holistic and original. Future studies can investigate and analyze the elements of the proposed smart city approach to collect more information about them from literature or smart city cases.

References

1. Korachi, Z., Bounabat, B.: Towards a platform for defining and evaluating digital strategies for building smart cities. 2019 3rd International Conference On Smart Grid And Smart Cities (ICSGSC) (2019). https://doi.org/10.1109/icsgsc.2019.00-22

2. Gokozan, H., Tastan, M., Sari, A.: Smart cities and management strategies. Chapter 8 In Book: 2017 Socio-Economic Strategies. ISBN: 978–3–330–06982–4 (2017)
3. Agbali, M., Trillo, C., Ibrahim, I., Arayici, Y., Fernando, T.: Are smart innovation ecosystems really seeking to meet citizens' needs? insights from the stakeholders' vision on smart city strategy implementation. Smart Cities **2**(2), 307–327 (2019). https://doi.org/10.3390/smartcities2020019
4. Afonso, R.A., dos Santos Brito, K., do Nascimento, C.H., Garcia, V.C., Álvaro, A: Brazilian smart cities. Proceedings of the 16th Annual International Conference on Digital Government Research—Dg.o '15 (2015). https://doi.org/10.1145/2757401.2757426.
5. Kuyper, T.: Smart City Strategy & Upscaling: Comparing Barcelona and Amsterdam. https://doi.org/10.13140/RG.2.2.24999.14242. Master Thesis, Msc. IT & Strategic Management (2016)
6. Angelidou, M.: Smart city policies: a spatial approach. Cities **41**(2014), S3–S11 (2014). https://doi.org/10.1016/j.cities.2014.06.007
7. Mora, L., Deakin, M., Aina, Y., Appio, F.: Smart City Development: ICT Innovation for Urban Sustainability. Encyclopedia of the UN Sustainable Development Goals, pp. 589–605 (2020). https://doi.org/10.1007/978-3-319-95717-3_27
8. Angelidou, M.: Smart cities: a conjuncture of four forces. www.elsevier.com/locate/cities. Cities 47 (2015) 95–106 (2015a). http://dx.doi.org/https://doi.org/10.1016/j.cities.2015.05.004
9. Bastidas, V., Bezbradica, M., Helfert, M.: Cities as enterprises: a comparison of smart city frameworks based on enterprise architecture requirements. Smart Cities, pp. 20–28 (2017). https://doi.org/10.1007/978-3-319-59513-9_3
10. Korachi, Z., Bounabat, B. (2018). Data driven maturity model for assessing smart cities. ACM International Conference Proceeding Series, pp. 140–147, 2nd International Conference on Smart Digital Environment, ICSDE'18, October 18–20, 2018, Rabat, Morocco; © 2018 Association for Computing Machinery; ACM ISBN 978–1–4503–6507–9/18/10; ENSIAS Rabat; Morocco. https://doi.org/10.1145/3289100.3289123
11. Korachi, Z., Bounabat, B.: Integrated methodological framework for digital transformation strategy building (IMFDS). Int. J. Adv. Comput. Sci. Appl. **10**(12) (2019). https://doi.org/10.14569/ijacsa.2019.0101234
12. Korachi, Z., Bounabat, B.: Towards a maturity model for digital strategy assessment. Adv. Intell. Syst. Comput. **1105**, 456–470 (2020). Springer, Cham. https://doi.org/10.1007/978-3-030-36674-2_47
13. Korachi, Z., Bounabat, B.: Towards a frame of reference for smart city strategy development and governance. J. Comput. Sci. **16**(10), 1451–1464 (2020). https://doi.org/10.3844/jcssp.2020.1451.1464
14. Oliveira, T., Oliver, M., Ramalhinho, H.: Challenges for connecting citizens and smart cities: ICT. E-Governance Blockchain. Sustain. **12**(7), 2926 (2020). https://doi.org/10.3390/su12072926
15. Darmawan, A., Siahaan, D., Susanto, T., Hoiriyah, Umam, B.: Identifying success factors in smart city readiness using a structure equation modelling approach. 2019 International Conference On Computer Science, Information Technology, And Electrical Engineering (ICOMITEE) (2019). https://doi.org/10.1109/icomitee.2019.8921312.
16. Dabeedooal, Y., Dindoyal, V., Allam, Z., Jones, D.: Smart tourism as a pillar for sustainable urban development: an alternate smart city strategy from mauritius. Smart Cities **2**(2), 153–162 (2019). https://doi.org/10.3390/smartcities2020011
17. Rotuna, C., Gheorghita, A., Zamfiroiu, A., Smada, D.: Smart city ecosystem using blockchain technology. Informatica Economica, **23**(4/2019), 41–50 (2019). https://doi.org/10.12948/issn14531305/23.4.2019.04.
18. Saba, D., Sahli, Y., Berbaoui, B., Maouedj, R.: Towards smart cities: challenges, components, and architectures. Toward Social Internet of Things (Siot): Enabling Technologies, Architectures And Applications, pp. 249–286 (2019). https://doi.org/10.1007/978-3-030-24513-9_15

19. Einola, S., Kohtamäki, M., Hietikko, H.: Open Strategy in a Smart City. Tchnology Innovation Management Review, September 2019 (Volume 9, Issue 9) (2019)
20. Aljowder, T., Ali, M., Kurnia, S.: Systematic literature review of the smart city maturity model. 2019 International Conference on Innovation and Intelligence for Informatics, Computing, and Technologies (3ICT) (2019). https://doi.org/10.1109/3ict.2019.8910321
21. Komninos, N., Bratsas, C., Kakderi, C., & Tsarchopoulos, P. (2015). Smart City Ontologies: Improving the effectiveness of smart city applications. Journal Of Smart Cities, 1(1). https://doi.org/10.18063/jsc.2015.01.001.
22. Korachi, Z., Bounabat, B.: Integrated methodological framework for smart city development. Proceedings of the International Conferences ICT, Society, and Human Beings 2019; Connected Smart Cities 2019; and Web Based Communities and Social Media (2019). https://doi.org/10.33965/csc2019_201908l030
23. Korachi, Z., Bounabat, B.: Towards a frame of reference for smart city strategy development and governance. J. Comput. Sci. **16**(10), 1451–1464 (2020). https://doi.org/10.3844/jcssp.2020.1451.1464
24. Dameri, R.P., Benevolo, C., Veglianti, E., Li, Y.: Understanding smart cities as a glocal strategy: a comparison between Italy and China. Technol. Forecast. Soc. Chang. (2018). https://doi.org/10.1016/j.techfore.2018.07.025
25. Kadhim, W.: Case study of Dubai as a Smart City. Int. J. Comput. Appl. **178**(40), 35–37 (2019). https://doi.org/10.5120/ijca2019919291
26. Chourabi, H., Nam, T., Walker, S., Gil-Garcia, J. R., Mellouli, S., Nahon, K., Scholl, H.J., et al.: Understanding smart cities: an integrative framework. 2012 45th Hawaii International Conference on System Sciences (2012). https://doi.org/10.1109/hicss.2012.615.
27. Petrolo, R., Loscrì, V., Mitton, N.: Towards a smart city based on cloud of things, a survey on the smart city vision and paradigms. Trans. Emerging Telecommun. Technol. **28**(1), e2931 (2015). https://doi.org/10.1002/ett.2931
28. Khatoun, R., Zeadally, S.: Smart cities: concepts, architectures, research oppor-tunities. Commun. ACM **59**(8), 46–57 (2016). https://doi.org/10.1145/2858789
29. Allam, Z., Newman, P.: Redefining the smart city: culture, metabolism and governance. Smart Cities **1**(1), 4–25 (2018). https://doi.org/10.3390/smartcities1010002
30. Kesswani, N., Kumar, S.: The smart-X model for smart cities. 2018 IEEE 42nd Annual Computer Software and Applications Conference (COMPSAC) (2018). https://doi.org/10.1109/compsac.2018.00112.
31. Giffinger, R.,Fertner, C.,Kramar, H., Kalasek, R., Pichler-Milanović, N., Meijers, E.: Smart cities—ranking of European medium-sized cities. [Online] Centre of Regional Science (SRF), Vienna University of Technology in October 2007 (2007). Available at: http://www.smart-cities.eu/download/smart_cities_final_report.pdf [Accessed 17 Jun 2019]
32. Dameri, R.P., Rosenthal-Sabroux, C.: Smart city and value creation. Progress IS, pp. 1–12 (2014). https://doi.org/10.1007/978-3-319-06160-3_1
33. Asri, N.A.M., Ibrahim, R., Jamel, S.: Designing a model for smart city through digital transformation. Int. J. Adv. Trends Comput. Sci. Eng. (2019). https://doi.org/10.30534/ijatcse/2019/6281.32019
34. Joshi, S., Saxena, S., Godbole, T., Shreya: Developing smart cities: an integrated framework. Procedia Comput. Sci. **93**, 902–909 (2016). https://doi.org/10.1016/j.procs.2016.07.258
35. Hämäläinen, M.: A framework for a smart city design: digital transformation in the Helsinki Smart City. Contribut. Manage. Sci. **63–86**,(2019). https://doi.org/10.1007/978-3-030-23604-5_5
36. Haller, S., Neuroni, A., Fraefel, M., Sakamura, K.: Perspectives on smart cities strategies. Proceedings of the 19th Annual International Conference on Digital Government Research Governance in the Data Age—Dgo '18 (2018). https://doi.org/10.1145/3209281.3209310
37. Maestre-Gongora, G., Bernal, W.: Conceptual model of information technology management for smart cities. J. Glob. Inf. Manag. **27**(2), 159–175 (2019). https://doi.org/10.4018/jgim.2019040109

38. Hämäläinen, M., Tyrväinen, P.: Improving smart city design: a conceptual model for governing complex smart city ecosystems. 31st Bled Econference: Digital Transformation: Meeting the Challenges (2018). https://doi.org/10.18690/978-961-286-170-4.17
39. Kumar, M.: Building Agile Data Driven Smart Cities (IDC: October 2015), White Paper, (Sponsored by EMC) (2015). http://docplayer.net/8696930-Building-agile-data-driven-smart-cities.html [Last Access 21/06/2020]
40. Guntur Alam, R., Ibrahim, H.: Cybersecurity strategy for smart city implementation. The International Archives of the Photogrammetry, Remote Sensing and Spatial Information Sciences, vol. XLII-4/W17, 2019. 4th International Conference on Smart Data and Smart Cities, 1–3 October 2019, Kuala Lumpur, Malaysia (2019). https://doi.org/10.5194/isprs-archives-XLII-4-W17-3-2019

Hyperspectral Data Preprocessing of the Northwestern Algeria Region

Zoulikha Mehalli, Ehlem Zigh, Abdelhamid Loukil, and Adda Ali Pacha

Abstract Hyperspectral images provide a rich source of information and allow us to broaden the field of applications like material identification, agriculture, and environmental studies. However, these images commonly suffer from atmospheric effects, sensor noise, etc. Preprocessing is a crucial task to reduce/remove these effects for better extraction of the desired information. The present paper aims to apply some preprocessing techniques on Djebel Meni zone, situated in the Northwestern of Algeria, to prepare reliable images for the next processing stages. The database used in this work is the Hyperion level L1R dataset. The preprocessing approach adopted in this work involves: bad band removal, i.e., removing the bands with no information, radiometric calibration, and atmospheric correction in order to handle the atmospheric effects, correcting the continuous vertical stripes and the abnormal pixels in bands. Also, we have applied dimensionality and noise reduction to deal with the huge volume of spectral data and the noise present in the Hyperion hyperspectral dataset. These preprocessing steps are an indispensable prerequisite for any further processing and accurate interpretation of spectra of different surface objects. Also this study recommends that QUAC atmospheric correction has a capability to compensate the atmospheric effects than IARR method for mineralogy studies

1 Introduction

Remote sensing measurement of the earth's surface is influenced by the atmosphere due to many causes like gases, sensor noise, and other illumination effects. For that,

Z. Mehalli (✉) · A. Ali Pacha
Coding and Information Security Laboratory, University of Science and Technology of Oran - Mohamed Boudiaf, Oran, Algeria

E. Zigh
LaRATIC laboratory. National Institute of Telecommunications and ICT of Oran, Oran, Algeria

A. Loukil
Department of Electronics, Faculty of Electrical Engineering, University of Science and Technology Oran, Oran, Algeria

M. Ben Ahmed et al. (eds.), *Networking, Intelligent Systems and Security*,
Smart Innovation, Systems and Technologies 237,
https://doi.org/10.1007/978-981-16-3637-0_45

it is crucial to reduce these effects in order to improve the quality of hyperspectral images. The preprocessing of these images could be considered as the first important step to ensure better further processing and to allow more accurate interpretation of objects spectra in images. Many scientific researches have been done in the field of hyperspectral images preprocessing [1–5].The preprocessing steps include generally bad band removal, geometric correction, radiometric calibration, atmospheric correction, and noise reduction. The geometric correction aims to correct the distortion in the acquired image caused by the earth rotation and curvature. Atmospheric corrections aim to match the image endmember spectra with the reference spectral libraries or ground data. They also aim to minimize the effect of atmospheric agents that alter radiance data. There are different atmospheric correction techniques divided into empirical approaches (Statistical Based) and atmospheric models (Physics Based).

The empirical approaches include quick atmospheric correction (QUAC) [6], empirical line correction (ELC) [7], internal average relative reflectance (IARR) [8], and flat field (FF) [9]. The atmospheric models are: ATmospheric REMoval Program (ATREM) [10], Atmospheric CORrection Now (ACORN) [11], Fast Line-of-sight Atmospheric Analysis of Spec-tral Hypercube (FLAASH) [6], and ATmospheric CORrection (ATCOR) [12].

To preprocess the Hyperion dataset of Djebel Meni (Northwestern Algeria), we have been inspired by [13] to apply an empirical approach quick atmospheric correction that determines atmospheric compensation parameters directly from the observed pixel spectra, without ancillary information. It generally produces reflectance spectra within the range of roughly+-15% of the physics-based approaches [14].

The next preprocessing operation is a vertical stripes removal to enhance the clarity of the images. After that, we have applied MNF technique to decrease data dimension while conserving the important information. This final pre-processing stage is recommended to ensure the low cost for downstream processing [15].

2 Study Area and Data Used

Djebel Meni is located in Mostaganem (Northwestern of Algeria). The area falls between latitudes 36°04’28.06” and 36°03’45.11” N and longitudes 0°23’15.02” and 0°31’08.59” E (Fig. 1). For this paper, data from EO-1 Hyperion satellite have been used. The EO-1 space-borne launched on November 21, 2000 by the National Aeronautics and Space Administration (NASA) for earth observation studies, and it orbits the earth in a sun-synchronous orbit altitude of 705 Km [16]. The Hyperion sensor has two spectrometers, the first operates in a visible near infrared region (VNIR), 0.355–1 um, having 70 bands and the second operates in short wave infrared region (SWIR), 0.9 um to 2.5 um, having 172 bands. So there are 242 wavebands with a sampling interval of 10 nm. The EO-1 hyperspectral image has a spatial resolution of 30m. The Hyperion hyperspectral data characteristics used in this paper are given in Table 1.

Fig. 1 Study area localization

Table 1 Summary of characteristics of the Hyperion hyperspectral data used in this study

Hyperion hyperspectral image of the region of Djebel Meni	
Acquisition date	17–12–2010
Spatial resolution	30 m
Spectral resolution	10 nm
Number of bands	242

3 Methodology

In this section, we described the steps of the adopted method for hyperspectral image preprocessing illustrated and resumed in flowdiagram (Fig. 2).

3.1 Bad Bands Removal

Hyperion hyperspectral data has 242 bands. The bands which do not have any pixel information are called zero bands, so we need to remove them. Concerning our image, bands 1 to 7 and 225 to 242 are not illuminated, bands 58 to 78 fall in the overlap region of the two spectrometers (VNIR, SWIR), i.e., bands 56, 57 and 77, 78. We need to remove the water vapor absorption bands too. These last are identified as bands 120 to 132,165 to 182 and 221 to 242 [17].The summary of removed bands is given in Table 2.

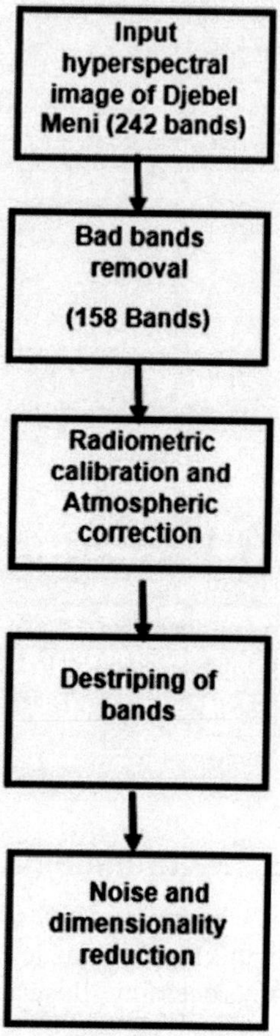

Fig. 2 Flowdiagram of the methodology

Table 2 Summary of removed bands

Bands removed	Reason
1–7	Not illuminated
58–78	Overlap region
120–132	Water vapor absorption band
165–182	Water vapor absorption band
185–187	Identified by Hyperion bad band list
221–224	Water vapor absorption band
225–242	Not illuminated

3.2 Radiometric Calibration and Atmospheric Correction

The electromagnetic signal captured by the Hyperion hyperspectral sensor is modified by the atmosphere between sensor and the ground. The atmospheric effects are scattering and absorption by gases such as carbon dioxide, ozone, water vapor, and other gases. The radiance recorded by the hyperspectral sensor is influenced by the atmosphere in two ways, the first way is by attenuating the energy illuminating the object of the earth and the second way is by adding the path radiance to the signal captured by the sensor. These two ways effects are represented mathematically as follows:

$$R_{\text{tot}} = \frac{P.E.T}{2\pi} + R_P \tag{1}$$

where R_{tot} is the total spectral radiance measured by the sensor, R_p is the path radiance, p is the reflectance of the object, T is the Transmitted energy, and E is the irradiance on object caused by directly reflected sunlight and diffused skylight.

The radiometric calibration is performed in order to remove the path radiance, and we need also to compensate the atmospheric attenuation effect. For that, we have used quick atmospheric correction method (QUAC) in order to get the true reflectance energy of the ground object. Quick atmospheric correction method determines the atmospheric compensation parameters directly from the observed pixel spectra in hyperspectral image. This method is based on the empirical finding that the mean spectrum of a collection of diverse material spectra, such as the endmember spectra in a scene, is not dependent on each scene. Quick atmospheric correction method is suitable for real-time applications than the first-principles methods because his faster computational speed. Furthermore, it performs a more approximate atmospheric correction than fast line-of-sight atmospheric analysis of spectral Hypercube (FLAASH) or other physics-based first-principles methods, generally producing reflectance spectra within the range of approximately 10 percent of the ground truth [13, 18]. QUAC also allows for any view angle or solar elevation angle, if a sensor does not have proper radiometric or wavelength calibration, or if the solar illumination intensity is unknown (with cloud decks, for example).

3.3 *Destriping of Band*

Among the problems associated with Hyperion hyperspectral images is the vertical stripes caused by calibration differences in Hyperion sensor array and temporal variations in the sensor response[19]. These stripes contain corrupted pixels that make the image unclear and will give a negative impact on further processing results[20]. We have proposed to remove these stripes by calculating the mean of every *n*th line and normalizing each line to its respective mean.

3.4 *Minimum Noise Fraction (MNF) Transform*

The minimum noise fraction transform is used to determine the inherent dimensionality of image data, to maximize the signal to noise ratio (SNR) of image data, and to minimize the computational requirements for subsequent processing [21]. The MNF can be processed as two consecutive data principal component transformations. The first is the conversion of the noise covariance matrix to an identity matrix, and the second is the principal component transformation of the noise whitened dataset maximizing the signal to noise ratio (SNR) and removing the noise from the acquired signal [22, 23]. The noise statistics are calculated using the shift difference method.

4 Results and Discussion

In the proposed methodology, Hyperion hyperspectral image of the region of Djebel Meni is of MxNxB size, where B is the number of bands in the image and MxN is the size of each band as shown in Table 3.

The true color image used as an input image is displayed in Fig. 3a and Fig. 4 depicts a Hyperspectral data cub

Only 158 of the 242 total hyperspectral bands were used because many of bands do not have any pixel information which are called zero bands. Accepted bands and noisy ones are shown in Fig. 5. Next radiometric calibration is applied to remove the path radiance effect from the acquired signal, and the result is shown in Fig. 3b. The quick atmospheric correction should be performed to handle the atmospheric effect (Fig. 3c).

A comparison of the reflectance spectral profiles of a randomly selected pixel in: the original image, after bad bands removal, after radiometric calibration, and after atmospheric corrections is shown in Fig. 6.

The operation of destriping should be performed on the Hyperion hyperspectral image of Djebel Meni because many bands suffer from this striping issue. The pixels at stripes can be replaced with either the mean values of neighboring columns, and the result of destriping for bands 2,3,4 is shown in Fig. 7

After applying destriping method on Hyperion data, minimum noise fraction (MNF) is applied to segregate the random noise and reliable estimation of data dimensionality from the signal information. The MNF eigenvalue graph is shown

Table 3 Data used characteristics

Hyperion hyperspectral image of the region of Djebel Meni	
Type	HDF EO-1 Hyperion
Columns (N)	256
Rows (M)	3296
Bands (B)	242

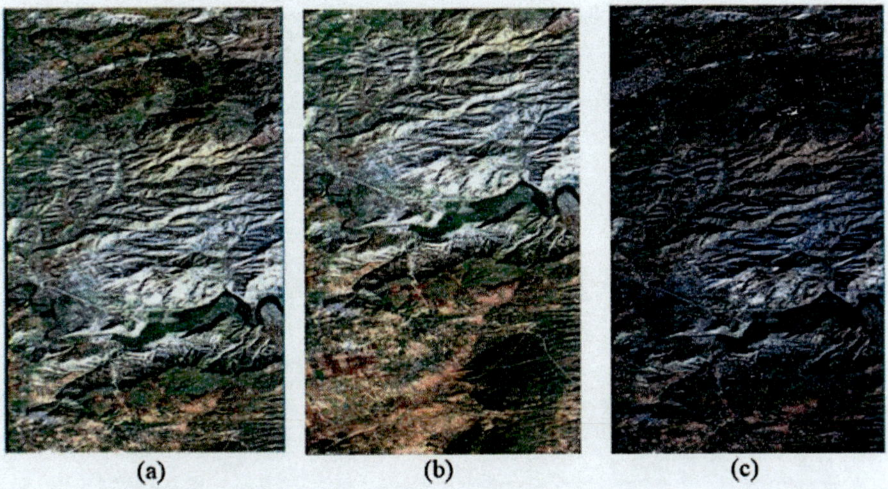

Fig. 3 Hyperspectral image, **a** original image, **b** image after radiometric calibration, **c** image after atmospheric correction

Fig. 4 Hyperion image cube

in Fig. 8. It represented each band with its corresponding eigenvalue. Bands with eigenvalues close to 1 are mostly noise.

Table 4 listed the first ten MNF bands selected with their corresponding eigenvalue, these bands contain the higher percentage of information.

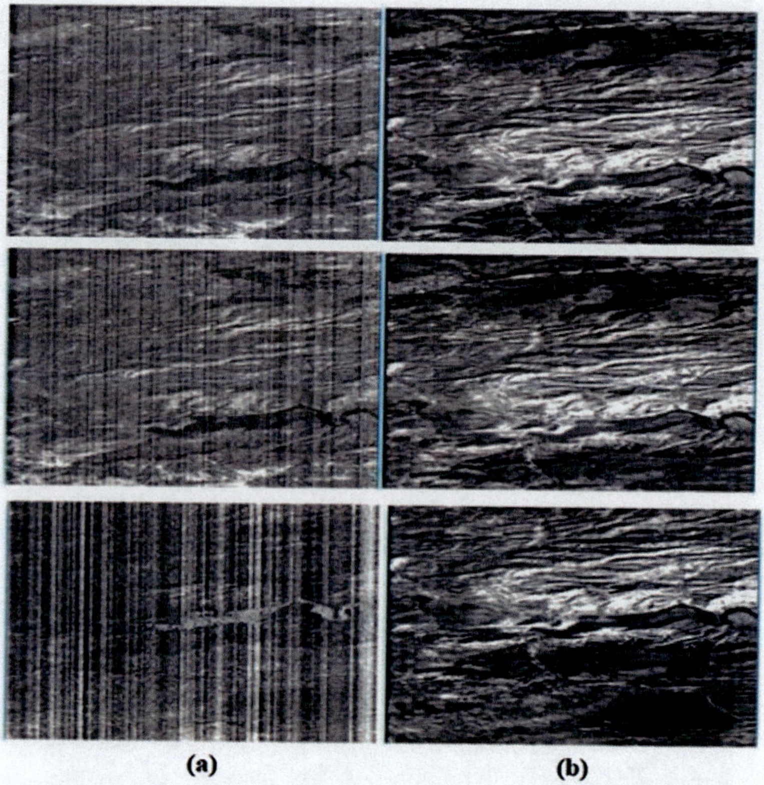

Fig. 5 Gray scale displaying Hyperion bands, **a** bands 179,180,123 removed, **b** bands 13,14.16 accepted

The results showed that the maximum noise fraction of an image has been reduced without losing information. Therefore, the overall proposed methodology is able to guarantee interesting preprocessed data for further processing or analysis.

5 Comparison of Efficient Techniques of Atmospheric Corrections on Hyperspectral Data for Mineralogy Studies

5.1 Methodology

In this section, we described the steps of the adopted method for a comparative analysis between internal average relative reflectance (IARR) atmospheric correction

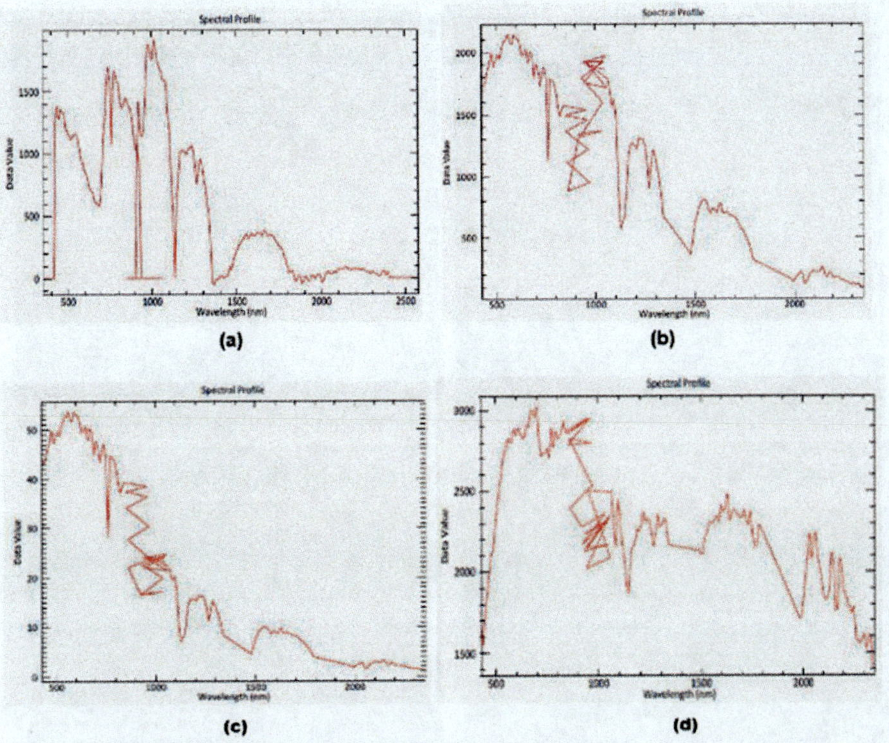

Fig. 6 Spectral profile of a randomly selected pixel, **a** original image, **b** after bad bands removal, **c** after radiometric calibration, **d** after atmospheric correction

and quick atmospheric correction (QUAC), illustrated and resumed in flowdiagram (Fig. 9).

5.1.1 Bad Bands Removal and Radiometric Calibration (Previously Explained in Sect. 3.1 and 3.2)

5.1.2 Internal Average Relative Reflectance (IARR).

IARR method divided each pixel in the image into a reflectance spectrum to generate relative reflectance. However, the reflectance spectrum for IARR is the mean spectrum of the complete image. It works best for arid areas with no vegetation [24, 25].

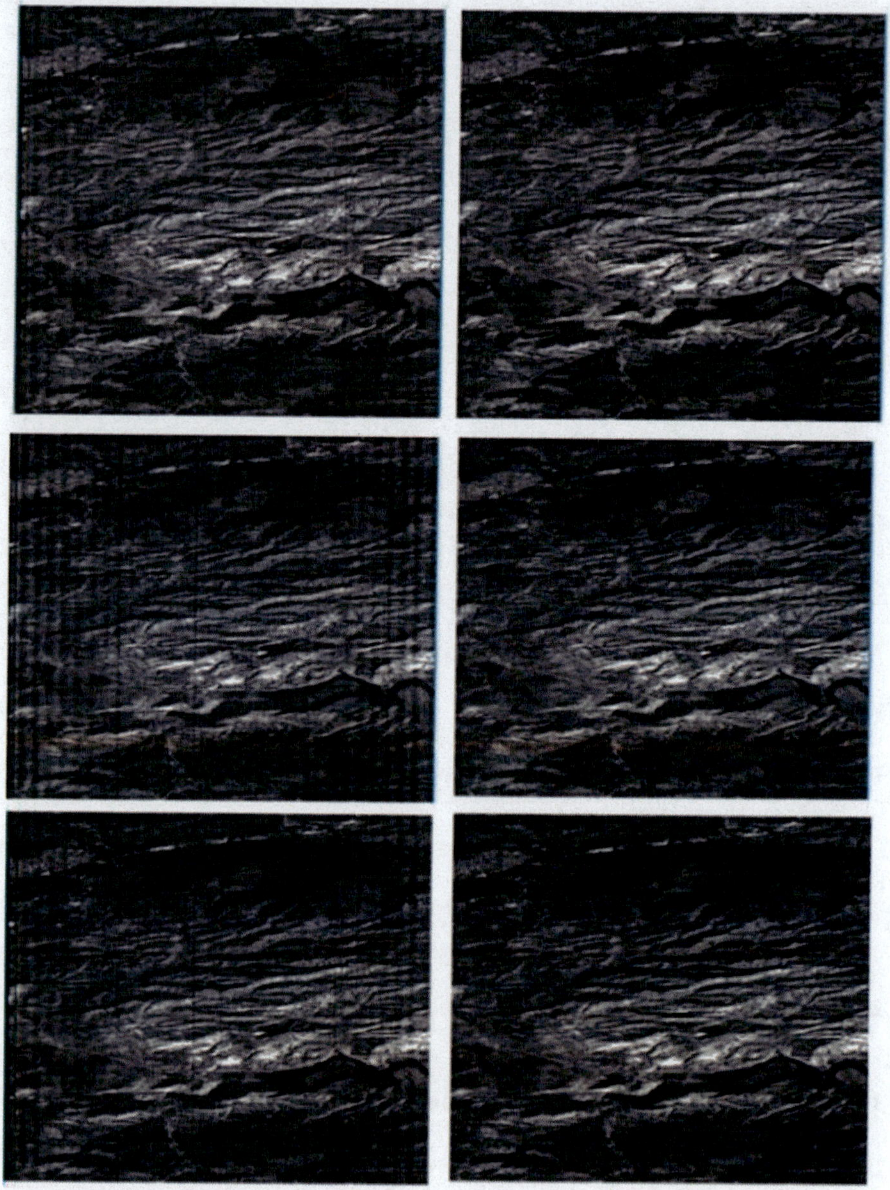

(a) (b)

Fig. 7 Destriping of bands 2, 3, 4, **a** bands before destriping, **b** bands after destriping

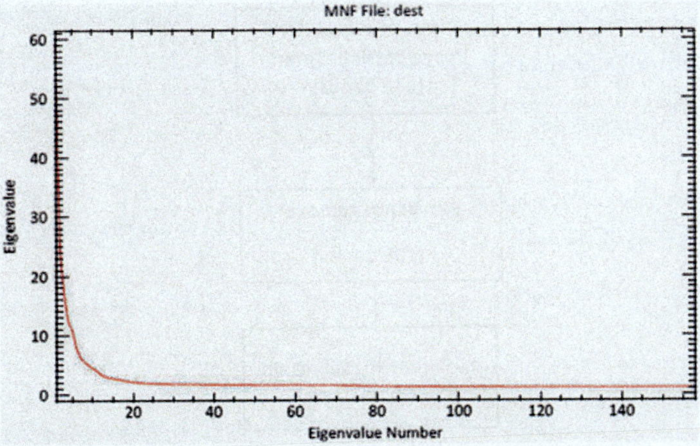

Fig. 8 MNF eigenvalue graph

Table 4 First ten MNF bands with their corresponding eigenvalue

MNF	Eigenvalue
1	59.3072
2	24.3788
3	16.9522
4	12.2546
5	10.7573
6	7.8286
7	6.1890
8	5.5210
9	4.8821
10	4.4532

5.1.3 Quick Atmospheric Correction (QUAC) (Previously Explained in Sect. 3.2)

5.1.4 Spectral Angle Mapper (SAM)

After applying IARR and QUAC atmospheric corrections, the spectral angle mapper algorithm is applied for IARR corrected image end QUAC corrected image.

The spectral angle mapper algorithm calculates the spectral similarity between the spectral signature of each pixel of the image and the spectral signatures of 21 clays minerals which are represented in Fig. 10 (5 spectral signatures of Illite, 8 spectral signatures of kaolinite and 8 spectral signatures of montmorillonite) introduced via the United States Geological Survey (USGS) spectral library [26]. These spectral

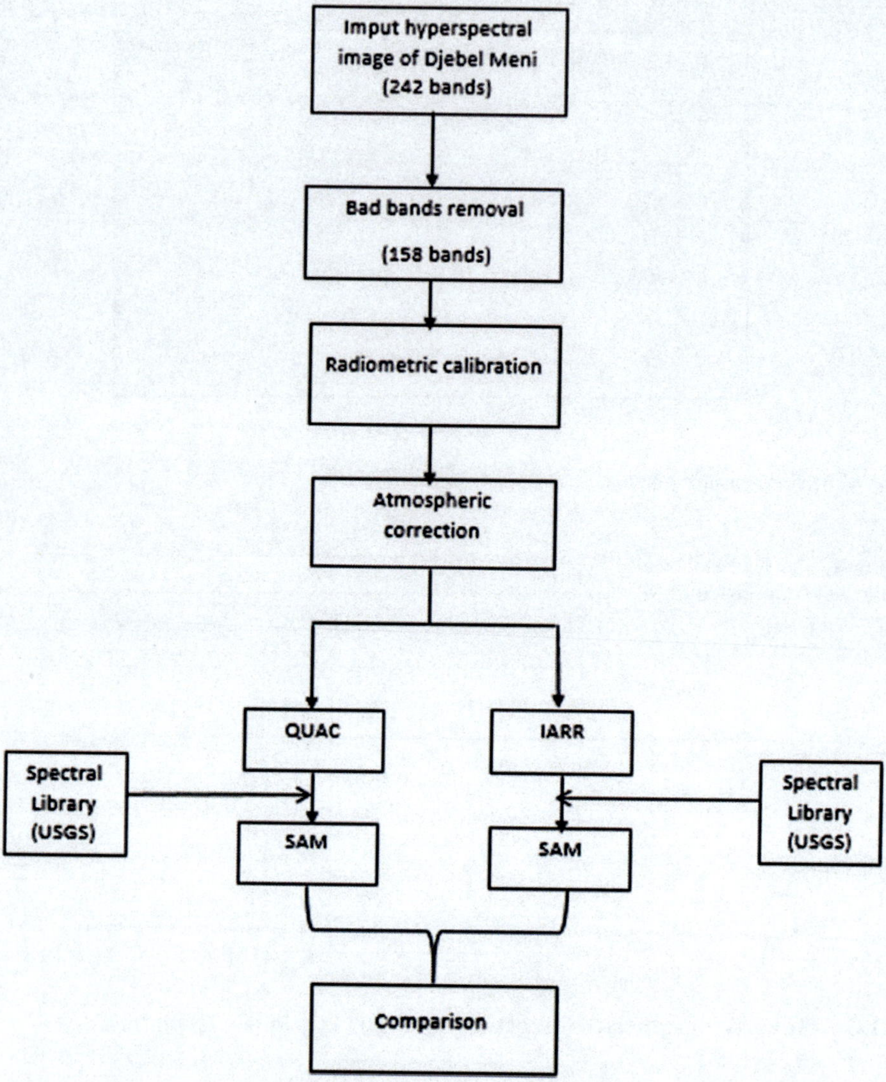

Fig. 9 Flowdiagram of the methodology

signatures of clay minerals were chosen due to the geological nature of the study area of Djebel Meni which is usually covered with this clay minerals[13].

SAM determines the similarity of an unknown spectrum t to a reference spectrum r by applying [27, 28]:

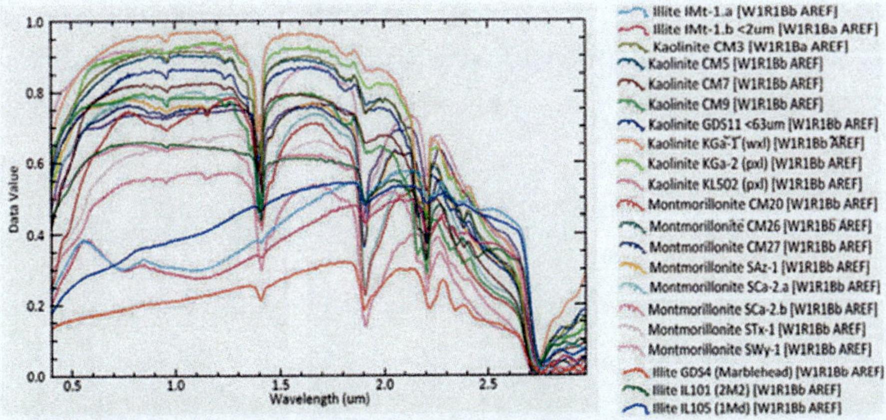

Fig. 10 Spectral signature of clay minerals from USGS spectral library

$$\alpha = \cos^{-1}\left(\frac{\sum_{i=1}^{nb} t_i r_i}{\left(\sum_{i=1}^{nb} t_i^2\right)^{0.5}\left(\sum_{i=1}^{nb} r_i^2\right)^{0.5}}\right) \tag{2}$$

where nb is the number of bands, ti is the test spectrum, and ri is the reference spectrum.

The SAM classification result is represented by a color-coded image that show the best SAM match at each pixel.

5.2 *Results and Discussion*

After the image bands have been resized to 158 bands (see Sect. 3.1), radiometric calibration is applied, and the result is shown previously in Fig. 3b.

Next the IARR and QUAC atmospheric corrections methods are performed, and the result of original image, QUAC, and IARR corrected images is shown in Fig. 11. Visual analysis of the result (Fig. 11) shows that there is no significant difference in the three images (original image, QUAC corrected image, IARR corrected image), so to compare and evaluate the differences between QUAC and IARR atmospheric corrections methods, we used the spectral angle mapper method that permits a rapid mapping by calculating the similarity between spectral signature of each pixel on the hyperspectral image and the spectral signatures of 21 clay minerals (illite, kaolinite, montmorillonite) of the USGS spectral library (this clay minerals were chosen because the Djebel Meni area is covered with it). The atmospheric correction method with which we identify more types of clay minerals is considered as the best.

Fig. 11 **a** Original image, **b** IARR corrected image, **c** QUAC corrected image

The result of SAM method applied to QUAC corrected image and IARR corrected image is illustrated in Fig. 12, and the histogram of Fig. 13 shows the clay minerals types identified with their number of pixel covered.

After analyzing the results of Figs. 12 and 13, QUAC atmospheric correction permits to identify 13 spectral signatures profils of illite, kaolinite, and montmorillonite clay minerals; but with IARR atmospheric correction, we identified only 10 spectral profils of illite et montmorillonite. Kaolinite is not identified, also with QUAC corrected image more pixels are classified than IARR corrected image as shown in Tables 5 and 6. So QUAC is rigorous method and shows better correction results, it has a capability to compensate the atmospheric effects than IARR method, and it performs a good approximate atmospheric correction to IARR.

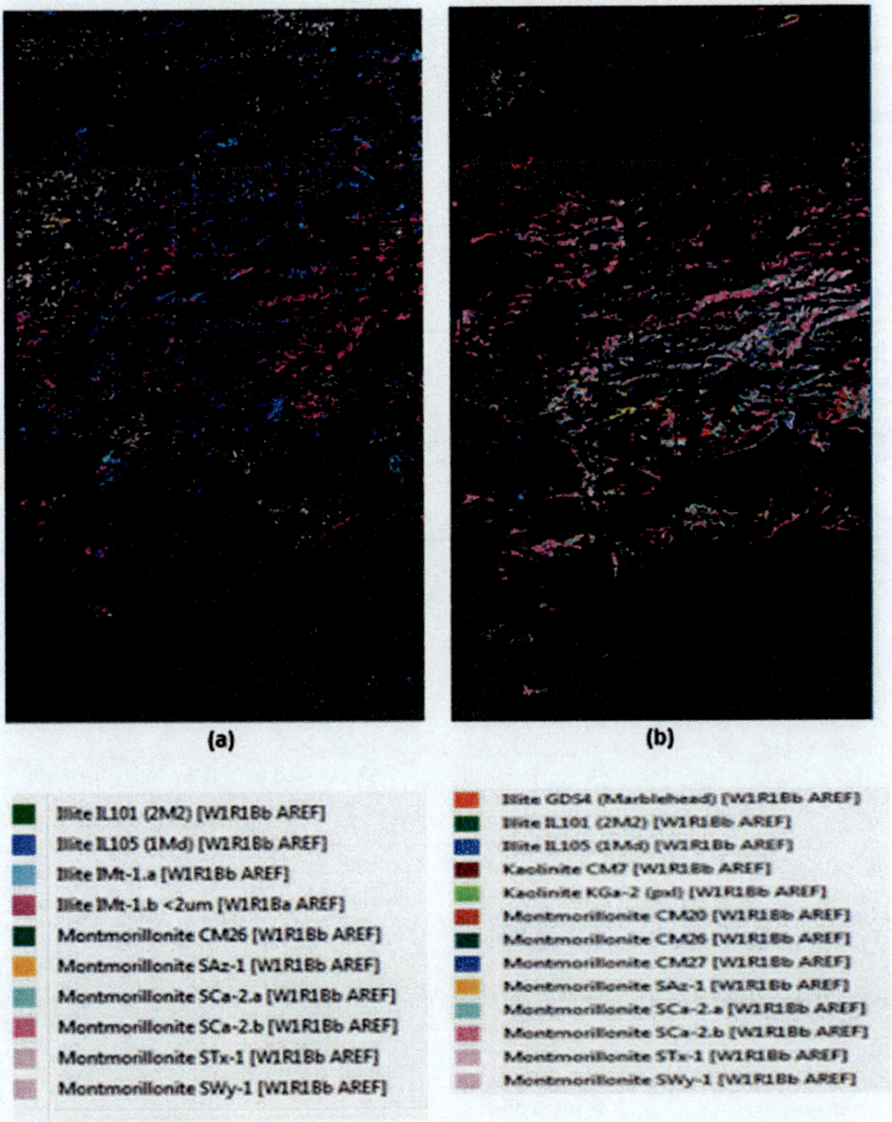

Fig. 12 **a** SAM applied to IARR corrected image, **b** SAM applied to QUAC corrected image

6 Conclusion

This scientific research aimed to propose a preprocessing scheme for the Hyperion dataset of the region of Djebel Meni (Northwestern Algeria). It includes four main steps, well-chosen to overcome input image drawbacks like geometric distortions, striping, low signal to noise ratio, and high dimensionality. Therefore, bad bands are

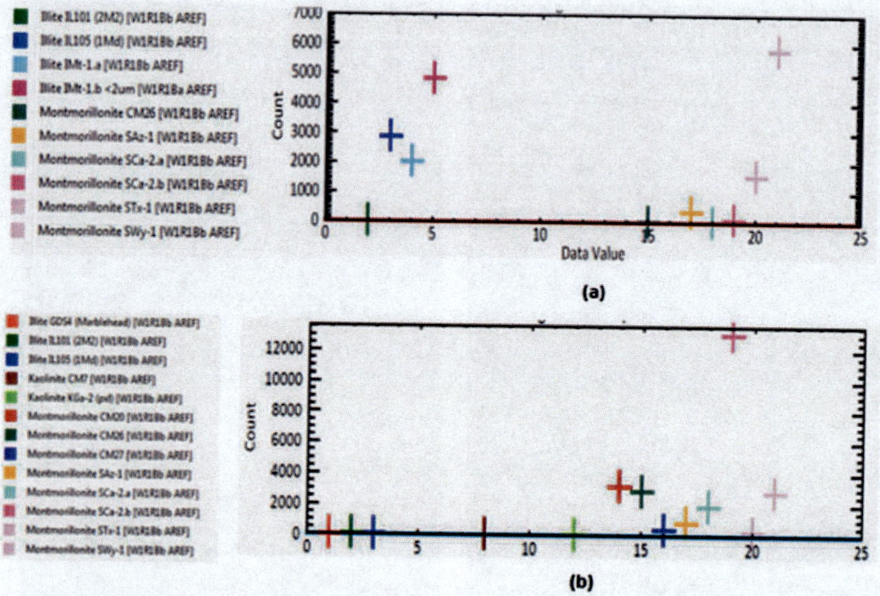

Fig. 13 Histogram of clay minerals identified. **a** SAM applied to IARR corrected image, **b** SAM applied to QUAC corrected image

Table 5 Capability of clay minerals identification

Situation		
	SAM applied to QUAC corrected image	SAM applied to IARR corrected image
Illite	Identified	Identified
Kaolinite	Identified	Not identified
Montmorillonite	Identified	Identified

Table 6 Comparison between SAM applied to QUAC and IARR corrected images

	Number of clay minerals profiles identified	Number of pixels classified
SAM applied to QUAC corrected image	13	25,601
SAM applied to IARR corrected image	10	17,513

firstly removed, only 158 of the 242 total Hyperion bands were used, than, radiometric calibration was performed to eliminate the path radiance effect from the acquired signal, after that, quick atmospheric correction (QUAC) is applied to compensate the effects of atmospheric absorption. This atmospheric absorption effect can lead to wrong interpretation and identification of objects because it influences the reflectance spectra .After removing atmospheric effects, destriping method is performed in order to correct the abnormal pixels of vertical stripes. Finally, to process and analyze the hyperspectral imagery with low computational cost, we have applied the minimum noise fraction transform to decrease the dimensions of the data; while conserving the important information, this method chooses the new components to maximize the signal to noise ratio (SNR) and orders them according to increasing image quality or decreasing noise. The obtained results show the high contribution of preprocessing proposed stages to enhance the quality of input image.

A methododology for a comparative analysis between internal average relative reflectance (IARR) atmospheric correction and quick atmospheric correction (QUAC) for mineralogy studies is also proposed. QUAC compensates the atmospheric effects more than IARR method in the field of mineral identification.

Comparison of other atmospheric correction will be an interesting perspective for this research, and it is highly recommended for the future.

References

1. Amigoa, J.M., Santosb, C.: Preprocessing of hyperspectral and multispectral images. Elsevier, pp. 37–53 (2020)
2. Jia, B., Wang, W., Ni, X., Lawrence, K.C., Zhuang, H., Yoon, S.C.,Gao, Z.: Essential processing methods of hyperspectral images of agricultural and food products. Elsevier, pp. 1–11 (2020)
3. Kale, K.V., Solankar, M.M., Nalawade, D.B., Dhumal, R.K., Gite, H.R.: A Research Review on Hyperspectral Data Processing and Analysis Algorithms, pp. 541–555, Springer (2017)
4. Tripathi, M.K., Govil, H.: Evaluation of Aviris-NG Hyperspectral Images for Mineral Identification and Mapping. Elsevier, pp. 1–10 (2019)
5. Gore, R., Mishra, A., Deshmukh, R.: Mineral mapping at lonar crater using remote sensing. J. Sci. Res. pp. 359–365 (2020)
6. Rani, N., Mandla, V.R., Singh, T.: Evaluation of atmospheric corrections on hyperspectral data with special reference to mineral mapping. Elsevier, –12 (2016)
7. Karpouzli, E., Malthus, T.: The empirical line method for the atmospheric correction of IKONOS imagery. Int. J. Remote Sens. pp. 1143–1150 (2003)
8. Tuominen, J, Lipping, T.: Atmospheric correction of hyperspectral data using combined empirical and model based method. In: Proceedings of the 7th European Association of Remote Sensing Laboratories Sig-imaging Spectroscopy Workshop (2011)
9. Kumar, M.V., Yarrakula, K.:Comparison of efficient tech-niques of hyper-spectral image preprocessing for mineralogy and vegetation studies (2017)
10. Thompson, D.R., Gao, B.C., Green, R.O., Roberts, D.A., Dennison, P.E.: Lundeen SR Atmospheric correction for global mapping spectroscopy: ATREM advances for the HyspIRI preparatory campaign. Remote Sens. Environ. **167**, 64–77 (2015)
11. Gao, B.C., Montes, M.J., Davis, C.O., Goetz, A.F.: Atmospheric correction algorithms for hyperspectral remote sensing data of land and ocean. Remote Sens. Environ. **113**, S17–S24 (2009)

12. Pflug, B., Main-Knorn, M.: Validation of atmospheric correction algorithm ATCOR. SPIE Proc. Lidar Radar Passive Atmos. Measure. II, **9242**(92420W), 1–8 (2014)
13. Zazi, L., Boutaleb, A., Guettouche, M.S.: Identification and mapping of clay minerals in the region of Djebel Meni (Northwestern Algeria) using hyperspectral imaging, EO-1 Hyperion sensor. Springer, 2–10 (2017)
14. Vignesh Kumar, M., Yarrakula, K.: Comparison of efficient techniques of hyper-spectral image preprocessing for mineralogy and vegetation studies. Indian J. Geo Marine Sci. pp. 1008–1021 (2017)
15. Wang, J., Chang, C.I.: Independent component analysis-based dimensionality reduction with applications in hyperspectral image analysis. IEEE Trans. Geosci. Remote. Sens. **44**(6), 1586–1600 (2006)
16. Pearlman, J., Carman, S., Lee, P., Liao, L., Segal, C.: Hyperion imaging spectrometer on the new millennium program Earth Orbiter-1 system. In Proceedings, International Symposium on Spectral Sensing Research (ISSSR), Systems and Sensors for the New Millennium, published on CD-ROM, International Society for Photogrammetry and Remote Sensing (ISPRS) (1999)
17. Datt, B., McVicar, T.R., Van Niel, T.G., Jupp, D.L.B., Pearlman, J.S.: Preprocessing eo-1 hyperion hyperspectral data to support the application of agricultural indexes. IEEE Trans. Geosci. Remote Sens. **41**(6), 1246–1259 (2003)
18. Bernstein, L.S., Adler-Golden, S.M., Jin, X., Gregor, B., Sundberg, R.L.: Quick atmospheric correction (QUAC) code for VNIR-SWIR spectral imagery: algorithm details. In Hyperspectral Image and Signal Processing (WHISPERS), 2012 4th Workshop on (pp. 1–4). IEEE (2012)
19. Acito, N., Diani, M., Corsini, G.: Subspace-based striping noise reduction in hyperspectral images. IEEE Trans. Geosci. Remote Sens. (2010)
20. Han, T., Goodenough, D.G., Dyk, A., Love, J.: "Detection and correction of abnormal pixels in Hyperion images," In IEEE International Geoscience and Remote Sensing Symposium, Toronto, Ont.,Canada, pp. 1327–1330
21. Shirmard, H., Farahbakhsh, E., Pour, A.B., Muslim, A.M., Müller, R.D., Chandra, R.: Integration of selective dimensionality reduction techniques for mineral exploration using ASTER satellite data. MDPI, pp. 1–29 (2020)
22. Phillips, R.D., Watson, L.T., Blinn, C.E., Wynne, R.H.: An adaptive noise reduction technique for improving the utility of hyperspectral data. In: Proceedings of the 17th William T. Pecora Memorial Remote Sensing Symposium, pp. 16–20 (2008)
23. Islam, M.R., Hossain, M.A., Ahmed, B.: Improved Subspace Detection Based on Minimum Noise Fraction and Mutual Information for Hyperspectral Image Classification. Springer, pp. 631–641 (2020)
24. Chakouri, M., Lhissou, R., El Harti, A., Maimouni, S., Adiri, Z.: Assessment of the image-based atmospheric correction of multispectral satellite images for geological mapping in arid and semi-arid regions. J. Preproof, pp. 1–33 (2020)
25. Merzah, Z.F., Jaber, H.S.: Assessment of Atmospheric Correction Methods for Hyperspectral Remote Sensing Imagery Using Geospatial Techniques. IOP Publishing, 1–7 (2020)
26. Ren, Z., Sun, L., Zhai, Q.: Improved k-means and spectral matching for hyperspectral mineral mapping. Elsevier, pp. 1–12 (2020)
27. Gopinath, G., Sasidharan, N., Surendran, U.: Landuse classification of hyperspectral data by spectral angle mapper and support vector machine in humid tropical region of India. Springer, pp. 1–9 (2020)
28. Govil, H., Mishra, G., Gill, N., Taloor, A., Diwan, P.: Mapping Hydrothermally Altered Minerals and Gossans using Hyperspectraldata in Eastern Kumaon Himalaya, India. Elsevier, pp. 1–7 (2021)

Smart Agriculture Solution Based on IoT and TVWS for Arid Regions of the Central African Republic

Edgard Ndassimba, Nadege Gladys Ndassimba, Ghislain Mervyl Kossingou, and Samuel Ouya

Abstract The Central African Republic is threatened by desertification. Its economy is mainly dependent on agriculture. However, in arid zones where rainfall is low, environmental factors have made it difficult to develop sustainable agriculture. The collapse of the level of agricultural production makes populations vulnerable to food insecurity. Today, there are several innovative solutions to monitor and automate field irrigation in these arid and semi-arid regions. This paper proposes a system for monitoring and automatic irrigation of fields in these areas. The proposed solution is based on the Internet of Things (IoT) using a raspberry pi 3 b + with temperature and soil moisture sensors, a submersible water pump and a relay. The system uses a TVWS network that offers broadband Internet to farmers allowing them to connect their smartphones via Wi-Fi. These farmers use a web interface to access the field temperature and soil moisture monitoring screen and then remotely activate or deactivate the irrigation system, although this can also be done automatically. The results of this experiment allow the implementation of an automatic irrigation system in arid and semi-arid regions with low rainfall. This solution would strengthen the Central African Republic's economy and help develop sustainable agriculture to meet the urgent needs of the population threatened by famine.

1 Introduction

Over the last ten years, several regions of Central African Republic, covering an area of about 623,000 km^2, have been threatened by desertification. The localities in the North East are practically devoid of water. Locally, the climate has changed: temperature has increased and rainfall has decreased [1]. Due to a long dry season, food is also no longer produced because the climatic conditions are not favorable for agriculture. One of the solutions to promote agriculture is the use of intelligent agriculture.

E. Ndassimba (✉) · N. G. Ndassimba · G. M. Kossingou · S. Ouya
Laboratory of Computing Networks and Telecommunications (LIRT) of Higher Polytechnic School At the University Cheikh Anta Diop (UCAD), Dakar, Senegal

M. Ben Ahmed et al. (eds.), *Networking, Intelligent Systems and Security*, Smart Innovation, Systems and Technologies 237,
https://doi.org/10.1007/978-981-16-3637-0_46

Several research works have been carried out on the use of the Internet of Things based on the Raspberry pi in the field of agriculture. The authors in [2] presented a soil quality monitoring system using wireless sensor nodes. Others in [3] developed intelligent monitoring and security devices based on IoT for agriculture. The results of this work have shown that with the Internet of Things (IoT), we can predict and analyze the parameters of the greenhouse effect to improve crop quality and productivity, and then ensure the safety of farms against rodents.

Our approach differs from existing uses. Firstly, because of the use of TV White Space to bring broadband internet to these arid areas, and secondly on how to monitor and control the irrigation system which can be done automatically by the system and manually remotely from a smartphone of the farmer connected to the TV White Space Wi-Fi network.

The rest of this paper is organized as follows. Section 2 is reserved for the state of the art. Section 3 presents the materials and methodology used. Section 4 presents the results and discussions about our solution, and finally Section 5 provides the conclusion.

2 State of the Art

2.1 *Impact of Desertification on Agriculture in the Central African Republic*

The concept of desertification is defined as land degradation in arid, semi-arid and dry sub-humid areas, often referred to simply as “drylands." It is believed to result from a combination of factors, including climate change and human activity. Arid zones occupy 41.3% of the Earth’s surface and are characterized by water scarcity [4, 5].

In the Central African Republic, climate change is perceptible over the entire national territory and deserves special attention from the main land users whose ecosystem functions need to be preserved in order to have an impact on productivity and farm incomes. Droughts are now frequent in the northern, northeastern and eastern regions, once renowned for their agricultural production. There is growing evidence that groundwater reserves are being depleted, leading to a sharp reduction in productivity in these areas [6].

Desertification is increasing every year, contributing to the disappearance of several million hectares and affecting farmers. This creates food insecurity, poverty and high-economic cost in the Central African Republic.

2.2 Climate-Smart Agriculture

Climate-smart agriculture (CSA) is an approach that helps guide the actions needed to transform and reorient agricultural systems to effectively support development and ensure food security in a changing climate. Climate-smart agriculture is one of the techniques that maximize agricultural yields through good management of inputs according to climate conditions [7]. The CSA has three main objectives: to sustainably increase agricultural productivity and incomes; to adapt and build resilience to climate change; and to reduce and/or eliminate greenhouse gas emissions wherever possible [8, 9]. Intelligent agriculture uses new technologies, such as satellite imagery and computers, satellite positioning systems such as GPS, also through the use of sensors that will collect useful information on soil condition, moisture content, mineral salt content, etc. and send this information to the farmer to take the necessary measures to ensure good production. Generally, irrigation is used to improve agricultural production in the face of climate change. Irrigation is the supply of water to crops by artificial means to enable agriculture in arid areas and to compensate for the effects of drought in semi-arid areas. Where traditional rain-fed agriculture is at high risk, irrigation can help to ensure stable production.

The Central African Republic has a hot and humid equatorial climate, characterized by two seasons: a rainy season that lasts from April to October, and a dry season, between November and March. Annual rainfall is higher in the Ubangi Valley (1780 mm) than in the central part (1300 m) and in the semi-arid northeastern and eastern areas (760 mm). The development of sustainable agriculture in the Central African Republic can help mitigate the secondary effects of the country's dire situation. Smart agriculture can lead to a growing economy at the micro and macro levels by increasing production [10].

Vulnerability to climate change in the Central African Republic and low capacity to adapt to its adverse effects pose serious threats to the management of ecosystems and agricultural resources and to sustainable development, hence the importance of using smart agriculture in desertification-affected regions.

2.3 The Internet of Things in Agriculture

The Internet of Things (IoT), refers to the set of infrastructures and technologies set up to make various objects work through an Internet connection. We then speak of connected objects. These objects can be controlled remotely, most often using a computer, a smartphone or a tablet.

Numerous research works have shown that the Internet of Things (IoT) can be used in several fields such as transport, health, home automation, agriculture, etc. In [11], the authors proposed an intelligent agricultural system (AgriSys) that can analyze an agricultural environment and intervene to maintain its suitability. The authors in [12] focused their work on the introduction of a Smart Drone for crop management where real-time data from the UAV, combined with IoT and Cloud Computing technologies, help to build a sustainable intelligent agriculture. The authors in [13] presented an intelligent solution, gCrop, to monitor the growth and development of leafy crops and to update the status in real time using IoT, image processing and machine learning technologies. In [14], the authors proposed adapted good practices to reduce the water footprint in agricultural crop fields with traditional methods. The combination of biochemistry and the Internet of Things contributes to improve the competitiveness of agricultural economic activities near cities and at the same time to avoid water crisis.

The results of this work have proved that the challenges of IoT are numerous. Among these challenges, agriculture is undergoing its digital transformation. Farmers can accurately control environmental parameters (air and soil humidity, temperature, etc.) recorded by sensors, and remotely control the irrigation of their fields for better productivity and profitability.

2.4 The TVWS in Agriculture

The TV White Space is a technology that allows the use of free frequencies from television to provide a broadband network in a given region. The free UHF frequency bands used are 470–790 MHz in Europe and 54–698 MHz in the United States. The use of White Space is based on secondary unlicensed dynamic spectrum access (DSA—Dynamic Spectrum Alliance) under the principle of non-detrimental interference for television operators operating in the area.

By revolutionizing traditional wireless broadband connectivity, the TV White Space is typically used to bring broadband Internet to rural areas with difficult access.

Several researchers have worked on the relevance of using TVWS. In [15], the authors showed the opportunity of vehicular communications on TV White Space in the presence of secondary users. The results show that there are opportunities for vehicular access even when a White-Fi network occupies the TVWS. In [16], the authors studied and adopted TV White Space technology as a rural telecommunication solution in Indonesia in relation to its performance. They concluded from a simulation that TV White Space is an appropriate technological alternative for rural conditions.

The results of this work proved the potential that TV White Space has offered for a wide range of innovative applications. For example, White Space TV can establish high-bandwidth links between a farmer's home Internet connections and an on-farm IoT base station with sensors for intelligent agriculture in arid rural areas.

3 Materials and Methodology

3.1 Architecture of Our Solution

The architecture of our solution includes:

- A Raspberry Pi 3 b + equipped with a 40-pin GPIO extension board and a GPIO cable mounted on a breadboard, and deployed on a RAB platform;
- A Relay 4 Module (VCC: 5 V positive supply voltage, GND: ground, IN1: relay control port);
- A DHT11 Temperature/Humidity Sensor (VCC: 5 V supply voltage, GND: ground, DATA: serial data port, single bus);
- Soil Moisture Sensor (VCC: 5 V supply voltage, GND: ground, GIS);
- A mini submersible pump (Motor Submersible Water Pump);
- Silicone tubing;
- 5 V power supply for the pump and Raspberry Pi;
- A TV White Space network with a Base Station (BS) and customers called Customer Premises Equipment (CPE);
- Internet connectivity.

The following figure shows the interconnection of the intelligent agriculture system test bed components (Fig. 1).

The following figures show the different elements of the Intelligent farming system test bed (Fig. 2).

4 relay module is a 4-channel 5 V relay interface card and each channel requires a driver current of 15–20 mA. It can be used to control various devices and equipment with high current. It is equipped with high current relays operating under AC250V

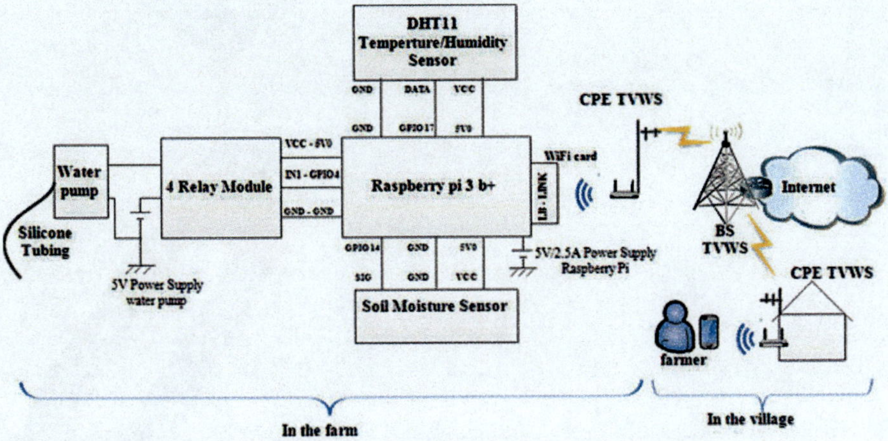

Fig. 1 Proposed solution architecture

Fig. 2 Relay module

10A or DC30V 10A. It has a standard interface that can be directly controlled by a microcontroller (Fig. 3).

The DHT11 digital temperature and humidity sensor are a composite sensor that contains a calibrated digital signal output of temperature and humidity. The application of dedicated digital module collection technology and temperature and humidity sensing technology ensures that the product has high reliability and excellent long-term stability. The sensor consists of a resistive wet component sensor and NTC temperature measuring device, and is connected to a high-performance 8-bit microcontroller (Fig. 4).

Fig. 3 DHT11 temperature/humidity sensor

Fig. 4 Soil moisture sensor

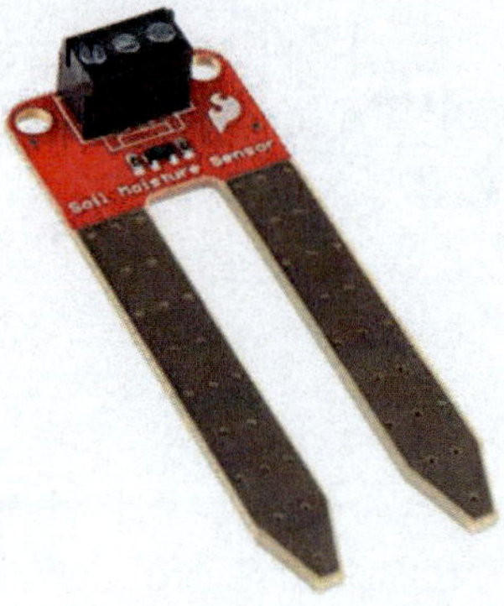

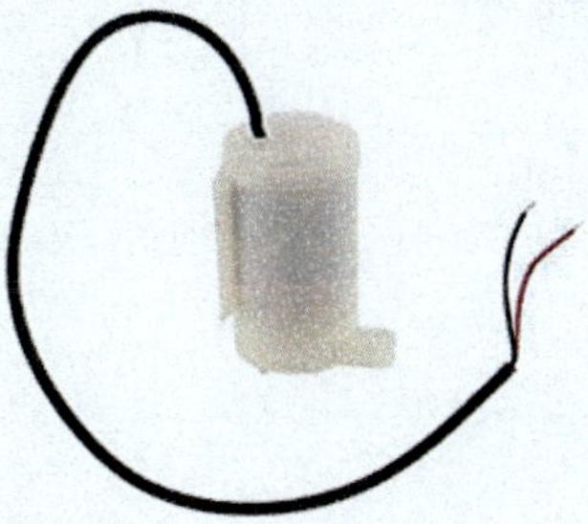

Fig. 5 Mini DC 3-6 V 120L/H brushless motor submersible water pump

The SparkFun soil moisture sensor is a simple escape to measure the moisture content of soil and similar materials. The two large exposed studs act as probes for the sensor, acting together as a variable resistance. The more water in the soil, the better will be the conductivity between the pads, resulting in lower resistance and higher SIG output (Fig. 5).

The mini water pump is made of plastic material and electronic components. It operates with a DC voltage of 2.5-6 V and can provide a flow rate of 80-120L/H with a power of 0.4–1.5 W (Fig. 6).

The raspberry pi is a motherboard of a mini-computer that can be connected to any device (mouse, keyboard…). This card is made to help study computers and to represent a means of learning computer programming in several languages (python, scratch…).

The raspberry pi 3 model B + includes a Broadcom BCM2837B0 64-bit quad-core ARM Cortex-A53 processor running at 1.4 GHz, a new CYW43455 supporting Dual-band 802.11ac Wi-Fi and Bluetooth version 4.2, and the support for power over Ethernet through an additional element.

The following figure shows the cabling of our test environment (Fig. 7).

Fig. 6 Raspberry Pi 3 model B +

Fig. 7 Testing environment

3.2 *How the Solution Works*

To operate, the mini submersible water pump must be completely submerged in a water tank in order to pump water. This pump is powered by a 5 V charger and connected to the relay. The pump is connected to the GPIO4 as a LED. The DHT11 sensor is connected to the GPIO17. The soil moisture sensor is connected to the GPIO14 as a button. For the SparkFun soil moisture sensor to work, we have connected the VCC and GND pins to the raspberry pi. We receive a SIG, which depends on the amount of water in the soil. When the soil moisture sensor gives the raspberry pi an input voltage of 1 or 0 depending on whether it is wet or dry, it behaves like a button pressed when wet, and a button released when dry. The pump will receive from the python written program of the raspberry pi via the relay an output voltage activated or deactivated like a LED that triggers watering and administers irrigation periods. The soil moisture sensor detects the condition of the soil (its dryness rate) in order to adjust the water supply accordingly to efficiently water the fields. The DHT11 sensor allows you to stop watering immediately when it begins to rain.

The system can operate in automatic mode, manual mode, and factory water mode.

In order to access our raspberry remotely via the TV White Space Wi-Fi broadband network and control the GPIOs, we need to make our card a Nginx web server capable of hosting the website and thus ensure communication with customers and respond to their requests using the HTTP network protocol. This allows them to consult in real time on a smartphone or computer via a dynamic web page of the web server that displays the ground parameters.

```
pi@raspberrypi: ~
root@raspberrypi:/home/pi# cd /var/www/html/IoT_Plant/
root@raspberrypi:/var/www/html/IoT_Plant# ls
eon  plant.html  Planty.py  README.md
root@raspberrypi:/var/www/html/IoT_Plant# python Planty.py
```

Fig. 8 Execution of the Planty.py python code

4 Results and Discussions

The results of the test bench are based on the libraries and dependencies of dedicated peripherals for the temperature/humidity sensor and the soil moisture sensor and GPIO raspberry pi that we downloaded and installed. We created our application with PubNub functions.

4.1 *Irrigation System Test Without Wi-Fi Connection*

To start the system, we executed the Planty.py python code by connecting to the Raspberry via SSH as shown in the following Fig. 8:

Figure 9 showed the values of temperature and humidity obtained on the computer connected by SSH to the raspberry showing the values of the two variables: Temp and humidity.

Figure 10 showed the automatic irrigation triggered by the Raspberry Pi when we remove the soil moisture sensor from the soil. This sensor emitted a voltage when wet and none when dry. Removing this sensor sent a signal that the soil is dry, which caused the pump that started the irrigation to start.

4.2 *Remote Control of the Irrigation System by Simulating the Connection to the TVWS Network*

To demonstrate the relevance of using TVWS technology for broadband Internet connectivity in rural areas for remote monitoring and control of the irrigation system, we used a Wi-Fi network card (LB-LINK) installed on the raspberry pi and connected to a Wi-Fi access point.

We used the web interface: http://192.168.43.140/IoT_Plant/plant.html for remote monitoring of field soil parameters and control of the irrigation system.

Figures 11 and 12 showed the graphs of the temperature and humidity data obtained in real time from the Nginx web server installed on the raspberry pi.

On the web interface, there are three buttons for remote control of the system.

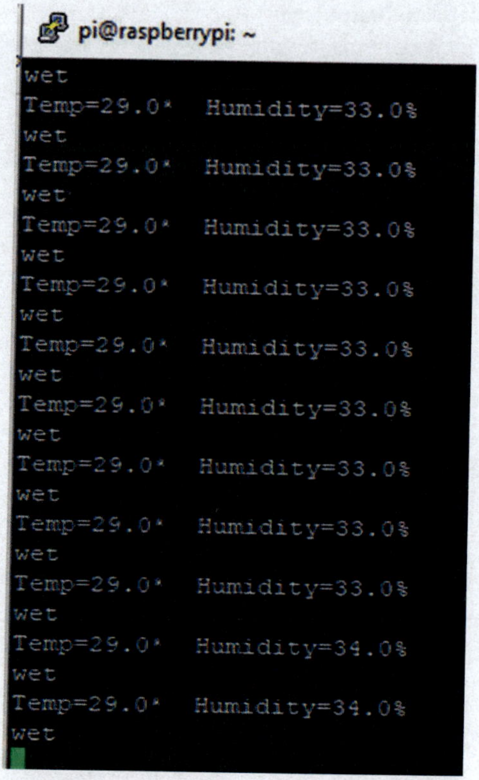

Fig. 9 Publication of temperature and humidity results

Fig. 10 Automatic start of the irrigation system

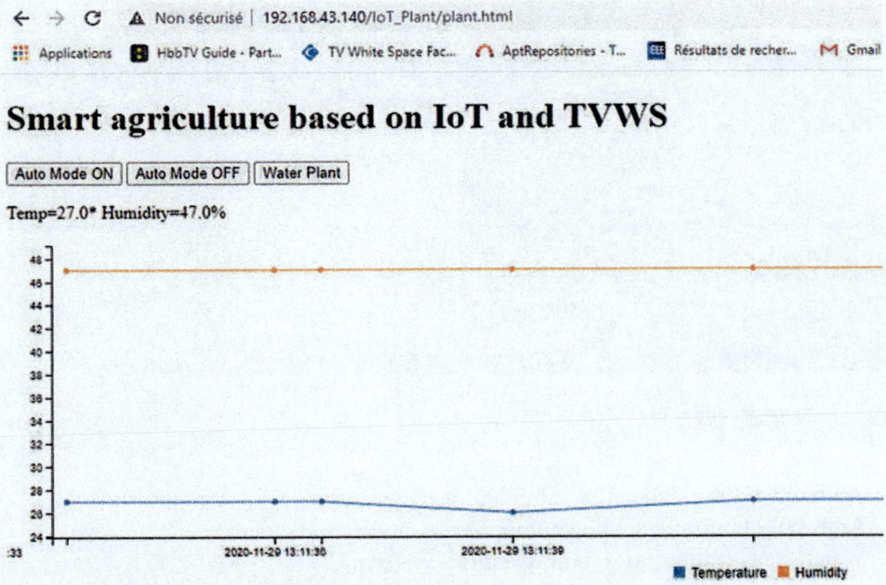

Fig. 11 Graph of data on a computer

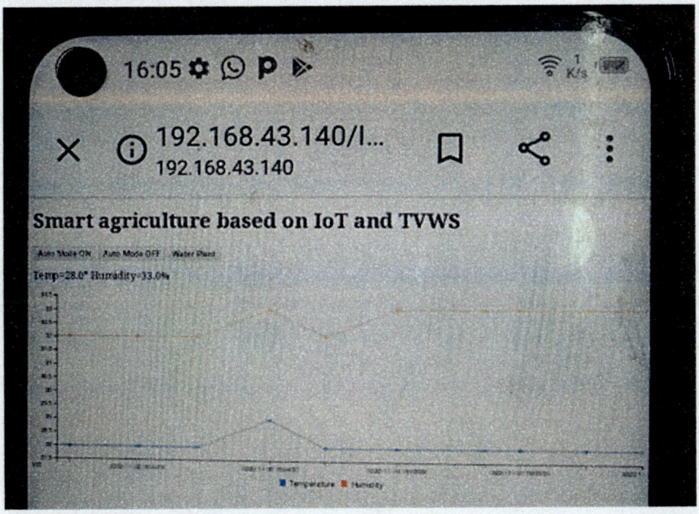

Fig. 12 Data graph on a smartphone

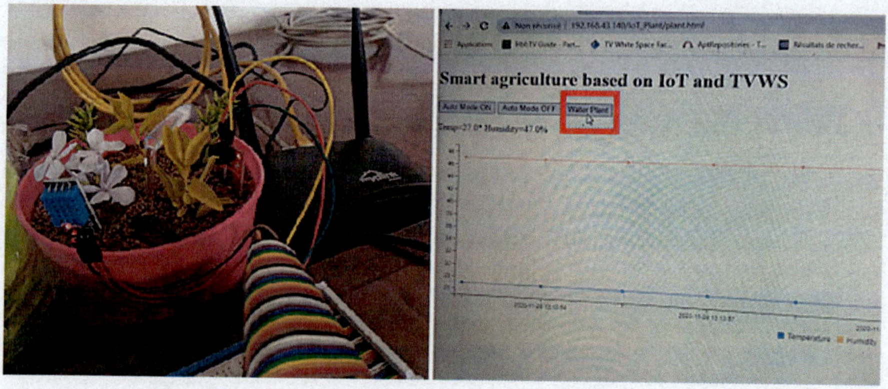

Fig. 13 Remote control of the irrigation system

- Auto ON button: when the automatic watering mode is activated.
- Auto OFF button: when the automatic watering mode is off.
- Water Plant Button: to remotely control watering.

Figure 13 showed the remote control of the irrigation system. Pressing the water plant button will remotely control the irrigation.

Contrary to the different types of networks used in Internet of Objects solutions (standard Wi-Fi IEEE 802.11, LoRa, GSM, etc.), TVWS technology is best suited to bring high-speed Internet to rural areas in relation to its technical and economic advantages. Its technical advantages are its extended range of 10 to 30 km; broadcasting over uneven areas, through obstacles, forests or buildings. The economic advantages concern the deployment of the technology at a lower cost.

In Central African Republic, unused television White Space (TVWS) would solve the problem of Internet connectivity to establish a farmer's home broadband Internet connection and the Internet of Things (IoT) deployed in the field through interconnected TVWS customer premium equipment (CPE) on a TVWS base station (BS).

5 Conclusion

In this work, we proposed a smart agriculture solution for the arid regions of the Central African Republic. This system was designed based on a raspberry pi 3 b + and a set of sensor networks connected to a TV White Space broadband network.

Intelligent farming practices have proven effective in several countries that have experienced droughts, and could be implemented in agricultural systems in countries that have experienced food crises due to droughts.

This solution will help farmers to remotely control the soil parameters of their fields and then trigger the irrigation system, although this can also be done automatically. The impact will be to increase production and solve the problems of food insecurity and climate change.

In future work, we will propose an independent mobile network infrastructure based on TVWS in rural areas of the Central African Republic using the Internet of Things with D-GSM in Osmocom and Freeswitch to send SMS/MMS alerts to farmers on the critical state of soil parameters (If the soil moisture level is below the normal threshold, the farmer receives an SMS notification and remote control of the automatic irrigation).

References

1. Forest Management and Deforestation in Central African Republic. http://www.ajer.org/papers/v5(04)/I0504079090.pdf, last accessed 2020/11/21
2. Shinde, D., Siddiqui, N.: "IOT Based environment change monitoring & controlling in greenhouse using WSN," *2018 International Conference on Information , Communication, Engineering and Technology (ICICET)*, pp. 1–5, Pune (2018)
3. Baranwal, T., Nitika and P.K. Pateriya, "Development of IoT based smart security and monitoring devices for agriculture," *2016 6th International Conference - Cloud System and Big Data Engineering (Confluence)*, pp. 597–602, Noida (2016)
4. United Nations Decade for Deserts and the Fight Against Desertification. https://www.un.org/en/events/desertification_decade/whynow.shtml, last accessed 2020/11/21
5. Trees, forests and land use in drylands The first global assessment. http://www.fao.org/3/a-i5905e.pdf, last accessed 2020/11/21
6. SOCIAL WATCH poverty eradication and gender justice. https://www.socialwatch.org/node/13981, last accessed 2020/11/22
7. Tenzin, S., Siyang, S., Pobkrut, T., Kerdcharoen, T.: "Low cost weather station for climate-smart agriculture," *2017 9th International Conference on Knowledge and Smart Technology (KST)*, pp. 172–177. Chonburi (2017)
8. Climate-Smart Agriculture. http://www.fao.org/climate-smart-agriculture/en/, last accessed 2020/11/24
9. Climate-smart agriculture for food security. https://www.researchgate.net/publication/273448307_Climate-smart_agriculture_for_food_security, last accessed 2020/11/24
10. Sustainable Agriculture in the Central African Republic. https://www.borgenmagazine.com/sustainable-agriculture-in-the-central-african-republic/, last accessed 2020/11/25
11. Abdullah, A., Al Enazi, S., Damaj, I.: "AgriSys: a smart and ubiquitous controlled-environment agriculture system," *2016 3rd MEC International Conference on Big Data and Smart City (ICBDSC)*, pp. 1–6, Muscat (2016)
12. Namani, S., Gonen, B.: "Smart agriculture based on iot and cloud computing," *2020 3rd International Conference on Information and Computer Technologies (ICICT)*, pp. 553–556, San Jose, CA, USA (2020)
13. Kumar, S., Chowdhary, G., Udutalapally, V., Das, D., Mohanty, S.P.: "gCrop: Internet-of-Leaf-Things (IoLT) for monitoring of the growth of crops in smart agriculture," *2019 IEEE International Symposium on Smart Electronic Systems (iSES) (Formerly iNiS)*, pp. 53–56, Rourkela, India (2019)
14. Larios, V.M., Michaelson, R., Virtanen, A., Talola, J., Maciel, R., Beltran, J.R.: "Best practices to develop smart agriculture to support food demand with the rapid urbanization trends in Latin America," *2019 IEEE International Smart Cities Conference (ISC2)*, pp. 555–558, Casablanca, Morocco (2019)

15. Arteaga, A., Céspedes, S., Azurdia-Meza, C.: Vehicular communications over TV white spaces in the presence of secondary users. IEEE Access **7**, 53496–53508 (2019)
16. Aji, L.S., Wibisono, G., Gunawan, D.: "The adoption of TV white space technology as a rural telecommunication solution in Indonesia," *2017 15th International Conference on Quality in Research (QiR) : International Symposium on Electrical and Computer Engineering*, pp. 479–484 Nusa Dua (2017)

Model-Driven Engineering: From SQL Relational Database to Column—Oriented Database in Big Data Context

Fatima Zahra Belkadi and Redouane Esbai

Abstract The growth of application architectures in all areas (e.g., astrology, meteorology, E-commerce, social network, etc.) has resulted in an exponential increase in data volumes, now measured in Petabytes. Managing these volumes of data has become a problem that relational databases are no longer able to handle because of the acidity properties. In response to this scaling up, new concepts have emerged such as NoSQL. In this paper, we show how to design and apply transformation rules to migrate from an SQL relational database to a big data solution within NoSQL. For this, we use the model driven architecture (MDA) and the transformation languages like as MOF 2.0 QVT (Meta-Object Facility 2.0 Query-View-Transformation) and Acceleo which define the meta-models for the development of transformation model. The transformation rules defined in this work can generate, from the class diagram, a CQL code for creation column-oriented NoSQL database.

1 Introduction

In recent years, the world of data storage is changing rapidly. New technologies and new actors are settling when the old ones make the move. This scientific revolution that has invaded the world of information, and the Internet has imposed new challenges on researchers in recent years and has led them to design new tools for specific storage and manipulation. The development of these tools is generating a growing interest among scientific and economic actors to offer them the possibility of managing all these masses of data with reasonable response times. Big data is correlated between four notions generally grouped under the acronym "4 V," namely: Volume, variety, velocity and variability [1].

Our focus in this paper is only on big data storage. Using relational databases prove to be inadequate for all applications, particularly ones involving large volumes of data. In this context, NoSQL databases offer new storage solutions in large-scale environments, replacing many traditional database management systems [2]. The

F. Z. Belkadi · R. Esbai (✉)
Mohammed First University, Oujda, Morocco

M. Ben Ahmed et al. (eds.), *Networking, Intelligent Systems and Security*,
Smart Innovation, Systems and Technologies 237,
https://doi.org/10.1007/978-981-16-3637-0_47

key feature of NoSQL databases is that they are schema-less, meaning that data can be inserted in the database without upfront schema definition. Nevertheless, there is still a need for a semantic data model to define how data will be structured and related in the database [3]; it is generally accepted that UML meets this requirement [4].

This paper aims to rethink the work presented in [5]. However, we develop the transformation rules using the MOF 2.0 QVT standard to generate a file which contains a code for creation a column-oriented NoSQL model [6]. Our approach includes UML modeling and automatic code generation using Acceleo with the aim to facilitate and accelerate the creation of column-oriented NoSQL database.

This paper is organized as follows: related works are presented in the second section, the third section defines the MDA approach, and the fourth section presents the NoSQL and its implementation as a database, column-oriented in this case. In the fifth section, we present the source and target meta-models. In the sixth section, we present the transformation process M2M and M2T from UML class diagram model to the column-oriented NoSQL database. The last section concludes this paper and presents some perspectives.

2 Related Works

Many researches on MDA and the process of transforming relational databases into a NoSQL model have been conducted in recent years. The most relevant are [3, 5–10]: Chevalier et al. [7] defined rules to transform a multidimensional model into NoSQL column-oriented and document-oriented models. The links between facts and dimensions have been converted using imbrications. Although the transformation process proposed by authors start from a multidimensional model, it contains facts, dimensions and one type of links only. Gwendal et al. [3] describe the transformation from an UML conceptual model into a graph databases via an intermediate graph meta-model. These transformation rules are specific to graph databases used as a framework for storing, managing and querying complex data with many connections.

Li et al. [8] propose a MDA approach to transform UML class diagram into HBase. After building the meta-models of UML class diagram and HBase, the authors have proposed mapping rules to realize the transformation from the conceptual level to the physical level. These rules are applicable to HBase only. Another works followed the same logic and have been the subject of a work Vajk et al. [9]. The authors propose a mapping from a relational model to document-oriented model using MongoDB.

The purpose of the work [10] presented by Abdelhedi et al. is to implement a conceptual model describing big data into NoSQL database and they choose to focus on column-oriented NoSQL model.

This paper aims to rethink and to complete the work presented by Abdelhedi et al. [5, 10], by applying the standard MOF 2.0 QVT and Acceleo to develop the transformation rules aiming at automatically generating the creation code of column-oriented NoSQL database. It is actually the only work for reaching this goal.

3 Model Driven Architecture (MDA) Approach

3.1 *The Transformations of MDA Model*

The MDA identifies several transformations during the development cycle [11]. It is possible to make three different types of transformations: CIM to PIM, PIM to PSM and PSM to code.

In this paper, we chose two types of transformation, we start with the transformation PIM to PSM using the approach by modeling. This type of transformation will allow us to automatically generate a column-oriented NoSQL model from an UML model. The second transformation is of type PSM to Code using the approach by template with Acceleo to develop the transformation rules aiming at automatically generating the creation code of column-oriented NoSQL database [12].

3.2 *The Elaborationist Approach*

The elaborationist approach is the one used in the present paper. The main advantage of MDA in the development of column-oriented NoSQL databases is the automation. This way, to demonstrate the automation support provides by our MDA approach, we are using the "Elaborationist approach" (see Fig. 1). With the elaborationist approach, the definition of the application is built up progressively as you progress through from PIM to PSM to code. When the PIM has created, the tool generates a skeleton or first-section PSM which the developer can then "elaborate" by adding more detail. Similarly, the final code is generated from PSM, and this can also be elaborated.

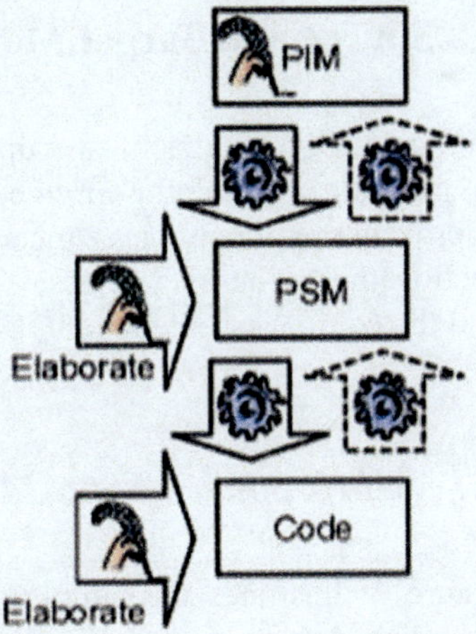

Fig. 1 Elaborationist approach [13]

4 Column-Oriented NoSQL Database

There are four basic types of NoSQL databases: key-value, document-oriented, column-oriented and graph-oriented [2]. In this paper, we choose to focus on column-oriented NoSQL model.

The column-oriented databases were originally created by Facebook to store messages (non-instant) between users [14]. It is a key-value database extension, because the column model is more evolved, it is called super-column or column-family that a line identifier can store a structured set of data. A column-family has the following characteristics: the data is sorted, associated, and can contain an array of columns of unlimited size.

The storage of column-oriented databases is by column and not by row. These bases can evolve over time, either in number of rows or in number of columns. In other words, and unlike a relational database where columns are static and present for each row, the column-oriented databases are dynamic and present only when needed.

In column-oriented databases such as Cassandra [15] or HBase [16], there are some additional concepts that are the column-family, which are a logical grouping of rows. In the relational world, this would be equivalent to a table. Cassandra offers an extension to the base model by adding an extra dimension called "Super Column" which itself contains other columns.

The concept of column-oriented databases is created by the big web actors, to meet the processing needs of large volumes of data precisely to manage large volumes of structured data. Often, these databases integrate a minimalist query system close to SQL called CQL [2].

In this paper, we choose the principle actor of column-oriented database such as Cassandra.

5 Source and Target Meta-Models

In our MDA approach, we opted for the modeling and template approaches to generate the column-oriented NoSQL database. As mentioned above, these approaches require a source meta-model and a target meta-model. We present in this section, the various meta-classes forming the UML class diagram source meta-model and the column-oriented NoSQL target meta-model.

5.1 UML Source Meta-Model

Figure 2 illustrates the simplified UML source meta-model based on packages including data types and classes. Those classes contain typed properties and they

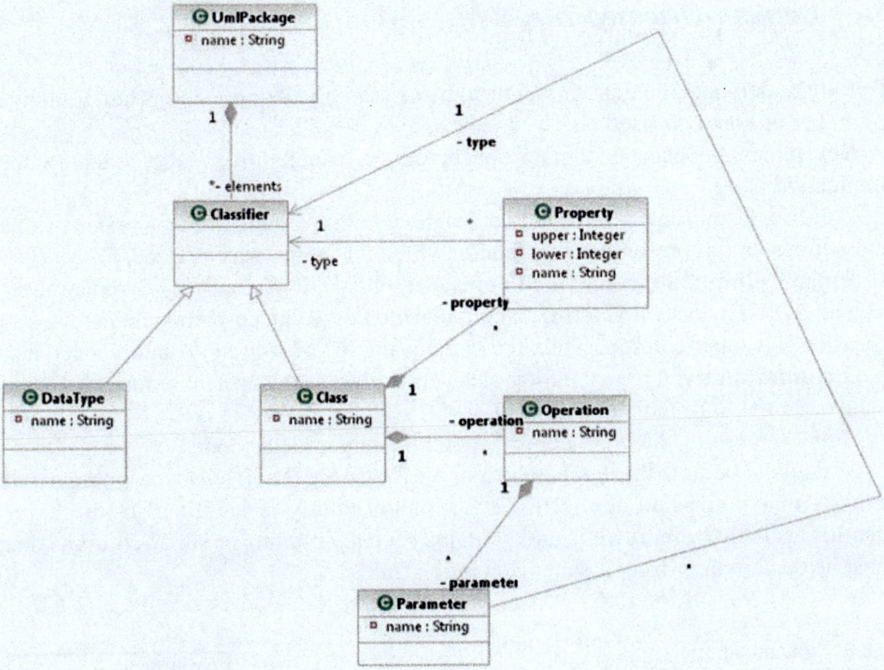

Fig. 2 Simplified UML source meta-model

are characterized by multiplicities (lower and upper). The classes are composed of operations with typed parameters.

UmlPackage: is the concept of UML package. This meta-class is connected to the meta-class Classifier.

Classifier: This is an abstract meta-class representing both the concept of UML class and the concept of data type.

Class: is the concept of UML class.

DataType: represents UML data type.

Operation: is used to express the concept of operations of a UML class.

Parameter: expresses the concept of parameters of an operation. These are of two types, Class or DataType. It explains the link between parameter meta-class and classifier meta-class.

Property: expresses the concept of properties of a UML class. These properties are represented by the multiplicity and meta-attributes upper and lower.

The works [17, 18] contain more details related to this section topic.

5.2 Column-Oriented Target Meta-Model

To fully understand the data model used by Cassandra [19], it is important to define a number of concepts used:

Keyspace: Appears as a namespace, this is usually the name given to the application.

Column: Represents a value, and it has three fields (see Fig. 3): its name, its value and a timestamp representing the date on which this value was inserted.

Super-Column: it is a list of columns (see Fig. 4), if you want to compare them with an SQL database, it is a row. It contains the key-value correspondence; the key identifies the super column, while the value is the list of columns that compose it.

Column-Family: it is a container of several columns or super-columns. Its notion is closer to the SQL table (see Fig. 5).

Figure 6 presents these concepts through the target meta-model.

By default, we store the database in a single Keyspace. This Keyspace is comprised of a set of column-families [20]. Each Column-family is identified by a unique identifier called "PrimaryKey" and contains several columns or super-columns that must be declared up front at schema creation time.

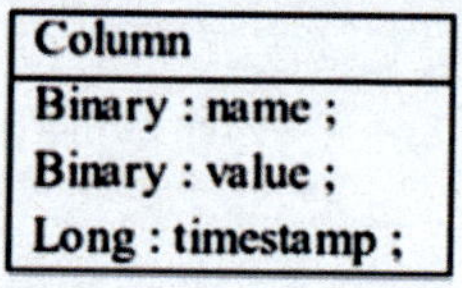

Fig. 3 Structure of the column element

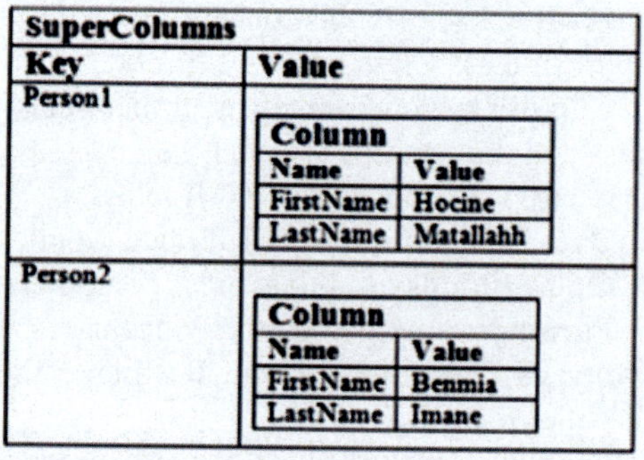

Fig. 4 Structure of the super-column element

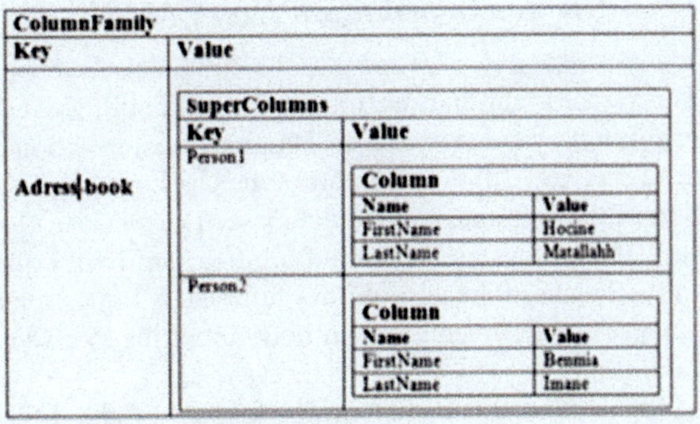

Fig. 5 Structure of the Column-Family element

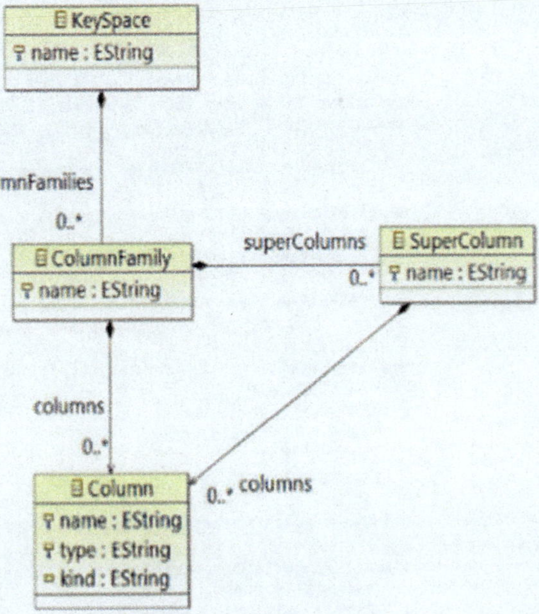

Fig. 6 Simplified column-oriented target meta-model

6 The Process of Transforming UML Source Model to Column—Oriented Target Code

We first developed ECORE models corresponding to our source and target meta-models. The development of many meta-models requires multiple model transformations. From these developed meta-models, M2M (Model to Model) and M2T (Model to Text) transformations are needed, to generate the code needed to create the column-oriented database. We have implemented the M2M transformation algorithm (see Sect. 6.1) using the QVT Operational Mappings language [21], and then the second M2T transformation is done with the Acceleo language [22] (see Sect. 6.2).

6.1 The Transformation Rules M2M

This transformation uses, in entry, a model of the UML type, and in output a model of column-oriented database. The first transformation rule establishes the correspondence between all the elements of the UML package and the element of the Keyspace type of the column-oriented database. The purpose of the second rule is to transform each UML class and association into a family of columns by creating the columns and references for each column-family. It is a question of transforming each property of these classes in column, without forgetting to give names and types to the various columns.

Figure 7 presents the principle part of the M2M transformation with QVT language.

```
UML2NoSQL.qvto
modeltype UML uses "http:///umlMM.ecore";
modeltype NoSQL uses "http://nosqlmm/1.0";
transformation UML2NoSQL(in umlModel: UML, out noSQLModel:NoSQL);
main() {
    umlModel.objects()[UmlPackage]->map UmlPackage2keySpace();
}
mapping UmlPackage::UmlPackage2keySpace () : KeySpace {
    name:= self.name;
    columnFamilies:=umlModel.objects()[Class] ->
            map classToColumnsFamily();
    columnFamilies+=umlModel.objects()[Association] ->
            map AssociationToColumnsFamily();
}
mapping Class::classToColumnsFamily():ColumnFamily{
 name:=self.name;
 columns:= object Column {
    name:= "id"+self.name;
    type:="uuid";
    kind:="PrimaryKey";
    };
 columns+= self._property -> map propertyToColumn();
}
mapping Property::propertyToColumn():Column{
    name:=self.name;
    type:= self.type.name;
}
mapping Association::AssociationToColumnsFamily():ColumnFamily{
 name:=self.name;
 var cs : Sequence (Column);
 self.links-> forEach(ls){
                cs+=object Column {
                    name:=ls.name+".Reference";
                    type:="uuid";
                    kind:="PrimaryKey";
                }
            };
 columns:=cs;
}
```

Fig. 7 M2M transformation with QVT from UML to NoSQL model

```
UML2NoSQL.mtl
[comment encoding = UTF-8 /]
[module UML2NoSQL('http://nosqlmm/1.0', 'http://umlMM.ecore')]

[template public generateElement(aKeySpace : KeySpace)]
[comment @main/]
[file (aKeySpace.name.concat('.cql'), false, 'UTF-8')]
create KEYSPACE  [aKeySpace.name/] WITH REPLICATION =
{ 'class' : 'SimpleStrategy', 'replication_factor' : 1 };

USE [aKeySpace.name/] ;

[for (cf:ColumnFamily|aKeySpace.columnFamilies)]
[ColumnFamilyToCQL(cf)/]
[/for]

[/file]
[/template]

[template public ColumnFamilyToCQL(cf: ColumnFamily)]
[let s:String='' ]
CREATE TABLE [cf.name/] (
[for (c:Column|cf.columns) separator(',')][c.name/] [c.type/] [/for],
PRIMARY KEY ([for (c:Column|cf.columns->
    select(f:Column|f.kind='PrimaryKey')) separator(',')]
        [c.name/]
[/for])
);
[/let]
[/template]
```

Fig. 8 M2T transformation with Acceleo to generate a CQL code

6.2 The Transformation Rules M2M

The transformation M2T toward the creation code of column-oriented database in Cassandra is realized with Acceleo transformation language, and the writing of the transformation rules itself does not present any problems in practice. It simply boils down to creating a text file where the transformation rules are written.

Figure 8 presents the transformation rules with Acceleo to generate a CQL file.

6.3 Result

To validate our transformation rules, we conducted several tests.

For example, we considered the class diagram composed by the classes department, employee and city (see Fig. 9).

After applying the transformation on the UML source model, we generated the column-oriented PSM target model (see Fig. 10).

Figure 10 shows the result after applying the transformation rules M2M.

Figure 11 illustrates the results of our M2T transformation. Our application generates a CQL code for a department management database on Cassandra platform.

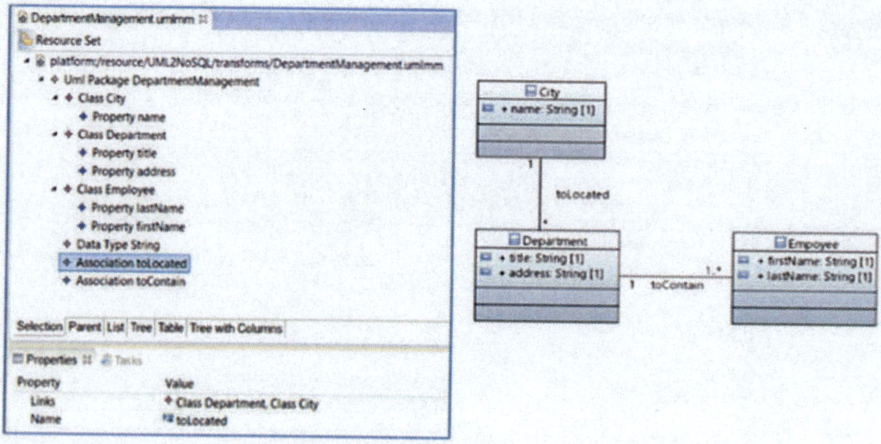

Fig. 9 UML source model: Class diagram EMF model and Class diagram instance model

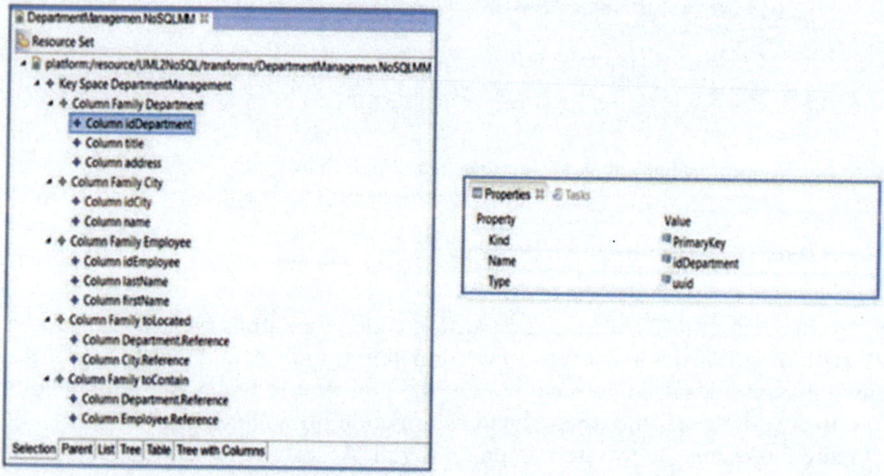

Fig. 10 Column-oriented cassandra PSM: resource set and their properties

7 Conclusion and Perspectives

In this paper, we have proposed an MDA approach to migrate UML class diagram representing a relational database to a column-oriented database. The transformations rules were developed using QVT to transform the class diagram into column-oriented model and then the automatic code generation using Acceleo with the goal to accelerate and makes easy the creation of NoSQL databases in Cassandra platform. In future, this work should be extended to allow the generation of other NoSQL

Fig. 11 CQL file generated

```
DepartmentManagement.cql
CREATE KEYSPACE  DepartmentManagement WITH REPLICATION
= { 'class' : 'SimpleStrategy', 'replication_factor' : 1 };

USE DepartmentManagement ;

CREATE TABLE Department (
idDepartment uuid ,
title text ,
address text ,
PRIMARY KEY (idDepartment)
);
CREATE TABLE City (
idCity uuid ,
name text ,
PRIMARY KEY (idCity)
);
CREATE TABLE Employee (
idEmployee uuid ,
lastName text ,
firstName text ,
PRIMARY KEY (idEmployee)
);
CREATE TABLE toLocated (
Department.Reference uuid ,
City.Reference uuid ,
PRIMARY KEY (Department.Reference,City.Reference)
);
CREATE TABLE toContain (
Department.Reference uuid ,
Employee.Reference uuid ,
PRIMARY KEY (Department.Reference,Employee.Reference)
);
```

solutions such as document-oriented and graph-oriented. Afterward we can consider integrating other big data platforms like HBase, Redis, Neo4j and others.

References

1. Chen, C.L.P., Zhang, C.: Data-intensive applications, challenges, techniques and technologies: a survey on big data. Inf. Sci. **275**, 314–347 (2014)
2. Cattell, R.: Scalable SQL and NoSQL data stores. ACM SIGMOD Rec. **39**(4), 12–27 (2011)
3. Gwendal, D., Gerson, S., Jordi, C.: UMLtoGraphDB: mapping conceptual schemas to graph databases. In: The 35th International Conference on Conceptual Modeling (ER) (2016)
4. Abello, A.: Big data design. In: Proc. of the ACM Eighteenth International Workshop on Data Warehousing and OLAP, Australia (2015)
5. Abdelhedi, F., Brahim,, A.A., Faten, A., Zurfluh, G.: MDA-based approach for NoSQL Databases Modelling, In: International Conference on Big Data Analytics and Knowledge Discovery (DaWaK 2017), Lyon, France, (28–31 Aug 2017)
6. OMG, XML Metadata Interchange (XMI), version 2.1.1, OMG (2007)
7. Chevalier,M., El Malki, M., Kopliku, A., Teste, O., Tournier, R. : Implementing multidimensional data warehouses into NoSQL. In: International Conference on Enterprise Information Systems (ICEIS 2015), Barcelona, Spain (2015)
8. Li, Y., Gu, P., Zhang, C.: Transforming UML Class Diagrams into HBase Based on Meta-model. Information Science, Electronics and Electrical Engineering (ISEEE) (2014)
9. Vajk, T., Feher, P., Fekete, K., Charaf, H.: Denormalizing data into schema-free databases. In: 4th International Conference CogInfoCom. pp. 747–752 (2013)
10. Abdelhedi, F., Brahim, A.A., Atigui, F., Zurfluh, G.: Big Data and knowledge management: how to implement conceptual models in NoSQL systems?. In: 8th International Conference on

Knowledge Discovery, Knowledge Engineering and Knowledge Management (IC3K 2016), Porto, Portugal, (9–11 Nov 2016)
11. Gotti, S., Mbarki, S.: IFVM bridge: a model driven IFML execution. Int. J. Online Biomed. Eng. (iJOE). **15**(4), 111–126 (2019)
12. Miller, J., Mukerji, J.: MDA Guide Version 1.0.1, OMG, (2003)
13. Papajorgjin, P., Pardalos, P.M.: Towards a model-centric approach for developing enterprise information systems. Enterprise Information Systems and Implementing It Infrastructures: Challenges and Issues. IGI Global; 1st edn. pp. 140–158 (2010)
14. Radoslava, S.K., , S.K., Nina, S., Petia, K., Nadejda, B.: Design and analysis of a relational database for behavioral experiments data processing. Int. J. Online Biomed. Eng. (iJOE). **14**(02) (2018), 117–132 (2019)
15. Apache Cassandra, http://cassandra.apache.org/
16. Apache HBase, https://hbase.apache.org/
17. Oualid, B., Saida, F., Amine, A., Mohamed, B.: Applying a model driven architecture approach: transforming CIM to PIM using UML. Int. J. Online Biomed. Eng. (iJOE). **14**(9), 170–181 (2018)
18. Arrhioui, K., Mbarki, S., Erramdani, M.: Applying CIM-to-PIM model transformation for development of emotional intelligence tests platform. Int. J. Online Biomed. Eng. (iJOE). **14**(8), 160–168 (2018)
19. Abadi, D., Boncz, P., Harizopoulos, S.: The design and implementation of modern column-oriented database systems". Found. Trends Databases **5**(3), 197–280 (2012)
20. Angadi, A.B., Angadi, A.B., Gull, K.C.: Growth of New databases & analysis of NOSQL datastores. Int. J. Adv. Res. Comput. Sci. Softw. Eng. **3**(6) (June 2013)
21. OMG, Meta Object Facility (MOF) 2.0 Query/View/Transformation, V1.1 (2011)
22. Acceleo, http://www.eclipse.org/acceleo
23. OMG, UML Infrastructure Final Adopted Specification, version 2.0, September 2003
24. OMG, Meta Object Facility (MOF), version 2.0, OMG (2006)

Data Lake Management Based on DLDS Approach

Mohamed Cherradi, Anass EL Haddadi, and Hayat Routaib

Abstract Over the past few years, big data is at the center of the concerns of actors in all fields of activity. The rapid growth of this massive data requires the question of its storage. Data lakes meet these storage needs, offering data storage without a predefined schema. In this context, a strategy for building a clear data catalog is fundamental for any organization that stores big data, helping to ensure the effective and efficient use of information. Setting up a data catalog in a data lake remains a complicated task and presents a major issue for data managers. However, the data catalog is still essential. This article presents the use of XML and JAXB technologies in the modeling of the data catalog by proposing an approach called DLDS (stands for Data Lake Description Service) and enables to build a central catalog file that allows the users to search, locate, understand and query different data sources stored in the lake.

1 Introduction

The term "data lake" was established by James Dixon founder and former CTO of Pentaho. According to Dixon, Data lake is very efficient compared to data marts, offering as an efficient solution to the problem of data silos linked to data marts: "If you think of a data mart as a store of bottled water, one packaged for easy consumption, the data lake is a great source of water in its natural state." [1] The emergence of the data lake concept in the last five decides has seen increasing interest compared to the data warehouse as shown in Fig. 1, which represents the number of times the terms "data lake" and "data warehouse" searched in the last five years on Google trends.

Data lake is a concept linked to the big data movement, refers to a centralized storage space of the new generation, which allows to store large amounts of data, whatever their format, without time limit and strict schema, this model described as

M. Cherradi (✉) · A. EL Haddadi · H. Routaib
Data Science and Competitive Intelligence Team (DSCI), Applied Science Laboratory ENSAH/UAE, AL Hoceima, Morocco

M. Ben Ahmed et al. (eds.), *Networking, Intelligent Systems and Security*,
Smart Innovation, Systems and Technologies 237,
https://doi.org/10.1007/978-981-16-3637-0_48

Fig. 1 "Data lake" queries in google trends

"schema on-read" [2], and the on-read model schema does not impose any structuring on the data, thus maintaining their original appearance. This flexibility thus ensures that the data is used either for analysis purposes in order to make effective decisions. The data lake can easily turn into a "data swamp" [3], and due to the absence of any enforced schema, the fact of not imposing a well-defined schema for the data during their ingestion presents an obvious risk of quality, reliability, trust, etc. In this context, data governance therefore appears to be one of the major challenges for ensuring the proper functioning of a data lake. He corresponds to a set of processes, rules, standards to ensure the effective and efficient use of information in the data lake. It defines the responsibilities ensuring the quality and security of data within an organization; data governance ingests in their process data catalog who is combined with governance and also ensures reliability of the data. A data lake governance provides assurance that the data is accessible, reliable and of high quality. In contrast, the data catalog authenticates the data stored in the lake using structured workflows.

The content of this paper is organized like this as follows: Identification of the related work is described in Sect. 2. Section 3 presents the challenges of data governance in data lake. Section 4 explains the main factors to consider in the design of the data lake to solve the issue of data swamp. Section 5 presents a formalization of the data lake and our proposition about the architecture of data lake adopted in our approach DLDS. Section 6 gives an overview about technical details associated with our approach. Section 7 presents the results and a critical study. Finally, Sect. 6 concludes the fruit of our work.

2 Related Work

Some people mistakenly think that a data lake is only version 2 of a data warehouse. Although in reality, the last two storage techniques are totally different. In fact, in the literature, there is an extremely large agreement in defining a data lake concept, in which Ref [4] defines data lake as *"big data repositories which store raw data and provide functionality for on-demand integration with the help of metadata descriptions."* On other hand, Ref [5] resumes a data lake like a "massive scalable storage repository that holds a vast amount of raw data in its native format (as is)." Then, it is clear that data lake uses a flat architecture that stores data in their native format, following Ref [6], "Each data entity in the lake is associated with a unique identifier and a set of extended metadata."

In the absence of metadata management, data lake can easily be changed into data swamps [3]. It is important to note that the management of data lakes to a large extent is based on metadata management systems. Indeed, metadata management is a crucial element and key component in the architecture of data lakes. In [7], the authors propose a generic model named MEDAL for managing the metadata of a data lake. This model adopts a modeling of the metadata system based on graphs.

Any data lake design must integrate a metadata storage strategy [8] to allow users to search, locate and understand the datasets that are available in the lake. In this context, our paper aims to propose a comprehensive data catalog that contains metadata about all assets that have been ingested into our data lake. The comprehensive data catalog created based on Java Architecture Xml Binding (JAXB) API, which makes it possible to match an XML document to a set of classes and vice versa via serialization/deserialization operations called marshaling/unmarshaling. The data catalog does not contain the data itself but rather metadata about the data, like its source, owner and other metadata if available.

3 Challenges of Data Catalog in Data Lake

It is no secret that the amount of data is growing exponentially, we often talk about big data. If we are talking about big data today, tomorrow we are going to talk about what we call "Huge Data." That creates the issue of storage. In this context, data lake appeared as an efficient and powerful solution for the storage of big data [9]. A data lake acts as a flexible repository that stores different types of data in their native format, exactly the way it comes in (as is) without any schema defined. One of the keys to this flexibility is the absence of a strict schema imposed on incoming flows [10]. Beyond the storage stage, one of the challenges of the data lake is to be able to facilitate access to data with the objective of carrying out advanced analyzes and meeting business needs. Seeing this observation, a data catalog appeared as one of the best ways to facilitate the management of data lakes in order to avoid their transformation into a data swamp.

One of the key components that must be taken into consideration to effectively manage a data lake is the incorporation of a data catalog [11], to enable business users to search, locate and understand the datasets contained in the lake. This is where the data catalog comes in. A data catalog is a descriptor center where users come to find the data they need to solve a business or technical problem at hand. The catalog contains only the descriptive metadata like the source, authors and title.

The data catalog provides the ability to query all assets stored in the lake. The catalog was also designed to provide a single source of information about the contents of the data lake. It presents an overview of the content contained in the data lake. It is interesting to note that there are several tools for building a data catalog [12–15]. Since most of these tools are paid solutions or are limited in terms of functionality, we have developed our own data catalog which ensures accessibility and an understanding of the different sources of the lakes, and it is indeed a centralized repository. We identify in our paper five major functionalities that must be provided in any data management system that favors the data catalog that ensures governance, such as

Data enrichment (DE) is one of the most important characteristics which has the role of supplementing the data, improving it and structuring it, so that it provides valuable information. Data enrichment is more than correcting errors it also improves data accuracy [16].

Data indexing (DI) consists of organizing a collection of data sources so that we can later easily find the one that interests us according to specific keywords. It is thanks to this functionality that we can simplify and speed up the operations of searching, sorting, joining, etc. It is very useful for structured and unstructured textual data [17].

Data versioning (DV) is a method of managing versions of the same data source. It consists of working directly on the lake's data source, keeping all the previous versions. And therefore, it also supports a continuous evolution of the data, in particular in their schema [18].

Number of descriptive variables (NDV) these variables consist of describing the data stored in the lakes. These variables are varied; the more we have a large number of variables, the more we have a clear vision of the content of the lake. It is very useful to organize the data in a synthetic way [19].

Data accessibility (DA) allows all users to access the different data sources that exist in the lakes and easily navigate through the location of the different data sources. Data accessibility is defined as the range in which the different data sources are available or easily and quickly retrieved [20].

Table 1 presents a comparative and synthetic study of some data management solution that ensures data governance based on data catalog. The main objective of this study is to give a global overview of the functionalities provided by our model compared to others. It appears clearly that our system is complete in terms of these six features that we have proposed (Table 2).

In our opinion, this lack of functionalities clearly demonstrates the complexity of design and implementation. But to design an effective system, it turns out that there are other features that can give great added value. In our approach, we have to concentrate only on the characteristics which seem necessary to us.

Table 1 Features provided by data catalog tools

Systems	DE	DI	DV	NDV	DA
Ckan [12]	✗	✓	✗	✗	✓
Collibra [21]	✗	✓	✓	✗	✓
Erwin [14]	✗	✓	✗	✗	✓
CoreDB [15]	✗	✓	✗	✗	✗
DLDS	✓	✓	✓	✓	✓

Table 2 Overview of the data lake resources

Documents	ID	Metadata
Document 1	Identifiant 1	Metadata 1
Document 2	Identifiant 2	Metadata 2
Document 3	Identifiant 3	Metadata 3
…	…	…
Document n	Identifiant n	Metadata n

4 What to Consider When Designing a Data Lake

Data lake design is a complex task, but necessary process. It involves a functional information system capable of managing all the data (structured, semi-structured and unstructured) of a company in one place, called a "data lake repository."

When designing a data lake, there are a lot of factors to consider in order to ensure that it can do what is required of it. Among these factors to be considered during the database design process, we cite:

1. **Metadata storage**: Metadata provides information about the structures that contain the data lake (it is data about data). A data lake design must include a metadata storage functionality to allow users to locate, search and understand the datasets in the lake. According to [22], the most significant keyword to express data lake is "metadata."
2. **Independent of fixed schema**: With the problems of increasing data volumes and the insufficiency of traditional methods, another approach was born, known by what is called "schema on–read," which allows for data to be inserted without applying any schema (upload data as it arrives without any transformations). With this type of approach, we do not talk about the extract transformation load (ETL) process, and on the other hand, we talk about extract load transformation (ELT). With the absence of a fixed schema (schema on-read), data lake can easily adapt to change, contrariwise schema on write. According to [22], an important keyword used to exprime data lake also is schema "on-read" (or "on-demand").
3. **Support for different data**: The main objective of a data lake is to create a centralized repository that supports a large amount of data sources in different types, whatever the format (structured, semi-structured or unstructured), accessible to a variety of end users like data scientist and data analyst. This flexibility enables organizations to store anything in raw format [23].
4. **Scalability and durability**: A data lake architecture designed to store different data sources for long periods of time. That makes data scalability and durability very important keys to design data lakes effectively, often traditional rdbms we face the limitations of scalability and durability due to their design [24].

A data lake offers other key factors, but in our article, we looked at the necessary elements that must be present in any data lake architecture in order to provide faster query results with low-cost storage.

5 Data Lake Architecture

5.1 Formalization of Our Approach

Before moving to the architecture of a data lake, it is essential to give it a formalization. A data lake has a flat architecture [25] for storing heterogeneous and voluminous data. Each data has a unique identifier and is informed by metadata, as you can see in Table 1.

In our approach, data lake DL is designed as a set of documents, like this:

DL = { document_1, document_2, …, document_n / document € Data Sources }

With: document = {(id, metadata) / id € ID and metadata € Metadata)}

metadata = ∪ (attrib(document))

It is interesting to note that the formalization of our data lake focuses only on description metadata, which allows you to identify, select the document and finally access it. They meet bibliographic objectives by adapting them according to our needs via our data lake descriptor, and they obviously act from our data catalog file.

To keep things simple and clear, for companies to effectively maximize the data stored in data lakes, they need to add context to their data based on policies that classify and identify information in the lake, in order to give an overview of the contents of the lakes. This cannot be achieved unless we have a global descriptor named in our paper **DLDS** (stands for Data Lake Descriptor Service) that can describe and index data. It also gives the ability to trust the data which it needs for its business value or to gain a competitive advantage. Figure 2 shows the different features of the data catalog.

In related research to data quality, most researchers stress the importance of data quality for the efficient construction of the data catalog. In addition to this, they agree that every company needs a data catalog to improve its use of its data.

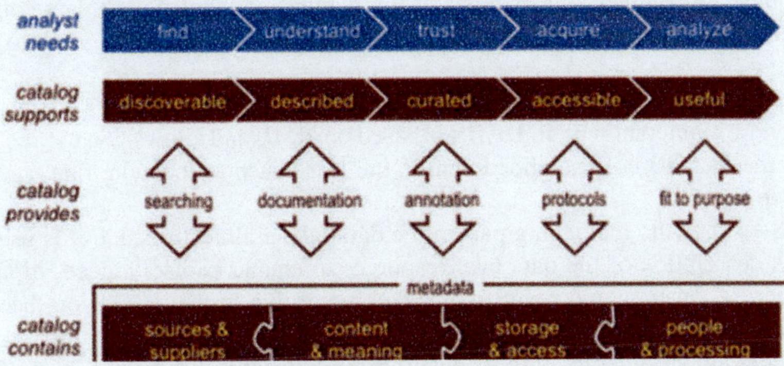

Fig. 2 Main features of the data catalog

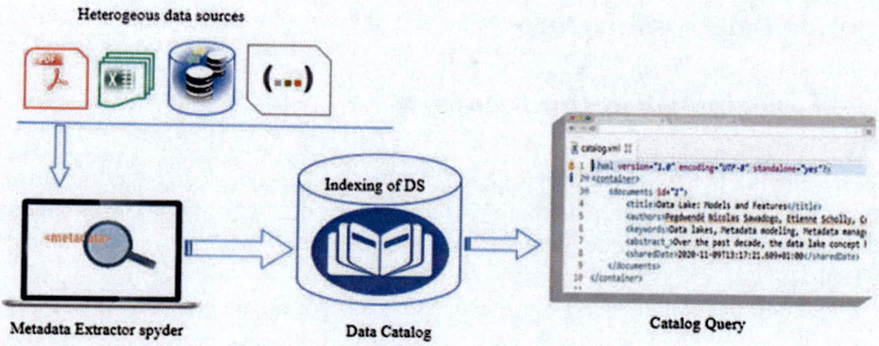

Fig. 3 Data lake architecture

5.2 *Proposed Architecture*

In this paper, we present a new architecture for management different data sources that exists in the lake via data catalog; as you see in Fig, 3, this architecture contains the main steps proposed by our approach, like (1) metadata extractor spyder, which is based on the extraction of the various descriptive attributes of each data source, while using specific API for each data source in order to guarantee the powerful extraction of metadata, (2) data catalog, which allows users to explore different data sources and understand their content via descriptive metadata and (3) catalog query, which allows to query the data stored in the data catalog according to the requested need.

6 Technical Details

In this article, we present a new methodology to build a data catalog to provide users with a guide to discover and understand the content of the different data sources in the lake, with the main objective of preventing data lake from turning into data swamp.

The proposed service presents an interface to users and shows how to use the data lake and how to interact with it. **DLDS** is based on XML and allows you to precisely describe the metadata of each document in the lakes such as location, title, authors and keywords.

DLDS ensures the messaging part in the data lake architecture, and it is used to provide a description of the data lake resources to enable its use. Indeed, to allow a customer to consume and perform analyzes according to their needs on the data sources that exist in the lake, and the latter needs a detailed description of the service before being able to interact with it. A DLDS provides this description in an XML document. DLDS plays an important role in the data lake architecture by providing

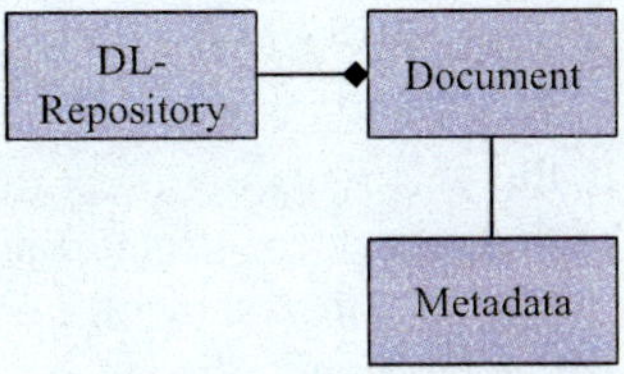

Fig. 4 Overview of the resources that make up the data lake

the description part: It contains all the information necessary to invoke the service it describes.

As described in Sect. 4.2, the data lake has two essential parts (data and metadata), architectured as shown in Fig. 4.

In order to efficiently organize all these documents stored in the lake and benefit from the wealth of these data, DLDS has been designed to meet this need. It is an XML document which describes the lake documents independently of any language. This shows the flexibility of our approach. Using DLDS as an xml-based data catalog, it allows tools from different systems, platforms and languages to use the content of a WSDL to generate code to consume different source data in the lake.

To respond to the approach we have designed, we based on the jaxb specification, as shown in Fig. 5.

The major objective of relying on this architecture is to facilitate the construction of the data catalog, by converting an object into an XML document. This document is linked to an XML schema which gives an overview of our data container, and it describes all the elements necessary to interact with our data lake. The description of this schema file is described below.

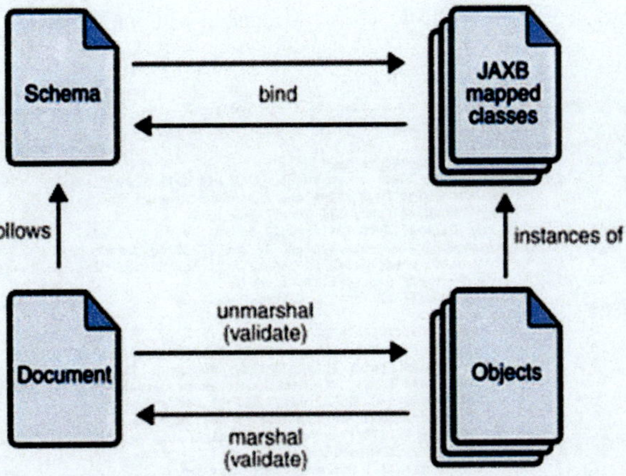

Fig. 5 Jaxb specification architecture

7 Results and Discussion

In this section, we present the result of our DLDS approach, which gives birth to a new approach for structuring the different data sources existing in the lake via data catalog. Figure 6 shows the data catalog we designed, menu with its associated XML schema to standardize it.

In fact, this schema defines the existing contract between users and our data lake. To interact with data lake, we will use the catalog that we have generated based on the descriptive metadata of each data source. Figure 7 shows an extract of the data catalog that we built.

This section also aims to present the critical aspect of our approach in the form of a discussion. It is true that our approach presents an original idea and makes a

```xml
<xs:element name="containerDL" type="containerDL"/>
<xs:element name="document" type="document"/>

<xs:complexType name="containerDL">
  <xs:sequence>
    <xs:element name="data-sources" type="document" minOccurs="0" maxOccurs="unbounded"/>
  </xs:sequence>
</xs:complexType>

<xs:complexType name="document">
  <xs:sequence>
    <xs:element name="title" type="xs:string" minOccurs="0"/>
    <xs:element name="authors" type="xs:string" minOccurs="0"/>
    <xs:element name="keywords" type="xs:string" minOccurs="0"/>
    <xs:element name="sharedDate" type="xs:dateTime" minOccurs="0"/>
    <xs:element name="updatedDate" type="xs:dateTime" minOccurs="0"/>
    <xs:element name="path" type="xs:string" minOccurs="0"/>
    <xs:element name="publisher" type="xs:string" minOccurs="0"/>
    <xs:element name="version" type="xs:string" minOccurs="0"/>
```

Fig. 6 Generated XML schema associated with the DLDS

```xml
data-catalog.xml   schema-catalog.xsd   ContainerDL.java   Document.java   Serialisation.java
<?xml version="1.0" encoding="UTF-8" standalone="yes"?>
<containerDL>
    <data-sources identifiant="1">
        <title>Data lake: a new ideology in big data era</title>
        <authors>Pwint Phyu Khine and Zhao Shun Wang</authors>
        <keywords>Data Lake, Big Data</keywords>
        <sharedDate>2018-06-18T12:00:17.534+01:00</sharedDate>
        <updatedDate>2018-06-18T12:00:17.536+01:00</updatedDate>
        <path>D:\Cherradi\Research-papers\Articles\Dl-New-Ideology-In-Big-Data-Era.pdf</path>
        <publisher>EDP Sciences</publisher>
        <version>original - not modified</version>
    </data-sources>
    <data-sources identifiant="2">
        <title>A Mapping Study about Data Lakes: An Improved Definition and Possible Architectures</title>
        <authors>Julia Couto, Olimar Borges, Duncan D. Ruiz, Rafael Prikladnicki and Sabrina Marczak</authors>
        <keywords>Data lakes, Big Data, Literature review, PRISMA, Hadoop.</keywords>
        <sharedDate>2019-07-18T12:00:17.535+01:00</sharedDate>
        <updatedDate>2019-07-18T12:00:17.537+01:00</updatedDate>
        <path>D:\Cherradi\Research-papers\Articles\Dl-New-Ideology-In-Big-Data-Era.pdf</path>
        <publisher>IJSEKE</publisher>
        <version>original - not modified</version>
    </data-sources>
    <data-sources identifiant="3">
    <data-sources identifiant="4">
    <data-sources identifiant="5">
</containerDL>
```

Fig. 7 Extract from the data catalog that we built

strong contribution to the scientific literature, especially in terms of the structuring of heterogeneous data stored in the lake via our approach called DLDS. But this does not prevent us from saying that our article is complete, on the other hand perhaps missing the following points, which can be considered as the subject of future research:

- Demonstration of metastore link to an available XML standards and semantic vocabularies.
- Connectors to storage space, such as: HDFS, Hive, Cassandra, MongoDB or others as needed, instead of an XML descriptor.
- Leveraging the power of machine learning to enrich metadata extraction process.
- etc.

8 Conclusion

Improving the quality of data organization via data catalog is becoming a brilliant technique in the world of heterogeneous data management. Indeed, the construction of the data catalog with a contract in the form of an XML schema considered to be a reference architecture to fully understand and interact with the different data sources exists in the lake.

This paper should not be interpreted as finalized. But rather it is the starting point or the birth of a new approach that deserves to be complemented by other work whose objective is to effectively manage data lake.

References

1. Dixon, J.: Pentaho, Hadoop, and Data Lakes | James Dixon's Blog (2010). https://jamesdixon.wordpress.com/2010/10/14/pentaho-hadoop-and-data-lakes/. Accessed 10 Feb 2021
2. Mathis, C.: Data lakes. Datenbank-Spektrum **17**, 1–5 (2017). https://doi.org/10.1007/s13222-017-0272-7
3. Suriarachchi, I., Plale, B.: Crossing Analytics Systems: A Case for Integrated Provenance in Data Lakes (2016)
4. Hai, R., Geisler, S., Quix, C.: Constance: an intelligent data lake system. In: Proceedings of the 2016 International Conference on Management of Data. Association for Computing Machinery, USA, pp. 2097–2100, New York, NY, (2016)
5. Miloslavskaya, N., Tolstoy, A.: Big data, fast data and data lake concepts. Procedia Comput. Sci. **88**, 300–305 (2016). https://doi.org/10.1016/j.procs.2016.07.439
6. Rangarajan, S., Liu, H., Wang, H., Wang, C.-L.: Scalable architecture for personalized health-care service recommendation using big data lake. In: Beheshti, A., Hashmi, M., Dong, H., Zhang, W.E. (eds.) Service Research and Innovation, pp. 65–79. Springer International Publishing, Cham (2018)
7. Scholly, E., Sawadogo, P.N., Favre, C., Ferey, E., Loudcher, S., Darmont, J.: Système de métadonnées d'un lac de données : modélisation et fonctionnalités (2019)
8. Sawadogo, P.N., Darmont, J.: On data lake architectures and metadata management. J. Intell. Inf. Syst. **56**, 1–24 (2021). https://doi.org/10.1007/s10844-020-00608-7
9. Khine, P., Wang, Z.: Data lake: a new ideology in big data era. ITM Web Conf. **17**, 03025 (2018). https://doi.org/10.1051/itmconf/20181703025

10. Sawadogo, P.N., Scholly, E., Favre, C., Ferey, E., Loudcher, S., Darmont, J.: Metadata Systems for Data Lakes: Models and Features (2019)
11. Chen, M.: Why Data Lakes Need a Data Catalog (2019). https://blogs.oracle.com/bigdata/why-data-lakes-need-a-data-catalog. Accessed 15 Feb 2021
12. ckan ckan. In: Data Cat. https://ckan.org/. Accessed 15 Feb 2021
13. Collibra Data Catalog on-demand demo. In: Data Manag. Data Cat. https://www.collibra.com/download/data-catalog-demo. Accessed 15 Feb 2021
14. Erwin Data Catalog Free Demo. In: Erwin Inc. https://erwin.com/erwin-data-catalog-free-demo/. Accessed 15 Feb 2021
15. Beheshti, A., Benatallah, B., Nouri, R., , V., Xiong, H., Zhao, X.: CoreDB: a Data Lake Service, pp. 2451–2454 (2017)
16. Azad, S., Wasimi, S., Ali, A.B.M.: Business Data Enrichment: Issues and Challenges, pp. 98–102 (2018)
17. Singh, K., Paneri, K., Pandey, A., Gupta, G., Sharma, G., Agarwal, P., Shroff, G.: Visual Bayesian Fusion to Navigate a Data Lake (2016)
18. Hellerstein, J.M., Sreekanti, V., Gonzalez, J.E., Dalton, J., Dey, A., Nag, S., Ramachandran, K., Arora, S., Bhattacharyya, A., Das, S., Donsky, M., Fierro, G., She, C., Steinbach, C., Subramanian, V., Sun, E.: Ground: A Data Context Service. 12
19. Yellapu, V.: Descriptive statistics. Int J. Acad Med **4**, 60 (2018). https://doi.org/10.4103/IJAM.IJAM_7_18
20. Pipino, L.L., Lee, Y.W., Wang, R.Y.: Data quality assessment. Commun ACM **45**, 211–218 (2002). https://doi.org/10.1145/505248.506010
21. Collibra Trusted data for your entire organization. In: Collibra. https://www.collibra.com/. Accessed 16 Feb 2021
22. Chihoub, H., Madera, C., Quix, C., Hai, R.: Architecture of Data Lakes. pp. 21–39 (2020)
23. Anne Laurent Dominique Laurent Cédrine Madera (2020) Data Lakes | Wiley Online Books. book/https://onlinelibrary.wiley.com/doi/10.1002/9781119720430. Accessed 16 Feb 2021
24. Bhawkar, A.: A Comparative Study to Analyze Scalability, Availability and Reliability of HBase and MongoDB (2018). In: ResearchGate. https://www.researchgate.net/publication/330675690_A_comparative_study_to_analyze_Scalability_Availability_and_Reliability_of_HBase_and_MongoDB. Accessed 16 Feb 2021
25. Alrehamy, H., Walker, C.: Personal Data Lake With Data Gravity Pull (2015)

Evaluation of Similarity Measures in Semantic Web Service Discovery

Mourad Fariss, Naoufal El Allali, Hakima Asaidi, and Mohamed Bellouki

Abstract The semantic Web service discovery is the process of finding services that could potentially meet the consumer requirements by choosing between several services. The matchmaking between the consumer request and the semantic Web services is the main task of any semantic Web service discovery mechanism. To make this, many works use similarity measures to choose a similar semantic Web service and the consumer's request. In this paper, we assess similarity measures to help us define their problems and determine which measure is most appropriate for Web service discovery approaches. For this matter, we used a test collection of semantic Web services with different domains. The results of our evaluation show the weakness of similarity measures to have an interesting efficiency.

1 Introduction

Service-oriented architecture (SOA) is an architectural style for designing distributed applications using functionality implemented by third-party providers. In a SOA, the service consumer satisfies its specific needs by using services offered by service providers. One concrete technology used for implementing SOA is Web service.

According to the W3C, a Web service is defined as "a software system designed to support interoperable machine-to-machine interaction over a network" [1]. Its interface can be described as a Web service description language (WSDL) document that contains structured information about the Web service's location, its offered operations and the input/output parameters. Interface descriptions (WSDL documents) enable Web services to be discovered, used by applications or other Web services and composed into new more complex Web services.

The Web service architecture consists of three entities: the service provider, the service registry and the service consumer. The service provider simply creates or offers the Web service. The service provider describes the Web service in a standard

M. Fariss (✉) · N. El Allali · H. Asaidi · M. Bellouki
Laboratory MASI, FPN, UMP, Nador, Morocco
e-mail: m.fariss@ump.ac.ma

M. Ben Ahmed et al. (eds.), *Networking, Intelligent Systems and Security*,
Smart Innovation, Systems and Technologies 237,
https://doi.org/10.1007/978-981-16-3637-0_49

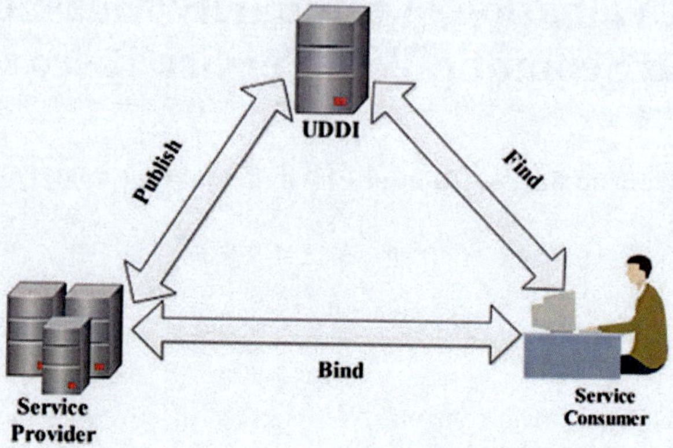

Fig. 1 Web service standard model

format, often in XML, and publishes it in a central service registry. The service registry contains additional information about the service provider, such as the address and contact details of the company providing the services, as well as technical details about the service. The service consumer extracts the information from the registry and uses the resulting service description to bind and call the Web service (Fig. 1).

The Universal Description Discovery and Integration (UDDI) registry was proposed for the publication of services. The service consumer can access this registry in order to find the best service that meets their needs. Since UDDI uses syntactic information and does not use semantic information, the search for the most appropriate service is limited in the sense that consumers cannot make requests asking for specific desirable properties such as quality of service parameters related to reliability, performance, security, response time, etc.

As long as several researches have adopted the semantic description of Web services [2, 3] [4], a new problem has appeared to measure the degree of similarity between the service consumer and the Web services registered in the database. In this paper, we propose an evaluation of the most used similarity measures in Web service discovery. Section 2 presents the problem of Web service discovery. Section 3 cites the related works for the similarity measures on the Web service discovery. Session 4 describes the similarity measures studied. The evaluation of the similarity measures studied is represented in Session 5. Finally, we present the conclusion and future works.

2 Web Service Discovery

The Web service discovery process is carried in three stages. Firstly, the service provider publishes the Web service in public repositories by registering the Web service description file written in WSDL [5]. In the second step, the service consumer sends a request with the requirements in predefined format to the Web service registry.

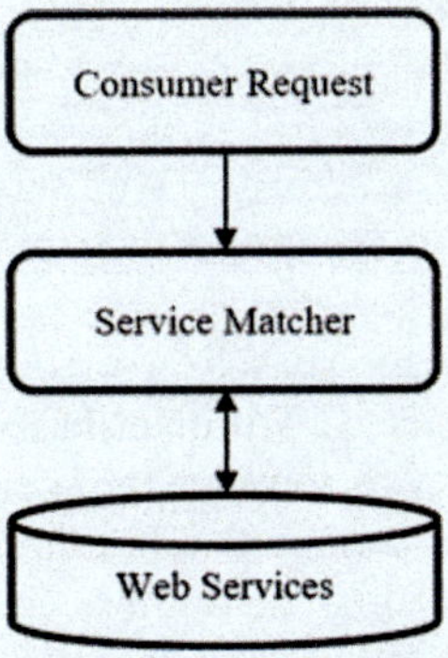

Fig. 2 Web service discovery. Since it has been observed that the search capabilities of the current discovery mechanisms are limited because they are mostly based on keyword matching, the service consumer searches the Web service in the UDDI register and submits the requirements with keywords. This requires a different mechanism, which includes the location of Web services based on the features they provide. Semantic technologies in Web service play an important role in the seamless integration of different services based on different terminologies

Web service matcher finds the corresponding Web service candidates to the consumer request. Finally, the selection and invocation of one of the retrieved Web services (Fig. 2).

The combination of the theory of semantic Web and Web services gives rise to what is known as semantic Web services. There are several approaches to add semantic information to services such as OWL-S [6], WSDL-S [7] and WSMO [8]. The term "ontology" is used in many fields such as artificial intelligence, software engineering, information architecture and many others. Ontology is a structural framework for organizing the representation of information or knowledge about the world or part of a world, which describes a set of concepts, its properties and the relationship between them. Ontology is a "formal specification of a shared conceptualization," providing a shared vocabulary, a taxonomy for a particular domain that defines objects, classes, its attributes and their relationships.

An ontology provides semantics to Web services and its resources and can greatly improve discovery mechanisms. To implement the vision of the semantic Web, the researchers proposed several languages, algorithms and architectures. Semantic Web services are part of the semantic Web because it uses markings that make the data machine readable [9]. Semantic Web services use standards such as OWL-S, WSDL-S, WSMO, OWLS-LR and others.

3 Related Works

Several works are concentrated on semantic Web service discovery using similarity measures [10–13]. In this section, we give some studies that use the three evaluated similarity measures.

The Web service discovery method based on semantic and clustering was proposed [14], and the similarity measure used a WordNet-based semantic similarity measurement. It uses the cosine similarity to compute the score of similarity on this method. The cosine measure was used to compute similarities and retrieve relevant WSDL service descriptions [15]. To do this, the authors create the service request vector according to the domain ontology and then project the description vectors and the request vector. Some researchers were present an approach for automated service discovery based on ontology [16]; this approach adds a filtering method based on logical reasoning before using the cosine similarity in the matching algorithm for the Web services filtered. In [17], Wu and Palmer's similarity was used to compute the similarity between the document word vectors obtained by analyzing the OWL-S Web service documents, before the LDA clustering in the Web service discovery process. The same measure of similarity was used to compute the similarity between words related to the requested query and service parameters on the proposed approach [18]. The Web service discovery approach combines the LDA clustering and the k-Medoids to reduces search space for Web service discovery. The Jaccard coefficient was used to calculate the similarity between Web services in a method to improve the Web service discovery using the user's experiences with similar queries [19]. Based on the experimental results of measuring the performance of similarity metrics for text information retrieval provided by Cohen et al. [20], the authors of [21] selected the top-performing ones to build the OWLS-MX variants on the proposed approach of Web service discovery. These symmetric token-based string similarity measures are the cosine similarity, the extended Jacquard similarity, the intentional loss of information and the Jensen–Shannon information divergence.

Researchers are still finding a difficulty to choose the most appropriate similarity measure to solve the Web service discovery problem. These measures aim to measure the similarity between the consumer request and the Web service available in the Web services registry. This paper provides a tool to evaluate and compare the similarity measures most used in the literature, to help researchers choose the most satisfactory of them for Web services defined in OWLS.

4 Similarity Measures

Semantic similarity between concepts is a method of measuring semantic similarity or the semantic distance between two concepts according to a given ontology. In other words, semantic similarity is used to identify concepts that have common "characteristics". Although a human does not know the formal definition of the connection between concepts, he can judge the relationship between them. For example, a young child may say that "apple" and "peach" are more related than "apple" and "tomatoes". These few concepts are interrelated, and their definition of structure is formally called the "is-a" hierarchy. Semantic similarity methods are used intensively for most applications of intelligent systems with knowledge-based and semantic information retrieval (identify an optimal match between query terms and documents) [22] [23], the ambiguity of the senses [24] and bioinformatics [10]. Semantic similarity and semantic relatedness [25] are two related words, but the semantic similarity is more specific than relatedness and can be considered as a type of semantic relatedness. For example, "Student" and "Teacher" are related terms, which are not similar. All similar concepts are related, but the reverse is not always true.

Semantic similarity and semantic distance are defined in the reverse direction. Let C_1 and C_2 be two concepts belonging to two different nodes n_1 and n_2 in a given ontology, and the distance between the nodes determines the similarity between these two concepts. Both n_1 and n_2 can be considered as an ontology that contains a set of synonymous concepts and consequently. Two terms are synonymous if they are in the same node, and their semantic similarity is maximized.

The use of ontologies to represent the concepts or terms (humans or computers) that characterize different communication sources is useful in making knowledge comprehensible. Furthermore, it is possible to use different ontologies to represent the concepts for each source of knowledge. Then, the mapping or comparing concepts based on the same or different ontologies ensures the exchange of knowledge between concepts. The mapping must find the similarity between terms or concepts based on domain-specific ontologies. The similarity between concepts or entities can be identified if they share common attributes or if they are related to other semantically related entities in an ontology [26].

The algorithm used to compute the similarity between two concepts, for as the Web service and the consumer request, is presented as follows:

Algorithm: Semantic Similarity Score Computation

Input: WSs and CR // *WSs is the Web services, and CR is the consumer request*
Output: Semantic Similarity Score

2. Convert CR to vector Vr.
3. **for** $i = 1$ to n **do**
//n is the number of web services
4. compute Measure_Similarity(V[i], Vr).
//Measure_Similarity is replaced by the measure tested
5. **end**

To calculate the similarity score between Web services and the consumer request, first, we convert the Web services and the customer request to a vector, and then, we calculate the similarity score according to the wanted similarity measure for each Web service.

In the following, the definition of the similarity measures evaluated in this paper.

4.1 The Cosine Similarity

In the process of semantic Web service discovery, it is very important to find out the semantic Web service which can meet the consumer's needs in specific contexts through the specific keyword query based on the keyword of the text. The cosine similarity algorithm uses the cosine of angle between two different vectors in the vector space to determine the difference on content between two vectors, it is mainly based on consumer's own preference and the level of difference between provided Web services, determine a semantic Web service that ultimately conforms to consumer's context, and then, feed the semantic Web service back to the consumer to meet the different needs of the consumer in different contexts (Fig. 3).

In two dimensions, the angle cosine of the vector a and the vector b is calculated as follows:

$$\begin{aligned}\cos(\alpha) &= \frac{a \cdot b}{\|a\| \times \|b\|} \\ &= \frac{(x_1, y_1) \cdot (x_2, y_2)}{\sqrt{x_1^2 + y_1^2} \times \sqrt{x_2^2 + y_2^2}} \\ &= \frac{x_1 x_2 + y_1 y_2}{\sqrt{x_1^2 + y_1^2} \times \sqrt{x_2^2 + y_2^2}}\end{aligned} \tag{1}$$

In the case of multidimensional, the angle cosine of the vector a and the vector b is calculated as follows:

$$\cos(\alpha) = \frac{\sum_{i=1}^{n} (x_i \times y_i)}{\sum_{i=1}^{n} (x_i)^2 \times \sum_{i=1}^{n} (y_i)^2} \tag{2}$$

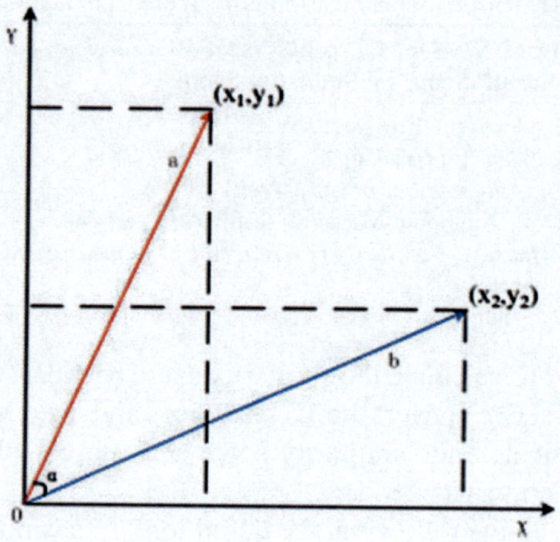

Fig. 3 Angle cosine of the vector **a** and the vector **b**

Since the cosine similarity algorithm focuses on the difference of vectors' directions, it is not sensitive to their size, so it is mainly used by consumers to determine whether the content of the Web service is interested.

4.2 Wu and Palmer's Similarity Measure

Path-based similarity measure usually utilizes the information of the shortest path between two concepts, of the generality or specificity of both concepts in ontology hierarchy and of their relationships with other concepts. Wu and Palmer [27] present a similarity measure finding the most specific common concept that subsumes both of the concepts being measured. The path length from the most specific shared concept is scaled by the sum of IS-A links from it to the compared two concepts.

$$S_{W\&P}(C_1, C_2) = 2N/(N_1 + N_2 + 2N) \tag{3}$$

where N_1 and N_2 are the distance that separates, respectively, the concept C_1 and C_2 from the specific common concept, and N is the distance which separates the closest common ancestor of C_1 and C_2 from the root node.

4.3 Jaccard Index

The Jaccard index, also known as the Jaccard similarity coefficient, is a statistic used for comparing the similarity and diversity of sample sets. The Jaccard coefficient measures similarity between sample sets and is defined as the number of common objects divided by the total number of objects, minus the number of common objects.

$$g^{(J)}(x_a, x_b) = x_a^T x_b/(\|x_a\|_2^2 + \|x_b\|_2^2 - x_a^T x_b) \tag{4}$$

5 Experimentation and Results

In this section, we will present the experimentation results done on the similarity measures evaluation, using the well-known service repository of OWLS-TC v4.0 test collection. There are 1083 semantic Web services written in OWL-S 1.1 and 42 requests. There are nine service domains: education, medical, food, travel, communication, economy, weapon, geography and simulation. Table 1 shows the details of OWLS-TC v4.0.

Table 1 Details of OWLS-TC v4.0

Domains rainfall data	Number of services	Number of requests
Education	286	6
Medical care	73	1
Food	34	1
Travel	197	6
Communication	59	2
Economy	395	12
Weapon	40	1
Geography	60	10
Simulation	16	3

Some services appear in more than one category. Therefore, the number of services is 1083 if we consider just the first occurrence of each service and 1140 if we consider repetitions across different categories.

To evaluate the three measures of similarity cited below, we used five consumer's requests and across about 1000 Web services. For each consumer, we calculated the degree of similarity with different similarity measures, and the execution time between each request of the Web services returned during the search. In order to judge the effectiveness of similarity measures, we compared the similarity measures with the number of Web service of each domain. The following Table 2 shows the samples chosen for the consumer request:

The following Table 3 represents some results of similarity measures for the consumer 1 request "*car_price_service*" with different services from all domains of dataset.

From the results of the similarity measurements tested, we can conclude the following points:

- The time to calculate similarity measures remains acceptable for all the similarity measures tested.
- We cannot say that there is an agreement between the different similarity measures, which will influence the Web service discovery process based on their similarity measures (Fig. 4).

Table 2 Consumer's request

Consumer request	Request name	Domains rainfall data
Consumer 1	car_price_service	Travel
Consumer 2	hospital_investigating_service	Medical care
Consumer 3	book_price_service	Education
Consumer 4	getLocationOfUSCity	Geography
Consumer 5	preparedfood_price_service	Food

Table 3 Similarity result for consumer 1

Web services	Domains rainfall data	Similarity measure					
		Cosine similarity	Time (μs)	Wu and Palmer's similarity	Time (μs)	Jaccard index	Time (μs)
AcademicBookNumberOrISBNSearch	Education	30	1	15.7	3	13.9	1
HealthInsurance_service	Medical care	73.7	1	10.4	3	56.3	2
corporation_apple_service	Food	11	3	30.8	1	66.4	1
_internalchange_EarthSystemservice	Travel	100	0	30.1	2	100	2
_filmvideomediaDiscoveryChannel_service	Communication	33.5	1	89.3	2	43.4	3
4wheeledcar_price_service	Economy	88.6	1	78.8	2	93.2	1
governmentweapon_funding_service	Weapon	23.4	2	12.4	4	28.4	1
GetCoordinatesOfAddress	Geography	100	1	88.4	2	23.4	0
greenLight_to_off	Simulation	34.5	2	45.7	3	67.7	1

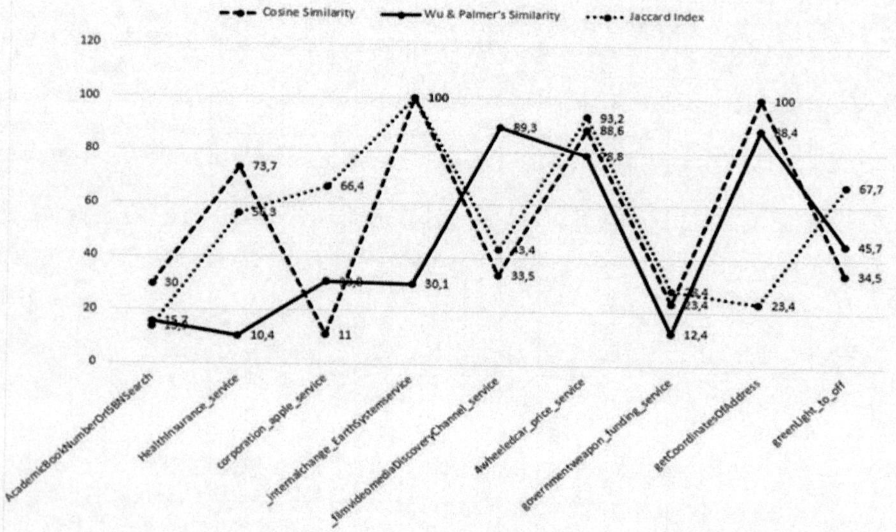

Fig. 4 Similarity measures of web services for consumer 1

- For example:
- For Consumer 1, the degree of similarity of the service "*HealthInsurance_service*" is 73.7% for the cosine similarity and 56.3% for the Jaccard index, but for the Wu & Palmer's Similarity, it is 10.4%
- For Consumer 1, the degree of similarity of the "*GetCoordinatesOfAddress*" service is 100% for cosine similarity and 88.4% for the Wu & Palmer's Similarity, but for the Jaccard index, it is 23.4%,

- The number of Web services discovered varies depending on the measure of similarity used, because of the degree of similarity of each service and the consumer's request (Fig. 5).

Fig. 5 Number of discovered services for consumer 1

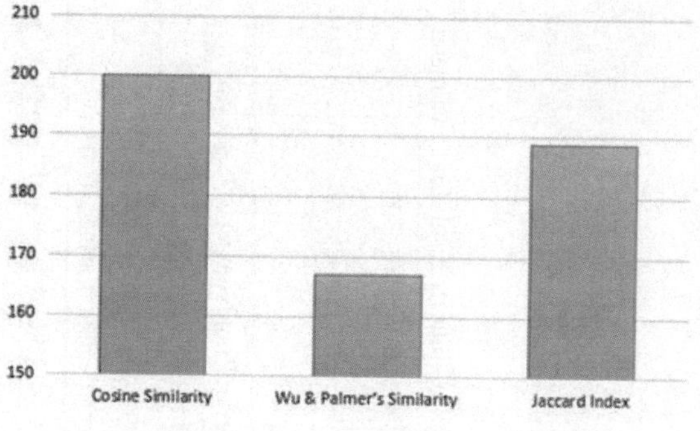

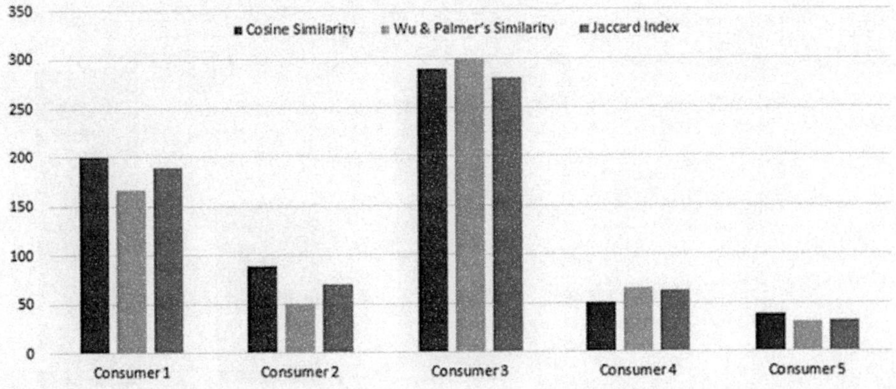

Fig. 6 Number of web services discovered for each consumer

- The results obtained with Consumer 1 for the discovered Web services based on similarity measures remain reliable for five different consumer's requests tested in different domains (Fig. 6).
- The results show that all the similarity measures tested do not give effective results. The similarity values obtained do not always meet the consumer's request. Furthermore, the Web services discovered using the tested similarity measures stay far away from the number of Web services in each domain.

To measure the accuracy of each similarity measure, we calculate the precision and the recall. Precision is the number of correct results divided by the number of all returned results. Recall is the number of correct results divided by the number of results that should have been returned as shown in the following equations:

$$\text{Precicion}(P) = \text{Correct Relevant Services Found}/\text{Total Relevant Services Found} \quad (5)$$

$$\text{Recall}(R) = \text{Correct Relevant Services Found}/\text{Relevant Services should be Found} \quad (6)$$

- The recall and the precision presented in Fig. 7 show the weakness of the three studied similarity measures, namely the cosine similarity, the Wu & Palmer's Similarity and the Jaccard index.

6 Conclusion and Perspectives

In this paper, we proposed an evaluation of the three most used similarity measures in the semantic Web service discovery. Indeed, our objective was to give the mean to compare and evaluate similarity measures with existing measures. The studied

Fig. 7 Comparison of recall and precision of similarity measures

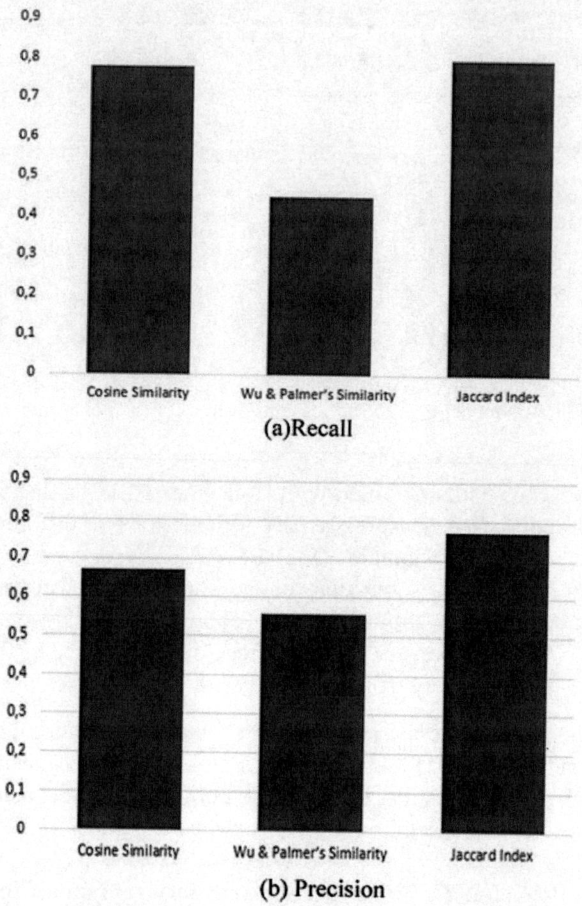

similarity measures show a weakness of decision either at the level of degree of similarity and also at the level of precision and recall in the Web service discovery.

In addition, we aim to

- Suggest solutions to solve the problem of similarity measures.
- Study other measures of similarity that exist and integrate them into this evaluation.
- Introduce the concept of quality of service at the task of the Web service discovery.

References

1. Moreau, J.J., Chinnici, R., Ryman, A., Weerawarana, S.: Web services description language (WSDL) version 2.0 part 1: core language. Candidate Recomm. **W3C**, 7 (2006)

2. Fethallah, H., Ismail, S.M., Mohamed, M., Zeyneb, T.: "An outranking model for web service discovery." Int. Conf. Math. Inf. Technol. (ICMIT) **2017**, 162–167 (2017)
3. Fariss, M., El Allali, N., Asaidi, H., Bellouki, M.: Review of Ontology Based Approaches for Web Service Discovery. Springer International Publishing (2019)
4. Malburg, L., Klein, P., Bergmann, R.: "Using Semantic Web Services for AI-Based Research in Industry 4.0," *arXiv Prepr. arXiv2007.03580* (2020)
5. Christensen, E., Curbera, F., Meredith, G., Weerawarana, S. et al.: "Web Services Description Language (WSDL) 1.1." Citeseer (2001)
6. Martin, D., et al.: "OWL-S: Semantic markup for web services," *W3C Memb. Submiss. 22*(4) (2004)
7. Akkiraju, R., Farrell, J., Miller, J.A., Nagarajan, M., Sheth, A.P., Verma, K.: "Web Service Semantics-wsdl-s," (2005)
8. Roman, D., et al.: Web service modeling ontology. Appl. Ontol. **1**(1), 77–106 (2005)
9. Malaimalavathani, M., Gowri, R.: "A survey on semantic web service discovery." Int. Conf. Inf. Commun. Embed. Syst. ICICES **2013**, 222–225 (2013)
10. Ehsani, R., Drabløs, F.: TopoICSim: A new semantic similarity measure based on gene ontology. BMC Bioinf. **17**(1), 1–14 (2016)
11. Wu, J., Chen, L., Zheng, Z., Lyu, M.R., Wu, Z.: Clustering web services to facilitate service discovery. Knowl. Inf. Syst. **38**(1), 207–229 (2014)
12. Alwasouf, A.A., Deepak, K.: "Research challenges of web service composition, software engineering," *Adv. Intell. Syst. Comput.*, vol. 731, 2019.
13. El Allali, N., Fariss, M., Asaidi, H., Bellouki, M.: "Semantic web services composition model using ant colony optimization," *4th Int. Conf. Intell. Comput. Data Sci. ICDS 2020* (2020)
14. Wen, T., Sheng, G., Li, Y., Guo, Q.: "Research on web service discovery with semantics and clustering," *Proc.—2011 6th IEEE Jt. Int. Inf. Technol. Artif. Intell. Conf. ITAIC 2011, 1*, 62–67 (2011)
15. Paliwal, A.V., Adam, N.R., Bornhövd, C.: "Web service discovery: adding semantics through service request expansion and latent semantic indexing," *Proc.—2007 IEEE Int. Conf. Serv. Comput. SCC 2007*, no. Scc, pp. 106–113 (2007)
16. Fang, M., Wang, D., Mi, Z., Obaidat, M.S.: Web service discovery utilizing logical reasoning and semantic similarity. Int. J. Commun. Syst. **31**(10), 1–13 (2018)
17. Zhao, H., Chen, J., Xu, L.: "Semantic web service discovery based on LDA clustering," In *Web Information Systems and Applications*, pp. 239–250 (2019)
18. Jalal, S., Yadav, D.K., Negi, C.S.: "Web service discovery with incorporation of web services clustering," *Int. J. Comput. Appl.*, 0(0), 1–12 (2019)
19. Nayak, R.: Data mining in Web services discovery and monitoring. Int. J. Web Serv. Res. **5**(1), 63–81 (2008)
20. Cohen, W.W., Ravikumar, P., Fienberg, S.E.: "A comparison of string distance metrics for name-matching tasks," *Proc. IJCAI-2003 Work. Inf. Integr. Web* (2003)
21. Klusch, M., Fries, B., Sycara, K.: "Automated semantic web service discovery with OWLS-MX," in *Proceedings of the Fifth International Joint Conference on Autonomous Agents and Multiagent Systems—AAMAS '06*, p. 915 (2006)
22. Budan, I.A., Graeme, H.: Evaluating wordnet-based measures of semantic distance. Comutational Linguist. **32**(1), 13–47 (2006)
23. Sim, K.M., Wong, P.T.: "Toward agency and ontology for web-based information retrieval," *IEEE Trans. Syst. Man, Cybern. Part C (Applications Rev.*, vol. 34, no. 3, pp. 257–269, 2004.
24. Patwardhan, S.: Incorporating Dictionary and Corpus Information into a Context Vector Measure of Semantic Relatedness. University of Minnesota, Duluth (2003)
25. Gracia, J., Mena, E.: "Web-based measure of semantic relatedness," in *International Conference on Web Information Systems Engineering*, pp. 136–150 (2008)

26. Li, Y., Bandar, Z.A., McLean, D.: An approach for measuring semantic similarity between words using multiple information sources. IEEE Trans. Knowl. Data Eng. **15**(4), 871–882 (2003)
27. Palmer, M.: "Verb semantics and lexical zhibiao W u," *32nd Annu. Meet. Assoc. Comput. Linguist.*, pp. 133–138 (1994)

Knowledge Discovery for Sustainability Enhancement Through Design for Relevance

Abla Chaouni Benabdellah, Asmaa Benghabrit, Imane Bouhaddou, and Kamar Zekhnini

Abstract Data is collected and accumulated at a dramatic pace from many different resources and services across a wide variety of fields, particularly for industrial companies. Hence, to capture long-term revenue, sustainable assessment and help humans extract valuable knowledge from the rapidly increasing amounts of digital data, companies must adopt a new generation of analytical theories and methods. A well-known fundamental task is the use of knowledge discovery in databases (KDD). In this respect, the aim of this paper is to adopt the KDD process to extract information from data that are generated through the use of different design for X techniques named Design for Relevance. Since we are looking to find a structure for sustainability enhancement in an unlabeled dataset related to collaborative product development, clustering is the most appropriate data mining (DM) task in our context. However, with the modified applications for various domains, several clustering algorithms have been provided. This multiplicity makes it difficult for researchers to define both the appropriate algorithms and the appropriate measures of validity. To deal with these issues, a related work focusing on comparing various clustering algorithms for real industrial datasets is presented. After that, for Design for Relevance dataset and by following the KDD process, several clustering algorithms were implemented and compared using several internal validity indices. In addition, we highlighted the best performing clustering algorithm that gives us successful clusters for the dataset to achieve improvement in sustainability.

1 Introduction

Over the past decade, scientific consensus on the importance of sustainability has tended to exist, both at the organizational level [1] and at the national and global level

---

A. C. Benabdellah (✉) · I. Bouhaddou · K. Zekhnini
L2M3S Laboratory, ENSAM, Moulay Ismail University, 50500 Meknes, Morocco

A. Benghabrit
LMAID Laboratory, ENSMR, Mohamed V University, 10070 Rabat, Morocco

M. Ben Ahmed et al. (eds.), *Networking, Intelligent Systems and Security*,
Smart Innovation, Systems and Technologies 237,
https://doi.org/10.1007/978-981-16-3637-0_50

[2, 3]. Organizations are not only supposed to be sustainable within this framework, but it is also in their interest to do so [4]. Hence, companies are currently in a state of change. In fact, additional criteria have been imposed on new products regarding consumer concerns about total cost of ownership, perceived efficiency, cost savings, long-term product support, and environmental effects. Different markets, digital demand, changing business environment, uncertainty and cost pressure, labor costs, are considered as the new challenges that companies have to face in order to stay competitive [5]. Thus, companies must commit to bringing new goods to market on a recurring basis to gain long-term sales and sustainable competitive advantage by knowing what consumers want [6] and considering a new generation of computational theories and tools to assist human in extracting useful information from the rapidly growing volumes of digital data.

Data produced by machines and computers, product lifecycle management (PLM) solutions, design for X techniques (DFX), supply chain processes, product creation, production planning systems, or quality and inventory management systems have reached an enormous volume of more than 1,000 Exabytes per day in line with the massive advancement and development of internet and online world technologies [7]. As a consequence, a well-known fundamental task for discovering useful knowledge from data and involves the evaluation and interpretation of the patterns to make decision is the use of knowledge discovery in databases (KDD).

The main objective of KDD is to extract high-level knowledge from low-level information, or in other words, to automatically process large quantities of raw data, identify the most significant and meaningful patterns, and present these as knowledge appropriate for achieving the user goals [8]. This process is mainly based on data mining step that is responsible for extracting patterns or generating models from the output of the data processing step and then feeding them into the decision-making step, which takes care of transforming its input into useful knowledge.

Traditional DM approaches have worked successfully in modeling variables of interest, and their core technologies have been identified and categorized into mining technologies for clustering, classification, and frequent patterns [9]. In this regard, three main considerations are target, data characteristics, and mining algorithm in choosing the applicable mining technologies for the problem to be solved by the KDD technology. However, since we are looking to find a structure for sustainability enhancement in an unlabeled dataset related to collaborative product development while considering different virtues X, clustering is the most appropriate method in our context. Indeed, without a priori knowledge of the structure of a database, only unsupervised classification can automatically detect the presence of relevant subgroups (or clusters).

Clustering is considered to be one of the most challenging tasks due to its unsupervised nature [10]. The numerous algorithms developed by the researchers over the years lead to different data clusters, including for the same algorithm, the selection of different parameters or the order of presentation of data objects may have a major impact on the final clustering partitions. However, with the vast number of surveys and comparative studies concerning the clustering algorithms, exploring the algorithm that cluster industrial sparse dataset still remains an open issue. Therefore, to deal with these issues, a related work focusing on comparing various clustering algorithms for real industrial datasets is presented. After that, by considering the Design for Relevance dataset which implements the sustainability concerns while considering different virtues in the product development and by following the KDD process, several clustering algorithms were implemented and compared using several internal validity indices. In addition, we highlighted the best performing clustering algorithm that gives us successful clusters for the dataset to achieve improvement in sustainability.

This paper is structured as follows; Sect. 2 presents the basic terminologies. Section 3 presents a related work over the past twenty years on research and review articles with a focus on comparative research. By providing a categorized framework, Sect. 4 provides a description of the considered clustering algorithms which will be evaluated properly. Section 5 detailed the KDD process for Design for Relevance dataset. Section 6 concludes the paper and discusses future research.

2 Preliminaries

The concepts of "Data Mining" (DM) and "Knowledge Discovery in Databases" (KDD) are confused by many researchers. More clearly, in addition to being used as one of the phases of the KDD system, some consider the DM as a synonym for KDD. In this section, key terms are defined to prepare the groundwork for the subsequent study.

2.1 Knowledge Discovery in Databases (KDD)

KDD has evolved, and continues to evolve, from the intersection of research in such fields as databases, machine learning, pattern recognition, statistics, artificial intelligence, and reasoning with uncertainty, knowledge acquisition for expert systems, data visualization, machine discovery [8], scientific discovery, information retrieval, and high-performance computing. Theories, algorithms, and methods from both of these areas are integrated into KDD software systems.

The literature analysis shows that there is no unified definition of KDD [8, 9, 12–14]. But generally, there is a common consensus that KDD is essentially an iterative and interactive process of discovering useful knowledge from a collection of data. More clearly, the goal of the KDD process is to transform data (large, multifaceted, stored on media that can be transmitted in different formats) into information. It is possible to convey this information in the form of general concepts that enrich the user's semantic field in relation to a query that concerns him. For decision making, they can be represented as a mathematical or logical model. Based on this, Fayyad et al. [11] define KDD as a process based on nine steps such as (Fig. 1):

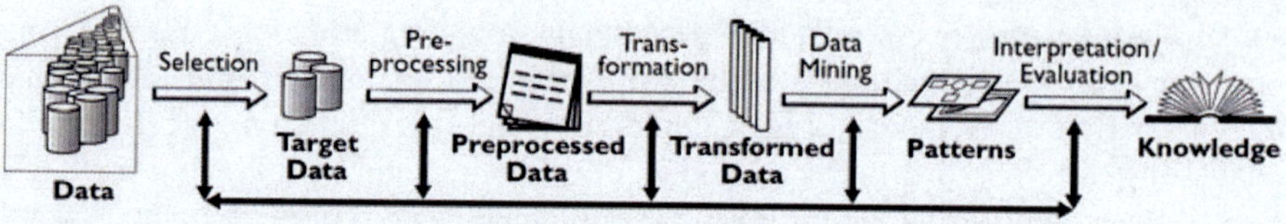

Fig. 1 Overview of the steps constituting the KDD process [11]

- Developing an understanding of the application domain and the related prior knowledge and, define the purpose of the KDD process from the customers point of view.
- Creating a target dataset.
- Data cleaning and preprocessing.
- Data reduction and projection
- Matching the goals of the KDD process to a particular data mining method.
- Exploratory analysis and model and hypothesis selection.
- Data mining.
- Interpreting mined patterns.
- Acting on the discovered knowledge.

The KDD process requires crucial steps. In [15], Yoong and Kerschberg claim that the discovery of information critically depends on how well a database is described and how the current and discovered knowledge is constantly generated. In this way, the phase description of the KDD process channel has a strong effect on the final mining results. For example, for mining, not all the attributes of the data are useful. For example, for mining, not all the attributes of the data are useful. The result is that Date Mining algorithms can find it difficult to find useful information if the selected attributes are unable to fully reflect the data characteristics [9]. However, the question arises, what is exactly data mining?

2.2 *Data Mining (DM)*

Fayad et al. in [11] define DM as a step in the KDD process involving the application of data analysis and the discovery of algorithms that generate a specific enumeration of patterns (or models) over the data. But any DM process needs a previous phase in data processing, also known as data warehousing (DW) [11]. DW refers to collecting and cleaning transactional data to make them available for online analysis and decision support. DW helps set the stage for KDD in two important ways: data cleaning and data access.

Data mining is the application, under human control, of low-level mining methods, which are in turn defined as algorithms designed to analyze data or to extract specific patterns (or patterns) from a large amount of data [16]. More clearly, DM focuses mostly on discovering knowledge in association with six basic tasks:

- **Clustering or segmentation** aims at grouping dataset (objects) into clusters such that similar objects are aggregated together in the same cluster while dissimilar ones should be belonging to different clusters. Furthermore, from an optimization perspective, the main goal of clustering is to maximize both the (internal) homogeneity within a cluster and the (external) heterogeneity among different clusters [17].
- **Classification** includes evaluating and assigning to a given class the attributes of a specific object. For example, to discern suspicious characters at an airport security check, identify a fraudulent transaction, or identify potential for a new service, classification may be used to split a customer base into best, average, and low-value customers.
- **Estimation** assigns an object some numeric value that is continually valued. Credit risk evaluation, for instance, is not simply a yes/no question; it may be some kind of scoring that assesses a propensity to default on a loan. As part of the classification process, estimation may be used (such as using an estimation model as part of a market segmentation process to guess the annual salary of a person).
- **Prediction** attempts to identify objects according to any potential actions predicted. By applying historical data where the classification is already defined to construct a model, classification and estimation can be used for prediction (this is called training). To predict future behavior, the model can then be applied to new data.
- **Affinity grouping** evaluates relationships or associations between data elements that demonstrate some kind of affinity between objects.
- **Definition** tries to explain what has been discovered or attempting to explain the results of the process of data mining.

On the basis of KDD, the landscape of business decision support is evolving with an increasing body of emerging applications, such as risk analysis, targeted marketing, customer retention, portfolio management, and brand loyalty [18]. In modeling variables of interest, traditional DM methods have proven to be successful, so that these variables can be projected in future scenarios and effective decisions made based on that forecast can be taken. More clearly, three key considerations in choosing the applicable mining technologies for the problem to be solved by the KDD technology: objective, characteristics of data, and mining algorithm. However, how to test a candidate model is one of the most critical aspects of DM problems, and obviously, this issue depends on the type of DM task at hand. Therefore, most of the DM issues can be considered as optimization issues, where the goal is to create a candidate model that optimizes those performance parameters which is a multi-objective concern in nature.

Since we are looking to find a structure for sustainability enhancement in an unlabeled dataset related to collaborative product development while considering different virtues X, clustering is the most appropriate method in our context [7].

Indeed, without a priori knowledge of the structure of a database, only unsupervised classification can automatically detect the presence of relevant subgroups (or clusters).

3 Related Works

The clustering publications have been rising over the past years, showing that researchers are paying more and more attention to this issue. Some researchers have improved clustering algorithms for a particular domain, while others have implemented new algorithms, or have researched and compared clustering algorithms. By using different databases such as Scopus, Elsevier, Emerald, Taylor & Francis, Springer, IEE, and Google Scholar, we have grouped the literature into four key categories: review and surveys, comparative studies; clustering methods dealing with a new algorithm, and finally clustering applications. According to Fig. 2 and without limitations constraints, twenty-three percent of publications focused on comparing various algorithms, while thirty-eight percent of publications applied clustering to several domains such as image processing, speech processing, information retrieval, web application processing.

Using the Newbert methodology [5, 19], this systematic methodology ensured that the analysis was thorough, leading to a large number of comparative studies being reviewed concerning to clustering algorithms for different domains. Among the 87 papers chosen in the literature review, 28 were eventually used for comparative studies to research these works in detail, while 10 were used for the application of the clustering approach in the industry field. As a result, some of the selected studies are presented in Table 1 using both the comparative studies carried out in the literature for all fields and the clustering algorithms used in the industry.

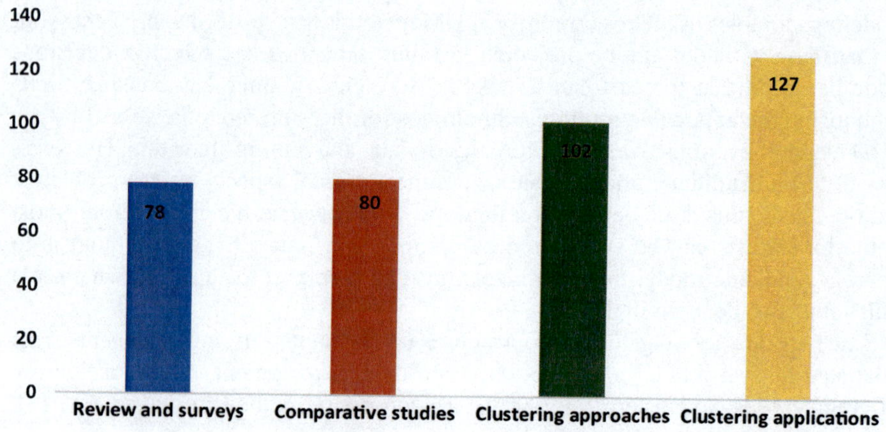

Fig. 2 Classification of clustering publications

Table 1 The literature review of clustering algorithms with respect to the considered classification

References		Description
[22]	Comparative studies	A detailed survey of current clustering algorithms in data mining with making a comparison among them according to their scores (merits). The authors identified the problems to be solved and the new directions in the future according to the application requirements in multimedia domain
[20]		A survey of clustering algorithms that require less amount of knowledge about the domain being clustered
[23]		The performance of several clustering algorithms is been computed with respect to several indices such as homogeneity, separation scores, silhouette width, redundant score, and WADP
[21]		A comparison of performance of various clustering algorithms based on size of dataset, time taken to form clusters, and the number of clusters formed
[24]	Both	A survey of different clustering algorithm (DBSCAN and K-means) for analyzing different financial datasets for a variety of applications: credit cards fraud detection, investment transactions, stock market, etc
[25]		A comparative study with three algorithms such as K-means, hierarchical agglomerative clustering, and SOM was used to modularize a packaging system by reducing the variety of packaging sizes
[17]		A comparative study of different clustering algorithms such as K-means, DBSCAN, hierarchical agglomerative algorithm, and SOM for analyzing four different real industrial dataset
[26]	Applications	Applied SOM to visualize and analyze large data bases of industrial systems such as forest industry. The goal was to cluster the pulp and paper mills of the world
[29]		The identification of high-profit, high-value and low-risk customers by the customer clustering which is a one of the data mining technique
[27]		A K-means clustering approach to determine and categorize the environmental risk zoning. The clustering result with the optimal clustering number is then used for the environmental risk zoning, and the zoning result is mapped using the geographic information system
[28]		The relationship between perceptions of clusters and innovation for firms operating through technicities that are accepted as innovative clusters. To test the propositions, a field survey using questionnaires was conducted
[30]		A survey on data mining in steel industries. Steel consists of alloy iron, carbon, and manganese with small amount of silicon, phosphorous, and Sulfur. Steel production stages: Heating, cooling, melting, solidification

With an in-depth study of these papers, we may state that several studies [20, 21] have extensively studied common and known algorithms such as K-means, DBSCAN, DENCLUE, K-NN, fuzzy k-means, and SOM to discuss their advantages and disadvantages, taking into account several factors that may affect the criterion when selecting a suitable clustering algorithm. While other studies [22, 23] have examined the clustering algorithm surveys based on various parameters, such as their score (merits), their problems solved, their applicability, their domain awareness, and also on the size of the dataset, the number of clusters, the type of dataset, the software used, the complexity of time, stability, etc. Researchers [24–30] have examined various algorithms such as K-means, DBSCAN, agglomerative hierarchical clustering, and SOM algorithm for cluster packaging and environmental risk, financial, female employees, consumer preferences, industrial hygiene, and forest industry datasets in further research and in particular in the field of industry.

However, there are some shortcomings of all these surveys and comparisons found in the literature, such as the characteristics of the algorithms are not well studied, no systematic empirical study was carried out to assess the value of one algorithm for a particular type of dataset over another. Thus, the only paper that deals with such issue is one of our previous work [10] which deals with various algorithms with different real industrial datasets. Therefore, overviewing and exploring the algorithms that determine the best clusters for sparse industrial dataset remains an open issue.

4 Methodology

Algorithms for clustering have a strong relationship with many fields, especially statistics and science. Different starting points and parameters typically lead to various clustering algorithm taxonomies. As a result, a large number of algorithms have been proposed in the literature. Depending on the approach considered in terms of data processing, clustering techniques can be divided into five categories [17]: Partitioning-based algorithms; hierarchy-based algorithms; density-based algorithms; grid-based algorithms; and model-based algorithms. However, based on the comparison realized in the paper [17] and due to the popularity, versatility, applicability to industrial datasets, we can claim that the chosen algorithms are: k-means algorithm, ward-distance agglomerative hierarchical algorithm, and self-organization map (SOM).

4.1 K-means

The K-means algorithm is the best-known approach, used and extended in the different communities dedicated to clustering. The principle is "natural," given the distribution of individuals X in the description space and a fixed number of groups, the objective is to minimize the dispersion of individuals relative to a set of prototypes representative of these groups. In other words, the first step is to randomly select k, where k equals the number of clusters that you chose. Data points representing the center of a cluster are called centroids [31]. By a two-step method called expectation–maximization, the key elements of the algorithm operate. Each data point is assigned to its closest centroid by the expectation stage. Then, for each cluster, the maximization step computes the average of all the points and sets the new centroid. The pseudo-code of the K-means algorithm is as follows:

Algorithm: K-means
1: Specify the number of k of clusters to assign
2: Randomly initialize k centroids
3: **repeat**
4: **expectation**: Assign each point to its closest centroid
5: **maximization**: Compute the new centroid (mean) of each cluster 6: **until** The centroid positions do not change

4.2 *Hierarchical*

By creating a hierarchy, hierarchical clustering decides cluster assignments. Either a bottom-up or a top-down approach implements this: the bottom-up approach is agglomerative clustering. Until all points have been grouped into a single cluster, it merges the two most similar points. In contrast, the top-down strategy is divisive clustering. It begins as one cluster with all points, and at each stage separates the least related clusters until only single data points remain [32]. A tree-based hierarchy of points called a dendrogram is generated by these techniques. The number of clusters (k) in hierarchical clustering is always predetermined by the user, similar to partitional clustering. Clusters are allocated by cutting the dendrogram at a given depth, resulting in smaller dendrograms in k groups. The pseudo-code of the hierarchical algorithm is as follows:

Algorithm: Hierarchical agglomerative
1: Compute the proximity matrix for data points
2: **repeat**
3: **combine** the two closest clusters into one cluster
4: **update** the proximity matrix 5: **until** there is only one cluster

4.3 *Self-Organizing Maps*

A self-organizing map (SOM) is an unsupervised competitive learning algorithm from an artificial neural network. It is a very popular nonlinear technique for reducing and visualizing data. When an observation is recognized, the activation of a neuron in the network, selected by a competition between neurons, has the effect of strengthening this neuron and inhibiting others (this is the rule of "Winner Takes All")[33]. Therefore, each neuron specializes during learning in recognizing a certain type of observation. The self-organizing map is made up of a set of neurons connected by topological links that form a two-dimensional grid. Each neuron is connected to n inputs (corresponding to the n dimensions of the representation space) according to n weights which form the prototype vector of the neuron. Neurons are also connected to their neighbors by topological links. The dataset is used to organize the map according to the topological constraints of the input space. Thus, a configuration between the input space and the network space is constructed; two nearby observations in entry space activate two nearby units on the map. The pseudo-code of the SOM algorithm is as follows:

Algorithm: Self-organizing maps

Select random values for the initial weight vector wj
repeat
    Select a sample from the input dataset
    Find the winning neuron for the sample input
    Adjust the weights of nearby neurons
until the feature map stops changing

---

5 Results and Discussion

In this section, we describe all the phases presented in the generic KDD pipeline (Fig. 3) to discover and extract useful knowledge from a collection of industrial data. More clearly, we are going to describe the considered dataset, the data-preprocessing

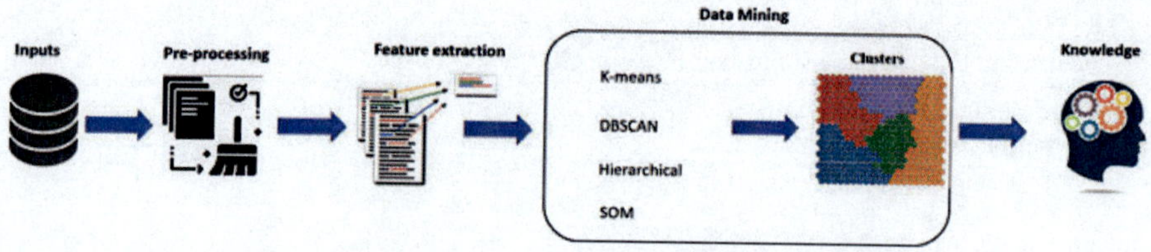

Fig. 3 KDD pipeline for clustering design for relevance dataset

and transformation tasks, the data mining process which is clustering as well as the useful knowledge which are the obtained clusters.

5.1 Dataset Collection

Manufacturing industries that rely on a higher degree of value-added manufacturing, such as the automotive industry, are increasingly trying to innovate in the face of rising sustainability challenges by bringing new products to market and providing more complete solutions that incorporate both product and service components [5]. However, it is a difficult mission to shift from a product-centric perspective to a solution-oriented perspective. It involves the use of modern sustainable business models to ensure design activities and to ensure sustainable assessment, from creation, through manufacturing, use, until the end-of-life of the product, meets consumer and societal needs. In this respect, the "design for" or DFX literature is reviewed by this new product development study to consolidate the existing body of expertise and to seek the future direction of the industry [7]. In terms of economics (dominated by supply chain design techniques), ecology (dominated by environmental design techniques), and social equity, DFX techniques can be put under the heading of sustainability [34].

However by analyzing the different DFX techniques which are most applicable during the design phase and by considering also the relationships and trade-offs between design decision across the main phases of product lifecycle while considering the IATF 16,949 quality attributes, the most relevant DFX in the automotive sector are Design for manufacture and assembly, Design for quality, Design for service; Design for safety; Design for supply chain and finally design for environment [7]. We shall from now, that the six DFX chosen in this paper will be named by Design for Relevance. The term relevance refers to the relation and interaction between DFX techniques and attributes of quality in a way that makes it beneficial to consider second when considering first. That implies that the two principles contribute to unity and cannot be isolated. More clearly, the Design for Relevance dataset contains 492 design factors (knowledge) and 29 modules.

The aim of clustering in Design for relevance is to assign related objects to the same cluster and to distinguish them from clusters containing modules that are not similar. We refer from this point onwards to objects as modules and present a group as Family Modules (MF). The basic concept is that the MF helps a project team to understand and design the design factors (knowledge) common to a group of modules at the same time. The efficacy of the designs importance will be improved by this technique. The dataset is presented in details in paper [7].

5.2 Data-Preprocessing and Transformation

Data-preprocessing is one of the most important steps in a data mining process. To improve their consistency, it deals with the preparation and transformation of the initial dataset. Generally, this phase is divided into four categories: data cleaning, data incorporation, transformation of data, and data reduction [11]. In our Design for Relevance dataset, we need the data cleaning and data transformation in order to improve the quality of our data, and we can start our clustering with no noise features and with numerical inputs dataset.

Data cleaning task consists of cleaning the information from noisy and inconsistent design factors. In this step, we first begin by unifying concepts that have the same thoughts but with different names. After that, and since our dataset is small, we define some noisy concepts with the aid of an expert and compute their neighborhood for each noisy point to decide whether it is a core point (the volume for each point is the same and the mass of neighborhood is variable). If yes, we start a cluster somewhere else around this point, we mark the point as an outlier. These two steps are repeated until all points are either allocated to a cluster or designated as an outlier [35].

Data transformation task deals with the conversion or aggregation of data into suitable forms for mining. The set of modules is described in the vector space model in this phase. This implies that the interactions between modules and design variables are determined using the tf scores [36] and are expressed in the incidence matrix of the binary module design factor [aij], where I represent the ith design variables associated with all modules and j represents the jth module. After doing this transformation; the only entries of our dataset are 0 and 1. Entry 1 indicates that the design factor of the ith is part of the jth module, while entry 0 indicates that the design factor of the ith is not part of the jth module. We note that with the aid of the designer interpreted as one expert or a team of experts, the binary assignment is achieved.

5.3 Data-Mining: Clustering

In several real-world cluster applications, one of the most important questions is how to decide a suitable number of clusters K. There is no easy way to know the number without a priori information, the only way to do it is by computing several internal indices such as CH index [37], Gamma index [38], DB index [39] and Hubert index[40] that allow us to choose the best cluster number [17]. The optimal number of clusters for each index is based on the maximum (or minimum) index value, the maximum (or minimum) index hierarchy difference, the maximum (or minimum) index hierarchy value, the maximum (or minimum) second index level difference, or the use of the critical value [41]. To compute the indices, user can request the indices one by one, by setting the argument index to the name of the index. Hence and according to the majority rules, 5 would be the best number of clusters for our dataset. In addition, if we look at the Hubert index, which is a graphical tool, the

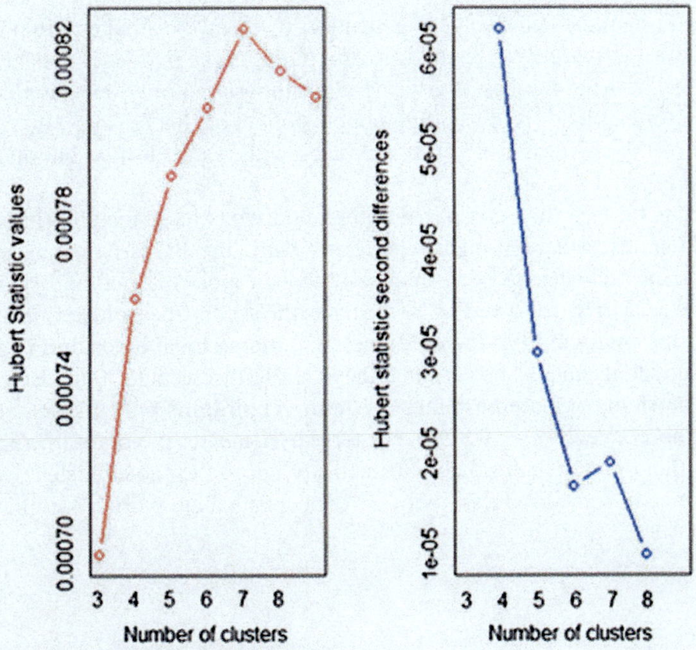

Fig. 4 Hubert index for design for relevance dataset

optimal number of clusters in the plot of index values against the number of clusters is defined by a significant knee. As the number of clusters ranges from minimum to maximum, this knee leads to a significant increase or significant decrease in the index. In other words, the relevant number of clusters is indicated by a large peak in the second differential value map. Hence, as shown in Fig. 4, the Hubert index confirms our purpose and propose 5 as the best number of clusters.

We can now run the selected three algorithms after selecting the required number of clusters in our data to compare, first, the best clustering algorithm in our case and, secondly, to define the groups of modules. However, we should consider the following points before beginning our comparative analysis:

- The SOM algorithm differs from other algorithms for clustering, especially from other artificial neural networks. Indeed, SOM is a common nonlinear dimension reduction and data visualization technique that does not provide clusters or groups [26]
- SOM uses a neighborhood function to preserve the topological properties of the input space.

Thus, an efficient method to grouping problem is based on the learning of a SOM. We use SOM to measure a set of reference vectors representing local means of data based on topological properties in the first step of the method. In the second point, the vectors obtained are clustered using the two standard K-means clustering methods and the agglomerative hierarchical algorithm to form the final partitioning. This method is most commonly referred to in the literature as two-level clustering algorithms [42]. Choosing the best clustering algorithm based only on a single measure for our dataset can lead to misleading conclusions. For this reason, Fig. 5 presents the results of the clustering of four candidates with regard to five internal validity measures presented above.

First, it can be seen that, compared to the remaining clustering algorithms except for the CH index, where k-means performs well, the SOM/K-means algorithm provides the best clustering performance based on most internal and stabilization indices. The SOM/HC followed by K-means is the second-best clustering algorithm in terms of internal validity. The agglomerative hierarchical algorithm is, thus, the worst. Moreover, it can also be seen that the SOM/K-means algorithm also generates connected clusters compared to other clustering algorithms [43]. In fact, according to Fig. 6, the connectivity of SOM clusters to K-means is approximately 15.97, suggesting that only 16% of related artifacts are not in the same cluster. This result is normal since in some modules, there are concepts that are partially similar to each

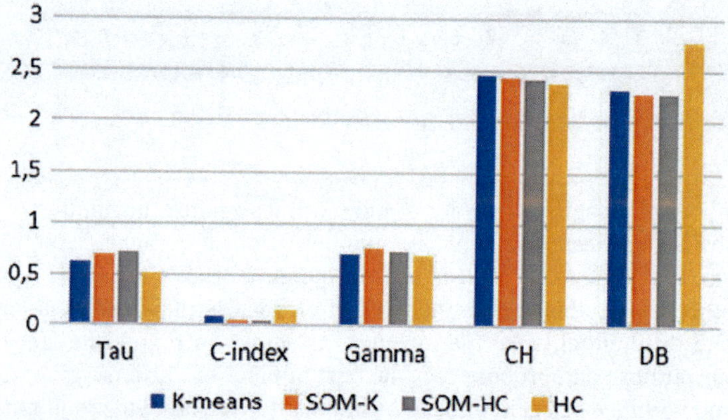

Fig. 5 Internal validation indices for the selected algorithms

Fig. 6 Connectivity index for the selected algorithms

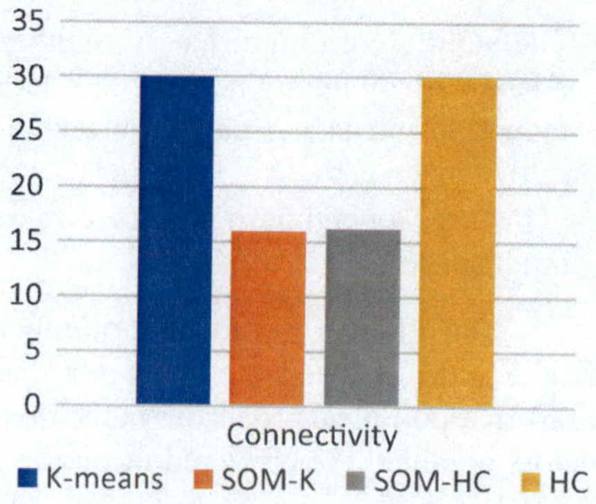

other but are in the same module. In terms of compactness and separation, it can be seen that not only compact clusters, but also well-separated clusters can often be generated by SOM/HC than K-means, which in turn is better than the Hierarchical algorithm. This is substantiated by indices of tau, Gamma, and DB performance.

Finally, and after analyzing the results of testing the clustering algorithms and evaluating them under different indices, we can remark that SOM combining with agglomerative hierarchical algorithm shows more accuracy in clustering most of the objects into their suitable clusters.

5.4 Knowledge for Sustainability Enhancement of the Design for Relevance

According to our previous comparison, SOM combined with hierarchical clustering algorithm indicates that the overall Design for Relevance can be accomplished in five self-contained modules:

- **Module Group 1** addresses logistics issues, it consists of packaging criteria, transport criteria, warehouse criteria, and Industrial footprint modules.
- **Module Group 2** addresses reverse logistics issues, it consists of recycling and disassembly criteria, reverse logistics criteria, and environmental issues modules.
- **Module Group 3** addresses service issues, it consists of product service system, service control, quality insurance, and service criteria modules.
- **Module Group 4** addresses feasibility issues, it consists of assembly features, assembly requirements, environmental requirements, environmental considerations, materials safety, materials selection, safety criteria, quality performance, service requirements, and human safety modules. Finally,
- **Module Group 5** addresses design issues, it consists of manufacture criteria, assembly criteria, assembly process, quality control, quality system, safety requirement, and manufacture planning and control modules. In designing these modules groups, the project team considers only the necessary and relevant design factors.

By so doing, the scope and degree of complexity of the design become more manageable and take a holonic perspective for sustainability assessment while developing a product from all its life cycle. For example, not only the widely used manufacturing design factors are considered by the design module community, but also across other modules, and other important and required design factors are included such as controlled parameters and control plan in quality control module; Assembly, transport and environment requirements and safety requirements, etc.). This simply signifies that the design module group is not an isolated activity and may not be designed in a vacuum. As a result, by achieving and integrating other important and necessary factors such as consistency, protection, climate, operation, assembly, and logistics, this approach improves the effectiveness as well as the sustainability consideration of the product design.

6 Conclusion

With the rapid growth in number and dimension of databases and database applications in business, administrative, industrial, and other fields, the automated extraction of information from these broad databases needs to be examined. These have become rich and reliable sources of information generation and verification due to knowledge extraction from databases, and knowledge discovery can be applied in software management, process querying, decision making, process control, and many other fields of interest. In addition to that, given the problems facing product manufacturing, which are becoming increasingly complex, companies need to consider complexity in both technological and other multidisciplinary fields. Manufacturing firms will not only need to implement flexible strategies to reap the gains in the future, but they must also successfully innovate and manage various problems with different X-factors to reach the sustainability assessment [7].

In this respect, this paper provides a comprehensive survey and intends to compare popular, flexible, and applicable clustering algorithms in industry field. More clearly, by considering the Design for Relevance dataset, we facilitate the interface and the collaboration between all project teams. However, to manage and organize a large number of design factors involved in the design of integrated DFX in an unlabeled dataset, we have used a clustering data mining task clustering. However, many researchers have developed and provided several clustering algorithms with updated implementations for different domains. As a result, finding suitable algorithms helps to organize information significantly and extract the correct queries from various database queries.

Through an exhaustive search, a related work was first presented in four categories, namely review and surveys; comparative studies; clustering approaches and clustering applications, according to twenty last year. After that, and based on the clustering categorization framework presented in paper [17], the most representative clustering algorithm of each category (excepted for the grid-based) has been implanted and compared for the Design for Relevance dataset. Following the KDD process, in the type of hierarchy, we have generated a taxonomy, i.e., a category of knowledge entities, according to the presumed relationships of the real-world entities they represent. SOM combined with hierarchical clustering offers an effective clustering algorithm for solving integrated DFX problems and achieving sustainable progress after comparing three commonly used algorithms (k-means, agglomerative hierarchical clustering with ward distance, and self-organizing SOM maps). This algorithm creates self-contained clusters that can be interpreted, changed, and applied more easily.

Furthermore, we have tried to reveal future directions for the researchers. While manufacturing processes are among the most controlled and supervised, this trend is being driven today by IoT growth. Data collected by manufacturing firms is increasing sharply, but software management tools, on the other hand, are typically unable to perform a detailed analysis of this information. In other words, KDD methods are not common, and businesses are still not leveraging the full potential of their collected data, especially about information acquisition. Thus, recent developments in innovative architectures and methods, such as IoT and DM, enable manufacturing management systems to rapidly evolve toward capacity forecasting, modeling, and optimization.

References

1. Salvioni, D.M., Gennari, F., Bosetti, L.: Sustainability and Convergence: The Future of Corporate Governance Systems? mdpi.com. https://doi.org/10.3390/su8111203
2. Clayton, T., Radcliffe, N.: Sustainability: A Systems Approach; Routledge, New York, NY, USA (2015)
3. Drexhage, J., Murphy, D.: Sustainable development: from Brundtland to Rio 2012. Background paper prepared for consideration by the High-Level Panel on Global Sustainability at its first meeting 19 September 2010. Adv. Appl. Sociol., **5**(12) (2015)
4. Zbuchea, A. Are customers rewarding responsible businesse...—Google Scholar
5. Benabdellah AC, Bouhaddou I, Benghabrit A, Benghabrit O (2019) A systematic review of design for X techniques from 1980 to 2018: concepts, applications, and perspectives. Int. J. Adv. Manuf. Technol., pp. 1–30. https://doi.org/10.1007/s00170-019-03418-6
6. Holmes, A., Moore, L., Sundsfjord, A., et al.: Understanding the mechanisms and drivers of antimicrobial resistance. Elsevier
7. Benabdellah, A.C., Benghabrit, A., Bouhaddou, I., Benghabrit, O.: Design for relevance concurrent engineering approach: integration of IATF 16949 requirements and design for X techniques. Res. Eng. Des. **31**, 323–351 (2020). https://doi.org/10.1007/s00163-020-00339-4
8. Matheus, C.J., Chan, P.K., Piatetsky-Shapiro, G.: Systems for knowledge discovery in databases. IEEE Trans Knowl Data Eng **5**, 903–913 (1993). https://doi.org/10.1109/69.250073
9. Tsai, C.-W., Lai, C.-F., Chiang, M.-C., Yang, L.T.: Data mining for internet of things: a survey. IEEE Commun. Surv. Tutorials **16**, 77–97 (2014). https://doi.org/10.1109/SURV.2013.103013.00206
10. Benabdellah, A.C., Benghabrit, A.: Science IB-P computer. A survey of clustering algorithms for an industrial context. Elsevier (2019)
11. Fayyad, U., Piatetsky-Shapiro, G.: Magazine PS-AI, 1996. From data mining to knowledge discovery in databases. aaai.org
12. Hilbert, M., Lopez, P.: The world's technological capacity to store, communicate, and compute information. Science **80**(332), 60–65 (2011). https://doi.org/10.1126/science.1200970
13. Cicardi, M., Aberer, W., Banerji, A., et al.: Classification, diagnosis, and approach to treatment for angioedema: consensus report from the hereditary angioedema international working group. Allergy **69**, 602–616 (2014)
14. Shen, W., Hao, Q., Yoon, H., et al.: Applications of agent-based systems in intelligent manufacturing: An updated review. Elsevier
15. Yoon, J.P., Kerschberg, L.: A framework for knowledge discovery and evolution in databases. IEEE Trans Knowl Data Eng **5**, 973–979 (1993). https://doi.org/10.1109/69.250080
16. Klösgen, W., data JŻ-A in knowledge discovery and, 1996. Knowledge discovery in databases terminology.dl.acm.org
17. Benabdellah, A.C., Benghabrit, A., Bouhaddou I (2019) A survey of clustering algorithms for an industrial context. In: Procedia Computer Science

18. Gamarra, C., Guerrero, J.: Energy EM-R and S, 2016. A knowledge discovery in databases approach for industrial microgrid planning. Elsevier
19. Newbert, S.L.: Empirical research on the resource-based view of the firm: an assessment and suggestions for future research. Strateg. Manag. J. **28**, 121–146 (2007)
20. Treshansky, A.: … RM-T for S, 2001. Overview of clustering algorithms. spiedigitallibrary.org
21. Chen, L., Ellis, S., Holsapple, C.: Supplier development: a knowledge management perspective. Knowl. Process. Manag. **22**, 250–269 (2015). https://doi.org/10.1002/kpm.1478
22. He, L., Wu, L.: Computers YC-A research of, 2007. Survey of Clustering Algorithms in Data Mining. en.cnki.com.cn
23. Chen, G., Jaradat, S., Banerjee, N., et al.: Evaluation and comparison of clustering algorithms in analyzing ES cell gene expression data. JSTOR
24. Cai, F., Le-Khac, N.-A., Kechadi, T.: Clustering Approaches for Financial Data Analysis: a Survey (2016)
25. Zhao, C., Johnsson, M., He, M.: Data mining with clustering algorithms to reduce packaging costs: a case study. Packag. Technol. Sci. **30**, 173–193 (2017). https://doi.org/10.1002/pts.2286
26. Simula, O., Vasara, P.: JV-IA of, 1999. The self-organizing map in industry analysis. books.google.com
27. Shi, W., Zeng, W.: Application of k-means clustering to environmental risk zoning of the chemical industrial area. Front. Environ. Sci. Eng. **8**, 117–127 (2014). https://doi.org/10.1007/s11783-013-0581-5
28. Yıldız, T.: Sciences ZA-P-S and B. Clustering and innovation concepts and innovative clusters: an application on technoparks in Turkey. Elsevier (2015)
29. Saraee, M., Moghimi, M.: AB the FIW on. Modeling batch annealing process using data mining techniques for cold rolled steel sheets.dl.acm.org (2011)
30. Umeshini, S., PSumathi, C.: A Survey on Data Mining in Steel Industries. pdfs.semanticscholar.org
31. MacQueen, J.: Some methods for classification and analysis of multivariate observations. Proc. Fifth Berkeley Symp. Math. Stat. Probab. **1**, 281–297 (1967)
32. Johnson, S.C.: Hierarchical clustering schemes. Psychometrika **32**, 241–254 (1967). https://doi.org/10.1007/BF02289588
33. Kohonen, T.: 1 The Self-Organizing Map (SOM) (2001)
34. Arnette, A.N., Brewer, B.L., Choal, T.: Design for sustainability (DFS): the intersection of supply chain and environment. J. Clean. Prod. **83**, 374–390 (2014)
35. Campello, R.J.G.B., Moulavi, D., Sander, J.: Density-Based Clustering Based on Hierarchical Density Estimates, pp. 160–172 (2013)
36. Aizawa, A.: An information-theoretic perspective of tf–idf measures. Inf. Process. Manage. **39**(1), 45–65 (2003)
37. Caliński, T.: JH-C in S. A dendrite method for cluster analysis. Taylor Fr (1974)
38. Baker, F.B., Hubert, L.J.: Measuring the power of hierarchical cluster analysis. J. Am. Stat. Assoc. **70**, 31–38 (1975). https://doi.org/10.1080/01621459.1975.10480256
39. Davies, D.: Analysis DB-I transactions on pattern. A cluster separation measure. ieeexplore.ieee.org (1979)
40. Statistical HL-J of the A. On the Kolmogorov-Smirnov test for normality with mean and variance unknown. amstat.tandfonline.com (1967)
41. Sajana, T., Rani, C.M.S.: KVN-I Journal of S. A survey on clustering techniques for big data mining. researchgate.net (2016)
42. Cabanes, G.: Maps YB-S-O. Learning the number of clusters in Self Organizing Map. intechopen.com (2010)
43. Xing, G., Wang, X., Zhang, Y., Lu, C., Pless, R., Gill, C.: Integrated coverage and connectivity configuration for energy conservation in sensor networks. ACM Trans, Sens. Netw. (TOSN) **1**(1), 36–72 (2005)

Location Finder Mobile Application Using Android and Google SpreadSheets

Adeosun Nehemiah Olufemi and Melike Sah

Abstract In this paper, we introduce a mobile application (app) called Location Finder that is built with Java using Android Studio and the Google Spreadsheets. This application works semantically as different locations can be automatically detected. It is a complete native Android application to find different locations such as restaurant, hotel, shops, bus stops, ATM, universities, hospitals, gas station and many more. It provides advanced mapping systems that will show different places of interest for app users. This paper discusses the system architecture and details of the design. One of the advantages of the proposed mobile app is that it is easy to develop and based on interoperable Google Spreadsheets platform.

1 Introduction

With the recent developments in the cellular world, the high-end mobile phones and PDAs are becoming pervasive and are being used in different application domain. Integration of the Web services and cellular domains leads to the new application domain, mobile Web services [1].

In this paper, we propose Location Finder which is a mobile Web application that allows people to explore the world around them by leveraging contexts that are meaningful to them. Someone that finds himself/herself in a city or town that he/she is not familiar with, say North Cyprus, for example, may decide to go to a restaurant and have no clue of how to find one. He/she may decide to open the mobile app and search for query like "Restaurant". The Location Finder mobile application (Fig. 1) combines an innovative interface and architecture to support ready exploration of rich information. The mobile application was built with Java using Android Studio,

A. N. Olufemi
Software Engineering Department, Near East University, Nicosia, North Cyprus
e-mail: 20195289@std.neu.edu.tr

M. Sah (✉)
Computer Engineering Department, Near East University, Nicosia, North Cyprus
e-mail: melike.sah@neu.edu.tr

M. Ben Ahmed et al. (eds.), *Networking, Intelligent Systems and Security*,
Smart Innovation, Systems and Technologies 237,
https://doi.org/10.1007/978-981-16-3637-0_51

Fig. 1 Interface of Location Finder mobile application; different features are supported by the app such as map, favorites and categories

and the Google Spreadsheets has been used for the backend. Further, the application supports the publication of new information in the backend: Public toilet, accident-prone area, etc., can be semantically published which will dynamically reflect on the mobile application. In these respects, Location Finder application uses technologies that provides an effective information exploration experience.

Another related work is mSpace Mobile [2, 3]. It is a semantic mobile application. The main goal of this semantic mobile app is to provide information about topics of chosen interest, based upon the location, as determined by an optional GPS receiver. It queries the mSpace Query Service (MQ), which is connected to RDF triplestore knowledge interfaces (MK). Simple Object Access Protocol (SOAP) and HTTP are used for the communication between the mSpace Mobile application and the mSpace Query Service. mSpace Mobile uses semantic Web technologies to explore information. The mSpace Mobile interface is designed to let users of small screen devices run complex queries through direct manipulation, and no typing is required. To this end, the application utilizes the primary features of the mSpace interaction model [4]. mSpace Mobile is an application that was developed for Windows phone users. Whereas in our work, Location Finder has been built and developed with Android Studio which will run on Android devices and based on interoperable Google SpreadSheets.

Other related works are AWARE [5], Momento [6], Device Analyser [7], MetricWire [8] and Google Timelines [9]. AWARE [5] is an open source platform for catching, deducing and creating context on mobile devices. It is an Android application which collects data from several sensors such as emails and messages which can be used to create context-aware software. Momento [6] is also another context-aware mobile application which explicitly asks the user his/her location, nearby persons and other information in order to provide context-aware data. Device Analyser [7] is a free data collection tool for Android devices, which can be used for ubiquitous computing. MetricWire [8] is an Android mobile application that is used for data collection from the user's mobile phone. Google Timelines [9] mobile application can also be utilized to track the location of the user.

2 System Architecture of Location Finder

Location Finder mobile application has been built with Java using Android Studio and Online Google Spreadsheet. The project is as simple as possible for better understanding. It contains few JAVA and XML files. The app has been built on Google standard Google map to ensure highest accuracy and portability and has a dynamic backend based on Online Google Spreadsheet, and this will help the administrator to update and maintain backend data without any additional server or database. In the following sub-sections, we explain the details of system components.

2.1 Google Spreadsheets for Storing Information of Places

First, information will be inserted to the Google Sheets as shown in Fig. 2. Google Sheets is a part of free, Web-based office software offered by Google Drive service. The service also includes Google Docs and Google Slides, a word processor and presentation program, respectively. It makes data to be created charts and graphs.

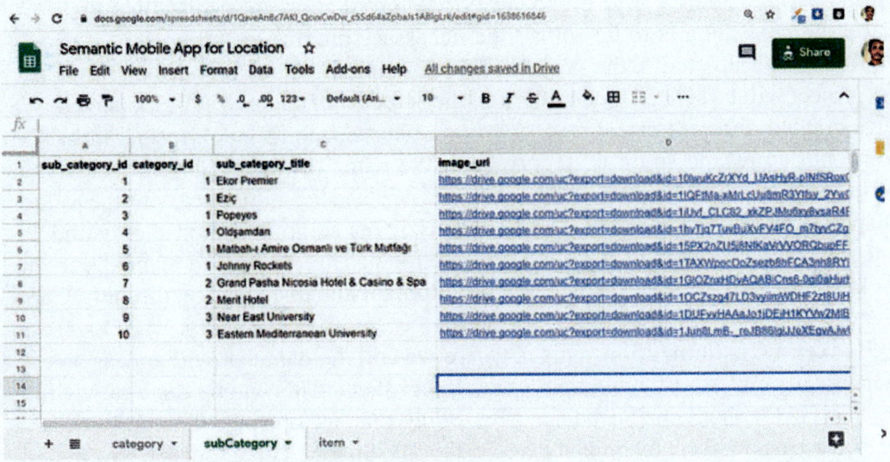

Fig. 2 Information is inserted into Google SpreadSheets

Built-in formulas, pivot tables and conditional formatting options save time and simplify common spreadsheet tasks.

2.2 Material Design and Optimization

Then, we design the UI/UX of the Location Finder. Material Design [10] is a visual language that synthesizes the classic principles of good design with the innovation of technology and science. It is very important to have a presentable user interface by using a well-pleasing design. To identify a set of UI component categories that frequently occur in Android apps, we referenced popular design tools and languages that expose component libraries such as Balsamiq [11] and Google's Material Design [12]. The Material Design has been maintained for Location Finder mobile application to decorate the user interface and to ensure the app is mobile friendly and fully optimized. All basic components are nicely decorated with unique looking and gorgeous color combination.

Optimization of code is essential to make the app run smoothly without lagging and that is why app has been developed with the code fully optimized. Every single implementation has been optimized for highest performance. Also, the code has been beautifully crafted and modularized to enable other developers to easily understand the code. Comments are used where necessary to describe certain line of code.

2.3 Android Studio Setup

This software is built purposely to develop Android apps. After installing this software, the next thing is to download the necessary plugins including Java IDE and SDK. In Android studio, the project folder is named. Once that has been selected, the project will be synced by Gradle, and the screenshot below will be displayed. We have created different categories for certain places such as hotel, restaurant, ATM point, bus stops and so on. Category takes the name of this places and displays it on the app in real-time mode. Subcategories are needed in order to put the values of specific places under each category, e.g., Ezic restaurant is a subcategory under a category called restaurant. All these information has been recorded manually in our backend (on Google Sheet) which will be automatically displayed on the app for app users.

Backend Customization. This is where we add the data that will appear on app dynamically. We start by creating a new Excel sheet. This can be done by visiting the link [13] with a Gmail account. After doing this, we open the Google Drive and then create a new sheet by going to New > Google Sheets. Three sheets were created, namely category, subcategory and items. The files in the folder contain the Java and XML files.

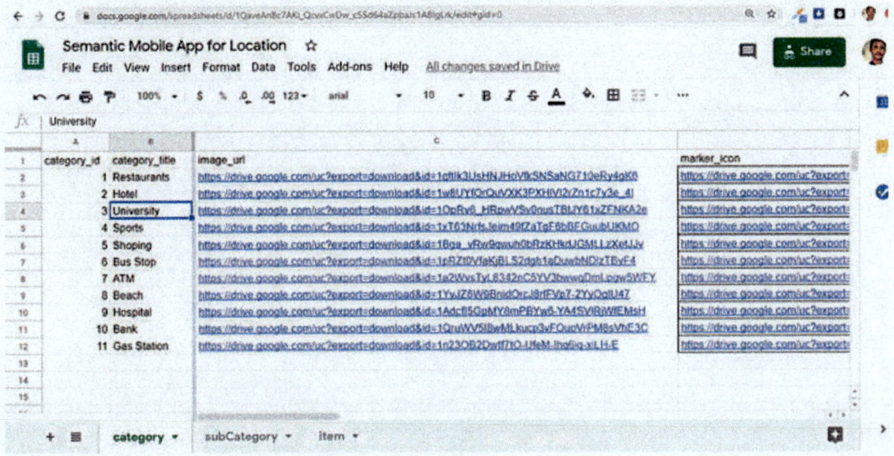

Fig. 3 Creating categories

Category Sheet Section. In this section, we have three columns, namely category_id (number must be unique), category_title (the name of category) and image_url (link at which image that will be displayed for each category is located) (Fig. 3).

Subcategory Sheet Section. This section consists of four columns, namely subcategory_id (a unique number for this column), category_id (the id of category in which subcategory will fall), sub_category_title (the title of subcategory) and image_url (link at which image that will be displayed for each subcategory is located) as shown in Fig. 4.

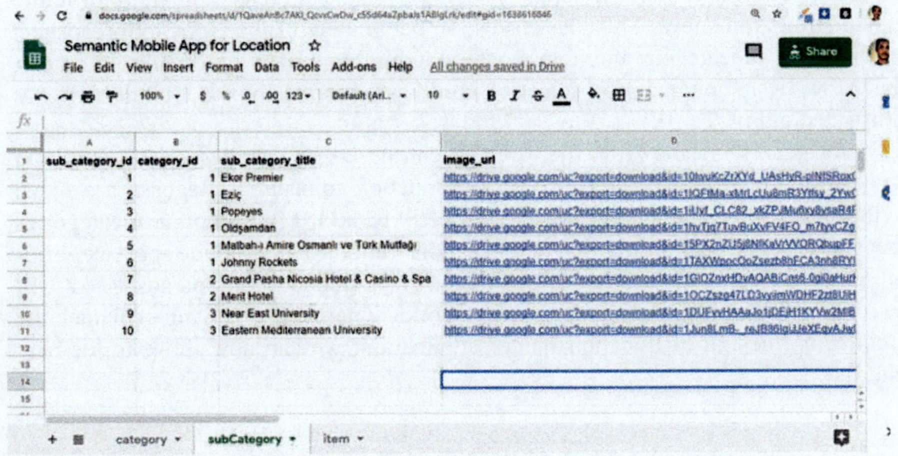

Fig. 4 Creating subcategories

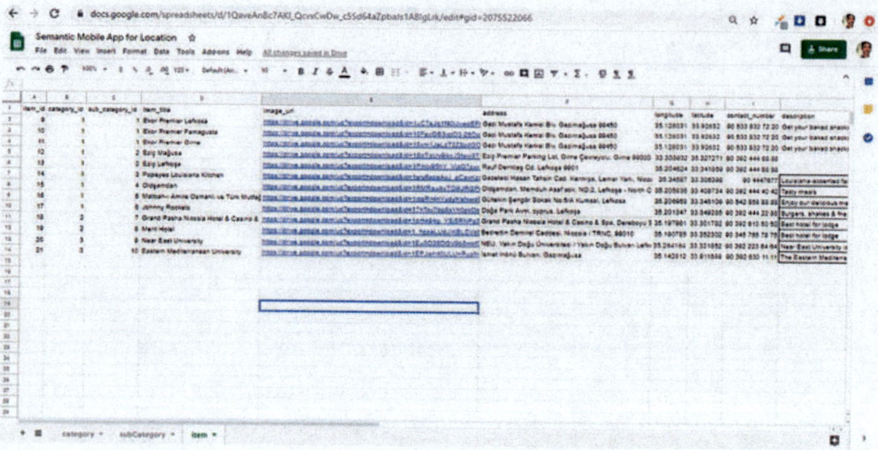

Fig. 5 Item page sheet page

Item Sheet Section. Lastly, we created a sheet named item, and this section is having eleven columns which are item_id (unique name for item), category_id (the column at which item will fall), sub_category_id (the subcategory id at which item will fall), item_title (name of item), image_url (item URL), address (business address of a particular location), longitude (distance from current location), latitude (distance from current location), contact_number (business contact of a particular location) and description (business description) shown in Fig. 5.

After creating the Google Sheet, then we save it and copy the sheet id from the URL as shown Fig. 6.

The marked id will be copied, and this is what will be used in Android Studio to display information on the app. A Java class has been created in Android Studio called HttpParams.java. This class takes the parameters of the Google Spreadsheet id, category, subcategory and items. After adding this project id to Android Studio, the APK file is generated by selecting Build > Generate Signed Bundle/APK, as illustrated in Fig. 7.

After generating the APK file, the app can be either installed on any Android device or published on Google Play Store for people to install. Categories are added to the spreadsheet, and more categories can still be added (the amount of categories that can be added are unlimited). For the demonstrations of our mobile app, categories presented in Fig. 8 are created. Once categories are added to the spreadsheet, it will be reflected automatically on the app as it works dynamically. Having explained the architecture design of Location Finder application, we can now move to the user interface.

Fig. 6 Project URL

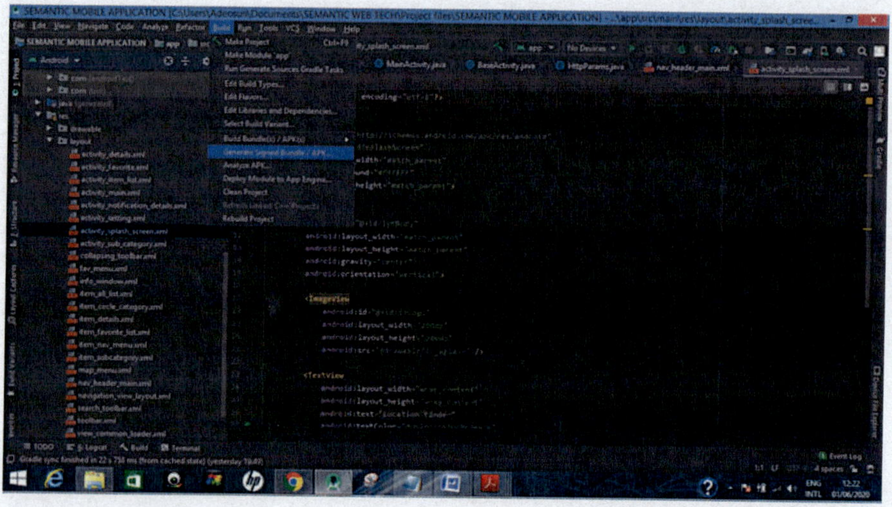

Fig. 7 Generating the .apk file

Fig. 8 Created subcategory hierarchy for demonstrations

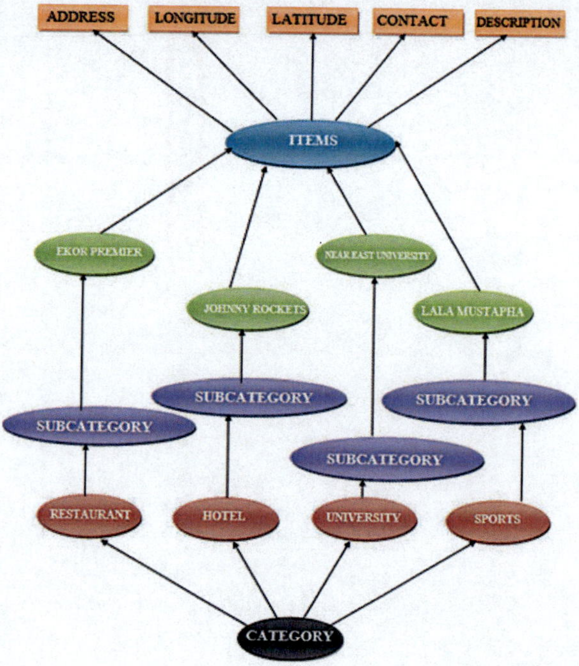

3 User Interface of Location Finder

The Location Finder app interface is designed to allow users of small screen devices run a search for a particular query like "Restaurant, Hotel and so on". To this end, the application utilizes the primary features of the app interaction model. The features of Location Finder app are as follows:

- App will automatically detect user current location.
- The contact number of each business can be easily dialled for more information about a particular location.
- It allows user to easily navigate to desired destination
- List of locations are already created.
- Users can save an item or a particular location of interest for further exploration in future.
- Categories of different items (i.e., names of available restaurants).

These user interactions are illustrated in Fig. 9. The user just opens up Location Finder app (Fig. 9a). The first screen that the user will see is the splash screen activity with a beautifully designed logo and a progress bar. User gives permission to app to access the user's current location (Fig. 9b). Then, user's location is detected (Fig. 9c). User then navigates using the navigation bar (Fig. 9d). User can set a radius setting in km, where the app looks for items of interest in this area (Fig. 9e). User just selected "Restaurant" from the list since s/he wants to find nearby restaurants (Fig. 9f). Ekor Premier has been chosen by the user (Fig. 9g). The user is trying to locate Ekor Premier in Famagusta using the map (Fig. 9h). A particular location can also be saved to easily access it next time as shown in Fig. 9i.

To summarize, Location Finder is a complete Android application to find different categories such as restaurant, hotel, popular places, shopping mall, ATM, hospital, fuel station, popular food item place, public toilet, accident-prone area and many more. Every categories also have a subcategory. The app has been built on standard Google Map to ensure highest accuracy and portability. The backend has been built dynamically based on Google Sheet. It will help us to update and maintain data without additional server or database. It provides advanced mapping systems that will show different places for a specific category. It would be easy to search nearby place and navigate in a user-friendly way. Initially, it will find the nearest points based on current user location. User will be able to find a particular location from a different place using built-in custom search facility. Our code is also reusable since it has been built with JAVA and Google SpreadSheets. In the next section, we also compare performance of Location Finder with other similar work in terms of execution time.

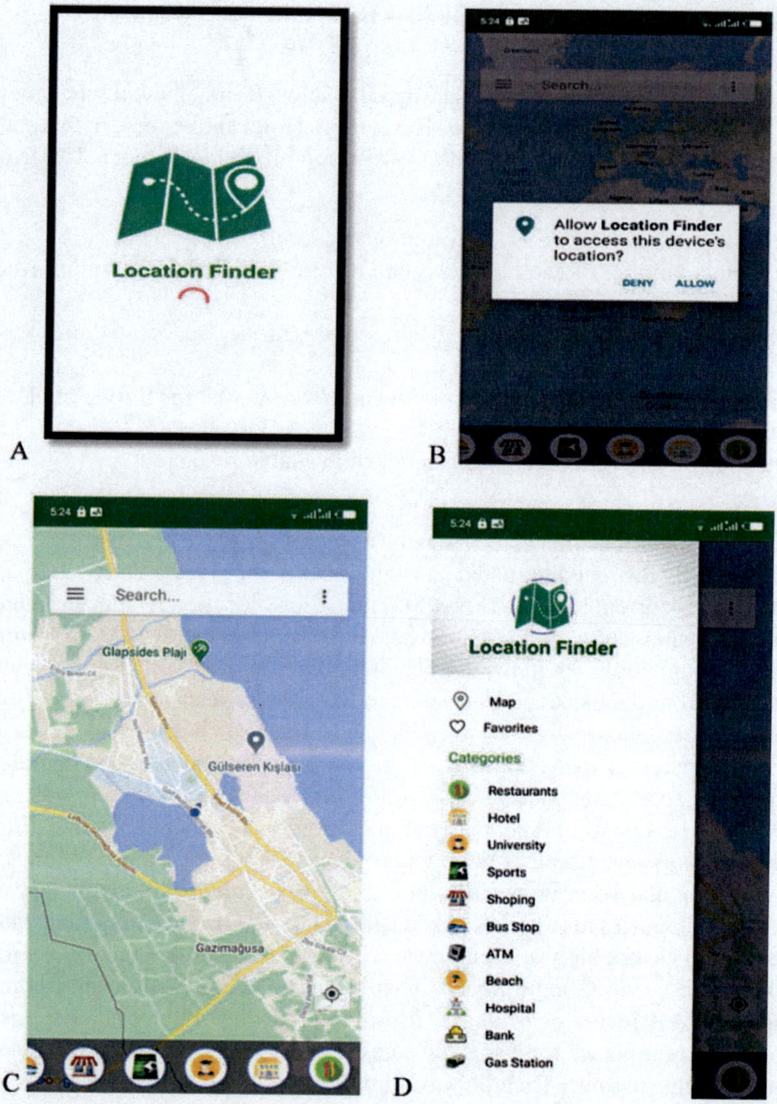

Fig. 9 Nine screens that explain how the app works. The user just opens up Location Finder app and is about to find a restaurant. **a** The first screen that user will see is the splash screen activity with a beautifully designed logo and a progress bar. **b** Gives permission to app to access the user's current location. **c** User's location has been detected. **d** User navigates using the navigation bar. **e** Settings screen. **f** User just selected "Restaurant" menu. **g** Ekor Premier has been chosen by the user. **h** The user is trying to locate Ekor Premier in Famagusta. **i** This shows that a particular location can be saved to easily access it next time

Fig. 9 (continued)

Fig. 9 (continued)

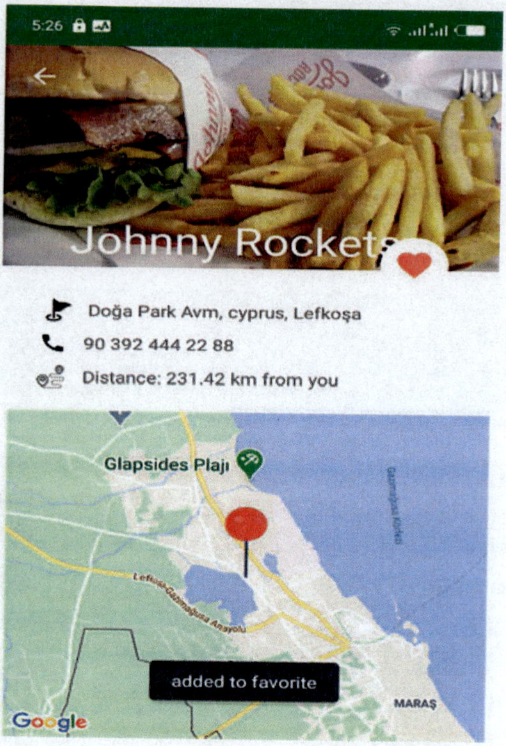

4 Evaluation and Comparison of Mobile Application Performances

To assess performance of the proposed mobile application, Location Finder, we test the response time using Android Profiler on Android Studio. Android Profiler is part of Android Studio 3.0 that provides real-time graphical display of CPU usage, memory and network usage for the running a mobile application. Furthermore, we compare the results of the Location Finder, with other mobile applications, namely AWARE [5], Device Analyser [6], Momento [7], MetricWire [8] and Google Timelines [9]. Experiments are tested in a mobile phone with the following technical details: 4 GB RAM, Android version 6 and CPU Octa-core.

For the experiment, we follow the guidelines from [14]. In particular, four test cases are defined: (1) Accessing an element from an array, (2) Iterate through an array, (3) Iterate through a matrix and (4) Calculation of Fibonacci number [14]. The test cases require input parameters, and three ranges are defined such as small, medium and large. For the small range, an arbitrary and small enough value is taken that contributes to changing the utilization of the memory and CPU. For the highest range, a rough upper boundary value is taken, after which the device has insufficient

Table 1 Comparative results for response time

Mobile applications	Platform	Response time	
		Wi-Fi (s)	4G (s)
AWARE [5]	Android/iOS	7.09	9.03
Device Analyser [6]	Android	15.04	16.11
Momento [7]	Android	8.91	9.02
MetricWire [8]	Android/iOS	10.19	10.34
Google Timeline [9]	Android/iOS	7.23	7.31
Proposed approach	Android	5.28	5.71

resources to perform the action. For the details of the experimental ranges and the settings, please refer to [14].

We compare performances of all mobile applications with the proposed mobile application. Results are shown in Table 1. Among the different mobile applications, Device Analyser has the highest response time for Wi-Fi and 4G network with 15.04 s and 16.11 s, respectively. Comparing to other applications, AWARE and Google Timelines achieve quicker response times. But among all mobile applications, Location Finder has the best response times with 5.28 s and 5.71 s for Wi-Fi and 4G network, respectively.

The good performance of Location Finder is attained for a number of reasons. Firstly, the mobile application has been developed on Android platform. Secondly, the proposed mobile application is fully optimized by getting rid of codes that are not adding any value, checking the application's efficiency, trying and testing the code, use profiling tools for monitoring, emphasizing on increasing app usability and focusing on the user interface. Having putting these into consideration, we are able to have a better response time which makes our application faster than the other ones. In our work, beyond identifying the location, the processing is kept to minimum. Thus, it can save battery life. On the other hand, many of the compared popular mobile applications run numerous background processes, data collection and sharing tools, which add extra processing time and increase the response time of the mobile applications.

5 Conclusions

We explained the proposed Location Finder mobile application that is based on Google SpreadSheets and Android devices. This application works semantically as different locations can be automatically detected. It is a complete native Android application to find different locations such as restaurant, hotel, shops, bus stops, ATM, universities, hospitals, gas station and many more. It provides advanced mapping systems that will show different places of interest for application users. Users can also adjust the search settings (categories, searching area radius) and can save a place for finding it next time. As a result, we provided an easy to use and reusable mobile application. Experiments on the performance of Location Finder also shows that the proposed approach provides much faster response times and better than other popular mobile applications in the domain.

References

1. Farley, P., Capp, M.: Mobile web services. BT Technol. J. **23**(2), 202–213 (2005)
2. Harris, C., et al.: mSpace: exploring the semantic web. In: A Technical Report in Support of the mSpace Software Framework, p. 98. IAM Group, University of Southampton, Southampton (2004)
3. Max, W., Alistair R., Daniel A., Alisdair O.: mSpace Mobile: A Mobile Application for the Semantic Web. IAM Research Group School of Electronics and Computer Science University of Southampton, SO17 1BJ. http://mspace.fm/
4. Schraefel, M.C., Karam, M., Zhao, S.: mSpace: interaction design for userdetermined, adaptable domain exploration in hypermedia. In: International Workshop on Adaptive Hypermedia. 2003. Nottingham, UK
5. Ferreira, D., Kostakos, V., Dey, A.K.: AWARE: mobile context instrumentation framework. Front. ICT, 20 April 2015. https://doi.org/10.3389/fict.2015.00006
6. Carter, S., Mankoff, J., Heer, J.: "/". In: Proceedings of the 2007 Conference on Human Factors in Computing Systems, 2007. https://doi.org/10.1145/1240624.1240644
7. Wagner, D.T., Rice, A., Beresford, A.R.: Device analyzer: understanding smartphone usage. In: International Conference on Mobile and Ubiquitous Systems: Computing, Networking and Services, 2014
8. MetricWire Mobile Application. Last accessed at 16th of February 2021. https://play.google.com/store/apps/details?id=com.metricwire.android3
9. Rodriguez, A.M., Tiberius, C., van Bree, R., Geradts, Z.: Google timeline accuracy assessment and error prediction, Forensic Sci. Res. **3**, 3, 240–255 (2018). https://doi.org/10.1080/20961790.2018.1509187
10. Material Design. https://material.io/design/
11. Balsamiq Studios: 2018. basalmiq. (2018). https://balsamiq.com/
12. Call-Em-All: 2018. Material-UI. (2018). https://material-ui-next.com/
13. Google Account: https://accounts.google.com
14. Andonoska, A., Jakimoski, K.: Performance Evaluation of Mobile Applications (2018). https://www.researchgate.net/publication/337437805

Sign Language Recognition with Quaternion Moment Invariants: A Comparative Study

Ilham El Ouariachi, Rachid Benouini, Khalid Zenkouar, Arsalane Zarghili, and Hakim El Fadili

Abstract In this paper, we aim to carry out a brief study of a new sets of Quaternion Discrete Moment Invariants (QDMI) in uniform lattice named: Quaternion Tchebichef Moment Invariants (QTMI), Quaternion Krawtchouk Moment Invariants (QKMI), and Quaternion Hahn Moment Invariants (QHMI), for hand gesture-based sign language recognition. For this purpose, we present briefly the principles of those moment invariants. Then, we test them on several datasets that contain challenging attributions, with regard to invariability and recognition under noise-free and noisy conditions. We conclude our paper by discussing the obtained results and the future works in this field.

1 Introduction

Sign language is a challenging task of research due to the complexity involved by gestures of signs. The basic component of sign language is hand gestures, aimed at communicating a message, by combining it with other body postures and facial expressions. Hand gesture-based sign language recognition requires solving a number of computer vision issues, including (i) hand gesture detection (ii) hand gesture feature extraction (iii) hand gesture recognition[1, 2].

In general, feature extraction algorithms for hand gestures-based sign language can be classified in two main types: dynamic and static algorithms [3]. In the case of static systems, different methods and algorithms have been proposed to extract features from hand gestures, including adaptive skin color model switching [4], hierarchical elastic graph matching algorithm [5], histogram of oriented gradient

I. El Ouariachi (✉) · R. Benouini · K. Zenkouar · A. Zarghili
Laboratory of Intelligent Systems and Application (LSIA), Faculty of Sciences and Technology, University Sidi Mohamed Ben Abdellah, Fez, Morocco
e-mail: ilham.elouariachi@usmba.ac.ma

H. El Fadili
Ecole Nationale des Sciences Appliquees of Fez, University Sidi Mohamed Ben Abdellah, Fez, Morocco

M. Ben Ahmed et al. (eds.), *Networking, Intelligent Systems and Security*, Smart Innovation, Systems and Technologies 237,
https://doi.org/10.1007/978-981-16-3637-0_52

[6], Gabor filters [4], scale invariant feature transform [7], and local binary pattern [8].

Despite the huge work that have been done in this area[3–8], feature extraction from hand gesture-based sign language remains unresolved issue due to appearance variations of the moving hand gesture, the complex background, and rotation scale and translation (RST) deformations.

Recently, with the availability of 3D sensors, such as Microsoft Kinect, and Asus Xtion [3], there has been an increasing concern in the research area related to hand gesture-based sign language. In fact, the RGB-D sensors allow to capture red, green, blue, and depth (RGB-D) informations from a scene at the same time [3].

Currently, image moments and moment invariants are one of the most active research areas in the fields of pattern recognition and computer vision [1, 9], and this is due to their properties that represent the information with minimum redundancy, robustness to different kinds of noise, and the invariability property against geometric deformations [10].

In that respect, various methods have been proposed for hand gesture-based sign language feature extraction using image moments and moment invariants [11–13]. However, the effectiveness of the moment invariants in hand gesture-based sign language recognition by using RGB-D images has not been amply examined, and only very few researches have been done with this interest.

Motivated by the search in moments and moment invariants area related to hand gesture-based sign language, we present a comparison between three Quaternion Discrete Moment Invariants(QDMI) in uniform lattice: Quaternion Tchebichef Moment Invariants(QTMI), Quaternion Krawtchouk Moment Invariants(QKMI), and Quaternion Hahn Moment Invariants(QHMI).

In this paper, we propose a new sets of moment invariants named: Quaternion hahn moment invariants(QHMI), we present a brief study of QDMI in the uniform lattice: QTMI, QKMI, and QHMI, also we evaluate and compare them using challenging hand gesture-based sign language databases.

The remainder of this paper is organized as follows: Sect. 2 we present a brief introduction of Quaternion algebra. Then, in Sect. 3, we introduce the principles of the three evaluated methods: QTMI, QKMI including the proposed QHMI. Section 4 reports the evaluation of the experiments result. Finally, Sect. 5 concludes this paper and projects some future works.

2 Quaternion Algebra

Quaternions have been introduced by Hamilton in 1843 [14]. The quaternion number has four parts, one real part and three imaginary parts. The formula of a quaternion q is defined as follows:

$$q = q_r + q_i i + q_j j + q_k k, \tag{1}$$

where q_r, q_i, q_j, q_k are real numbers and i, j, k are complex operators, which obey the following rules:

$$i^2 = j^2 = k^2 = -1, ij = -ji = k, jk = -kj = i, ki = -ik = j. \tag{2}$$

Let $f(x, y)$ be an RGB-D image function defined in cartesian coordinate system, using the quaternion representation, each pixel can be represented as a quaternion [1] as following:

$$f(x, y) = f_D(x, y) + f_R(x, y)i + f_G(x, y)j + f_B(x, y)k, \tag{3}$$

with $f_D(x, y)$, $f_R(x, y)$, $f_G(x, y)$ and $f_B(x, y)$ are, respectively, the depth, red, green, and blue components of the pixel (x, y).

3 Evaluated Quaternion Moment Invariants Methods

3.1 *Quaternion Tchebichef Moment Invariants*

Elouariachi et al. [1] proposed a robust hand gesture recognition system using a new set of Quaternion Tchebichef Moment Invariants. The authors derived directly the proposed invariants from their orthogonal moments, based on the algebraic properties of the discrete Tchebichef polynomials. The proposed method based on quaternion algebra is suggested to process the four components holistically, for a robust and efficient hand gesture recognition system.

Let $f^d(x, y)$ be the deformed version of the original image $f(x, y)$, defined by:

$$f^d(x, y) = f(a_{11}x + a_{12}y, a_{21}x + a_{22}y). \tag{4}$$

The Quaternion Tchebichef Moment Invariants $\mathrm{QTMI}_{n,m}^{RST}$ of a deformed image $f^d(x, y)$, which is rotation, scaling, and translation invariance is defined as:

$$\begin{aligned}\mathrm{QTMI}_{n,m}^{RST} &= \sum_{i=0}^{n}\sum_{j=0}^{m}\sum_{s=0}^{i}\sum_{t=0}^{j}\sum_{u=0}^{i+j-s-t}\sum_{v=0}^{s+t}(-1)^{j-t}\binom{i}{s}\binom{j}{t} \times A_{n,i}A_{m,j} \\ &\times B_{i+j-s-t,u}B_{s+t,v} \times (\lambda_f)^{-\frac{i+j+2}{2}}(\cos\theta_f)^{i+t-s}(\sin\theta_f)^{j-t+s}QTM_{u,v}^t.\end{aligned} \tag{5}$$

Along this paper, it will be noted as QTMI.

3.2 Quaternion Krawtchouk Moment Invariants

The Krawtchouk polynomials were introduced by Mikhail Kravchuk in [15]. The classical Krawtchouk polynomials of the nth order are defined in terms of the hypergeometric function ${}_2F_1(.)$ as follows:

$$k_n(x; p, N) = {}_2F_1(-n, -x; -N; 1/p). \tag{6}$$

Similarly, to the Quaternion Tchebichef Moment Invariants, we can define the following $\text{QKMI}^{\text{RST}}_{n,m}$ which is rotation, scaling, and translation invariance as follows:

$$\text{QKMI}^{\text{RST}}_{n,m} = \sum_{i=0}^{n}\sum_{j=0}^{m}\sum_{s=0}^{i}\sum_{t=0}^{j}\sum_{u=0}^{i+j-s-t}\sum_{v=0}^{s+t}(-1)^{j-t}\binom{i}{s}\binom{j}{t} \times C_{n,i}C_{m,j} \times D_{i+j-s-t,u}D_{s+t,v} \times (\lambda_f)^{-\frac{i+j+2}{2}}(\cos\theta_f)^{i+t-s}(\sin\theta_f)^{j-t+s}QKM^t_{u,v}, \tag{7}$$

Along this paper it will be noted as QKMI.

3.3 Quaternion Hahn Moment Invariants

The Hahn polynomials have been firstly introduced in the field of image analysis by Zhu et al. in [16]. Similarly to Tchebichef and Krawtchouk polynomials, we can define the $n-th$ order Hahn polynimials in terms of the hypergeometric function ${}_3F_2(.)$ as follows:

$$h_n^{(a,b)}(x; N) = \frac{(-1)^n(b+1)_n(N-n)_n}{n!} {}_3F_2(-n, -x, n+1+a+b; b+1, 1-N; 1), \tag{8}$$

We can define the following $QHMI^{RST}_{n,m}$ which is rotation, scaling, and translation invariance as follows:

$$\text{QHMI}^{\text{RST}}_{n,m} = \sum_{i=0}^{n}\sum_{j=0}^{m}\sum_{s=0}^{i}\sum_{t=0}^{j}\sum_{u=0}^{i+j-s-t}\sum_{v=0}^{s+t}(-1)^{j-t}\binom{i}{s}\binom{j}{t} \times E_{n,i}E_{m,j} \times F_{i+j-s-t,u}F_{s+t,v} \times (\lambda_f)^{-\frac{i+j+2}{2}}(\cos\theta_f)^{i+t-s}(\sin\theta_f)^{j-t+s}QHM^t_{u,v}, \tag{9}$$

Along this paper it will be noted as QHMI.

4 Experimental Results

In this experimental study, three real hand gesture-based sign language databases, namely HKU [17], NTU [3], and ASL[18] are used to evaluate the three sets of discrete quaternion moment invariants. The first one contains 10 gestures with 20 different poses from 5 subjects. Therefore, there are a total of 1000 cases, each of which consists of a pair of color and depth information. The NTU contains 1000 cases of 10 signs from 10 subjects. The third database captures about 65000 samples of 24 signs (English letters expect j and z) from 5 subjects. The hands are located and segmented using the hand-wrist belt for the second dataset and depth thresholding for the others.

All the experiments are conducted on a personal computer equipped with an Intel(R) Core i7 2.1 GHz and 4 GB of memory. And all the algorithms are coded in Matlab 8.5.

4.1 *Invariability Property*

The task of sign language recognition is highly challenging, due to the geometric deformations performed by the hand, during the execution of the sign. To solve this issue, an analysis of the three moment invariants with respect to the variation of image geometric deformation, translation, scale, and rotation is presented.

This section is intended to test the performance of the proposed QTMI, QKMI, and QHMI quaternion discrete orthogonal moment invariants using RGB-D images. We evaluate the invariability property under various geometric deformations and noise conditions. For this, we use a sign language image, whose size is 150 × 150 pixels [17].

Let RE be the relative error between two sets of quaternion discrete orthogonal moment invariants corresponding to the original and transformed image as:

$$\mathrm{RE}(f, f^t) = \frac{||\mathrm{QDMI}(f) - \mathrm{QDMI}(f^t)||}{||\mathrm{QDMI}(f)||},$$

where f, f^t, $\mathrm{QDMI}(f)$ *and* $QDMI(f^t)$ represent, respectively, the original and the transformed images, as well as the quaternion discrete orthogonal moment invariants of the original image and the transformed one. The testing protocol is then constructed as follows: the test image is scaled with the factor $\lambda \in 0.7, 1.2$, and rotated from 0 to 360, with an increment 10; then the test image was translated by $(\Delta x, \Delta y) = (4, 4)$. Accordingly, the obtained relative error for the rotation, scaling, and translation transformations are shown, respectively, in Fig. 1a–c.

In the second part of this experiments, we will investigate the invariability property of the proposed QDMI in the presence of noisy effects. In fact, the previous hand

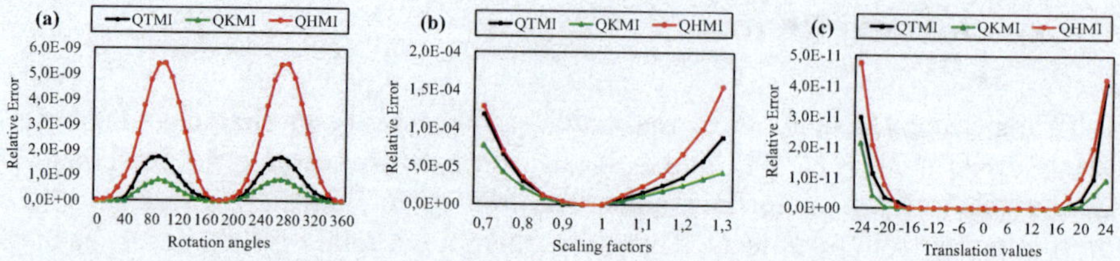

Fig. 1 Relative errors of the proposed QTMI, QKMI, and QHMI against: **a** Rotation, **b** Scaling and **c** Translation transformations

gesture test image is corrupted with different kinds of noise. First, the test image was corrupted by Gaussian noise with standard deviations varying from 0 to 5. Then, affected by salt-and-pepper noise with a density in the range 0–5%. Finally, degraded by speckle noise with standard deviation starting from 0 to 5. Consequently, the obtained results for noise invariance are summarized in Fig. 2. It is important note that the parameters of Krawtchouk and Hahn polynomials are restricted to $p = 0.5$, $a = 0$ and $b = 0$. In addition, moment invariants up to the order $(n, m \leq 3)$ are used to construct the feature vector.

As it can be seen from the Figs. 1 and 2, QKMI perform better than the QTMI and QHMI, not only in the case of geometric transformations (scale, rotation and translation), but also in the presence of noisy effects (salt-and-pepper, Gaussian and speckle).

According to the obtained results in the two figures, it is clear that the relative error is very low indicating that the proposed set of QDMI remain unchangeable under geometric transformations and in the presence of the noisy conditions. Therefore, we can conclude that these new sets of QDMI could be useful in pattern recognition tasks, especially in hand gesture-based sign language area, which suffer from various challenging transformations and noisy effects.

4.2 Recognition Results Under Noise-Free Conditions

The goal of this experiment is to evaluate the performance of the proposed descriptors without noisy effects in the task of sign language recognition. We carry out a series of experiments on the three databases as follow:

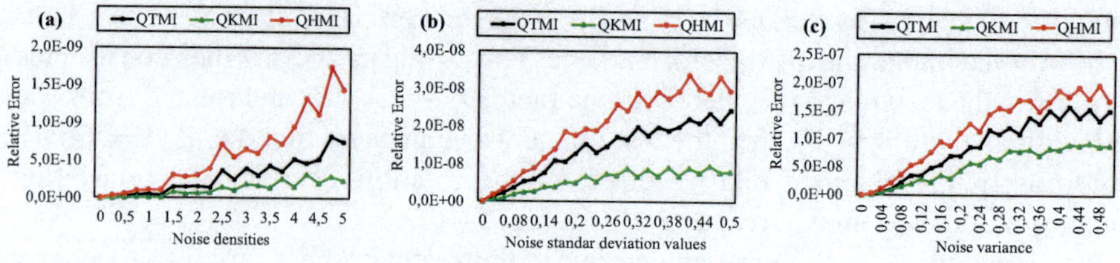

Fig. 2 Relative errors of the proposed QTMI, QKMI, and QHMI for different types of noise: **a** salt-and-pepper, **b** Gaussian and **c** speckle

Table 1 Comparative analysis of recognition accuracy (%) on HKU, NTU and ASL, by using the studied methods: QTMI, QKMI and QHMI, with a varying number of neighbors k

		Different distance metric *k*						
Databases	Methods	$k = 1$	$k = 2$	$k = 3$	$k = 4$	$k = 5$	$k = 6$	Average
	QTMI	71	71.5	72	72,9	73	72.8	72.2
	QKMI	70.2	70.9	71.3	72	72.7	72.3	71.57
NTU	QHMI	54.2	57.9	63.1	62.1	62.2	62.4	60.32
	QTMI	74.3	75	76	76.8	77,1	77	76
	QKMI	70.9	71.5	72,3	73.9	75	74.8	73.1
HKU	QHMI	55.9	56.6	56	56.6	58.8	59.4	57.27
	QTMI	80,6	81	81.9	82	82.7	82.1	81.72
	QKMI	75.7	76.5	77.3	78.6	79.1	78.2	77.6
ASL	QHMI	57.8	61.7	64.5	66.7	68.2	69.5	64.7

Experiment 1: We first evaluate the recognition accuracy of the studied QDMI, with a chosen number of coefficients in the feature vector (16 coefficients), and using k-NN as classifier with different values of the neighborhood parameter k, with 5-folds cross validation.

Experiment 2: We evaluate the recognition accuracy using different number of coefficients in the feature vector, with k =5 for the k-NN classifier.

The results of experiment 1 can be found in Table 1, where each row in Table 1 present the recognition accuracies of our moment invariants, for different values of the neighborhood parameter *k*; columns from three to ten present the recognition accuracies of different parameters *k*. The last column is the average recognition accuracies of each methods for all the chosen values of *k*.

As one can see from Table 1, the proposed QTMI demonstrates a superiority in terms of image classification accuracy for all databases compared with QKMI and QHMI. Therefore, this experiment approves the discrimination capability of our methods even though with the simplest classifier.

Generally, the utilization of moment invariants up to a higher order leads to a higher image description power. However, it has been proved that only a small number of moment invariants coefficients can produce accurate classification [10]. In this connection, we will investigate the influence of the number of moment invariants coefficient in the feature vector. The results of experiment 2 are shown in Table 2. Each row in Table 2 presents the recognition accuracies with variation in the number of coefficients in the feature vector, of the used methods; each column represents the recognition accuracies for a specific number of coefficients in the feature vector. The last column is the average accuracy of each methods for all the numbers of coefficients in the feature vector up to 64 coefficients.

Obviously, we can notice from the Table 2 that the augmentation in the number of coefficients has a good impact on the recognition accuracy for all the methods. In addition, as remarked in the experiment 1, the QTMI gives the best results among

Table 2 Comparative analysis of recognition accuracy (%) on HKU, NTU, and ASL, by using the studied methods: QTMI, QKMI, and QHMI with a varying number coefficient in the feature vector

Databases	Methods	N of coefficients in the feature vector						
		9	16	25	36	49	64	Average
NTU	QTMI	72.4	73	74.4	76.8	78.4	79.5	75.75
	QKMI	72	72.7	73.2	75.7	77.9	78.4	74.98
	QHMI	63.8	62.2	59.9	62.2	62.8	61.3	62.03
HKU	QTMI	69.3	77.1	85.1	84.1	85.3	91.4	82.05
	QKMI	73.4	75	77.9	80.8	82.6	88.7	79.73
	QHMI	59.5	58.8	58.4	55.6	52.2	57.8	57.05
ASL	QTMI	75	82,7	89.7	91.4	91.3	91.2	86.88
	QKMI	77.6	79.1	80.7	82.9	85.2	88.4	82.32
	QHMI	61.42	68.2	60.3	57.3	52.5	58.9	59.77

the proposed methods. This obtained results are justified by the fact that the QTMI are global shape descriptors [19], and they extract features from the whole sign gesture image. In the contrary, QKMI is a local descriptor, since it is computed with emphasis on a specific region of an image [15]. The proposed QHMI gives the lowest results compared with both QTMI and QKMI, due to its computational complexity and numerical instability in the calculation of polynomial values [20]. Finally, in this work, we choose to use for the rest of experiments, a feature vector of 16 elements [1].

4.3 Recognition Results Under Noisy Conditions

It is important to inspect the capability of our descriptors to remain robust against different sources of noise. In fact, the three databases used in this paper provide a real examples of the problems that can be faced in sign language recognition applications using depth cameras, such as depth noise and the presence of the arm and other parts of the body [3]. Moreover, we test our QDMI in the presence of other kinds of noise, salt-and-pepper and Gaussian noise.

Each image of the three datasets is corrupted by salt-and-pepper noise with noise densities varying from 1% to 5% with 1% increments, as well as the other type of noise: Gaussian with densities changing from 0.1 to 0.5 with 0.1 increments.

The performance of those moment invariants on the datasets is presented on Table 3, for the two types of noises, where each row of a dataset corresponds to the recognition accuracies for the different densities, and each column corresponds to recognition accuracies of all methods for a specific density. The last column is the average recognition accuracies of each method for all densities.

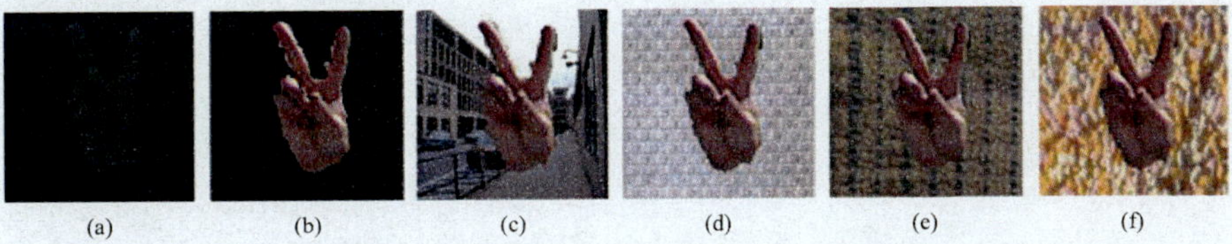

Fig. 3 Examples of hands gesture image, **a** original depth image, **b** segmented color image and different background images from: **c** Vistex, **d** Brodatz, **e** Outex and Amsterdam database **f**

Considering the presented results in Tables 3, it is obvious that the best average recognition rate for the three datasets was obtained by QTMI, and it is closely followed by the proposed QKMI.

4.4 Recognition Results Under Complex Background

In this experiment, we have considered four color texture image databases, namely VisTex [21], Brodatz [22], Outex [23], and Amsterdam [24], to generate four variations of color background on each of the testing hand gesture datasets HKU, NTU, and ASL. In fact, we have selected 50 different images from the three texture databases. These selected images have been randomly used as complex background for each sign image. And therefore, we have created four additional testing cases for every hand gesture-based sign language database (Fig. 3).

Table 4 summarizes, respectively, the obtained recognition results on the three testing datasets, HKU, NTU, and ASL, for different texture background, by using the studied quaternion moment invariant, QTMI, QKMI, and QHMI. According to the presented results in Table 4, we can see that only the proposed QTMI and QKMI are appropriate to be used to classify the sign gestures with complex background, with an average recognition rate above 70%. However, the proposed QHMI cannot obtain a higher average recognition rate, more than 60%. It is important to note that the QTMI and QKMI obtain quite similar recognition rate for many testing cases, showing high robustness against complex backgrounds.

Finally, we can deduce the importance incorporating depth information in feature extraction process, and we can conclude that QTMI and QKMI could be very effective for the applications involving depth and color information.

Table 3 Comparative results of recognition accuracy (%) on HKU, NTU, and ASL affected by different salt-and-pepper and Gaussian noise standard deviations, using the studied methods: QTMI, QKMI, and QHMI

Databases	Methods	Noise free	Salt-and-Pepper					Average	Gaussian					Average
			1%	2%	3%	4%	5%		0.01	0.02	0.03	0.04	0.05	
NTU	QTMI	73	72.1	71.8	70.9	70.2	69.5	71.25	71.9	70.8	70.4	69.9	69.2	70.86
	QKMI	72.7	71.9	70.5	70.1	70.6	66.8	70.43	71.1	70.4	70.5	70.1	69.8	70.76
	QHMI	62.2	61.9	60.8	60.3	59.7	59	60.65	61.9	60.8	60.3	59.7	59	60.65
HKU	QTMI	77.1	75.4	74.7	73.3	70.8	70	73.55	76.8	75.7	73	72.3	70.5	74.23
	QKMI	75	73.1	72.8	71.2	70.7	70.3	72.18	71	69.8	69.5	68.4	67.6	70.21
	QHMI	58.8	57.6	57.3	56.9	55.5	54.7	56.8	57.6	57.3	56.9	55.5	54.7	56.8
ASL	QTMI	82.7	80.4	78.8	76.7	74.2	73	77.63	81.2	80.7	78.6	77	76.4	79.43
	QKMI	79.1	78.2	76.3	75.8	74.2	73.7	76.21	77.6	76.8	75.5	75.1	75	76.51
	QHMI	68.2	67.7	65.3	64.1	63	62.4	65.11	67.7	65.3	64.1	63	62.4	65.11

Table 4 Influence of different complex backgrounds on the hand gesture-based sign language recognition accuracy (%) for HKU, NTU and ASL, by using the studied methods: QTMI, QKMI, and QHMI

Databases	Methods	Different types of complex background					Average
		Uniforme background	VisTex	Brodatz	Outex	Amsterdam	
NTU	QTMI	73	71.3	74.3	70.5	73.9	72.6
	QKMI	72.7	70.5	73.2	69.8	71.6	71.56
	QHMI	62.2	56.1	54.5	64.4	65.8	60.6
HKU	QTMI	77.1	76	74.3	75.7	72.1	75.04
	QKMI	75	74.8	73.4	74	72.6	73.96
	QHMI	58.8	46.1	44.9	63.6	60.5	54.78
ASL	QTMI	82.7	80.5	81.7	79.4	78.3	80.52
	QKMI	79.1	77.6	78.4	76.3	75	77.28
	QHMI	68.2	51.6	49.2	64.6	64.3	59.58

5 Conclusion

In this paper, we carried out a comparative study between a three Quaternion Moment Invariants in uniform lattice for hand gesture-based sign language recognition. The experimental results with the sets of moment invariants show that the QTMI performs well in terms of accuracy and robustness followed by the QKMI. There are many interesting ways to extend this work in the future. First, the KNN classifier is simple and could be replaced with a more sophisticated classifiers. Second, it would be interesting to automate the system of hand gesture-based sign language and use it for dynamic hand. Finally, we are interested to adopt the three proposed QDMI on more challenging situations such as luminosity changes, occultation of the target, and many others issues.

References

1. Elouariachi, I., Benouini, R., Zenkouar, K., Zarghili, A.: Robust hand gesture recognition system based on a new set of quaternion Tchebichef moment invariants. Pattern Anal. Appl. 1–17 (2020)
2. Elouariachi, I., Benouini, R., Zenkouar, K., Zarghili, A., El Fadili, H.: Explicit quaternion krawtchouk moment invariants for finger-spelling sign language recognition. In: 2020 28th European Signal Processing Conference (EUSIPCO), pp. 620–624. IEEE (2021, January)
3. Ren, Z., Yuan, J., Meng, J., Zhang, Z.: Robust part-based hand gesture recognition using kinect sensor. IEEE Trans. Multimed. **15**, 1110–1120 (2013)
4. Huang, D.Y., Hu, W.C., Chang, S.H.: Gabor filter-based hand-pose angle estimation for hand gesture recognition under varying illumination. Expert Syst. Appl. **38**(5), 6031–6042 (2011)
5. Li, Y.T., Wachs, J.P.: HEGM: a hierarchical elastic graph matching for hand gesture recognition. Pattern Recognit. **47**(1), 80–88 (2014)

6. Lin, J., Ding, Y.: A temporal hand gesture recognition system based on hog and motion trajectory. Optik **124**(24), 6795–6798 (2013)
7. Patil, S.B., Sinha, G.R.: Distinctive feature extraction for Indian Sign Language (ISL) gesture using scale invariant feature Transform (SIFT). J. Inst. Eng. (India): Ser. B **98**(1), 19–26 (2017)
8. Zhang, F., Liu, Y., Zou, C., & Wang, Y.: Hand gesture recognition based on HOG-LBP feature. In: 2018 IEEE International Instrumentation and Measurement Technology Conference (I2MTC), pp. 1–6. IEEE (2018, May)
9. Benouini, R., Batioua, I., Elouariachi, I., Zenkouar, K., Zarghili, A.: Explicit separable two dimensional moment invariants for object recognition. Procedia Comput. Sci. **148**, 409–417 (2019)
10. Flusser, J., Suk, T., Zitov, B.: 2D and 3D image analysis by moments. Wiley, Hoboken (2016)
11. Jadooki, S., Mohamad, D., Saba, T., Almazyad, A.S., Rehman, A.: Fused features mining for depth-based hand gesture recognition to classify blind human communication. Neural Comput. Appl. **28**(11), 3285–3294 (2017)
12. Hu, Y.: Finger spelling recognition using depth information and support vector machine. Multimedia Tools Appl. **77**(21), 29043–29057 (2018)
13. Gallo, L., Placitelli, A.P.: View-independent hand posture recognition from single depth images using PCA and Flusser moments. In: 2012 eighth international conference on signal image technology and internet based systems, pp. 898–904. IEEE (2012, November)
14. Hamilton, W.R.: Elements of quaternions. Longmans, Green, & Company (1866)
15. Krawtchouk, M.: On interpolation by means of orthogonal polynomials. Memoirs Agric. Inst. Kyiv **4**, 21–28 (1929)
16. Zhou, J., Shu, H., Zhu, H., Toumoulin, C., & Luo, L.: Image analysis by discrete orthogonal Hahn moments. In International Conference Image Analysis and Recognition, pp. 524–531. Springer, Berlin, Heidelberg (2005, September)
17. Wang, C., Liu, Z., Chan, S.C.: Superpixel-based hand gesture recognition with kinect depth camera. IEEE Trans. Multimedia **17**(1), 29–39 (2014)
18. Pugeault, N., Bowden, R.: Spelling it out: Real-time ASL fingerspelling recognition. In: 2011 IEEE International conference on computer vision workshops (ICCV workshops), pp. 1114–1119. IEEE (2011, November)
19. Mukundan, R., Ong, S.H., Lee, P.A.: Image analysis by Tchebichef moments. IEEE Trans. Image Process. **10**(9), 1357–1364 (2001)
20. Karakasis, E.G., Papakostas, G.A., Koulouriotis, D.E., Tourassis, V.D.: Generalized dual Hahn moment invariants. Pattern Recogn. **46**(7), 1998–2014 (2013)
21. VisionTexture (VisTex). http://vismod.media.mit.edu/vismod/imagery/VisionTexture/Images/Reference/
22. Colored Brodatz (CBT). http://multibandtexture.recherche.usherbrooke.ca/colored%20_brodatz.html
23. Outex texture (Outex). http://lagis-vi.univ-lille1.fr/datasets/outex.html
24. Amsterdam Library of Textures (Amsterdam). http://aloi.science.uva.nl/public_alot/

Virtual Spider for Real-Time Finding Things Close to Pedestrians

Souhail Elkaissi and Azedine Boulmakoul

Abstract The active growth of technology gives a big push to the way we collect the spatial data either by surveying which means collect the data by surveyors and then use it to create highly precise maps, they calculate the precise position of points, distances, and angles through geometry. In this paper, we consider the collection of timely spatial data by a method based on spiders' behavior. The data can be collected by remote sensing which uses satellites orbiting the Earth to capture information of the surface and atmosphere. One of the major challenges in collecting spatial data is to get it as fast as possible and to be able to scale contingent on the amount and the size of the data. In this matter, we see that messages brokers like Kafka can be very useful, thus it gives the ability to provide a real-time architecture for a real-time streaming data, it is scalable, durable, and a fault-tolerant published-subscribe messaging system. The other major challenge in collecting spatial data is to be able to manipulate this data with a smoother and faster way. Yet geography is a natural data domain for graphs and graph databases. Geometries and topologies could simply be drawn on graph databases like Neo4j. With their ability of expressing geo data intuitive way, graph databases could be used from calculating routes between locations in an abstract network such as a road or rail network, airspace network, or logistical network to spatial operations such as find all points of interest in a bounded area.

1 Introduction

With the growth of technology, the location awareness became more and more fluid, this is due to the usage of connected objects around us, like GPS devices, smartphones, sensors, etc. These devices collect a lot of information about individual persons, communities, and the eco-system that we live in. It collects what each individual does, or where he goes, when he eats his breakfast, his heartbeat rating, etc. These collected data can be very massive in terms of field diversity and in term of size, so it should

S. Elkaissi · A. Boulmakoul (✉)
LIM/IOS, FSTM, Hassan II University of Casablanca, Casablanca, Morocco

M. Ben Ahmed et al. (eds.), *Networking, Intelligent Systems and Security*,
Smart Innovation, Systems and Technologies 237,
https://doi.org/10.1007/978-981-16-3637-0_53

be stored in a scalable format where the data can be manipulated and transformed smoothly [16–18]. That said graph databases provide an excellent infrastructure to link diverse data. With easy expression of entities and relationships between data, graph databases make it easier for programmers, users, and machines to understand the data and find insights. This deeper level of understanding is vital for successful machine learning initiatives, where context-based machine learning is becoming important for feature engineering, machine-based reasoning, and inferencing [1, 2, 10]. These collected data can be used to enhance the quality and comfort of the pedestrian on the pedestrian walkaway. We can help to increase the safety of the pedestrian by minimizing the number of eventual accidents that could happen, and also, it will help to increase the walkability of the surrounding. We could find out ways or keys of amelioration of pedestrian walkway. The data allows to understand how the pedestrian walks. To sum up, the pedestrian study helps to increase the sustainability of the walking area. In the paper, we have created a spatial data pipeline based on a virtual spider. Let us imagine that it exists an intelligent virtual spider that can travel on streets or avenues or even cities and it is conscious of its position, and it is aware of all the elements that are in its circular area of radius r such as stores, avenues, highways, institutions. The spider will go through long straight streets and try to collect all the possible information around. All the collected information is spatial data, which is a geometry like points, lines or polygons that have a meaningful semantic match of a real and a physical object. At the end of its journey, the spider could give us the point of interest based on a criterion. This kind of systems (like the spider) is extremely useful nowadays, and it could be such a relief in situations of finding extremely fast some points of interest like, for example, the time when it is safe to go on the pedestrian walkway [14, 15]. Trajectory-based human mobility data analysis research has largely focused on the trajectories of people and vehicles, driven by the fact that geographic information science has traditionally supported spatial information from moving objects [5–9, 11–13]. Currently, the interest of spatiotemporal data analysis for road safety is a real challenge for modern cities [3, 4].

The rest of this article is organized as follows; the following section describes the main components of the proposed system, namely the graph database and the messaging broker. Section 3 details the implementation of the proposed architecture. The article ends with a conclusion and offers further work in Sect. 4.

2 System Components

2.1 Graph Database

Graph databases are purpose-built to store and navigate relationships. Relationships are first-class citizens in graph databases, and most of the value of graph databases

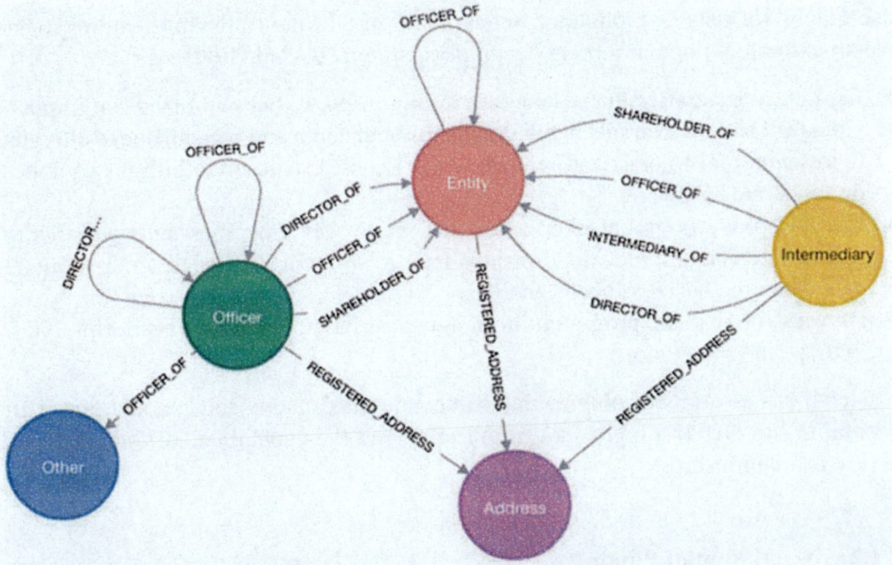

Fig. 1 Graph of different nodes and edges (Neo4j (https://neo4j.com/))

is derived from these relationships. Graph databases use nodes to store data entities and edges to store relationships between entities. An edge always has a start node, end node, type, and direction, and an edge can describe parent–child relationships, actions, ownership, and the like. There is no limit to the number and kind of relationships a node can have (Fig. 1).

A graph in a graph database can be traversed along specific edge types or across the entire graph. In graph databases, traversing the joins or relationships is amazingly fast because the relationships between nodes are not calculated at query times but are persisted in the database. Graph databases have advantages for use cases such as social networking, recommendation engines, and fraud detection, when you need to create relationships between data and quickly query these relationships.

2.1.1 Neo4j Graph Database

Neo4j is an open-source, NoSQL, native graph database that provides an ACID-compliant transactional backend for your applications. The source code, written in Java and Scala, is available for free on GitHub or as a user-friendly desktop application download. Neo4j is referred to as a native graph database because it **efficiently implements the property graph model down to the storage level.** This means that the data is stored exactly as your whiteboard is, and the database uses pointers to navigate and traverse the graph. In contrast to graph processing or in-memory libraries, Neo4j also provides full database characteristics, including ACID transaction compliance, cluster support, and runtime failover—making it suitable to use graphs for data in production scenarios. Some of the following features make Neo4j extremely popular among developers, architects, and DBAs:

- **Cipher**, a declarative query language similar to SQL, but optimized for graphs.
- **Constant time traversals** in big graphs for both depth and breadth due to efficient representation of nodes and relationships. Enables scale-up to billions of nodes on moderate hardware.
- **Flexible** property graph schema that can adapt over time, making it possible to materialize and add new relationships later to shortcut and speed up the domain data when the business needs change.
- **Drivers** for popular programming languages, including Java, JavaScript,.NET, Python, and many more.

Neo4j has a variety of plugins that can extend its capacity and can add powerful capabilities to Neo4j. Plugins are meant to extend the capabilities of the database, nodes, or relationships.

2.1.2 Neo4j Spatial Plugin

Neo4j spatial is a plugin utility for Neo4j that facilitates the enabling of spatial operations on data. In particular, you can add spatial indexes to already located data and perform spatial operations on the data like searching for data within specified regions or within a specified distance of a point of interest. In addition, classes are provided to expose the data to GeoTools and thereby to GeoTools-enabled applications like GeoServer and uDig.

2.2 Kafka Message Broker

Kafka[1] technically speaking is an event streaming platform which is the practice of capturing data in real time from event sources like databases, sensors, mobile devices, cloud services, and software applications in the form of streams of events; storing these event streams durably for later retrieval; manipulating, processing, and reacting to the event streams in real time as well as retrospectively; routing the event streams to different destination technologies as needed. Event streaming thus ensures a continuous flow and interpretation of data so that the right information is at the right place, at the right time.

---

[1] https://kafka.apache.org/.

2.3 Geographical Information System

A geographic information system (GIS) is a conceptualized framework that provides the ability to capture and analyze spatial and geographic data. GIS applications (or GIS apps) are computer-based tools that allow the user to create interactive queries (user-created searches), store and edit spatial and non-spatial data, analyze spatial information output, and visually share the results of these operations by presenting them as maps. Modern GIS technologies use digital information for which various digitized data creation methods are used. The most common method of data creation is digitization, where a hard copy map or survey plan is transferred into a digital medium through the use of a CAD program and geo-referencing capabilities. With the wide availability of ortho-rectified imagery (from satellites, aircraft, and UAVs), heads-up digitizing is becoming the main avenue through which geographic data is extracted. Heads-up digitizing involves the tracing of geographic data directly on top of the aerial imagery instead of by the traditional method of tracing the geographic form on a separate digitizing tablet (heads-down digitizing). Geoprocessing is a GIS operation used to manipulate spatial data. A typical geoprocessing operation takes an input dataset, performs an operation on that dataset, and returns the result of the operation as an output dataset. Common geoprocessing operations include geographic feature overlay, feature selection and analysis, topology processing, raster processing, and data conversion. Geoprocessing allows for definition, management, and analysis of information used to form decisions.

3 System Design

In this part, we will talk about the design and architecture of our system (Fig. 2).

We will go and explain in details every section in the following sections.

3.1 Collected Data

We are interested into gathering as much data as possible from many sources at once and bind that data together and store it into one database. The first part of the system is the data gathering (Fig. 3).

The spider will go around pedestrian walkways to collect all kind of useful information. Each of this information above is sent by the spider using a Kafka producer to a specific topic in the Kafka broker.

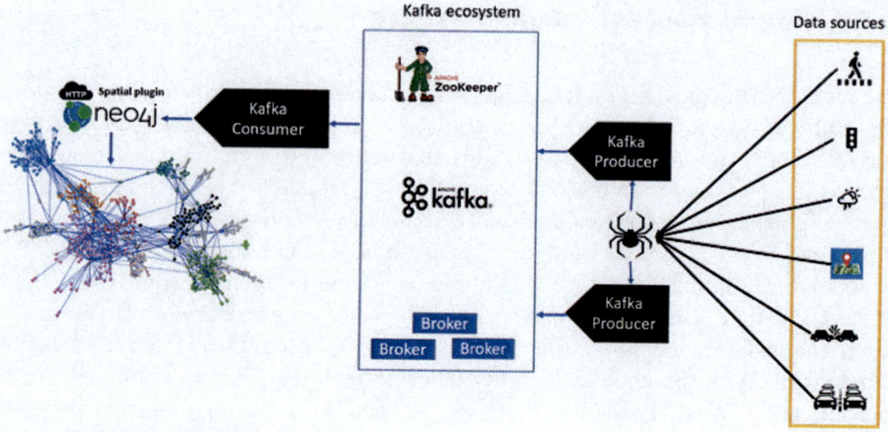

Fig. 2 System architecture of spider's gathering data

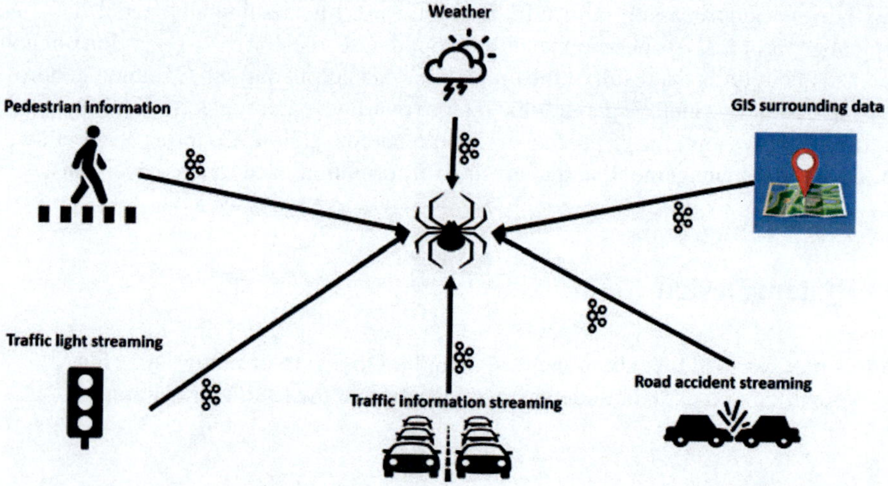

Fig. 3 All data sources of the spider

3.1.1 GIS Surrounding Data

We got an intelligent spider that can detect its position and be aware of all the elements that are in a circular area of radius R. We can expand the radius R to collect a mush GIS data as possible for a big surrounding area. The spider can move freely over streets ups and downs. At each point in its path beside or on pedestrian walkway, the spider collects all the spatial data on a circular area of a radius R, indeed the spatial data can be any kind of geometries such as points, polygons, or lines. Every geometry is an abstraction of real and physical places or locations that may be a point of interest. You can see in red circles on the next images the points that are beside

Ibn Rochd Avenue. Also, in the next images, we can see in purple the polygons close to Ibn Rochd Avenue (Figs. 4, 5).

For example, if the spider is at the point with the coordinates (33.987181, −6.852403) on the road of the Ibn Rochd Avenue and the radius is 70 m, the spider will collect the following points (Fig. 4).

- Burger kings Rabat,
- **Laboratoire Ibn Nafiss,**

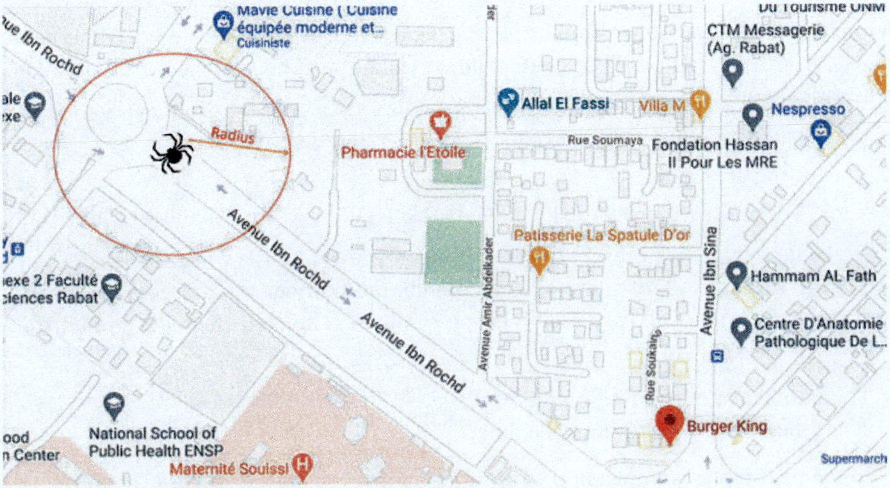

Fig. 4 Example of closest points to spider

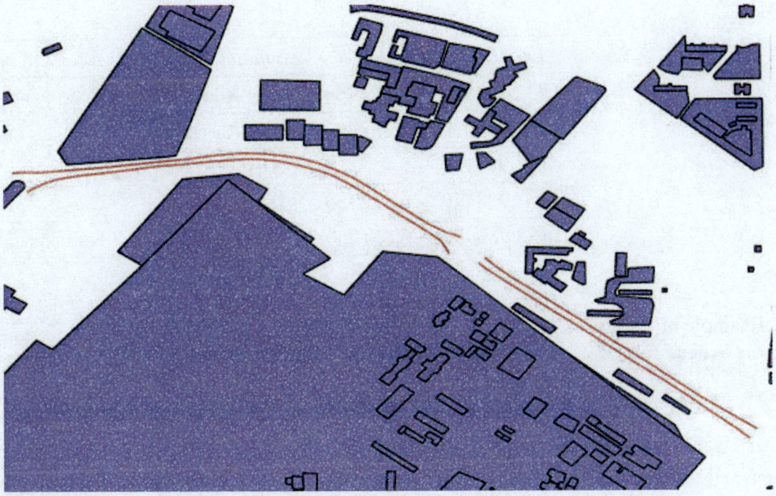

Fig. 5 Polygons beside Ibn Rochd Avenue

- Café Sokaina,
- Latiere Anwar,
- Creches maternelle les P'tits Explorateurs.

All these GIS information helps us to measure the sustainability of the walking area.

3.1.2 Road Accidents Data

While the spider traveling from road and pedestrian walkways, it detects the accidents closer to its radius. The accidents data (thus it can be retrieved from the local authorities) contains (see Table 1):

- localization of the accident,
- time of the accident,
- type of the accident (accident between two cars, or between a car and a pedestrian)

In the next table, we present a slice of data retrieved by the spider (Table 2).

Traffic flow data. Also, the spider collects the number of cars going on every road and pedestrian walkway beside going beside it. The collected data contains:

- The name of the road.
- The number of cars going on the road.
- The time of the collection of the data.

In the next table, we present a slice of data retrieved by the spider (Table 3).

Traffic light data. The spider collects the traffic light state on the roads beside the pedestrian walkway. The state of the traffic light nearby the pedestrian every minute

Table 1 Example table of road accidents data

Localization	Time	Accident type
(33.988849, −6.824750)	2019-01-19 22 h	Cars
(33.988039, −6.852750)	2019-03-25 04 h	Cars
(33.988129, −6.854112)	2019-05-11 14 h	Humans
(33.988479, −6.858350)	2019-5-09 13 h	Cars
(33.988409, −6.851250)	2019-11-02 20 h	Humans

Table 2 Example of traffic flow data on Avenue Ibn Rochd

Name	Number of cars	Time
Avenue Ibn Rochd	15	2019-01-10 21 h 10 min
Avenue Ibn Rochd	3	2019-01-10 21 h 12 min
Avenue Ibn Rochd	21	2019-01-10 21 h 14 min
Avenue Ibn Rochd	09	2019-01-10 21h16min
Avenue Ibn Rochd	29	2019-03-10 21 h 20 min

Table 3 Example table of traffic lights on Avenue Ibn Rochd crossroads

Road name	Time	Is red
Avenue Ibn Rochd	2019-01-10 19 h 53 min	True
Avenue Ibn Rochd	2019-01-10 21 h 10 min	True
Avenue Ma El Aynane	2019-01-10 19 h 53 min	False
Avenue Ma El Aynane	2019-01-10 19 h 55 min	True
Avenue Ibn Rochd	2019-01-10 19 h 55 min	False

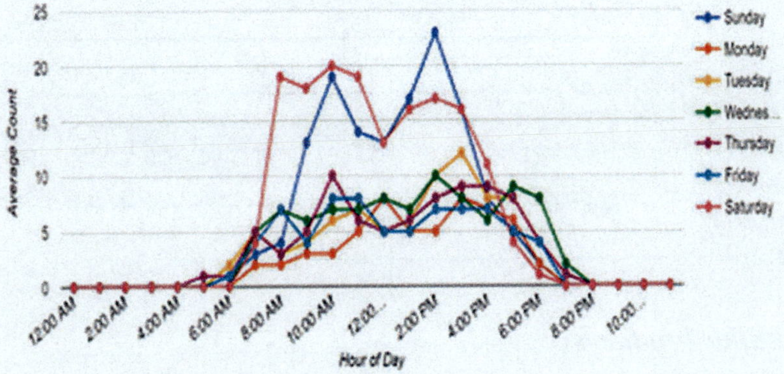

Fig. 6 Example of pedestrian count

can be red or green. In the next table, we present a slice of data retrieved by the spider.

Pedestrian information. Here, the spider collects the number of pedestrians walking on the pedestrian walkway (Fig. 6). The information collects contains:

- The number of pedestrians walking.
- The time the spider retrieves the data.

As you can see in Fig. 6, on Sunday there are more people outside walking on the pedestrian walkaway at 2 p.m. Also, we can see that there are more people walking on the pedestrian walkway between 7 a.m. and 5 p.m. over all the days of week.

Weather. The spider collects the information about the weather state on its localization, beside the pedestrian walkway. The information contains:

- The localization of the spider.
- The time of the collection of the data.
- The degree in Celsius
- Whether it is raining or not.

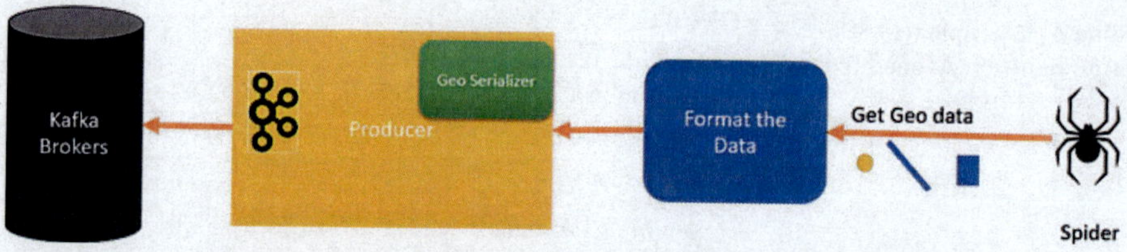

Fig. 7 Producing spatial data into Kafka broker

Fig. 8 Kafka consumer

3.2 *Kafka Producer*

While the spider collects the spatial data, there is a data formatter that processes these data, deletes the unwanted information from the geometry's attributes, and deletes the wrong or inappropriate ones. After the steps of formatting and processing have been finished, the Kafka producers send the data to the Kafka topics using a Geo serializer (Figs. 7, 8).

3.3 *Kafka Consumer*

The spatial Neo4j Kafka consumers listen to the brokers on the servers. If there is any messaging event coming from the tracked partitions, it gets all these messages and tries to deserialize it is using the Geo deserializer been implemented (Fig. 8). After getting and deserializing this data, the consumer sends the data to Neo4j instance graph databases. We use a plugin in the Neo4j database called neo4j spatial, and it helps to enable geographic capabilities on your data and complement the spatial functions that come with graph database Neo4j.

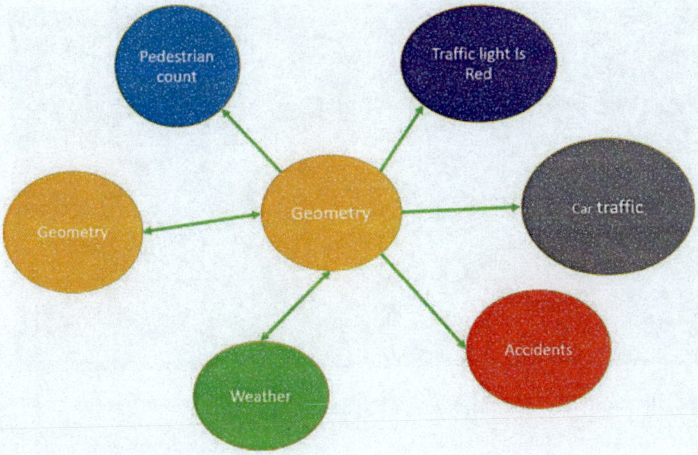

Fig. 9 Data representation in the graph

3.4 Neo4j Graph in the Database

All the collected data is sent to neo4j database, and in this section, we will see how the data it's been modeled in the graph database (Fig. 9).

Relationship types in our database. As you can see in Fig. 9, the main node is the geometry it can be a point (stores…) and lines (highway, pedestrian walkway…), polygons (building…):

- The connection between two geometries is called IS_NEARBY.
- The connection between a geometry node and accidents node is called ACCIDENT_HAPPENED, and it contains the property time of the accident.
- The connection between a geometry and car traffic node is called HAD_CAR_TRAFFIC, and it contains the property time.
- The connection between the node geometry and the traffic light is red node, it is called HAD_TRAFFIC_LIGHT, and it contains the property time.
- The connection between the node geometry and pedestrian count node, it is called HAD_PEDESTRIAN, and it contains the property time.

Note that the property time is a representation of the time with days, hours, and minutes, this choice was made in purpose to be able to know what happens in or beside every geometry in our database.

3.5 Find Things Close to Pedestrian

Now that the database contains the geo-temporal-spatial information that we want, we can make specific requests to the database using cipher (Fig. 10).

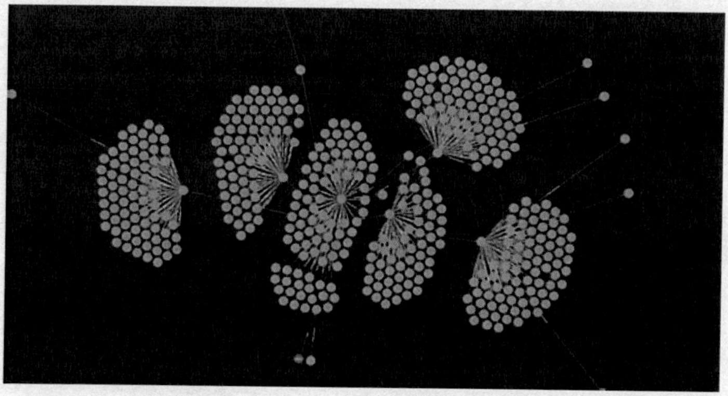

Fig. 10 Neo4j database after the insertion of the data

Examples of finding things close to pedestrian. To find places close to pedestrian in Ibn Rochd avenue at 11 h 00 a.m., we could formulate this cipher query:

```
Match (:Geometry {Name: 'Ibn Rochd'})-[:IS_NEARBY {time:
'11:00'}]->(place:Geometry) RETURN place.name
```

To find the number of accidents that occurs at Ibn rochd avenue in a raining weather, we could use the following cipher query:

```
Match (accidents:Accidents)<-[:ACCIDENT_HAPPENED]-(:Geometry
{Name: 'Ibn Rochd'})-[:HAD_WEATHER]->(weather:Weather) Where
weather.State = 'Rain' RETURN accidents.number
```

4 Conclusion

In this work, we presented the first development of a system for collecting opportunistic spatial data by a method based on spiders' behavior. The main challenge in collecting spatial data is to be able to obtain it as quickly as possible depending on the quantity and size of the data. In this construction, message brokers like Kafka are extremely useful, giving the possibility of providing the real-time architecture for real-time data delivery. Kafka is scalable and offers a fault-tolerant subscription messaging system. The developed system allows the storage of the collected spatial data in a spider network defined by a Neo4j graph. This persistence will be used later for analytical purposes. The specialization of the proposed solution to smart city needs is scheduled in our future work.

Acknowledgments This work was partially funded by Ministry of Equipment, Transport, Logistics and Water–Kingdom of Morocco, The National Road Safety Agency (NARSA), and National

Center for Scientific and Technical Research (CNRST). Road Safety Research Program# An intelligent reactive abductive system and intuitionist fuzzy logical reasoning for dangerousness of driver-pedestrians interactions analysis: Development of new pedestrians' exposure to risk of road accident measures.

References

1. Boulmakoul, A., Fazekas, Z., Karim, L., Gáspár, P., Cherradi, G.: Fuzzy similarities for road environment-type detection by a connected vehicle from traffic sign probabilistic data. Procedia Comput. Sci. **170**, 59–66 (2020). ISSN 1877-0509
2. Boulmakoul, A., Karim, L., El Bouziri, A., Lbath, A.: A system architecture for heterogeneous moving-object trajectory metamodel using generic sensors: tracking airport security case study. IEEE Syst. J. **9**(1), 283–291 (2015)
3. Coulton, C.J., Jennings, M.Z., Chan, T.: How big is my neighborhood? Individual and contextual effects on perceptions of neighborhood scale. Am. J. Community Psychol. **51**, 140–150 (2013)
4. Erkan, I.: Cognitive analysis of pedestrians walking while using a mobile phone. J. Cogn. Sci. **18**(3), 301–319 (2017). https://doi.org/10.17791/jcs.2017.18.3.301
5. Georgiou, H., et al.: Moving Objects Analytics: Survey on Future Location &Trajectory Prediction Methods, Technical Report. arXiv:1807.04639 (2018)
6. Gómez, L.I., Kuijpers, B., Vaisman, A.A.: Analytical queries on semantic trajectories using graph databases. Trans. GIS **23**(5). (John Wiley & Sons Ltd)
7. Güting, R.H., de Almeida, V.T.: Ding Z Modeling and querying moving objects in networks. VLDB J **15**(2), 165–190 (2006)
8. Güting, R.H., Schneider, M.: Moving objects databases. Morgan Kaufmann, San Francisco, CA (2005)
9. Gómez, L.I., Kuijpers, B., Vaisman, A.A.: Analytical queries on semantic trajectories using graph databases. Trans. GIS. **23**, 1078–1101. https://doi.org/10.1111/tgis.12556
10. Maguerra, S., Boulmakoul, A., Karim, L., et al.: Towards a reactive system for managing big trajectory data. J Ambient Intell. Human Comput. **11**, 3895–3906 (2020). https://doi.org/10.1007/s12652-019-01625-3
11. Open Geospatial Consortium, Inc. OGC KML documentation. http://www.opengeospatial.org/standards/kml/ (2012)
12. Parent, C., Spaccapietra, S., Renso, C. Andrienko, G., Andrienko, N., Bogorny, V., Yan, Z.: Semantic trajectories modeling and analysis. ACM Comput. Surv. **45**(4), 42:1–42:32 (2013)
13. Popa, I.S., Zeitouni, K., Oria, V., Kharrat, A.: Spatiotemporal compression of trajectories in road networks. GeoInformatica **19**(1), 117–145 (2015)
14. Qi, F., Du, F.: Trajectory data analyses for pedestrian space-time activity study. J. Vis. Exp. **72**, 50130 (2013)
15. Vecchio, P., Secundo, G., Maruccia, Y., Passiante, G.: A system dynamic approach for the smart mobility of people: Implications in the age of big data. Technol. Forecast. Soc. Change **149**, 119771 (2019)
16. Yoon H., Zheng Y., Xie X., Woo W.: Smart itinerary recommendation based on user-generated GPS trajectories. In: Yu, Z., Liscano, R., Chen, G., Zhang, D., Zhou, X. (eds) Ubiquitous Intelligence and Computing. UIC 2010. Lecture Notes in Computer Science, vol. 6406. Springer, Berlin, Heidelberg. https://doi.org/10.1007/978-3-642-16355-5 5 (2010)
17. Zheng, Y., Zhou, X., et al. :Computing with Spatial Trajectories. Springer, Berlin (2011). ISBN 978-1-4614-1629-6
18. Zhong, R.Y., Huang, G.Q., Shulin, L., Dai, Q.Y., Xu, C., Zhang, T.: A big data approach for logistics trajectory discovery from RFID-enabled production data. Int. J. Prod. Econ. **165**, 260–272 (2015)

Evaluating the Impact of Oversampling on Arabic L1 and L2 Readability Prediction Performances

Naoual Nassiri, Abdelhak Lakhouaja, and Violetta Cavalli-Sforza

Abstract Most Arabic educational corpora, which are used to elaborate readability prediction models, suffer from an unbalanced distribution of texts among difficulty levels. We argue that readability prediction using machine learning (ML) methods should be addressed through a balanced learning corpus. In this work, we address, in a first experiment, the problem of imbalanced data by clustering classes, an approach adopted by several state-of-the art studies. We then present the results of a second experiment in which we adopted an oversampling technique on a unbalanced corpus in order to train the models on balanced data. This experiment was carried out on four corpora, including three dedicated to Arabic as a foreign language (L2) and one for Arabic as a first language (L1). The results show that balanced data give a significant improvement on readability prediction model performances.

1 Introduction

Readability represents the complexity of vocabulary and syntax in a text. It plays an important role in the process of providing students and readers with materials at an appropriate difficulty level. Research in this field began first with the use of mathematical formulas, known as traditional approaches. These scores are a simple way of measuring whether written information is likely to be understood by the intended reader. They produce an index that can indicate approximately the level of education a person will need to be able to read a text easily.

N. Nassiri (✉) · A. Lakhouaja
Department of Computer Science, Faculty of Sciences, University Mohamed First, Oujda, Morocco

V. Cavalli-Sforza
School of Science and Engineering, AI Akhawayn University, Ifrane, Morocco
e-mail: v.cavallisforza@aui.ma

M. Ben Ahmed et al. (eds.), *Networking, Intelligent Systems and Security*, Smart Innovation, Systems and Technologies 237,
https://doi.org/10.1007/978-981-16-3637-0_54

Later, researchers used machine learning (ML) approaches to automatically predict the difficulty level of a given text. The approach is a supervised ML process that consists of applying a classification algorithm to text features in order to generate models from a set of pre-classified training texts. The models are then used to classify new texts. These approaches have shown better results compared to traditional techniques, but this performance is highly correlated to the learning corpus.

Most Arabic readability studies suffer from the problem of unbalanced data. Unbalanced data means that at least one class is under-represented or over-represented compared to the others, thus perturbing learning algorithms and making them unable to recognize and classify minority instances. This phenomenon occurs in several application areas, such as fraud detection, text categorization, and medical diagnostics, and represents a challenge for the data mining community.

The task of automatically measuring text readability is frequently faced with challenges such as the lack of data and the imbalance of available data sets. ML algorithms, applied to automatically predict readability, tend to produce unsatisfactory classifiers when confronted with unbalanced data sets. If the text whose difficulty we aim to measure belongs to the minority class, it is generally incorrectly classified. The main question asked when analyzing such a data is: how can we balance the data set to overcome this phenomenon? In order to address this issue, we will present in this paper a set of experiments that are based on balancing the learning data in order to build automatic readability prediction models on a balanced data set. A number of techniques can be adopted to address this problem such as:

1. Class clustering.
2. Transformation of the learning algorithm.
3. Sampling methods such as oversampling and under-sampling.

In this paper, we are interested in data sampling methods and particularly "oversampling." These approaches are widely used and have given encouraging results in cases where the data are unbalanced.

The rest of this paper is organized as follows. We review in Section 2 a set of studies on Arabic automatic readability measurement, and we highlight the weaknesses of these approaches. Details on different unbalanced data classification techniques are given in Sect. 3. Section 4 presents the tools, the data, and the process that we adopted for this study. The tests and the results are discussed in Sect. 5. Finally, Sect. 6 presents a conclusion and some future directions to further improve the obtained results.

2 Related Work

Readability assessment is considered, in the majority of research, as a classification problem. This process is carried out in three main steps:

1. Assembling a reference corpus: The quality and the quantity of the texts composing the reference corpus are important.

2. Transforming corpus texts into feature vectors: The choice of the features to be extracted is an essential step in the construction of the prediction model. These must have a correlation with the level to be predicted.
3. Applying a classification algorithm on the data.

For Arabic as a foreign language (L2), most of the studies carried out following this process are conducted using unbalanced corpora. In 2014, Forsyth described in his thesis [8], a study is based on a corpus comprising 179 texts. The corpus texts were annotated with difficulty levels based on the Interagency Language Roundtable (ILR) organization scale [7]. The scale is used to measure language proficiency by concerned entities in the U.S. federal government. This scale assesses language skills on levels ranging from 0 to 5. Levels 0+, 1+, 2+, 3+, or 4+ are used when a person's skills far exceed those of a given level but are insufficient to reach the next level. Forsyth's corpus is annotated with five difficulty levels (1, 1+, 2, 2+, and 3) using this scale. The corpus contains an unbalanced distribution of texts between the different levels as given in Table 1. Based on cross-validation, he reported a maximum F-score value of 0.52 and 0.78, respectively, for five and three classes (obtained by grouping levels 1 and 1+ and levels 2+ and 3). The gain in F-score obtained by using three classes is due to the combination of the levels, and this leads to a relative balance in the data.

In 2015, Saddiki et al. carried out a study [13] in which they gathered a corpus of 251 texts distributed over five ILR difficulty levels, as given in Table 1. They reported a maximum accuracy value of 59.76% and a maximum F-score of 0.58 using five classes, versus a maximum accuracy of 73.31% and a maximum F-score of 0.73 using three classes. In 2018, Nassiri et al. [12] collected a corpus of 230 texts (Table 1) and reported a maximum accuracy value of 89.56% using a five-way classification when testing the models on the training data.

Studies on Arabic as a first language (L1) started in 2010 with Al-Khalifa and Al-Ajlan [2], who used, for the development of their tool, a corpus collected manually from reading books of elementary, intermediate and secondary school curricula in Saudi Arabia. Their corpus was composed of 150 texts distributed in a balanced way between three classes (50 text per level). They reported an accuracy rate of 77.77%.

In 2014, Al-Tamimi et al. [4] collected a corpus of ten levels containing a total of 1,196 texts gathered from the Jordanian curriculum. They then re-annotated the

Table 1 State-of-the-art readability corpora

Level	Forsyth [8]	Saddiki et al. [13]	Nassiri et al. [12]
1	20	33	27
1+	14	28	19
2	80	91	87
2+	40	63	62
3	25	63	35
Total	179	251	230

corpus with three classes (easy, medium, and difficult) and achieved an average accuracy of 83.23%.

Finally, in 2018, Saddiki et al. conducted a study [14] to predict the readability of L1 texts based on a corpus developed by Al Khalil et al. [3]. This corpus is composed of 27,688 texts divided into four difficulty levels. The first three levels have been extracted from UAE textbooks, and the fourth level contains novels. Their best accuracy result is 94.8%.

Most of the studies we presented in this section used class clustering to address the problem of unbalanced data. Unfortunately, only Saddiki et al. [14] adopted an approach in which they limited the length of texts by splitting them into approximately equal sizes, thus increasing the number of texts in a class. In the remainder of this paper, we will use both class clustering and oversampling techniques to address this phenomenon.

3 Classification and Class Imbalance

Most existing classification methods are not suitable for use with the minority class when the class is extremely unbalanced. This problem has become a challenge for many researchers, as it is present in many real-world applications. To deal with this challenge, two approaches are widely used:

1. Under-Sampling the learning data, and
2. Oversampling the learning data.

3.1 Under-Sampling

Under-sampling consists in balancing the data set by decreasing the number of instances of the majority class. Among the best-known algorithms of this category of approaches, we cite:

- **Random Under-sampling** [9]: It involves randomly drawing samples from the majority class, with or without replacement. However, it can increase the variance of the classifier and can potentially eliminate useful or important samples.
- **Tomek Links** [15]: It removes unwanted overlap between classes where majority class links are removed until all pairs of closest neighbors, at minimum distance, are of the same class.

3.2 Oversampling

Oversampling consists in balancing the data set by artificially increasing the number of instances of the minority class. Among the best-known oversampling algorithms, we have:

- **Random Oversampling** [9]: It involves supplementing the training data with multiple copies of some instances of the minority class.
- **SMOTE (Synthetic Minority Over-sampling Technique)** [6]: Rather than reproducing minority observations, SMOTE creates a user-selected number of synthetic observations on segments between items close to the minority class. SMOTE introduces the synthetic examples randomly on the line connecting the minority class point concerned and one of its closest K neighbors (the number of neighbors 'k' is a user-defined parameter). The principle is based on the difference between a minority instance and its closest neighbor, this difference is multiplied by a random number between 0 and 1, this causes the selection of a random point along the line segment between two minority instances. For a sufficient number of synthetic instances, the SMOTE algorithm consists of:
  - Selecting a minority instance A,
  - Selecting one instance B among the closest neighbors,
  - Selecting a random weight between 0 and 1 (W) to create the new synthetic instance C, and
  - Applying, for each attribute, the formula:

$$attValue\_C = attValue\_A + (attValue\_B - attValue\_A) * W$$

4 Resources and Methodology

In this study, we adopted two approaches to develop readability prediction models based on balanced data sets:

1. The first approach consists of redistributing the data into three classes: easy, medium, and difficult. These three levels are obtained by grouping adjacent levels together.
2. The second approach consists of using the SMOTE technique to re-balance the learning data.

In this section, we will present the data on which we evaluated our approaches, the tools we used, and the process we adopted in each approach.

4.1 Data

In this section, we will present three of the most used corpora for L2 readability measurement and one corpus dedicated to L1 called MoSAR [10].

- **L2 readability corpora**: For Arabic as L2, most of the published research is performed using corpora from the GLOSS[1] platform, whose content was developed by the Defense Language Institute Foreign Language Center (DLIFLC[2]), considered one of the best foreign language schools. The platform offers thousands of lessons in dozens of languages for independent learners to strengthen their listening skills and reading. MSA texts in GLOSS are annotated with five difficulty levels using the ILR scale, described earlier. We collected two corpora from this source, namely:

  - **GLOSS-Reading**, which consists of 271 reading texts.
  - **GLOSS-Listening**, which comprises 227 listening texts.

  We have also used the texts available online by Aljazeera[3] to form the third L2 corpus. The Aljazeera-learning Web site, for learning Arabic, presents texts for educational purposes. The texts are annotated with five levels of difficulty, namely: Beginner1, Beginner2, Intermediate1, Intermediate2, and Advanced. The gathered corpus contains a total of 321 texts.
  In order to facilitate the representation of the results, we relabeled the three corpora levels into Level 1, Level 2, ..., Level 5. Note that this new labels meaning varies from one corpus to another. Table 2 presents the distribution of these three corpora across the five difficulty levels.
  The statistics given in Table 2 allow us to appreciate the great imbalance in the distribution of corpora texts between the five difficulty levels. For Aljazeera-learning, for example, we have 220 texts in level 2 while we have only 8 texts in level 5. Similarly for the GLOSS-listening corpus, we have 18 texts for level 2 compared to 65 texts for level 3. It is this problem of imbalance that we will try to solve in the remainder of this paper.
- **L1 readability corpus**: For L1, we used the modern standard Arabic readability (**MoSAR**) corpus. MoSAR is composed of a set of texts collected from textbooks used in the six Moroccan primary school levels and annotated according to these six levels. MoSAR consists of 602 texts as given in Table 3.

[1] https://gloss.dliflc.edu/.

[2] https://www.dliflc.edu/global-language-online-support-system-gloss/.

[3] https://learning.aljazeera.net.

Table 2 L2 corpora texts distributed by five difficulty levels

	GLOSS-reading	GLOSS-listening	Aljazeera-learning
Level 1	34	43	12
Level 2	29	18	220
Level 3	95	65	54
Level 4	68	62	27
Level 5	45	39	8
Total	271	227	321

Table 3 MoSAR text distribution by levels

Level	Number of books	Number of texts
1	1	51
2	2	139
3	1	60
4	2	136
5	1	86
6	1	130
Total	8	602

4.2 Morphological Annotation

For the morphological annotation of the four corpora, we used **Alkhalil-Toolkit** which is a morphological analysis and disambiguation system composed of several independent tools. For this task, we used specifically:

1. **Alkhalil-Lemmatizer** [5], which assigns a unique lemma to each word taking into account its context, and
2. **Alkhalil-PoS-Tagger** [1], which offers a very rich set of basic tags providing syntactic information for each word of the analyzed text.

4.3 Methodology

The first experiment that we will present consists of building predictive models on five classes (unbalanced corpora) then on three classes data sets distributed as given in Table 4.

The experimental process consists of three steps, as follows (see Fig. 1):

1. **Morphological analysis**: The input of this phase is a text from one of the corpora. The result of this phase is a file (for each text) annotated with information such as PoS, lemma, and diacritical signs.

Table 4 Corpora distribution over three levels

Level	GLOSS-reading	GLOSS-listening	Aljazeera-learning	MoSAR
Easy	63	61	232	190
Medium	95	65	54	196
Difficult	113	101	35	216
Total	271	227	321	602

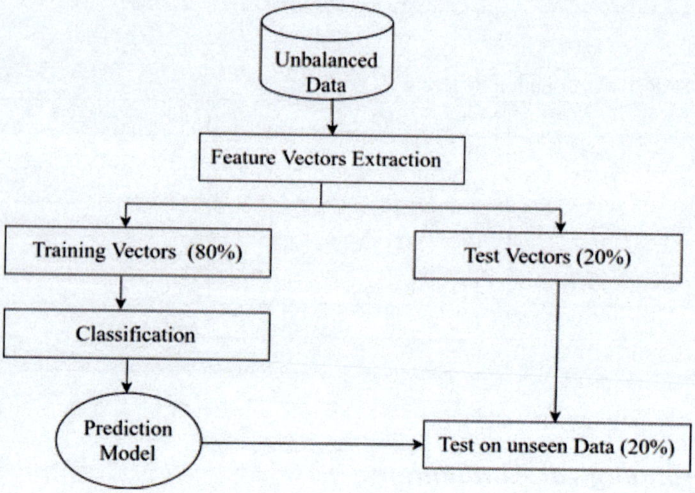

Fig. 1 Model generation process on unbalanced data

2. **Feature extraction**: In this step, we extract and calculate a list of features [11]. For each corpus file, we obtain a feature vector that we will use to prepare the input file for the classification phase.
3. **Classification**: We apply a classification algorithm on 80% of the generated vectors (training data), randomly selected, in order to build a prediction model. The obtained model is tested on the remaining 20% of the generated vectors.

The second experiment consists of regenerating the models on balanced learning vectors. To do this, we proceed as follows (see Fig. 2).

1. Apply the SMOTE technique on the learning vectors (the 80%) to balance them,
2. Use the balanced vectors set to train and build a prediction model, and
3. Test the obtained model on the remaining 20% unseen data.

This second experiment is applied only on texts distributed among five levels, since the objective is to evaluate the impact of balancing the training data on the original corpus distribution.

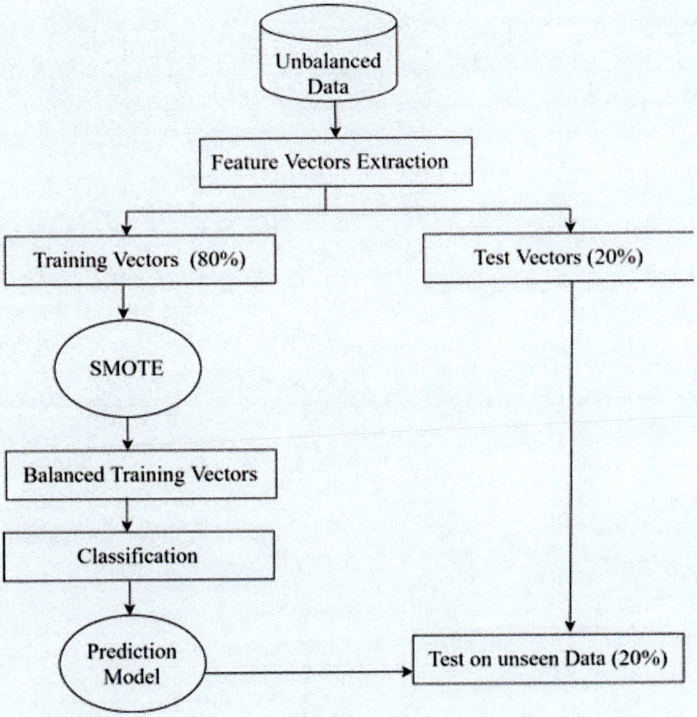

Fig. 2 Model generation process on balanced data

5 Results and Discussions

In this section, we will present and discuss the results of the experiments for which we have already presented the processes in Sect. 4.3.

5.1 L2 Prediction Models

Table 5 presents the results obtained using a five-class classification on the three corpora. GLOSS-reading achieved an overall accuracy of 60%. This result hides how accuracy values vary from one class to another; for example, the model was able to predict level 4 with an accuracy of 25%, while it achieved an accuracy of 83.33% for level 1. For the GLOSS-listening corpus, we obtained a total accuracy of 56.52%, and for Aljazeera-learning, we obtained 89.23% as accuracy value.

Merging the classes and switching to a three-class classification, allowed us to obtain the results presented in Table 6. For this second experiment, classification accuracy for the GLOSS-reading corpus increased from 60 to 70%, for GLOSS-

Table 5 Test results on L2 corpora using five levels

	GLOSS-reading (%)	GLOSS-listening (%)	Aljazeera-learning (%)
Level 1	83.33	77.77	–
Level 2	66.66	16.66	95.65
Level 3	80	75	66.66
Level 4	25	45.45	88.88
Level 5	66.66	50	0
Total	60	56.52	89.23

Table 6 Test results on L2 corpora using three levels

	GLOSS-reading (%)	GLOSS-listening (%)	Aljazeera-learning (%)
Easy	90	85.71	94.28
Medium	68.75	76.92	72.72
Difficult	61.11	70.58	83.33
Total	70	75	88

Table 7 Test results on L2 corpora after oversampling

	GLOSS-reading (%)	GLOSS-listening (%)	Aljazeera-learning (%)
Level 1	83.33	88.88	–
Level 2	87.5	33.33	97.82
Level 3	60	75	100
Level 4	31.25	45.45	100
Level 5	77.77	62.5	100
Total	60	63.04	98.46

listening, it increased from 56.52 to 75%, and for the Aljazeera-learning corpus, it decreased slightly from 89.23 to 88%.

The improvements obtained through the application of this first approach based on merging classes are very important and are encouraging, but it should be noted that sometimes we need to classify the texts according to more detailed levels of difficulty (school grade levels for example), so we should try to improve the results of the models based on five classes as well. To do so, we present in Table 7 the results of applying the SMOTE technique on the learning vectors of our three corpora.

The comparison of the results of Tables 5 and 7 leads to the conclusion that the use of oversampling techniques has improved the classification results both in terms of total accuracy and in terms of the accuracy of each level independently.

Table 8 MoSAR test results before and after oversampling

	Before SMOTE		After SMOTE
	Six levels (%)	Three levels (%)	Six levels (%)
1	0	61.76	25
2	84.21		84.21
3	55.55	67.64	66.66
4	77.77		55.55
5	25	75	37.5
6	91.66		83.33
Total	67.21	68	67.21

5.2 *L1 Prediction Models*

Table 8 presents the results obtained using five classes and three classes classifications and those obtained when applying the SMOTE technique on MoSAR corpus. For MoSAR, the total accuracy of 67.21% with five classes was increased to 68% using three classes. If we compare the five classes results with those obtained using the SMOTE technique, we can conclude that the use of sampling techniques improves the classification results at the individual class level, while keeping the same maximum value of 67.21% overall.

6 Conclusions and Futures Works

This paper proposes a readability prediction approach of MSA texts based on ML techniques. The main objective of the study is to deal with unbalanced data sets when training readability prediction models. For these, we experimented with two balancing approaches: the first one consists of grouping adjacent levels together and the second one consists of applying an oversampling technique on the training data. These approaches are evaluated on four corpora, of which three are intended for L2 learners and one for L1 learners.

Using the four corpora with their original unbalanced distributions, we get an accuracy value of 60, 56.52, 89.23, and 67.21%, while using three classes instead of five gave corresponding values of 70, 75, 88, and 68%. The application of the SMOTE oversampling technique enabled us to achieve accuracy values of 60, 63.04, 98.46, and 67.21% for the same corpora, respectively, thus maintaining or improving the performance previously achieved with five classes and, in some cases, even with three classes. This is a promising result when it is important to allow finer discrimination of text difficulty.

In our future work, we aim to further improve the results of our models by using alternative data balancing techniques and making changes to the learning algorithms to support the unbalanced situations. We aim also to perform a n-fold cross-validation in order the validate the obtained results.

References

1. Ababou, N., Mazroui, A.: A hybrid Arabic POS tagging for simple and compound morphosyntactic tags. Int. J. Speech Technol. **19**(2), 289–302 (Jun 2016). http://orcid.org/10.1007/s10772-015-9302-8
2. Al-Khalifa, H., Al-Ajlan, A.: Automatic readability measurements of the Arabic text: An exploratory study. Arab. J. Sci. Eng. **35**, 103–124 (12 2010)
3. Al Khalil, M., Saddiki, H., Habash, N., Alfalasi, L.: A leveled reading corpus of Modern Standard Arabic. In: Proceedings of the Eleventh International Conference on Language Resources and Evaluation (LREC 2018). European Language Resources Association (ELRA), Miyazaki, Japan (May 2018), https://www.aclweb.org/anthology/L18-1366
4. Al-Tamimi, A.K., Jaradat, M., Aljarrah, N., Ghanim, S.: Aari: Automatic Arabic readability index. Int. Arab J. Inf. Technol. **11**, 370–378 (07 2014)
5. Boudchiche, M., Mazroui, A.: A hybrid approach for Arabic lemmatization. Int. J. Speech Technol. **22**(3), 563–573 (2019). http://orcid.org/10.1007/s10772-018-9528-3
6. Chawla, N., Bowyer, K., Hall, L., Kegelmeyer, W.: Smote: Synthetic minority over-sampling technique. J. Artif. Intell. Res. (JAIR) **16**, 321–357 (06 2002). https://doi.org/10.1613/jair.953
7. Clark, J.L.D., Clifford, R.T.: The fsi/ilr/actfl proficiency scales and testing techniques: development, current status, and needed research. Stud. Second Lang. Acquisition **10**(2), 129–147 (1988). http://orcid.org/10.1017/S0272263100007270
8. Forsyth, J.N.: Automatic readability detection for modern standard Arabic. Brigham Young University (2014)
9. Japkowicz, N., Stephen, S.: The class imbalance problem: A systematic study. Intelligent data analysis **6**(5), 429–449 (2002)
10. Nassiri, N., Cavalli-Sforza, V., Lakhouaja, A.: Mosar: Modern standard Arabic readability corpus for l1 learners. In: Proceedings of the 4th International Conference on Big Data and Internet of Things. BDIoT'19, Association for Computing Machinery, New York, NY, USA (2019). https://doi.org/10.1145/3372938.3372961
11. Nassiri, N., Lakhouaja, A., Cavalli-Sforza, V.: Arabic readability assessment for foreign language learners. In: Silberztein, M., Atigui, F., Kornyshova, E., Métais, E., Meziane, F. (eds.) Natural Language Processing and Information Systems. pp. 480–488. Springer International Publishing, Cham (2018)
12. Nassiri, N., Lakhouaja, A., Cavalli-Sforza, V.: Modern standard arabic readability prediction. In: Lachkar, A., Bouzoubaa, K., Mazroui, A., Hamdani, A., Lekhouaja, A. (eds.) Arabic Language Processing: From Theory to Practice, pp. 120–133. Springer International Publishing, Cham (2018)
13. Saddiki, H., Bouzoubaa, K., Cavalli-Sforza, V.: Text readability for Arabic as a foreign language. pp. 1–8 (11 2015). https://doi.org/10.1109/AICCSA.2015.7507232
14. Saddiki, H., Habash, N., Cavalli-Sforza, V., Al Khalil, M.: Feature optimization for predicting readability of Arabic L1 and L2. In: Proceedings of the 5th Workshop on Natural Language Processing Techniques for Educational Applications, pp. 20–29. Association for Computational Linguistics, Melbourne, Australia (Jul 2018). https://doi.org/10.18653/v1/W18-3703, https://www.aclweb.org/anthology/W18-3703
15. Tomek, I., et al.: An experiment with the edited nearest-nieghbor rule. IEEE Trans. Syst. Man Cybern. **SMC-6**(6), 448–452 (1976). https://doi.org/10.1109/TSMC.1976.4309523

An Enhanced Social Spider Colony Optimization for Global Optimization

Farouq Zitouni, Saad Harous, and Ramdane Maamri

Abstract An improved variant of the social spider optimization algorithm is introduced. It is inspired by the hunting and mating behaviors of spiders in nature. We call it enhanced social spider colony optimization (ESSCO). The performance of ESSCO is evaluated using the benchmark CEC 2020. To validate the proposed algorithm, the obtained statistical results are compared to eleven recent state-of-the-art metaheuristic algorithms. The comparative study shows the competitiveness of ESSCO in finding efficient solutions to the considered test functions.

1 Introduction

Over millions of years of evolution, many intelligent behaviors have arisen in nature. Biological agents (e.g., insects, birds, mammals, and fishes) have perpetually exhibited adaptability, self-learning, robustness, and efficiency to solve complex tasks [4, 14, 18, 42]. Researchers started to mimic natural systems during the last decades to develop several solutions to address difficult problems. Therefore, special attention has been given to metaheuristic algorithms. Metaheuristics are nature-inspired

F. Zitouni (✉)
Department of Computer Science, Kasdi Merbah University, Ouargla, Algeria

LIRE Laboratory, Abdelhamid Mehri University, Constantine, Algeria
e-mail: farouq.zitouni@univ-constantine2.dz

S. Harous
Department of Computer Science and Software Engineering UAE University, Abu Dhabi, United Arab Emirates
e-mail: harous@uaeu.ac.ae

R. Maamri
Department of Computer Science, Abdelhamid Mehri University, LIRE Laboratory, Abdelhamid Mehri University, Constantine, Algeria
e-mail: ramdane.maamri@univ-constantine2.dz

M. Ben Ahmed et al. (eds.), *Networking, Intelligent Systems and Security*,
Smart Innovation, Systems and Technologies 237,
https://doi.org/10.1007/978-981-16-3637-0_55

algorithms: i.e., they mimic biological and physical phenomena. They provide near-optimal solutions in a reasonable amount of time [46, 57, 62, 84, 89]. We identify three classes of metaheuristics: (i) algorithms that are based on Darwinian principles, (ii) algorithms that are based on laws of physics and chemistry, and (iii) algorithms that are based on swarm intelligence. Table 1 summarizes popular algorithms in each class.

Metaheuristic algorithms are grouped into two prominent families: (i) algorithms that use one agent (i.e., individual) and (ii) algorithms that use many agents (i.e., population). In the first family, the individual swarms in the search space for a given number of iterations [70, 79]. Its final position is considered a solution to the optimization problem. In the second family, several agents swarm together in the search space for a given number of iterations [32, 76]. The position of the best individual is considered a solution to the optimization problem. Algorithms of the second family generally outperform the first family's algorithms concerning the optimality of obtained solutions [86]. The No-Free-Lunch theorem [35, 88] has declared the nonexistence of a metaheuristic algorithm that can efficiently solve all optimization problems. That is to say, some algorithms that show good performance on a given problem might exhibit poor performance on another one. However, the averaged performance of all metaheuristics on all optimization problems is equal.

All metaheuristic algorithms share two common steps: exploration and exploitation of the search space. In the exploration step, the algorithm tries to find promising areas [30]. In the exploitation step, the algorithm attempts to investigate a given region [11]. In metaheuristics, efficient balancing between these two steps would lead the algorithm to find near-optimal solutions [12].

The main contribution of the paper is the improvement of the metaheuristic algorithm proposed in [19] for global optimization. The enhancements are summarized as follows.

- Adding a hunting phase to the basic algorithm;
- Using the blend crossover [25, 26] for generating new offsprings;
- The enhanced algorithm has four parameters to set: number of insects, number of male spiders, number of female spiders, and an insect's probability of getting stuck.

The remainder of the paper is organized as follows. Section 2 presents the biological inspiration of ESSCO, its mathematical model, and algorithms. Section 3 summarizes the obtained numerical results and discusses the performance of ESSCO. Section 4 concludes the paper with a conclusion and some future work.

Table 1 Popular metaheuristic algorithms for global optimization

Class 1	Class 2	Class 3
Genetic algorithm [33, 63, 78]	Simulated annealing [44]	Particle swarm optimization [43, 71]
Evolution strategies [9]	Thermodynamic laws [17]	Ant colony optimization [20, 21]
Evolutionary programming [27, 91]	Gravitation [64, 82]	Artificial bee colony [8, 37]
Genetic programming [45]	Big bang–big crunch[24]	Grey wolf optimizer [54]
Differential evolution [75]	Charged system [41]	Bat algorithm [87]
Biogeography-based optimization [73]	Central force [28]	Whale optimization algorithm [53]
Quantum-inspired evolutionary algorithm [77]	Chemical reaction [2]	Dragonfly algorithm [52]
	Black hole [31]	Dolphin echolocation [39]
	Ray [40]	Fruit fly optimization [61]
	Small-world [22]	Krill herd [29]
	Galaxy-based [69]	Bird mating optimizer [5]
	General relativity theory [55]	Hunting search [60]
	Spherical search [47]	Firefly algorithm [85]
	Solar system algorithm [92]	Dolphin partner optimization [72]
		Cuckoo search [90]
		Bee collecting pollen algorithm [49]
		Marriage in honey bees [1, 83]
		Monkey search [58]
		Termite [65]
		Fish swarm algorithm [48]
		Tunicate swarm algorithm [38]

2 Enhanced Social Spider Colony Optimization

2.1 *Inspiration*

Spiders belong to the class of arachnids. They have eight legs, and the body is divided into two parts. All spiders are predators. Some species are active hunters, whereas other species weave webs to capture prey. Some spiders inject venom into their victims to kill and eat them. Other spiders immobilize the prey by making silk wrappers around it. According to their social behavior, spiders can be classified into solitary and social spiders [50]. Solitary spiders weave webs and live there lonely. Social spiders share the same web and have social relationships (building and maintaining the web, hunting, and mating).

A colony of spiders is composed of males and females. Usually, the percentage of females is high (some studies assume that the percentage of males can barely attain 30% of the population size [6]). There are two forms of interactions between social spiders: direct (i.e., body contact) or indirect (i.e., vibrations transmitted through the web) [59]. Thus, the shared web is used as a medium of communication. Intensities of vibrations are used to encode several messages: e.g., size of captured insects, nature of neighbors. Vibrations' intensities depend on the weight and the distance of the source that has initiated them [68]. Social spiders have two main behaviors: hunting and mating [23, 81].

- The hunting behavior can be summarized as follows. When an insect is trapped in the web, it tries to escape, generating vibrations through the web. When spiders sense them, they move toward the source of vibrations.
- The mating behavior can be summarized as follows. The role of males is to fertilize females. Males are either dominant or non-dominant. Dominant males are attracted to the closest females. Non-dominant males move toward the center of the male population to get females. Females exhibit either an attraction or repulsion to males. This attitude is an answer to the intensity of received vibrations. It depends on the weight and the distance of males that provoked them.

2.2 *Mathematical Model and Optimization Algorithm*

We enhance the metaheuristic algorithm proposed in work [19], for global optimization. It mimics the hunting and mating behaviors of social spiders. We name it enhanced social spider colony optimization (ESSCO). We suppose a search space of dimension D and an optimization function f to be minimized. We assume three sets E, M, and F of N_E insects, N_M male spiders, and N_F female spiders, respectively:

$$\begin{cases} E = \{e_1, e_2, \ldots, e_{N_E}\} \\ M = \{m_1, m_2, \ldots, m_{N_M}\} \\ F = \{f_1, f_2, \ldots, f_{N_F}\} \end{cases}$$

To describe the behaviors of insects and spiders in ESSCO, we adopt the following assumptions.

1. An insect moves randomly in the search space according to a Lévy flight distribution [7].
2. An insect can be trapped in the web according to a given probability.
3. When a spider feels a trapped insect, it moves to its location.
4. A dominant male chooses the closest female for mating.
5. A non-dominant male does not mate. It moves in the direction of the center of males.
6. A female either has attraction or repulsion for a given male.
7. When two spiders mate, a new spider is generated, and the worst spider in the current population is replaced. The gender of the new spider is the same as the replaced one.

Algorithms 1, 2, and 3 outline the steps of ESSCO. Their instructions are explained in Sects. 2.2.1, 2.2.2, and 2.2.3, respectively.

2.2.1 Algorithm 1

- Lines 1 to 3: For each insect $e_i \in E$, we generate a random location $X^{\langle e_i \rangle}$ using Eq. 1.
- Lines 4 to 6: For each male spider $m_i \in M$, we generate a random location $X^{\langle m_i \rangle}$ using Eq. 1.
- Lines 7 to 9: For each female spider $f_i \in F$, we generate a random location $X^{\langle f_i \rangle}$ using Eq. 1.
- Lines 10 to 13: We iterate the algorithm for a given number of iterations. In each iteration, we run the hunting (Algorithm 2) and mating (Algorithm 3) behaviors.
- Line 14: We get X_m^{best}: i.e., the best solution among male spiders.
- Line 15: We get X_f^{best}: i.e., the best solution among female spiders.
- Line 16: We get X^*: i.e., the best solution among male and female spiders.

$$X^{\langle y \rangle} = (\alpha_1 \times (x_{\max}^1 - x_{\min}^1) + x_{\min}^1, \ldots, \alpha_D \times (x_{\max}^D - x_{\min}^D) + x_{\min}^D) \quad (1)$$

where

y : might be e_i, m_i, or f_i.

$\alpha_1, \ldots, \alpha_D$: uniformly distributed random numbers between 0 and 1.

Algorithm 1: The Enhanced Social Spider Colony Optimization.

Input: D is the dimension of the search space.
Input: $\Delta^1 = [x^1_{\min}, x^1_{\max}], \ldots, \Delta^D = [x^D_{\min}, x^D_{\max}]$ are the input domains.
Input: f is the objective function to be minimized.
Input: $E = \{e_1, e_2, \ldots, e_{N_E}\}$: i.e., the set of insects ($N_E = |E|$).
Input: $M = \{m_1, m_2, \ldots, m_{N_M}\}$: i.e., the set of mal spiders ($N_M = |M|$).
Input: $F = \{f_1, f_2, \ldots, f_{N_F}\}$: i.e., the set of female spiders ($N_F = |F|$).
Input: ρ is the probability of an insect to get trapped.

foreach $e_i \in E$ **do**
  Compute $X^{\langle e_i \rangle} = (x_1^{\langle e_i \rangle}, \ldots, x_D^{\langle e_i \rangle})$ using Equation 1;
end
foreach $m_i \in M$ **do**
  Compute $X^{\langle m_i \rangle} = (x_1^{\langle m_i \rangle}, \ldots, x_D^{\langle m_i \rangle})$ using Equation 1;
end
foreach $f_i \in F$ **do**
  Compute $X^{\langle f_i \rangle} = (x_1^{\langle f_i \rangle}, \ldots, x_D^{\langle f_i \rangle})$ using Equation 1;
end
for $t \leftarrow 1$ **to** $Iter_{\max}$ **do**
  /* Hunting behavior */
  Run Algorithm 2;
  /* Mating behavior */
  Run Algorithm 3;
end
$X^{\text{best}}_m \leftarrow \underset{i \in \{1,\ldots,N_M\}}{\operatorname{argmin}} \{f(X^{\langle m_i \rangle})\}$;
$X^{\text{best}}_f \leftarrow \underset{i \in \{1,\ldots,N_F\}}{\operatorname{argmin}} \{f(X^{\langle f_i \rangle})\}$;
Return $X^* = \underset{i \in \{m,f\}}{\operatorname{argmin}} \{f(X^{\text{best}}_i)\}$ as best solution;

2.2.2 Algorithm 2

- Lines 1 to 14: This loop translates the swarming behavior of insects $e_i \in E$ in the search space.
- Line 2: For each insect $e_i \in E$, we generate a random number ρ_i between 0 and 1.
- Lines 3 to 5: If ρ_i is greater than or equal to ρ, it means that the considered insect e_i is not trapped in the web.
- Line 4: We update location $X^{\langle e_i \rangle}$ using Eqs. 2 and 3 [51]. If the values of $X^{\langle e_i \rangle}$ are out of permitted ranges, they are adjusted.

$$X^{\langle e_i \rangle} = X^{\langle e_i \rangle} + \left[\frac{u_1}{(v_1)^{1/\beta}}, \ldots, \frac{u_D}{(v_D)^{1/\beta}} \right] \qquad (2)$$

$$\begin{cases} u_i \sim \mathfrak{N}(0, \sigma^2)\ ,\ \sigma = \left(\frac{\Gamma(1+\beta)\sin(\frac{\pi\beta}{2})}{\Gamma(\frac{1+\beta}{2})\times\beta\times 2^{\frac{\beta-1}{2}}} \right)^{\frac{1}{\beta}} & ,\ i \in \{1, \ldots, D\} \\ v_i \sim \mathfrak{N}(0, \sigma'^2)\ ,\ \sigma' = 1 & ,\ i \in \{1, \ldots, D\} \end{cases} \tag{3}$$

- Lines 6 to 13: If ρ_i is less than ρ, it means that the considered insect e_i is trapped in the web.
- Lines 7 to 9 and 10 to 12: Each male/female spider $m_i \in M / f_i \in F$ moves in the direction of the trapped insect.
- Lines 8 and 11: We update location $X^{\langle m_i \rangle}/X^{\langle f_i \rangle}$ using Eq. 4. If the values of $X^{\langle m_i \rangle}/X^{\langle f_i \rangle}$ are out of permitted ranges, they are adjusted.

$$X^{\langle y \rangle} = X^{\langle y \rangle} + \frac{e^{-\delta \| X^{\langle e_i \rangle} - X^{\langle y \rangle} \|^2}}{f(X^{\langle e_i \rangle})} (X^{\langle e_i \rangle} - X^{\langle y \rangle}) \tag{4}$$

where

Γ : is the gamma function [3].
$\mathfrak{N}(\mu, \sigma)$: is the normal distribution of mean μ and standard deviation σ.
β : is the power law index $1 \leq \beta \leq 2$.
y : might be m_j or f_j.
δ : is a coefficient that defines the attenuation of vibrations.

Algorithm 2: The hunting behavior of ESSCO.

Input: D is the dimension of the search space.
Input: $\Delta^1 = [x^1_{\min}, x^1_{\max}], \ldots, \Delta^D = [x^D_{\min}, x^D_{\max}]$ are the input domains.
Input: f is the objective function to be minimized.
Input: $E = \{e_1, e_2, \ldots, e_{N_E}\}$: i.e., the set of insects ($N_E = |E|$).
Input: $M = \{m_1, m_2, \ldots, m_{N_M}\}$: i.e., the set of mal spiders ($N_M = |M|$).
Input: $F = \{f_1, f_2, \ldots, f_{N_F}\}$: i.e., the set of female spiders ($N_F = |F|$).
Input: ρ is the probability of an insect to get trapped.

1 **for** $i \leftarrow 1$ **to** N_E **do**
2   Generate a random number ρ_i between 0 and 1;
3   **if** ($\rho_i \geq \rho$) **then**
4   Update $X^{\langle e_i \rangle}$ using Equations 2 and 3. Adjust its values;
5   **end**
6   **else**
7   **for** $j \leftarrow 1$ **to** N_M **do**
8   Update $X^{\langle m_j \rangle}$ using Equation 4. Adjust its values;
9   **end**
10   **for** $j \leftarrow 1$ **to** N_F **do**
11   Update $X^{\langle f_j \rangle}$ using Equation 4. Adjust its values;
12   **end**
13   **end**
14 **end**

Algorithm 3: The mating behavior of ESSCO.

Input: D is the dimension of the search space.
Input: $\Delta^1 = [x^1_{\min}, x^1_{\max}], \ldots, \Delta^D = [x^D_{\min}, x^D_{\max}]$ are the input domains.
Input: f is the objective function to be minimized.
Input: $E = \{e_1, e_2, \ldots, e_{N_E}\}$: i.e., the set of insects ($N_E = |E|$).
Input: $M = \{m_1, m_2, \ldots, m_{N_M}\}$: i.e., the set of mal spiders ($N_M = |M|$).
Input: $F = \{f_1, f_2, \ldots, f_{N_F}\}$: i.e., the set of female spiders ($N_F = |F|$).
Input: ρ is the probability of an insect to get trapped.

/* Movement of male spiders */
1 **for** $i \leftarrow 1$ **to** N_M **do**
2 | Compute the weight $w^{\langle m_i \rangle}$ of male m_i using Equations 5 and 6;
3 **end**
4 Compute $\tilde{M}$ the median of weights $w^{\langle m_i \rangle}$, where $i \in \{1, \ldots, N_M\}$;
5 **for** $i \leftarrow 1$ **to** N_M **do**
6 | **if** ($w^{\langle m_i \rangle} \geq \tilde{M}$) **then**
7 | | Update $X^{\langle m_i \rangle}$ using Equation 7. Adjust its values;
8 | **end**
9 | **else**
10 | | Update $X^{\langle m_i \rangle}$ using Equation 8. Adjust its values;
11 | **end**
12 **end**

/* Movement of female spiders */
13 **for** $i \leftarrow 1$ **to** N_F **do**
14 | Update $X^{\langle f_i \rangle}$ using Equation 9. Adjust its values;
15 **end**

/* Mating of males and females */
16 **for** $i \leftarrow 1$ **to** N_M **do**
17 | **if** ($w^{\langle m_i \rangle} \geq \tilde{M}$) **then**
18 | | Choose the closest female $X^{\langle f \rangle}$;
19 | | Generate a new spider $X^{\langle \text{new} \rangle}$ using Equations 10, 11 and 12. Adjust its values;
20 | | Replace the worst male or female spider;
21 | **end**
22 **end**

2.2.3 Algorithm 3

- Lines 1 to 3: For each male $m_i \in M$, we compute its weight using Eqs. 5 and 6.

$$w^{\langle m_i \rangle} = \frac{F(X^{\langle m_i \rangle})}{\sum_{j=1}^{N_M} F(X^{\langle m_j \rangle})} \tag{5}$$

$$F(X^{\langle m_i \rangle}) = \frac{f(X^{\langle m_i \rangle}) - \mu - \sigma}{\sigma} \tag{6}$$

- Line 4: We compute the median $\tilde{M}$ of weights $w^{\langle m_i \rangle}$, where $i \in \{1, \ldots, N_M\}$.
- Lines 5 to 12: This loop translates the swarming behavior of male spiders $m_i \in M$ in the search space.
- Lines 6 to 8: If value $w^{\langle m_i \rangle}$ is greater than or equal to $\tilde{M}$, it means that male m_i is dominant.
- Line 7: We update location $X^{\langle m_i \rangle}$ using Eq. 7. If the values of $X^{\langle m_i \rangle}$ are out of permitted ranges, they are adjusted.

$$\begin{cases} X^{\langle m_i \rangle} = X^{\langle m_i \rangle} + \alpha_1\, T1 + \alpha_2\, T2 \\ T1 = w^{\langle \text{nf} \rangle}\, e^{-\|X^{\langle \text{nf} \rangle} - X^{\langle m_i \rangle}\|^2}\, (X^{\langle \text{nf} \rangle} - X^{\langle m_i \rangle}) \\ T2 = (\alpha_3 - 0.5) \end{cases} \tag{7}$$

- Lines 9 to 11: If value $w^{\langle m_i \rangle}$ is less than $\tilde{M}$, it means that male m_i is non-dominant.
- Line 10: We update location $X^{\langle m_i \rangle}$ using Eq. 8. If the values of $X^{\langle m_i \rangle}$ are out of permitted ranges, they are adjusted.

$$X^{\langle m_i \rangle} = X^{\langle m_i \rangle} + \alpha_1 \left(\frac{\sum_{j=1}^{N_M} (X^{\langle m_j \rangle} \times w^{\langle m_j \rangle})}{\sum_{j=1}^{N_M} w^{\langle m_j \rangle}} - X^{\langle m_i \rangle} \right) \tag{8}$$

- Lines 13 to 15: For each female $f_i \in F$, we compute its preference (i.e., attraction or repulsion) to males. In other words, we update location $X^{\langle f_i \rangle}$ using Eq. 9. If the values of $X^{\langle f_i \rangle}$ are out of permitted ranges, they are adjusted.

$$\begin{cases} X^{\langle f_i \rangle} = X^{\langle f_i \rangle} + (-1)^b\, \alpha_1\, T_1 + (-1)^b\, \alpha_2\, T_2 + \alpha_3\, T_3 \\ T_1 = w^{\langle \text{nb} \rangle}\, e^{-\|X^{\langle \text{nb} \rangle} - X^{\langle m_i \rangle}\|^2}\, (X^{\langle \text{nb} \rangle} - X^{\langle m_i \rangle}) \\ T_2 = w^{\langle \text{gb} \rangle}\, e^{-\|X^{\langle \text{gb} \rangle} - X^{\langle m_i \rangle}\|^2}\, (X^{\langle \text{gb} \rangle} - X^{\langle m_i \rangle}) \\ T_3 = (\alpha_4 - 0.5) \\ b \sim \mathfrak{B}(0.5) \end{cases} \tag{9}$$

- Lines 16 to 22: For each dominant male $m_i \in M$, we choose the closest female and generate a new spider using Eqs. 10, 11, and 12 [25, 26]. Then, we replace the worst male or female spider in the current population. The gender of the new spider is the same as the replaced one.

$$X^{\langle \text{new} \rangle} = \begin{bmatrix} \alpha_1 (f_2(x_1^{\langle m \rangle}, x_1^{\langle f \rangle}) - f_1(x_1^{\langle m \rangle}, x_1^{\langle f \rangle})) + f_1(x_1^{\langle m \rangle}, x_1^{\langle f \rangle}) \\ \vdots \\ \alpha_D (f_2(x_D^{\langle m \rangle}, x_D^{\langle f \rangle}) - f_1(x_D^{\langle m \rangle}, x_D^{\langle f \rangle})) + f_1(x_D^{\langle m \rangle}, x_D^{\langle f \rangle}) \end{bmatrix}^{\mathsf{T}} \tag{10}$$

$$f_1(x_1, x_2) = \min\{x_1, x_2\} - \alpha(\max\{x_1, x_2\} - \min\{x_1, x_2\}) \tag{11}$$

$$f_2(x_1, x_2) = \max\{x_1, x_2\} + \alpha(\max\{x_1, x_2\} - \min\{x_1, x_2\}) \tag{12}$$

where

μ : is the mean of values $f(X^{\langle m_i \rangle})$, where $i \in \{1, \ldots, N_M\}$.
σ : is the standard deviation of values $f(X^{\langle m_i \rangle})$, where $i \in \{1, \ldots, N_M\}$.
$\alpha_1, \ldots, \alpha_D$: uniformly distributed random numbers between 0 and 1.
$X^{\langle \text{nf} \rangle}$: location of the nearest female spider.
$\mathfrak{B}(0.5)$: is the Bernoulli distribution that has probability of success 0.5.
$X^{\langle \text{nb} \rangle}$: location of the nearest spider that has the best weight.
$X^{\langle \text{gb} \rangle}$: location of the best spider in the current population.

3 Experimental Results and Discussion

The benchmark CEC 2020 [66] is used to assess the performance of ESSCO. The obtained numerical results are compared to eleven state-of-the-art metaheuristic algorithms. All the experiments were performed on a personal computer (16 GB RAM, CPU of Intel(R) Core(TM) i7-9750H CPU @ 2.60GHz 2.59 GHz, and Windows 10 OS) using the Java programming language. The benchmark CEC 2020 comprises ten test functions: one is unimodal, three are multimodal, three are hybrid, and three are composite. All the functions are created employing 14 basic test functions, which are: High Conditioned Elliptic, Bent Cigar, Discus, Rosenbrock, Ackley, Weierstrass, Griewank, Rastrigin, Modified Schwefel, Happy Cat, HGBat, Expanded Griewank plus Rosenbrock, Expanded Schaffer, and Lunacek bi-Rastrigin. More specifications on these functions can be found in [66]. Tables 2, 3, 4, and 5 summarize the obtained outcomes of the comparative study for the benchmark CEC 2020. The next 11 state-of-the-art metaheuristic algorithms are studied.

- Improving Cuckoo Search: Incorporating changes for CEC 2017 and CEC 2020 Benchmark Problems (CSsin) [66].
- A Multi-Population Exploration-only Exploitation-only Hybrid on CEC 2020 Single Objective Bound Constrained Problems (MP-EEH) [13].
- Ranked Archive Differential Evolution with Selective Pressure for CEC 2020 Numerical Optimization (RASP-SHADE) [74].
- Improved Multi-operator Differential Evolution Algorithm for Solving Unconstrained Problems (IMODE) [67].
- DISH XX Solving CEC2020 Single Objective Bound Constrained Numerical Optimization Benchmark (DISH-XX) [80].
- Evaluating the Performance of Adaptive Gaining-Sharing Knowledge Based Algorithm on CEC 2020 Benchmark Problems (AGSK) [56].
- Differential Evolution Algorithm for Single Objective Bound Constrained Optimization: Algorithm j2020 (j2020) [15].
- Eigenvector Crossover in jDE100 Algorithm (jDE100e) [16].

Table 2 Mean, SD, and performance (P) obtained in 30 independent runs by ESSCO, CSsin, MP-EEH, and RASP-SHADE on 20-D CEC 2020 problem suite

Function	ESSCO		CSsin			RASP-SHADE			MP-EEH		
	Mean	SD	Mean	SD	P	Mean	SD	P	Mean	SD	P
F_1	0.00E+00	0.00E+00	9.33E+09	2.53E+09	+	0.00E+00	0.00E+00	=	0.00E+00	0.00E+00	=
F_2	3.00E−01	1.87E−01	9.83E+01	8.33E+01	+	1.70E+02	9.42E+01	+	1.38E−01	4.54E−02	–
F_3	1.72E+01	2.55E+00	2.55E+01	2.27E+00	–	2.33E+01	6.13E+00	+	2.05E+01	1.89E−01	–
F_4	8.02E−02	6.23E−02	0.00E+00	0.00E+00	–	4.25E−01	1.41E−01	+	4.53E−01	4.18E−02	–
F_5	6.57E+00	3.85E+00	1.16E+02	6.34E+01	+	2.36E+02	7.80E+01	+	1.41E+00	1.25E+00	–
F_6	1.59E−01	2.83E−02	6.72E−01	8.22E+00	+	4.27E+01	5.23E+01	+	1.71E−01	5.74E−02	+
F_7	4.82E−01	1.53E−01	2.62E+00	2.26E+00	+	7.77E+01	6.52E+01	+	7.27E−01	2.40E−01	+
F_8	9.07E+01	5.30E+00	9.89E+01	5.59E+00	+	8.00E+01	4.07E+01	+	1.00E+02	0.00E+00	–
F_9	1.02E+02	3.09E+00	1.03E+02	1.82E+01	+	9.67E+01	1.83E+01	+	3.42E+02	3.51E+01	+
F_{10}	4.00E+02	6.91E−01	3.99E+02	2.44E−01	–	4.01E+02	3.78E+00	+	4.14E+02	1.14E−13	–
+/–/=			7/3/0			9/0/1			3/6/1		

Table 3 Mean, SD, and performance (P) obtained in 30 independent runs by ESSCO, IMODE, DISH-XX, and AGSK on 20-D CEC 2020 problem suite

Function	ESSCO		IMODE			DISH-XX			AGSK		
	Mean	SD	Mean	SD	P	Mean	SD	P	Mean	SD	P
F_1	0.00E+00	0.00E+00	0.00E+00	0.00E+00	=	0.00E+00	0.00E+00	=	0.00E+00	0.00E+00	=
F_2	3.00E−01	1.87E−01	5.13E−01	7.13E−01	+	8.67E+01	1.11E+02	+	2.68E+03	1.60E+02	+
F_3	1.72E+01	2.55E+00	2.05E+01	1.26E−01	–	2.13E+01	3.74E+00	+	7.37E+01	5.25E+00	+
F_4	8.02E−02	6.23E−02	0.00E+00	0.00E+00	–	2.47E−04	1.35E−03	–	5.37E+00	4.25E−01	+
F_5	6.57E+00	3.85E+00	1.09E+01	4.33E+00	+	5.63E+01	6.63E+01	+	2.44E+02	3.97E+01	+
F_6	1.59E−01	2.83E−02	3.02E−01	8.17E−02	+	1.50E+01	3.57E+01	+	3.35E+00	2.17E+00	+
F_7	4.82E−01	1.53E−01	5.24E−01	1.64E−01	+	5.09E+00	6.42E+00	+	5.86E+01	1.09E+01	+
F_8	9.07E+01	5.30E+00	8.40E+01	1.89E+01	+	1.00E+02	1.39E−13	–	1.00E+02	0.00E+00	–
F_9	1.02E+02	3.09E+00	9.67E+01	1.83E+01	+	4.05E+02	2.50E+00	–	4.39E+02	2.95E+01	+
F_{10}	4.00E+02	6.91E−01	4.00E+02	6.18E−01	–	4.14E+02	2.54E−02	–	4.14E+02	8.87E−03	–
+/–/=			6/3/1			5/4/1			7/2/1		

Table 4 Mean, SD, and performance (P) obtained in 30 independent runs by ESSCO, j2020, jDE100e, and OLSHADE on 20-D CEC 2020 problem suite

Function	ESSCO		j2020			jDE100e			OLSHADE		
	Mean	SD	Mean	SD	P	Mean	SD	P	Mean	SD	P
F_1	0.00E+00	0.00E+00	0.00E+00	0.00E+00	=	0.00E+00	0.00E+00	=	0.00E+00	0.00E+00	=
F_2	3.00E−01	1.87E−01	2.60E−02	2.47E−02	–	1.48E+00	1.50E+00	+	1.15E+02	7.62E+01	+
F_3	1.72E+01	2.55E+00	1.44E+01	9.29E+00	+	2.10E+01	4.09E−01	–	2.52E+01	7.63E+00	+
F_4	8.02E−02	6.23E−02	1.80E−01	7.84E−02	+	3.47E−01	8.04E−02	+	1.01E+00	1.22E+00	+
F_5	6.57E+00	3.85E+00	7.78E+01	5.75E+01	+	2.37E+00	8.49E−01	–	1.78E+01	4.14E+01	+
F_6	1.59E−01	2.83E−02	1.92E−01	1.01E−01	+	1.15E−01	3.31E−02	+	5.17E−01	0.00E+00	–
F_7	4.82E−01	1.53E−01	1.98E+00	4.02E+00	+	2.14E−01	1.14E−01	–	8.42E−01	1.61E−01	+
F_8	9.07E+01	5.30E+00	9.27E+01	2.21E+01	+	1.00E+02	0.00E+00	–	1.00E+02	7.01E−01	–
F_9	1.02E+02	3.09E+00	3.39E+02	1.28E+02	+	4.05E+02	1.74E+00	–	1.10E+02	3.06E+01	+
F_{10}	4.00E+02	6.91E−01	3.99E+02	4.02E−02	–	4.13E+02	2.78E+00	+	4.10E+02	6.15E+00	+
+/–/=			7/2/1			4/5/1			7/2/1		

Table 5 Mean, SD, and performance (P) obtained in 30 independent runs by ESSCO, mpm-LSHADE, and SOMA-CL on 20-D CEC 2020 problem suite

Function	ESSCO		mpm-LSHADE			SOMA-CL		
	Mean	SD	Mean	SD	P	Mean	SD	P
F_1	0.00E+00	0.00E+00	0.00E+00	0.00E+00	=	0.00E+00	0.00E+00	=
F_2	3.00E−01	1.87E−01	3.97E−02	2.12E−02	−	7.36E+00	2.15E+01	+
F_3	1.72E+01	2.55E+00	2.04E+01	4.67E−13	−	2.14E+01	8.06E−01	−
F_4	8.02E−02	6.23E−02	4.97E−01	4.23E−02	−	1.05E+00	1.83E−01	+
F_5	6.57E+00	3.85E+00	1.38E+00	1.45E+00	−	1.45E+02	8.01E+01	+
F_6	1.59E−01	2.83E−02	2.05E−01	4.71E−02	+	2.71E−01	8.35E−02	+
F_7	4.82E−01	1.53E−01	5.10E−01	1.19E−01	−	9.22E+00	8.93E+00	+
F_8	9.07E+01	5.30E+00	1.00E+02	7.47E−13	−	9.91E+01	3.75E+00	−
F_9	1.02E+02	3.09E+00	4.01E+02	6.68E−01	−	3.99E+02	4.54E+01	+
F_{10}	4.00E+02	6.91E−01	4.14E+02	2.74E−04	−	4.71E+02	3.23E+01	+
+/−/=			1/8/1			7/2/1		

- Large Initial Population and Neighborhood Search incoporated in LSHADE to solve CEC2020 Benchmark Problems (OLSHADE) [10].
- Multi-population Modified L-SHADE for Single Objective Bound Constrained Optimization (mpm-LSHADE) [34].
- SOMA-CL for Competition on Single Objective Bound Constrained Numerical Optimization Benchmark (SOMA-CL) [36].

The parameters of ESSCO are $|N_E| = 50$, $|N_M| = 24$, $|N_F| = 56$, and $\rho = 0.5$. The search space dimension is set to 20. The maximum function evaluation is fixed to 1,000,000 for 30 different runs. The range of search space is $[-100, 100]^d$. The same settings and evaluation criteria in [66] are adapted in this work. The default values for different metaheuristic algorithms' parameters are used (as reported in the corresponding published papers). The mean, standard deviation (SD), and obtained performance (P) of each metaheuristic are given. Signs '+', '−', and '=' designate, respectively, that the performance of ESSCO is better/worst/equal than/to the considered algorithm. The statistical results of different metaheuristics are shown in Tables 2, 3, 4, and 5.

Based on the result given in the last row of Tables 2, 3, 4, and 5, the performance of ESSCO is better than CSsin, MP-EEH, RASP-SHADE, IMODE, DISH-XX, AGSK, j2020, jDE100e, OLSHADE, mpm-LSHADE, and SOMA-CL on 7, 9, 3, 6, 5, 7, 7, 4, 7, 1, and 7 functions, respectively. However, the performance of CSsin, RASP-SHADE, IMODE, DISH-XX, AGSK, j2020, jDE100e, OLSHADE, mpm-LSHADE, and SOMA-CL is better than ESSCO on 3, 6, 3, 4, 2, 2, 5, 2, 8, and 2 functions, respectively.

4 Conclusion

This work improves the optimization algorithm proposed in [19]. It is based on the hunting and mating behaviors of social spiders. Enhanced social spider colony optimization is evaluated on ten test functions of the benchmark CEC 2020. The numerical results are compared to the 11 state-of-the-art metaheuristic algorithms and show the competitiveness of ESSCO in finding efficient solutions. In future work, we plan to combine ESSCO with quantum genetic algorithms.

Acknowledgements This research work is supported by UAEU Grant: 31T102-UPAR-1-2017.

References

1. Abbass, H.A.: Mbo: Marriage in honey bees optimization-a haplometrosis polygynous swarming approach. In: Proceedings of the 2001 Congress on Evolutionary Computation (IEEE Cat. No. 01TH8546). vol. 1, pp. 207–214. IEEE (2001)

2. Alatas, B.: Acroa: artificial chemical reaction optimization algorithm for global optimization. Expert Syst. Appl. **38**(10), 13170–13180 (2011)
3. Artin, E.: The Gamma Function. Courier Dover Publications (2015)
4. Ashby, W.R.: Principles of the self-organizing system. In: Facets of Systems Science, pp. 521–536. Springer (1991)
5. Askarzadeh, A., Rezazadeh, A.: A new heuristic optimization algorithm for modeling of proton exchange membrane fuel cell: bird mating optimizer. Int. J. Energy Res. **37**(10), 1196–1204 (2013)
6. Aviles, L.: Sex-ratio bias and possible group selection in the social spider Anelosimus eximius. Am. Nat. **128**(1), 1–12 (1986)
7. Barthelemy, P., Bertolotti, J., Wiersma, D.S.: A lévy flight for light. Nature **453**(7194), 495–498 (2008)
8. Basturk, B.: An artificial bee colony (abc) algorithm for numeric function optimization. In: IEEE Swarm Intelligence Symposium, Indianapolis, IN, USA, 2006 (2006)
9. Bergmann, H.W.: Optimization: Methods and Applications, Possibilities and Limitations: Proceedings of an International Seminar Organized by Deutsche Forschungsanstalt Für Luft-und Raumfahrt (DLR), Bonn, June 1989, vol. 47. Springer Science & Business Media (2012)
10. Biswas, P.P., Suganthan, P.N.: Large initial population and neighborhood search incorporated in lshade to solve cec2020 benchmark problems. In: 2020 IEEE Congress on Evolutionary Computation (CEC), pp. 1–7. IEEE (2020)
11. Blum, C., Roli, A.: Metaheuristics in combinatorial optimization: overview and conceptual comparison. ACM Comput. Surv. (CSUR) **35**(3), 268–308 (2003)
12. Blum, C., Roli, A.: Hybrid metaheuristics: an introduction. In: Hybrid Metaheuristics, pp. 1–30. Springer, Berlin (2008)
13. Bolufé-Röhler, A., Chen, S.: A multi-population exploration-only exploitation-only hybrid on cec-2020 single objective bound constrained problems. In: 2020 IEEE Congress on Evolutionary Computation (CEC), pp. 1–8. IEEE (2020)
14. Borenstein, Y., Moraglio, A.: Theory and Principled Methods for the Design of Metaheuristics. Springer, Berlin (2014)
15. Brest, J., Maučec, M.S., Bošković, B.: Differential evolution algorithm for single objective bound-constrained optimization: algorithm j2020. In: 2020 IEEE Congress on Evolutionary Computation (CEC), pp. 1–8. IEEE (2020)
16. Bujok, P., Kolenovsky, P., Janisch, V.: Eigenvector crossover in jde100 algorithm. In: 2020 IEEE Congress on Evolutionary Computation (CEC), pp. 1–6. IEEE (2020)
17. Černỳ, V.: Thermodynamical approach to the traveling salesman problem: an efficient simulation algorithm. J. Opt. Theor. Appl. **45**(1), 41–51 (1985)
18. Ciarleglio, M.I.: Modular abstract self-learning tabu search (masts): Metaheuristic search theory and practice (2008)
19. Cuevas, E., Cienfuegos, M., ZaldíVar, D., Pérez-Cisneros, M.: A swarm optimization algorithm inspired in the behavior of the social-spider. Expert Syst. Appl. **40**(16), 6374–6384 (2013)
20. Dorigo, M., Birattari, M.: Ant colony optimization. In: Sammut, C., Webb, G. I. (Eds.) Encyclopedia of Machine Learning and Data Mining, pp. 56–59. Springer US, Boston, MA (2017), ISBN:978-1-4899-7687-1. https://doi.org/10.1007/978-1-4899-7687-1_22.
21. Dorigo, M., Di Caro, G.: Ant colony optimization: a new meta-heuristic. In: Proceedings of the 1999 congress on evolutionary computation-CEC99 (Cat. No. 99TH8406). vol. 2, pp. 1470–1477. IEEE (1999)
22. Du, H., Wu, X., Zhuang, J.: Small-world optimization algorithm for function optimization. In: International Conference on Natural Computation, pp. 264–273. Springer, Berlin (2006)
23. Elias, D.O., Andrade, M.C., Kasumovic, M.M.: Dynamic population structure and the evolution of spider mating systems. In: Advances in Insect Physiology, vol. 41, pp. 65–114. Elsevier (2011)
24. Erol, O.K., Eksin, I.: A new optimization method: big bang-big crunch. Adv. Eng. Softw. **37**(2), 106–111 (2006)

25. Eshelman, L.J.: Crossover operator biases: exploiting the population distribution. In: Proceedings of International Conference on Genetic Algorithms, 1997 (1997)
26. Eshelman, L.J., Schaffer, J.D.: Real-coded genetic algorithms and interval-schemata. In: Foundations of Genetic Algorithms, vol. 2, pp. 187–202. Elsevier (1993)
27. Fogel, D.B.: Artificial intelligence through simulated evolution. Wiley-IEEE Press (1998)
28. Formato, R.: Central force optimization: a new metaheuristic with applications in applied electromagnetics. prog electromagn res **77**: 425–491 (2007)
29. Gandomi, A.H., Alavi, A.H.: Krill herd: a new bio-inspired optimization algorithm. Commun. Nonlinear Sci. Numer. Simul. **17**(12), 4831–4845 (2012)
30. Glover, F.W., Kochenberger, G.A.: Handbook of metaheuristics, vol. 57. Springer Science & Business Media (2006)
31. Hatamlou, A.: Black hole: a new heuristic optimization approach for data clustering. Inf. Sci. **222**, 175–184 (2013)
32. Helbig, M., Engelbrecht, A.P.: Population-based metaheuristics for continuous boundary-constrained dynamic multi-objective optimisation problems. Swarm Evol. Comput. **14**, 31–47 (2014)
33. Holland, J.H.: Genetic algorithms. Sci. Am. **267**(1), 66–73 (1992)
34. Jou, Y.C., Wang, S.Y., Yeh, J.F., Chiang, T.C.: Multi-population modified l-shade for single objective bound constrained optimization. In: 2020 IEEE Congress on Evolutionary Computation (CEC), pp. 1–8. IEEE (2020)
35. Joyce, T., Herrmann, J.M.: A review of no free lunch theorems, and their implications for metaheuristic optimisation. In: Nature-inspired algorithms and applied optimization, pp. 27–51. Springer, Berlin (2018)
36. Kadavy, T., Pluhacek, M., Viktorin, A., Senkerik, R.: Soma-cl for competition on single objective bound constrained numerical optimization benchmark: a competition entry on single objective bound constrained numerical optimization at the genetic and evolutionary computation conference (gecco) 2020. In: Proceedings of the 2020 Genetic and Evolutionary Computation Conference Companion, pp. 9–10 (2020)
37. Karaboga, D., Basturk, B.: A powerful and efficient algorithm for numerical function optimization: artificial bee colony (ABC) algorithm. J. Glob. Optim. **39**(3), 459–471 (2007)
38. Kaur, S., Awasthi, L.K., Sangal, A., Dhiman, G.: Tunicate swarm algorithm: a new bio-inspired based metaheuristic paradigm for global optimization. Eng. Appl. Artif. Intell. **90**, 103541 (2020)
39. Kaveh, A., Farhoudi, N.: A new optimization method: Dolphin echolocation. Adv. Eng. Softw. **59**, 53–70 (2013)
40. Kaveh, A., Khayatazad, M.: A new meta-heuristic method: ray optimization. Comput. Struct. **112**, 283–294 (2012)
41. Kaveh, A., Talatahari, S.: A novel heuristic optimization method: charged system search. Acta Mechanica **213**(3–4), 267–289 (2010)
42. Keller, E.F.: Organisms, machines, and thunderstorms: a history of self-organization, part two. complexity, emergence, and stable attractors. Hist. Stud. Natural Sci. **39**(1), 1–31 (2009)
43. Kennedy, J., et al.: Encyclopedia of machine learning. Particle Swarm Optim. 760–766 (2010)
44. Kirkpatrick, S., Gelatt, C.D., Vecchi, M.P.: Optimization by simulated annealing. Science **220**(4598), 671–680 (1983)
45. Koza, J.R., Koza, J.R.: Genetic programming: on the programming of computers by means of natural selection, vol. 1. MIT press (1992)
46. Koziel, S., Yang, X.S.: Computational optimization, methods and algorithms, vol. 356. Springer, Berlin (2011)
47. Kumar, A., Misra, R.K., Singh, D., Mishra, S., Das, S.: The spherical search algorithm for bound-constrained global optimization problems. Appl. Soft Comput. **85**, 105734 (2019)
48. Li, X.: A new intelligent optimization-artificial fish swarm algorithm. Doctor thesis, Zhejiang University of Zhejiang, China (2003)
49. Lu, X., Zhou, Y.: A novel global convergence algorithm: bee collecting pollen algorithm. In: International Conference on Intelligent Computing, pp. 518–525. Springer, Berlin (2008)

50. Lubin, Y., Bilde, T.: The evolution of sociality in spiders. Adv. Study Behavior **37**, 83–145 (2007)
51. Mantegna, R.N.: Fast, accurate algorithm for numerical simulation of levy stable stochastic processes. Phys. Rev. E **49**(5), 4677 (1994)
52. Mirjalili, S.: Dragonfly algorithm: a new meta-heuristic optimization technique for solving single-objective, discrete, and multi-objective problems. Neural Comput. Appl. **27**(4), 1053–1073 (2016)
53. Mirjalili, S., Lewis, A.: The whale optimization algorithm. Adv. Eng. Softw. **95**, 51–67 (2016)
54. Mirjalili, S., Mirjalili, S.M., Lewis, A.: Grey wolf optimizer. Adv. Eng. Softw. **69**, 46–61 (2014)
55. Moghaddam, F.F., Moghaddam, R.F., Cheriet, M.: Curved space optimization: a random search based on general relativity theory. arXiv preprint arXiv:1208.2214 (2012)
56. Mohamed, A.W., Hadi, A.A., Mohamed, A.K., Awad, N.H.: Evaluating the performance of adaptive gainingsharing knowledge based algorithm on cec 2020 benchmark problems. In: 2020 IEEE Congress on Evolutionary Computation (CEC), pp. 1–8. IEEE (2020)
57. Molina, J., Rudnick, H.: Transmission expansion plan: Ordinal and metaheuristic multiobjective optimization. In: 2011 IEEE Trondheim PowerTech, pp. 1–6. IEEE (2011)
58. Mucherino, A., Seref, O.: Monkey search: a novel metaheuristic search for global optimization. In: AIP Conference Proceedings, vol. 953, pp. 162–173. AIP (2007)
59. Murata, K., Tanaka, K.: Spatial interaction between spiders and prey insects: horizontal and vertical distribution in a paddy field. Acta arachnologica **53**(2), 75–86 (2004)
60. Oftadeh, R., Mahjoob, M., Shariatpanahi, M.: A novel meta-heuristic optimization algorithm inspired by group hunting of animals: hunting search. Comput. Math. Appl. **60**(7), 2087–2098 (2010)
61. Pan, W.T.: A new fruit fly optimization algorithm: taking the financial distress model as an example. Knowl.-Based Syst. **26**, 69–74 (2012)
62. Puchinger, J., Raidl, G.R.: Combining metaheuristics and exact algorithms in combinatorial optimization: a survey and classification. In: International work-conference on the interplay between natural and artificial computation, pp. 41–53. Springer, Berlin (2005)
63. Rajeev, S., Krishnamoorthy, C.: Discrete optimization of structures using genetic algorithms. J. Struct. Eng. **118**(5), 1233–1250 (1992)
64. Rashedi, E., Nezamabadi-Pour, H., Saryazdi, S.: GSA: a gravitational search algorithm. Inf. Sci. **179**(13), 2232–2248 (2009)
65. Roth, M., Wicker, S.: Termite: A swarm intelligent routing algorithm for mobile wireless ad-hoc networks. In: Stigmergic Optimization, pp. 155–184. Springer (2006)
66. Salgotra, R., Singh, U., Saha, S., Gandomi, A.H.: Improving cuckoo search: incorporating changes for CEC 2017 and CEC 2020 benchmark problems. In: 2020 IEEE Congress on Evolutionary Computation (CEC), pp. 1–7. IEEE (2020)
67. Sallam, K.M., Elsayed, S.M., Chakrabortty, R.K., Ryan, M.J.: Improved multi-operator differential evolution algorithm for solving unconstrained problems. In: 2020 IEEE Congress on Evolutionary Computation (CEC), pp. 1–8. IEEE (2020)
68. Salomon, M., Sponarski, C., Larocque, A., Avilés, L.: Social organization of the colonial spider leucauge sp. in the neotropics: vertical stratification within colonies. J. Arachnology **38**(3), 446–451 (2010)
69. Shah-Hosseini, H.: Principal components analysis by the galaxy-based search algorithm: a novel metaheuristic for continuous optimisation. Int. J. Comput. Sci. Eng. **6**(1–2), 132–140 (2011)
70. Shi, J., Zhang, Q.: A new cooperative framework for parallel trajectory-based metaheuristics. App. Soft Comput. **65**, 374–386 (2018)
71. Shi, Y., Eberhart, R.C.: Empirical study of particle swarm optimization. In: Proceedings of the 1999 Congress on Evolutionary Computation-CEC99 (Cat. No. 99TH8406), vol. 3, pp. 1945–1950. IEEE (1999)
72. Shiqin, Y., Jianjun, J., Guangxing, Y.: A dolphin partner optimization. In: 2009 WRI Global Congress on Intelligent Systems, vol. 1, pp. 124–128. IEEE (2009)

73. Simon, D.: Biogeography-based optimization. IEEE Trans. Evol. Comput. **12**(6), 702–713 (2008)
74. Stanovov, V., Akhmedova, S., Semenkin, E.: Ranked archive differential evolution with selective pressure for CEC 2020 numerical optimization. In: 2020 IEEE Congress on Evolutionary Computation (CEC), pp. 1–7. IEEE (2020)
75. Storn, R., Price, K.: Differential evolution-a simple and efficient heuristic for global optimization over continuous spaces. J. Glob. Optim. **11**(4), 341–359 (1997)
76. Talbi, E.G., Jourdan, L., Garcia-Nieto, J., Alba, E.: Comparison of population based metaheuristics for feature selection: Application to microarray data classification. In: 2008 IEEE/ACS International Conference on Computer Systems and Applications, pp. 45–52. IEEE (2008)
77. Talbi, H., Draa, A.: A new real-coded quantum-inspired evolutionary algorithm for continuous optimization. Appl. Soft Comput. **61**, 765–791 (2017)
78. Tang, K.S., Man, K.F., Kwong, S., He, Q.: Genetic algorithms and their applications. IEEE Signal Process. Mag. **13**(6), 22–37 (1996)
79. Van Laarhoven, P.J., Aarts, E.H.: Simulated annealing. In: Simulated Annealing: Theory and Applications, pp. 7–15. Springer, Berlin (1987)
80. Viktorin, A., Senkerik, R., Pluhacek, M., Kadavy, T., Zamuda, A.: Dish-xx solving cec2020 single objective bound constrained numerical optimization benchmark. In: 2020 IEEE Congress on Evolutionary Computation (CEC), pp. 1–8. IEEE (2020)
81. Vollrath, F., Rohde-Arndt, D.: Prey capture and feeding in the social spider Anelosimus eximius. Zeitschrift für Tierpsychologie **61**(4), 334–340 (1983)
82. Webster, B., Philip, J., Bernhard, A.: Local search optimization algorithm based on natural principles of gravitation, ike'03, las vegas, Nevada, USA (2003, June)
83. Yang, C., Tu, X., Chen, J.: Algorithm of marriage in honey bees optimization based on the wolf pack search. In: The 2007 International Conference on Intelligent Pervasive Computing (IPC 2007), pp. 462–467. IEEE (2007)
84. Yang, X.S.: Engineering Optimization: An Introduction with Metaheuristic Applications. Wiley, Hoboken (2010)
85. Yang, X.S.: Firefly algorithm, stochastic test functions and design optimisation. arXiv preprint arXiv:1003.1409 (2010)
86. Yang, X.S.: Nature-Inspired Metaheuristic Algorithms. Luniver Press (2010)
87. Yang, X.S.: A new metaheuristic bat-inspired algorithm. In: Nature Inspired Cooperative Strategies for Optimization (NICSO 2010), pp. 65–74. Springer, Berlin (2010)
88. Yang, X.S.: Swarm-based metaheuristic algorithms and no-free-lunch theorems. Theor. New Appl. Swarm Intell. **9**, 1–16 (2012)
89. Yang, X.S.: Optimization and metaheuristic algorithms in engineering. In Metaheuristics in Water, Geotechnical and Transport Engineering, pp. 1–23 (2013)
90. Yang, X.S., Deb, S.: Cuckoo search via lévy flights. In: 2009 World Congress on Nature & Biologically Inspired Computing (NaBIC), pp. 210–214. IEEE (2009)
91. Yao, X., Liu, Y., Lin, G.: Evolutionary programming made faster. IEEE Trans. Evol. Comput. **3**(2), 82–102 (1999)
92. Zitouni, F., Harous, S., Maamri, R.: The solar system algorithm: a novel metaheuristic method for global optimization. IEEE Access (2020)

Data Processing on Distributed Systems Storage Challenges

Mohamed Eddoujaji, Hassan Samadi, and Mohamed Bohorma

Abstract Hadoop is an open-source software framework for storing data and launching applications on standard machine clusters. This solution provides massive storage space for all types of data, tremendous processing power, and the ability to support virtually any amount of work. Based on Java, this framework is part of the Apache project, sponsored by Apache Software Foundation [Hadoop official site. http://hadoop.apache.org/]. Thanks to the MapReduce framework, it can handle huge amounts of data. Rather than having to move the data to a network for processing, MapReduce allows you to directly move the processing software to the data. No one will be able to neglect the results realized by the Hadoop technique nor to question its power and its performance to manage important volumetrics of the data of any type: structured, semi-structured, or unstructured [https://www.lebigdata.fr/hadoop]. But it does not shine by its processing speed when datasets are smaller and even smaller. This problem, the "small files" problem has been well-defined by the Hadoop community, and also by the researchers, the majority of the proposed solutions and concepts only deal with the performance and the pressure exerted on the NameNode memory; some new approaches have been proposed as the casi-random, grouping of small heterogeneous files and in different formats, this solves the problem of memory because the number of metadata has been considerably reduced; the real challenge companies face is the performance of the Hadoop cluster when processing a very large number of small files.

1 Introduction and Background

Hadoop was designed to efficiently manage effectively large files, especially when traditional systems are facing limitations to analyze this new data dimension of data caused by its exponential growth. However, Hadoop is not deployed to handle only large files! The heterogeneity and diversity of information from multiple sources

M. Eddoujaji (✉) · H. Samadi · M. Bohorma (✉)
Abdelmalek Essaâdi University, National School of Applied Sciences of Tangier, Tangier, Morocco

M. Ben Ahmed et al. (eds.), *Networking, Intelligent Systems and Security*,
Smart Innovation, Systems and Technologies 237,
https://doi.org/10.1007/978-981-16-3637-0_56

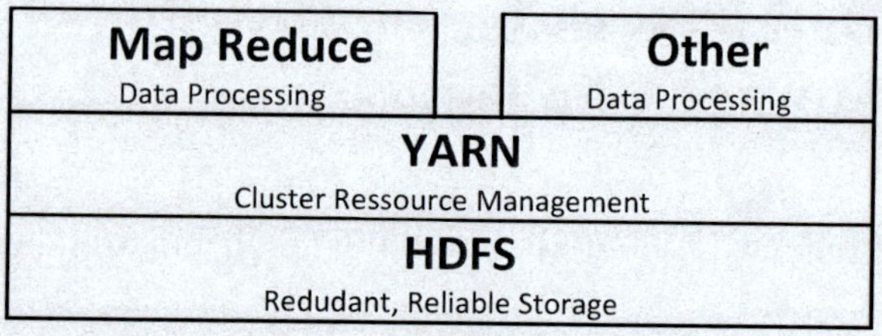

Fig. 1 Hadoop 2.0 main components

(intelligent devices, IOTs, Internet users, log files, and security events) have become the normal flow of Hadoop architectures.

In today's world, most domains permanently and constantly generate a very large amount of information in the form of small files. Multiple domains store and analyze millions and millions of small files, such as analysis for multimedia data mining [3], astronomy [4], meteorology [5], signal recognition [6], climatology [7, 8], energy, and e-learning. [9], without forgetting the astronomical information processed by social networks; Facebook stores more than 350 million images every day [10].

In biology, the human genome generates up to 30 million files which do not exceed on average 190 KB [11] (Fig. 1).

1.1 HDFS—Reliable Storage

Due to its massive capacity and reliability, HDFS is a storage system that is very suitable for Big Data. In combination with YARN, this system increases the data management capabilities of the HDFS Hadoop cluster and thus enables the efficient processing of Big Data. Among its main features, there is the possibility of storing terabytes, even petabytes of data [12].

The system is capable of handling thousands of nodes without the intervention of an operator. It allows simultaneous benefits of parallel computing and distributed computing [33]. After a modification, it allows to easily restore the previous version of a data.

HDFS can be launched on commodity hardware, which makes it very tolerant of errors. Each piece of data is stored in several places and can be retrieved under any circumstances. In the same way, this replication makes it possible to fight against the potential corruption of the data (Fig. 2).

2 Paper Sections

The following parts in this paper will be organized as follows:

Chapter 3 describes the related work and some of the previous results.

Chapter 3 describes our proposed approach.

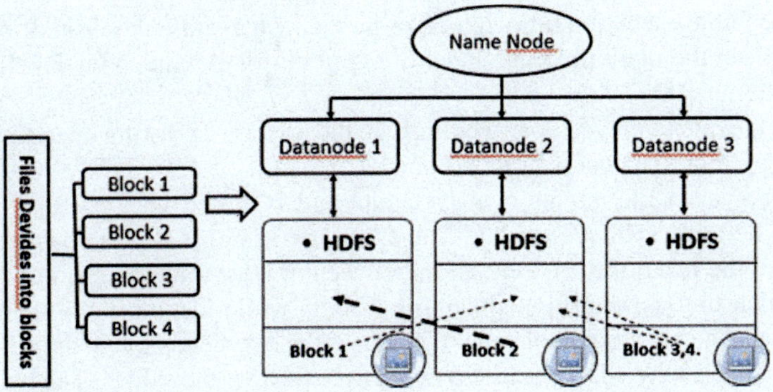

Fig. 2 Storage and replication of blocks in HDFS

Chapter 4 presents our experimental works and results.
Finally, and to conclude this research,
Chapter 5 for conclusion and expectation.

3 Related Work

3.1 Reminder

In most data-intensive applications, we mainly work on two types of file systems, the distributed file systems and the parallel file systems [33].

The distributed file system type is broadly used in online services, and one of the unmistakable examples involves Amazon Simple Storage Service known as (Amazon S3), Google's file system, and of course, the open-source Hadoop file system [13].

Parallel file system type is mainly built targeting high compute processing (HPC) applications that run-on large-scale clusters and require highly extendable and scalable storage solutions that can perform simultaneous I/O operations [2]. Some illustrations of parallel file systems are covered by Sun's LustreFS [14], IBM's Spectrum Scale based on GPFS [15], and open-source Parallel Virtual FS (PVFS) [14].

3.2 Motivation

HDFS has been designed mainly to be able to manage sizable files, but not small ones. That is the reason why it may face some issues when asked to manage a big amount of small files. For example, if we have around 600,000 small files which have sizes that vary between 1 and 10 KB were stored into HDFS, the process led to observe the following phenomena [16]:

Unacceptable execution time. It took more than 7 h to store these small-sized files into HDFS, in the main time in a local file system, for example, "ext3," the storing time was about 600 s.

The usage rate of the memory was high. In the process of storing operations, AND in an idle system, the occupation was at 63%.

In HDFS, each file has its own metadata and is stored with multiple replicas, three as a default replication value. Managing metadata in HDFS consumes much time due to the need of the cooperation between at least three nodes. For small file I/O, so much time is spent in managing metadata while just a little is spent in data transferring. The big amount of small files raises the overhead of metadata operations in HDFS. This is why HDFS needs so much time to store these files. The NameNode manages and stores the metadata, the DataNodes preserve fragmented information, and all these data are loaded on the physical memory for exploitation.

As a result, the more number of small files the more memory usage rate increases. We have to adopt an optimized approach to build a middleware on HDFS so as to satisfy the application demands of small files IO performance, and we should consider the file access patterns of special applications [17, 18].

4 Existing Solutions

4.1 HAR Files

Hadoop itself provides a tool Archives Hadoop (HAR) [3, 19], and it can conduct a small file in file processing. It adds an index file to an archive and also provides convenience for the MapReduce operation, but there are still shortcomings. Due to the small file in the access to HAR based on the two index files to find which leads to client relying on the HAR file to access small files, and the merged process takes a long time, which is less efficient than reading small files from the HDFS. You cannot modify The HAR file created by this method, instead, you have to recreate the HAR file if you want to make changes (add or delete content on it) [20] (Fig. 3).

Fig. 3 HAR file layout

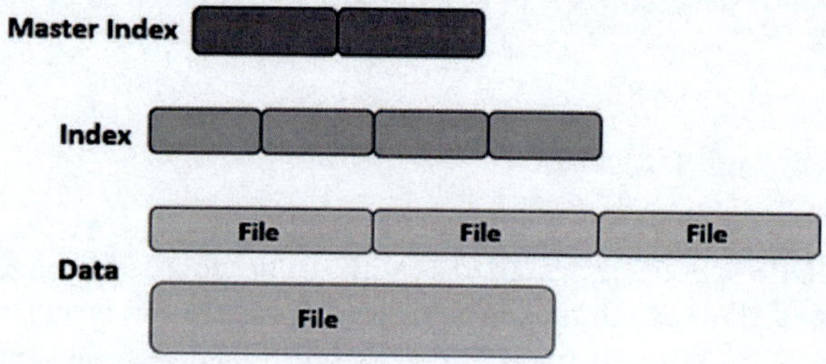

5 SequenceFiles

In the case of SequenceFile, we have no solution to list all the keys of a file, except reading through the entire file. (MapFiles files, which look like SequenceFiles with sorted keys, keep a partial index, so they cannot list all their keys either—see the diagram.) [21, 22].

SequenceFile is a binary format program compatible with Hadoop API [3, 23], whose data structure is composed of a series of key/binary values, the small files can be stored in a single unit. The principle is simple, and it consists of merging and grouping a large number of small files into one large file. It provides good support and high scalability and performance for local MapReduce data management [24] (Fig. 4).

Unlike the HAR files, the SequenceFile supports compression, and they are more suitable for MapReduce tasks as they are splittable [3, 23], so mappers can operate on chunks independently. However, converting into a SequenceFile can be a time-consuming task, and it has a poor performance during random read access.

To improve the metadata management, Mohd Abdul Ahada, Ranjit Biswasa [10] merged small files into a single larger file, using the HAR through a MapReduce task. The small files are referenced with an added index layer (Master index, Index) delivered with the archive, to retain the separation, and to keep the original structure of files.

C. Vorapongkitipunet al. [3, 11] proposed an improved approach of the HAR technique, by introducing a single index instead of the two-level indexes. Their new indexing mechanism aims to improve the metadata management as well as the performance during file access without changing the implemented HDFS architecture.

Rattanaopas and Kaewkeeree. [25], and Mir and Ahmed [26] proposed to combine files using the SequenceFile method. Their approach reduces memory consumption on the NameNode but it did not show how much the read and write performances are impacted.

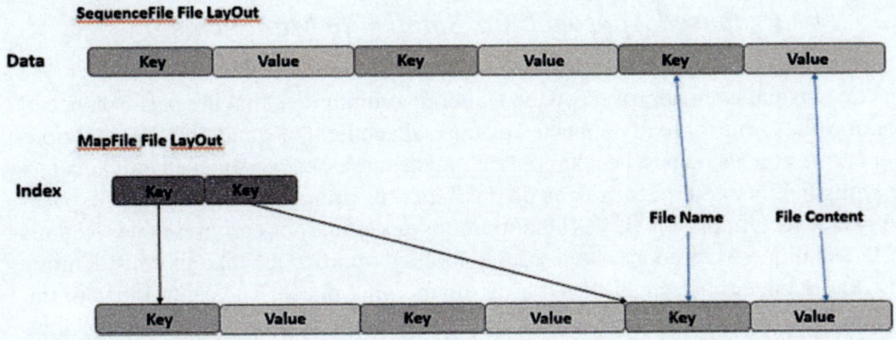

Fig. 4 SequenceFile layout

Zheng et al. [32] proposed merging related small files according to WebGIS application, which improved the storage efficiency and HDFS metadata management; however, the results are limited by the scene.

Mir and Ahmed [7] proposed a modification of the existing HAR. They used a hashing concept based on the sha256 as a key. This can improve the reliability and the scalability of the metadata management, also the reading access time is greatly reduced, but it takes more time to create the NHAR archives compared to the HAR mechanism.

Niazi and Ronström [27] proposed a scheme for combining small file, merging, prefetching the related small files which improve the storage and access efficiency of small files but do not give an appropriate solution for independent small files.

6 Proposed Work

6.1 Little Reminder

Hadoop is designed for powerful management of large files, unlike the management of large volumes of small files where the technique always requires improvements, in fact, Hadoop passes each small file to a map () function, which creates a large number of mappers, and therefore, the solution is not very effective for this kind of traffic. For example, the 100,000 files smaller than 2Mo will need 10,000 mappers, which is very inefficient, which could be a problem throughout the system.

The main objective of these improvements and new approaches is on the one hand to solve this problem, and on the other hand, it is to accelerate the execution of operations (reading and writing) Hadoop. This solution is to combine small files into larger files, and this will reduce the number of map () functions that are executed and thus significantly improve performance [28, 32].

6.2 The Proposed Approach for Small File Management

It is known and even approved by the Hadoop communities that the performance of Hadoop platforms is greatly impacted by the management of small files. Les solutions précédentes ou les recherches faites sur ce sujet, surpassent ce souci en empaquetant de petits fichiers hétérogènes dans de dossiers plus grands. Cette manière de packaging et le facteur principale des améliorations des temps des écritures et des lectures de la technique Mapreduce, none of the adopted approaches take in consideration how to organize those small files during the merging phase. The central idea of our approach is to start the organization and management of files when starting flows containing small files, combined if relevant, that will be combined with different other client streams into blocks based on their relevance; moreover, they are arranged in an efficient way as the files with most fetch probability appear always on top. This efficiency and performance capabilities can be reached utilizing "**Hadoop File Server Analyzer**" as shown in "Fig. 6."

The foundational idea of our approach is to consolidate files from different clients that contain series of small files, combined by relevance, and unify them through a merge process, to be stored optimally before the current SFA connection closes. This realization was carried out in our previous work, see "Fig. 6" as the main task of the "**Small File Analyzer server.**" In the present research, we have enhanced the suggested SFA method, to be able to handle new modules and make it possible for us to acknowledge other parameters within the merge process [3].

We have introduced a sorting process which can work as an unrelated module in the SFA server. The compressor module is an extra add-on that has enabled us to apply a new compression layer on top of the merged files that are no longer in use, or barely solicited, so that we can have a huge storage capacity advantage. Additionally, we have used a "**prefetch and caching**" technique to improve the overall performance when reading identical small files [3].

6.3 Operational Mode Explanation

The operational mode of our idea can be divided into three major phases:

File combining, File mapping, Prefetching and caching

6.4 File Combining

Conventionally, NameNode stores metadata of files and blocks.

File metadata consists of the following entries:

file name, file permissions, owner and group, access time, modification time, file length, replication.

Block metadata consists of the following information:

information on all the blocks containing the data of the file, location of these blocks.

Thus, in conventional approaches, storing a large number of small files consumes a large amount of memory of the "NameNode" machine.

In order to optimize the memory consumption needed to store the NameNode metadata, several small files will be merged and combined into a single file that we can call a "combined file."

As a result, NameNode will store the metadata of a single large file instead of keeping the metadata of several small files.

In addition, this technique will ensure reading and grouping all similar files together.

Below is a small description of how MapReduce works for the "combined file" which will be called the MapReduce combiner:

In the Map phase, the mappers generate key/value pairs.

During the shuffle/sort phase, these pairs are distributed and ordered on one or more nodes depending on the value of the key.

During the Reduce phase, one or more reducers aggregate the key/value pairs according to the value of the key.

6.5 File Mapping

The second phase consists of analyzing these files based on their size (how these small files are used), then put them in appropriate groups in MapFile. As we mentioned above, the MapFile is another file packaging format developed by Hadoop based on the indexation of SequenceFile (Figs. 5 and 6).

When a request for a particular file is executed, a request is sent by the HDFS client to the NameNode machine to obtain the metadata of the requested file.

In the proposed approach, instead of processing file metadata directly, like a traditional Hadoop processing, the NameNode will opt for metadata processing of the combined files based on a mapping file.

This mapping file consists of:

the file name, file length, file offset, block number of the combined file.

So we call a "mapping record" each entry in the mapping file.

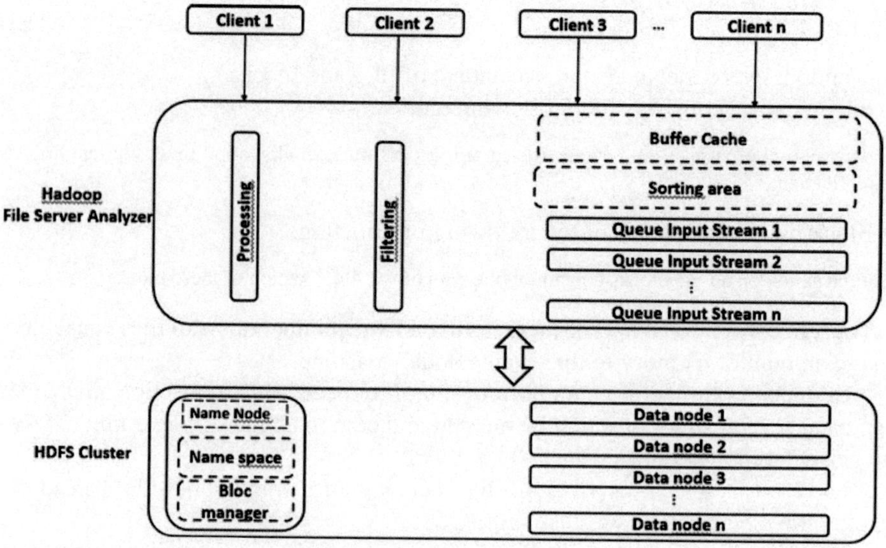

Fig. 5 Hadoop file server analyzer

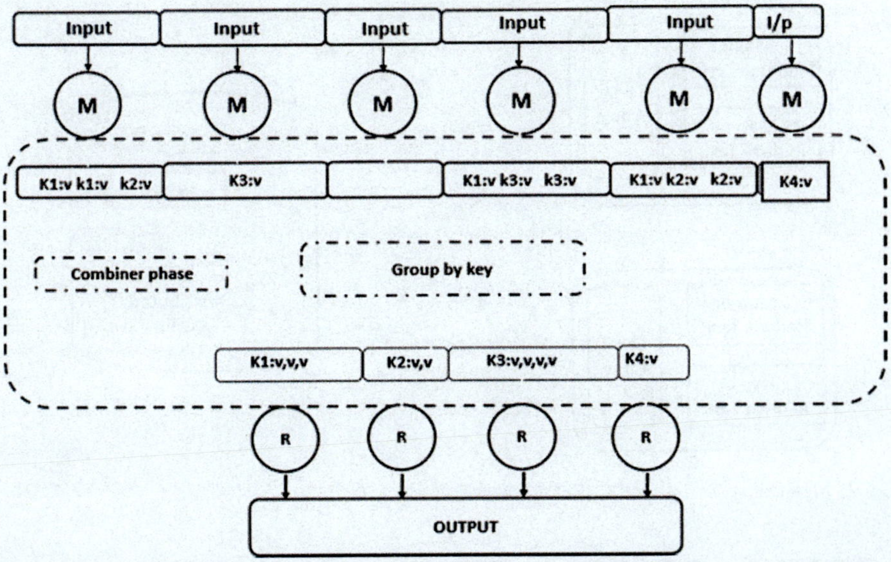

Fig. 6 Combiner phase

6.6 Prefetching and Caching

The proposed approach reduces load on NameNode by using prefetching and caching techniques that are generally used as storage optimization techniques [23]. Prefetching conceals visible I/O cost and improves access time by utilizing correlations between files and fetching data into cache prior to they are requested [29].

The study of the behavior of data processing is not limited to the study of the consumption of the memory and the processors, but also depends on storage parameters, and we can include here two parameters used in the optimization of storage: prefetching and caching [30].

Prefetching conceals visible I/O costs and improves file access time by retrieving data files in the cache before they are requested [31]. This means that access to the metadata will be acquired directly from the cache instead of the disk volumes.

This greatly improves file access times (Figs. 7 and 8).

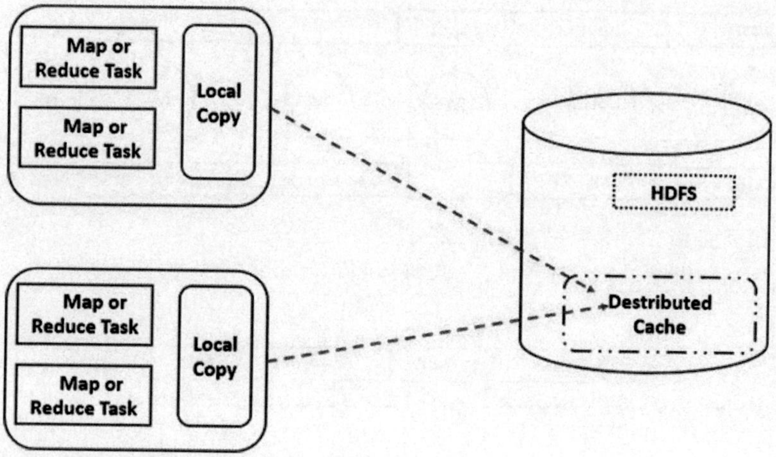

Fig. 7 MetaData caching

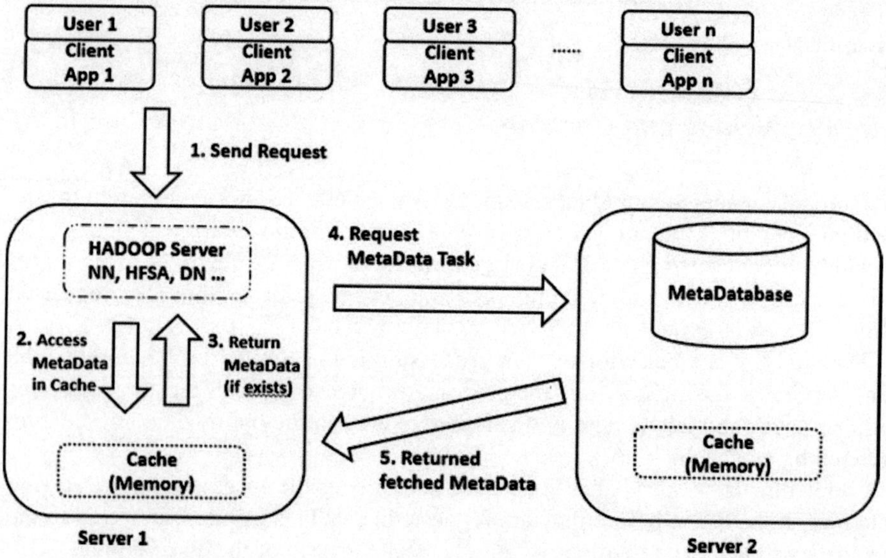

Fig. 8 Prefetching-based MetaData

7 Experimental

7.1 Experimental Environment

The suggested implementation in this paper can be contrasted to the original HDFS in terms of NameNode memory usage along with MapReduce job performance, during sequential, and selective file access. Our simulation was conducted using Hadoop 2.4.0, with a cluster of nodes each node consisting of the following specs:

1. Node x
2. 3.10 GHz clock rate
3. 8 GB of memory
4. 1 Gbit Ethernet—Network interface controller
5. 4 Data Nodes

All nodes offer a 500 GB hard drive and are deployed over Ubuntu server 14.04 distribution. The replication rate will be preserved as the default 3, and the block size of HDFS is picked as 64 MB. Pilot datasets are mostly auto-generated, and we also have the ability to use public datasets on [27].

7.2 Phases Algorithms

Input:

```
  Block_Size: Defined block size
  CombinedFile: Name of combined file.
  smallFile_List: list of small files
  smallFile_Dir: Directory containing small files
```

Output:

```
  File_Index: index of CombinedFile.
  Initialize smallFile_Map
  if smallFile_List is provided as input
  for each file fl in File_List
  compute code(fl.name)
  add fl to smallFile_Map with key: code,
  values:name(fl), length(fl)
  end for
  else if smallFile_Dir is given as input
  for each file f in the directory tree at
smallFile_Dir
  compute code(fl.name)
  add f to sF_Map with key: code,
  values:name(f), length(fl)
  end for
  end if
  if CombinedFile exists
  open CombinedFile for append
  LocalIndex= LocalIndex(CombinedFile );
  blockNum = length(combinedFile)/Block_Size
  cur_offset = length(CombinedFile) % Block_Size
```

```
else
create CombinedFile; open CombinedFile for write
Initialize File_Index; set cur_offset= 0, blockNum=0
end if
for each file with code fl in smallFile_ap in order
of
increasing length(fl)
name = name(fl), ln=length(fl); combinedIn=CombinedFile
if cur_offset + ln > Block_Size
blockNum++,
start = blockNum*Block_Size, end= start + ln,
curr_offset = ln ++
else if cur_offset + ln = Block_Size
start= blockNum*Block_Size + cur_offset, end= start
+ ln,
curr_offset=0, blockNum++
else
start= blockNum*Block_Size + cur_offset,
end = start + ln,
curr_offset=end ++
endif
append f to CombinedFile
Insert key fl into LocalIndex with name, ln, start,
end, combinedIn as values
end for
close CombinedFile
return File_Index.
```

7.3 Results

The amount of files generated by size is presented in Fig. 10; the total number of files is 20,000, and the size of files range is from 1 KB to 4.35 MB (Fig. 9).

The workloads for measuring the time taken for read and write operations are a subset of the workload used for memory usage experiment, containing the above datasets (Fig. 11; Table 1).

As shown in the figure, we can conclude that memory consumption using the approach is 20–30% lower than that consumed by the original HDFS. Indeed, the NameNode, for original HDFS, stores file and bloc metadata for each file.

This means that by increasing the number of files stored the memory consumption increases. On the other hand, as for the proposed approach, the NameNode stores only the metadata of each small file. For the block metadata, the NameNode stores them as a single combined file and not for every single small file, which explains the reduction of the memory used by the proposed approach.

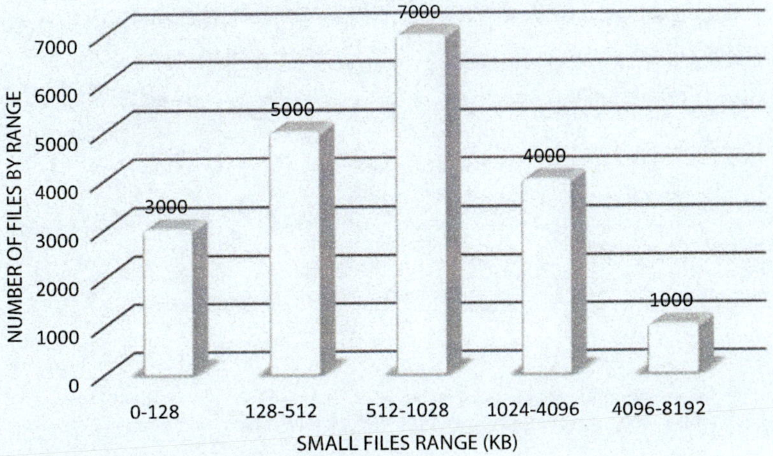

Fig. 9 Distribution of file sizes in our experiment

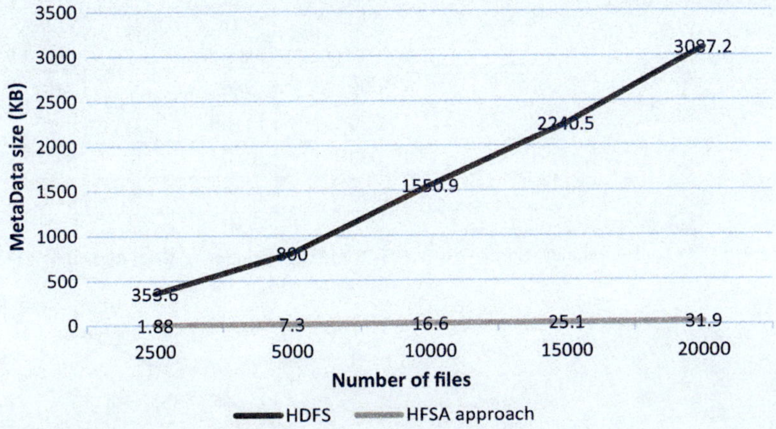

Fig. 10 Memory consumption

7.4 Performances Comparison

7.4.1 Writing Test

Results of writing time are shown in this Fig. 12.

In the following, we performed MapReduce jobs on five datasets. As shown in the diagram, when the number of files is important, the gain in writing performance becomes more important using our approach. For example, we have a performance of 9% for 1000 files beside 36% for 20,000 files.

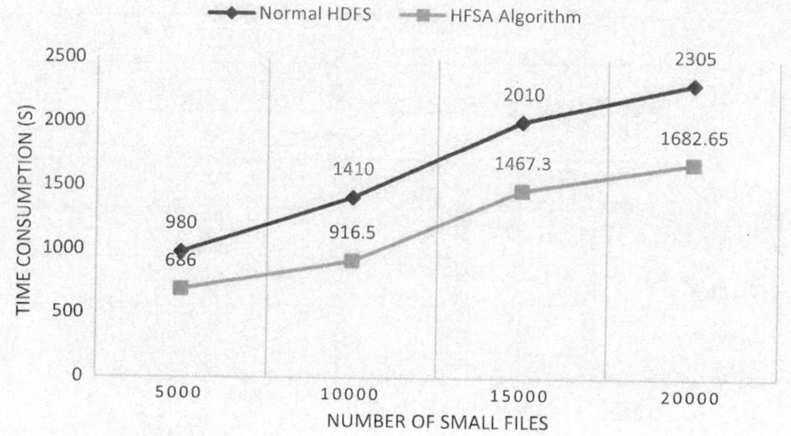

Fig. 11 Memory usage by NodeName

Table 1 Comparison of the NameNode memory usage

Dataset #	Number of small files	Time consumption (s)	
		Normal HDFS	HFSA algorithm
1	5000	980	268
2	10,000	1410	390
3	15,000	2010	530
4	20,000	2305	620

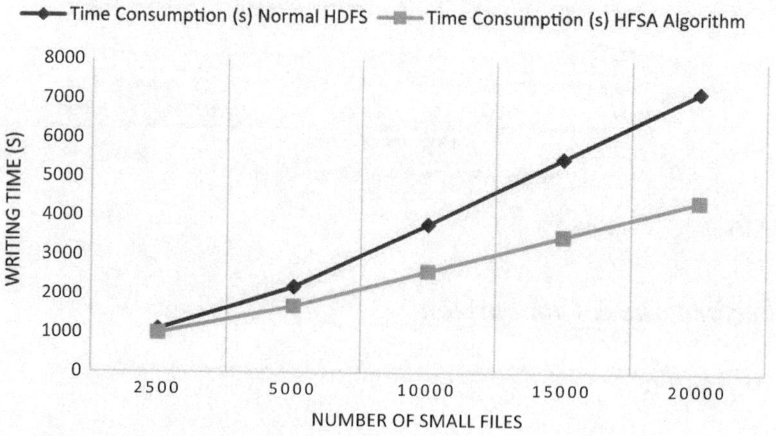

Fig. 12 Performance evaluation: writing time

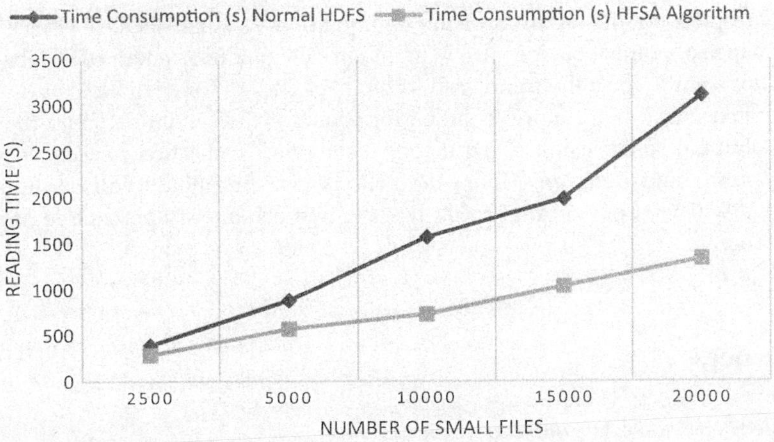

Fig. 13 Performance evaluation: reading time

Through the above comparison, it is proved that our approach can correctly enhance the effectivity of file writing.

7.4.2 Reading Test

The average sequence reading time of HFSA is 788,94 s, and the average read time of the original HDFS is 1586,58 s. The comparison shows that the average reading speed of SFS is 1.36 times of HDFS, 13.03 times of HAR.

Applying our approach, we had a performance of around 33% for writing process and more than 50% for reading (Fig. 13).

8 Conclusion and Future Works

In this paper, we described, in a detailed way, our approach and solution to address Hadoop technology defects related to distributed storage of large volumes of small files.

The Hadoop Server File Analyzer supports the combination of a set of files into MapFile and then categorizes them.

This technique greatly improved the write and read performance of the classic Hadoop system and also greatly reduced the RAM consumption on the DataNode.

Several researches and several scenarios have been launched to meet the same need and to improve the technique, such as HAR and NHAR or other technologies such as SPARC and STORM but each proposed solution, and each approach developed does not only respond to a very specific need.

If we opted for the solution of the combination of small files, it is to gain in performance of course; but it is also to maintain a cleaner file system HDFS, because we do not want to have thousands and millions of checks for each file!

The next phase is to improve the performance of the search on small files in a huge volume of data; using graph theory techniques in the first phase, especially the ***Dijkstra's*** and ***Bellman–Moore algorithms***, this first phase will be the initial basis that will feed our ***A* algorithm*** that we will use in our approach to artificial intelligence.

References

1. Hadoop official site. http://hadoop.apache.org/
2. https://www.lebigdata.fr/hadoop
3. Achandair, O., Elmahouti, M., Khoulji, S., Kerkeb, M.L.: Improving Small File Management in Hadoop, pp. 1–14 (2017).
4. Bende, S., Shedge, R.: Dealing with Small files problem in hadoop distributed file system. Procedia Comput. Sci. **79**, 1001–1012 (December 2016)
5. Cai, X., Chen, C., Liang, Y.: An optimization strategy of massive small files storage based on HDFS. In: 2018 Joint International Advanced Engineering and Technology Research Conference (JIAET 2018) (2018)
6. Niazi, S., Ronström, M., Haridi, S., Dowling, J.: Size Matters: Improving the Performance of Small Files in Hadoop. Middleware'18. ACM, Rennes, France (2018)
7. Mir, M.A., Ahmed, J.: An Optimal Solution for Small File Problem in Hadoop. Int. J. Adv. Res. Comput. Sci. (2017)
8. Alange, N., Mathur, A.: Small sized file storage problems in hadoop distributed file system. In: Second International Conference on Smart Systems and Inventive Technology (ICSSIT 2019), IEEE Xplore (2019)
9. Archid, A.S., Mangala, C.N.: Improving Hadoop Performance in Handling Small Files. Int. J. Eng. Res. Technol. (IJERT) (2016)
10. Ahada, M.A., Biswasa, R.: Architecture for Efficiently Storing Small Size Files in Hadoop. Procedia Comput. Sci. **132**, 1626–1635 (2018)
11. Vorapongkitipun, C., Nupairoj, N.: Improving performance of smallfile accessing in hadoop. In: IEEE International Conference on Computer Science and Software Engineering (JCSSE), pp. 200–205 (2014)
12. Sheoran, S., Sethia, D., Saran, H.: Optimized MapFile based storage of small files in hadoop. In: ACM International Symposium on Cluster, Cloud and Grid Computing
13. https://searchstorage.techtarget.com/definition/parallel-file-system
14. Carns, P.H., Ligon III, W.B., Ross, R.B., Thakur, R.: Pvfs: A parallel file system for linux clusters. In: Proceedings of the 4th Annual Linux Showcase and Conference, pp. 317–327. USENIX Association.
15. Alange, N., Mathur, A.: Small sized file storage problems in hadoop distributed file system. In: 2019 International Conference on Smart Systems and Inventive Technology (ICSSIT). IEEE (2019)
16. https://dataottam.com/2016/09/09/3-solutions-for-big-datas-small-files-problem/
17. Bende, S., Shedge, R.: Dealing with small files problem in hadoop distributed file system. In: 7th International Conference on Communication, Computing and Virtualization (2016)
18. Implementing WebGIS on Hadoop: A Case Study of Improving Small File I/O Performance on HDFS
19. Wang, K., Yang, Y., Qiu, X., Gao, Z.: MOSM: An approach for efficient string massive small files on Hadoop. In: International Conference on Big Data Analysis (ICBDA), IEEE (2017)

20. Huang, L., Liu, J., Meng, W.: A review of various optimization schemes of small files storage on Hadoop. In: Joint International Advanced Engineering and Technology Research Conference (JIAET 2018) (2018)
21. Tchaye-Kondi, J., Zhai, Y., Lin K.J., Tao, W., Yang, K.: Hadoop perfect file: a fast access container for small files with direct in disc metadata access. IEEE
22. Ciritoglu, H.E., Saber, T., Buda, T.S., Murphy, J., Thorpe, C.: Towardsa better replica management for hadoop distributed file system. IEEE Big Data Congress '18At: San Francisco (2018)
23. Cheng, W., Zhou, M., Tong, B., Zhu, J.: Optimizing Small File Storage Process of the HDFS Which Based on the Indexing Mechanism. In: 2nd IEEE International Conference on Cloud Computing and Big Data Analysis (2017)
24. Venkataramanachary, V., Reveron, E., Shi, W.: Storage and rack sensitive replica placement algorithm for distributed platform with data as files. In: 2020 12th International Conference on Communication Systems & Networks (COMSNETS) (2020)
25. Rattanaopas, K., Kaewkeeree, S.: Improving Hadoop MapReduce Performance with Data Compression: A Study Using Wordcount Job. IEEE (2017)
26. El-Sayed, T., Badawy, M., El-Sayed, A.: SFSAN approach for solving the problem of small files in Hadoop. In: 2018 13th International Conference on Computer Engineering and Systems (ICCES) (2018)
27. Niazi, S., Ronström, M.: Size Matters: Improving the Performance of Small Files in Hadoop. In: The 19th International Middleware Conference
28. Climate Data Online, available from National Centers for Environmental Information at https://www.ncdc.noaa.gov/cdo-web/datasets
29. Merla, P.R., Liang, Y.: Data analysis using hadoop MapReduce environment. IEEE
30. Tao, W., Zhai, Y., Tchaye-Kondi, J.: LHF: A New Archive based Approach to Acclerate Massive Small Files Access Performance in HDFS. EasyChair Preprint n°. 773 (2017)
31. Shah, A., Padole, M.: Optimization of hadoop MapReduce model in cloud computing environment. IEEE (2019)
32. Zheng, T., Guo, W., Fan, G.: A method to improve the performance for storing massive small files in Hadoop. In: The 7th International Conference on Computer Engineering and Networks (CENet2017) Shanghai (2017)
33. https://arxiv.org/ftp/arxiv/papers/1904/1904.03997.pdf

COVID-19 Pandemic

Data-Based Automatic Covid-19 Rumors Detection in Social Networks

Bolaji Bamiro and Ismail Assayad

Abstract Social media is one of the largest sources of propagating information; however, it is also a home ground for rumors and misinformation. The recent extraordinary event in 2019, the COVID-19 global pandemic, has spurred a web of misinformation due to its sudden rise and global widespread. False rumors can be very dangerous; therefore, there is a need to tackle the problem of detecting and mitigating false rumors. In this paper, we propose a framework to automatically detect rumor on the individual and network level. We analyzed a large dataset to evaluate different machine learning models. We discovered how all our methods used contributed positively to the precision score but at the expense of higher runtime. The results contributed greatly to the classification of individual tweets as the dataset for the classification task was updated continuously, thereby increasing the number of training examples hourly.

1 Introduction

In our world today, we have economic, technological, and social systems built with high complexity to help the human society. However, these systems can be highly unpredictable during extraordinary and unprecedented events. The most recent global pandemic called COVID-19 started gaining attention from late December 2019 and has affected the world greatly with more than 45 million cumulative worldwide cases[1] of infection currently [1]. During these shocking periods, cooperation is crucial to mitigate the impact of the pandemic on the collective well-being of the public.

[1] https://coronavirus.jhu.edu/map.html

B. Bamiro (✉)
African Institute for Mathematical Sciences, Mbour, Senegal
e-mail: bolaji.r.bamiro@aims-senegal.org

I. Assayad
LIMSAD Faculty of Sciences and ENSEM, University Hassan II of Casablanca, Casablanca, Morocco

M. Ben Ahmed et al. (eds.), *Networking, Intelligent Systems and Security*,
Smart Innovation, Systems and Technologies 237,
https://doi.org/10.1007/978-981-16-3637-0_57

Social media, which is a complex society that aids in global communication and cooperation, has, however, become one of the major sources of information noise and fake news. 'Fake news spreads faster and more easily than this virus, and is just as dangerous.' were the words of Dr. Tedros Adhanom Ghebreyesus at the World Health Organization, Munich Security Conference, February 15, 2020 [2]. The waves of unreliable information being spread may have a hazardous impact on the global response to slow down the pandemic [3]. Most of the fake news is harmful and problematic as they reach out to thousands of followers. The possible effects are widespread fear [4], wrong advice that may lead to the encouragement of risky behavior, and contribution to the loss of life during the pandemic [5, 6].

There are recognized organizations that have dealt with rumors such as the International Fact-Checking Network (IFCN) [7], the World Health Organization (WHO), and United Nations Office (UN).

This paper aims to achieve the goal by designing a framework that can effectively detect rumor over time by analyzing tweets on twitter. Twitter is one of the largest social media platforms [8]; therefore, we obtained the dataset from the platform. The contributions made in this paper are as follows:

- evaluate methods and models to detect rumors with high precision using a large dataset obtained from Twitter;
- addition of image analysis to text analysis.
- designing a unified framework that detects rumor effectively and efficiently in real time.

2 Related Works

Research on rumor detection has been receiving a lot of attention across different disciplines for a while now [9, 10]. New approaches have been arising to tackle the problem of fake news specifically on social media using computational methods. These methods have been shown by [11–13] to be efficient not just in solving the rumor detection problem but also in identification of such fake news on time [14]. Some of the methods used are machine learning, n-gram analysis, and deep learning models to develop detection and mitigation tools for classification of news [14, 15]. Some take this further to apply several tools for higher precision [13]. Much previous research approaches these problems by analyzing a large number of tweets during the COVID-19 epidemic to analyze the reliability of the news on social media that poses serious threat amidst the epidemic [10, 16]. Another approach that has been used to study social media news is fake images, especially on COVID-19 [17]. Few studies have investigated the reliability of images on social media. Such methods have been used by [18] by analyzing a large number of tweets to characterize the images based on its social reputation and influence pattern using machine learning algorithms.

Most studies make use of response information: agreement, denial, enquiry, and comment in their rumor detection model, and have shown good performance improvement [19]. Text content analysis is also an important method that has been employed by most previous studies on rumor detection. It includes all the post text and user responses. Deceptive information usually has a common content style that differs from that of the truth, and researchers explore the sentiment of the users toward the candidate rumors such as [19]. It is important to note that although textual content analysis is quite important in rumor detection, many studies point that just is not sufficient [20]. Visual features (images or videos) are also an important indicator for rumor detection and have been shown [17, 21]. Rumors are sometimes propagated using fake images which usually provokes user responses.

Network-based rumor detection is very useful because it involves construction of extensible networks to indirectly collect possible rumor propagation information. Many studies have utilized this method such as [13, 18, 22]. Knowledge base (KB) has also been shown to be quite important for detecting fake news. This involves using known truth about a situation. Some studies on employing KB in the past such as [23]. Very few previous studies have, however, been designed for real-time rumor detection systems [19, 24]. This paper aims to develop a framework for a practical rumor detection system that uses available information and models by collectively involving major factors which are text analysis, knowledge base, deep learning, natural language processing (NLP), network analysis, and visual context (Images).

3 Background

The definition of rumor is quite ambiguous and inconsistent as various publications have different definitions for it. Having a solid definition is crucial toward making a well-informed classification of news by understanding its properties. However, this paper will emphasize on these two definitions.

3.1 A Rumor is a Statement Whose Truth Value is Either True, False or Unverified [15]

The definition generally means that the truth value of a rumor is uncertain. The main problem arises when the rumor is false, and this is often referred to as false or fake news.

3.2 A Rumor is Defined as Unreliable Information that is Easily Transmissible, Often Questioned [25, 26]

For further clarity on this definition, we emphasize on the properties of a rumor which are unreliable information, easily transmissible, often questioned. Rumors cannot be relied upon because its truth value is uncertain and controversial due to lack of evidence. Rumors easily transmit from one person/channel to another. Also, study from [27] shows that false rumor spreads wider and faster than true news. Rumors cause people to express skepticism or disbelieve, that is, verification, correction, and enquiry [13].

4 Problem Statement

This paper aims to solve the rumor detection problem: A post p is defined as a set of i connected news $N = \{n_1, n_2, \ldots, n_i\}$ where n_1 is the initial news in which other posts spanned from. We define a network of posts from a social media platform where the nodes are the messages while the edges are the similarities between the two nodes. Each node n_i has its unique attributes that represent its composition. This consists of the user id, post id, user name, followers count, friend count, source, post creation time, and accompanying visuals (images). Given these attributes, the rumor detection function takes as input the post p, the set of connected news N together with the attributes to return an output {True, False} that determines whether that post is a rumor or not.

5 Real-Time Computational Problem

For each tweet from the tweet stream containing the text and image information, we extract its attributes as a set $\{t_1, t_2, \ldots t_i\}$. For the rumor detection problem, we aim to predict whether the tweet is a rumor or not using the attributes of each tweet.

6 Methodology

According to Fig. 1, there are four phases involved in this paper's framework for rumor detection. The methods used in this section are a modification and improvement of the paper [13]. These phases are as follows.

(A) Extraction and storage of tweets;
(B) Classification of tweets;
(C) Clustering of tweets and;

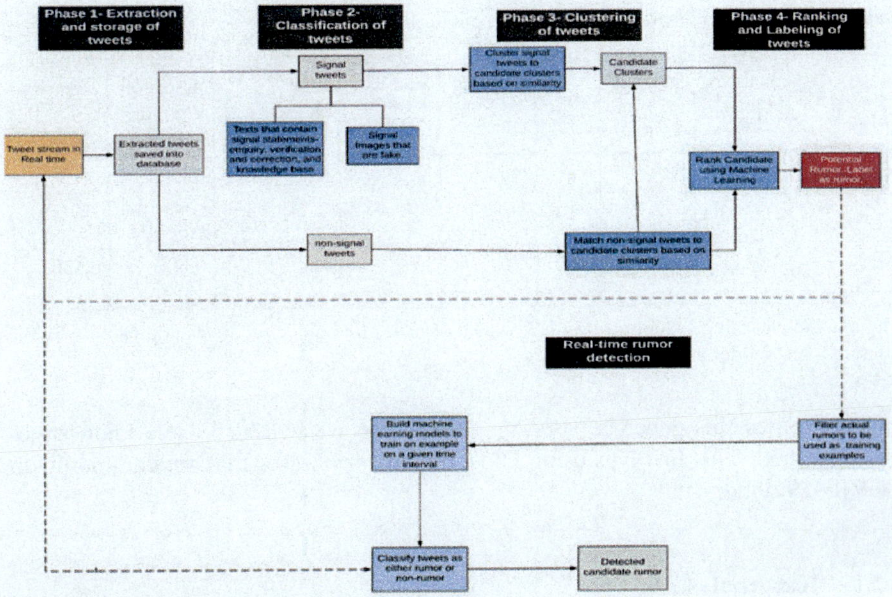

Fig. 1 Framework showing the different phases of rumor detection

(D) Ranking and labeling of tweets.

6.1 Extraction and Storage of Tweets

The goal of this paper is to detect rumor early, and hence, extraction and storage of tweets are quite important. The tweets are streamed using a python library called Tweepy and cleaned to remove stop words, links, and special characters, and also extract mentions and hashtags. The tweets are then stored in a MySQL database.

6.2 Classification of Tweets

The second phase involves classifying the tweets into signal and non-signal tweets. In this paper, signal tweets are tweets that contain unreliable information, verification questions, corrections, enquiry, and fake images. Basically, a tweet conveying information usually contains a piece of knowledge, ideas, opinions, thoughts, objectives, preferences, or recommendations.

Verification/confirmation questions have been found to be good signals for rumors [13] and also visual attributes [17]. Therefore, this phase will explore text and image

Table 1 Method and result comparison with enquiring minds [13]

	Text analysis	Image analysis	NLP	Deep learning	Network analysis	Knowledge base	Average precision without ML	Average precision with ML
Enquiring minds [13]	Yes	No	Yes	No	Yes	No	0.246	0.474
Our method	Yes	Yes	Yes	Yes	Yes	Yes	0.602	0.638

The machine learning model used in [13] was a decision tree model and was compared with the random forest model used in this study

analysis. Since this paper is based on COVID-19, we will also use a knowledge-based method. This involves using known words or phrases that are common with COVID-19 rumors.

6.2.1 Text Analysis

At this stage, we want to extract the signal tweets based on regular expressions. The signal tweets will be obtained by identifying the verification and correction tweets as used in [13]. We will also add known fake websites identified by Wikipedia[2] and WHO Mythbusters as shown in Table 1. We make use of the spaCy python library to match the tweets to the phases.

6.2.2 Image Analysis

This project aims to use the visual attributes of the tweet as one of the factors to detect rumor. We approach this by using three stages of analysis for the images. At the first stage, the image metadata is analyzed to detect software signatures. It is the fastest and simplest method to classify images. However, image metadata analysis is unreliable because there are existing programs that can alter it such as Microsoft Paint. An image without an altered metadata will contain the name of the software used for the editing, for example, Adobe Photoshop. The second stage makes use of the error level analysis (ELA) and local binary pattern histogram (LBPH) methods. ELA basically detects areas in an image where the compression levels are different. After the ELA, the image is passed into the local binary pattern algorithm. LBPH algorithm is actually used for face recognition and detection, but, in this paper, it is useful for generating histograms and comparing them. At the third and final stage, the image is reshaped to 100px x 100px image. This aspect involves deep learning. We used the pre-trained model VGG16 and also added a CNN model. Then, these

[2] https://en.wikipedia.org/wiki/Fake_news.

10,000 pixels with RGB values will be fed into the input layer of the multilayer perceptron network. Output layer contains two neurons: one for fake images and one for real images. Therefore, based on the neuron outputs, we determine whether the images are real or fake.

6.2.3 Clustering of Tweets

The third phase involves clustering of tweets to highlight the candidate rumors. Usually, a rumor tweet is retweeted by other users or recreated similarly to the original tweet. This is why clustering of similar tweets is quite important. However, to reduce the computational costs, memory, and time used by other clustering algorithms, we treat the rumor clusters as a network. This method can be quite efficient for this phase as shown in [13]. We define a network where the nodes represent the tweets while the edges represent the similarity. Nodes with high similarity are connected. We define this network as an undirected graph to analyze the connected components, that is, a path connects every pair of nodes in the graph. We measure the similarity between two tweets t_1 and t_2 using Jaccard coefficient. The Jaccard coefficient between t_1 and t_2 can be measured using:

$$J(t_1, t_2) = \frac{|N \text{ gram } (t1) \cap N \text{ gram } (t2)|}{|N \text{ gram } (t1) \cup N \text{ gram } (t2)|} \quad (1)$$

where N is 1-g of tweets t_1 and t_2.

Jaccard distance is commonly used to measure similarity between two texts [13]. The similarity values range from 0 to 1 and values tending to 1 mean a higher similarity. However, computing these similarities for each pair of tweets may be time consuming; therefore, we make use of the MinHash algorithm [28]. The threshold set for high similarity is at 0.7, that is 70 % similarity between the pair of tweets. After clustering the signal tweets, we also add the non-signal tweets to the network also using Jaccard similarity, however, with a threshold of 60 %.

6.2.4 Ranking and Labeling of Tweets

At this phase, each tweet has a degree centrality score. However, tweets with high degree centrality may not be rumors. Therefore, this phase applies machine learning to rank the tweets. We extract features from the candidate rumors that may contribute to predicting whether the candidate is a rumor. Some of these features were used in [13]. The following are the features used were Twitter_id, follower's count, location, source, Is_verified, friends count, retweet count, favorites count, reply, Is_protected, sentimental analysis, degree centrality score, Tweet length, signal tweet ratio, subjectivity of text, average tweet length ratio, retweet ratio, image reliability, hashtag ratio, and mentions Ratio.

6.2.5 Experimental Materials and Structure

Dataset

The initial data set used is the COVID-19 tweets selected randomly from February 2020 to October 2020. The total amount of data collected for labeling was 79,856. The data set used to train the images was obtained from MICC-2000 [29]. It consists of 2000 images, 700 of which are tampered and 1300 originals.

Ground Truth

The dataset collected was then labeled for training. The labels were assigned according to the definitions given in Sect. 3, and also, some tweets had to be confirmed by web search. The reliability achieved a Cohen's Kappa score of 0.75.

Evaluation Metric

We divided the dataset labeled into train and validation sets. The validation set contains 13,987 tweets. Then different machine learning models were used to rank the test set. The evaluation of the model will be based on its top N rumor candidates where N is varied.

$$\text{Precision} = \frac{\text{TP}}{\text{TP} + \text{TN}} \tag{2}$$

The detection time and batch size are also taken into consideration.

Baseline Method

The baseline methods consist of the framework provided without machine learning and Image analysis.

Text analysis (verification and correction only): This involves using only text analysis at the classification of tweets into signal tweets. We evaluate the efficiency of this method without including the visual attributes and knowledge base using the rank of the output from the machine learning models.

Without machine learning: This involves using only the degree centrality method to rank the candidate rumor. At phase 4, the different machine learning models are skipped and evaluated for efficiency. This also includes the omission of the CNN model at the image analysis stage. Therefore, the rumor detection algorithm based on this method outputs the rank of the clusters without any machine learning involved in the process.

Variants

To improve on the baseline methods, we introduce three variants. These variants will enable us to understand the effectiveness of the method. The variants are as follows:

Text (verification and correction only) and Image Analysis: For this variant, we use verification and correction, and image analysis to classify the tweets into signal tweets.

Text analysis (verification and correction, and knowledge base): For this variant, we use verification and correction, and knowledge base without image analysis to classify the tweets into signal tweets.

Text (verification and correction, and knowledge base) and Image Analysis: For this variant, we use verification and correction, knowledge-based method, and image analysis to classification of tweets into signal tweets. This is our method, and it is a collation of all the methods in the framework. We evaluate the efficiency of this method without including the visual attributes using the rank of the output from the machine learning models.

Machine learning: This variant involves using various machine learning models to rank the rumor candidates.

Algorithm 1: Ranking Clustered Tweets

```
Input: Document term, Tweets
Output: Ranked signal tweets
Initialize: N
for text in document term do
 | s := signal tweets                      (get signal tweets)
end
for id, text1 in tweets do
   for id, text2 in s do
     | J := Jaccard (text1, text2)         (calculate Jaccard Coefficient)
     |   if J > a                          (set Coefficient threshold)
     |     | append to dataframe N
     |   end if
     end
end
 D := Degree centrality(text)
Rank dataframe N based on D in descending order
```

Result and Discussions

Precision of Methods

The precision of these methods is evaluated using 10 min of tweets collected on October 27, 2020. This dataset consists of 13,437 tweets with 213 images in total. The precision value only takes into account the top 100 tweets ranked by without machine learning (degree centrality score is used) and with machine learning (CatBoost model) for the baseline and variants methods, respectively. The results obtained show that the collation of our methods (text (verification and correction + knowledge base) and image analysis) detected more signal tweets and candidate rumor than the other method with higher precision with machine learning ranking. The precision of our method outperformed other methods with a precision of 0.65 with machine learning. The results also showed that the signal tweets and candidate rumor detected using our method is much larger than using the baseline method.

Ranking Candidate Rumor

After the clustering of the tweets, we use different ranking methods to rank the candidate rumors. The baseline ranking method is ranking based on the degree centrality score while our method is based on using machine learning models. Among all machine learning models, we selected a logistic, tree and boosting model-random forest, logistic regression, and CatBoost model. For the machine learning models, we use the 20 statistical features described in the methodology section. We trained the models and tested their performances for comparison. A tenfold cross-validation was carried out to obtain the average performance. These graphs show a general reduction in precision as N increases. However, the CatBoost model outperforms other ranking methods except the Text + Image analysis method where the degree centrality ranking method performs the best. Logistic regression, however, does not perform well which may be due to overfitting.

The text + knowledge base method performs best at N = 10 and 20 at with an average precision value of 0.95 but decreases gradually as N tends to 100. Our method shows an improvement in most methods especially for the CatBoost model but its value decreases steadily with increase in N.

Early Detection

It is very useful to detect rumors early; therefore, early detection is a key objective in this paper. The run time was measured between the baseline method-text analysis only and variants to determine how early the method can detect the rumor. The results showed that as the number of tweets increases, the run time increases much faster in the variant methods as compared to the baseline method. However, the number of signal tweets detected also increases which improve the precision. This difference because of the time taken to get each image and classify as rumor or non-rumor. The higher the number of tweets, the higher the number of images, and hence, the higher the run time.

Efficiency of Real-Time Detection Framework

Using the above result, we develop an application that detects streaming tweets in real time while the dataset used for the prediction is appended continuously using the top 10 candidate rumor detected by the text (verification and correction + knowledge base) and image analysis hourly. The real-time rumor detection application predicts an average of 4.04 rumors per second. The web application built using Flask detects an average of 38 rumors every 8 s.

6.3 Discussion

In this paper, we built a framework to take advantage of the text and visual attributes of a tweet to classify the tweet as a rumor. It improves the verification and correction method by including other known regular expressions associated with the problem and publicly declared fake new websites. We went further to use different ranking methods to rank clustered rumors using complex network algorithms. We extracted 20 features from the rumors to train models for prediction. We observed that the top features that have a high impact on the ranking are sentiment analysis, location, and friends count using the CatBoost model. CatBoost has shown to be very effective in ranking the candidate rumor because it outperforms other algorithms. This is very useful because it gives us an idea of the features of tweets needed for real-time detection and the best model that can deliver the highest precision. The precision can be improved with higher number of training examples. Our method, however, takes much longer to run as compared to the baseline method because of the addition of the image component. Therefore, we have to decide if early detection is a price to pay for higher precision. The real-time detection component, however, solves this problem as it predicts the tweet as they stream in.

7 Conclusion

False rumors can be very dangerous especially during a pandemic. Rumor super spreaders are taking the COVID-19 pandemic to confuse social media users. This is why it is very important to detect rumors as early as possible. The World Health Organization is working hard to dispute many false rumors and has provided some information. Using these details, we built a framework to detect rumors. The approach used is quite efficient with machine learning because it yields high precision. The real-time detection model detects 4.04 rumors per second using training examples appended continuously from the approach. This approach can be improved upon by reducing the analysis run time for our method.

Acknowledgements Special thanks go out to the African Institute for Mathematical Sciences (AIMS) and LIMSAD for their support toward this paper.

References

1. W. H. Organization et al.:Coronavirus disease 2019 (covid-19): situation report, 103 (2020)
2. Zarocostas, J.: How to fight an infodemic. The Lancet **395**, 676 (2020)
3. Anderson, J., Rainie, L.: The Future of Truth and Misinformation Online, vol. 19. Pew Research Center (2017)
4. Latif, S., Usman, M., Manzoor, S., Iqbal, W., Qadir, J., Tyson, G., Castro, I., Razi, A., Boulos, M.N.K., Weller, A., et al.: Leveraging data science to combat covid-19: a comprehensive review (2020)
5. Tasnim, S., Hossain, M.M., Mazumder, H.: Impact of rumors or misinformation on coronavirus disease (covid-19) in social media (2020)
6. Hossain, M.S., Muhammad, G., Alamri, A.: Smart healthcare monitoring: a voice pathology detection paradigm for smart cities. Multimedia Syst. **25**(5), 565–575 (2019)
7. Perrin, C.: Climate feedback accredited by the international fact-checking network at poynter. Clim. Feedback **24** (2017)
8. Kouzy, R., Abi Jaoude, J., Kraitem, A., El Alam, M. B., Karam, B., Adib, E., Zarka, J.. Traboulsi, C., Akl, E.W., Baddour, K.: Coronavirus goes viral: quantifying the covid-19 misinformation epidemic on twitter. Cureus **12** (2020)
9. Li, Q., Zhang, Q., Si, L., Liu, Y.: Rumor detection on social media: Datasets, methods and opportunities. arXiv preprint arXiv:1911.07199 (2019)
10. Shahi, G.K., Dirkson, A., Majchrzak, T.A.: An exploratory study of covid-19 misinformation on twitter. arXiv preprint arXiv:2005.05710 (2020)
11. Ahmed, H., Traore, I., Saad, S.: Detection of online fake news using n-gram analysis and machine learning techniques. In: International Conference on Intelligent, Secure, and Dependable Systems in Distributed and Cloud Environments. Springer, Berlin, pp. 127–138 (2017)
12. Bharadwaj, A., Ashar, B.: Source based fake news classification using machine learning. Int. J. Innov. Res. Sci. Eng. Technol. 2320–6710 (2020)
13. Zhao, Z., Resnick, P., Mei, Q.: Enquiring minds: early detection of rumors in social media from enquiry posts. In: Proceedings of the 24th International Conference on World Wide Web, pp. 1395–1405 (2015)
14. Liu, Y., Wu, Y.-F.B.: Early detection of fake news on social media through propagation path classification with recurrent and convolutional networks. In: Thirty-second AAAI conference on artificial intelligence (2018)
15. Qazvinian, V., Rosengren, E., Radev, D., Mei, Q.: Rumor has it: identifying misinformation in microblogs. In: Proceedings of the 2011 Conference on Empirical 15 Methods in Natural Language Processing, pp. 1589–1599 (2011)
16. Al-Rakhami, M.S., Al-Amri, A.M.: Lies kill, facts save: detecting covid-19 misinformation in twitter. IEEE Access **8**, 155961–155970 (2020)
17. Jin, Z., Cao, J., Zhang, Y., Zhou, J., Tian, Q.: Novel visual and statistical image features for microblogs news verification. IEEE Trans. Multimedia **19**, 598–608 (2016)
18. Gupta, A., Lamba, H., Kumaraguru, P., Joshi, A.: Faking sandy: characterizing and identifying fake images on twitter during hurricane sandy. In: Proceedings of the 22nd International Conference on World Wide Web, pp. 729–736 (2013)
19. Liu, X., Nourbakhsh, A., Li, Q., Fang, R., Shah, S.: Real-time rumor debunking on twitter. In: Proceedings of the 24th ACM International on Conference on Information and Knowledge Management, pp. 1867–1870 (2015)
20. Chua, A.Y., Banerjee, S.: Linguistic predictors of rumor veracity on the internet. In: Proceedings of the International MultiConference of Engineers and Computer Scientists, vol. 1, p. 387 (2016)
21. Wang, Y., Ma, F., Jin, Z., Yuan, Y., Xun, G., Jha, K., Su, L., Gao, J.: Eann: event adversarial neural networks for multi-modal fake news detection. In: Proceedings of the 24th acm sigkdd International Conference on Knowledge Discovery & Data Mining (2018), pp. 849–857

22. Wu, K., Yang, S., Zhu, K.Q.: False rumors detection on sina weibo by propagation structures. In: 2015 IEEE 31st International Conference on Data Engineering (IEEE, 2015), pp. 651–662
23. Hassan, N., Arslan, F., Li, C., Tremayne, M.: Toward automated fact-checking: detecting check-worthy factual claims by claimbuster. In: Proceedings of the 23rd ACM SIGKDD International Conference on Knowledge Discovery and Data Mining (2017), pp. 1803–1812
24. Liu, X., Li, Q., Nourbakhsh, A., Fang, R., Thomas, M., Anderson, K., Kociuba, R., Vedder, M., Pomerville, S., Wudali, R., et al.: Reuters tracer: a large scale system of detecting & verifying real-time news events from twitter. In: Proceedings of the 25th ACM International on Conference on Information and Knowledge Management (2016), pp. 207–216
25. DiFonzo, N., Bordia, P.: Rumor Psychology: Social and Organizational Approaches. American Psychological Association (2007)
26. Bugge, J.: Rumour has it: a practice guide to working with rumours. Communicating with Disaster Affected Communities (CDAC) (2017)
27. Vosoughi, S.: Automatic detection and verification of rumors on twitter. Ph.D. thesis, Massachusetts Institute of Technology (2015)
28. Wu, W., Li, B., Chen, L., Gao, J., Zhang, C.: A review for weighted minhash algorithms. IEEE Trans. Knowl. Data Eng. (2020)
29. Amerini, I., Ballan, L., Caldelli, R., Del Bimbo, A., Serra, G.: A sift-based forensic method for copy–move attack detection and transformation recovery. IEEE Trans. Inf. Forensics Secur. **6**, 1099–1110 (2011)

Security and Privacy Protection in the e-Health System: Remote Monitoring of COVID-19 Patients as a Use Case

Mounira Sassi and Mohamed Abid

Abstract The Internet of Things (IoT) is characterized by heterogeneous technologies, which contribute to the provision of innovative services in various fields of application. Among these applications, we find the field of e-Health which provides a huge amount of data that served the health of patients remotely and in real time but also medical records, health monitoring and emergency response. e-Health systems require low latency and delay which is not guaranteed since data are transferred to the cloud and then back to the application, which can seriously affect performance. Also, COVID-19 pandemic has accelerated the need of remote monitoring of patients to reduce chances of infection among physicians and healthcare workers. To this end, Fog computing has emerged, where cloud computing is extended to the edge of the network to reduce latency and network congestion. This large amount of data is downloaded and stored on remote public cloud servers to which users cannot be fully trusted, especially when we are dealing with sensitive data like health data. In this scenario, meeting the confidentiality and patient privacy requirements becomes urgent for a large deployment of cloud systems. In this context, we offer a solution to secure the personal data of the e-Health system and protect the privacy of patients in an IoT–Fog–cloud environment while being based on cryptographic techniques especially CP-ABE and the blockchain paradigm. The results obtained are satisfactory, which allowed us to deduce that the solutions are protected against the most known attacks in IoT–Fog–cloud systems.

M. Sassi (✉)
Laboratory Hatem Bettaher Irescomtah, Faculty of sciences of Gabes,
University of Gabes, Gabes, Tunisia

M. Abid
Laboratory Hatem Bettaher Irescomtah, National School of Engineering of Gabes,
University of Gabes, Gabes, Tunisia
e-mail: mohamed.abid@enig.rnu.tn

M. Ben Ahmed et al. (eds.), *Networking, Intelligent Systems and Security*,
Smart Innovation, Systems and Technologies 237,
https://doi.org/10.1007/978-981-16-3637-0_58

1 Introduction

Countries around the world have been affected by the COVID-19 pandemic since December 2019, and the health care systems are rapidly adapting to the increasing demand. E-Health systems offer remote patient monitoring and share of information between physicians. Hence, it helps facilitate and improve the prevention, diagnosis and treatment of patients at a distance. Indeed, health data is collected by sensors and then transmitted through the Internet to the cloud for consultation, evaluation and recommendations by professionals.

According to the World Health Organization (WHO) [1], "COVID-19 is the disease caused by a new coronavirus, SARS-CoV-2." It especially infects the respiratory system of patients. Some people who have contracted COVID-19, regardless of their condition, continue to experience symptoms, including fatigue and respiratory or neurological symptoms. Therefore, doctors use intelligent equipments which collect the measurements of a patient at home and sends them to the Fog. The latter can be a treatment center installed in the hospital. Then, the Fog sends this data to the cloud storage service for consultation by doctors. This process helps professionals understand the behavior of this pandemic and gives them a hint about its evolution.

Despite the importance of e-Health system and their good results, it is necessary to protect the confidentiality of the data, to secure the sharing and to protect the private life of the patients. However, the implementation of treatment and storage in Fog and cloud to store and process sensitive data raises many security issues (waste, leakage or theft). Consequently, to use a model based on IoT–Fog–cloud architecture, a reinforcement of the security measures is mandatory. Thus, the confidentiality, integrity and access control of stored data are among the major challenges raised by external storage. To overcome the challenges mentioned above, cryptography techniques are widely adopted to secure sensitive data.

In this paper, a new solution to secure e-Health applications by exchanging data confidentially and protecting patient privacy in an IoT–Fog–cloud architecture is proposed. Our system can offer these basic functionalities:

Achieve a hard authentication and secure key sharing between the oximetry characterized by a limited resource in memory and computation and the Fog. This allows confidential transfer of data between the two entities.

Apply a public key (one-to-many) encryption scheme for secure cloud storage and data sharing between a group of physicians. This scheme allows the implementation of access control according to their attributes.

Combine cryptography technologies and blockchain to strengthen the management of decentralized access control, keep the traceability of data traffic, and obtain a level of anonymity offered by the blockchain.

Our system can effectively resist against the most well-known attacks in IoT and against tampering of control messages.

The rest of the article is organized as follows. The related works on securing e-Health systems are discussed in Sect. 2. We present the basic knowledge (preliminaries) in Sect. 3. We describe the secure data sharing e-Health system that protects patient privacy based on CP-ABE encryption and blockchain in IoT- Fog-cloud architecture in Sect. 4. We provide security and performance analysis in Sect. 5. Section 6 concludes the article.

2 Related Works

There are many research works focusing in securing IoT application especially for *ambient-assisted living* (AAL) application and e-Health system. Some researchers used public key identity-based security or on lightweight cryptographic primitives [2], such as one-way hash function and XOR operations. Others concentrated on securing access to data, and many research used blockchain to secure the e-Health system.

Chaudhari and Palve [3] developed a system that provides mutual authentication between all system components (human body sensor, handheld device and server/cloud). The generated data is encrypted using RSA.

Wang et al. [4] proposed a scheme based on a fully homomorphic design for the protection of privacy and the treatment of data in the e-Health framework. The proposed architecture consists in performing a transmission mode for the electronic health record. This mode ensures diagnostic of the patient based on encrypted records residing in the cloud by a remote physician without being decrypted.

Bethencourt et al. [5] presented a system to achieve complex access control over encrypted data based on encryption policy attributes. The access policy is embedded in ciphertext. Attributes are used to describe the credentials of a user. A set S of descriptive attributes is used to identify the private keys. When a party wishes to encrypt a message through an access tree structure of a policy, his/her private keys must satisfy to be able to decrypt. The CP-ABE scheme includes four main algorithms: initialization, encryption, decryption and generation of the secret key.

Another scheme based on attribute encryption called "CCA" for architectures that integrate the Fog as an outsourced decryption. It is proposed by Zu et al. [6] in order to protect data in cloud computing. The main idea, to achieve this schema, is to allow the decryptor to have the ability to verify the validity of the ciphertext. The public key used is non-transformable and the type of attribute-based encryption used is OD-ABE (*attribute-based encryption with outsourced decryption*).

Wang [7] proposed a secure data sharing scheme to ensure the anonymity and identity confidentiality of data owners. Symmetric encryption, search-able encryption and attribute-based encryption techniques are used to keep data outsourced to the cloud secure. Due to the risk of breach and compromise of patient data, medical organizations have a hard time adopting cloud-stored services. Moreover, the existing authorization models follow a patient-centered approach. Guo et al. [4] proposed a CP-DABKS schema (*ciphertext-policy decryptable attribute-based keyword search*) which allows an authorized user to decrypt data in a supposedly completely insecure network. The architecture of this schema includes the four other components: *KDG (key general center)* and the *data center* are the data owners. The data center lies down the keywords and the access structure linked to the data. The third data receiver element is the *data consumer*. It has a set of attributes, and a generated trap is used to identify it in order to have the capacity to decrypt the data. Finally, the *cloud server* plays the role, in this diagram, of a storage base for the data sent by the data sender and a verification of the satisfaction of the access structure of a secret key received by the data receiver.

Blockchain attracts attention in several academic and industrial fields [8]. It is a technology that was first described in 1991 when researchers Stuart Haber and W. Scott Stornetta introduced a computer solution, allowing digital documents to be time-stamped and therefore never backdated or altered [9]. It is based on cryptographic techniques: hash functions and asymmetric encryption. This technology was at the origin of the Bitcoin "electronic money" paradigm described in the article by Nakamoto [10] in 2009. Blockchain is an innovation in storage. "It allows information to be stored securely (each writing is authenticated, irreversible and replicated) with decentralized control (there is no central authority which would control the content of the database)" [11]. This technology has been used in several areas of the Internet of Objects and often used as a means to secure data as in the paper of Gupta et al. [12] who proposed a model to guarantee the security of transmitted data and received by nodes of an Internet of Things network and control access to data [13].

Blockchain is seen as a solution to make secure financial transactions without an authority. It is also used in several areas with the aim of decentralizing security and relieving the authorities. For example, the vehicular area integrates this technology to solve security problems. We mention the article by Yao et al. [14] who have proposed a BLA (*blockchain-assisted lightweight anonymous authentication mechanism*) to achieve inter-center authentication that allows a vehicle to decide to re-authenticate in another location. At the same time, they used the blockchain to eliminate communications between vehicles and service managers (SM), which considerably reduces the communication time.

In recent years, to overcome security issues in e-Health systems, several solutions are based on the blockchain to achieve *personal health information (PHI)* sharing with security and privacy preservation due to its advantages of immutability. C. Nguyen et al. [13] proposed a new framework (architecture) for offloading and sharing of *electronic health records (EHR)* that combines blockchain and the decentralized *interplanetary file system (IPFS)* on a mobile cloud platform. In particular, they created a reliable access control mechanism using smart contracts to ensure secure sharing of electronic health records (EDRs) between different patients and medical providers. In addition, a data sharing protocol is designed to manage user access to the system. Zhang et al. [15] built two types of blockchain (the private blockchain and the consortium blockchain) to share PHI on secure health and maintain confidentiality. The private blockchain is responsible for storing the PHI, while the consortium blockchain keeps its records at secure indexes. Next, we give the basic knowledge that we use in the design of the new solution.

3 Preliminaries

To properly design our solutions, we must have prior knowledge of certain cryptographic tools. Thus, this section is devoted for a generality on mathematical notions where we refer to the books "Cryptography and Computer Security" [16], "Introduction to Modern Cryptography" by Katz and Lindel [17] and course by Ballet and Bonecaze [18].

Access Structure

Definition 1 $\{P1, P2, ..., Pn\}$ is a set of parties. A collection A $\subseteq 2^{\{P1,P2,...,Pn\}}$ is monotone if $\forall B, C : if\, B \in A$ and $B \in C$ then $C \in A$. An access structure (respectively, monotone access structure) is a collection (respectively, monotone collection) A of non-empty subsets of $\{P1, P2, ..., Pn\}$, i.e., A $\subseteq 2^{\{P1,P2,...,Pn\}}$ $\varnothing$. The sets in A are called the authorized sets, and the sets not in A are called the unauthorized sets [5].

Security Model for CP-ABE

The ABE security model is based on the following functions:
Install: The challenger executes the configuration algorithm and gives the public parameters, PK to the opponent.
Phase 1: The adversary creates repeated private keys corresponding to the sets of attributes $S1$,. . ., Sq1.
Challenge: The opponent submits two messages of equal length $M0$ and $M1$.

Moreover, the adversary gives a challenge access structure A * such that none of the sets $S1$,. . ., Sq1 satisfy the access structure. The challenger flips a random coin b and digits Mb under A *. The CT * ciphertext is given to the opponent.
Phase 2: Phase 1 is repeated with the restriction that none of the attribute sets Sq1 + 1,. . ., Sq satisfy the access structure corresponding to the challenge.
Supposition: The opponent makes a supposition b 'of b. The advantage of an adversary A in this game is defined as $Pr[b' = b] - \frac{1}{2}$

4 Secure Storage and Sharing of E-Health Data

4.1 An Overview of Our Proposed Schemes

In this section, we present our contribution to secure the sharing and storage of data and preserve the privacy of patients in the e-Health system. Figure 1 shows the different components of our architecture.

Connected objects (Oximeter): They generate the data (oxygen level in the blood) of patients remotely and send them in real time to the Fog (which in turn is responsible

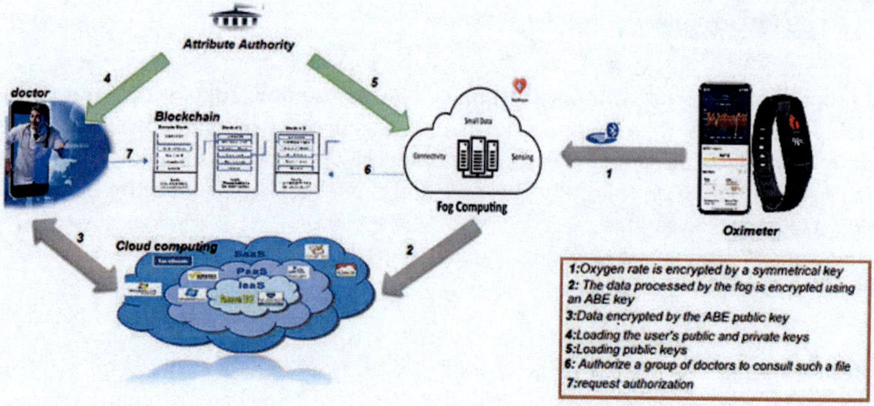

Fig. 1 Global architecture of our contribution

for processing them and sending them to the cloud in order to be stored and consulted by doctors) in a secure manner using symmetric encryption after an authentication phase and the exchange of a secret key.

Proxy/Fog Computing: It is an intermediary between health sensors and the cloud. It offers storage, computation and analysis services. The Fog first decrypts the data sent by the sensors through an anonymity proxy. He analyzes them. In case of emergency, it sends an alert message to the ambulance. Using attribute-based encryption, the Fog encrypts data and sends it to the cloud storage server where it will be saved.

Cloud: An infrastructure used in our system to store encrypted data and share it for legal users. Attributes Authority: The attributes authority manages all attributes and generates, according to the identity of the users, the set of key pairs and grants them access privileges to end users by providing them with their secret keys according to their attributes.

Users: Doctors and caregivers are the consumers of data. They request access to data according to their attributes from cloud servers. Only users who have the attributes satisfying the access policies can decrypt the data. Doctors can also add diagnostics and recommendations to share with colleagues. This data is encrypted by ABE and stored in the cloud.

Blockchain: It is a decentralized base of trust. It is used to ensure access control management while ensuring data integrity and traceability of transactions made over an insecure network.

We need a public key infrastructure (PKI) for entity identity verification for blockchain operation and digital signatures. Table 1 shows the notations used to describe the CP-ABE scheme and their meanings.

4.2 Modeling of Our Blockchain

We first present our records on the blockchain in the form of a token presenting a pseudo transaction.

Table 1 Notation

Notation	Description
Pks	The set of public attribute keys
SKs	The set of secret attribute keys
A	List of attributes of a user
CT	Ciphertexts (data encrypted by ABE)

The blockchain, in our construction, is used as a distributed, persistent and tamper-proof book. It manages access control messages. In addition, one of the advantages of using the blockchain is denying access to data by the cloud. A record (contained in a block) in our distributed database is presented in the form of an access authorization token (designated authorisation) on blockchain, equivalent to a pseudo crypto-currency.

- Authorization (idx, @ cl, @ gr, @pr): This is a record in the blockchain. It allows to specify an authorization by the Fog (the entity that encrypts the data by the ABE encryption, which signs this token) with a blockchain @pr address so that a group of doctors with address @gr can access data stored in the @cl cloud (blockchain address of the storage provider). The cloud identifies data by the idx index. This idx is calculated by the Proxy/Fog as a sequence idx = HMAC (CT).

We define two transactions carried out in the blockchain:

- GenAutorization (idx, @ cl, @ gr, @pr): It is the source transaction which is generated by the owners of the data. Once the idx value is calculated, Fog computing broadcasts this transaction to transfer this idx to the group of authorized consultants.
- RequestAuthorization (authorization (idx, @ cl, @ gr, @pr), @ pr, @ rq): This transaction is used to load authorization within the blockchain from an actor's account to the applicant's account. The requestor uses their @rq address and sends a request to the storage provider (@cl). The cloud sends this request to the Fog (@pr). The token circulation processes to obtain authorization to access data are illustrated in Fig. 2:

- 1: RequestAuthorization (authorization (idx, @ cl, @ gr, @pr), @ rq, @ cl): The requestor (@rq) transfers a transaction to the cloud site (@cl) in order to obtain the permission to download data.
- 2: Authorization request (authorization (idx, @ cl, @ gr, @pr), @ cl, @ pr): The cloud transfers the transaction from the user request to the Fog (@pr).
- 3: RequestAuthorization (authorization (idx, @ cl, @ gr, @pr), @ pr, @ rq): The Fog verifies the requester authorization and transfers the transaction to the doctor site.

Fig. 2 Token circulation processes in blockchain

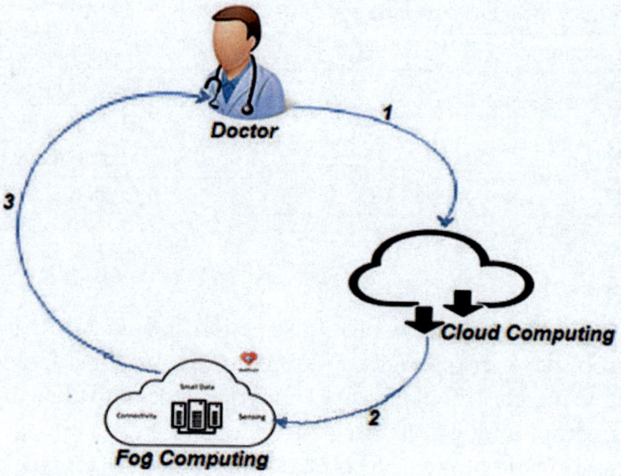

4.3 *The Proposed ABE Encryption Algorithm*

Let $G0$ and $G1$ be two bilinear groups of prime order p, g is a generator point of $G0$, and also, e is a bilinear map defined by: $G0 \times G0 \rightarrow G1$.

initialization: Setup()

The algorithm selects the groups $G0$ and $G1$ of order p and generator g of $G0$, and then, it chooses two random numbers α and β in Zp and produces the keys:

$$\text{PK} = (G0, G1, g, h = g^{\beta}, Y = e(g, g)^{\alpha}).$$
$$\text{MSK} = (\alpha, \beta)$$

Message encryptions: Encry (M, T, PK)

The encryption algorithm encrypts a message M under the access structure T. It first chooses a polynomial qx for each node in the tree T in a descending manner, from the root node R. Indeed, for each node x of T, we define the degree dx of the polynomial qx by dx = $kx - 1$ with kx presents the threshold of this node.

Starting with the root node R, the algorithm chooses a random $s \in$ Zp and defines qR (0) = s. Then, it chooses dR at random in order to define it completely. For any node x of the structure, this algorithm defines qx (0) = qparent (x) (index (x)), and also, it chooses the degree dx in a random way.

We put the subset X of the leaf nodes of T, by giving the access structure to the tree T and building the ciphertext:

$$\text{CT} = \{T,\ E_1 = MY^s\ , E_2 = h^s,\ \forall i \in X:\ E_i = g_i^q(0),\ E' = f(\text{attributes}(i)^{q_i(0)})\}$$

Private key generation: GenerKeyS (AU, MSK)

The algorithm takes an MSK master key and a set of AU attributes as input and generates a secret key that identifies with that set. We also pose a function:

$f: 0, 1 * \rightarrow$ G0 maps any attribute described as a binary string to a random group element. First, the algorithm selects random $r \in$ Zp, then a random rj $\in$ Zp for each attribute $j \in S$. Then, it calculates the key as:

$$\mathrm{SK} = (\mathrm{D} = g^{\frac{(\alpha+r)}{\beta}}, \mathrm{D} = g^{r}.E2, \forall j \in AU,\ D_j = g^{r}.f(j)^{r_j}, \mathrm{D}j = g^{r_j}).$$

Decryptions

It is a recursive algorithm. To facilitate the calculation, we present the simple form of decryption and improve it later.

First, we start to define the partial algorithm DecryNode (CT, SK, x) which takes as input a ciphertext CT, a secret key SK which is associated with a set of attributes AU, and with a node x of T the access shaft.

If node x is a leaf node, then we set i = attribute (x) and therefore

- If $i \in AU$, then

$$\mathrm{DecryNode(CT, SK, x)} = \frac{e(D_i, Cx)}{e(D_i', Cx)}$$
$$= e(g, g)^{rqx(0)}$$

- If i is not in AU, then DecryptNode (CT, SK, x) = $\perp$.

  After that, we move on to recursion. This amounts to saying that when x is a leaf node the DecryNode (CT, SK, x) algorithm proceeds as follows: For all Z nodes children of X, it calls DecryNode (CT, SK, z) and stores the output as as long as Fz. Let Sx be an arbitrary set of child nodes of size kx such that Fz different from $\perp$ (if Fz equal to $\perp$, then child node Fz is not satisfied then returns $\perp$), so we calculate:

$$F_x = \prod F_Z^{\Delta_i, S'x(0)},\ \mathrm{i = index(Z), S'x = \{index(Z)} : Z \in Sx\}$$
$$F_x = \prod (e(g, g)^{r.q_{parent}(z)(\mathrm{index}(z))})^{\Delta_i, S'x(0)}$$
$$F_x = \prod e(g, g)^{(r.q_x(i))^{\Delta_i, S_x'(0)}}$$
$$F_x = \prod e(g, g)^{(r.q_x(0))},\ \text{(polynomialinterpolation)}$$

4.4 Secure Data Storage and Sharing Scheme

In order to ensure effective access control of sensitive recording and protect patient privacy, we offer a system based specifically on symmetric encrypt, CP-ABE encrypt and blockchain.

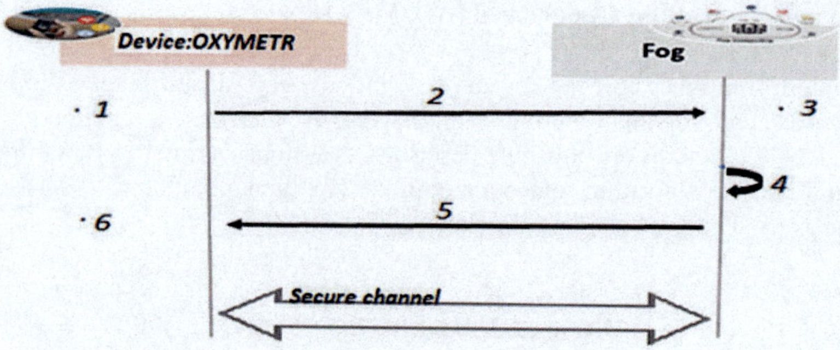

Fig. 3 Authentication phase and key exchange between the device and Fog

Generation of the public key PKs: Initialization

The attributes authority (AA) is responsible for generating the public attribute keys and transferring them to the Fog and/or doctors for later use if necessary. It executes the Setup() algorithm.

Data Generation Phase

Figure 3 illustrates the different steps for authenticating and sharing the secure key in order to obtain a secure channel to transfer data between the data generating devices and the Fog.

1. First the device selects a secure random a_d and calculates the value $R_d = a_d$.G.
2. Then, the device signs an identity id_d and encrypts the value R_d and the identity id_d by the public key of Fog PK_Fog. Then, it sends the information to the Fog: $E_{PK_{Fog}}(id_d) \parallel E_{SK_{d}evice}(id_d) \parallel E_{PK_{Fog}}(R_d) \parallel E_{SK_{d}evice}(R_d)$.
3. On its part, upon receipt of the message, the Fog decrypts and verifies the message in order to obtain the information id_d necessary for authentication and R_d necessary for the calculation of the symmetric key. Then, it performs a signature verification function.
4. If the signatures received are correct. Then, the device authenticated successfully. The Fog in turn selects a secure random value a_F and calculates $R_F = a_F.G$. Finally, it calculates the symmetric common key $S_K = R_d.a_F$.
5. The Fog encrypts and signs the R_F value and sends the message to the device: $E_{PK_{d}evice}(R_F) \parallel E_{SK_{Fog}}(R_F)$.
6. The device decrypts the $E_{PK_{d}evice}(R_F)$ message and verifies the validity of the signature. Finally, it calculates the symmetric common key $S_K = R_F.a_d$.

Note that the public parameters are as follows: - G the generator point. - The public keys of Fog PK_Fog and device PK_divice.

The secure channel is ready to transmit the data generated by the sensors.

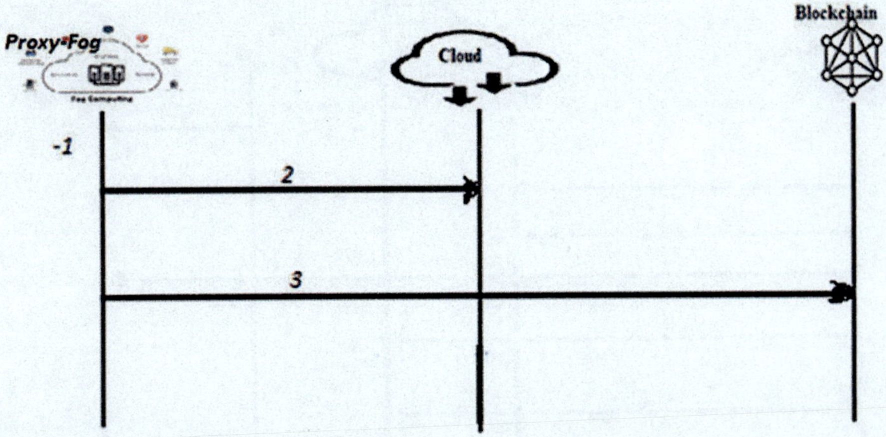

Fig. 4 Data logging phase

Data logging phase

Healthcare devices run the E_{S_K}(M) algorithm by encrypting the data and sending it to the Fogs. Once received, the latter executes the algorithm: Encry (M, T, PK) → CT, it calculates the data identifier "idx" = Hash (CT) and transfers the ciphertext to the storage provider where it is stored, and simultaneously, the proxy-Fog broadcasts the transaction:

GenAutorization (idx, @ cl, @ gr, @pr).

The steps of this phase are shown in Fig. 4
1: The Fog calculates the CT = Encry (M, T, PK) and idx = HMAC (CT)
2: The Fog sends CT, idx to the cloud
3: At the same time, the Fog broadcasts

GenAutorization (idx, @ cl, @ gr, @pr).

Access authorization phase

Authorization is given by the data signature (Fog). Indeed, the Fog generates an authorization (idx, @ cl, @ gr, @pr) which is used to authorize a group to access its data in the cloud. If a user wants to view data, it defuses a transaction to the site to the cloud which transmits it to the Fog. Then, the data owner checks the authorization right of this user and broadcasts the

RequestAutorization (authorization, @ pr, @ rq).

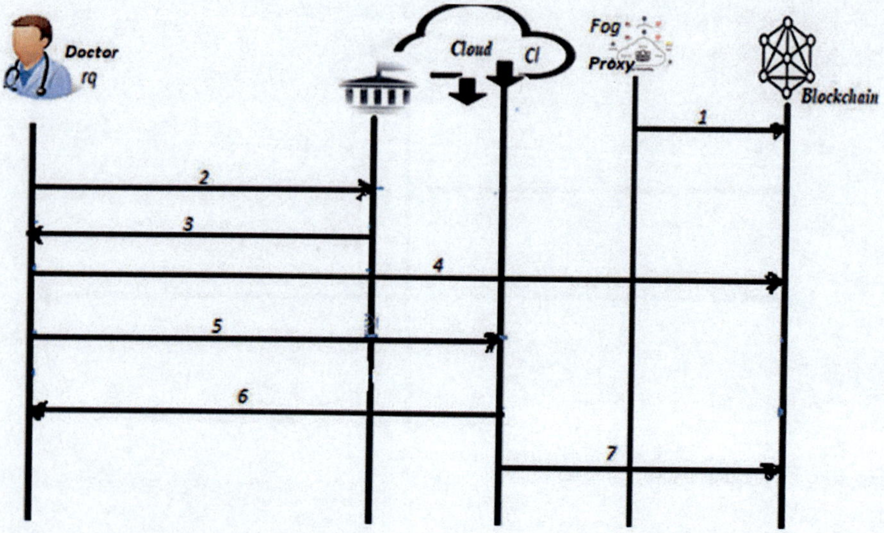

Fig. 5 Authorization and data access phase

Access authorization phase

When a doctor receives authorization to access data, first of all, he/she authenticates to the cloud with his/her professional card which defines his/her attributes. If the authentication is successful, the attribute authority executes the *GenerKeyS (AU, MSK)→ SK* attribute key generation algorithm. The output of this algorithm is transferred to the requestor in a secure manner. Also, the requestor broadcasts a *DemAutorization transaction (authorization, @ rq, @ cl)* to transfer to the cloud. The storage service sends the requester the encrypted text that is identified by the idx. It then broadcasts a DemAutorization transaction (authorization, @ cl, @ pr) in order to inform the Fog that its data has been consulted.

The doctor uses his/her secret ABE key and retrieves the data in clear. Figure 5 summarizes the authorization and data access phase:

1. RequestAuthorization (authorization, @ pr, @ rq).
2. The doctor looks for his/her secret ABE key from the attribute authority.
3. The authority securely sends the secret key to the requester.
4. RequestAuthorization (authorization, @ rq, @ cl)
5. The consultant sends a request to consult the data.
6. The cloud sends the encrypted CT text to the doctor.
7. The cloud broadcasts an $Ack_{D}emAutorization$ (authorization, @ cl, @ pr).

5 Security And Performance Analysis

In this section, we present the security and performance analysis of our new solution.

5.1 *Security Analysis*

Unlike traditional communication security and privacy protection, our cryptography scheme ensures security and privacy. We presented a formal safety analysis and validated it formally with the AVISPA simulator [19]. The symmetrical scheme proves through AVISPA that it is safe. The security model is presented as follows:

Suppose we have a polynomial probabilistic adversary A can break our scheme with a significant advantage AdvA = ε. We will show that we can, then, build a B simulator that can solve the DBDH problem with a significant advantage.

Simulator B will use A to find the solution to the DBDH problem. For the demonstration of the security of the diagram, we estimate the advantage of simulator B.

If $\mu = 1$, the ciphertext gives no information about Υ. So P $[\Upsilon' = \Upsilon | \ \mu = 1] = \frac{1}{2}$. The decision of B will be based on the result of A, if Υ'different from Υ, then B conclude that $\mu' = 1$ and if $\Upsilon' = \Upsilon$, B will choose $\mu' = 0$.

When $\mu = 0$, the advantage of A is ε. By definition P $[\Upsilon' = \Upsilon \mid \mu = 0] = \varepsilon + \frac{1}{2}$. B selects $\mu' = 0$ and $\Upsilon' = \Upsilon$ then P $[\mu' = \mu \mid \mu = 0] = \varepsilon + \frac{1}{2}$. Finally, the general advantage of B: AdvB = ε / 2. And since ε is assumed not to be negligible, then ε / 2 is also not negligible.

5.2 *Performance Analysis*

To analyze the performance of our solution, we implemented:

an anonymity proxy and attribute-based encryption by Fog computing and secure data storage in the cloud.

A blockchain to achieve decentralized access control message management and fine-grained access control is achieved through the encryption scheme based on attributes conjugated by the Fog.

To check that our scheme achieves its objectives, we analyze the performance of our implementation by modifying each time the number of attributes $N = \{2, 3, 4, 6, 8\}$, $N = \{10, 20, 30, 40, 50, 60, 70\}$ which are considered to be representative for real-world ranges for attribute-based encryption.

Figure. 6 shows the total execution time (ABE encryption + Symmetric encryption). The ABE scheme encryption time is a function of the numbers of attributes in the access structure. The results are considered good since the encryption time is done slightly as the number of attributes increases, but at a certain level, the execution time remains stable. On the other hand, the symmetric encryption time is considered

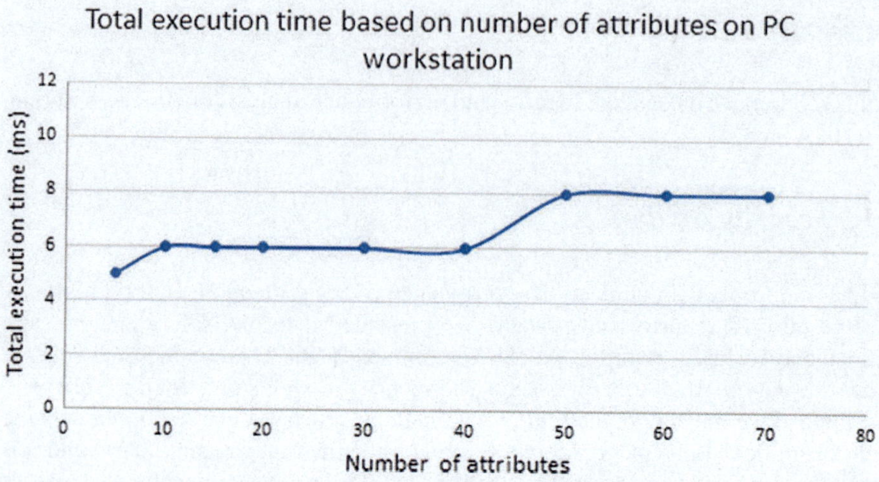

Fig. 6 Total execution time based on number of attributes on PC workstation

to be zero (between the device and the Fog server). Indeed this time has no effect on the total execution time. This allows us to say that our proposal respects the real-time constraint.

6 Conclusion

Through this article, we have proposed a solution to secure e-Health applications by exchanging data confidentially and protecting patient privacy in an IoT–Fog–cloud architecture. Our solution uses symmetric encryption and asymmetric encryption (ABE) techniques. It integrates the blockchain in order to strengthen security at the level of data access control management. In addition, our proposal ensures integrity and keeps track of data sharing. In order to move to a fully distributed architecture, we can integrate the smart contract for the execution of the encryption and decryption algorithms, and we can also strengthen our model by using machine learning to secure the cloud computing environment and detect the Man-In-The-Middle (MITM) attack in a network of connected objects for an upcoming job.

References

1. https://apps.who.int/iris/handle/10665/331421
2. Li, X., Niu, J., Karuppiah, M., Kumari, S., Wu, F.: Secure and efficient two-factor user authentication scheme with user anonymity for network based e-health care applications. J. Med. Syst. 40(12), 268 (2016)

3. Anitha, G., Ismail, M., Lakshmanaprabu, S.K.: Identification and characterisation of choroidal neovascularisation using e-Health data through an optimal classifier in Electronic Government. Int. J. **16**(1–2) (2020)
4. Wang, X., Bai, L., Yang, Q., Wang, L., Jiang, F.: A dual privacy-preservation scheme for cloud-based eHealth systems. J. Inf. Secur. Appl. **132–138** (2019)
5. Bethencourt, J., Sahai, A., Waters, B.: Encryption, Ciphertext-Policy Attribute-Based: IEEE Symposium on Security and Privacy (SP '07), p. 2007. France, May, Berkeley (2007)
6. Zuo, C., Shao, J., Wei, G., Xie, M., Ji, M.: CCA-secure ABE with outsourced decryption for fog computing. Future Gener. Comput. Syst. **78**(2), 730–738 (January 2018)
7. Wang, H.: Anonymous data sharing scheme in public cloud and its application in E-health record. In: IEEEaccess May 22, 2018, date of current version June 19 (2018)
8. Liu, Q., Zou, X.: Research on trust mechanism of cooperation innovation with big data processing based on blockchain. EURASIP J. Wirel. Commun. Network. **2019**, Article number: 26 (2019)
9. https://www.binance.vision/fr/blockchain/history-of-blockchain
10. Nakamoto, S.: Bitcoin : A peer-to-peer electronic cash system (2009). https://doi.org/10.1007/11823285_121
11. Genestier, P., Letondeur, L., Zouarhi, S., Prola, A., Temerson, J.: Blockchains et smart contracts: des perspectives pour lInternet des objets (IoT) et pour l'e-santé. Annales des Mines - Réalités industrielles, août **2017**(3), 70–73 (2017). http://orcid.org/10.3917/rindu1.173.0070
12. Gupta, Y., Shorey, R., Kulkarni, D., Tew, J.: The applicability of blockchain in the internet of things. In: 2018 10th International Conference on Communication Systems Networks (COMSNETS), pages 561–564 (2018)
13. Nguyen, C., Pathirana, N., Ding, M., Seneviratne, A.: Blockchain for Secure EHRs Sharing of Mobile Cloud Based E-Health Systems in ieeeAccess May 17, 2019, date of current version June 4 (2019)
14. Yao, Y., Chang, X., Misić, J., Misić, V.B., Li, L.: BLA: Blockchain-Assisted Lightweight Anonymous Authentication for Distributed Vehicular Fog Services. IEEE Internet Things J. Citation information https://doi.org/10.1109/JIOT.2019.2892009
15. Zhang, A., Li, X.: Towards Secure and Privacy-Preserving Data Sharing in e-Health Systems via Consortium Blockchain, Springer Science+Business Media, LLC, part of Springer Nature (2018)
16. Dumont, R.: Cryptographie et Sécurité informatique. http://www.montefiore.ulg.ac.be/~dumont/pdf/crypto.pdf
17. http://www.enseignement.polytechnique.fr/informatique/INF550/Cours1011/INF550-2010-7-print.pdf
18. Zhang, P., Chen, Z., Liu, J.K., Kaitai, L., Hongwei, L.: An efficient access control scheme with outsourcing capability and attribute update for fog computing. Future Gener. Comput. Syst. 78, 753–762 (2018)
19. The Avispa-Project http://www.avispa-project.org/

Forecasting COVID-19 Cases in Morocco: A Deep Learning Approach

Mustapha Hankar, Marouane Birjali, and Abderrahim Beni-Hssane

Abstract The world is severely affected by the COVID-19 pandemic, caused by the SARS-CoV-2 virus. So far, more than 108 million confirmed cases have been recorded, and 2.3 million deaths (according to Statistica data platform). This has created a calamitous situation around the world and fears that the disease will affect everyone in future. Deep learning algorithms could be an effective solution to track COVID-19, predict its growth, and design strategies and policies to manage its spread. Our work applies a mathematical model to analyze and predict the propagation of coronavirus in Morocco by using deep learning techniques applied on time series data. In all tested models, long short-term memory (LSTM) model showed a better performance on predicting daily confirmed cases. The forecasting is based on history of daily confirmed cases recorded from March 2, 2020, the day the first case appeared in Morocco, until February 10, 2020.

1 Introduction

In the last days of December 2019, the novel coronavirus, of an unknown origin, first time appeared in Wuhan, a province in China. Health officials are still tracing the exact source of this new virus; early hypotheses thought it may be linked to a seafood market in Wuhan [1]. After then, it was noticed that some people who visited the market have developed viral pneumonia caused by the new coronavirus [2]. A study that came out on January 25, 2020, notes that the individual with the first reported case became ill on December 1, 2019, and had no link to the seafood market [3]. Investigations are ongoing as to how this virus originated and spread. It appears after the person has been exposed to the virus for the first time that many symptoms are showing up within 14 days of the first exposure to the virus, including fever, dry cough, fatigue, breathing difficulties, and loss of smell and taste.

M. Hankar · M. Birjali (✉) · A. Beni-Hssane
LAROSERI Laboratory, Computer Science Department, Faculty of Sciences, University of Chouaib Doukkali, El Jadida, Morocco

M. Ben Ahmed et al. (eds.), *Networking, Intelligent Systems and Security*,
Smart Innovation, Systems and Technologies 237,
https://doi.org/10.1007/978-981-16-3637-0_59

COVID-19 mainly spreads through the air when people are close to each other long enough, primarily via small droplets or aerosols, as an infected person breathes, coughs, sneezes, or speaks [4]. In some cases, people who do not show any symptoms, or asymptomatic patients, remain infectious to others with a transmission rate equal to that of symptomatic people [5].

Amid a pandemic that has taken many lives so far, threatens the lives of others in the world, we are obligated to act, as researchers in machine learning and its real-world applications, which COVID-19 is one of the biggest actual challenges, in order to collaborate in the solution process. Machine learning algorithms can be deployed very effectively to track coronavirus disease and predict epidemic growth. This could help decision makers to design strategies and policies to manage its spread. In this work, we built a mathematical model to analyze and predict the growth of this pandemic. A deep learning model using feedforward LSTM neural network has been applied to predict COVID-19 cases in Morocco on time series data. The proposed model based its predictions on the history of daily confirmed cases, as a training phase, which have been recorded from the start of the pandemic in March 2, 2020, to February 20, 2020.

After training LSTM model on time series data, we tested it within a period of 60 days to assess the accuracy of the model and compared the obtained results with other applied models such as auto-regressive integrated moving averages (Auto-ARIMA), K-nearest neighbor (KNN) regressor, random forest regressor (RFR), and Prophet.

2 Related Works

Recently, deep learning techniques have been serving the medical industry [6, 7], bringing with them the new technology and its revolutionary solutions that are changing the shape of health care. Deep learning provides the healthcare industry with the ability to analyze large datasets at exceptional speeds and make accurate model.

Fang et al. [8] investigated the effect of early recommended or mandatory measures on reducing the crowd infection percentage, using a crowd flow model.

Hu et al. [9] developed a modified stacked auto-encoder for modeling the transmission dynamics of the epidemics. Using this framework, they forecasted the cumulative confirmed cases of COVID-19 across China from January 20, 2020, to April 20, 2020.

Roosa et al. [10] used phenomenological models that have been validated during previous outbreaks to generate and assess short-term forecasts of the cumulative number of confirmed reported cases in Hubei Province, the epicenter of the epidemic, and for the overall trajectory in China, excluding the province of Hubei. They collected daily report of cumulative confirmed cases for the 2019-nCoV outbreak for each Chinese province from the National Health Commission of China. They provided 5, 10, and 15 days forecasts for five consecutive days, with quantified uncertainty based on a generalized logistic model.

Liu and colleagues [11] used early reported case data and built a model to predict the cumulative COVID-19 cases in China. The key features of their model are the timing of implementation of major public policies restricting social movement, the identification and isolation of unreported cases, and the impact of asymptomatic infectious cases.

In [12], Peng et al. analyzed the COVID-19 epidemic in China using dynamical modeling. Using the public data of National Health Commission of China from January 20th to February 9th, 2020, they estimated key epidemic parameters and made predictions on the inflection point and possible ending time for 5 different regions.

In [13], Remuzzi analyzed the COVID-19 situation in Italy and mentioned if the Italian outbreak follows a similar trend as in Hubei Province, China, the number of newly infected patients could start to decrease within 3–4 days, departing from the exponential trend, but stated this cannot currently be predicted because of differences between social distancing measures and the capacity to quickly build dedicated facilities in China.

In [14], Ayyoubzadeh et al. implemented linear regression and LSTM models to predict the number of COVID-19 cases. They used tenfold cross-validation for evaluation, and root-mean-squared error (RMSE) was used as the performance metric.

In [15], Canadian researchers developed a forecasting model to predict COVID-19 outbreak using state-of-the-art deep learning models such as LSTM. They evaluated the key features to predict the trends and possible stopping time of the current COVID-19 pandemic in Canada and around the world.

3 Data Description

On March 11, 2020, the World Health Organization (WHO) declared COVID-19 as a pandemic, pointing to over 118,000 confirmed cases of coronavirus in over 110 countries and territories around the world at that time. The data used in this study was collected by many sources including the World Health Organization, Worldometers, and Johns Hopkins University, sourced from data delivered by the Moroccan Ministry of Health. The dataset is in a CSV format taken from the link: https://github.com/datasets/covid-19. It is maintained by the team at Johns Hopkins University Center for Systems Science and Engineering (CSSE) who have been doing a great public service from an early point by collecting data from around the world. They have cleaned and normalized data and made it easy for further processing and analysis, arranging dates and consolidating several files into normalized time series. The dataset is located in the data folder in a CSV file format. The team has been recording and updating all the daily cases in the world since January 22, 2020.

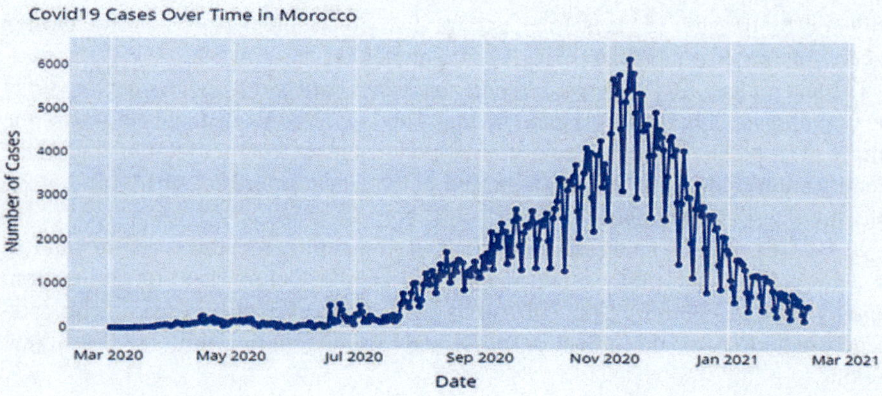

Fig. 1 Daily cases over time

The file contains six columns: cumulative confirmed cases, cumulative fatalities, dates of recording these cases, recovered cases, region/country, and finally province/state. Since we are working on Moroccan data, we filtered it based on country column to get the cases recorded in Morocco from March 2, 2020, to February 10, 2021. Since we are interested in daily confirmed cases only, which is not found in the dataset, we had to code a Python script to compute confirmed cases per day indexed by dates and then feed it to the algorithms.

As mentioned above, we transformed the original given data into univariate time series format of recorded confirmed cases. The values of the single-columned data frame are the number of cases per day, indexed by date/time. The plotting of daily cases in the dataset is shown in Fig. 1.

It is noticeable from the plot above (Fig. 1), COVID-19 daily cases are likely stabilized by the beginning of March with a small margin because of the strict measures taken by the health authorities. By the end of July, as the authorities start easing these measures, the cases began to increase exponentially this time due to the increase in population movements and travels during summer. After November 12, the cases started obviously to decrease, which could be a result of suppressing the virus by taken measures, or tested cases have obviously declined.

4 LSTM Model for Forecasting

Hochreiter and Sherstinsky [16, 17] published a theoretical and experimental works on the subject of LSTM networks and reported astounding results across a wide variety of application domains, especially on a sequential data. The impact of the LSTM network has been observable in natural language processing domains, like speech-to-text transcription, machine translation, and other applications [18].

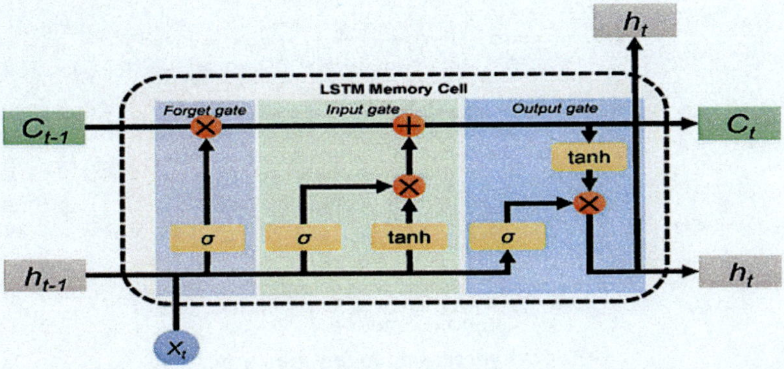

Fig. 2 LSTM cell architecture

LSTM is the type of recurrent neural networks (RNNs) that have feedback looping, meaning they are able to maintain information over time. They can process not only single data points, but also entire sequences of data such as speech or video, much applicable in time series data [19]. A LSTM unit is composed of a cell, an input gate, an output gate, and a forget gate (Fig. 2). The cell remembers values over arbitrary time intervals, and the three gates regulate the flow of information into and out of the cell [16]. All RNNs keep information in their memory over short period of time, because the gradient of its loss function fades exponentially [20]. Therefore, it could be difficult to train standard RNNs to solve problems that require learning long-term temporal dependencies like time series data. LSTM is a designed RNN architecture to address the vanishing gradient problem [21]. The reason we chose to implement this method is that LSTM units include a memory cell that can maintain information in memory for long period of time. A set of gates is used to control when information enters the memory, when it is output, and when it is forgotten.

In the equations below, the variables represent vectors. Matrices W_q{\displaystyle W_{q}} and U_q{\displaystyle U_{q}} contain, respectively, the weights of the input and recurrent connections, where the subscript q {\displaystyle _{q}} can either be the input gate i {\displaystyle i}, output gate o{\displaystyle o}, the forget gate f{\displaystyle f}, or the memory cell c{\displaystyle c}, depending on the activation function being calculated. In this section, we are using a "vector notation."

The equations for the forward pass of a LSTM unit with a forget gate are defined [22]:

$$f_t = \sigma_g\left(W_f x_t + U_f h_{t-1} + b_f\right) \quad (1)$$

$$i_t = \sigma_g(W_i x_t + U_i h_{t-1} + b_i) \quad (2)$$

$$o_t = \sigma_g(W_o x_t + U_o h_{t-1} + b_o) \quad (3)$$

$$\tilde{c}_t = \sigma_c(W_c x_t + U_c h_{t-1} + b_c) \tag{4}$$

$$c_t = f_t \circ c_{t-1} + i_t \circ \tilde{c}_t \tag{5}$$

$$h_t = o_t \circ \sigma_h(c_t) \tag{6}$$

- Activation functions used:

σ_g : sigmoid function.
σ_c : hyperbolic tangent function.
σ_h : hyperbolic tangent function.

- Variables used:

$x_t \in \mathbb{R}^d$: input vector to the LSTM unit
$f_t \in \mathbb{R}^h$: forget gates activation vector
$i_t \in \mathbb{R}^h$: input/update gates activation vector
$o_t \in \mathbb{R}^h$: output gates activation vector
$h_t \in \mathbb{R}^h$: hidden state vector also known as output vector of the LSTM unit
$\tilde{c}_t \in \mathbb{R}^h$: cell input activation vector
$c_t \in \mathbb{R}^h$: cell state vector
$W \in \mathbb{R}^{h\times d}$, $U \in \mathbb{R}^{h\times h}$ and $b \in \mathbb{R}^h$: weight matrices and bias vector parameters which need to be learned during training

RNN architectures using LSTM cells can be trained in a supervised way on training sequences, using an optimization algorithm, like gradient descent, combined with backpropagation through time to compute the gradients needed during the optimization process, in order to change each weight of the LSTM network in proportion to the derivative of the error (at the output layer of the LSTM network) with respect to corresponding weight [23]. Figure 3 shows the structure of a neural network using LSTM units.

The reason we proposed to use LSTM neural network goes to the nature of data. Since we are dealing with COVID-19 cases as time series values, we prioritized implementing this method over other techniques such as random forest regressor (RFR), which is rarely applied on times series data. On the other side, LSTM model showed, among other models, hopeful results and performance in predicting based on two essential metrics: root-mean-squared-error (RMSE) metric and mean absolute percentage error (MAPE) (Fig. 4).

Before getting to modeling section, it is a common practice to separate the available data into two main portions: **training** and **test** data (or validation data), where

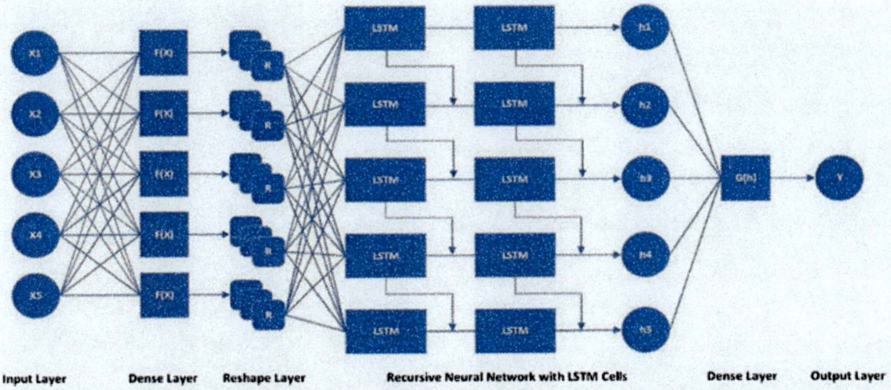

Fig. 3 Feedforward LSTM network structure

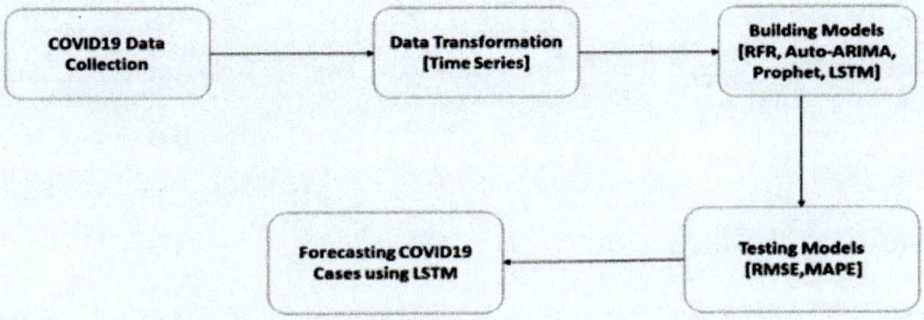

Fig. 4 Data pipeline

the training data is used to estimate the parameters of a forecasting method and the test data is used to evaluate its accuracy and estimate the loss function. Because the test data is not used in determining the forecasts, it should provide a reliable indication of how the model will likely forecast on new data. After splitting the data, we standardize the values with a MinMax scaler and then reshape the inputs in the right shape. In time series problem, we predict a future value in a time T based on a period of time T–N with T is the number time steps to be chosen as hyperparameter. We obtained good results by taking $N = 60$ days. Thus, the training inputs have to be in a three-dimensional shape (training inputs, time steps, and number of features) before beginning the training.

LSTM network is set to be trained over 300 epochs on more than 80% of the dataset and tested on a period of 60 days (20% of the dataset). The screenshot below, taken from the code source, shows the architecture of the trained feedforward LSTM network (Fig. 5).

The number of hidden layers, the dropout rate, and the optimization method to minimize the errors are essential hyperparameters to fine-tune in order to achieve hopeful results and performance of a deep learning model. In our case, the model contains three LSTM layers with a dropout rate of 0.4 each, dense layer to output

```
Model: "sequential"
_________________________________________________________________
Layer (type)                 Output Shape              Param #
=================================================================
lstm (LSTM)                  (None, 60, 64)            16896
_________________________________________________________________
dropout (Dropout)            (None, 60, 64)            0
_________________________________________________________________
lstm_1 (LSTM)                (None, 60, 64)            33024
_________________________________________________________________
dropout_1 (Dropout)          (None, 60, 64)            0
_________________________________________________________________
lstm_2 (LSTM)                (None, 64)                33024
_________________________________________________________________
dropout_2 (Dropout)          (None, 64)                0
_________________________________________________________________
dense (Dense)                (None, 1)                 65
=================================================================
Total params: 83,009
Trainable params: 83,009
Non-trainable params: 0
```

Fig. 5 LSTM model architecture summary

the forecasting results, and the "adam" optimizer given its best results compared to "rmsprop," for example.

5 Results and Discussion

The part of evaluating the model is a deductive part in our work. Therefore, the choice of a metric to evaluate the model matters and gives an insight about its performance on testing data and how the model will perform on new data. The dataset contains 294 records. We left a portion of 80% for training the model and 20% to test it. Since we used the metrics to compare the performance of an LSTM model with other models, we evaluated the models by two common methods.

5.1 Root-Mean-Squared Errors

RMSE is a method proposed in 2005 by the statistician Rob J. Hyndman [24] as the measure of a forecast accuracy. RMSE metric computes the residuals between predicted values and observed knowing that a forecast error is simply defined by the equation: et $= \mathrm{y} - \hat{y}_i$, where y is the true value and $(\hat{y})$ is the predicted value.

Accuracy measures that are based only on the error et are, therefore, scale dependent and cannot be used to make comparisons between series that involve different units. RMSE method is one of the two most commonly used scale-dependent measures. It is resulted by the formula:

$$\text{RMSE} = \sqrt{\frac{\sum_{i=1}^{N} (y_i - \hat{y}_i)^2}{N}}$$

5.2 Mean Absolute Percentage Error

MAPE is a measure of prediction accuracy of a forecasting method in statistics, such as time series data in our case, also used as a loss function for regression problems in machine learning [24]. Percentage errors have the advantage of being scale independent and frequently used to compare forecast performance across different datasets. MAPE metric usually expresses the accuracy as a ratio defined by the formula:

$$\text{MAPE} = \frac{1}{n}\sum_{t=1}^{n}\left|\frac{Y_t - F_t}{Y_t}\right|,$$

where Y_t is the actual value and F_t is the forecast value. MAPE is also sometimes reported as a percentage, which is the above equation multiplied by 100 making it a percentage error: the absolute difference between Y_t and F_t divided by the actual value Y_t summed for every forecasted point in time and divided by the number of fitted points n.

Considering the size of the dataset, which is likely small in this case, the model took an estimated time of 406 s in training over 300 epochs. As we can see in Fig. 6, the loss function began to minimize the errors in the first fifty epochs, after then the

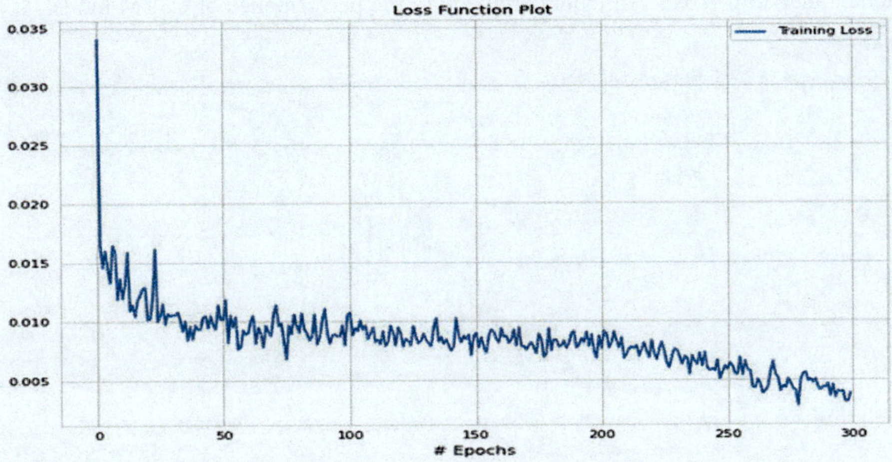

Fig. 6 Loss function plot over 300 epochs

Table 1 Forecast errors for the tested models

Model	RMSE	MAPE (%)
LSTM	357.90	29.31
Prophet	412.022	37.01
Auto-ARIMA	1699.47	215.87
Random forest regressor	977.53	83.41

cost function slowly decreases to the end. As the loss function decreases, the model on the other way increases its accuracy leading us to get a better outcome.

The results showed a better performance of LSTM model compared to other models like Prophet (Facebook forecasting algorithm), Auto-ARIMA, and random forest regressor. Table 1 shows the comparative performance of the four tested models based on two metrics.

Based on the results above, we chose to forecast COVID-19 daily cases on testing data using LSTM model (357.90 of RMSE), which outperforms other models by minimizing the loss function. When compared to bidirectional LSTM neural network architecture, the results of the latter were much closer to feedforward LSTM model than other models. In Fig. 6, we plot the whole data frame segmented into training set (more than 80% of the dataset) and testing set (almost 20%) to see how the model performs versus actual COVID-19 cases. As the chart shows, LSTM model accuracy did not reach the best wanted results, but it is very obvious that the model recognizes the trend within data and learned the overall pattern from the previous cases of training set. We also noticed that the performance of LSTM model increases when adding more data to the dataset. Meanwhile, RFR model and Auto-ARIMA model performances diminish.

To compare the presented forecasting results from the graph above, we tested other models on the same test set; Fig. 7 illustrates the predictions of Prophet model, RFR model, and Auto-ARIMA model compared to the performance of LSTM model. It

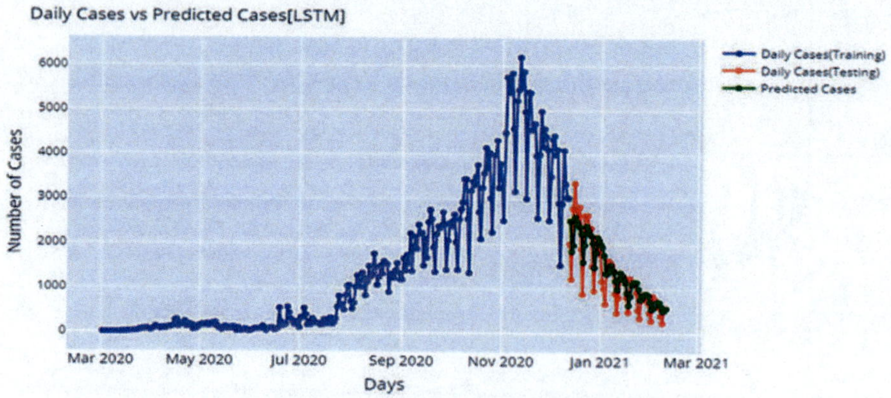

Fig. 7 LSTM predictions

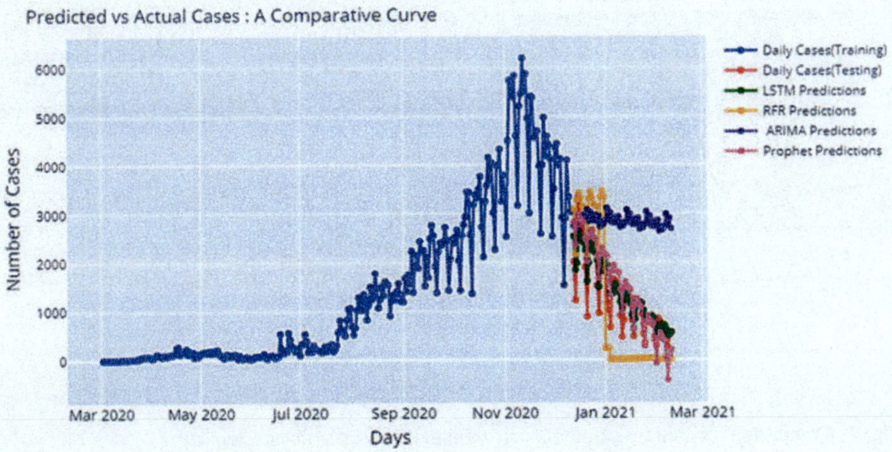

Fig. 8 Comparing predictions of tested models with the actual cases

is observed that Prophet's performance is more likely to learn trend from data than RFR and Auto-ARIMA models. The results of the latter are the worst among all models which is shown in Table 1 and Fig. 8.

The LSTM model showed a good performance in the training phase, because the loss function was at its lowest level by increasing the number of epochs to 300. This may lead to overfitting due to small amount of training data. However, a low training error might indicate that LSTM model can extract the pattern from data, which is obvious in our predictions (Fig. 6). Therefore, we certainly assume that the model could lead to better results if we have more training data. We also notice that the same proposed model showed good results on Bitcoin time series data predicting the daily stock prices, since we trained it on a sufficient amount of data. And yet, despite the small size of the dataset, LSTM model outperformed other models on this task (Fig. 9).

We assumed before getting into this study that the data provided till date is not big enough to train the model, meaning that our findings will not be at the very good level, but remain hopeful showing at least the trending behavior of how the coronavirus spreads over time, which is a helper factor to anticipate the future of its growth and give insights to health officials leading them to take actions and slow down the propagation of the virus, preventing vulnerable people from unbearable consequences. Due to measures taken during quarantine for more than three months, the curve of COVID-19 cases was likely stable and the virus propagation was almost controllable, but shutting down the economy and holding people in their places are not the ultimate solutions. It could actually be the problem itself.

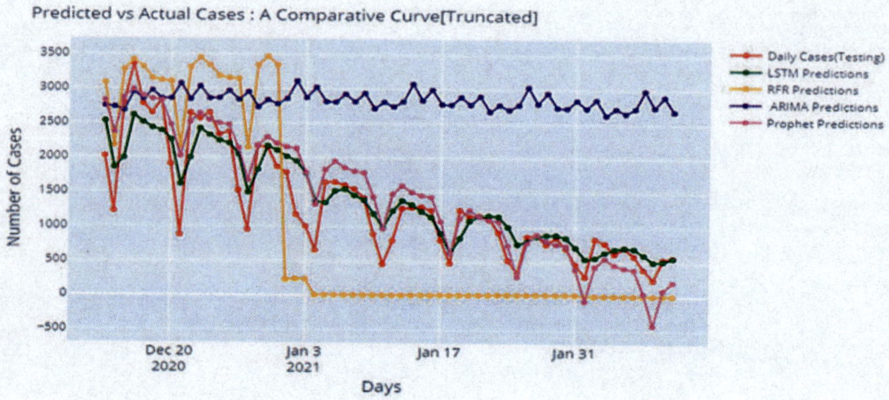

Fig. 9 Comparing predictions of the models with daily cases [truncated chart]

6 Conclusion

Considering the serious situation of recording thousands of COVID-19 daily cases in Morocco lately, an early prediction and anticipation of the virus transmission could help decision makers to take preventive actions to slow down its growth. This chapter is a contribution to solve this problem by implementing machine learning and statistical models. The results show that LSTM model yielded a hopeful accuracy score and a minimum root-mean-squared error.

References

1. Zhu, N., Zhang, D., Wang, W., Li, X., Yang, B., Song, J., Zhao, X., Huang, B., Shi, W., Lu, R., Niu, P., Zhan, F., Ma, X., Wang, D., Xu, W., Wu, G., Gao, G.F., Tan, W.: A novel coronavirus from patients with pneumonia in China, 2019. N. Engl. J. Med. (2020). https://doi.org/10.1056/nejmoa2001017
2. Zu, Z.Y., Di Jiang, M., Xu, P.P., Chen, W., Ni, Q.Q., Lu, G.M., Zhang, L.J.: Coronavirus disease 2019 (COVID-19): a perspective from China. Radiology (2020). https://doi.org/10.1148/radiol.2020200490
3. Huang, C., Wang, Y., Li, X., Ren, L., Zhao, J., Hu, Y., Zhang, L., Fan, G., Xu, J., Gu, X., Cheng, Z., Yu, T., Xia, J., Wei, Y., Wu, W., Xie, X., Yin, W., Li, H., Liu, M., Xiao, Y., Gao, H., Guo, L., Xie, J., Wang, G., Jiang, R., Gao, Z., Jin, Q., Wang, J., Cao, B.: Clinical features of patients infected with 2019 novel coronavirus in Wuhan, China. Lancet (2020). https://doi.org/10.1016/S0140-6736(20)30183-5
4. Karia, R., Gupta, I., Khandait, H., Yadav, A., Yadav, A.: COVID-19 and its modes of transmission. SN Compr. Clin. Med. (2020). https://doi.org/10.1007/s42399-020-00498-4
5. Oran, D.P., Topol, E.J.: Prevalence of asymptomatic SARS-CoV-2 infection: a narrative review. Ann. Intern. Med. (2020). https://doi.org/10.7326/M20-3012
6. Alhussein, M., Muhammad, G.: Voice pathology detection using deep learning on mobile healthcare framework. IEEE Access (2018). https://doi.org/10.1109/ACCESS.2018.2856238

7. Yuan, W., Li, C., Guan, D., Han, G., Khattak, A.M.: Socialized healthcare service recommendation using deep learning. Neural Comput. Appl. (2018). https://doi.org/10.1007/s00521-018-3394-4
8. Fang, Z., Huang, Z., Li, X., Zhang, J., Lv, W., Zhuang, L., Xu, X., Huang, N.: How many infections of COVID-19 there will be in the "Diamond Princess" predicted by a virus transmission model based on the simulation of crowd flow. ArXiv (2020)
9. Hu, Z., Ge, Q., Li, S., Jin, L., Xiong, M.: Artificial intelligence forecasting of COVID-19 in China. ArXiv (2020)
10. Roosa, K., Lee, Y., Luo, R., Kirpich, A., Rothenberg, R., Hyman, J.M., Yan, P., Chowell, G.: Real-time forecasts of the COVID-19 epidemic in China from February 5th to February 24th, 2020. Infect. Dis. Model. (2020). https://doi.org/10.1016/j.idm.2020.02.002
11. Liu, Z., Magal, P., Seydi, O., Webb, G.: Predicting the cumulative number of cases for the COVID-19 epidemic in China from early data. Math. Biosci. Eng. (2020). https://doi.org/10.3934/MBE.2020172
12. Peng, L., Yang, W., Zhang, D., Zhuge, C., Hong, L.: Epidemic analysis of COVID-19 in China by dynamical modeling. ArXiv (2020). https://doi.org/10.1101/2020.02.16.20023465
13. Remuzzi, A., Remuzzi, G.: COVID-19 and Italy: what next? Lancet (2020). https://doi.org/10.1016/S0140-6736(20)30627-9
14. Sajadi, M.M., Habibzadeh, P., Vintzileos, A., Shokouhi, S., Miralles-Wilhelm, F., Amoroso, A.: Temperature and latitude analysis to predict potential spread and seasonality for COVID-19. SSRN Electron. J. (2020). https://doi.org/10.2139/ssrn.3550308
15. Chimmula, V.K.R., Zhang, L.: Time series forecasting of COVID-19 transmission in Canada using LSTM networks. Chaos Solitons Fractals (2020). https://doi.org/10.1016/j.chaos.2020.109864
16. Hochreiter, S., Schmidhuber, J.: Long short-term memory. Neural Comput. (1997). https://doi.org/10.1162/neco.1997.9.8.1735
17. Sherstinsky, A.: Fundamentals of recurrent neural network (RNN) and long short-term memory (LSTM) network. Phys. D Nonlinear Phenom. (2020). https://doi.org/10.1016/j.physd.2019.132306
18. Lin, H.W., Tegmark, M.: Critical behavior in physics and probabilistic formal languages. Entropy (2017). https://doi.org/10.3390/e19070299
19. Karevan, Z., Suykens, J.A.K.: Transductive LSTM for time-series prediction: an application to weather forecasting. Neural Netw. (2020). https://doi.org/10.1016/j.neunet.2019.12.030
20. Bai, S., Kolter, J.Z., Koltun, V.: An empirical evaluation of generic convolutional and recurrent networks for sequence modeling. ArXiv (2018)
21. Graves, A., Schmidhuber, J.: Framewise phoneme classification with bidirectional LSTM and other neural network architectures. Neural Netw. (2005). https://doi.org/10.1016/j.neunet.2005.06.042
22. Gers, F.A., Schmidhuber, J., Cummins, F.: Learning to forget: continual prediction with LSTM. Neural Comput. (2000). https://doi.org/10.1162/089976600300015015
23. Kolen, J.F., Kremer, S.C.: Gradient flow in recurrent nets: the difficulty of learning long term dependencies. In: A Field Guide to Dynamical Recurrent Networks (2010). https://doi.org/10.1109/9780470544037.ch14
24. Hyndman, R.J., Koehler, A.B.: Another look at measures of forecast accuracy. Int. J. Forecast. (2006). https://doi.org/10.1016/j.ijforecast.2006.03.001

The Impact of COVID-19 on Parkinson's Disease Patients from Social Networks

Hanane Grissette and El Habib Nfaoui

Abstract Multiple examples of unsafe and incorrect treatment recommendations shared everyday. The challenge, however, to provide efficient, credible, and quick relevant access to reliable insight. Providing computational tools for Parkinson's Disease (PD) using a set of data-objects that contain medical information is very desirable for alleviating the symptoms that can help to discover the risk of this disease at an early stage. In this paper, we propose an automatic CNN-clustering aspect-based identification method for drug mentions, events, treatments from daily PD narratives digests. Therefore, a BiLSTM-based Parkinson classifier is developed regarding both varied emotional states and common senses reasoning, which further used to seek the impactful COVID-19 insights. The embedding strategy characterized polar facts through concept-level distributed biomedical representation associated with real-world entities, which are operated to quantifying the emotional state of the speaker context in which aspect are extracted. We conduct comparisons with neural networks state-of-art algorithms and biomedical distributed systems. Finally, as a result, the classifier achieves an accuracy of 85.3%, and facets of this study may used in many health-related concerns such as: Analyzing change in health status, unexpected situations or medical conditions, and outcome or effectiveness of a treatment.

1 Introduction

Having multiple voices who can relate to a similar situation, or who have experienced similar circumstances, always garner greater persuasion than that of a single brand.[1] Understanding emotions is the full-stack study that aims at recognizing, interpreting, processing, and simulating human emotions and affects. Nowadays, affective

[1] https://www.pwc.com/us/en/industries/health-industries/library/health-care-social-media.html.

H. Grissette (✉) · E. H. Nfaoui
LISAC Laboratory, Faculty of Sciences, Sidi Mohamed Ben Abdellah University, FEZ, Morocco
e-mail: hanane.grissette@usmba.ac.ma

M. Ben Ahmed et al. (eds.), *Networking, Intelligent Systems and Security*,
Smart Innovation, Systems and Technologies 237,
https://doi.org/10.1007/978-981-16-3637-0_60

computing (AC) and sentiment analysis (SA) have been considered as significant emerging approach used to discriminate fine grain information regarding the emotional state of patients. Indeed, instead of just detecting the polarity of given document [1], they are used to interpret the emotional state of patients, detect misunderstanding of drug related-information or side effects, and ensure a competitive edge to better understanding patients' experiences in a given condition [2].

Parkinson's Disease (PD) is a condition that can qualify a person for social security disability benefits. It is second common emotions-related disorders that affect an estimated 7–10 millions people and families worldwide. Few works have been provided to distil sentiment conveyed towards a drug/treatments on social networks whereby distinguish impactful facts degrees regarding PD-related drug-aspects. In previous work [3], authors proved the ability of based-neural network model to probe what kind of based-treatment target may result in enhanced model performance by detecting genuine sentiment in polar facts. Indeed, numerous of serious factors increase the failure rate in detecting $harmful$ and $non-harmful$ patients' notes regarding related-medication targets [1]. Noticeably, many of them may fail to retrieve the correct impact due to the inability to define complex medical components in text.

Each post may cite or refer to a drug reaction or/and misuse, which may lead to be categorized to harmful impact or beneficial reaction, where beneficial adverse reactions widely detected as harmful components [4]. This study is an affective distillation regarding Parkinson's disease related drug-targets, which further may used for fine-tuning tasks of many health-related concerns such as defining change in health status or unexpected situations/medical conditions, and monitoring outcome or effectiveness of a treatment. In addition, it is a powered neural network model for detecting polar medical statements, which further investigate what kind of based-treatment target may result in improved emotion parkinson's model performance. Technically, it consists of mining and personalizing various changeable emotional state toward specific objects/subjects regarding various aspect of PD's patients, which further used to track the impact of social media messages in daily PD patients' lives regarding given aspects and related-medical contexts such the case of COVID'19 pandemic. At this and (1) Firstly, we investigate a based-phrase transition method between social media messages and formal description on medical anthologies such as MedLine, life science journals, and online data systems. Then, (2) An automatic CNN-clustering-based model regarding PD-aspects for given targets, e.g., drugs mentions, events, physical treatments, or healthcare organizations at large. Therefore, (3) a BiLSTM-based emotional parkinson classifier is developed for the evaluation fine-tuning tasks. The main contributions of this paper can be summarized as follow: First, a embedding-based conceptualization that relies on various sub-steps such as medical concept transition-based normalization, the later is proposed for disambiguating medical-related expressions and concepts. Second, an affective distinction knowledge and common senses reasoning for improved sentiment inference.

The rest of paper is organized as follows: Sect. 2 briefly overviews of sentiment analysis and affective computing related works regarding healthcare context. Section 3 introduces the proposed method and knowledge base that we extended in this paper, and describe the whole architecture. In Sect 4, experimental results are presented and discussed. Finally, Sect. 5 concludes the paper and presents future perspectives.

2 Sentiment Analysis and Modelling Language

2.1 Sentiment Analysis and Affective Computing

In a broader scope, sentiment analysis (SA) and affective computing (AC) are allowing the investigation and comprehension of the relation between human emotions and health services as well as application of assistive and useful technologies in the medical domain. Recognizing and modeling patient perception by extracting affective information from related-medical text is of critical importance. Due to the diversity and complexity of human language, it is been mandatory to prepare taxonomy or ontology to capture concepts with various granularities in every domain. First initiatives in this era was: providing a knowledge-based methods that aim at building vocabulary and understanding language used to describe medication-related experiences, drugs issues and other related-therapies topics.

From the literature, efficient methods assume the existing of annotated lexicons regarding various aspects of analysis. Indeed, many lexicons have been annotated in term of sentiment for both public and depend-domain. they differ in annotation schemata: multinomial values (e.g., surprised, fear, joy) or continuous value as sentiment quantification that means extract the positiveness or negativeness parameters of probabilistic model or generative model as well. Existing large scale knowledge bases including Freebase [5], SenticNet [6], and Probase [7]. Most prior studies focused on exploring an existing or customized lexicons regarding depend-context, such as medical and pharmaceutical context.

Typically, neural network brought many success to enhancing these corpora capabilities, for example [6] proposed a SenticNet sub-symbolic and symbolic AI that automatically discover conceptual primitives from text and link them to common sense concepts and named entities in a new three-level knowledge representation for sentiment analysis. Other very related work is [8] that provide for affective common sense knowledge acquisition for sentiment analysis. Sentiment analysis has been known as a novel that allows a new form of sentiment-annotation regarding various aspects of analysis such as attention, motivation and emotions, namely aspect-based sentiment analysis (AbSA). Existing AbSA classifiers do not meet medical requirements. However, various lexicons and vocabularies are defined to identify very complicated medication concepts, e.g., adverse drug reactions (ADRs) or drug descriptions. For example, authors in [9] present deep neural network (DNN) model that utilizes the chemical, biological, and biomedical information of drugs to detect ADRs. Most of existing models aimed to fulfil two main purposes: (i) identifying the potential ADRs of drugs, (ii) defining online criteria/characteristics of drug reactions, and (ii) predicting the possible/unknown ADRs of a drug. Traditional approaches widely used medical concept extraction system such as ADRMine [5] that uses conditional random fields (CRFs), they used a variety of features, including a novel feature for modeling words' semantic similarities.

Parkinson's Disease Social Analysis Background Reports pour daily into health-care communities and micro-blogs at a staggering rate range from drug uses, side effects of some treatments to potential adverse drug reactions. Parkinson's disease (PD) is the second most important age-related disorder, after Alzheimer's disease, with a prevalence ranging from 41 per 100,000 in the fourth decade of life to over 1900 per 100,000 in people over 80 years of age. Emotional dysregulation is an essential dimension that may occur in several psychiatric and neurologic disorders. In most cases, it has been focused on clinical characteristics of emotional state variations in bipolar disorder in Parkinson's disease [10]. In both pathologies, the emotional intensity variability involves important diagnostic and therapeutic issues.

Few data mining and natural language processing techniques have been proposed in the PD context, and less efficient. Nowadays, advanced machine learning get researchers and professionals' attention and result in good results in quantifying short-term dynamics of Parkinson's Disease using self-reported symptom data from social network. Where [11] proved the power of machine learning in a Parkinson's Disease digital biomarker dataset construction (NNC) methodology that discriminates patient motor status. Other contributions used ensemble techniques to efficiently enhance models performance. For example, [12] proposed hybrid intelligent system for the prediction of PD progression using noise removal, clustering and prediction methods, an adaptive neuro-fuzzy inference system (ANFIS) and support vector regression (SVR) for prediction of PD progression. This study aims to perform an efficient neural networks-solution that attempts a public concerns and impact of COVID-19 pandemic on Parkinson's disease patients.

2.2 *Medical Conceptualization*

Patients and health consumers are storming intentionally their medications' experiences and their related-treatment opinions that describe all the incredibly complex processes happening in real-time treatment in a given condition. Patients self-reports on social networks frequently capture varied elements ranging from medical issues, product accessibility issues to potential side effects. Deep learning based neural networks have widely attracted many researchers' attention on normalization, matching and classification tasks by exploiting their ability to learn more under distributed representations.

Embedding approaches are the most accurate methods that used for constructing vector representations of words and documents [13]. The problem is that these algorithms got in low-medical entities recognition recall which further require intervention of both formal external medical knowledge and real-world examples to learn natural medical concepts patterns. Recently, researches paid great attention to conquer these limitations by many alternatives: (i) extending embeddings on large amounts of data or depend-domain corpora, namely semisupervised feature-based methods [14], (ii) fusing weighted distributed features, (iii) manifold regularization for enhancing the performance of sentiment classification of medication-related online examples. In this study, we aim at imitating and retrieving the corresponding of real-word related drug entities in the formal medical ontologies by adopting a distributed representation of two databases PubMed and MIMIC III Clinical notes as described in Table 1. To summarize, we aim at developing an approach that takes unstructured data as input, constructing the embedding matrix by incorporating biomedical vocabulary, and discriminating drug reaction multi-word expressions.

3 Proposed Approach

In this section, we will introduce the proposed methodology that consists of medical conceptualization and affective-aspect analysis model.

3.1 Text Vectorization and Embedding

The exploitation of such overwhelming unstructured, rich, and high-dimensional data on social media from patients self-reports is of critical importance. Technically, each post encoded as vector of the individual document to be input to the neural network model. Moreover, medication-related text has a wide variety of medical entities and medical facts, it requires a well-encoding approach to successfully leverage the real meaning and conveyed sentiment toward a given medical aspects such as drug entity, treatment, condition...etc.

However, the situation is complex, related-medication text depends on various aspects. The popular SA approaches ignore many types of information, e.g., sexual and racial information for medical oriented-analysis that certainly affects the emotion computing process, but using the same information for personalizing patient's behavior is not. Moreover, patient self-reports contains various related-medical components range from drug/treatment entities, adverse drug events/reactions to potential unsuspected adverse drug reaction [15]. The existing techniques for both sentiment analysis and affective computing are not able to efficiently extract those concepts and topics due to limited resources in this case. Otherwise, probabilistic models could be used to extract low-dimensional topics from document collections, such models without any human knowledge often produce topics that are not interpretable. To address these problems, we approach this step by incorporating real-world knowledge encoded by entities embedding to automatically learn medical representation from an external biomedical corpora EU-ADR [16] and ADRMINE [5] Lexicon in a unified model. As depicted in Fig. 2, incorporating depend-domain medical knowledge support two main functionalities in this stage:

- **Stage (1)**: Build an embedding representation: since biomedical knowledge is multi-relational data, we seek to represent knowledge as triple fact, e.g., (Drug name, disease, association). We aim at incorporating medical knowledge into embedding vectors space. Indeed, for this purpose, we combine two annotated corpora, (1) EU-ADR corpus, The EU-ADR corpus has been annotated for drugs, disorders, genes and their inter-relationships. Moreover, for each of the drug–disorder, drug–target, and target–disorder relations three experts have annotated a set of 100 abstracts, (2) ADRMINE Corpus is results of supervised sequence labeling CRF classifier is established that extracts mentions of ADR and indications from inputs, adverse drug reactions (ADRs) training data consisted of 15,717 annotated tweets. Indeed, each annotation includes cluster ID, semantic type (i.e., ADR, indication, Drug interaction, Beneficial effect, other), drug name and corresponding UMLS ID that means ADR Lexicon compile an exhaustive list of ADR concepts and their corresponding UMLS IDs. Moreover, it includes concepts from SIDER, a subset of CHV (Consumer health vocabulary) and COSTART (The Coding Symbols for a Thesaurus of Adverse Reaction Terms).
- **Stage (2)**: Define a distance measure for unknown entities. In fact, we want to rely on a measurement to be internal semantic representation for stream data and map to some previously sentiment-polarized vectors. We would be able to identify exact duplicates, making practically online rich information useful for tracking new issues, thoughts and even probable diseases.

The incorporating knowledge technique is inspired by previous work [17], this latter aim at defining fact-oriented knowledge graph embedding that automatically captures relations between entities in knowledge base. However, this knowledge space is embedded into low-dimensional continuous vector space while new properties have been created and preserved. Generally, each entity is treated as a new triple with variable: value format. e.g. [c_1, Drug, aspirin, C0004057]. We combined this latter with EU-ADR corpus to link drug with probable disease and ADR in the space. Each time, annotation may have many attributes, while we project them into the same format, then we copy all the entities triples of each document that shows up in our training vocabulary. Thus, each term outvocabulary is convolutionally aggregated to the closest entities in the embedding matrix, we multiply frequencies of the same term in both text documents. Finally, we defined a distance between two entities by using soft cosine similarity measure as shown in the formula below Eq. 1.

$$f_i = \text{cosSimilarity}(, c_i, E) \tag{1}$$

In such case, we need to consider the semantic meaning that conveyed regarding medical aspects in texts, where similar entity meanings may attribute varied facets of sentiments. A similarity metric that gives higher scores forc_i in documents belonging to the same topic and lower scores when comparing documents from different topics.

Since, neural network is a method based on nonlinear information processing, typically we use continuous BOW for building an embedding vector space regarding related-medical concepts. The obtained embedded vectors trained regarding preserved ADRMINE parameters and context features where the context is defined with seven features including the current token t_i, the three preceding ($t_{i-3}, t_{i-2}, t_{i-1}$), and three following tokens ($t_{i+3}, t_{i+2}, t_{i+1}$), in the input. Moreover, these samples were passing by a set of normalization and processing steps to be able for our neural inference model. Indeed, every single tweet including includes spelling, correction, lemmatization, and tokenization. Our dataset consists of a separate document that saves the life of correlate entities contained regarding medical and pharma objects.

Convolutional neural network provides an efficient mechanism for aggregating information at a higher level of abstraction; we exploit convolutional learning to learn data properties and tackle ambiguities types through common semantics and contextual information.Considering a window of words $[w_i, w_{i+1}, \ldots, w_{i+k-1}, w_{i+k}]$, the concatenated vector of the ith window is then:

$$S_i = [w_i, w_{i+1}, \ldots, w_{i+k-1}, w_{i+k}] \in \mathbb{R}^{k*d} \tag{2}$$

The convolution filter is applied to each window, resulting in scalar values r_i, each for the ith window:

$$r_i = g(x_i * u) \in \mathbb{R} \tag{3}$$

In practice one typically applies more filters, $u_1, \ldots, u_l$, which can then be represented as a vector multiplied by a matrix U and with an addition of a bias term b:

$$r_i = g(x_i * u + b) \tag{4}$$

with $r_i \in \mathbb{R}^l$, $x_i \in \mathbb{R}^{k.d*l}$ and $b \in \mathbb{R}l$

CNN features are also great at learning relevant features from unlabelled data and got huge success in many unsupervised learning case study. CNN-based clustering method use these feature to be input to K-mean clustering and parameterized manifold learning. It is of extracting the structural representation by polar medical facts and non-polar facts. This is because of the need to distinct false positives and negatives usually obtained by baselines.

3.2 Common Sense Detection

Accurate emotions analysis approach relies on the accuracy of vocabulary and the way we define emotions regarding related-medication concepts (Drugs, ADRs, and diseases), events, and facts. Indeed, patients self-reports may refer to various concepts in different ways regarding various context. Not only surface analysis of the text is required, but also common sense analysis based knowledge approach is needed. To bridge the cognitive and affective gap between word-level natural language data and the concept-level sentiments conveyed by them, affective common sense knowledge is needed [17]. For this purpose, a conceptualization technique is involved in discovering the conceptual primitives of each entity by means of contextual embedding. Each entity may belong to many concepts regarding clusters we preserved from first learning in previous stage c_{drug} =treatment, doctor, ADR, indication.

People frequently express their opinions with particular backgrounds and set of morals aspects, such as ethics, spirituality, and positionality. In this window, it is widely accepted that before patient or his family make a decision such a straightforwardly information about drugs and adverse drug reactions is learned. Further, we are willing to put this working hypothesis to the test of rational discourse, believing that other persons acting on a rational basis will agree. Thus, the weighing and balancing of potential risks and benefits becomes an essential component of the reasoning process in applying the principles. Moreover, related-medication mining involves systematizing and defining concepts-related meanings. Indeed, it seeks to resolve questions of human perception by defining affective concepts information. Patient perception is main aspect to define what a person is permitted to do in a specific situation or a particular domain of action. Technically, we aim at extending the following assumption regarding these common senses: "a set of affective concepts correlate with affective words and affective common sense knowledge consists of information that people usually take for ambiguous status, hence, normally leave unstated." Especially, It is concerned the computational treatment representing the affective meaning of given input that allows a new form. Recognizing emotional information requires the extraction of meaningful patterns from the gathered data first otherwise, CNN-based model is calculated.

In this stage, a distinction between means and effect is performed. Meaning, a distinction between word-level natural language data and the concept-level sentiments conveyed by them is also required. Affective common sense, in fact, is not a kind of knowledge that we can find in formal knowledge such as Wikipedia, but it consists in all the basic relationships among words, concepts, phrases, emotions, and thoughts that allow people to communicate with each other and face everyday life problems and experiences. For this reasons, we chose to use SenticNet [6] as prior knowledge that is seminal domain-independent knowledge base which is constructed for concept-based sentiment analysis through a multidisciplinary approach, namely sentic computing. Sentics is affective semantics where are operated to extract the affective concept-based meaning. Thus, a pre-trained embedding from ConceptNet is basically added to seek the concept-based affective information and collect such kind of knowledge through label sequential rules (LSR), crowd sourcing, and GWAP techniques. Indeed, sentic computing is a multidisciplinary approach to opinion mining and sentiment analysis at the crossroads between affective computing and common sense computing, which exploits both computer and social sciences to better recognize, interpret and process opinions and sentiments over the Web. It provides the cognitive and affective information associated to concepts extracted from opinionated text by means of a semantic parser.

4 Experiments and Output Results

The performance of adopting existing SA method to the medical situations and case studies can be summarized as follow: (1) sentiment analysis systems are able to perform sentiment analysis toward a given entity fairly well, but poorly on clarifying sentiment towards medical targets, (2) they got in low recall in term of distinction multi-word expressions that may refer to an adverse drug reaction. The paper investigates the challenges of considering biomedical aspects through sentiment tagging task. An automatic approach to generate sentimental based-aspect concerning drug reaction multi-word expressions toward varied related medication contexts, it considered as domain-specific sentiment lexicon by considering the relationship between the sentiment of both words features and medical concepts features. From our evaluation on large Twitter data set, we proved the efficiently of our features representation of drug reaction, which is dedicated to matching expressions from everyday patient self-reports.

In order to understand the difference of deriving features from various source data, we choose to utilize a predefined corpora trained on a different corpus. A results from [7] assumes that is better than using a subselection of the resources and delivered an unified online corpus for emotion quantification. Technically, deep neural networks have achieved great success in enhancing model reliability. Authors in [17] provide a novel mechanism to create medical distributed N-grams by enhancing convolutional representation, which is applied for featuring text regarding medical setting and clarifying contextual sentiment in a given target. Correlation between Knowledge, experience and common Sense are assessted through this study. Each time, a sentiment value is contributed to each vector. We use two benchmarks for model development: (1) **lex1**: ConceptNet as a representation of commonsense knowledge [2] (2) **lex2**: SenticNet.[3]

Since patients perceptions of drug-related knowledge are usually considered empty of content and untruthful, this application of emotional state comes into focus of understanding the unique features and polar facts that provide the context for the case. Therefore, obtaining the relevant and accurate facts is an essential component of this approach to decision making. Noticeably, we got a great changes and shift in patient statements on everyday shared conversations on the pandimic period. Table 4 shows a comparison of positives and negatives statements on COVID-19 period and Before the pandemic period. Where we used a parkisons datasets collected for previous studies on the year of 2019.

Emotional and common senses detection performance are assessed through experiments on varied online datasets (Facebook, Twitter, Parkinson Forum), as summarized in Table 2. An extensive evaluation of different features, including medical corpora, ML-algorithms, have been performed. As shown in Table 2, a sample of PD-related posts (dataset can be found in this link[4]) was collected from the

[2] https://ttic.uchicago.edu/~kgimpel/commonsense.html.

[3] https://sentic.net/downloads/.

[4] https://github.com/hananeGrissette/Datasets-for-online-biomedical-WSD.

Table 1 Biomedical corpora and medical ontologies statistics used for biomedical distributed representation

Sources	Documents	Sentences	Tokens
PubMed	28,714,373	181,634,210	4,354,171,148
MIMIC III Clinical notes	2,083,180	41,674,775	539,006,967

Table 2 Summarize online datasets from varied platforms used for both training and model developments

Plateform source	#Posts	Keywords used
Twitter	256,703	Parkinson, disorder, seizure, Chloroquine, Corona, Virus, Remdesivir, Disease, infectious, treatments, COVID-19
Facebook	49,572	COVID-19,Chloroquine, Corona, Virus, Remdesivir, disease, infectious, Parkinson, disorder, seizure, treatments
PD forum	30,748	Chloroquine, COVID-19 Corona, Virus, Remdesivir, disease, infectious, treatments

online healthcare community of Parkinson's Disease and normalized to enrich the vocabulary. For twitter, we collect more than 25000 tweets in the $Parkinson$ and $COVID-19$ contexts that have been prepared to be input to the neural classifier for defining medical concepts and then re-define distributed representation for unrelated items of natural medical concepts cited in real-life patients narratives. A based-keywords crawling system is created for the collection of Twitter posts. In this study, we focused on distracting drug reaction in the COVID19 contexts. We used a list of related-COVID-19 Keywords, e.g., $Corona$, and $Chloroquine$. Thus, we are interested in getting information attributed by : ['id','created_at', 'source', 'original_text', 'retweet_count', 'ADR', 'original_author', 'hashtags', 'user_mentions', 'place', 'place_coord_boundaries']. The Table 2 summarizes statistics of raw data grouped in terms of some related-drugs keywords in different slice of time. Twitter Data relies on relies on large volumes of unlabeled data, and thus diminishing the need for memorable positive and negative statement regarding each based-target class is assessed through the experiments,whereby, is considered as an automatic supervised learning problem regarding emotion information. The based CNN-clustering are one or more axes-matter. For fine-grained analysis, the evaluation may also applied on varied axes such as: age-axis and gender-axis that allows us to show peaks of positiveness of PD's patients.

We conducted many experimentation derived from the application of the proposed method based on hybrid medical corpora based-text and concepts, a minute conceptualization is released. As illustrated in Table 3, the BiLSTM-based Parkinson's classifier outperforms other neural network algorithms regardless the sentiment lexicon and the medical knowledge used. The support vector machine (SVM) classi-

Table 3 Experiments results overview on different platforms data using sentiment lexicons discussed above

Dataset	Algorithm	Sentiment	Medical knowledge/ADRs	Accuracy %
Twitter	BiLSTM	Lex1	PubMED + clinical notes MIMIC III + EU-ADR	0.71
	BiLSTM	Lex1 +lex2	PubMED+ EU-ADR+ ADRMINE	**0.81**
	LSTM	Lex1+lex2	PubMED + EU-ADR+ ADRMINE	0.73
	SVM	Lex1+lex2	PubMED + EU-ADR+ ADRMINE	0.61
	stacked-LSTM	Lex1+lex2	PubMED + EU-ADR+ ADRMINE	0.73
Facebook	BiLSTM	Lex1	PubMED + clinical notes MIMIC III + EU-ADR	0.71
	BiLSTM	Lex1 +lex2	PubMED+ EU-ADR+ ADRMINE	**0.79**
	LSTM	Lex1+lex2	PubMED + EU-ADR+ ADRMINE	0.71
	SVM	Lex1+lex2	PubMED + EU-ADR+ ADRMINE	0.59
	stacked-LSTM	Lex1+lex2	PubMED + EU-ADR+ ADRMINE	0.71
PD forum	BiLSTM	Lex1	PubMED + clinical notes MIMIC III + EU-ADR	0.76
	BiLSTM	Lex1 +lex2	PubMED+ EU-ADR+ ADRMINE	**0.85**
	LSTM	Lex1+lex2	PubMED + EU-ADR+ ADRMINE	0.71
	SVM	Lex1+lex2	PubMED + EU-ADR+ ADRMINE	0.68
	Stacked-LSTM	Lex1+lex2	PubMED + EU-ADR+ ADRMINE	**0.80**

fier has got acceptable results in classifying polar facts and non-polar facts. The based stacked-LSTM and BiLSTM model consistently improved the sentiment classification performance, but is efficient when we exploit proposed configuration on PD post from forum due to the post's length(it contains more details and clear drug-related descriptions). We also conducted an evaluation on different dataset from Facebook, which is collected in a previous study. The got in low results than other baselines in term of entity recognition recall, which reflect on model performance (Table 4).

However, it deserves to be noted that if we replace tokens with n-grams and train on small datasets, which is CNN-clustering based architecture improved the representations over the obtained biomedical distributed representations on top of those features, then we may get whopping 1.8 bumps in the accuracy and it boosts accuracy to over 87%. Thus, we end up by learning some deeper representations and new multi-word expressions vectors are inserted in the vocabulary each time.

Table 4 Percentage of sentiment terms (positive and negative) extracted before the COVID-19 period and in the pandemic period

Sources	Before COVID-19		In COVID-19 period	
	Positive (%)	Negative (%)	Positive (%)	Negative (%)
PD's forum	30	10	20	35
Twitter	33	16	28	47
	43	13	30	50
	51	19	37	43
Facebook	15	17	25	45
	32	20	40	52

5 Conclusion

This article is intended to be brief introduction to the use of neural networks to efficiently leverage patient emotions regarding various affective aspects. We proposed an automatic CNN-clustering aspect-based identification method for drug mentions, events, treatments from daily PD narratives digests. The experiments proved emotional Parkinson classifier ability to translate varied facets of sentiment and seek the impactful COVID-19 insights from generated narratives. The study of what is morally right by patient in given condition and what is not, is our perspectives. We aim at defining an neural network approach based on set of morals aspects in which the model rely on variables that can be shown to substitute for morals aspects regarding the emotion quantity. It also involved to provide a proper standard of care that avoids or minimizes the risk of harm that is supported not only by our commonly held moral convictions, but by the laws of society as well.

References

1. Grissette, H., Nfaoui, E.H.: Drug reaction discriminator within encoder-decoder neural network model: Covid-19 pandemic case study. In: 2020 Seventh International Conference on Social Networks Analysis, Management and Security (SNAMS), pages 1–7 (2020)
2. Grissette, H., Nfaoui, E.H.: A conditional sentiment analysis model for the embedding patient self-report experiences on social media. In: Advances in Intelligent Systems and Computing (2019)
3. Grissette, H., Nfaoui, E.H.: The impact of social media messages on parkinson's disease treatment: detecting genuine sentiment in patient notes. In: Book Series Lecture Notes in Computational Vision and Biomechanics. SPRINGER International Work Conference on Bioinspired Intelligence (IWOBI 2020) (2021)
4. Grissette, H., Nfaoui, E.H.: Daily life patients sentiment analysis model based on well-encoded embedding vocabulary for related-medication text. In: Proceedings of the 2019 IEEE/ACM International Conference on Advances in Social Networks Analysis and Mining, ASONAM 2019 (2019)

5. Nikfarjam, A., Sarker, A., O'Connor, K., Ginn, R., Gonzalez, G.: Pharmacovigilance from social media: Mining adverse drug reaction mentions using sequence labeling with word embedding cluster features. J. Am. Med. Inf. Assoc. (2015)
6. Cambria, E., Li, Y., Xing, F.Z., Poria, S., Kwok, K.: SenticNet 6: ensemble application of symbolic and subsymbolic AI for sentiment analysis. In: International Conference on Information and Knowledge Management, Proceedings (2020)
7. Wu, W., Li, H., Wang, H., Zhu, K.Q.: Probase: a probabilistic taxonomy for text understanding. In: Proceedings of the ACM SIGMOD International Conference on Management of Data (2012)
8. Cambria, E., Xia, Y., Hussain, A.: Affective common sense knowledge acquisition for sentiment analysis. In: Proceedings of the Eighth International Conference on Language Resources and Evaluation (LREC'12), pages 3580–3585, Istanbul, Turkey. European Language Resources Association (ELRA) (2012)
9. Shiang Wang, C., Ju Lin, P., Lan Cheng, C., Hua Tai, S., Kao Yang, Y.H., Hsien Chiang, J.: Detecting potential adverse drug reactions using a deep neural network model. J. Med. Internet Res. (2019)
10. Grover, S., Somaiya, M., Kumar, S., Avasthi, A.: Psychiatric Aspects of Parkinson's Disease (2015)
11. Tsoulos, I.G., Mitsi, G., Stavrakoudis, A., Papapetropoulos, S.: Application of machine learning in a parkinson's disease digital biomarker dataset using neural network construction (NNC) methodology discriminates patient motor status. Front, ICT (2019)
12. Nilashi, M., Ibrahim, O., Ahani, A.: Accuracy improvement for predicting Parkinson's disease progression. Sci. Rep. (2016)
13. Pennington, J., Socher, R., Manning, C.D.: GloVe: global vectors for word representation. In: Proceedings of the Conference EMNLP 2014—2014 Conference on Empirical Methods in Natural Language Processing (2014)
14. van Engelen, J.E., Hoos, H.H.: A survey on semi-supervised learning. Mach. Learn. (2020)
15. Nikfarjam, A.: Health Information Extraction from Social Media. ProQuest Dissertations and Theses (2016)
16. van Mulligen, E.M., Fourrier-Reglat, A., Gurwitz, D., Molokhia, M., Nieto, A., Trifiro, G., Kors, J.A., Furlong, L.I.: The EU-ADR corpus: annotated drugs, diseases, targets, and their relationships. J. Biomed. Inf. (2012)
17. Grissette, H., Nfaoui, E.H.: Enhancing convolution-based sentiment extractor via dubbed N-gram embedding-related drug vocabulary. Netw. Model. Anal. Health Inf. Bioinf. **9**(1), 42 (2020)

Missing Data Analysis in the Healthcare Field: COVID-19 Case Study

Hayat Bihri, Sara Hsaini, Rachid Nejjari, Salma Azzouzi, and My El Hassan Charaf

Abstract Nowadays, data is becoming incredibly important to manage in a variety of domains, especially healthcare. A large volume of information is collected through a lot of ways such as connected objects, datasets, records, or doctor's notes. This information can help clinicians in preventing diagnosis errors and reduce treatment complexity. However, data is not always available and reliable due to missing values and outliers which lead to a loss of a significant amount of information. In this paper, we suggest a model to deal with missing values of our system during the diagnosis of COVID-19 pandemic. The system aims to enhance the physical distancing and activities limitations related to the outbreak and then providing to the medical staff the necessary information to make the right decision.

1 Introduction

In the last century, the development of technology and innovation related to the field of the Internet of Things (IoT) has contributed enormously to improve several sectors like buildings, transportation, and health [1].

The treatment of collected information from these devices has a very important role in predicting the future and taking the right decision principally if such data is complete. However, this is not always the case, as this information is plagued by missing values and biased data.

In the healthcare domain, preventing diseases and the early detection of a complication regarding the patient's situation may save lives and avoid the worst, and we can find many datasets in the field such as the electronic health records (EHRs) that contain information about patients (habits, prescriptions of medication, medical history, the doctors' diagnosis, nurses' notes, etc.) [2]. But, one of the common problems which may affect the validity of the clinical results and decrease the precision of medical research remains the missing of data [3]. In fact, healthcare data analytics depend mainly on the availability of information and many others factors such as

H. Bihri (✉) · S. Hsaini · R. Nejjari · S. Azzouzi · M. E. H. Charaf
Faculty of Sciences, Ibn Tofail University, Kenitra, Morocco

M. Ben Ahmed et al. (eds.), *Networking, Intelligent Systems and Security*,
Smart Innovation, Systems and Technologies 237,
https://doi.org/10.1007/978-981-16-3637-0_61

disconnection of devices, a deterioration of network connection, or a failure in the equipment can lead to bias data or loss of precision [4].

In the current circumstance of the COVID-19 pandemic, and due to the severity of the virus in addition to the risk of transmission and contamination, our idea is to promote the physical distancing and to minimize the contact of the health staff with potential patients who could probably be affected by the virus. Thus, we need to deal with information delivered through the monitoring control, principally if they are not complete or biased, which can affect the final result and compromise the validity of the diagnosis.

To avoid this problem, the missing data must be treated and addressed correctly by the application of the appropriate method that suits the most with the issue of the study, and then facilitate extraction of useful information from the data collected in order to assist the decision making.

Furthermore, the use of the IOT technology was through sensors such as telemedicine tools in order to collect data and to offer then a monitoring control using smartphones, webcam-enabled computers, or robots of temperature measurement [5, 6].

In this article, we review the different techniques proposed in the literature to deal with the missing data treatment in the healthcare domain. Then, we describe our proposed system to protect both patients and the medical staff from the contamination by the virus in the context of COVID-19 pandemic, and how to manage missing data, resulting from data collection through different remote diagnosis tools. The study is in line with the idea detailed in [7]. The aim is to improve our previous works to monitor and control the spread of COVID-19 disease.

The paper is organized as follows: Sect. 2 gives some basic concepts related to missing data and prediction functions. Then, we introduce, respectively, in the Sects. 3 and 4 the problematic statement and some related works. Afterward, we present our contribution to manage missing data issues and the workflow of the proposed approach in the context of COVID-19 in Sect. 5. Finally, we give some conclusions and future works.

2 Preliminaries

Generally, we distinguish structured data which is an organized collection structured in rows and columns for storage and unstructured data which is not organized in predefined manner or model, and the processing of such data requires a specialized programs and methods to extract and analyze useful information [8, 9].

Furthermore, we can collect data using manually form for entering data, or by means of measuring features from monitoring systems.

2.1 *Missing Data and Data Collection Mechanisms*

The data collected using monitoring systems are usually missed or biased. Indeed, the type of missing data can indicate the appropriate approach to deal with the issue. These are three categories [10–12]:

- Missing completely at random (MCAR): can result independently from both the observed variables and the unobserved variables, and occur entirely in a random manner. For example, in the case of temperature measurement using an electronic device. The data can be missed if the device is running out of battery.
- Missing at random (MAR): occurs when the missing data is not unplanned. However, the missed data can be predictable from variables with complete information. For example, the measurement of the temperature failed with children due to the lack of cooperation of young people.
- Not missing at random (NMAR): In this case, the probability of the variable that's missing is directly related to the reason that is looking for or requested by the study. For example, patients with fever resist to the temperature measurement for fear of being diagnosed positive.

2.2 *Prediction Function*

Prediction function or imputation technique is a method used to predict incomplete information in a dataset. Therefore, several approaches exist to deal with missing values issue and can be divided into two main categories [13]:

- Deletion-based techniques which ignore incomplete data, we can distinguish:
  - List wise deletion that consists of deletion of the incomplete rows completely which can lead to a loss of a quantity of information;
  - Pairwise deletion: This technique maintains the maximum of information and requires the type MCAR of values that are missing.
- Recovering-based Techniques that aims to recover missing values by estimating them, through application of a variety of techniques such as:
  - Single imputation technique: In this technique, the missing entry is replaced by a calculated value using an appropriate formula or equation based on variable's observed;
  - Mean imputation: is one method of the single imputation techniques that consists of the replacement of the missing value by the mean of the values observed for this variable;
  - Last observation carried forward (LOCF): A kind of deterministic single imputation technique that aims to replace the missing value by the last observed one;

- Non-response weighting approach is developed to address unit non-response by creation and application of columns of weights to responses items;
- Multiple Imputations consists of three phases: the imputation phase, the analysis phase, and the pooling phase. The technique is based on the creation of m replacement for the missing value using m arbitrary constant;
- Full Information Maximum Likelihood (FIML) is based on the ignorance of non-response items and considers only the observed data. The results provided by FIML are optimal in a MAR context.

3 Problematic Statement

Over time, the management of health crises and problems has proven to be very difficult for both health workers and Nations, especially when it comes to contagious diseases or pandemics.

In December 2019, a new kind of corona virus (COVID-19) has been discovered in Wuhan. The disease is then spreading quickly to healthy persons having close contact with infected ones [14]. The virus causes severe respiratory problems which increases significantly intensive care unit admission and leads to a high mortality rate [15].

Therefore, in order to reduce contact between people, which lead to the rapid propagation of the virus a set of measures have been taken by the government, in order to break the chains of transmission of the infection, such as physical distancing measures, school closures, travel restrictions, and activities limitations [16]. Even more, the medical staff needs to adopt telemedicine as a safe strategy to restrict contact with infected patients. Indeed, technology can help to exchange information 24/24 and 7/7 using smartphones or IOT components providing thus a real situation of the patient and allow the health staff to control remotely infected persons [17]. However, the use of sensors and telemedicine tools to collect data can be faced to missing data issues.

In this paper, we suggest our approach to manage outliers and missing data in order to help medical staff to make the appropriate healthcare policy decisions when the knowledge is not available. The model as designed could be used to diagnosis remotely the COVID-19 patients.

4 Related Works

To deal with missing data especially in healthcare domain, many methods of treatment are available. They are different but they can help to predict missing items in a specific situation.

In this context, in [13] the authors tackled the issue of principle missing data treatments. According to their study, they conclude that the dominated approach used to address the missing data problem was deletion-based technique (36.7%).

The work [18] proposes an example of missing data in the work-family conflict (WFC) context. The author purposes four approaches to deal with missing data (he CWFC scores in this article), which are the following methods: Multiple Imputations analysis, linear regression models, and logistic regression.

Moreover, the authors in [19] conducted a study on 20 pregnant responding to a set of criteria, and they are in the end of their first trimester. The physical activity and heart data of the samples were collected and transmitted through a gateway device (Smartphone, PC) to make health decisions. Most of data analysis was performed to extract useful information regarding to maternal heart rate. In this context, the authors suggest an approach, based on imputation, to handle missing data issues that occur when the sensor is incapable to provide data.

Another work [20] tackles two predictive approaches: The single imputation technique as a suitable choice when the missing value is not informative. The second one is the multiple imputations that it is useful for a complete observation.

In [21], the authors highlight the benefit of the data and the electronic health records available in healthcare centers which bring important opportunities for advancing patient care and population health. The basic idea in [22] is to handle the problem of missing data often occurred in the case of a sensor failure or the network device problems, for example. To this end, the authors propose a new approach entitled: a Dynamic Adaptive Network-Based Fuzzy Inference System (D-ANFIS). In this study, the collected data was divided in two groups: complete data that be used to train the proposed method and incomplete data to fill the missing values.

Furthermore, the authors in [23] describe the principal methods to handle missing values in multivariate Data Analysis which can manage and treat missing data according to the nature of information, especially categorical continuous, mixed or structured variables. In this context, they introduce these methods: principal component analysis (PCA), multiple correspondence analysis (MCA), factorial analysis for mixed data (FAMD), and multiple factor analysis (MFA).

According to [24], the authors are mainly focused on the prediction of cerebral infarction risk since this is a fatal disease in the region of the study. To deal with the issue, the authors propose a new convolutional neural network-based multimodal disease risk prediction (CNN-MDRP) for both structured and unstructured data collected for their study. Another work [25] proposes a deep study to understand the origin of a chronic and complex disease called the inflammatory bowel disease (IBD). For this purpose, the authors introduce a new imputation method centered on latent-based analysis combined with patients clustering in order to face the issue of data missingness. The method as described allows to improve the design of treatment strategies and also to develop the predictive models of prognosis.

On the other hand, the authors in [26] emphasize the importance of handling missing data encountered in the field of clinical research. To deal appropriately with the issue, they purpose to follow four mean steps particularly: (1) trying to reduce the rate of missing data in the data collection stage; (2) performing a data diagnostic to understand the mechanism of missing value; (3) handling missing data by application of the appropriate method and finally (4) proceeding to the analyze of sensitivity when required.

Moreover, the authors in [27] highlight the importance of data quality when establishing statistics. In this context, the authors suggest to deal with missing values using cumulative linear regression as kind of imputation algorithm. The idea is to cumulate the imputed variables in order to estimate the missing values in the next incomplete variable. By applying the method on five datasets, the results obtained revealed that the performances differ according to the size of data, missing proportion, and the type of mechanisms used for this purpose.

In the next section, we describe our prototype to remotely monitor the patients' diagnosis then we explain how to tackle missing values issue.

5 COVID-19 Diagnosis Monitoring

In what follows, we tackle the missing data issue in the e-health domain and particularly in the COVID-19 context.

5.1 Architecture

Figure 1 describes the architecture of our diagnosis monitoring system.

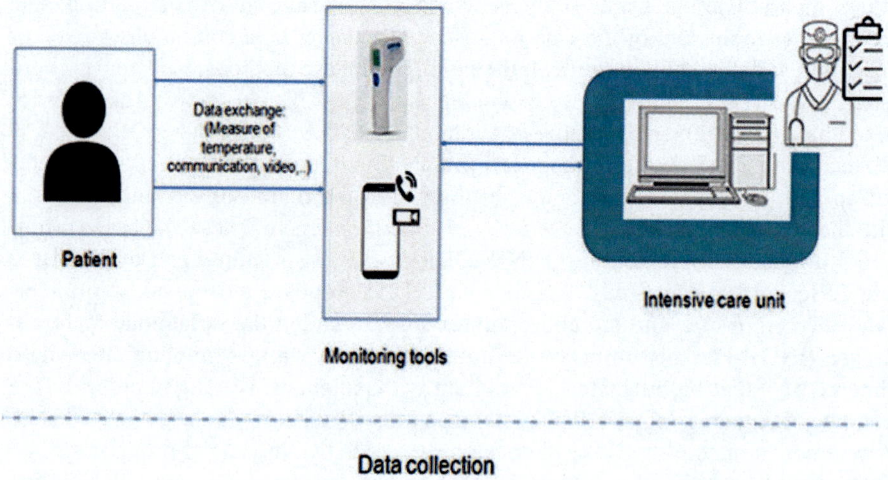

Fig. 1 Diagnosis monitoring architecture

In order to achieve the physical distancing recommended by the World Health Organization (WHO) in the context of COVID-19, we propose a diagnosis monitoring architecture that can be used to diagnose patients remotely without having to move to the medical unit and have direct contact with the doctor or other person of the medical staff.

In fact, the system proposes to diagnose cases probably infected with the virus through the use of several tools able to transmit information about the patients' condition to the medical unit.

Therefore, data is exchanged between the patient and intensive care unit using some remote monitoring tools. However, the data provided could become unavailable or incomplete due to failure in the data collection step or during the data transmission. This can occur when the connection to the medical unit server is broken down or interrupted. Otherwise, if the device doesn't work correctly or is disconnected from the network.

A scenario that we propose for the remote diagnosis of COVID-19 is the one used in Morocco. In fact, the patient who doubts or has symptoms such as fever, respiratory symptoms, or digestive symptom, such as lack of appetite, diarrhea, vomiting, and abdominal pain, can use this system to contact the intensive care unit. Afterward, a variety of accurate questions based on a checklist of common symptoms are communicated to the patient through our system to determine and identify whether the patient needs to seek immediate medical attention or not. Indeed, the application of such measure can contribute effectively to deal with the widespread of the disease by decreasing the high transmission rate, and then limit the spread of the virus. We suggest in the next subsection our prototype to face such situation and to handle missing values using the mean imputation approach.

5.2 *Missing Data Handling: Prototype*

Figure 2 describes the processes of making a decision using our diagnosis monitoring system. In fact, the proposed system aims to automate collection of data related to people likely be infected by COVID-19. Afterward, the system ensures the transfer of information to the intensive care unit. Thus, it will guarantee also the individuals distancing and protecting the medical staff from any probable contamination.

Subsequently, data is treated and stored in the appropriate server in order to be analyzed and to extract useful information necessary for making suitable decision.

Therefore, the system is subdivided into four main steps:

- Data collection phase;
- Data prediction phase;
- Data processing phase;
- Decision-making phase.

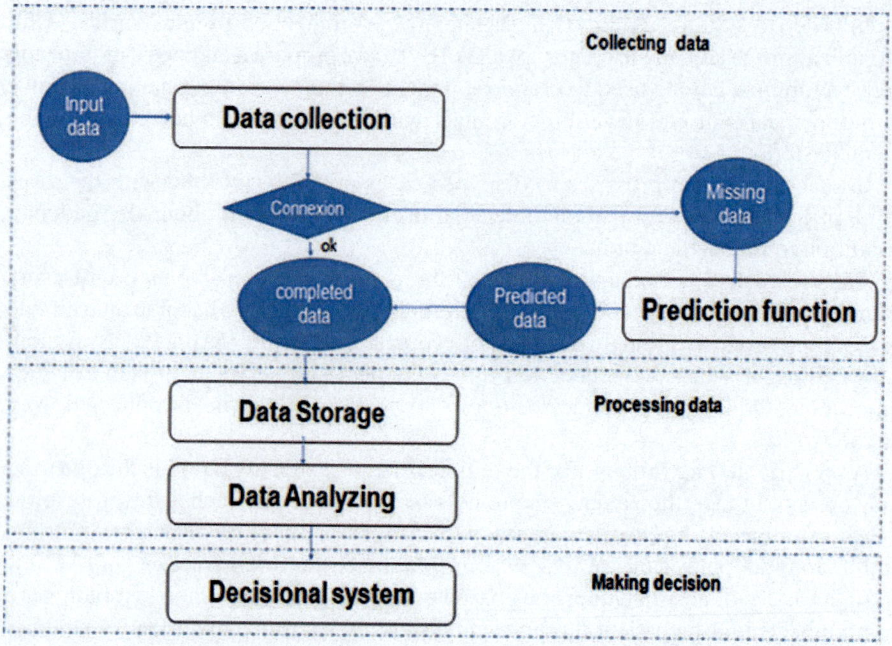

Fig. 2 Missing data handling prototype

5.2.1 Data Collection

This step aims to collect data using various tools for information exchanging such as smartphones, webcam-enabled computers, smart thermometers, etc. These data sources must provide a real situation of the patient and collect information that will be transmitted to the medical unit for treatment. We distinguish two types of data:

- ***Complete information*** refers to data measured and collected correctly. The observed entries are then automatically redirected for storage in a local server.
- ***Incomplete data*** refers to data collected through the monitoring tools. In such situations, data is usually biased or loosed due to the devices, or if the network connection is broken down for example.

5.2.2 Data Prediction

As described above, the input data is sent to the intensive unit care server to be treated. It refers to the information collected using monitoring tools (medicine sensors and mobile device). The data provided is then sorted, and two groups are distinguished: missing data and complete data.

For complete data, it will be redirected automatically to the next phase: Data processing. However, missing data need to be treated before being processed. In a single imputation method, missing data is filled in by some means and the resulting completed data set is used for inference [28]. In our case, we suggest replacing missing values using the mean of the available cases using the mean imputation approach as a kind of single imputation method. Even if the variability in the data is reduced, which can lead to underestimate standard deviations as well as variance estimates, the method is still easy to use and gives sufficiently interesting results in our case study.

5.2.3 Data Processing

The estimated data as well as complete data are both redirected to the storage handling phase. The system proceeds at this stage to the following operations:

- Data storage: In this case, we need to use the appropriate hardware/software tools to ensure data storage for both complete and predicted data. In fact, we need to take into account the types of data collected (recordings, video file...) as well as the way such data increase to better size the required capacity for the storage needs.
- Data analysis: It aims to analyze the data collected using the appropriate data analysis tools. This will help to improve the decision making in the next steps.

5.2.4 Making Decision

The objective is to give a screening of the patient state to the health staff and clinician using reporting and dashboard. The goal is to help them taking the right decision and to improve their prevention's strategy.

To meet the specific needs of physicians, the key performance indicators (KPIs) and useful information to be displayed must be carefully defined in collaboration with the medical professionals. Therefore, the implementation of this real-time monitoring system reduces the time and effort required to search and interpret information which will improve significantly the medical decision making.

5.3 Discussion

Missing data can present a significant risk of drawing erroneous conclusions from clinical studies. It is common practice to impute missing values, but this can only provide an approximate result to the actual expected outcome.

Many recent research works are devoted to handle missing data in the healthcare domain and to avoid the problems described previously. The review of some articles tackling different approaches used in the field reveals a variety of techniques to deal with this issue such as deletion-based techniques and recovering-based techniques.

Therefore, the use of an inappropriate method according to the missing item can bias results of the study. Hence, the identification of the suitable method depends mainly on whether the data is missing completely at random (MCAR), missing at random (MAR), or not missing at random (NMAR) as explained previously.

In the prototype proposed in this paper and according to the reasons leading to incomplete data, we consider MCAR as the appropriate type of missing variable. In addition, we opt for the use of mean imputation in order to predict missing values from the observed one. Whereas we exclude the use of techniques based on deletion due to the loss of data that occurs when such type of techniques is applied.

Moreover, even if mean, median or mode imputations are simples and easy to implement, we are aware that such techniques could underestimate variance and ignore the relationship with other variables. Therefore, further investigations need to be done and deeply analyzed to understand such relations and to enhance our model in order to obtain reliable results.

Furthermore, even if data security remains outside the scope of this paper, it is still highly recommended to take into consideration such aspects during the design of the application. In fact, the importance of data security in healthcare is becoming increasingly critical and nowadays, it is imperative for healthcare organizations to understand the risks that they could encounter in order to ensure protection against online threats.

On the other hand, the first experiments show that our proposition lack performance and could be limited if the data collected is not sufficient to execute the logic of the imputation function. In this context, many studies confirm, using the simulation of missing values, that a significant drop in performance is observed even in cases where only a third of records were missing or incomplete. Thus, it is strongly recommended to use large datasets to deal with this problem and promote prediction of missing items.

Finally, the completeness of the data should be assessed using a monitoring system that provides reports to the entire study team on a regular basis. These reports can be used then to improve the conduct of the study.

6 Conclusion

According to the current situation in the world related to the COVID-19 pandemic, we aim in this study to support the physical distancing by minimizing the contact of the health staff. To this end, we describe our monitoring prototype to deal with such situation and to perform patients' diagnosis remotely in order to reduce contamination risks.

However, healthcare analytics depends mainly on the availability of data and presence of missing values during data collection stage which can lead to bias or loss of precision. In this context, we suggest in this paper to use a prediction technique to avoid loss of data and by the way to predict missing information in order to take the right and valid decision. The method proposed for prediction is the mean imputation technique used in the collection stage to fill our dataset by estimated values.

As prospects, we plan to conduct more experimental studies regarding the performance of our prototype in other medical cases such as mammographic mass and hepatitis datasets. We will also enhance our model in the future to take other imputation methods, especially the multiple imputation method.

References

1. Balakrishnan, S.M., Sangaiah, A.K.: Aspect oriented modeling of missing data imputation for internet of things (IoT) based healthcare infrastructure. Elsevier (2018)
2. Wells, B.J., et al.: Strategies for handling missing data in electronic health record derived data. eGEMs (Generating Evidence & Methods to Improve Patient Outcomes), **1**(3), 7 (2013). https://doi.org/10.13063/2327-9214.1035
3. Donders, A.R.T., et al.: Review: a gentle introduction to imputation of missing values. J. Clin. Epidemiol. **59**(10), 1087–1091 (2006). https://doi.org/10.1016/j.jclinepi.2006.01.014
4. Ebada, A., Shehab, A., El-henawy, I.: Healthcare analysis in smart big data analytics: reviews. Challenges Recommendations (2019). https://doi.org/10.1007/978-3-030-01560-2_2
5. Yang, G.Z., et al.: Combating COVID-19-the role of robotics in managing public health and infectious diseases. Sci. Robot. **5**(40), 1–3 (2020). https://doi.org/10.1126/scirobotics.abb5589
6. Engla, N.E.W., Journal, N.D.: New England J. 1489–1491 (2010)
7. Hsaini, S., Bihri, H., Azzouzi, S., El Hassan Charaf, M.: Contact-tracing approaches to fight COVID-19 pandemic: limits and ethical challenges. In: 2020 IEEE 2nd International Conference on Electronics, Control, Optimization and Computer Science. ICECOCS 2020, (2020)
8. Wang, Y., et al.: United States Patent 2(12) (2016). Interfacial nanofibril composite for selective alkane vapor detection. Patent No.: US 10,151,720 B2. Date of Patent: 11 Dec 2018
9. Park, M.: United States Patent 1(12) (2010). Method and apparatus for adjusting color of image. Patent No.: US 7,852,533 B2. Date of Patent: 14 Dec 2010
10. Salgado, C.M., Azevedo, C., Proença, H., Vieira, S.M.: Missing data. In: Secondary Analysis of Electronic Health Records. Springer, Cham (2016)
11. Haldorai, A., Ramu, A., Mohanram, S., Onn, C.C.: EAI International Conference on Big Data Innovation for Sustainable Cognitive Computing (2018)
12. Sterne, J.A.C., et al.: Multiple imputation for missing data in epidemiological and clinical research: potential and pitfalls. BMC **339**(July), 157–160 (2009). https://doi.org/10.1136/bmj.b2393
13. Lang, K.M., Little, T.D.: Principled missing data treatments. Prev. Sci. **19**, 284–294 (2018). https://doi.org/10.1007/s11121-016-0644-5
14. Heymann, D., Shindo, N.: COVID-19: what is next for public health? Lancet 395 (2020). https://doi.org/10.1016/S0140-6736(20)30374-3
15. Huang, C., Wang, Y., Li, X., et al.: Clinical features of patients infected with 2019 novel coronavirus in Wuhan, China. Lancet **395**(10223), 497–506 (2020)
16. Huang, C., Wang, Y., Li, X., et al.: Clinical features of patients infected with 2019 novel coronavirus in Wuhan, China. Lancet **395**, (2020). https://doi.org/10.1016/S0140-6736(20)30183-5
17. Kiesha, P., Yang, L., Timothy, R., et al.: The effect of control strategies to reduce social mixing on outcomes of the COVID-19 epidemic in Wuhan, China: a modelling study. Lancet Public Health **5**, (2020). https://doi.org/10.1016/S2468-2667(20)30073-6
18. Nguyen, C.D., Strazdins, L., Nicholson, J.M., Cooklin, A.R.: Impact of missing data strategies in studies of parental employment and health: missing items, missing waves, and missing mothers. Soc. Sci. Med. **209**, 160–168 (2018)

19. Azimi, I., Pahikkala, T., Rahmani, A.M., et al.: Missing data resilient decision-making for healthcare IoT through personalization: a case study on maternal health. Future Gener. Comput. Syst. **96**, 297–308
20. Josse, J., Prost, N., Scornet, E., Varoquaux, G.: On the consistency of supervised learning with missing values. arXiv preprint arXiv:1902.06931 (2019)
21. Stiglic, G., Kocbek, P., Fijacko, N., Sheikh, A., Pajnkihar, M.: Challenges associated with missing data in electronic health records: a case study of a risk prediction model for diabetes using data from Slovenian primary care. Health Inform. J. **25**(3), 951–959 (2019)
22. Turabieh, H., Mafarja, M., Mirjalili, S.: Dynamic adaptive network-based fuzzy inference system (D-ANFIS) for the imputation of missing data for internet of medical things applications. IEEE Internet Things J. **6**(6), 9316–9325 (2019)
23. Josse, J., Husson, F.: missMDA: a package for handling missing values in multivariate data analysis. J. Stat. Softw. **70**(i01), (2016)
24. Chen, M., Hao, Y., Hwang, K., Wang, L., Wang, L.: Disease prediction by machine learning over big data from healthcare communities. IEEE Access **5**, 8869–8879 (2017). https://doi.org/10.1109/ACCESS.2017.2694446
25. Abedi, V., et al.: Latent-based imputation of laboratory measures from electronic health records: case for complex diseases. bioRxiv, pp. 1–13 (2018). https://doi.org/10.1101/275743
26. Papageorgiou, G., et al.: Statistical primer: how to deal with missing data in scientific research? Interact. Cardiovasc. Thorac. Surg. **27**(2), 153–158 (2018). https://doi.org/10.1093/icvts/ivy102
27. Mostafa, S.M.: Imputing missing values using cumulative linear regression. CAAI Trans. Intell. Technol. **4**(3), 182–200 (2019). https://doi.org/10.1049/trit.2019.0032
28. Jamshidian, M., Mata, M.: Advances in analysis of mean and covariance structure when data are incomplete. In: Handbook of Latent Variable and Elated Models, pp. 21–44 (2007). https://doi.org/10.1016/B978-044452044-9/50005-7

An Analysis of the Content in Social Networks During COVID-19 Pandemic

Mironela Pirnau

Abstract During the COVID-19 pandemic, Internet and SN technologies are an effective resource for disease surveillance and a good way to communicate to prevent disease outbreaks. In December 2019, the frequency of the words COVID-19, SARS-CoV-2, and pandemic was very low in online environment, being only few posts informing that, "the mysterious coronavirus in China could spread." After March 1, 2020, there have been numerous research projects that analyze the flows of messages in social networks in order to perform real-time analyses, to follow the trends of the pandemic evolution, to identify new disease outbreaks, and to elaborate better predictions. In this context, this study analyzes the posts collected during [August–September 2020], on the Twitter network, that contain the word "COVID-19," written both in Romanian and English. For the Romanian language posts, we obtained a dictionary of the words used, for which it was calculated their occurrence frequency in the multitude of tweets collected and pre-processed. The frequency of words for non-noisy messages was identified from the multitude of words in the obtained dictionary. For the equivalent of these words in English, we obtained the probability density of words in the extracted and pre-processed posts written in English on Twitter. This study also identifies the percentage of similarity between tweets that contain words with a high frequency of apparition. The similarity for the collected and pre-processed tweets that have "ro." in the filed called Language has been computed making use of Levenshtein algorithm. These calculations are intended to quickly help find the relevant posts related to the situation generated by the COVID-19 pandemic. It is well known that the costs of analyzing data from social networks are very low compared to the costs involved in analyzing data from the centers of government agencies; therefore, the proposed method may be useful.

M. Pirnau (✉)
Faculty of Informatics, Titu Maiorescu University, Bucharest, Romania
e-mail: mironela.pirnau@prof.utm.ro

M. Ben Ahmed et al. (eds.), *Networking, Intelligent Systems and Security*,
Smart Innovation, Systems and Technologies 237,
https://doi.org/10.1007/978-981-16-3637-0_62

1 Introduction

Social networks are widely used by people not only to share news, states of minds, thoughts, photos, etc. but also to manage disasters by informing, helping, saving, and monitoring health. The results obtained by processing the huge quantities of data that circulate in social networks allow identifying and solving several major problems that people have. In the present, Big data systems have come to enable multiple processing of data [1]. Thus, by means of the obtained information, Big data systems contribute to the management of risk in climate changes [2], to the analysis of the traffic in big locations [3], to the personalization of health care [4], to early cancer detection [5], to solving some social problems by building houses using 3D printer technology [6], to social media analysis [7], and to developing methods that contribute to identify the feelings and emotions of people [8]. Social networks have contributed to the creation and continuous development of massive amounts of data at a relatively low cost. There is a lot of research that analyzes large amounts of data and highlights both the role of computing tools and the manner of storing them [9, 10]. Big data in media communications and Internet technology have led to the development of alternative solutions to help people in case of natural disasters [11]. There are research studies that monitor different types of disasters based on the data extracted from social platforms that have highlighted real solutions [12–17]. By extension, these studies can also be used for the situation generated by the pandemic, which the whole of humanity is facing. The relevance of the words [17–19] can greatly help the analysis of data processed in the online environment, in a disaster situation. This study analyzes data collected in August–September 2020 from the Twitter network [20], to find the most common words relevant to the situation of COVID-19. In this respect, the study consists of (1) Introduction, (2) Related Works, (3) presentation of the data set used, (4) the results obtained, and (5) Discussions and Conclusions. Because, throughout this period, it has been a proven fact that the Public Health District Authority cannot handle the communication with the people who need its help, an automatic system for processing and correct interpretation of email messages received during the COVID-19 crisis would contribute to the decrease in the response and intervention time. In this sense, this study tries to prove that the identification of relevant words in online communication can contribute to the efficient development of a demand–response system.

2 Related Works

People have become more and more concerned about the exponential evolution of illness cases, about the deaths caused by them, as well as about the severe repercussions of the COVID-19 pandemic's evolution on the daily life. There are numerous studies that analyze the magnitude of disasters and their consequences [12–14, 21–24]. At the same time, there are studies that show an exponential pattern of increase in the case of the number of messages during the first interval after a sudden catastrophe [24], but there are also studies that analyze the behavior of news consumers on social networks [25–27]. Some research [28], based on analyzing the data taken from social networks, characterizes users in terms of the use of controversial terms during COVID-19 crisis. Knowing the most frequently used words in posts referring to a certain category of events enables the determination of the logical conditions for searching the events corresponding to emergencies [29–31]. Analyzing the data collected from social networks in order to identify emergencies, it is essential to establish the vocabulary used for a regional search, as shown in the studies [13, 29, 31, 32]. According to the studies [30, 33], the analysis of information with high frequency of occurrence from the social networks contributes to the rapid decrease in the effects of disasters. Online platforms allow the understanding of social discussions, of the manner in which people cope with this unprecedented global crisis. Informative studies on posts from social networks should be used by state institutions for developing intelligent online communication systems.

3 Data Set

Taking into account the rapid evolution of the COVID-19 pandemic, as well as the problems that have arisen regarding the people's concern about the significant increase in cases of illness and death caused by the infection of the population with Sarv-Cov_2, we have extracted tweets from the Twitter network during [August–September 2020] using the topic COVID-19 as a selective filter. The data have been extracted using the software developed in the research [20] and then cleaned to be consistent in the processing process. An important task of fundamental natural language processing (NLP) is lemmatization. Natural language processing (NLP) has the role of contributing to the recognition of speech and natural language. This is the most common text pre-processing technique used in natural language processing (NLP) and machine learning. Because lemmatization involves deriving the meaning of a word from a dictionary, it is time-consuming, but it is the simplest method. There are lemmatizers that are based on the use of a vocabulary and a morphological analysis of words. These work well for simple flexible forms, but for large compound words, it is necessary to use a rule-based system for machine learning from an annotated corpus [34–38]. Statistical processing of natural language is largely based on machine learning. From the collected messages, only the ones that have the language field completed with "en" and "ro" were used. The messages were cleaned and prepared for processing [39–41]. Many messages are created and posted by robots [42–44], automated accounts that enhance certain discussion topics, in contrast to the posts that focus on the public health problems, so that the human operators should find it difficult to manage such a situation. In the previous context, the noisy tweets were removed [45], but the hardest task was to rewrite the tweets that had been written using diacritics (ş, ţ, ă, î, â). After all this processes, 811 unique tweets written in Romanian and 43,727 unique tweets written in English were obtained for processing.

The pre-processing procedure and the actual processing procedure of these tweets were performed using the PHP 7 programming environment. MariaDB was used for the data management.

4 Results

Using regular expression, a database including all the words used for writing the 811 posts in Romanian was generated. Using my own application written in PHP for content parsing, only the words in Romanian were extracted. Thus, only 1103 words were saved. In the collected tweets, it was noticed the fact that, even if the language used for writing the tweets was “ro,” the users also operate with words originated from other languages when writing messages and words. The following were not taken into account: names of places, people’s names and surnames, names of institutions, prepositions, and articles (definite and indefinite). In Table 1, the average number of words used in one post was determined. The average value of 14 words, significant for the posts related to COVID-19 pandemic, indicates the fact that the average number of words used is enough to convey a real state or situation.

In the dictionary of words, only the ones with the occurrence frequency of at least 5 times in the analyzed posts were selected. Vector V_{ro} contains the words (written in Romanian language) that occur at least 5 times in posts, and vector F_{ro} contains the number of occurrences corresponding to vector V_{ro}.

$V_{\text{ro}}\{\} = \{$activ; afectat; ajutor; analize; antibiotic; anticorp; anti-covid; aparat; apel; azil; bilant; boala; bolnav; cadre; cauza; cazuri; central; centru; confirmat; contact; contra; control; convalescent; coronavirus; decedat; deces; depistat; diagnostic; disparut; donatori; doneaza; echipamente; epidemia; fals; pozitive; focar; forma; grav; gripa; imbolnavit; imun; infectat; informeaza; ingrijiri; inregistrat; intelege; laborator; localitate; lume; masca; medic; medicament; merg; moarte; mondial; mor; negativ; oameni; pacient; pandemia; plasma; pneumonia; negativ; post-covid; post-pandemic; pre-covid; pulmonary; raportat; rapus; reconfirmat; restrictii; rezultat; risc; scolile; sever; sicriu; situatia; spital; test; tragedie; transfer; tratament; tratarea; tratez; urgenta; vaccin; virus;. The English translation for the terms written above is as follows: active; affected; help; analysis; antibiotic; antibody; anti-covid; camera; call; asylum; balance sheet; disease; sick; frames; cause;

Table 1 Determination of an average number of words from the analyzed posts written in Romanian

Analyzed elements	Found values	Number of used characters	Average number of characters
Tweets	811	78,261	96.49
Words	1103	7378	6.68
		Average number of tweets words =	14.42

cases; central; center; confirmed; contact; against; control; convalescent; coronavirus; deceased; death; diagnosed; diagnosis; missing; donors; donate; equipment; epidemic; fake; false positives; outbreak; form; serious; flu; ill; immune; infected; inform; care; registered; understand; laboratory; locality; world; mask; doctor; drug; go; death; world; die; negative; people; patient; pandemic; plasma; pneumonia; negative; post-covid; post-pandemic; positive; pre-covid; pulmonary; reported; killed; reconfirmed; restrictions; result; risk; schools; severe; coffin; situation; hospital; test; tragedy; transfer; treatment; treating; treats; emergency; vaccine; virus}.

$F_{ro}\{\} = \{17; 6; 7; 8; 5; 5; 16; 9; 6; 5; 14; 18; 10; 8; 27; 87; 14; 5; 25; 5; 16; 6; 9; 55; 8; 32; 7; 7; 5; 5; 5; 5; 10; 7; 5; 6; 20; 8; 8; 13; 5; 54; 8; 5; 12; 5; 7; 45; 5; 15; 14; 7; 6; 5; 8; 34; 8; 10; 8; 15; 9; 6; 25; 7; 6; 20; 6; 5; 14; 6; 7; 5; 5; 7; 13; 6; 6; 23; 21; 62; 6; 5; 8; 5; 6; 5; 47; 69\}$.

For the number of words in vector V_{ro}, the number of occurrences is represented by the sum of elements of vector F_{ro}, meaning the 88 words that are used 1220 times in the 811 unique selected and cleaned tweets. The occurrence probability of the words in vector V_{ro} within the 811 posts may be calculated according to the following Eq. (1).

$$P = \frac{\text{No posts}}{\sum_1^n \text{Fro}} \tag{1}$$

where No posts is 811, n represents the number of words in vector V_{ro} with the value of 88, and F contains the number of occurrences of these words. Thus, $P =$ 66.48%. This value indicates the fact that vector V_{ro} of the obtained words provides an occurrence possibility of more than 20%, which demonstrates that these words are relevant for the tweets analyzed in the context of COVID-19 pandemic.

If we represent vectors V_{ro} and F_{ro} graphically, Fig. 1 is obtained, namely the Pareto graph. It indicates that the words with an occurrence frequency more than 50% are Vrelevant {cazuri; virus; test; coronavirus; infectat; vaccin; localitate; mor; deces; cauza; confirmat; positiv/pozitiv; situatia; spital; forma; negatic; boala;}/corresponding to the English words {cases; virus; test; coronavirus; infected; vaccine; locality; die; death; cause; confirmed; positive; situation; hospital; form; negative; disease;}.

For these words, Table 2 indicates the distribution of occurrence frequencies. The Pareto graph can be seen in Fig. 2.

Figure 2 highlights that both the groups of words {cazuri; virus; test; coronavirus; infectat;} and the group of words {vaccin; localitate; mor; deces; cauza; confirmat; pozitiv; situatia; spital; forma; negativ; boala} have the occurrence frequency of 50% in the analyzed posts.

Similarly, for the unique collected tweets written in English, vector V_{en} was determined. It contains the same number of words as vector V_{ro}, but in English, and their occurrence frequency was determined.

In Table 3, only the words written in English with the occurrence frequency of over 2% in tweets were kept.

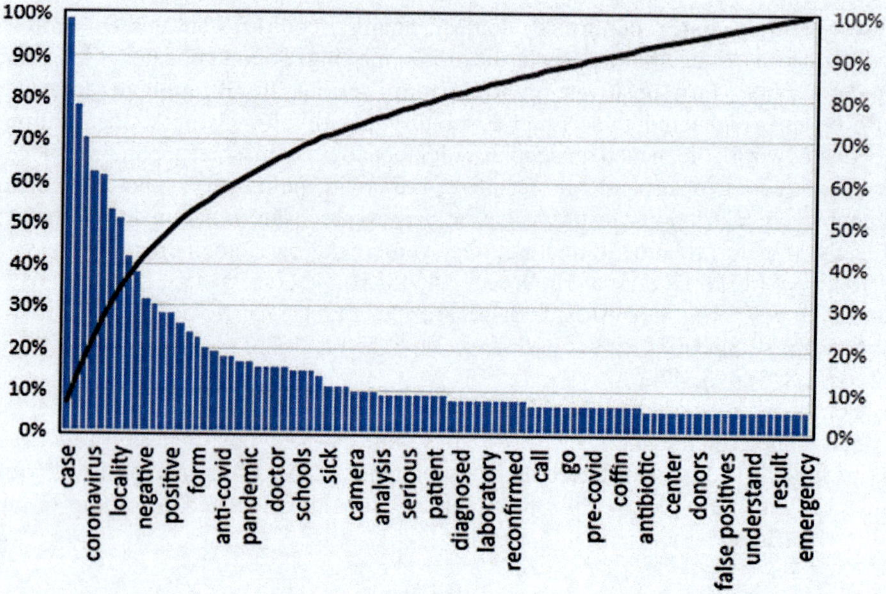

Fig. 1 Distribution of V_{ro} words in the analyzed posts

Table 2 Occurrence frequency in tweets for $V_{relevant}$

No current	Words in "ro"	Number of occurrences	Frequency of occurrence in the dictionary (%)	Frequency of occurrence in tweets (%)	Frequency of occurrence in the key word set (%)
1	Cazuri	87	7.89	10.73	98.86
2	Virus	69	6.26	8.51	78.41
3	Test	62	5.62	7.64	70.45
4	Coronavirus	55	4.99	6.78	62.50
5	Infectat	54	4.90	6.66	61.36
6	Vaccin	47	4.26	5.80	53.41
7	Localitate	45	4.08	5.55	51.14
8	Mor	34	3.08	4.19	38.64
9	Deces	32	2.90	3.95	36.36
10	Cauza	27	2.45	3.33	30.68
11	Confirmat	25	2.27	3.08	28.41
12	Pozitiv	25	2.27	3.08	28.41
13	Situatia	23	2.09	2.84	26.14
14	Spital	21	1.90	2.59	23.86
15	Forma	20	1.81	2.47	22.73
16	Negativ	20	1.81	2.47	22.73
17	Boala	18	1.63	2.22	20.45

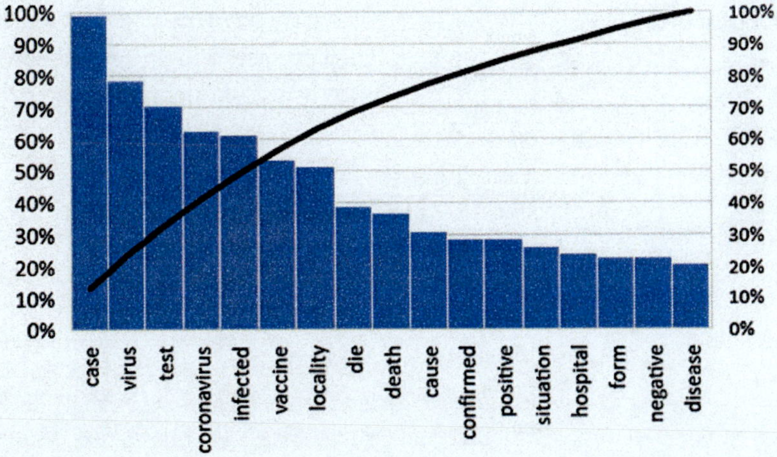

Fig. 2 Distribution of V_{relevant} words in the analyzed posts

Table 3 Words in V_{EN} with the frequency of occurrence in tweets > 2%

Words in English	Number of occurrences	The frequency of occurrence from 43,727 tweets (%)
Test	4171	9.54
Cases	3122	7.14
People	2364	5.41
Death	2010	4.60
Virus	1778	4.07
Help	1627	3.72
Form	1597	3.65
Positive	1533	3.51
Report	1387	3.17
Pandemic	1274	2.91
Situation	1273	2.91
Disease	1175	2.69
Vaccine	1168	2.67
World	1168	2.67
School	1122	2.57
Coronavirus	1011	2.31
Died	944	2.16
Risk	941	2.15
Mask	911	2.08

Table 4 Statistics indicators

Mean	39.06
Standard error	4.95
Median	32.00
Mode	25.00
Standard deviation	20.41
Sample variance	416.68
Kurtosis	0.13
Skewness	0.97
Range	69.00
Minimum	18.00
Maximum	87.00
Sum	664.00
Count	17.00
Confidence level (95.0%)	10.50

For the words in Table 2, the main statistical indicators were determined in Table 4. The value of the Kurtosis peakedness parameter indicates the fact that the curve is flatter than the normal one. The value of skewness parameter shows that the right side of the average is asymmetric.

One can notice that mode, the statistic indicator, is 25, which corresponds to the words of the group formed by "confirmat; pozitiv."

The distribution function of continuous probability was calculated based on the statistical indicators (the average and the standard deviation), and the group of words "vaccine, locality, die, death" has the greatest occurrence density, as seen in Fig. 3.

Table 5 is created by intersecting the set of values in Tables 2 and 3. It indicates the fact that there is a number of words with high occurrence frequency for the posts written both in English and in Romanian.

Because Pearson's r correlation coefficient is a dimensionless index between −1.0 and 1.0 that indicates the extent of the linear relationship between two data sets,

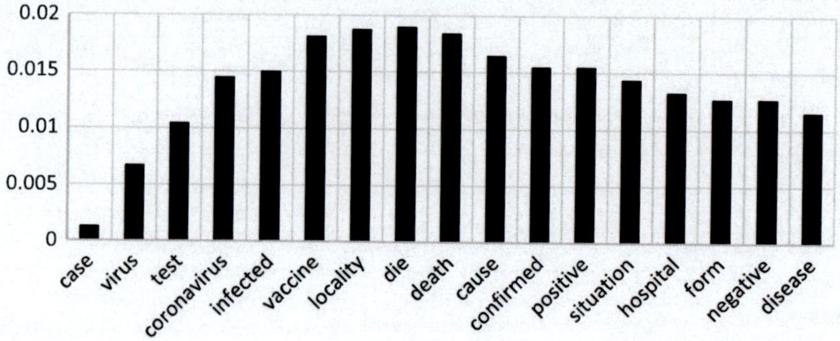

Fig. 3 Probability density for V_{relevant}

Table 5 Words with high occurrence frequency

English word	Romanian word	Frequency of English word (%)	Frequency of Romanian word (%)
Disease	Boala	2.69	2.22
Form	Forma	3.65	2.47
Hospital	Spital	2.08	2.59
Situation	Situatia	2.91	2.84
Positive	Pozitiv	3.51	3.08
Death	Deces	4.60	3.95
Die	Mor	2.16	4.19
Vaccine	Vaccin	2.67	5.80
Coronavirus	Coronavirus	2.31	6.78
Test	Test	9.54	7.64
Virus	Virus	4.07	8.51
Cases	Cazuri	7.14	10.73

we have noticed that the correlation between the Frequency of English word and the Frequency of Romanian word for data in Table 5 is 0.609, which means that there is a moderate to good correlation among the words found. The determined coefficient was calculated by Eq. (2) for Pearson's r correlation coefficient:

$$r = \frac{\sum(x - \bar{x})(y - \bar{y})}{\sqrt{\sum(x - \overline{x})^2 \sum(y - \overline{y})^2}} \quad (2)$$

where x and y are Frequency of English word and Frequency of Romanian word, respectively (in Table 5). For the unique tweets in Romanian, the similarity of contents was calculated for the posts, using the Levenshtein algorithm [46–48]. Levenshtein distance is a string metric that measures the difference between two sequences (Levenshtein, 1966) and represents the minimum number of operations, so that string X can be converted into string Y. Consider strings Xa $= x_1x_2..x_i$ and Yb $= y_1y_2..y_j$, where X and Y are sets of tweets referring to the same subject. If we define $D\,[a, b]$ the minimum number of operations by which Xa can be converted into Yb, then $D\,[i, j]$ is the Levenshtein editing distance looked for. Dynamic programming is the method by which this algorithm is implemented.

Thus, it was noticed that 190 posts have a similarity ranging between 50 and 90%, which represents a percent of 23,42% from the analyzed tweets, as seen in Fig. 4. The tweets with similarity over 90% were not considered because their content varied depending on the punctuation characters. Thus, they were considered informational noise. In 2017, Twitter's decision to double the number of characters from 140 to 280 allows users enough space to express their thoughts in their posts. Basically, identifying the similarity of tweets becomes relevant, because the user no longer has to delete words.

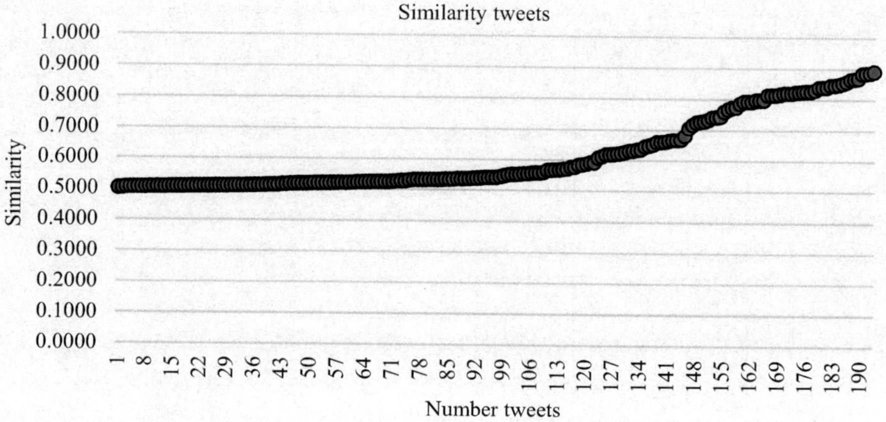

Fig. 4 Variation in content similarity

5 Discussions and Conclusions

Social networks play a vital role in the real-world events, including those incidents which happen in critical periods such as earthquakes, hurricanes, epidemics, and pandemics. It is a well-known fact that the social network messages have both positive and negative effects regarding the media coverage or the excessive publicity of disasters. During disasters, the messages from social networks could be used successfully by authorities for a more efficient management of the actual calamity. The same messages also represent a tool for malicious persons to spread false news. The study of social networks provides informative data that helps identifying the manner used by people to cope with a disaster situation. If, for the COVID-19 period, these data were replaced with real messages received by empowered bodies—governments—an automatic system could be built. This system could significantly contribute to diminishing the waiting time when receiving a reply from these types of institutions. Moreover, the empirical data from Table 3 indicate that a dictionary with common terms, regardless of the language it uses, could be used to implement an efficient call–response system, which would replace the human factor when communicating with the authorities that are overwhelmed by the situation created by the 2020 pandemic. The ICT systems, which use such dictionaries for their "communication" with people, must be highly complex to function efficiently in unexpected disaster conditions.

Acknowledgements I thank Prof. H.N. Teodorescu for the suggestions on this research and for correcting several preliminary versions of this chapter.

References

1. Avci, C., Tekinerdogan, B., Athanasiadis, I.N.: Software architectures for big data: a systematic literature review. Big Data Anal. **5**(1), 1–53 (2020). https://doi.org/10.1186/s41044-020-00045-1
2. Guo, H.D., Zhang, L., Zhu, L.W.: Earth observation big data for climate change research. Adv. Clim. Chang. Res. **6**(2), 108–117 (2015)
3. Zhao, P., Hu, H.: Geographical patterns of traffic congestion in growing megacities: big data analytics from Beijing. Cities **92**, 164–174 (2019)
4. Tan, C., Sun, L., Liu, K.: Big data architecture for pervasive healthcare: a literature review. In: Proceedings of the Twenty-Third European Conference on Information Systems, pp. 26–29. Münster, Germany (2015)
5. Fitzgerald, R.C.: Big data is crucial to the early detection of cancer. Nat. Med. **26**(1), 19–20 (2020)
6. Moustafa, K.: Make good use of big data: a home for everyone, Elsevier public health emergency collection. Cities **107**, (2020)
7. Kramer, A., Guillory, J., Hancock, J.: Experimental evidence of massive scale emotional contagion through social networks. PNAS **111**(24), 8788–8790 (2014)
8. Banerjee, S., Jenamani, M., Pratihar, D.K.: A survey on influence maximization in a social network. Knowl. Inf. Syst. **62**, 3417–3455 (2020)
9. Yue, Y.: Scale adaptation of text sentiment analysis algorithm in big data environment: Twitter as data source. In: Atiquzzaman, M., Yen, N., Xu, Z. (eds.) Big Data Analytics for Cyber-Physical System in Smart City. BDCPS 2019. Advances in Intelligent Systems and Computing, vol. 1117, pp. 629–634. Springer, Singapore (2019)
10. Badaoui, F., Amar, A., Ait Hassou, L., et al.: Dimensionality reduction and class prediction algorithm with application to microarray big data. J. Big Data **4**, 32 (2017)
11. Teodorescu, H.N.L., Pirnau, M.: In: Muhammad Nazrul Islam (ed.) Cap 6: ICT for Early Assessing the Disaster Amplitude, for Relief Planning, and for Resilience Improvement (2020). e-ISBN: 9781785619977
12. Shan, S., Zhao, F.R., Wei, Y., Liu, M.: Disaster management 2.0: a real-time disaster damage assessment model based on mobile social media data—A case study of Weibo (Chinese Twitter). Saf. Sci. **115**, 393–413 (2019)
13. Teodorescu, H.N.L.: Using analytics and social media for monitoring and mitigation of social disasters. Procedia Eng. **107C**, 325–334 (2015)
14. Pirnau, M.: Tool for monitoring web sites for emergency-related posts and post analysis. In: Proceedings of the 8th Speech Technology and Human-Computer Dialogue (SpeD), pp. 1–6. Bucharest, Romania, 14–17 Oct (2015).
15. Wang, B., Zhuang, J.: Crisis information distribution on Twitter: a content analysis of tweets during hurricane sandy. Nat. Hazards **89**(1), 161–181 (2017)
16. Eriksson, M., Olsson, E.K.: Facebook and Twitter in crisis communication: a comparative study of crisis communication professionals and citizens. J. Contingencies Crisis Manage. **24**(4), 198–208 (2016)
17. Laylavi, F., Rajabifard, A., Kalantari, M.: Event relatedness assessment of Twitter messages for emergency response. Inf. Process. Manage. **53**(1), 266–280 (2017)
18. Banujan, K., Banage Kumara, T.G.S., Paik, I.: Twitter and online news analytics for enhancing post-natural disaster management activities. In: Proceedings of the 9th International Conference on Awareness Science and Technology (iCAST), pp. 302–307. Fukuoka (2018)
19. Takahashi, B., Tandoc, E.C., Carmichael, C.: Communicating on Twitter during a disaster: an analysis of tweets during typhoon Haiyan in the Philippines. Comput. Hum. Behav. **50**, 392–398 (2015)
20. Teodorescu, H.N.L., Pirnau, M.: Analysis of requirements for SN monitoring applications in disasters—a case study. In: Proceedings of the 8th International Conference on Electronics, Computers and Artificial Intelligence (ECAI), pp. 1–6. Ploiesti, Romania (2016)

21. Ahmed, W., Bath, P.A., Sbaffi, L., Demartini, G.: Novel insights into views towards H1N1 during the 2009 pandemic: a thematic analysis of Twitter data. Health Inf. Libr. J. **36**, 60–72 (2019)
22. Asadzadeh, A., Kötter, T., Salehi, P., Birkmann, J.: Operationalizing a concept: the systematic review of composite indicator building for measuring community disaster resilience. Int. J. Disaster Risk Reduction **25**, 147–162 (2017)
23. Teodorescu, H.N.L., Saharia, N.: A semantic analyzer for detecting attitudes on SNs. In: Proceedings of the International Conference on Communications (COMM), pp. 47–50. Bucharest, Romania (2016)
24. Teodorescu, H.N.L.: On the responses of social networks' to external events. In: Proceedings of the 7th International Conference on Electronics, Computers and Artificial Intelligence, pp. 13–18. Bucharest, Romania (2015)
25. Gottfried, J., Shearer, E.: News use across social media platforms 2016. White Paper, 26. Pew Research Center (2016)
26. Gupta, A., Lamba, H., Kumaraguru, P., Joshi, A.: Faking sandy: characterizing and identifying fake images on twitter during hurricane sandy. In WWW'13 Proceedings of the 22nd International Conference on World Wide Web, pp. 729–736 (2013)
27. Allcott, H., Gentzkow, M.: Social media and fake news in the 2016 election. J. Econ. Perspect. **31**(2), 211–236 (2017)
28. Lyu, H., Chen, L., Wang, Y., Luo, J.: Sense and sensibility: characterizing social media users regarding the use of controversial terms for COVID-19. IEEE Trans. Big Data (2020)
29. Teodorescu, H.N.L., Bolea, S.C.: On the algorithmic role of synonyms and keywords in analytics for catastrophic events. In: Proceedings of the 8th International Conference on Electronics, Computers and Artificial Intelligence, ECAI, pp. 1–6. Ploiesti, Romania (2016)
30. Teodorescu, H.N.L.: Emergency-related, social network time series: description and analysis. In: Rojas, I., Pomares, H. (eds.) Time Series Analysis and Forecasting. Contributions to Statistics, pp. 205–215. Springer, Cham (2016)
31. Bolea, S.C.: Vocabulary, synonyms and sentiments of hazard-related posts on social networks. In: Proceedings of the 8th Conference Speech Technology and Human-Computer Dialogue (SpeD), pp. 1–6. Bucharest, Romania (2015)
32. Bolea, S.C.: Language processes and related statistics in the posts associated to disasters on social networks. Int. J. Comput. Commun. Control **11**(5), 602–612 (2016)
33. Teodorescu, H.N.L.: Survey of IC&T in disaster mitigation and disaster situation management, Chapter 1. In: Teodorescu, H.-N., Kirschenbaum, A., Cojocaru, S., Bruderlein, C. (eds.), Improving Disaster Resilience and Mitigation—IT Means and Tools. NATO Science for Peace and Security Series—C, pp. 3–22. Springer, Dordrecht (2014)
34. Kanis, J., Skorkovská, L.: Comparison of different lemmatization approaches through the means of information retrieval performance. In: Proceedings of the 13th International Conference on Text, Speech and Dialogue TSD'10, pp. 93–100 (2010)
35. Ferrucci, D., Lally, A.: UIMA: an architectural approach to unstructured information processing in the corporate research environment. Nat. Lang. Eng. **10**(3–4), 327–348 (2004)
36. Jacobs, P.S.: Joining statistics with NLP for text categorization. In: Proceedings of the Third Conference on Applied Natural Language Processing, pp. 178–185 (1992)
37. Jivani, A.G.: A comparative study of stemming algorithms. Int. J Comp Tech. Appl **2**, 1930–1938 (2011)
38. Ingason, A.K., Helgadóttir, S., Loftsson, H., Rögnvaldsson, E.: A mixed method lemmatization algorithm using a hierarchy of linguistic identities (HOLI). In: Raante, A., Nordström, B. (eds.), Advances in Natural Language Processing. Lecture Notes in Computer Science, vol. 5221, pp. 205–216. Springer, Berlin (2008)
39. Krouska, A., Troussas, C., Virvou, M.: The effect of preprocessing techniques on Twitter sentiment analysis. In: Proceedings of the International Conference on Information, Intelligence, Systems & Applications, pp. 13–15. Chalkidiki, Greece (2016)
40. Babanejad, N., Agrawal, A., An, A., Papagelis, M.: A comprehensive analysis of preprocessing for word representation learning in affective tasks. In: Proceedings of the 58th Annual Meeting of the Association for Computational Linguistics, pp. 5799–5810 (2020)

41. Camacho-Collados, J., Pilehvar, M.T.: On the role of text preprocessing in neural network architectures: an evaluation study on text categorization and sentiment analysis. In: Proceedings of the 2018 EMNLP Workshop BlackboxNLP: Analyzing and Interpreting Neural Networks for NLP, pp. 40–46. Association for Computational Linguistics (2018)
42. Davis, C.A., Varol, O., Ferrara, E., Flammini, A., Menczer, F.: BotOrNot: a system to evaluate social bots, a system to evaluate social bots. In: Proceedings of the 25th International Conference Companion on World Wide Web, pp. 273–274 (2016)
43. Ferrara, E.: COVID-19 on Twitter: Bots, Conspiracies, and Social Media Activism. arXiv preprint arXiv:2004.09531 (2020)
44. Metaxas, P., Finn, S.T.: The infamous#Pizzagate conspiracy theory: Insight from a Twitter Trails investigation. Wellesley College Faculty Research and Scholarship (2017)
45. Teodorescu, H.N.L.: Social signals and the ENR index—noise of searches on SN with keyword-based logic conditions. In: Proceedings of the International Symposium on Signals, Circuits and Systems. Iasi, Romania (2015)
46. Aouragh, S.I.: Adaptating Levenshtein distance to contextual spelling correction. Int. J. Comput. Sci. Appl. **12**(1), 127–133 (2015)
47. Kobzdej, P.: Parallel application of Levenshtein's distance to establish similarity between strings. Front. Artif. Intell. Appl. **12**(4) (2003)
48. Rani, S.; Singh, J.: Enhancing Levenshtein's edit distance algorithm for evaluating document similarity. In: Communications in Computer and Information Science, pp. 72–80. Springer, Singapore (2018)

Author Index

M. Ben Ahmed et al. (eds.), *Networking, Intelligent Systems and Security*, Smart Innovation, Systems and Technologies 237,
https://doi.org/10.1007/978-981-16-3637-0